Nanostructure Science and Technology

Series Editor

David J. Lockwood, FRSC
National Research Council of Canada
Ottawa, ON, Canada

D1809975

More information about this series at http://www.springer.com/series/6331

Hiromi Yamashita · Hexing Li
Editors

Core-Shell and Yolk-Shell Nanocatalysts

 Springer

Editors
Hiromi Yamashita
Division of Materials Science
Osaka University
Suita, Japan

Hexing Li
Department of Chemistry
Shanghai Normal University
Shanghai, China

ISSN 1571-5744 ISSN 2197-7976 (electronic)
Nanostructure Science and Technology
ISBN 978-981-16-0465-2 ISBN 978-981-16-0463-8 (eBook)
https://doi.org/10.1007/978-981-16-0463-8

This Springer imprint is published by the registered company Springer Nature Singapore Pte Ltd.
The registered company address is: 152 Beach Road, #21-01/04 Gateway East, Singapore 189721, Singapore

Preface

This book introduces recent progress in preparation and application of core–shell and yolk–shell structures for attractive design of catalyst materials. Core–shell nanostructures with active core particles covered directly with an inert shell can perform as highly active and selective catalysts with long lifetimes. Yolk-shell nanostructures consisting of catalytically active core particles encapsulated by hollow materials are an emerging class of nanomaterials. The enclosed void space is expected to be useful for encapsulation and compartmentation of guest molecules, and the outer shell acts as a physical barrier to protect the guest molecules from the surrounding environment. Furthermore, the tunability and functionality in the core and the shell regions can offer new catalytic properties, rendering them attractive platform materials for the design of heterogeneous catalysts. This book describes the recent development of such unique nanostructures to design effective catalysts which can lead to new chemical processes. It provides an excellent guide for design and application of core–shell and yolk–shell structured catalysts for a wide range of readers working on design of attractive catalysts, photocatalysts, and electrocatalysts for energy, environmental, and green chemical processes.

Osaka, Japan   Hiromi Yamashita
Shanghai, China   Hexing Li

Contents

About the Editors

Hiromi Yamashita has been a professor at Osaka University since 2004. He also has been a professor of Elements Strategy Initiative for Catalysts & Batteries (Kyoto University) since 2012. He received Ph.D. degree from Kyoto University (supervisor: Prof. S. Yoshida) in 1987. He was an assistant professor of Tohoku University (with Profs. A. Tomita, T. Kyotani), an associate professor of Osaka Prefecture University (with Prof. M. Anpo), and an invited professor of University Pierre and Marie Curie (with Prof. M. Che). He was also visiting research fellows of the Pennsylvania State University (with Prof. L. R. Radovic), the University of Texas at Austin (with Prof. M. A. Fox), California Institute of Technology (with Prof. M.E. Davis). He is the editor of Applied Catalysis B, the associate editors of Journal of Photochemistry and Photobiology C, the president of Catalysis Society of Japan (2019–2020), the president of Asia and Pacific Association of Catalysis Societies (2019–2022), and the Member of Academia Europea. He received awards from several societies such as Catalysis Society of Japan, the Japan Petroleum Institute, Japanese Photochemistry Association, the Japan Institute of Metals and Materials, Japan Society of Coordination Chemistry. His research interests include the design of single-site photocatalysts and nanostructured catalysts.

Hexing Li received his Ph.D. degree from Fudan University in 1998 (Advised by Academician Jingfa Deng). He has been a professor at Shanghai Normal University since 1999. He is an invited professor of Shanghai University and a Ph.D. student advisor of East China Normal University and East China Science and Technology University. He is now an associate editor of Applied Catalysis B: Environmental. In addition, he is a director of both Chinese Education Ministry Key Lab on Resource Chemistry and International Joint Lab co-established by Shanghai Normal University, Princeton University, and National University of Singapore. Meanwhile, he is also a vice-director of Chinese Photo-chemistry and Photocatalysis Professional Committee. His research interests include the design of metal, organometal and semiconductor nanomaterials for both thermo- and photocatalysis. He received Outstanding Youth Funding from NSFC and the first class of Nature Science Award from Chinese Education Ministry and Shanghai Local Government.

Chapter 1
Introduction

Yasutaka Kuwahara and Hiromi Yamashita

1.1 Introduction

Catalysis is an old but still growing field of science, which now accounts for over 90% of chemical processes and the production of 60% of all chemicals worldwide. In order to meet the urgent demand for high-performance catalysts with excellent activity, selectivity and stability, catalysis, especially heterogeneous catalysis, has undergone innovative advances accompanied with great progress in nanoscience and nanotechnology over the past decades. Thanks to rapid advances in synthesis chemistry, nanomaterials with well-defined sizes, shapes, structures, crystal facets, and compositions are currently available [1–3], which have provided many opportunities and possibilities for developing advanced catalysts. In particular, nanoparticles (NPs) have received considerable research attention due to their superior catalytic properties [4–7]. Different from conventional bulk catalysts, NP catalysts usually offer a significantly increased catalytic activity due to the high surface-to-volume ratio as well as a large fraction of active atoms with dangling bonds exposed to the surface. In addition, other unique properties of NPs such as variation in electronic state density and surface/lattice distortions could also affect the catalytic performance. Such property of NP catalysts offers solutions to the traditional demands for catalysts to be more efficient and more selective for particular catalytic reactions that will transform raw materials into targeted valuable chemicals, such as pharmaceuticals and fuels. Furthermore, these nanomaterials could also offer the advantages in terms of atom efficiency (i.e., reaction rates per an atom of active metal) and the amount of

Y. Kuwahara · H. Yamashita (✉)
Division of Materials and Manufacturing, Science, Graduate School of Engineering, Osaka University, Suita 565-0871, Osaka, Japan
e-mail: yamashita@mat.eng.osaka-u.ac.jp

Y. Kuwahara
e-mail: kuwahara@mat.eng.osaka-u.ac.jp

© The Author(s), under exclusive license to Springer Nature Singapore Pte Ltd. 2021
H. Yamashita and H. Li (eds.), *Core-Shell and Yolk-Shell Nanocatalysts*,
Nanostructure Science and Technology,
https://doi.org/10.1007/978-981-16-0463-8_1

metal used as well as waste reduction, hence contributing to the development of green, sustainable and economically viable chemical processes, which are currently regarded as the most desirable yet challenging areas in chemistry.

Among the variety of NP catalysts, core–shell NPs and yolk–shell NPs have recently emerged as a new type of important composite nanomaterial, because of their unique structural features and physicochemical properties The number of research papers published annually in the field of core–shell and yolk–shell catalysts has increased significantly (Fig. 1.1). (1) "core–shell structure" is broadly defined as a composite nanomaterial constructed with a single core material uniformly/partially surrounded by a secondary shell (layer) material and (2) "yolk–shell structure" (also known as rattle-type structure) generally represents a composite nanomaterial composed of a single movable core material inside a hollow shell material, which possesses a void space between the core and the shell regions. More detailed classifications have been made in some review papers according to their nano-configurations, such as "sandwiched structure" standing for multiple metal NPs embedded between the core and the shell boundaries, and "interior-decorated structures" standing for multiple metal NPs decorated on an interior surface of a hollow shell material. "Multi-core@shell structure" composed of multiple metal NP cores embedded within a secondary shell matrix and "core@multi-shell structure" composed of a single core material coated by multiple shell layers are also regarded as one class of "core–shell structures". Similarly, "multi-core@hollow structure" composed of multiple movable metal NP cores inside a hollow shell material and "core@multi-shell structure" composed of a single movable metal NP core encapsulated by multiple hollow

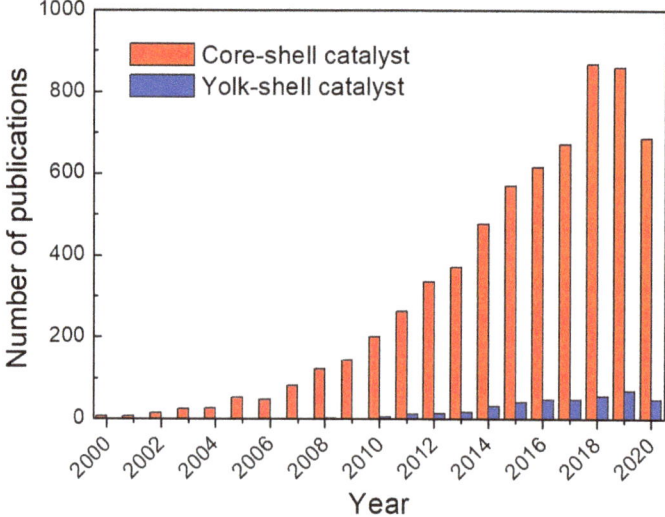

Fig. 1.1 The number of publications in the last 20 years found on Scopus for the entry "Core–shell catalyst" and "Yolk–shell catalyst" (data collected from Scopus database in August, 2020)

shell materials are usually categorized into "yolk–shell structure" as well (Fig. 1.2) [8–12].

Until now, a numerous number of work has been performed to design and construct different nanocomposites with core–shell and yolk–shell structures. Quest for new catalytic applications, performance improvement, and discovery of new functionalities (i.e., developing multifunctional catalysts) in catalysis field has still motivated a continuous development in NP research. This book is intended for the readers to overview current trends in catalysis field involving core–shell and yolk–shell structured NPs by introducing recent research results from renowned researchers all over the world.

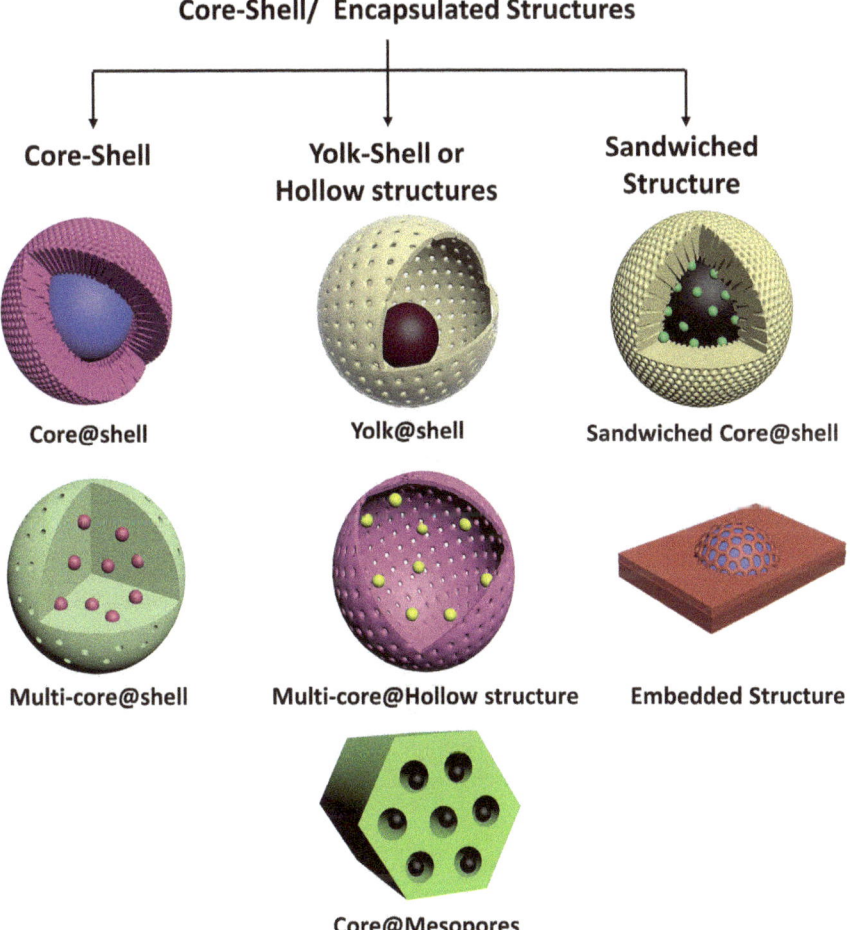

Fig. 1.2 Schematic of types of core–shell structures based on morphology. Reprinted with permission from Chem. Soc. Rev. (2020) 49, 2937, Ref. [9]. Copyright 2020 Royal Society of Chemistry

1.2   Core–Shell Structures for Catalysis

Integration of different materials such as noble metals (e.g., Pt, Pd, Au, and Ag, etc.) and metal oxides (e.g., SiO_2, TiO_2, CeO_2, and ZrO_2, etc.) into a single nanostructure has recently become one of the hottest research topics in catalysis, photo- and electrocatalysis. One possible method to accomplish this objective is to construct "core–shell structure" through the coating of core NPs with one or more shells (layers) of other materials, as illustrated in Fig. 1.3.

A range of core–shell NPs with tailorable catalytic properties and functionalities can be produced by rationally tuning the cores and the shells of such materials [8–11, 13]. For example, thermal stability or dispersibility can be improved by growing protective shells around the cores. Covering the core NP with shell materials possessing specific nanoarchitectures, such as micro/meso-porosity and high surface area, etc., endows the core material with a superior catalytic activity or a size-selective reactivity (i.e., molecular-sieving effect) together with an improved thermal stability [14, 15]. In addition, novel optical, magnetic, and electronic functionalities can be imparted to the core NP through the formation of a core–shell structure (e.g., magnetically separable NP catalysts can be produced by introducing ferromagnetic Fe_3O_4 particle as a core).

Core–shell structures not only show combined functions of individual materials, but also bring unique collective and synergetic catalytic properties compared with single-component materials, depending on the interactions between the cores and the shells. For example, it has been discovered that synergistic interactions between the core and the shell can change the properties of the original core particles. There are three major effects ever known (in most cases, it is a combined one) that play important roles in deciding the catalytic, photocatalytic, and electrocatalytic activity of such core–shell NP catalysts (Fig. 1.3): (a) "ligand effect", originating from the interaction at the interface between the core and the shell, which affects the charge transfer

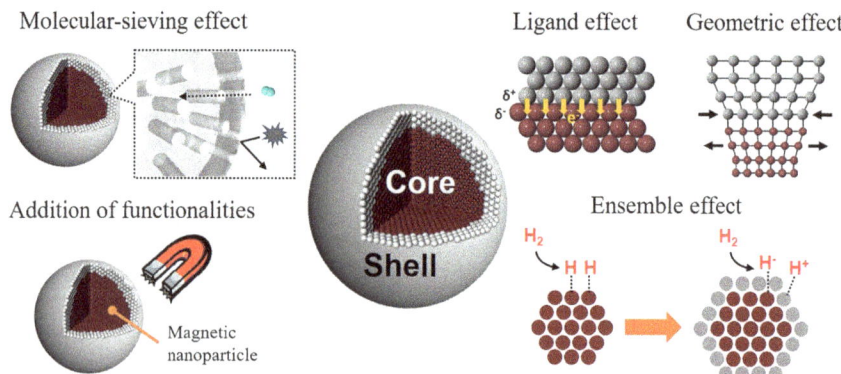

Fig. 1.3 A schematic illustration of "core–shell structure" particle and its functionalities

between the components and alters the electronic state. This "ligand effect" is significantly strengthened in core–shell structures, because the interfacial area between the inner cores and the outer shell is geometrically maximized; (b) "ensemble effect", affecting adsorption behavior of reactant molecules or alters reaction mechanism on the surface. This effect is brought about by the presence of secondary atoms which change the atomic arrangement and aggregation state of primary active metal atoms; (c) "geometric effect", originating from the three-dimensional structural constraints (e.g., lattice strain and surface strain) due to the heterojunction between the core and the shell materials, which alters electronic state and reactivity of the surface atoms. These effects can be manipulated by controlling the chemical compositions and relative sizes of the core and the shell, which give us opportunities and possibilities to precisely tune the catalytic properties of core–shell NPs [9].

As mentioned above, the properties of the core–shell NP catalysts can be modified by changing the constituting materials, the core/shell ratio, as well as the interface between the core and the shell. A precise control of the size of the core particle, the shell thickness, and the porosity in the core or the shell regions is important to materialize high-performance catalysts [13, 14, 16]. To this end, various synthetic methods for preparing different classes of core–shell NPs have been developed so far. Metal NP@metal oxide core–shell nanostructures are one of the simplest motifs in two-component systems [17–19]. The synthetic strategies for metal NP@metal oxide core–shell nanostructures are generally classified into three groups: (1) two-step methods based on seeded growth mechanism, involving the pre-preparation of metal NPs as the seeds, followed by formation of metal oxide shells; (2) one-pot methods, involving simultaneous formation of both metal cores and metal oxide shells; and (3) partial oxidation of the outside layer of metal NPs or selective oxidation of the shell of binary metal NPs [10, 17–20]. Many types of metal NP@metal oxide core–shell nanostructures have been successfully achieved by various strategies, most of which are solution-based synthetic processes. Besides, core–shell structured composite nanomaterials constructed in different combinations, such as metal NP@metal core–shell nanostructures, metal NP@MOF (MOF: metal-organic-framework) core–shell nanostructures, and metal NP@carbon core–shell nanostructures, have also been developed by unique synthetic methodologies.

Multifunctional core–shell NP catalysts are designed usually depending on the final application. For designing a catalyst for chemoselective transformation reactions that will transform raw materials into targeted valuable chemicals, such as pharmaceuticals, metal@metal oxide or metal@metal core–shell NPs exhibit excellent product selectivity due to the synergism at the interface between the core and the shell components (Chaps. 2, 3, 4, 5 and 9). Metal NPs encapsulated by porous MOF shells have recently emerged as new motifs which provide multiple synergistic effects in heterogeneous catalysis (e.g., molecular-sieving effect, one-pot multifunctional catalysis, and enhanced catalytic stability) (Chap. 6). For realizing the selective synthesis of valuable chemicals by syngas conversion or CO_2 hydrogenation, zeolites-based core–shell structured catalysts are effective because of the spatial confinement effect, the target product selectivity, and anti-sintering property

(Chaps. 7, 8, 11 and 12). Core–shell structures of photocatalysts composed of semi-conductor photocatalysts or plasmonic NPs act as promising photocatalysts for CO_2 reduction, H_2 generation and photocatalytic degradation of toxic compounds, due to efficient light absorption, adsorption and activation of reactant molecules, efficient photogenerated charge transfer, formation of Z-scheme electron transfer pathway (Chaps. 13 and 14). Optical property and photocatalytic property of plasmonic metal NPs are precisely tunable by the core–shell nanostructure formation, which leads to the fabrication of high-performance plasmonic photocatalysts (Chaps. 15, 16 and 17). Core–shell bimetallic NPs also exhibit excellent performance as electrocatalysts for hydrogen evolution reaction, oxygen evolution reaction, nitrogen reduction reaction, CO_2 reduction reaction, and methanol oxidation reaction, etc., due to the synergistic effect between core and shell materials (Chaps. 18, 19, 20 and 21).

1.3   Yolk–Shell Structures for Catalysis

Yolk–shell structure (also known as rattle-type structure) is a new class of special core–shell structures with a characteristic core@void@shell configuration (Fig. 1.4), which has recently gained considerable attention in chemistry field due to their complex hierarchical nanostructures and intriguing physical properties, such as low

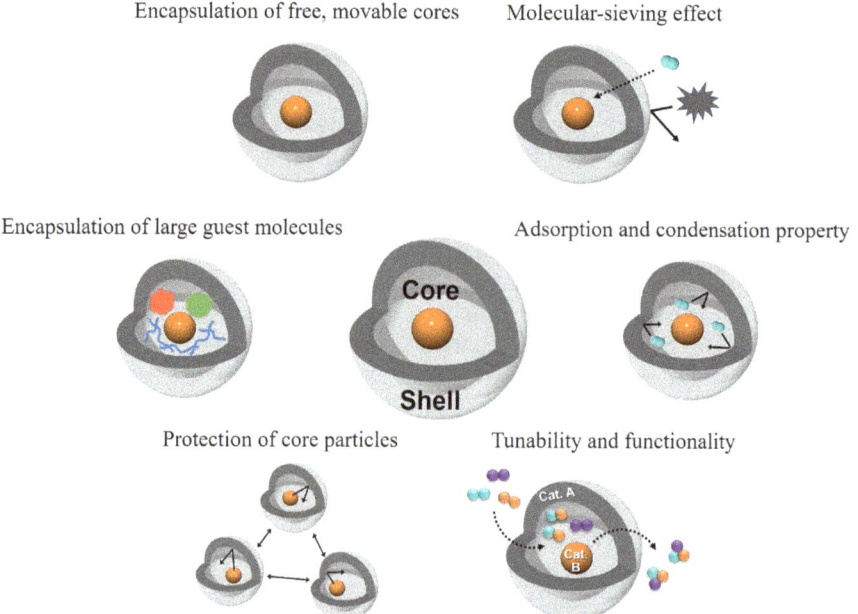

Fig. 1.4 A schematic illustration of "yolk–shell structure" particle and its functionalities

density, high surface area, and interstitial hollow spaces. Research over the last decade has proved that the distinct properties, which mainly come from the presence of an interior void space, are advantageous for a range of applications, including catalysis, drug/gene-delivery, bioimaging, and energy- and gas-storage [21–28].

Recently, catalytically active yolk–shell structured composite nanomaterials have been designed and synthesized by encapsulating catalytically active components, such as metal NPs, metal oxide NPs, and metal complexes, as the core, which have been proven to show outstanding catalytic performances for various catalytic reactions due to the following interesting physical and chemical properties (Fig. 1.4) [8, 19, 21, 22, 29, 30]: (1) encapsulation of free, movable cores within confined nanospace, which offers a homogeneous environment in heterogeneous catalysis; (2) encapsulation and compartmentation of large guest molecules, such as metal oxide particles, magnetic particles, quantum dot particles, fluorophores, and functional polymers, within the interior void space, which can add further functionalities; (3) protection of the encapsulated core particles from surrounding environments, leaching, sintering and aggregation by the outer shell, which can stabilize the core materials and elongate their lifetime; (4) molecular-sieving effect for reactant molecules, originating from permeable porous shells, which endows size-selective reactivity in catalysis; (5) adsorption and condensation property toward reactant molecules inside the cavity space, which enable reactant molecules to efficiently react on catalytically active core particles, thus accelerate catalytic reactions; (6) tunability and functionality in the core and the shell regions. For example, manipulation of the nanostructure/morphology of the core and the shell can tune the diffusion rates of the reactants, thus affects the reactivity. Furthermore, incorporation of secondary active components either in the hollow or in the shell regions is possible for achieving cooperative effects of the two catalysts or for constructing cascade reaction (namely, one-pot or domino reaction) systems. On the basis of these intriguing properties, yolk–shell nanostructured materials encapsulating catalytically active components as cores have recently been regarded as a promising platform for designing multi-functionalized heterogeneous catalysts and for realizing industrially applicable nanocatalyst systems [8, 19, 21, 22, 29, 30].

To date, a number of yolk–shell structured materials with configurations, compositions, sizes, and shapes have extensively been developed. Yolk–shell structured materials have been designed and engineered with a variety of chemical compositions, including metal@polymer, metal@metal oxide, metal@carbon, metal@MOF, metal oxide@metal oxide, and metal oxide@carbon, etc. In addition, yolk–shell nanostructures composed of multiple cores or multiple shells have also been fabricated and studied. Thus, yolk–shell structures developed so far are diverse in configurations, compositions, sizes, and shapes, hence they can be produced by a variety of techniques. Commonly, the synthesis methods of yolk–shell structured composite nanomaterials can be simply classified into three categories: (1) hard-/soft-templating methods; (2) template-free synthesis (e.g., Kirkendall effect, Galvanic replacement, Ostwald ripening, etc.); and (3) ship-in-bottle approach [10, 19–21, 30–38]. In the former case, such as hard-/soft-templating methods, the procedures normally involve the pre-preparation of core particles, coating the core particles with multiple shells (inner sacrificial shells and outer stable shells), and selective removal of the inner shells by post-treatments, therefore the synthesis procedures become tedious and time-consuming. In the case of template-free synthesis, it is not necessary to coat the core particles, therefore the procedures are much simpler.

A variety of yolk–shell structured catalysts with tuned physicochemical properties and multi-functionalities have been designed, synthesized, and studied in chemical reactions that requires excellent activity, high selectivity to desirable products, and long-term stability under severe reaction environments. For example, catalytic NPs (e.g., Pd NPs, Pt NPs) are usually entrapped in a surfactant/ligand-free state inside the void space, hence they commonly show a higher catalytic activity compared with surface-capped counterparts (Chaps. 22, 23 and 25). Hollow particles possessing permeable porous shells can act as a "nanoreactor" that can selectively catalyze the transformation of a specific molecule due to the molecular-sieving effect (Chap. 25). The large void space in the hollow particles has recently been used for loading functional molecules, such as biological molecules and functional polymers, which allow creation of unprecedented catalytic environments within the limited nanospace (Chaps. 24, 25 and 26). The encapsulation of active metal NPs inside the hollow MOF cavity can help to enhance the catalyst performance in terms of size- and chemo-selectivity and catalyst stability for recycling (Chap. 29). Furthermore, yolk–shell structures encapsulating semiconductor NPs and plasmonic NPs can be used in the field of photocatalysis, including photocatalytic pollutant degradation, hydrogen evolution reaction and carbon dioxide reduction due to efficient light absorption, adsorption and activation of reactant molecules, and separation of photogenerated charge carriers (Chaps. 27, 28, 30 and 31).

1.4   Outline of This Book

Based on their significance in catalysis, this book is dedicated to a comprehensive review on recent progress in the construction of core–shell/yolk–shell nanostructures and their applications in catalysts, photocatalysis, and electrocatalysis. This book is composed of four parts, "Core–Shell for Catalysis", "Core–Shell for Photocatalysis and Electrocatalysis", "Yolk–Shell for Catalysis", and "Yolk–Shell for Photocatalysis and Electrocatalysis".

In Part I (Core–Shell for Catalysis), a variety of core–shell structured catalysts for liquid-phase and gas-phase catalytic applications, including oxidative esterification, selective hydrogenation, CO_2 hydrogenation, conversions of syngas and methane, upgrading of C1-3 feedstock chemicals and hydrocarbons, and NOx removal, etc., that are recently developed are reviewed. In Part II (Core–Shell for Photocatalysis and Electrocatalysis), recent studies of core–shell structured catalysts for photocatalytic and electrocatalytic applications, including CO_2 reduction, H_2 production, degradation of toxic organic pollutant in water, HER, OER, as well as nitrogen reduction reaction, CO_2 reduction reaction, and methanol oxidation reaction, etc., are comprehensively summarized. In Part III (Yolk-shell for Catalysis), efficient catalytic reactions such as selective hydrogenation reactions and CO_2 transformation reactions using a variety of yolk–shell structured catalysts encapsulating active metal NPs are summarized. In Part IV (Yolk-Shell for Photocatalysis and Electrocatalysis), a range of potential photocatalytic and electrocatalytic applications of yolk–shell structures containing semiconductor or plasmonic NPs, including photocatalytic degradation of pollutants, HER, OER, photocatalytic CO_2 reduction, etc., are overviewed. The key synthesis approaches and general procedures for constructing each of these nanostructures with desirable configurations, compositions, sizes, and shapes are briefly mentioned in each chapter. Furthermore, the relationships between the core/yolk–shell nanostructures and catalytic performances, as well as the roles of core/yolk–shell nanostructures in deciding the catalytic, photocatalytic and electrocatalytic activity, are discussed.

We hope that this book will be useful to readers in various fields who take practical and scientific interests in the design, synthesis and application of core/yolk–shell nanostructure materials.

References

1. Zeng H-C (2013) Integrated nanocatalysts. Acc Chem Res 46(2):226–235
2. Goesmann H, Feldmann C (2010) Nanoparticulate functional materials. Angew Chem Int Ed 49(8):1362–1395
3. Zhuang Z, Peng Q, Li Y (2011) Controlled synthesis of semiconductor nanostructures in the liquid phase. Chem Soc Rev 40(11):5492–5513
4. Roucoux A, Schulz J, Patin H (2002) Reduced transition metal colloids: a novel family of reusable catalysts? Chem Rev 102:3757–3778
5. Schmid G (1992) Large clusters and colloids. Metals in the embryonic state. Chem Rev 92:1709–1727
6. Wilcoxon JP, Abrams BL (2006) Synthesis, structure and properties of metal nanoclusters. Chem Soc Rev 35(11):1162 1194
7. Tao AR, Habas S, Yang P (2008) Shape control of colloidal metal nanocrystals. Small 4(3):310–325
8. Gawande MB, Goswami A, Asefa T, Guo H, Biradar AV, Peng DL, Zboril R, Varma RS (2015) Core-shell nanoparticles: synthesis and applications in catalysis and electrocatalysis. Chem Soc Rev 44(21):7540–7590
9. Das S, Perez-Ramirez J, Gong J, Dewangan N, Hidajat K, Gates BC, Kawi S (2020) Core-shell structured catalysts for thermocatalytic, photocatalytic, and electrocatalytic conversion of co2. Chem Soc Rev 49(10):2937–3004
10. El-Toni AM, Habila MA, Labis JP, ALOthman ZA, Alhoshan M, Elzatahry AA, Zhang F (2016) Design, synthesis and applications of core-shell, hollow core, and nanorattle multifunctional nanostructures. Nanoscale 8(5):2510–2531
11. Ghosh Chaudhuri R, Paria S (2012) Core/shell nanoparticles: classes, properties, synthesis mechanisms, characterization, and applications. Chem Rev 112(4):2373–2433
12. Purbia R, Paria S (2015) Yolk/shell nanoparticles: classifications, synthesis, properties, and applications. Nanoscale 7(47):19789–19873
13. Mitsudome T, Kaneda K (2013) Advanced core-shell nanoparticle catalysts for efficient organic transformations. ChemCatChem 5(7):1681–1691
14. Zhang Q, Lee I, Joo J-B, Zaera F, Yin Y (2013) Core-shell nanostructured catalysts. Acc Chem Res 46(8):1816–1824

15. Joo SH, Park JY, Tsung CK, Yamada Y, Yang P, Somorjai GA (2009) Thermally stable pt/mesoporous silica core-shell nanocatalysts for high-temperature reactions. Nat Mater 8(2):126–131
16. Zhong C-J, Maye MM (2001) Core-shell assembled nanoparticles as catalysts. Adv Mater 13(19):1507–1511
17. Liu S, Bai S-Q, Zheng Y, Shah KW, Han M-Y (2012) Composite metal-oxide nanocatalysts. ChemCatChem 4(10):1462–1484
18. Sun H, He J, Wang J, Zhang SY, Liu C, Sritharan T, Mhaisalkar S, Han MY, Wang D, Chen H (2013) Investigating the multiple roles of polyvinylpyrrolidone for a general methodology of oxide encapsulation. J Am Chem Soc 135(24):9099–9110
19. Li G, Tang Z (2014) Noble metal nanoparticle@metal oxide core/yolk-shell nanostructures as catalysts: recent progress and perspective. Nanoscale 6(8):3995–4011
20. Liu R, Priestley RD (2016) Rational design and fabrication of core–shell nanoparticles through a one-step/pot strategy. J Mater Chem A 4(18):6680–6692
21. Li Y, Shi J (2014) Hollow-structured mesoporous materials: chemical synthesis, functionalization and applications. Adv Mater 26(20):3176–3205
22. Liu J, Qiao SZ, Chen JS, Lou XW, Xing X, Lu GQ (2011) Yolk/shell nanoparticles: new platforms for nanoreactors, drug delivery and lithium-ion batteries. Chem Commun 47(47):12578–12591
23. Tang F, Li L, Chen D (2012) Mesoporous silica nanoparticles: synthesis, biocompatibility and drug delivery. Adv Mater 24(12):1504–1534
24. Wang Z, Zhou L, Lou XW (2012) Metal oxide hollow nanostructures for lithium-ion batteries. Adv Mater 24(14):1903–1911
25. Li X, Yang Y, Yang Q (2013) Organo-functionalized silica hollow nanospheres: synthesis and catalytic application. J Mater Chem A 1(5):1525–1535
26. Chen Y, Meng Q, Wu M, Wang S, Xu P, Chen H, Li Y, Zhang L, Wang L, Shi J (2014) Hollow mesoporous organosilica nanoparticles: a generic intelligent framework-hybridization approach for biomedicine. J Am Chem Soc 136(46):16326–16334
27. Vaz B, Salgueirino V, Perez-Lorenzo M, Correa-Duarte MA (2015) Enhancing the exploitation of functional nanomaterials through spatial confinement: the case of inorganic submicrometer capsules. Langmuir 31(32):8745–8755
28. Croissant JG, Cattoen X, Wong MC, Durand JO, Khashab NM (2015) Syntheses and applications of periodic mesoporous organosilica nanoparticles. Nanoscale 7(48):20318–20334
29. Perez-Lorenzo M, Vaz B, Salgueirino V, Correa-Duarte MA (2013) Hollow-shelled nanoreactors endowed with high catalytic activity. Chem Eur J 19(37):12196–12211
30. Lee J, Kim SM, Lee IS (2014) Functionalization of hollow nanoparticles for nanoreactor applications. Nano Today 9(5):631–667
31. Yin Y, Rioux RM, Erdonmez CK, Hughes S, Somorjai GA, Alivisatos AP (2004) Formation of hollow nanocrystals through the nanoscale Kirkendall effect. Science 304:711–714
32. Fan HJ, Gosele U, Zacharias M (2007) Formation of nanotubes and hollow nanoparticles based on Kirkendall and diffusion processes: a review. Small 3(10):1660–1671
33. Zhang Q, Wang W, Goebl J, Yin Y (2009) Self-templated synthesis of hollow nanostructures. Nano Today 4(6):494–507
34. Wu XJ, Xu D (2010) Soft template synthesis of yolk/silica shell particles. Adv Mater 22(13):1516–1520
35. Wong YJ, Zhu L, Teo WS, Tan YW, Yang Y, Wang C, Chen H (2011) Revisiting the stober method: inhomogeneity in silica shells. J Am Chem Soc 133(30):11422–11425
36. Fang X, Zhao X, Fang W, Chen C, Zheng N (2013) Self-templating synthesis of hollow mesoporous silica and their applications in catalysis and drug delivery. Nanoscale 5(6):2205–2218
37. Qiao ZA, Huo Q, Chi M, Veith GM, Binder AJ, Dai S (2012) A "ship-in-a-bottle" approach to synthesis of polymer dots@silica or polymer dots@carbon core-shell nanospheres. Adv Mater 24(45):6017–6021
38. Anderson BD, Tracy JB (2014) Nanoparticle conversion chemistry: Kirkendall effect, galvanic exchange, and anion exchange. Nanoscale 6(21):12195–12216

Part I
Core-Shell for Catalysis

Chapter 2
Aerobic Oxidative Esterification of Aldehydes with Alcohols by Gold–Nickel Oxide Nanoparticle Catalysts with a Core–Shell Structure

Ken Suzuki

2.1 Introduction

Esterification, one of the most fundamental transformations in organic synthesis, is widely used in laboratories and industries [1]. Esterification of aldehydes with alcohols is an attractive method for the synthesis of esters because aldehydes are readily available raw materials on a commercial scale. Although several facile and selective esterification reactions have been reported [2], the development of a catalytic method for the direct oxidative esterification of aldehydes with alcohols under mild and neutral conditions in the presence of molecular oxygen as the terminal oxidant is highly desirable for both economic and environmental aspects.

Since Haruta et al. discovered that Au nanoparticles can catalyze aerobic oxidation reactions [3], Au-catalyzed oxidation reactions have been widely investigated. Efforts are being directed at achieving highly selective oxidation using molecular oxygen [4]. Several Au-nanoparticle-based catalysts for the aerobic esterification of aldehydes [5] or alcohols [6] have been reported. In this chapter, we report a highly selective and efficient catalytic method for the oxidative esterification of aldehydes with alcohols that employs supported gold–nickel oxide (Au–NiO$_x$) nanoparticles as the catalyst and molecular oxygen as the terminal oxidant (Scheme 2.1) [7].

As an example, the aerobic catalytic esterification of methacrolein **1a** with methanol to form methyl methacrylate (MMA; **2a**) was investigated under neutral conditions. The monomer MMA is mainly used to produce acrylic plastics such as poly(methyl methacrylate) (PMMA) and other polymer dispersions used in paints and coatings. MMA can be manufactured in numerous ways from C_2–C_4 hydrocarbon feedstocks [8]. Currently, MMA is mainly produced via the acetone cyanohydrin

K. Suzuki (✉)
Chemistry & Chemical Process Laboratory, Asahi Kasei Corporation, 2767-11 Niihama, Shionasu, Kojima, Kurashiki, Okayama 711-8510, Japan
e-mail: suzuki.kd@om.asahi-kasei.co.jp

© The Author(s), under exclusive license to Springer Nature Singapore Pte Ltd. 2021
H. Yamashita and H. Li (eds.), *Core-Shell and Yolk-Shell Nanocatalysts*,
Nanostructure Science and Technology,
https://doi.org/10.1007/978-981-16-0463-8_2

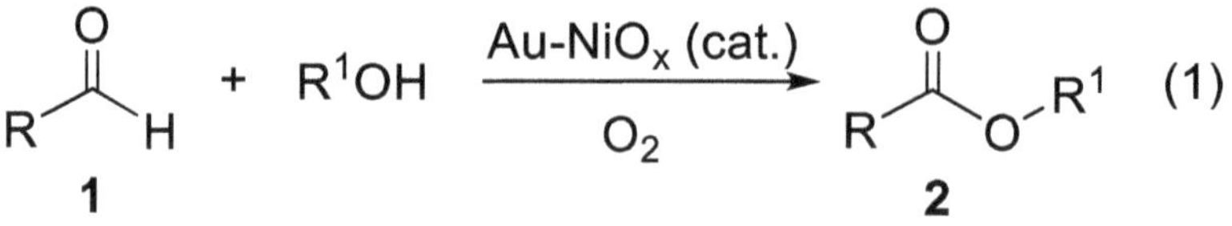

$$R\overset{\text{O}}{\underset{\text{H}}{\|}}\quad + \quad R^1OH \quad \xrightarrow[O_2]{\text{Au-NiO}_x \text{ (cat.)}} \quad R\overset{\text{O}}{\underset{\text{O}^{\smallsmile R^1}}{\|}} \quad (1)$$

1 **2**

Scheme 2.1 Aerobic oxidative esterification of aldehydes with alcohols by Au-NiOx catalyst

process, but there are problems in handling the resulting ammonium bisulfate waste and toxic hydrogen cyanide. Some manufacturers use isobutene or *tert*-butanol as the starting material, which is sequentially oxidized first to methacrolein and then to methacrylic acid, which in turn is esterified with methanol. Recently, an environmentally benign procedure based on the use of molecular oxygen and a Pd–Pb catalyst has been developed for the direct oxidative esterification of methacrolein with methanol to yield MMA[9]. This work was an important milestone in the aerobic oxidative esterification of aldehydes, as it put forth a clean and efficient method of forming carboxylic esters. However, the existing synthetic methods still suffer from several disadvantages; methods for successful catalytic oxidative esterification are limited as selective oxidation of methacrolein is extremely difficult because of the instability of α,β-unsaturated aldehydes. Therefore, the development of an efficient and highly selective catalytic system based on the above reaction remains a challenge.

2.2 Au–NiO$_x$-Catalyzed Oxidative Esterification of Aldehydes in Alcohols with Molecular Oxygen

Table 2.1 summarizes the activity of various catalysts used in the aerobic esterification of **1a** with methanol. The activity of previously reported Pd catalysts was investigated first [9]. When 2.5 wt% Pd/SiO$_2$–Al$_2$O$_3$ was used, **2a** could not be obtained in satisfactory yields (entry 1) because the decarboxylation of **1a** resulted in the formation of large amounts of propylene and CO$_2$ as by-products. When oxidative esterification was carried out by the addition of Pb(OAc)$_2$ to the reaction mixture, decarboxylation was inhibited, and the selectivity to **2a** improved to 84% (entry 2). The active species in the above reaction was found to be the intermetallic compound Pd$_3$Pb$_1$ [9]. During oxidative esterification in the presence of methanol, the excess methanol is oxidized to form methyl formate (MF) as a by-product (0.2 mol of MF per mole of MMA). The turnover number (TON) of the catalyst, defined as the total number of moles of the product **2a** formed per mole of the Pd-catalyst, was determined to be 61. Attempts to carry out esterification of **1a** to **2a** in the presence of other Pd-based catalysts were unsuccessful. We then turned our attention to nickel oxide. Nickel peroxide is known to be highly oxidizing and can stoichiometrically oxidize various alcohols [10]. The catalytic aerobic oxidation of alcohols was possible after the recent development of catalysts such as Ni–Al hydrotalcite and nanosized NiO$_2$ powder [11]. In the field of electronic materials, research is being conducted on

Table 2.1 Catalytic activity for aerobic oxidative esterification of methacrolein 1a with methanol[a]

Entry	Catalyst	Conversion of aldehyde 1 (%)[b]	Selectivity for ester 2 (%)[b]
1c	Pd/SiO$_2$–Al$_2$O3	20	40
2c	PdPb/SiO$_2$–Al$_2$O3	34	84
3	Au–NiO$_x$/SiO$_2$–Al$_2$O$_3$–MgO	58	98
4	Au–NiO$_x$/SiO$_2$–Al$_2$O$_3$	63	97
5	Au/SiO$_2$–Al$_2$O$_3$–MgO	14	91
6	Au/SiO$_2$–Al$_2$O$_3$	17	79
7	AuNi/SiO$_2$–Al$_2$O$_3$–MgO	12	89

[a]Reaction conditions: 1a (15 mmol), catalyst (Au: 0.1 mol%) in methanol (10 mL), O$_2$ (O$_2$/N$_2$ = 7:93 v/v, 3 MPa) at 60 °C for 2 h. [b]Determined by GC analysis using an internal standard. [c]Pd-based catalyst (Pd: 0.5 mol%)

NiO–M (M: Ni, Pd, Pt, Au, Ag, Cu) composite film to quicken the light-absorption response of Ni oxide film used as an electrochromic material. The metal doped into Ni oxide is supposed to act as a positive hole and improves the speed of oxidative coloring by converting Ni oxide into a higher oxidation state [12]. Nickel peroxide was also found to participate in oxidative esterification of aldehydes with alcohols [13]. We examined the relationship between the chemical form and reactivity and developed a new catalytic system of composite nanoparticles composed of NiO and Au active species.

Au and NiO were supported on SiO_2–Al_2O_3–MgO (average particle size of 60 μm) by co-precipitation. The amounts of Au and Ni in the supported nanoparticle were determined to be 0.9 and 1.1 wt%, respectively, by inductively coupled plasma-atomic emission spectroscopy (ICP-AES). Thus, the reaction of **1a** in the presence of Au–NiO_x/SiO_2–Al_2O_3–MgO in methanol at 60 °C under an oxygen–nitrogen mixture (7:93 (v/v), 3 MPa, outside flammability limits) for 2 h gave **2a** with 98% selectivity and 58% conversion (entry 3). Based on the moles of MMA formed per mole of the Au catalyst, the TON of the supported nanoparticle catalyst was determined to be 621, and its activity was approximately 10 times that of the Pd–Pb catalyst. Moreover, reduced by-product (MF) formation was observed in this case (0.007 mol of MF formed per mole of MMA). Oxidative esterification was found to proceed with high efficiency even when SiO_2–Al_2O_3 was used as carriers (entry 4). The catalyst supported with only Au nanoparticles showed lower activity and selectivity than the supported Au–NiO_x catalyst (entries 5 and 6). The activity and selectivity of the Au–Ni catalyst, prepared by reduction of the Au–NiO_x catalyst under H2 atmosphere at 400 °C for 3 h, was greatly decreased (entry 7). When 1 wt% Au–1 wt% MO_x/SiO_2–Al_2O_3–MgO (M: Cr, Mn, Fe, Co, Cu, Zn, Ga, Ge, Nb, In, Sn, Nb, Ta, and Pb) was used in oxidative esterification, similar effects to that of the Au–NiO_x catalyst could not be obtained. The oxidative esterification activity of the Au–NiO_x catalyst showed a strong dependence on the Au and NiO composition in the supported nanoparticle. The maximum activity was observed for 20 mol% of Au (Fig. 2.1).

Next, oxidative esterification of various aldehydes and alcohols was carried out by using the Au–NiO_x catalyst (Table 2.2). When oxidative esterification of **1a** was carried out in the presence of Au–NiO_x/SiO_2–Al_2O_3–MgO in methanol at 80 °C, under an oxygen–nitrogen mixture (7:93 (v/v), 3 MPa) for 1 h, **2a** was obtained with 98% selectivity and 62% conversion (entry 1). The conversion of benzaldehyde **1b** to methyl benzoate **2c** was highly efficient (entry 3). When oxidative esterification of **1a** and **1b** was carried out using ethanol in place of methanol, the corresponding esters (**2b** and **2d**) were obtained with high selectivity but with lower conversion efficiencies than those in the case of using methanol (entries 2 and 4).

Fig. 2.1 Yield of methyl methacrylate **2a** for oxidative esterification of methacrolein **1a** in methanol over catalysts with various Au/Ni compositions

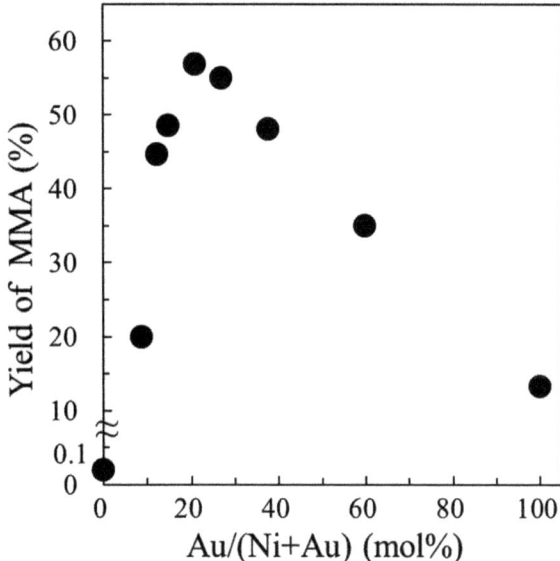

Table 2.2 Au–NiO$_x$-catalyzed aerobic oxidative esterification of aldehydes with alcohols[a]

Entry	Aldehyde	Alcohol	Product	Conversion (%)[b]/selectivity (%)[b]
1	![1a] **1a**	Methanol	![2a] **2a**	62/98
2	![1a] **1a**	Ethanol	![2b] **2b**	11/97
3	![1b] **1b**	Methanol	![2c] **2c**	61/97
4	![1b] **1b**	Ethanol	![2d] **2d**	10/97

[a]Reaction conditions: aldehyde (15 mmol), Au–NiO$_x$/SiO$_2$–Al$_2$O$_3$–MgO (Au: 0.1 mol%) in alcohol (10 mL), O$_2$ (O$_2$/N$_2$ = 7:93 v/v, 3 MPa) at 80 °C for 1 h. [b]Determined by GC analysis using an internal standard

2.3   Characterization of Catalyst and Reaction Mechanism

Spherical particles of the Au–NiO$_x$ catalyst that are uniformly distributed on the carrier can be seen in the transmission electron microscopy (TEM) images (Fig. 2.2). The particles have a diameter of 2–3 nm (number-average particle diameter: 3.0 nm). High-magnification images revealed a lattice of Au (111) particles with a d-spacing of 2.36 Å. Elemental analysis of individual particles by energy-dispersive X-ray (EDX) spectroscopy showed the presence of Ni and Au in the particles. The average Ni/Au atomic ratio of the nanoparticles was 0.82 (100 units used for calculation). As shown in Fig. 2.3, EDX analysis was performed on the scanning transmission electron microscopy (STEM) image of the nanoparticles. The results showed that the Ni/Au atomic ratio was 0.73 at the center of the particle (measurement point 1) but 2.95 at the edge of the particle (measurement point 2). Trace amounts of Ni were detected in areas that did not contain the particle (measurement point 3). Based on the composition profile observed in the direction of the scan, Ni appears to be more widely distributed than Au. Thus, Ni was distributed on the Au particles as well as around the edges of the particles. Hence, the nanoparticles were assumed that the surface of the Au particles is covered by Ni without alloy formation. However, TEM/STEM images of the Ni shell around the Au particles could not be obtained.

A broad diffraction peak attributable to Au^0 was observed in X-ray diffraction (XRD) patterns. The absence of diffraction peaks due to Ni suggested that Ni existed as a noncrystalline phase. The Au 4f and Ni 2p XPS spectra confirmed the oxidation states of Au and Ni to be 0 and +2, respectively.

When the variation in the electronically excited state was examined using ultraviolet–visible (UV–vis) spectroscopy, no surface plasmon absorption peak, as observed in the case of the Au catalyst, originated from the Au nanoparticles (~530 nm) in the

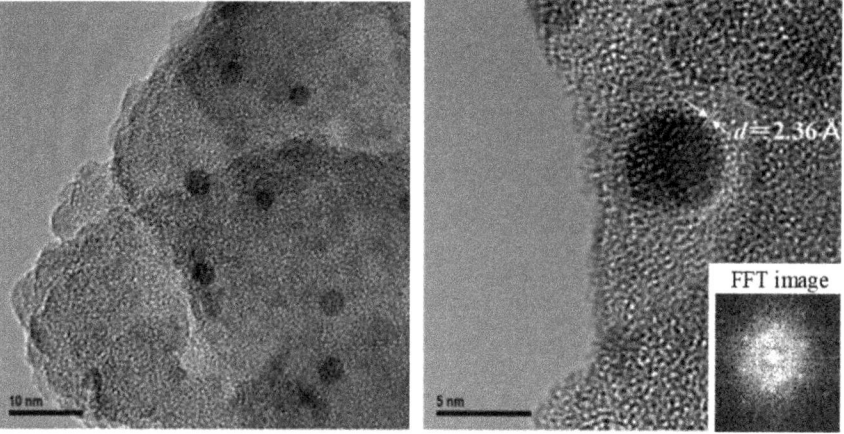

Fig. 2.2 Typical transmission electron microscopy images of Au–NiO$_x$/SiO$_2$–Al$_2$O$_3$–MgO catalyst

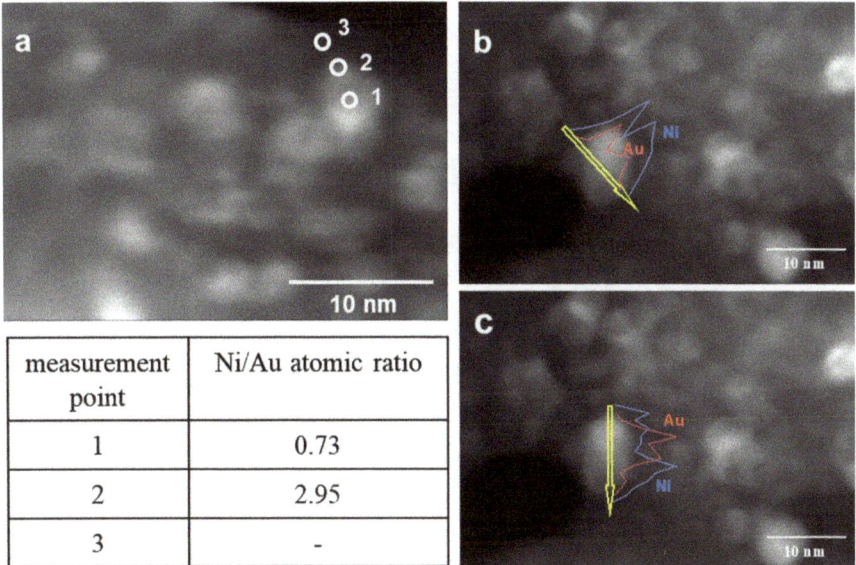

measurement point	Ni/Au atomic ratio
1	0.73
2	2.95
3	-

Fig. 2.3 Scanning transmission electron microscopy–energy-dispersive X-ray compositional analyses (**a**; point analyses, **b**, **c**; line analyses) of Au–NiO$_x$/SiO$_2$–Al$_2$O$_3$–MgO catalyst

case of the Au–NiO$_x$ catalyst. The Au–NiO$_x$ catalyst was brown in color and showed a broad absorption peak in the wavelength region 200–800 nm. The spectrum pattern and color of the catalyst were similar to those of NiO$_x$/SiO$_2$–Al$_2$O$_3$–MgO, synthesized by the oxidation of NiO/SiO$_2$–Al$_2$O$_3$–MgO using NaOCl. Thus, it can be deduced that the surface electronic state of the Au–NiO$_x$ catalyst differs from that of the Au-only catalyst as Ni was present in a highly oxidized state.

The representative Fourier transform-infrared (FT-IR) spectra of CO adsorbed on the catalysts were investigated. The Au catalyst showed an intense band attributed to Au^0–CO, at 2058 cm^{-1} [14]. In contrast, the Au–NiO$_x$ catalyst only showed a weak signal attributed to Ni^{2+}–CO, at 2111 cm^{-1} [15]. No peak corresponding to Au^0–CO could be observed.

Based on these results, the Au–NiO$_x$ nanoparticle was assumed to have a core–shell structure with Au nanoparticle at the core with its surface covered by highly oxidized NiO$_x$ (Fig. 2.4).

2.4 Industrial Catalyst and Catalytic Life Test

Stability and long life are the key requirements of an effective catalyst. We precisely controlled the distribution of Au–NiO$_x$ nanoparticles in the catalyst to decrease any loss of metals and to achieve high activity. Loss of metals from catalyst can occur

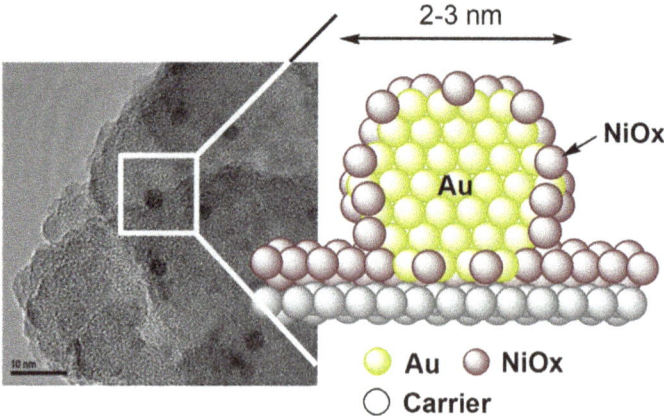

Fig. 2.4 Typical transmission electron microscopy images of Au–NiO$_x$ nanoparticle catalysts and proposed structure of nanoparticles

because of detachment or abrasion under reaction. Figure 2.5 shows the results of linear analysis on a particle cross section of a sample obtained by embedding Au–NiO$_x$/SiO$_2$–Al$_2$O$_3$–MgO in a resin followed by polishing using electron probe microanalysis (EPMA). The Au–NiO$_x$ layer was sharply distributed in a region within a

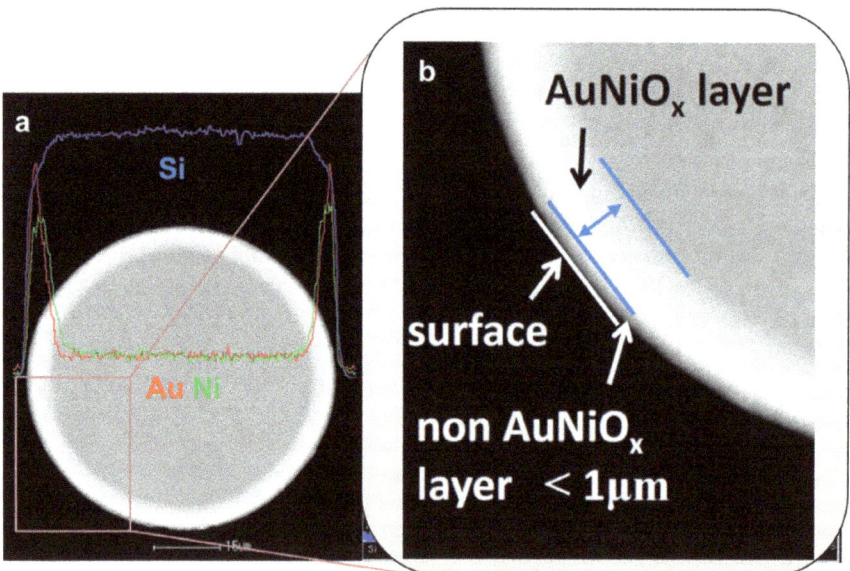

Fig. 2.5 Electron probe microanalysis spectra and elemental mapping measured for a single particle of Au–NiO$_x$/SiO$_2$–Al$_2$O$_3$–MgO

10-μm depth from the catalyst surface layer, and Au–NiO$_x$ was shifted by submicron from the surface of the catalyst to inside.

Details on the chemical state of Ni were investigated by using two-crystal high-resolution X-ray fluorescence (HRXRF) spectroscopy analysis. HRXRF has an extremely high energy-resolution capacity, and the chemical state can be analyzed from the energy state (chemical shift) or shape of the obtained spectrum. In particular, in the k spectrum of *3d* transition metal elements, a chemical shift or change in shape occurs because of a change in the valance or electronic state; a difference in the chemical state can be considered even though the valence is the same. The HRXRF results showed that Ni from the Au–NiO$_x$ catalyst can be considered to be in a high-spin bivalent state. The difference from the Ni Kα spectrum clearly shows that its chemical state is different from single NiO. NiO/SiO$_2$–Al$_2$O$_3$–MgO also showed a spectrum different from that for NiO powder. Thus, the Ni component of the Au–NiO$_x$ catalyst may form a composite with Au. In addition, free NiO may form a composite oxide or solid solution by reacting with metal component in the carrier.

The catalytic life of Au–NiO$_x$/SiO$_2$–Al$_2$O$_3$–MgO was evaluated by using a continuous-flow reaction apparatus (Fig. 2.6). When the conversion efficiency of the reaction was maintained at approximately 60%, **2a** was obtained with high selectivity (96–97%). No decrease in the catalyst activity or selectivity was observed over a period of 1000 h. Furthermore, Au leaching was negligible during prolonged reactions. The concentration of metal in the reaction mixture was determined to be less than 2.5 ppb by ICP-AES. TEM observation of the catalyst after the reaction showed

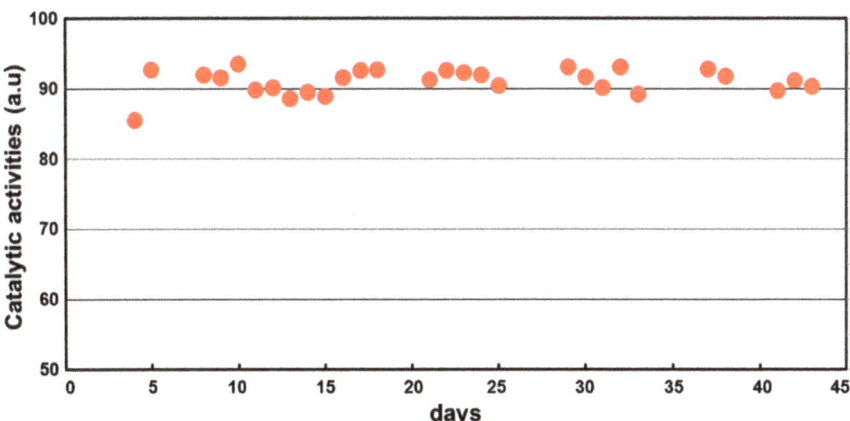

Fig. 2.6 Catalytic life test using a continuous-flow reaction apparatus (Reaction Conditions: The catalyst was packed in a 1.2-L stirring-type stainless-steel continuous reactor provided with a catalyst separator. A 37 wt% solution of **1a** in methanol was then continuously fed into the reactor at 0.6 L/h, and the reaction was carried out by blowing air in the reactor at 80 °C and 0.5 MPa until the exiting O$_2$ concentration became 4 vol%. Then, the mixture was stirred. The reaction product was continuously extracted from the reactor outlet by overflow; the reactivity was investigated by GC analysis, and the products were identified by GC–MS)

no sintering of the Au–NiO$_x$ nanoparticles. The TEM/STEM-EDX, UV–vis, and FT-IR results confirmed that the core–shell structure of Au–NiO$_x$ was preserved.

Evaluation of the catalyst life of Au/SiO$_2$–Al$_2$O$_3$–MgO showed that the activity decreased over time. After a reaction for 1000 h, the Au catalyst showed widening of the pore diameter of the carrier and sintering of Au. In the case of Au–NiO$_x$/SiO$_2$–Al$_2$O$_3$–MgO, NiO$_x$ was found to be present on the carrier unlike the Au–NiO$_x$ nanoparticles, and HRXRF confirmed that NiO$_x$ formed a composite compound or solid solution by reacting with the metal component of the carrier. The chemical stability of the carrier was assumed to be increased because of its action in stabilizing the Si–Al bridge structure. Stabilization of the carrier structure and anchor effect of NiO controlled the growth of the Au–NiO$_x$ nanoparticles and thus made it possible to greatly increase the catalyst life compared to that of the monometallic Au catalyst.

2.5 Practical Applicability

The practical applicability of this catalytic system was verified in a 100,000 ton/year MMA production plant in 2008. Thus, isobutene was oxidized in the gaseous phase using a Mo–Bi catalyst to synthesize methacrolein. Subsequent oxidative esterification of methacrolein in the presence of methanol using the Au–NiO$_x$ catalyst produced MMA. The block flow diagram of oxidative esterification process for MMA is shown in Fig. 2.7. This process confirmed the high selectivity, high activity, and long life of the Au–NiO$_x$ catalyst. This catalyst would help in saving energy and resources, in addition to being highly economical (Fig. 2.8).

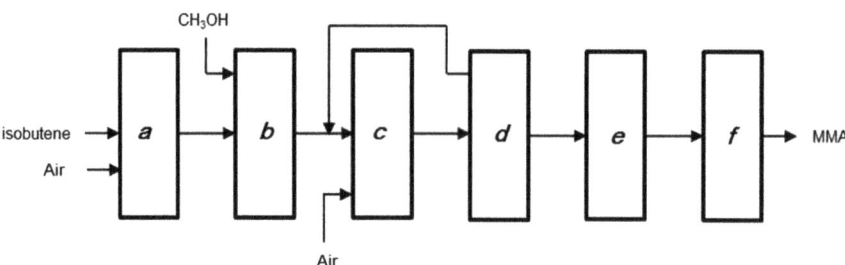

Fig. 2.7 The block flow diagram of oxidative esterification process for MMA. **a** Isobutene oxidation for methacrolein synthesis. **b** Methacrolein absorption by methanol. **c** Oxidative esterification for MMA synthesis. **d** Recovery column for unreacted methacrolein and methanol, and removal of low boiling point products. **e** High boiling point products separation tower. **f** MMA purification tower

Fig. 2.8 MMA production plant using Au–NiO$_x$ nanoparticle catalysts

2.6 Summary and Outlook

Oxidative esterification of aldehydes with alcohols proceeds with high efficiency in the presence of molecular oxygen, on supported gold–nickel oxide (Au–NiO$_x$) nanoparticle catalysts. The method is environmentally benign because it requires only molecular oxygen as the terminal oxidant and gives water as the side product. The Au–NiO$_x$ nanoparticles have a core–shell structure, with the Au nanoparticles at the core and the surface covered by highly oxidized NiO$_x$. This strategy provides an efficient and environmentally benign method for the synthesis of esters. The oxidation will be particularly important for exploring further aerobic catalytic oxidations.

References

1. Otera J (2003) Esterification: methods, reaction and applications. Wiley-VCH, Weinheim, Germany
2. Dehydrogenaton (a) Murahashi S-I, Naota T, Ito K, Maeda Y, Taki H (1987) J Org Chem 52:4319–4327. Oxidation with hydrogen peroxide (b) Gopinath R, Patel BK (2000) Org Lett 2:577–579. (c) Wu X-F, Darcel C (2009) Eur J Org Chem 1144–1147. (d) Gopinath R, Barkakaty B, Talukdar B, Patel BK (2003) J Org Chem 68:2944–2947. Oxidation with TBHP (e) Hashmi ASK, Lothschuetz C, Ackermann M, Doepp R, Anantharaman S, Marchetti B, Bertagnolli H, Rominger F (2010) Chem Eur J 16:8012–8019. Oxidation with benzyl chloride (f) Liu C, Tang S, Zheng L, Liu D, Zhang H, Lei A (2012) Angew Chem Int Ed 51:5662–5666. Reviews (g) Ekoue-Kovi K, Wolf C (2008) Chem Eur J 14:6302–6315

3. (a) Haruta M, Kobayashi T, Sano H, Yamada N (1987) Chem Lett 405–408. (b) Haruta M, Yamada N, Kobayashi T, Iijima S (1989) J Catal 115:301–309
4. Reviews (a) Hashmi ASK, Hutchings GJ (2006) Angew Chem Int Ed 45:7896–7936. (b) Arcadi A (2008) Chem Rev 108:3266–3325. (c) Li Z, Brouwer C, He C (2008) Chem Rev 108:3239–3265. (d) Pina CD, Falletta E, Prati L, Rossi M (2008) Chem Soc Rev 37:2077–2095. (e) Corma A, Garcia H (2008) Chem Soc Rev 37:2096–2126
5. (a) Marsden C, Taarning E, Hansen D, Johansen L, Klitgaard SK, Egeblad K, Christensen CH (2008) Green Chem 10:168–170. (b) Fristrup P, Johansen LB, Christensen CH (2008) Chem Commun 2750–2752. (c) Su F-Z, Ni J, Sun H, Cao Y, He H-Y, Fan K-N (2008) Chem Eur J 14:7131–7135. (d) Xu B, Liu X, Haubrich J, Friend CM (2009) Nat Chem 2:61–65
6. (a) Hayashi T, Inagaki T, Itayama N, Baba H (2006) Catal Today 117:210–213. (b) Nielsen IS, Taarning E, Egeblad K, Madsen R, Christensen CH (2007) Catal Lett 116:35–40. (c) Oliveira RL, Kiyohara PK, Rossi LM (2009) Green Chem 11:1366–1370. (d) Casanova O, Iborra S, Corma A (2009) J Catal 265:109–116. (e) Miyamura H, Yasukawa T, Kobayashi S (2010) Green Chem 12:776–778. (f) Costa VV, Estrada M, Demidova Y, Prosvirin I, Kriventsov V, Cotta RF, Fuentes S, Simakov A, Gusevskaya EV (2012) J Catal 292:148–156. (g) Kotionova T, Lee C, Miedziak P, Dummer NF, Willock DJ, Carley AF, Morgan DJ, Knight DW, Taylor SH, Hutchings G (2012) J Catal Lett 142:1114–1120
7. Suzuki K, Yamaguchi T, Matsushita K, Iitsuka C, Miura J, Akaogi T, Ishida H (2013) ACS Catal 3:1845–1849
8. Nagai K (2001) Appl Catal A 221:367–377
9. (a) Yamamatsu S, Yamaguchi T, Yokota K, Nagano O, Chono M, Aoshima A (2010) Catal Surv Asia 14:124–131. (b) Diao Y, Yan R, Zhang S, Yang P, Li Z, Wang L, Dong H (2009) J Mol Catal A Chem 303:35–42. (c) Wang B, Sun W, Zhu J, Ran W, Chen S (2012) Ind Eng Chem Res 51:15004–15010
10. Nakagawa K, Konaka R, Nakata T (1962) J Org Chem 27:1597–1601
11. (a) Choudary BM, Kantam ML, Rahman A, Reddy ChV, Rao KK (2005) Angew Chem Int Ed 40:763–766. (b) Ji H, Wang T, Zhang M, She Y, Wang L (2005) Appl Catal A 282:25–30
12. Ferreira FF, Fantini MCA (2003) J Phys D Appl Phys 36:2386–2392
13. The reaction of **1a** (15 mmol) in the presence of $NiO_2 \cdot nH_2O$ (0.3 g) in methanol at 80 °C under an oxygen-nitrogen mixture (7:93 (v/v), 3 MPa) for 2 h gave a trace of **2a** (1.5 μmol)
14. Mihaylov M, Knoezinger H, Hadjiivanov K, Gates BC (2007) Chem Ing Tech 79:795–806
15. Estrella Platero E, Scarano D, Zecchina A, Meneghini G, De Franceschi R (1996) Surf Sci 350:113–122

Chapter 3
Core–Shell Nanostructured Catalysts for Chemoselective Hydrogenations

Takato Mitsudome

3.1 Introduction

Selective hydrogenation is an important aspect of fine chemical synthesis. However, the chemoselective hydrogenation of targeted functional groups in the presence of other reducible groups, for example, alkene moieties, remains a significant challenge. Recent advances in nanoengineering have enabled the precise synthesis of metal nanoparticle (NP) catalysts [1–5]. Sophisticated metal NP catalysts have been produced with controlled particle size and shape, tuned metal electronic states, and modulated metal–support interactions, and have been used to improve the selectivity of various chemoselective hydrogenations. This section presents the recent progress in core–shell metal NP catalysts for chemoselective hydrogenations. These focused core–shell metal NPs exhibit high catalytic performance in the chemoselective hydrogenations of nitrostyrene, epoxides, unsaturated carbonyl compounds, and alkynes while preserving alkene groups, thereby overcoming the limitations of conventional metal NP catalysts. Furthermore, a green synthetic method for core–shell NP catalysts and their catalytic performances have also been described.

T. Mitsudome (✉)
Department of Materials Engineering Science, Graduate School of Engineering Science, Osaka University, 1-3 Machikaneyama, Toyonaka 560-8531, Osaka, Japan
e-mail: mitsudom@cheng.es.osaka-u.ac.jp

3.2 Design of Core-Metal-NP/Shell-Metal-Oxide Catalyst to Maximize Metal–Support Interaction

Rational control of metal-support interaction in the design of heterogeneous catalysts (e.g., metal oxide-supported metal NP catalysts) is one of the most important factors affecting catalyst performance. A core–shell morphology comprising an active metal NP core and metal oxide shell maximizes the metal NP-support interface area. Our research demonstrated that maximizing metal-support interaction via encapsulation of metal NPs with metal can lead to high chemoselective hydrogenation activity and selectivity. The resulting silver (Ag) and gold (Au) NPs wrapped by cerium oxide (CeO_2) outperformed conventional metal oxide-supported metal NP catalysts in various chemoselective hydrogenations of nitrostyrene, unsaturated carbonyl compounds, epoxides, and alkynes while retaining the C=C bonds, which overcomes the limitations of conventional metal NP catalysts.

3.3 Core-Ag NP/Shell-CeO_2 Catalyst (Ag@CeO_2) for Chemoselective Hydrogenation of Nitrostyrenes and Epoxides

Aniline derivatives are valuable synthetic intermediates in the production of agrochemicals, pharmaceuticals, and dyes [6]. The reduction of an aromatic nitro compound is the most straightforward method to synthesize the corresponding aniline [7]. However, targeted reduction of the nitro functionality of a mother nitro compound containing other reducible functionalities (e.g., C=C bonds) is difficult [8]. The chemoselective reduction of nitro compounds to the corresponding anilines in the presence of C=C bonds was conducted via stoichiometric reactions with excess amounts of Fe, Sn, Zn, and NaS_2O_4 [9], which produced large amounts of harmful waste with low atom-efficiencies. More practical and environmentally friendly approaches have been proposed using efficient catalysts for the chemoselective reduction of nitroaromatics with C=C double bonds [10]. To date, TiO_2-supported gold NPs (Au/TiO_2) may be the best-supported metal NP catalyst for the chemoselective reduction of nitro functionalities, where 3-nitrostyrene was reduced to the desired 3-vinylaniline with 98.5% conversion and 95.9% selectivity [11]. However, conventional metal NP catalysts, including Au/TiO_2, are associated with successive hydrogenation of C=C double bonds in vinylaniline to ethylaniline at a high conversion level, thereby gradually decreasing the selectivity for vinylaniline over an extended reaction time. Therefore, a new strategy for the design of a catalyst for the chemoselective hydrogenation of nitrostyrene with complete suppression of C=C double bond hydrogenation is required.

Ag NPs on a basic hydrotalcite support (Ag/HT) have been previously applied to catalyze the chemoselective reduction of nitrostyrenes [12] and epoxides [13, 14] to the corresponding anilines and alkenes using alcohols or CO/H_2O as a reducing

Fig. 3.1 Schematic of
Ag/HT-catalyzed
chemoselective reductions
with alcohols or CO/H$_2$O.
Reprinted with permission
from Ref. [15]. Copyright
2012 Wiley–VCH

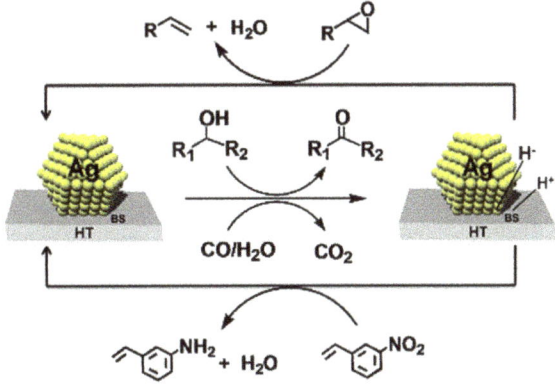

reagent, where complete retention of the reducible C=C bonds was achieved. During the reduction reaction, polar hydrogen species, namely the hydride and protons, were formed in situ at the Ag NPs − HT interface via cooperative catalysis between the Ag NPs and the basic sites (BS) of HT. The polar hydrogen species were exclusively active for the reduction of the polar functional groups, and inactive for the non-polar C=C bonds (Fig. 3.1), thereby enabling high chemoselectivity. However, the use of H$_2$ instead of alcohols or CO/H$_2$O in the Ag-catalyst system cause reduction in the polar groups (nitro and epoxide) and the non-polar C=C bonds. This non-selectivity was attributed to the formation of non-polar hydrogen species via the homolytic cleavage of H$_2$ at the surface of the Ag NPs, which were active for C=C bond reduction (Fig. 3.2a). Therefore, more efficient Ag catalysts suitable for use with H$_2$ are needed. Specifically, the number of interface sites between the Ag NPs and basic sites of the support (active toward polar groups) should be increased while

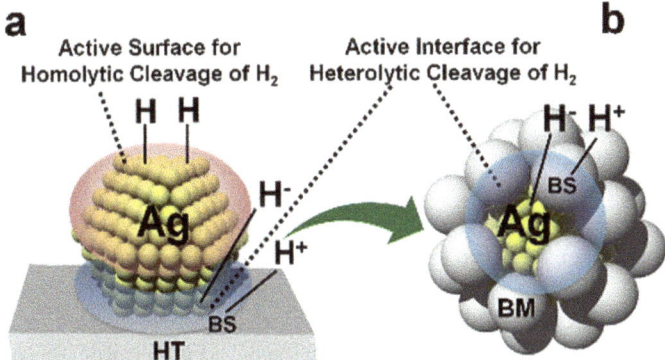

Fig. 3.2 Design of core–shell nanocomposite catalyst for chemoselective hydrogenation. **a** Ag/HT reacting with H$_2$ to form polar and non-polar hydrogen species. **b** Ag NPs encapsulated by a basic material (BM) (Ag@BM) reacting with H$_2$ for the exclusive formation of polar hydrogen species. BS represents a basic site. Reprinted with permission from Ref. [15]. Copyright 2012 Wiley–VCH

simultaneously decreasing the number of bare Ag NP surface sites (active toward C=C bonds).

A core–shell Ag@BM nanocomposite comprising Ag NPs encapsulated by basic metal oxides (BM) was designed to perform complete chemoselective reduction (Fig. 3.2b). The Ag@BM structure maximized the Ag NPs—BM interface area— while minimizing the area of bare Ag NPs. This enabled the exclusive formation of heterolytically cleaved hydrogen species, which relied on the combined action of the Ag NPs and basic sites of BM. Further, the unfavorable formation of homolytically cleaved hydrogen species on the bare Ag NPs was minimized. The Ag-hydride and proton species allowed for complete chemoselective reduction of polar functionalities while retaining the C=C bonds.

The facile synthesis of a core-Ag NP/shell-CeO$_2$ nanocomposite (Ag@CeO$_2$) based on this catalyst design concept was proposed, which involved the reverse micelle technique and the redox reaction between Ag (I) and Ce (III) [15]. The redox reaction between Ag (I) and Ce (III) to Ag (0) and Ce (IV) occurred spontaneously to form Ag@CeO$_2$ in a single step. The method did not require the use of additional reductants, and yielded small NPs (<10 nm) suitable for catalytically active metals. Further, the nanoporous shell facilitated the diffusion of reactants into the active core metal. This preparation method was straightforward compared to previously reported methods for the synthesis of core–shell NPs, which often require multiple steps and result in the formation of larger metal NPs (>10 nm). Scanning electron microscopy (SEM), transmission electron microscopy (TEM), energy-dispersive X-ray spectroscopy (EDS), and X-ray absorption fine structure (XAFS) analyses of Ag@CeO$_2$ revealed core-Ag NPs (mean diameter = 10 nm) encapsulated in a shell of assembled spherical CeO$_2$ NPs (diameter = 3–5 nm) (Fig. 3.3).

Ag@CeO$_2$ exhibited excellent selectivity for the chemoselective hydrogenation of 3-nitrostyrene (1) to 3-aminostyrene (2) (98% yield) without reduction of the C=C bond to 3-ethylaniline (3) under 6 atm of H$_2$ at 110 °C (Fig. 3.4a). Unlike previously reported catalysts, the C=C bond of 2 remained intact, even during a prolonged reaction time after complete conversion of 1. In contrast to the Ag@CeO$_2$ catalyst, non-encapsulated conventional Ag NPs supported on CeO$_2$ (Ag/CeO$_2$) exhibited low chemoselectivity toward 2 due to successive hydrogenation of 2 to 3 (Fig. 3.4b).

The advantage of the core–shell structure of Ag@CeO$_2$ was clearly demonstrated during hydrogenation testing with styrene under the same conditions used for the hydrogenation of nitrostyrene (Scheme 3.1). Ag@CeO$_2$ did not hydrogenate the C=C bond of styrene at all, whereas Ag/CeO$_2$ did, thereby proving that Ag@CeO$_2$ exclusively hydrogenated nitro groups did not exhibit activity for the C=C bond.

Ag@CeO$_2$ also performed well in the hydrogenation of other nitro compounds to form the corresponding aniline derivatives with high yields with a selectivity of >99%, while retaining C=C double bonds (Table 3.1). Furthermore, the Ag@CeO$_2$ particles were easily recovered after use via simple filtration from the reaction mixture, and the high chemoselectivity was maintained after repeated use (Table 3.1, entries 2 and 3). Inductively coupled plasma-atomic emission spectrometry (ICP-AES) analysis of the filtrate revealed the absence of Ag species, while the TEM and XAFS analyses revealed that the core–shell morphology was maintained. Further,

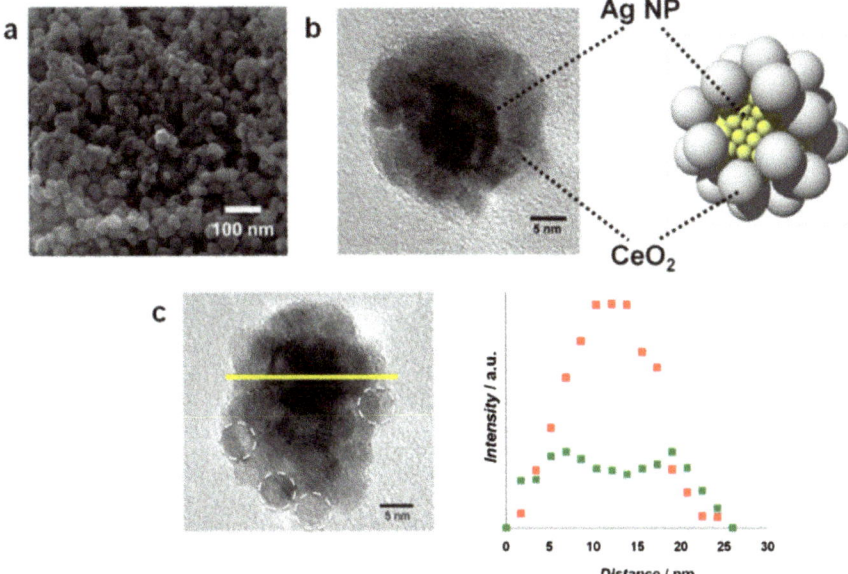

Fig. 3.3 **a** SEM image, **b** HRTEM image, and **c** line-scan STEM-EDS of Ag@CeO$_2$ (Ag = red; Ce = green). Reprinted with permission from Ref. [15]. Copyright 2012 Wiley–VCH

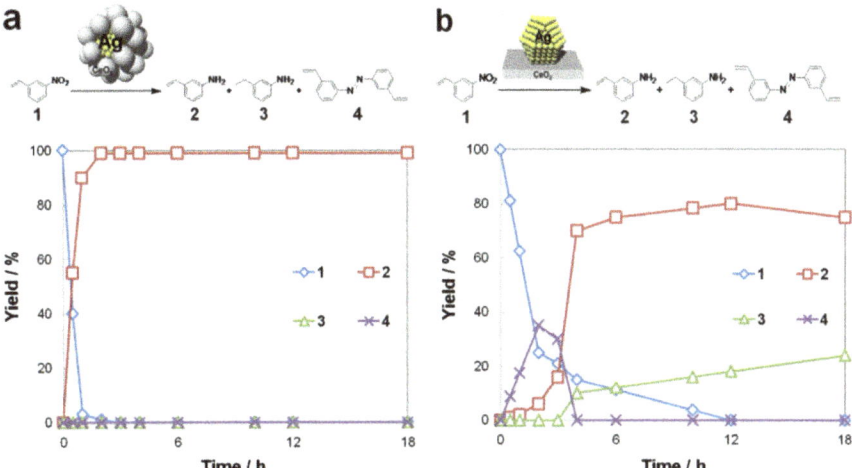

Fig. 3.4 Hydrogenation of 3-nitrostyrene over time using **a** Ag@CeO$_2$ and **b** Ag/CeO$_2$. Reprinted with permission from Ref. [15]. Copyright 2012 Wiley–VCH

Scheme 3.1 Hydrogenation activity test of Au@CeO$_2$ and Ag/CeO$_2$ using styrene

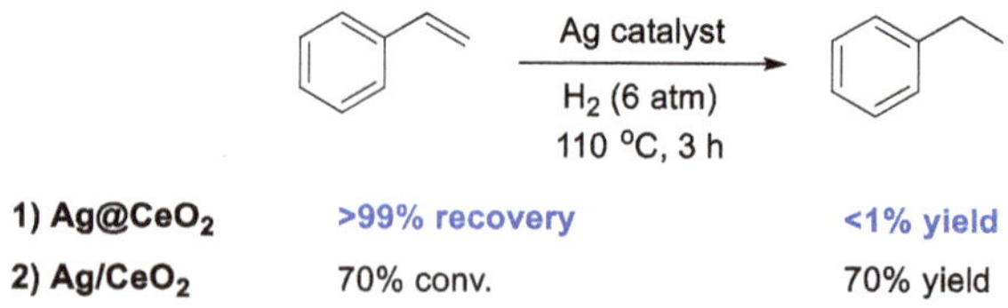

	>99% recovery	<1% yield
1) Ag@CeO$_2$	>99% recovery	<1% yield
2) Ag/CeO$_2$	70% conv.	70% yield

Table 3.1 Chemoselective hydrogenation of nitro compounds using Ag@CeO$_2$

Entry	Substrate	Product	Time (h)	Yield (%)	Sel. (%)
1	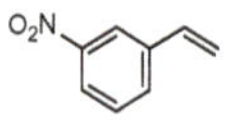	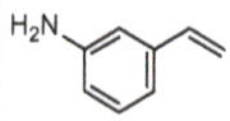	6	98	>99
2[a]	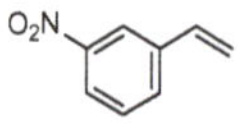	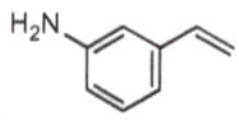	6	97	>99
3[b]	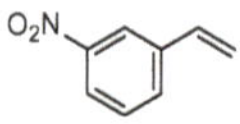	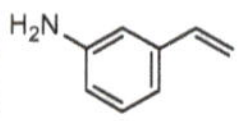	6	97	>99
4	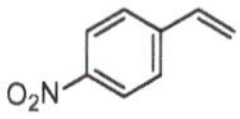	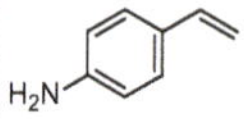	6	98	>99
5[c]	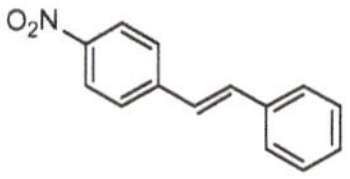	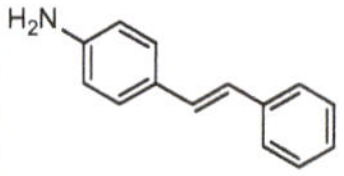	24	97	>99
6[c]	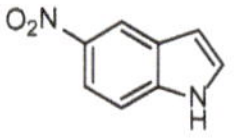	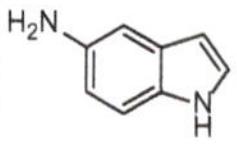	24	95	>99

Reaction conditions: Ag@CeO$_2$ (25 mg), substrate (0.5 mmol), dodecane (5 mL), H$_2$ (6 atm), 110 °C. [a]Reuse 1. [b]Reuse 2. [c]150 °C

the size and oxidation state of the Ag NPs in the core did not change after the reaction. These findings demonstrated the high durability of Ag@CeO$_2$.

Ag@CeO$_2$ also promoted the unique deoxygenation of epoxides to give the corresponding alkenes. This reaction has attracted attention in both organic synthesis and biological chemistry, as it plays a role in the protection-deprotection cycle of carbon–carbon double bonds [16–18] and in the production of vitamin K in the human body [19, 20]. Generally, stoichiometric deoxygenation of epoxides is conducted using expensive and toxic reagents such as low-valent metals, iodides, phosphines, and silane compounds. However, these systems involve tedious work-up procedures, air- and moisture-sensitive reaction conditions, low atom efficiency, and low catalytic activities. Therefore, there remains a need for highly efficient and environmentally friendly catalytic deoxygenation of epoxides.

Various epoxides, including aromatic, aliphatic, and alicyclic epoxides, have been successfully converted to the corresponding alkene with >99% selectivity (Table 3.2).

Table 3.2 Selective hydrogenation of epoxides to alkenes using Ag@CeO$_2$[a]

Entry	Substrate	Product	Time (h)	Yield (%)[b]	Sel. (%)[b]
1			6	97	>99
2			6	95	>99
3			12	98	>99
4			12	96	>99 E:Z = 3:1
5			24	96	> 99
6			24	95	>99
7			24	94	>99
8			6	96	>99

[a]Reaction conditions: Ag@CeO$_2$ (25 mg), substrate (0.5 mmol), toluene (5 mL), 110 °C, H$_2$ (6 atm).
[b]Determined by GC and LC using an internal standard

This is the first report on Ag-catalyzed chemoselective deoxygenation of epoxides to alkenes using H$_2$.

3.4 Highly Dispersed Ag@CeO$_2$ on Solid Support for Chemoselective Hydrogenation of Unsaturated Aldehydes

Ag@CeO$_2$ is an agglomerated powder, which results in partial covering of the pores between the CeO$_2$ NPs in the shell. This pore shielding inhibits diffusion of the substrate to the Ag NPs in the core, leaving some catalytically active Ag NP species ineffective. This agglomeration was prevented when the Ag@CeO$_2$ particles were highly dispersed on inorganic supports [21]. Ag@CeO$_2$ was successfully dispersed on a CeO$_2$ support (Ag@CeO$_2$-D), as visualized using SEM with elemental

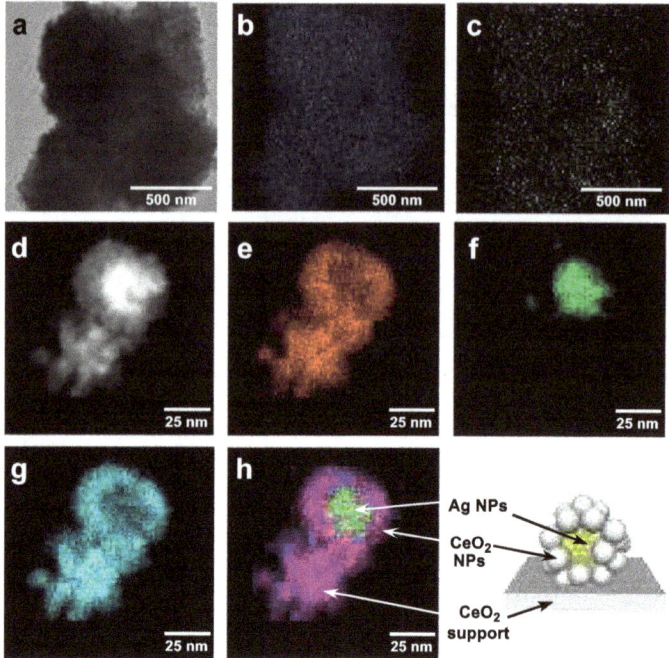

Fig. 3.5 a STEM images of Ag@CeO$_2$-D and corresponding elemental maps of **b** Ce and **c** Ag in the sample. **d** HAADF-STEM image of Ag@CeO$_2$-D and the corresponding elemental maps of **e** O, **f** Ag, **g** Ce, and **h** their overlap in the sample. Reprinted with permission from Ref. [21]. Copyright 2013 Wiley–VCH

mapping (Fig. 3.5a–c). A representative high-angle annular dark-field (HAADF) image and corresponding elemental maps revealed immobilized Ag@CeO$_2$ on the CeO$_2$ support, where the original core–shell structure of the catalyst was maintained (Fig. 3.5d–h).

The applicability of this Ag@CeO$_2$ design strategy for exclusive hydrogenation of polar functional groups while retaining alkene groups was extended by using Ag@CeO$_2$-D for the selective hydrogenation of unsaturated aldehydes to the corresponding unsaturated alcohols. This is an important reaction as unsaturated alcohol products are highly useful intermediates in fragrances and pharmaceutical products. Ag@CeO$_2$-D exhibited excellent selectivity for the desired unsaturated alcohols, where various aldehydes, including terpenes, aliphatic, aromatic α,β-unsaturated aldehydes, and unconjugated aldehydes, were efficiently converted to allylic alcohols with high selectivity (Fig. 3.6). Furthermore, the high dispersion of the agglomerated Ag@CeO$_2$ greatly enhanced its catalytic activity, and a sixfold increase in turnover frequency was achieved. Ag@CeO$_2$-D performed well under gram-scale reaction conditions to produce a high yield of unsaturated alcohols, thereby demonstrating the high stability of Ag@CeO$_2$-D. These results clearly demonstrated the potential of core–shell NP Ag@CeO$_2$ and this catalyst design strategy for the chemoselective

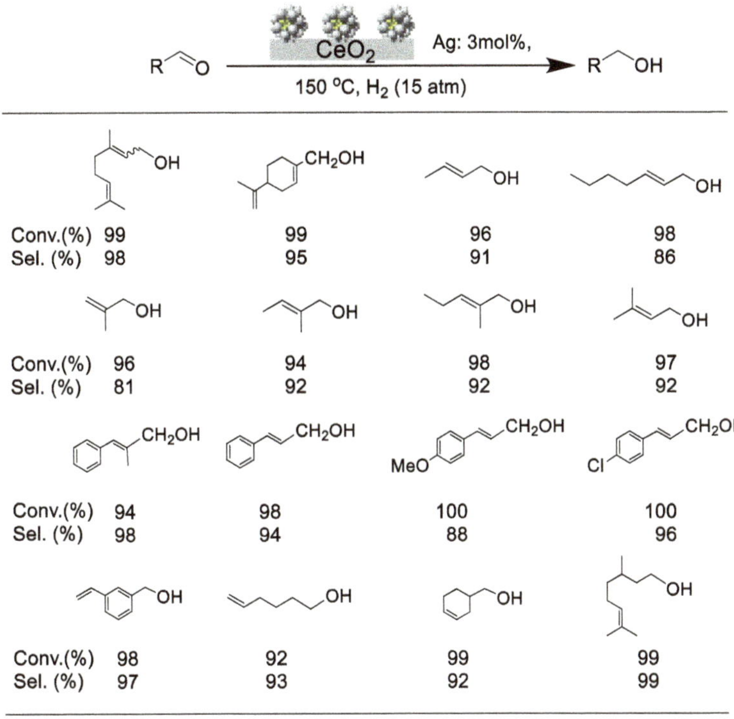

Fig. 3.6 Chemoselective hydrogenations of unsaturated aldehydes using Ag@CeO$_2$-D

hydrogenation of polar functional groups while maintaining C=C bonds. This design concept maximized the interface interaction between the Ag NPs and basic sites of CeO$_2$, which played a key role in inducing the heterolytic cleavage of H$_2$.

3.5 Core-Au NP/Shell-CeO$_2$ Catalyst (Au@CeO$_2$) for Selective Semihydrogenation of Alkynes

Core–shell metal NP catalysts have been synthesized via the seeded-growth method [22–25], galvanic replacement method [26–28], and decomposition method [29–31]. These methods allow for precise synthesis of core–shell NP catalysts, but are complicated and often require multiple time-consuming steps. In contrast, the redox-coprecipitation method used to produce the Ag@CeO$_2$ catalyst allowed for facile synthesis of core–shell metal NPs, where a redox reaction between the core and shell precursors facilitated spontaneous formation of the core–shell metal NPs in a single step. Redox-coprecipitation in a reverse micelle solution can be applied to the synthesis of Au@CeO$_2$ with a AuNPs core and CeO$_2$ NPs (mean diameter = 2 nm) assembled in a shell structure (Fig. 3.7) [32].

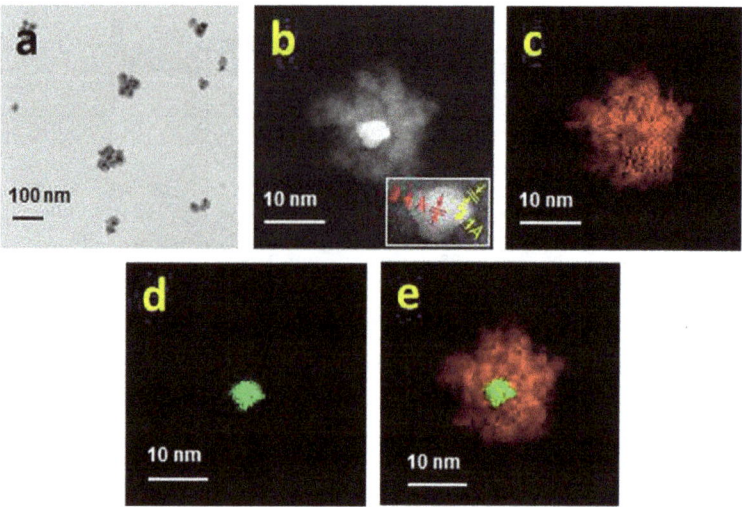

Fig. 3.7 **a** TEM and **b** HAADF-STEM images of Au@CeO$_2$ (inset: HRTEM showing lattice fringes), with corresponding elemental maps of **c** Ce, **d** Au, and **e** their overlap in the sample. Reprinted with permission from Ref. [32]. Copyright 2015 American Chemical Society

The catalytic potential of Au@CeO$_2$ was investigated for the semihydrogenation of alkynes for the synthesis of (Z)-alkenes. These products are important building blocks of fine chemicals, including bioactive molecules, flavors, and natural products. Lindlar catalyst, namely Pb(OAc)$_2$-treated Pd/CaCO$_3$ with large amounts of quinoline, has been widely used in these reactions [33]. However, this method has several drawbacks, including the use of a toxic Pb salt and large amount of quinoline to suppress the overhydrogenation of alkenes. Furthermore, the use of Lindlar catalysts to treat terminal alkynes results in rapid overhydrogenation of terminal alkene products. Therefore, the development of a sustainable and environmentally friendly Pb-free semihydrogenation alternative has attracted significant attention.

Au NP catalyst systems with toxic hydride reagents are often required to achieve high selectivity for alkenes [34–36]. Thus, the combination of Au NPs with hydride is vital for achieving high selectivity for alkenes. However, a previous study found that Au or Ag NPs and basic metal oxides cooperatively dissociate H$_2$ to polar hydrogen species, namely H$^{\delta+}$ and H$^{\delta-}$, which straddle the interfacial perimeter sites [15, 21, 37]. Therefore, the catalytic potential of Au@CeO$_2$ was investigated for semihydrogenation of alkynes, where the core–shell structure maximized the Au NP-CeO$_2$ interfacial sites to facilitate heterolytic dissociation of H$_2$ into polar hydrogen species. The resulting polar hydrogen species favored alkynes over alkenes, thus enabling highly selective semihydrogenation of alkynes.

The catalytic potential of Au@CeO$_2$ was investigated for the hydrogenation of phenylacetylene at room temperature under 30 atm H$_2$. The time profile (Fig. 3.8a) indicated a 89% yield of styrene using Au@CeO$_2$ with >99% selectivity after 12 h, where the high selectivity was maintained over a prolonged reaction time after full

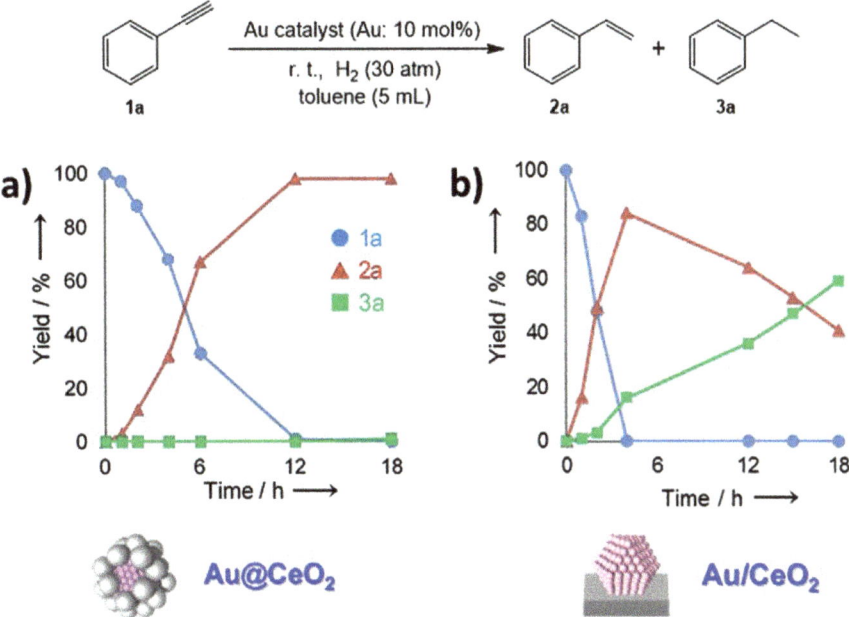

Fig. 3.8 Phenylacetylene hydrogenation over time using **a** Au@CeO₂ and **b** Au/CeO₂. Reprinted with permission from Ref. [32]. Copyright 2015 American Chemical Society

consumption of phenylacetylene. In contrast, conventional CeO₂-supported Au NPs (Au/CeO₂) without the core–shell structure exhibited a lower styrene yield, and the selectivity gradually decreased as overhydrogenation of styrene occurred to form ethylbenzene (Fig. 3.8b). The differences between the performance of Au@CeO₂ and Au/CeO₂ clearly demonstrated the key role of the core–shell structure in achieving highly selective semihydrogenation.

The substrate applicability of Au@CeO₂ for semihydrogenation is summarized in Table 3.3, where aromatic and aliphatic terminal alkynes were transformed to the corresponding alkenes with >99% selectivity at a high conversion level. This excellent selectivity for terminal alkenes is unique and valuable, as conventional catalysts such as the Lindlar catalyst causes rapid overhydrogenation of terminal alkenes to alkanes at a high conversion level. Consequently, the hydrogen uptake of these conventional processes must be closely monitored to prevent the formation of alkanes. Various internal alkynes also yielded alkenes with >99% selectivity, while reducible moieties, including halogen, methoxy, benzyl, cyano, hydroxyl, and ester groups, were completely intact under these conditions. These findings demonstrated that semihydrogenation of alkynes using Au@CeO₂ is a promising method for the synthesis of various functionalized alkenes from alkynes. This is the first example of employing an Au catalyst for the selective semihydrogenation of alkynes at ambient temperature under additive-free conditions.

Table 3.3 Selective semihydrogenation of alkynes using Au@CeO$_2$[a]

Substrate	Time (h)	Yield (%)[b]	Substrate	Time (h)	Yield (%)[b]
(phenylacetylene)	12	98	(cyclohexylacetylene)	12	72
Cl—(4-chlorophenylacetylene)	10	96	NC~~~(nitrile alkyne)	20	97
(benzyl propargyl)	45	99	(isoprenol alkyne) HO	7	>99
(benzyl ether alkyne) O~	24	95	—CO$_2$Et	18	>99
N (pyridyl alkyne)	24	99	—CO$_2$Et	20	>99
(decyne chain)	24	95	HO (alkyne)	15	93

[a]Reaction conditions: Ag@CeO$_2$ (Au: 10–16 mol%), substrate (0.5 mmol), toluene (5 mL), rt, H$_2$ (25–50 atm). [b]Determined by GC and LC using an internal standard

3.6 Environmentally Friendly One-Step Synthesis of Hydrotalcite-Supported Au@CeO$_2$ in Water

The redox-coprecipitation method in a reverse micelle solution is a useful method for the single synthesis of core–shell NP catalysts (Fig. 3.9a). However, the unavoidable use of surfactants, organic solvents, and additional bases leads to economic and

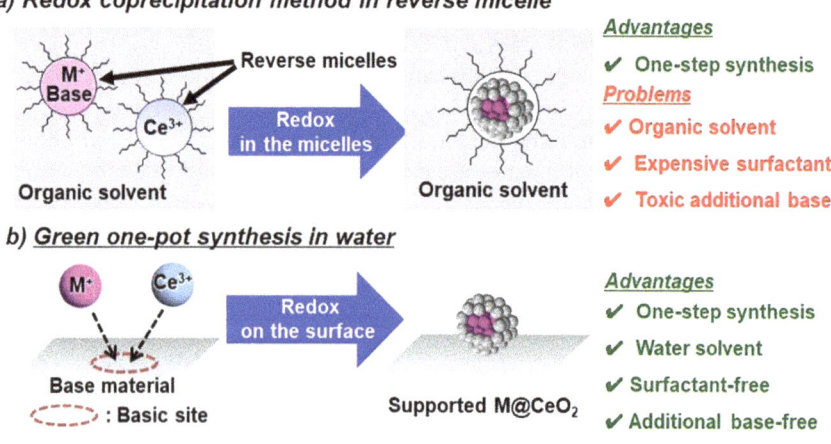

Fig. 3.9 Comparison of the synthetic methods of core–shell NPs

environmental issues, where the large amounts of waste and high energy consumption have limited wider implementation of core–shell NPs. Alternatively, core–shell NPs have been produced using a novel and environmentally friendly strategy without the use of organic reagents under neutral conditions (Fig. 3.9b) [38]. The core and shell metal precursors were simply mixed in water in the presence of metal oxides with basic properties, which led to one-step fabrication of core–shell NPs without the use of reductants, surfactants, organic solvents, or additional toxic bases. The basic sites on the metal oxides served as a nanoscale reaction site to promote the redox reaction between the core and shell precursors, leading to core–shell NPs highly dispersed on the solid surface.

Core–shell NPs were synthesized using $HAuCl_4$, $Ce(NO_3)_3$, and hydrotalcite (HT: $Mg_6Al_2(CO_3)(OH)_{16}\cdot4(H_2O)$) as the core, shell precursors, and basic metal oxides, respectively. Specifically, a mixed aqueous solution of $HAuCl_4$ and $Ce(NO_3)_3$ was added dropwise to an aqueous suspension of HT, yielding HT-supported core-Au NP/Shell-CeO_2 (Au@CeO_2/HT) in a single step. Microscope images revealed small core–shell NPs (Fig. 3.10) suitable for catalytic applications due to immediate immobilization of the in situ generated core–shell NPs on the metal oxide without overgrowth. Furthermore, the metal oxide-supported core–shell NPs were retrieved via simple filtration, and the conventional calcination process to remove residual surfactants was not necessary. Au@CeO_2/HT was a highly efficient and reusable heterogeneous catalyst in a series of highly chemoselective hydrogenation reactions of unsaturated aldehydes, alkynes, and epoxides, where the desired products were obtained with over 99% selectivity while preserving the C=C bonds under additive-free conditions. Au@CeO_2/HT also performed well in gram-scale reactions (Scheme 3.2), where 20 mmol of the substrate was chemoselectively hydrogenated with an excellent yield of the desired product. Moreover, the Au@CeO_2/HT catalyst was recovered easily via simple filtration and reused without loss of catalytic activity or selectivity. Overall, the synthesis and catalytic performance of Au@CeO_2/HT allowed for a lower environmental impact and reduced energy consumption throughout the process, including catalyst preparation, reaction, separation, and reuse.

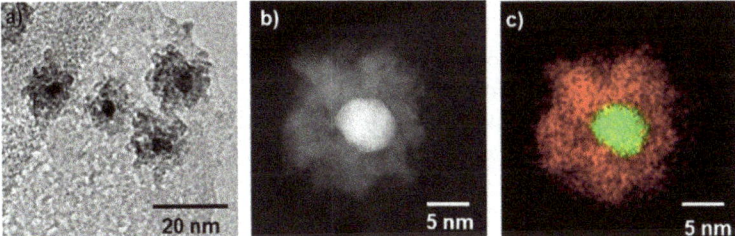

Fig. 3.10 **a** TEM and **b** HAADF-STEM image of Au@CeO_2/HT (inset: HR-TEM showing the lattice fringes) with corresponding elemental mapping images of **c** Au (green) and Ce (red)

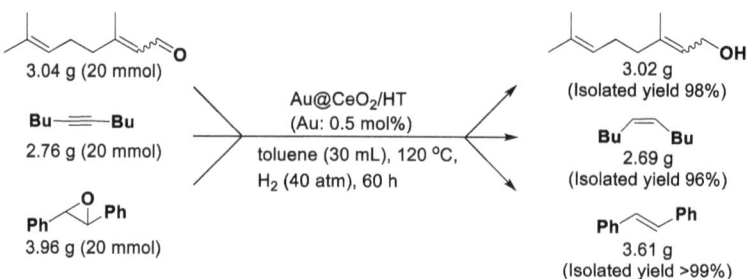

Scheme 3.2 Gram-scale reactions using Au@CeO₂/HT

3.7 Summary

This section introduced core–shell nanostructured catalysts for chemoselective hydrogenation. The core–shell metal NP catalysts comprised an Ag or Au NPs core and a shell of assembled CeO_2 NPs. The catalysts exhibited high potential for durability against aggregation, as well as excellent chemoselectivity for hydrogenations of various functional groups such as nitro, epoxide, aldehyde, and alkynes while preserving C=C bonds.

The core–shell metal NPs included a maximized active interfacial area between the core metal and the shell support for enhanced catalytic performance. Specifically, the Ag or Au@CeO₂ structure was designed to maximize the area of the interface between the Ag or AuNPs and CeO_2, while minimizing the area of bare Ag or Au NPs. This enabled the exclusive formation of heterolytically cleaved hydrogen species via combined action of the Ag or Au NPs and the basic sites of CeO_2. Further, the unfavorable formation of homolytically cleaved hydrogen species on the bare Ag or Au NPs was suppressed. The resulting metal-hydride and proton species facilitated complete chemoselective hydrogenation of the polar functionalities while retaining the nonpolar C=C bonds. Key aspects of these newly developed core–shell NP catalysts included: (i) high catalytic activity and chemoselectivity, (ii) high atom efficiency, (iii) reusability, and (iv) broad substrate scope.

Redox-coprecipitation in a reverse micelle solution was straightforward method for the one-step synthesis of core–shell NP catalysts. This method did not require the use of additional reductants. Further, the small size of the NPs was suitable for a catalytically active metal and a nanoporous shell, which facilitated the diffusion of reactants into the active core metal. This redox-coprecipitation preparation method allowed for novel environmentally friendly one-step synthesis of small core–shell NPs in water, where no surfactants, organic solvents, reductants, or additives were required. The resulting core–shell NP catalyst exhibited high catalytic performance and reusability. Consequently, high environmental compatibility and energy efficiency were achieved throughout the process, including catalyst preparation, reaction, separation, and reuse. This simple and powerful method is expected to facilitate the clean synthesis of other core–shell NPs.

References

1. Tauster SJ, Fung SC, Baker RT, Horsley JA (1981) Strong interactions in supported-metal catalysts. Science 211:1121–1125
2. Schubert MM, Hackenberg S, van Veen AC, Muhler M, Plzak V, Behm RJ (2001) CO oxidation over supported gold catalysts—"inert" and "active" support materials and their role for the oxygen supply during reaction. J Catal 197:113–122
3. Bell AT (2003) The impact of nanoscience on heterogeneous catalysis. Science 299:1688–1691
4. Tian N, Zhou ZY, Sun SG, Ding Y, Wang ZL (2007) Synthesis of tetrahexahedral platinum nanocrystals with high-index facets and high electro-oxidation activity. Science 316:732–735
5. Somorjai GA, Park JY (2008) Molecular factors of catalytic selectivity. Angew Chem Int Ed 47:9212–9228
6. Booth G (2002) Ullmanns encyclopedia of industrial chemistry. Wiley-VCH Verlag, Weinheim, Germany
7. Blaser HU, Siegrist U, Steiner H, Studer M (2001) In: Sheldon RA, van Bekkum H (eds) Fine chemicals through heterogeneous catalysis. Wiley-VCH, Weinheim, p 389
8. Houben-Weyl (1980) Methoden der Organischen Chemie. Thieme Verlag Stuttgart, 4/1c, p 511
9. Downing RS, Kunkeler PJ, van Bekkum H (1997) Catalytic syntheses of aromatic amines. Catal Today 37:121–136
10. Blaser HU, Steiner H, Studer M (2009) Selective catalytic hydrogenation of functionalized nitroarenes: an update. ChemCatChem 1:210–221
11. Corma A, Serna P (2006) Chemoselective hydrogenation of nitro compounds with supported gold catalysts. Science 313:332–334
12. Mikami Y, Noujima A, Mitsudome T, Mizugaki T, Jitsukawa K, Kaneda K (2010) Highly chemoselective reduction of nitroaromatic compounds using a hydrotalcite-supported silver-nanoparticle catalyst under a CO atmosphere. Chem Lett 39:223–225
13. Mikami Y, Noujima A, Mitsudome T, Mizugaki T, Jitsukawa K, Kaneda K (2010) Selective deoxygenation of styrene oxides under a CO atmosphere using silver nanoparticle catalyst. Tetrahedron Lett 51:5466–5468
14. Mitsudome T, Noujima A, Mikami Y, Mizugaki T, Jitsukawa K, Kaneda K (2010) Supported gold and silver nanoparticles for catalytic deoxygenation of epoxides into alkenes. Angew Chem Int Ed 49:5545–5548
15. Mitsudome T, Mikami Y, Matoba M, Mizugaki T, Jitsukawa K, Kaneda K (2012) Design of a silver-cerium dioxide core-shell nanocomposite catalyst for chemoselective reduction reactions. Angew Chem Int Ed 51:136–139
16. Corey EJ, Su WG (1987) Total synthesis of a C15 ginkgolide, (.+-.)-bilobalide. J Am Chem Soc 109:7534–7536
17. Kraus GA, Thomas PJ (1988) Synthesis of 7,7,8-trideuteriated trichothecenes. J Org Chem 53:1395–1397
18. Johnson WS, Plummer MS, Reddy SP, Bartlett WR (1993) The fluorine atom as a cation-stabilizing auxiliary in biomimetic polyene cyclizations. 4. Total synthesis of dl-ß-amyrin. J Am Chem Soc 115:515–521
19. Silverman RB (1981) Model studies for a molecular mechanism of action of oral anticoagulants. J Am Chem Soc 103:3910–3915
20. Preusch PC, Suttie JW (1983) A chemical model for the mechanism of vitamin K epoxide reductase. J Org Chem 48:3301–3305
21. Mitsudome T, Matoba M, Mizugaki T, Jitsukawa K, Kaneda K (2013) Core-shell AgNP@CeO$_2$ nanocomposite catalyst for highly chemoselective reductions of unsaturated aldehydes. Chem Eur J 19:5255–5258
22. Zhang J, Tang Y, Lee K, Ouyang M (2010) Nonepitaxial growth of hybrid core-shell nanostructures with large lattice mismatches. Science 327:1634–1638
23. Serpell CJ, Cookson J, Ozkaya D, Beer PD (2011) Core@shell bimetallic nanoparticle synthesis via anion coordination. Nat Chem 3:478–483

24. Zhang S, Hao Y, Su D, Doan-Nguyen VVT, Wu Y, Li J, Sun S, Murray CB (2014) Monodisperse core/shell Ni/FePt nanoparticles and their conversion to Ni/Pt to catalyze oxygen reduction. J Am Chem Soc 136:15921–15924
25. Lai J, Shafi KVPM, Ulman A, Loos K, Popovitz-Biro R, Lee Y, Vogt T, Estournes C (2005) One-step synthesis of core(Cr)/shell(γ-Fe$_2$O$_3$) nanoparticles. J Am Chem Soc 127:5730–5731
26. Lee WR, Kim MG, Choi JR, Park JL, Ko SJ, Oh SJ, Cheon J (2005) Redox−transmetalation process as a generalized synthetic strategy for core−shell magnetic nanoparticles. J Am Chem Soc 127:16090–16097
27. Yan JM, Zhang XB, Akita T, Haruta M, Xu Q (2010) One-step seeding growth of magnetically recyclable Au@Co core−shell nanoparticles: highly efficient catalyst for hydrolytic dehydrogenation of ammonia borane. J Am Chem Soc 132:5326–5327
28. Park S, Yoon D, Bang S, Kim J, Baik H, Yang H, Lee K (2015) Formation of a Cu@RhRu core–shell concave nanooctahedron via Ru-assisted extraction of Rh from the Cu matrix and its excellent electrocatalytic activity toward the oxygen evolution reaction. Nanoscale 7:15065–15069
29. Cargnello M, Wieder NL, Montini T, Gorte RJ, Fornasiero P (2010) Synthesis of dispersible Pd@CeO$_2$ core−shell Nanostructures by self-assembly. J Am Chem Soc 132:1402–1409
30. Qu F, Wang Y, Liu J, Wen S, Chen Y, Ruan S (2014) Fe$_3$O$_4$–NiO core–shell composites: hydrothermal synthesis and toluene sensing properties. Mater Lett 132:167–170
31. Lim Y, Kim SK, Lee SC, Choi J, Nahm KS, Yoo SJ, Kim P (2014) One-step synthesis of carbon-supported Pd@Pt/C core–shell nanoparticles as oxygen reduction electrocatalysts and their enhanced activity and stability. Nanoscale 6:4038–4042
32. Mitsudome T, Yamamoto M, Maeno Z, Mizugaki T, Jitsukawa K, Kaneda K (2015) One-step synthesis of core-gold/shell-ceria nanomaterial and its catalysis for highly selective semihydrogenation of alkynes. J Am Chem Soc 137:13452–13455
33. Lindlar H (1952) Ein neuer katalysator fur selektive hydrierungen. Helv Chim Acta 35:446–450
34. Yan M, Jin T, Ishikawa Y, Minato T, Fujita T, Chen LY, Bao M, Asao N, Chen MW, Yamamoto Y (2012) Nanoporous gold catalyst for highly selective semihydrogenation of alkynes: remarkable effect of amine additives. J Am Chem Soc 134:17536–17542
35. Vasilikogiannaki E, Titilas I, Vassilikogiannakis G, Stratakis M (2015) cis-Semihydrogenation of alkynes with amine borane complexes catalyzed by gold nanoparticles under mild conditions. Chem Commun 51:2384–2387
36. Wagh YS, Asao N (2015) Selective Transfer semihydrogenation of alkynes with nanoporous gold catalysts. J Org Chem 80:847–851
37. Noujima A, Mitsudome T, Mizugaki T, Jitsukawa K, Kaneda K (2011) Selective deoxygenation of epoxides to alkenes with molecular hydrogen using a hydrotalcite-supported gold catalyst: a concerted effect between gold nanoparticles and basic sites on a support. Angew Chem Int Ed 50:2986–2989
38. Urayama T, Mitsudome T, Maeno Z, Mizugaki T, Jitsukawa K, Kaneda K (2016) Green, multi–gram one–step synthesis of core–shell nanocomposites in water and their catalytic application to chemoselective hydrogenations. Chem Eur J 22:17962–17966

Chapter 4
Functions and Applications of Core–Shell Materials in Hydrogenation-Related Processes

Minghua Qiao, Hexing Li, and Baoning Zong

4.1 Introduction

Heterogeneous catalysts are usually prepared by the impregnation method, which gives rise to a fairly random size and spatial distribution of the active components on the catalyst surface. Due to the unrestricted migration and coalescence of the metal atoms on the surface, it is difficult for catalysts in this structure to avoid the loss of the active surface area and fast catalyst deactivation during the reaction. A catalyst design concept that shows promise to circumvent these limitations is to construct the catalyst in a core–shell structure. Typical core–shell catalysts contain an active metal core and a porous shell. The core–shell structure affords heterogeneous catalysts new structural and electronic characters, which endow the catalysts with new functions and enormous possibilities in hydrogenation-related processes. It is expected that the shell is capable of protecting the active sites in the core and improving the catalytic stability. The pore size in the shell can be utilized to control

M. Qiao (✉)
Department of Chemistry and Shanghai Key Laboratory of Molecular Catalysis and Innovative Materials, Collaborative Innovation Center of Chemistry for Energy Materials, Fudan University, Shanghai 200438, People's Republic of China
e-mail: mhqiao@fudan.edu.cn

H. Li
Education Ministry Key Laboratory of Resource Chemistry and Shanghai Key Laboratory of Rare Earth Functional Materials, Shanghai Normal University, Shanghai 200234, People's Republic of China
e-mail: hexing-li@shnu.edu.cn

B. Zong
State Key Laboratory of Catalytic Materials and Chemical Engineering, Research Institute of Petroleum Processing, SINOPEC, Beijing 100083, People's Republic of China
e-mail: zongbn.ripp@sinopec.com

© The Author(s), under exclusive license to Springer Nature Singapore Pte Ltd. 2021
H. Yamashita and H. Li (eds.), *Core-Shell and Yolk-Shell Nanocatalysts*,
Nanostructure Science and Technology,
https://doi.org/10.1007/978-981-16-0463-8_4

the entrance of the reactant molecules with different sizes, which is conducive to size-selective or regio-selective catalysis. The shell thickness can influence the diffusion of the reactants/products, thus changing the product distribution. By adjusting the hydrophilicity of the shell, the dispersion in the water phase or organic phase and the interaction with the reactant/product of the core–shell catalysts can be changed, making the catalysts versatile in water- or organic-phase reactions. When the core is not used as the active sites, it is usually made of magnetic materials to facilitate catalyst separation, or made of relatively inexpensive metals to act as a geometric or electronic modifier to promote the metal overlayer.

As the research about the core–shell catalysts continues, and with the development of in situ and operando techniques such as environmental transmission electron microscopy (ETEM), near-ambient pressure X-ray photoelectron spectroscopy (NAP XPS), and X-ray absorption spectroscopy (XAS), aside from the respective roles of the shell or the core, the core–shell interaction has been unveiled in some cases, which deepens the interpretation of the catalytic performance of the core–shell catalysts [1, 2]. Since dynamic compositional and/or structural evolution is likely to occur when the core and shell materials can react with each other or the core–shell material is bimetallic, care must be taken when correlating the structure–performance relationship using the static composition/structure. Moreover, the applications of the core–shell materials in measuring the hydrogen spillover distance [3] and as Raman signal enhancer [4] have been demonstrated, which are instructive to the establishment of more comprehensive hydrogenation mechanisms.

This chapter addresses studies mainly in this decade on the respective roles of the shell and the core of the core–shell catalysts in hydrogenation reactions, the dynamic compositional/structural changes of the core–shell materials under reaction conditions, and the applications of the core–shell materials in new hydrogenation-related processes. CO/CO_2 methanation to CH_4 is used to exemplify the advantages of the core–shell catalysts in challenging high-temperature and highly exothermic hydrogenation reactions as compared to the conventional catalysts. In the end, the implications and challenges of the core–shell catalysts in hydrogenation reactions are discussed.

4.2 The Functions of the Shell

Generally, the core–shell catalysts are prepared by coating the catalytically active metal cores with porous materials as shells, which protect the active sites in the core from leaching, sintering, and from being poisoned. Moreover, the properties of the shell, such as pore size, shell thickness, and hydrophilicity, are closely related to the performances of the core–shell catalysts in terms of activity and selectivity.

4.2.1 Effect of Pore Size

Zhang and co-workers produced Pd/SiO$_2$ core by loading Pd nanoparticles (NPs) on non-porous SiO$_2$ microspheres, and then coating the core with a perfect ZIF-8 shell by electrostatic induction, thus fabricating the Pd/SiO$_2$@ZIF-8 catalyst. The ZIF-8 shell with a pore size of 4.0–4.2 Å has two main functions. First, it can selectively allow reactants of appropriate sizes to pass through the shell and react on the core. Second, it can protect the Pd NPs on the core from leaching and from being poisoned. In the hydrogenation of 1-hexene, cyclohexene, and cyclooctene (Fig. 4.1), the conversion increased as the molecular size decreased. Intentionally adding 100 ppm of triphenylmethyl mercaptan to the reaction did not lower the conversion of 1-hexene. In addition, after four runs, the Pd loading on the core–shell catalyst minimally decreased. In contrast, the Pd/SiO$_2$ catalyst lost a significant amount of Pd [5]. Similarly, Yip and co-workers prepared the Pd/ZIF-8@ZIF-8 catalyst for olefin hydrogenation. The conversion of 1-hexene was higher than that of cyclooctene, and the small pore size of the ZIF-8 shell protected the Pd NPs from thiophene poisoning, showing the attractive size selectivity and anti-poisoning properties of the core–shell catalyst [6].

To geometrically direct the hydrogenation product of 3-methylcrotonaldehyde to prenol, the Pt@ZIF-8 catalyst with the pore size of ZIF-8 equivalent to the size of 3-methylcrotonaldehyde was prepared [7]. The narrow pores forced 3-methylcrotonaldehyde to linearly approach Pt. Therefore, the C=C bond in the middle of the 3-methylcrotonaldehyde molecule did not interact with Pt, while the C=O group at the terminal was easy to adsorb and be further hydrogenated to prenol. As a result, the Pt@ZIF-8 catalyst gave >84% selectivity to prenol even at >90% conversion. The Pt@ZIF-8 catalyst was also tested for the hydrogenation of acrolein. For acrolein, since the C=C and C=O groups are located at both ends, the C=C bond adsorption was not inhibited on Pt@ZIF-8. Thus, the selectivity to allyl alcohol was comparable on the supported Pt-0.47/ZIF-8 catalyst and on the Pt-1.16@ZIF-8 catalyst.

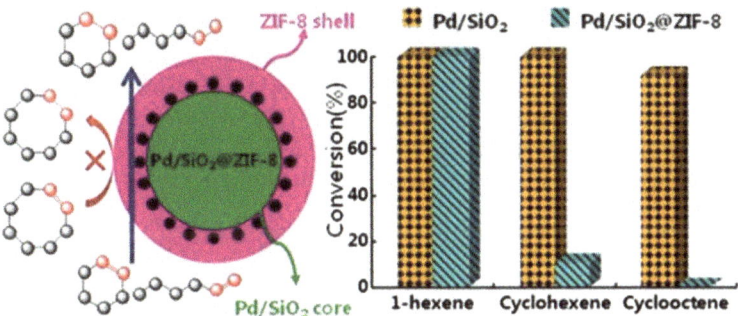

Fig. 4.1 The size-selective effect of Pd/SiO$_2$@ZIF-8 in olefin hydrogenation. Reprinted from Ref. [5]. Copyright 2014 American Chemical Society

Noticing that there was a lack of systematic comparison of the activities of the core–shell catalysts with different porosities, Semagina and co-workers studied the accessibility of metal sites in the $Pd@SiO_2$ catalysts prepared with and without porogen. In the first case, PVP was used as both a stabilizer of Pd and a potential porogen of SiO_2. It turned out that the activity of the resulting $Pd@SiO_2$ was extremely low due to the difficulty in accessing Pd. In the second case, the high surface-area $Pd@SiO_2$ prepared by separately introducing PVP and CTAB as additional porogen was active, with identical TOF as traditional catalysts while higher sintering resistance. However, even in highly porous catalysts, the authors found that two-thirds of the surface of Pd was blocked [8]. Thus, the high porosity of the shell must be ensured for efficient mass transfer within the core–shell catalyst.

Liu and co-workers prepared the $Pd@mSiO_2$ catalysts composed of a Pd core and a mesoporous SiO_2 shell with controllable pore size from 1.97 to 2.73 nm by changing the alkyl length of the surfactant $C_n TAB$ ($n = 14$–18) [9]. In nitrobenzene hydrogenation, the larger pore size of the catalyst was more conducive to the transportation of the reactants, and the smaller pore size was conducive to product selectivity. After five cycles, the $Pd@mSiO_2$ catalyst was more active and metal-loaded than supported catalysts and commercial Pd/C catalysts. The $Pd@mSiO_2$ catalyst may solve the problem of aggregation and sintering of active metals without hampering the transport of reactants.

4.2.2 Effect of Shell Thickness

4.2.2.1 Oxide Shell

Wang and co-workers prepared the $Fe_3O_4@SiO_2$ catalysts with different shell thicknesses to study the effect of the SiO_2 shell on the distribution of Fischer–Tropsch synthesis (FTS) products [10]. The inert microporous SiO_2 shell improved the activity and stability by protecting the core from cracking and sintering due to the structural restriction and separation effect. As the thickness of the shell increased, the product distribution shifted to C_1–C_4 gaseous hydrocarbons, which might be due to the diffusion restriction on the products. Although the selectivity of C_{5+} hydrocarbons was limited to 10%, the water–gas shift (WGS) activity was also lowered, which favorably increased the productivity of hydrocarbons. The micropores in the SiO_2 shell were likely to hinder more the diffusion of CO than H_2, thus increasing the partial pressure of H_2 on the core and consequently inhibiting the WGS reaction and the production of CO_2.

Xu et al. used skeletal Co as the core to prepare a Co@HZSM-5 catalyst via hydrothermal synthesis [11]. The thickness of the zeolite shell was tuned by the hydrothermal time. With the increase in the zeolite shell thickness, the catalytic activity first increased and then declined in FTS. The volcanic change in the activity was attributed to the existence of an acidic HZSM-5 shell, which is conducive to the cracking of long-chain hydrocarbons into short-chain hydrocarbons that are easy to desorb, so the activity increased first. Further increasing the zeolite shell thickness might hinder the syngas from entering the catalyst, and at the same time hinder the out-diffusion of the product, thus reducing the catalytic activity. At a suitable zeolite shell thickness, the long-chain hydrocarbons were cracked completely, leading to a high selectivity for the gasoline fraction.

Sun et al. synthesized the Raney Fe@HZSM-5 catalysts with different shell thicknesses via a one-pot strategy using FeAl alloy both as the Fe precursor and as the Al source (Fig. 4.2) [12]. The shell thickness was varied from 1.3 to 9.7 μm by changing the crystallization temperature. The core–shell catalyst with a thinner zeolite shell showed lower steady-state activity, indicating that there was wax deposition on the core, which blocked the active sites. There were long-chain products that had not been completely cracked, which also signified that the zeolite shell was not thick enough to completely crack long-chain hydrocarbons. On the other hand, too thick zeolite shell was adverse to the catalytic activity. Moreover, because the degree of catalytic cracking was too high, the product distribution shifted to short-chain hydrocarbons. On the core–shell catalyst with suitable shell thickness, the steady-state conversion was as high as 92.4%, and the product selectivity in the gasoline fraction reached a maximum of 71.1% [13].

Yip and co-workers prepared a core–shell zeolite catalyst consisting of a Pd/ZSM-5 core and a Silicalite-1 (S-1) shell (Pd/ZSM-5@S-1) by secondary crystallization of S-1 on Pd/ZSM-5 [14]. In liquid phase olefin hydrogenation, although both 1-hexene and cyclohexene can enter the 10-membered ring channel of the Pd/ZSM-5@S-1 catalyst, the conversion of the former (87%) was much higher than that of the latter (13%). With the increase of the S-1 layer thickness, the Pd/ZSM-5@S-1 catalysts showed increased conversion of 1-hexene over 1-heptene, even though both are linear olefins with similar kinetic diameters that can enter the MFI framework. The authors proved that due to the faster mass transfer rate, there is a strong correlation between the thickness of the S-1 shell and the conversion of light olefins. This work provides an example of how to change and improve the selectivity on zeolitic catalysts without relying on the typical size exclusion mechanism.

Fig. 4.2 Illustration of the formation process of the Raney Fe@HZSM-5 catalyst. **a** The starting FeAl alloy; **b** in situ dealumination of the FeAl alloy by TPAOH in the zeolite synthesis solution; **c** nucleation of HZSM-5 on Raney Fe; **d** removal of TPAOH by calcination in air; **e** reduction in H₂/Ar to restore the metallic Fe core [12]

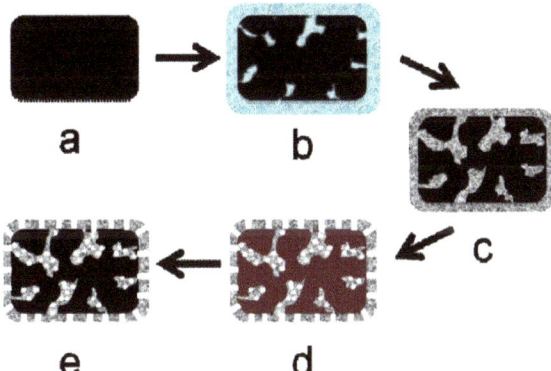

4.2.2.2 Metal Shell

Reducing the shell thickness to a few atomic layers while retaining the catalytic performance of the bulk catalyst reduces the consumption of precious metals. However, in addition to the synthesis challenges, since the structure often changes in a reactive environment, the few-layer core–shell catalysts may exhibit different performance over time. Therefore, it is necessary to gain an understanding of the behavior of few-layer core–shell catalysts in thermal and reaction environments.

Humphrey and co-workers synthesized Au@Rh NPs with adjustable shell thickness using a microwave-assisted heating method. Au@Rh NPs with the shell as thin as 2–4 Rh monolayers were prepared. The core–shell catalyst was more effective than pure Rh NPs per mol of Rh in gas-phase hydrogenation of cyclohexene at room temperature [15].

Pan and Yang reported that among the Rh, Pd, and Pd@Rh catalysts, when the average thickness of the Rh shell was about two atomic layers, the Pd@Rh catalyst showed the highest selectivity for CO_2 methanation. The thickening of the Rh shell during the high-temperature process was observed, which changed the selectivity of CH_4 due to the disappearance of the ligand effect of the core on the thick Rh shell. Therefore, the selectivity at high temperatures was similar to that of the Rh catalyst. Density functional theory (DFT) calculations showed that the dissociation of CO on the (111) surface of the Rh mono- or bi-layer on Pd was energetically more favorable than on pure Rh or Pd, which is beneficial for the hydrogenation of CO_2 to hydrocarbons [16].

Zhou and co-workers investigated the structural stability of small Pd@Pt octahedral NPs and the effect of shell thickness on the catalytic performance in the hydrogenation of chloronitrobenzene [17]. The 6–8 nm Pd@Pt NPs were prepared by a sequential reduction method, which could controllably synthesize Pd@Pt NPs with 1–4 Pt atomic layers. The Pd@Pt NPs with one atomic Pt layer showed excellent structural stability and high catalytic stability during cycles of p-chloronitrobenzene hydrogenation (Fig. 4.3). The Al_2O_3-supported Pd@Pt NPs showed the superior catalytic performance to Pt, Pd, and their physical mixtures. DFT calculations suggested that the unexpected structural stability of the Pd@Pt NPs with thin Pt

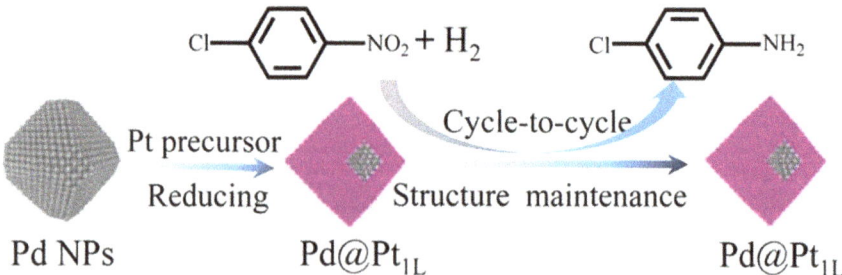

Fig. 4.3 Synthesis of Pd@Pt core–shell NPs with one atomic Pt layer and their stability during p-CNB hydrogenation reaction. Adapted from Ref. [17]. Copyright 2015 American Chemical Society

shells and their catalytic stability could be attributed to the strong binding of Pt with the reactants/products. The enhanced catalytic performance of the Pd@Pt octahedral NPs might originate from the core–shell interaction, which modulated the electronic state of surface Pt atoms.

Enantioselective hydrogenation over heterogeneous catalysts is an ideal method for the synthesis of chiral compounds in pesticides and drugs. So far, the supported precious metal catalysts have been the most extensively studied in heterogeneous enantioselective hydrogenation. However, the interaction between the prochiral molecules and the surface of the support, though weak, is detrimental to enantioselectivity. Chen and co-workers, through a simple strategy based on the reduction of Pd^I carbonyl complex, prepared a P25 TiO_2@Pd catalyst with a thin Pd shell ~1.0 nm thick. By shielding the negative interaction from the support, the P25@Pd catalyst with unique electronic properties of Pd exhibited higher activity and enantioselectivity in enantioselective hydrogenation of acetophenone than the Pd/P25 catalyst prepared by the impregnation method and the unsupported Pd black catalyst [18].

4.2.3 Hydrophilicity

Wei and co-workers prepared monodispersed mesoporous SiO_2 nanospheres (MSN) as the support for Pt NPs. The Pt/MSN particles were then used as the core and coated with a uniform mesoporous SiO_2 layer by a two-phase stratification method. The resulting dendritic mesoporous hierarchical core–shell nanostructures were hydrophobically modified by methyl groups. During the aqueous two-phase hydrogenation of nitrobenzene, the catalyst was located at the oil–water interface. Within 2 h without stirring, nitrobenzene was completely converted into aniline. For comparison, under the same conditions, the conversions on Pt/C and un-methylated catalyst were only 33% and 39%, respectively [19].

Pang and co-workers developed carbon nanocomposite catalysts (Ni@NCFs) with unique hydrophobicity from PVP/nickel nitrate through a one-step pyrolysis process [20]. Their structure and hydrophobicity could be adjusted by the annealing temperature. A positive correlation between the catalytic performance and hydrophobicity of the catalyst was identified. Especially for the Ni@NCF-700 catalyst, the superhydrophobic N-doped graphene shell acted as a "solid ligand" to modify the Ni core and exhibited a high affinity for organic substrates in the water phase and high catalytic performance in the one-step synthesis of aniline and N-heterocyclic aromatic compounds via the hydrogen transfer reaction. The superhydrophobic Ni@NCF-700 repelled the water by-product, thereby reducing the possible adverse effect of water on the catalyst. In addition, water can be used as the solvent, which satisfies the criterion of green chemistry.

In order to solve the limitations of conventional hydrophobic/hydrophilic catalysts in the water phase hydrogenation of organic compounds, Yang and co-workers synthesized a core–shell structured catalyst, in which Pd-supported fluorine-modified SiO_2 sphere acted as a hydrophobic core and the mesoporous SiO_2 as the hydrophilic shell, as inspired by the structure of natural enzymes [21]. The hydrophobic core–hydrophilic shell catalyst was well dispersed in water, while the hydrophobic core could adsorb the hydrophobic reactants from water to the active sites, making the catalyst more active than the Pd/SiO$_2$ catalyst in the water phase C=C bond hydrogenation in a series of acrylates. After the product was extracted with diethyl ether, the catalyst dispersed in water was used directly in the next cycle after removing the upper organic layer, which reduced catalyst loss during recycling. This hydrophobic core-hydrophilic shell strategy opens up new possibilities for designing effective catalysts for water phase organic reactions.

Xie and co-workers prepared a multifunctional nanomaterial (Fe$_3$O$_4$@SiO$_2$@C$_X$@NH$_2$) that contained a magnetic core, a silica protective interlayer, and an amphiphilic silica shell (C$_X$). After being loaded with Ru NPs, the catalyst was used for the hydrogenation of α-pinene. The new amphiphilic catalyst, acting as a solid foaming agent, increased the gas–liquid–solid three-phase interface and consequently accelerated the reaction. Under mild conditions (40 °C, 1 MPa H$_2$, 3 h), 99.9% α-pinene conversion and 98.9% *cis*-pinane selectivity were obtained [22].

4.3 Interaction Between Core and Shell

4.3.1 Effect of Shell Thickness on Core

Mehta and co-workers studied the effect of carbon shell thickness on the hydrogenation performance of the Pd@C material with a defined core size, as pulsed molecular beam experiments on carbon-free and carbon-containing Pd NPs showed that the latter significantly promoted the diffusion of subsurface hydrogen [23]. Interestingly, when the thickness of the shell increased, the H/Pd ratio was greatly improved, and the hydrogen-induced lattice expansion also increased. Meanwhile, the Pd 4d band center moved toward higher binding energy (BE) relative to that of bare Pd NPs. For the interaction between metal and hydrogen, the position of the d-band center relative to the Fermi level and the coupling matrix element between the adsorbate and the metal are the most important parameters that determine the reactivity of the metal toward hydrogen. The coupling matrix elements depend on the energy gap between the interacting atoms. Therefore, the shift of the d-band center away from the Fermi energy resulted in a stronger interaction between Pd and H (Fig. 4.4). This discovery may be useful in many hydrogen-related applications, such as hydrogen storage and catalysis, which require increased Pd–H interactions.

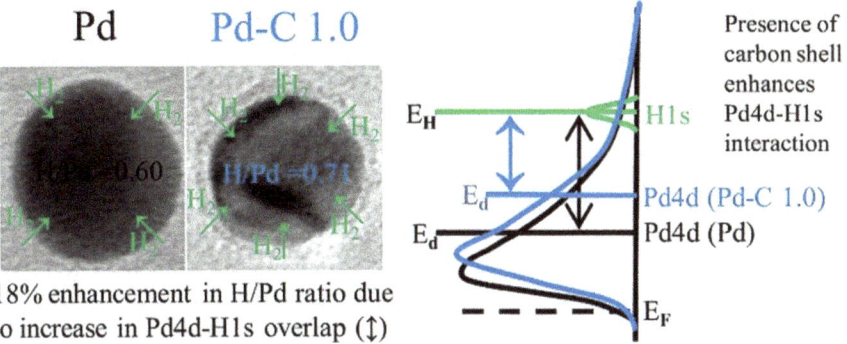

Fig. 4.4 Effect of carbon shell on the interaction between Pd and H. Reprinted from Ref. [23]. Copyright 2015 American Chemical Society

4.3.2 Effect of Core on Shell

Transition metal NPs have high catalytic activity in many reactions, so it is necessary to develop new methods for stabilizing tiny particles. The stabilization effect can be achieved by coating metal NPs with a carbon layer similar to graphene to form the metal@C nanocomposites. Interestingly, although the metal particles were encapsulated in the carbon shell, this core–shell material still had catalytic activity in many hydrogenation reactions. To explain this phenomenon, Erokhin and coworkers synthesized carbon-coated Ni and Fe NPs, which contained metal cores of about 5 nm in size and were wrapped in several layers of graphene-like carbon [24]. Ni@C and Fe@C gave high conversion in phenylacetylene hydrogenation above 150 °C and 300 °C, respectively. Fe@C displayed a high styrene selectivity of 86% with 99% phenylacetylene conversion at 300 °C. The authors ruled out the possibility of the presence of metal NPs on the outside of the pristine core–shell catalyst. They also confirmed that the possibility of carbon shell cracking during the reaction was minimal. Theoretical calculations further proved that due to the existence of space and structural defects on the graphene overlayer and/or the presence of transition metals in the subsurface layer, the carbon shells gained the ability to activate H_2 through dissociative adsorption. This work shows that for metal@C catalysts, the role of graphene shells as active sites in the hydrogenation reactions cannot be ruled out.

Ni is vulnerable to ambient oxidation, which is adverse to its catalytic activity and stability. Surface science studies showed that the sub-monolayer Ni coated on polycrystalline Au foil can resist oxidation. Therefore, Vinod and co-workers synthesized Au@Ni NPs [25]. By changing the concentration of the Ni precursor, the thickness of the Ni shell was adjusted from 2 to 8 nm. NAP XPS revealed that at a Ni shell thickness of about 2 nm, the shell exhibited oxidation resistance. At high temperatures, the thin Ni(OOH) shell was decomposed to metallic Ni, while the thick Ni(OOH) shell maintained its oxidation state. In partial hydrogenation of phenylacetylene, under mild reaction conditions, the Au@Ni catalyst with a shell thickness of about 2 nm was more active than the monometallic Ni or Au NPs and the core–shell catalyst with a thicker Ni shell, which was attributed to electron donation from the Au core to the thin Ni shell.

Román-Leshkov and co-workers demonstrated that compared with pure Pt NPs, atomically thin Pt shells coated on titanium tungsten carbide and titanium tungsten nitride cores, TiWC@Pt and TiWN@Pt, respectively, exhibited high sintering-resistance, enhanced CO tolerance, and increased selectivity in partial hydrogenation of acetylene [26]. They found that the electronic structure of the Pt shell on the TiWC and TiWN cores had changed significantly. The unoccupied Pt $5d_{3/2}$ state increased, which was attributed to the widening of the d-band. However, the Pt–Pt distance was almost the same as that of pure Pt NPs. Therefore, these electronic modifications were attributed to the ligand effect originated from the hybridization of the Pt and W d states rather than the strain effect. The DFT calculations proved the broadening of the Pt d-band accompanied by the downshift of the d-band center. These results demonstrate the importance of detailed physical characterization of the structure for the accurate elucidation of material properties.

4.4 Thermal Evolution of Core–Shell Material

Most studies on core–shell catalysts are focused on their catalytic performances. Moreover, the interpretations on the catalytic performances of the core–shell catalysts are primarily based on their static structures before or after a reaction, which is valid only when the core–shell structure remains intact throughout the reaction. However, at elevated temperatures or in the presence of the reactants, atom redistribution may occur. To obtain such dynamic structural information is challenging, as advanced in situ or operando techniques are required. Though only limited works had addressed this issue, the findings are enlightening and beneficial for the understanding of the thermal stability, activity/selectivity change, and the mechanism underlying the deactivation of the core–shell catalysts.

4.4.1 Monometallic Core–Shell Material

The thermal stability of the Pd@SiO$_2$ nanostructure was investigated by Baaziz and co-workers using ETEM under atmospheric pressure and electron tomography (ET) [27]. In the process of heating in gas at atmospheric pressure, the Pd core with the original shape of octahedron or icosahedron was tracked. In a reducing H$_2$ environment up to 400 °C, there was a strong dependence of the shape and facet transformation on the initial structure of the Pd particles. The octahedral single-crystalline Pd NPs were less stable than the icosahedral polycrystalline Pd NPs. For the former, the diffusion of Pd from the core to the outer surface of SiO$_2$ caused the gradual shrinkage of the core. The latter did not show morphology/faceting change, because in this case, due to the large number of crystal defects in the particles, the diffusion of atoms within the particles was favored against the diffusion to the SiO_2 shell.

The authors further studied the thermal behaviors of the Pd@SiO_2 material under reductive and oxidative conditions at temperatures up to 1000 °C, which was found to be closely related to the thermal response of the SiO_2 shell. Under H_2, the SiO_2 shell lost its porous structure and became densified, which blocked the Pd core by the thick shell. In the air, the porosity of the SiO_2 shell was maintained. The diffusion of Pd from the core to the outer surface of the SiO_2 shell increased with the temperature. At 850 °C, all Pd atoms were discharged to the outside of the SiO_2 shell.

4.4.2 Bimetallic Core–Shell Material

van Blaaderen and co-workers used in situ TEM and in situ EXAFS to study the thermally driven atom redistribution in single-crystalline Au@Ag nanorods coated with protective mesoporous SiO_2 layer [28]. They found that the increase in the Ag content resulted in slower metal redistribution, which was the opposite of the dependence of the melting temperature on the Au/Ag ratio. In addition, size-dependent alloying was found, where a reduction in particle size resulted in a lower alloying temperature.

Gao and co-workers found that the Ni–Au bimetallic catalyst had high selectivity for CO production during CO_2 hydrogenation. Before and after the reaction, the Ni–Au catalyst showed the structure of Ni core coated by a complete ultra-thin Au shell. However, the inactive Au shell surface could not account for this catalytic performance, as Au is inactive for this reaction. Through direct ETEM visualization combined with a variety of in situ techniques, including XAS, IR, and theoretical simulations, the authors revealed the formation of transient reconstructed alloy surface caused by CO adsorption during the reaction [29]. The discovery of this dynamic structural transition shows the importance of understanding the reaction mechanism beyond the static model for bimetallic core–shell catalyst.

4.5 Applications of Core–Shell Materials

4.5.1 Measurement of Hydrogen Spillover Distance

Hydrogen spillover is an important phenomenon that affects the catalytic performance of certain hydrogenation catalysts and the hydrogen storage capacity of certain materials. However, it is a challenging task to confirm the occurrence of hydrogen spillover and measure the distance of hydrogen spillover.

Zhan and Zeng noticed that the stability of ZIF-8 (based on Zn^{2+}) was different from that of ZIF-67 (based on Co^{2+}) in the presence of H_2 or atomic hydrogen. Both ZIF-8 and ZIF-67 maintained their structure in a H_2 environment up to 300 °C. On the contrary, because atomic H is highly active, the ZIF-67 skeleton was easily degraded by exposure to atomic H at lower temperatures (for example, 180 °C), while the ZIF-8 skeleton remained almost intact. On the basis of these facts, they designed a series of ZIF-67@ZIF-8 nanocubes with the ZIF-8 shell thickness varied from 0 to 50 nm (Fig. 4.5) [3]. The Pt NPs were then loaded on the outer surface to generate atomic H. Only when H atoms dissociated on Pt penetrate the inert ZIF-8 shell, does the decomposition of the ZIF-67 core occur. Therefore, in ZIF-67@ZIF-8/Pt, the ZIF-8 shell worked as a ruler to measure the travel distance of H atoms, while the ZIF-67 core acted as the terminator for H atoms. By determining the extent of ZIF-67 degradation through in situ gravimetric measurement under flowing H_2 in combination with ex situ morphology/structure characterizations, one can qualitatively measure the concentration of atomic H arriving at the ZIF-67 core and hence the spillover distance of H atoms. In this way, the authors confirmed that the ZIF-67@ZIF-8/Pt composite with a ZIF-8 shell thickness of 20 nm or 50 nm maintained their structural integrity in H_2 flow at 240 °C for 4 h, indicating that almost no H atoms reached the ZIF-67 core. When the ZIF-8 shell was as thin as 5 nm, H atoms penetrated the ZIF-8 layer even at 220 °C for 4 h, causing the hydrogenolysis of the ZIF-67 core. In addition, the authors proved that CO_2 hydrogenation on Co metal could also be used to track the spillover of H atoms on ZIF-8 under high pressure.

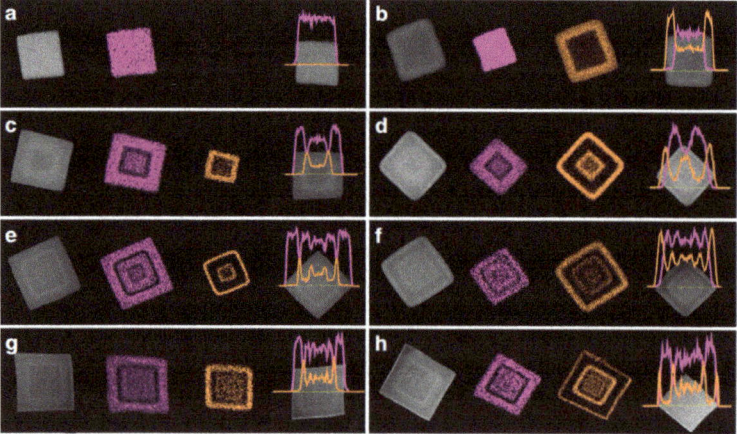

Fig. 4.5 EDX elemental mapping and line scanning of ZIF-67 and nano-Matryoshka-structured ZIFs. **a** ZIF-67, **b** ZIF-67@ZIF-8, **c** tri-layered ZIF, **d** tetra-layered ZIF, **e** penta-layered ZIF, **f** hexa-layered ZIF, **g** hepta-layered ZIF, and **h** octa-layered ZIF. Color code: purple represents cobalt; brown represents zinc. Reprinted from Ref. [3]

4.5.2 Enhancement of Raman Signal

Vibrational spectroscopy can provide molecular fingerprints of surface and gaseous species present during the catalytic reaction, and can be used under a wide range of experimental conditions, which is suitable for a molecular-level understanding of the actual structure–performance relationship. Although the detection limit of traditional Raman spectroscopy is low, surface-enhanced Raman spectroscopy (SERS) using signal enhancement technology can detect single molecules. By using Au or Ag NPs that exhibit local surface plasmon resonance (LSPR) effect when irradiated with light of right frequency, a strong electromagnetic field can be induced near the surface of the precious metal NPs. This makes SERS an extremely suitable tool for surface study, as it only enhances the signal of the species on or near the catalyst surface. In order to apply SERS to in situ and operando heterogeneous catalysis, Tian and co-workers adopted a thin layer of dielectric oxide (such as SiO_2) to stabilize and physically isolate the Au and Ag NPs. This technique, so-called shell-isolated nanoparticle-enhanced Raman spectroscopy (SHINERS) [30], is rapidly developing into a very useful and practical method for the in situ study of catalytic reaction.

By using in situ SERS and SHINERS, Chen and co-workers studied the roles of the Pt–Au and Pt–oxide–Au interfaces in H_2 activation and p-nitrothiophenol hydrogenation at the molecular level [4]. The Pt–Au and Pt–oxide–Au interfaces were made by the synthesis of Pt-on-Au and Pt-on-SHINs nanocomposites. Direct spectroscopic evidence showed that the H atoms generated on Pt diffused to Au through the Pt–Au and Pt–TiO_2–Au interfaces. But this diffusion pathway was blocked at the Pt–SiO_2–Au interface. This discrepancy led to different reaction pathways and product selectivity on Pt-on-Au and Pt-on-SHINs nanocomposites. In addition, the pinhole-free-SHINs shielded the influence of the core on the reaction, which can be used as a promising platform for the in situ study of heterogeneous catalysis.

However, the biggest limitation of such SHINs used in catalysis is their moderate thermal stability (~450 °C). Therefore, studies have been limited to precious metal catalysts, which can be easily reduced without high temperature. Non-precious metal catalysts are more widely used, but their reduction requires harsher temperature treatment. For example, the reduction of Ni requires high temperature, which would destroy the SiO_2 shell of the Au@SiO_2 SHINs, making the SHINERS study on such metals highly challenging. Weckhuysen and co-workers investigated various methods to prepare active Ni catalysts on Au@SiO_2 for the in situ SHINERS study [31]. They found that spark ablation could deposit metallic Ni NPs directly on the Au@SiO_2 SHINs. In the acetylene adsorption experiment, the Ni-acetylene species that has not been reported before were detected. This species disappeared after hydrogenation, and the Raman bands associated with ethylene and ethylidine on Ni were identified, thus confirming that hydrogenation reactions can be studied in situ on the Ni-based catalysts using the SHINERS method. This work opens the avenue to the elucidation of the reaction mechanism on non-precious metal catalysts including Fe, Co, Ni, and Cu that are preferred in industrial catalysis using the in situ SHINERS technique.

4.5.3 Magnetic Core–Shell Catalyst on Magnetic Reactor

The rapid development of general synthesis strategies for magnetic core–shell materials with controlled size, composition, and structure provides huge possibilities for the preparation of magnetic heterogeneous catalysts [32, 33]. Regardless of these impressive advances, it is surprising that the application of magnetic core–shell catalysts is quite primitive, that is, their magnetic properties are only used to help their separation. A magnetically stabilized bed (MSB) reactor is a kind of reactor that confines the magnetic catalyst by a magnetic field. As the strength of the magnetic field increases, the MSB displays three operating modes: particle scattering mode, chain mode, and magnetic coagulation mode. When the MSB is operated in the chain mode, the particles are oriented along the axis of the MSB in a chain-like manner. The voidages between the chains are uniform, and there is no gas bypassing and solid backmixing, thus rendering good contact with liquid or gas fluid. Fine particles can be used without high pressure drop [34]. Therefore, the MSB operating in the chain mode provides a good opportunity to develop reactors that are suitable for magnetic core–shell catalysts.

Zong and co-workers developed a magnetic $NiFe_2O_4@Al_2O_3$ material with sufficient mechanical strength by coating Al_2O_3 on magnetic $NiFe_2O_4$ spinel. Then, a magnetic $Pd/NiFe_2O_4@Al_2O_3$ catalyst with sufficient saturation magnetization was synthesized. The catalytic performance of the magnetic core–shell catalyst for acetylene hydrogenation in an MSB reactor under different operation conditions was studied. Under optimal reaction conditions, the acetylene conversion and ethylene selectivity were about 100% and 84%, respectively [35]. For the partial hydrogenation of benzene on the magnetic $Ru/Fe_3O_4@Al_2O_3$ catalyst in the MSB reactor, Fu and co-workers found that the catalyst operating in the chain mode exhibited higher benzene conversion and cyclohexene selectivity than in the particle scattering mode [36]. The "magnetic core–shell catalyst-on-magnetic reactor" strategy can find applications in energy- and environment-related reactions, and will stimulate the design of new synthetic methods for magnetic core–shell catalysts.

4.5.4 CO/CO₂ Methanation

CO/CO_2 methanation has gained more and more interest as a way to store surplus renewable energy in the form of CH_4, which is easy to store, transport, and use in existing industrial infrastructure. However, the highly exothermic nature of the methanation reactions requires catalysts that are stable under severe reaction conditions. In principle, the core–shell catalysts are suitable for this kind of reaction, because the catalyst structure can limit the sintering of the core metal [37]. Zhu and co-workers demonstrated that the $Ni@SiO_2$ catalyst gave a high CO conversion of 99.0% and a CH_4 yield of 89.8% in CO methanation. The $Ni@SiO_2$ catalyst exhibited good catalytic stability in 100 h on stream, which is better than the supported

Ni/SiO$_2$ catalyst. The improvement in the catalytic performance of the Ni@SiO$_2$ catalyst was proposed as the strong interaction between the Ni core and the SiO$_2$ shell that effectively inhibits the growth of Ni and the deposition of coke [38].

Güttel and co-workers studied the effect of the structure of the iron catalyst on the catalytic performance in CO$_2$ hydrogenation, paying special attention to the coking resistance [39]. Bare α-Fe$_2$O$_3$, SiO$_2$-supported iron oxide, and a core–shell catalyst with nano-sized Fe core embedded in a SiO$_2$ shell (15Fe@SiO$_2$) were prepared. The bare α-Fe$_2$O$_3$ catalyst and the impregnated catalyst showed relatively high activity at 400 °C and 1 bar, but suffered severe deposition of coke. In contrast, the core–shell catalyst had higher resistance to coke formation, thermal sintering, and particle attrition. The amount of carbon deposition on the used core–shell catalyst was about 300 times less than that on the bare α-Fe$_2$O$_3$ catalyst, and 32 times less than that on the impregnated catalyst. However, the activity of the core–shell catalyst was low and should be improved.

The authors also explored the applicability of Co-based core–shell catalysts in CO$_x$ methanation [37]. The catalyst consisted of Co NPs as the core, wrapped in an amorphous mesoporous SiO$_2$ shell (Co@mSiO$_2$). By comparing the Co@mSiO$_2$ catalyst with the supported Co/mSiO$_2$ catalyst, the core–shell catalyst achieved higher CH$_4$ selectivity in CO$_2$ methanation, which was attributed to the higher chance of readsorption and subsequent hydrogenation of the CO intermediate. For CO methanation, the core–shell catalyst exhibited rapid temperature-dependent deactivation due to coking and possible pore blockage at temperatures above 350 °C. Nevertheless, the Co@mSiO$_2$ catalyst was thermally more stable than the supported catalyst. In simultaneous CO/CO$_2$ methanation, the Co@mSiO$_2$ catalyst reacted flexibly upon varying the composition of the feed gas, making the catalyst a promising candidate to convert various process gases in the steel industry with a wide range of CO/CO$_2$ ratio into CH$_4$.

The limitation of heat transport has an adverse effect on the yield and selectivity and the service life of the catalyst due to deactivation. Considering that heat transfer through the solid catalyst largely depends on the thermal conductivity of the support, Lee and co-workers developed a method of interfacial hydrothermal oxidation of Al metal particles in a Ni salt aqueous solution to produce a core–shell microstructure consisting of a high thermal conductivity Al metal core densely covered with a NiAl-layered double hydroxide shell (Al@NiAl-LDH) [40]. The reduction of Al@NiAl-LDH extracted Ni atoms from the NiAl-LDH shell, generating finely dispersed Ni NPs on the Al@Al$_2$O$_3$ support (Al@Al$_2$O$_3$/Ni). The thermal conductivity of Al metal is more than one order of magnitude higher than that of Al$_2$O$_3$, SiO$_2$, TiO$_2$, etc., which significantly increased the thermal conductivity of the catalyst. Compared with the Ni/Al$_2$O$_3$ catalyst prepared by the impregnation method, the Al@Al$_2$O$_3$/Ni catalyst showed higher turnover frequency (TOF) and lower CO selectivity in CO$_2$ methanation, manifesting the excellent inherent catalytic performance of the Ni/Al$_2$O$_3$@Al catalyst.

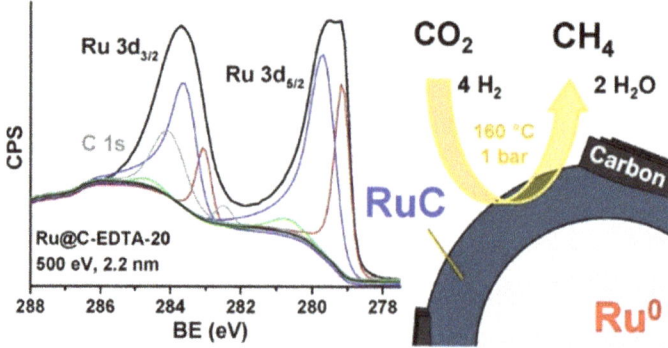

Fig. 4.6 Determination of the Ru@RuC structure and its excellent activity in CO_2 methanation. Reprinted from Ref. [41]. Copyright 2019 American Chemical Society

Ru is a highly active metal for CO_2 methanation at low temperatures. However, the space–time yield (STY) of CH_4 reported so far was too low for industrial applications. Corma and co-workers synthesized a ruthenium carbide catalyst with a core–shell structure formed by a metallic ruthenium core and a ruthenium carbide shell (labeled as Ru@C) through a gentle hydrothermal method using EDTA or glucose as a carbon source [41]. It was determined that although metallic Ru was dominant in the inner layer of the material, ruthenium carbide was located on the upper surface layer (Fig. 4.6). The Ru@C catalyst could activate CO_2 and H_2, and showed high activity for CO_2 methanation at low temperature (160–200 °C). Under the conditions of atmospheric pressure and feed rate of 8.3 ml g^{-1} s^{-1}, the STY at 160 °C was 3.5 μmol_{CH4} s^{-1} $g_{cat.}^{-1}$, and increased to 13.8. μmol_{CH4} s^{-1} $g_{cat.}^{-1}$ at 200 °C. Based on catalytic studies and isotopic $^{13}CO/^{12}CO_2/H_2$ experiments, the active sites responsible for the high activity were related to the surface ruthenium carbide (RuC) species, which activated CO_2 and converted it into CH_4 through a direct CO_2 hydrogenation mechanism. The catalyst also showed good stability with CH_4 selectivity higher than 99.9%, which makes the Ru@C-EDTA and Ru@C-glucose catalysts promising for the Sabatier reaction.

4.6 Summary and Outlook

The advent of core–shell materials has opened up enormous possibilities for creating heterogeneous catalysts with new properties and new functions. The core–shell materials have been extensively developed and applied in many fields and can serve as promising catalysts in a variety of hydrogenation-related processes, which show superior performances as compared to conventional catalysts, especially in size- or shape-dependent reactions and high-temperature reactions. Advanced in situ and operando techniques together with theoretical calculations are powerful tools to reveal compositional/structural changes and catalytic mechanisms of the core–shell catalysts, which can accelerate the development of highly active and selective core–shell hydrogenation catalysts. However, the industrial application of the core–shell catalysts in hydrogenation processes has not yet been achieved. It is crucial to develop synthetic methods that are capable of preparing uniform and well-defined core–shell catalysts in a facile and low-cost manner to facilitate their large-scale commercialization. When designing the core–shell catalysts, care should be taken to balance the confinement effect of the shell and the accessibility of the core to attain a high utility of the core atoms. For bimetallic core–shell catalysts, there are good chances to develop new catalysts with enhanced activity and/or selectivity on the basis of the strain effect or the ligand effect. For this purpose, the atomic layer deposition (ALD) technique, which can accurately control the shell composition and thickness [42], is promising to prepare core–shell catalysts with a defined structure and hence tailor-made catalytic performance.

Acknowledgements The support from the National Key R&D program of China (2018YFB0604501) and the Science and Technology Commission of Shanghai Municipality (19DZ2270100) are cordially acknowledged.

References

1. Bugaev L, Guda AA, Lomachenko KA, Shapovalov VV, Lazzarini A, Vitillo JG, Bugaev LA, Groppo E, Pellegrini R, Soldatov AV, van Bokhoven JA, Lamberti C (2017) Core-shell structure of palladium hydride nanoparticles revealed by combined X-ray absorption spectroscopy and X-ray diffraction. J Phys Chem C 121:18202–18213
2. Bugaev LA, Usoltsev OA, Lazzarini A, Lomachenko KA, Guda AA, Pellegrini R, Carosso M, Vitillo JG, Groppo E, van Bokhoven JA, Soldatov AV, Lamberti C (2018) Time-resolved operando studies of carbon supported Pd nanoparticles under hydrogenation reactions by X-ray diffraction and absorption. Faraday Discuss 208:187–205
3. Zhan GW, Zeng HC (2018) Hydrogen spillover through Matryoshka-type (ZIFs@)$_{n-1}$ZIFs nanocubes. Nat Commun 9:3778
4. Zhang H, Zhang XG, Wei J, Wang C, Chen S, Sun HL, Wang YH, Chen BH, Yang ZL, Wu DY, Li JF, Tian ZQ (2017) Revealing the role of interfacial properties on catalytic behaviors by in situ surface-enhanced Raman spectroscopy. J Am Chem Soc 139:10339–10346
5. Lin L, Zhang T, Zhang XF, Liu HO, Yeung KL, Qiu JS (2014) New Pd/SiO$_2$@ZIF-8 core-shell catalyst with selective, antipoisoning, and antileaching properties for the hydrogenation of alkenes. Ind Eng Chem Res 53:10906–10913
6. Yin H, Choi J, Yip ACK (2016) Anti-poisoning core−shell metal/ZIF-8 catalyst for selective alkene hydrogenation. Catal Today 265:203–209
7. Lan XC, Huang N, Wang JF, Wang TF (2017) Geometric effect in the highly selective hydrogenation of 3-methylcrotonaldehyde over Pt@ZIF-8 core−shell catalysts. Catal Sci Technol 7:2601–2608
8. Habibi AH, Hayes RE, Semagina N (2018) Bringing attention to metal (un)availability in encapsulated catalysts. Catal Sci Technol 8:798–805
9. Lv MX, Yu ST, Liu SW, Li L, Yu HL, Wu Q, Pang JH, Liu YX, Xie CX, Liu Y (2019) One-pot synthesis of stable Pd@mSiO$_2$ core−shell nanospheres with controlled pore structure and their application to the hydrogenation reaction. Dalton Trans 48:7015–7024

10. Tian ZP, Wang CG, Si Z, Wang YC, Chen LG, Liu QY, Zhang Q, Xu Y, Ma LL (2018) Product distributions of Fischer-Tropsch synthesis over core-shell catalysts: the effects of diverse shell thickness. Chem Sel 3:12415–12423

11. Xu K, Cheng Y, Sun B, Pei Y, Yan SR, Qiao MH, Zhang XX, Zong BN (2015) Fischer-Tropsch synthesis over skeletal Co@HZSM-5 core-shell catalysts. Acta Phys Chim Sin 31:1137–1144

12. Sun B, Yu GB, Lin J, Xu K, Pei Y, Yan SR, Qiao MH, Fan KN, Zhang XX, Zong BN (2012) A highly selective Raney Fe@HZSM-5 Fischer−Tropsch synthesis catalyst for gasoline production: one-pot synthesis and unexpected effect of zeolites. Catal Sci Technol 2:1625–1629

13. Sun B (2013) PhD dissertation, Shanghai

14. Jia XC, Jeong Y, Baik H, Choi J, Yip ACK (2018) Increasing resolution of selectivity in alkene hydrogenation via diffusion length in core-shell MFI zeolite. Catal Today 314:94–100

15. Garía S, Anderson RM, Celio H, Dahal N, Dolocan A, Zhou JP, Humphrey SM (2013) Microwave synthesis of Au−Rh core−shell nanoparticles and implications of the shell thickness in hydrogenation catalysis. Chem Commun 49:4241–4243

16. Pan YT, Yang H (2017) Rhodium-on-palladium nanocatalysts for selective methanation of carbon dioxide. ChemNanoMat 3:639–645

17. Zhang PP, Hu YB, Li BH, Zhang QJ, Zhou C, Yu HB, Zhang XJ, Chen L, Eichhorn B, Zhou SH (2015) Kinetically stabilized Pd@Pt core-shell octahedral nanoparticles with thin Pt layers for enhanced catalytic hydrogenation performance. ACS Catal 5:1335–1343

18. Gao XY, He LL, Xu JT, Chen XY, He HY (2019) Facile synthesis of P25@Pd core-shell catalyst with ultrathin Pd shell and improved catalytic performance in heterogeneous enantioselective hydrogenation of acetophenone. Catalysts 9:513

19. Wei J, Zou LK, Li YL, Zhang XM (2019) Synthesis of core−shell-structured mesoporous silica nanospheres with dual-pores for biphasic catalysis. New J Chem 43:5833–5838

20. Pang SF, Zhang YJ, Su Q, Liu FF, Xie X, Duan ZY, Zhou F, Zhang P, Wang YB (2020) Super-hydrophobic nickel/carbon core−shell nanocomposites for the hydrogen transfer reactions of nitrobenzene and N-heterocycle. Green Chem 22:1996–2010

21. Hao YJ, Jiao X, Zou HB, Yang HQ, Liu J (2017) Growing a hydrophilic nanoporous shell on a hydrophobic catalyst interface for aqueous reactions with high reaction efficiency and in situ catalyst recycling. J Mater Chem A 5:16162–16170

22. Wu FZ, Yu FL, Yuan B, Xie CX, Yu ST (2019) Highly selective and recyclable hydrogenation of α-pinene catalyzed by ruthenium nanoparticles loaded on amphiphilic core−shell magnetic nanomaterials. Appl Organomet Chem 33:e5165

23. Singh V, Mehta BR, Sengar SK, Kulriya PK, Khan SA, Shivaprasad SM (2015) Enhanced hydrogenation properties of size selected Pd-C core-shell nanoparticles; Effect of carbon shell thickness. J Phys Chem C 119:14455–14460

24. Erokhin AV, Lokteva ES, Yermakov AY, Boukhvalov DW, Maslakov KI, Golubina EV, Uimin MA (2014) Phenylacetylene hydrogenation on Fe@C and Ni@C core−shell nanoparticles: about intrinsic activity of graphene-like carbon layer in H_2 activation. Carbon 74:291–301

25. Bharathan VA, Jain R, Gopinath CS, Vinod CP (2017) Diverse reactivity trends of Ni surfaces in Au@Ni core−shell nanoparticles probed by near ambient pressure (NAP) XPS. Catal Sci Technol 7:4489–4498

26. Garg A, Goncalves DS, Liu YS, Wang ZS, Wang LX, Yoo JS, Kolpak A, Rioux RM, Zanchet D, Román-Leshkov Y (2019) Impact of transition metal carbide and nitride supports on the electronic structure of thin platinum overlayers. ACS Catal 9:7090–7098

27. Baaziz W, Bahri M, Gay AS, Chaumonnot A, Uzio D, Valette S, Hirlimann C, Ersen O (2018) Thermal behavior of Pd@SiO_2 nanostructures in various gas environments: a combined 3D and in situ TEM approach. Nanoscale 10:20178–20188

28. van der Hoeven JES, Welling TAJ, Silva TAG, van den Reijen JE, La Fontaine C, Carrier X, Louis C, van Blaaderen A, de Jongh PE (2018) In situ observation of atomic redistribution in alloying gold-silver nanorods. ACS Nano 12:8467–8476

29. Zhang XB, Han SB, Zhu BE, Zhang GH, Li XY, Gao Y , Wu ZX, Yang B, Liu YF, Baaziz W, Ersen O, Gu M, Miller JT, Liu W (2020) Reversible loss of core–shell structure for Ni–Au bimetallic nanoparticles during CO_2 hydrogenation. Nat Catal. https://doi.org/10.1038/s41929-020-0440-2

30. Li JF, Huang YF, Ding Y, Yang ZL, Li SB, Zhou XS, Fan FR, Zhang W, Zhou ZY, Wu DY, Ren B, Wang ZL, Tian ZQ (2010) Shell-isolated nanoparticle-enhanced Raman spectroscopy. Nature 464:392–395
31. Wondergem CS, Kromwijk JJG, Slagter M, Vrijburg WL, Hensen EJM, Monai M, Vogt C, Weckhuysen BM (2020) In situ shell-isolated nanoparticle-enhanced Raman spectroscopy of nickel-catalyzed hydrogenation reactions. ChemPhysChem 21:625–632
32. Vono LLR, Damasceno CC, Matos JR, Jardim RF, Landers R, Masunaga SH, Rossi LM (2018) Separation technology meets green chemistry: development of magnetically recoverable catalyst supports containing silica, ceria, and titania. Pure Appl Chem 90:133–141
33. Zhang QQ, Yang XY, Guan JQ (2019) Applications of magnetic nanomaterials in heterogeneous catalysis. ACS Appl Nano Mater 2:4681–4697
34. Cheng M, Xie WH, Zong BN, Sun B, Qiao MH (2013) When magnetic catalyst meets magnetic reactor: etherification of FCC light gasoline as an example. Sci Rep 3:1973
35. Dong MH, Pan ZY, Peng Y, Meng XK, Mu XH, Zong BN, Zhang JL (2008) Selective acetylene hydrogenation over core–shell magnetic Pd-supported catalysts in a magnetically stabilized bed. AIChE J 54:1358–1364
36. He TT, Mu SL, Fu QT, Liu CG (2018) Synthesis of Fe_3O_4 core/alumina shell nanospheres for partial hydrogenation of benzene. Mater Sci Eng 292:012019
37. Ilsemann J, Straß-Eifert A, Friedland J, Kiewidt L, Thöming J, Bäumer M, Güttel R (2019) Cobalt@silica core-shell catalysts for hydrogenation of CO/CO_2 mixtures to methane. ChemCatChem 11:4884–4893
38. Han Y, Wen B, Zhu MY (2017) Core-shell structured $Ni@SiO_2$ catalysts exhibiting excellent catalytic performance for syngas methanation reactions. Catalysts 7:21
39. Kirchner J, Zambrzycki C, Kureti S, Güttel R (2020) CO_2 methanation on Fe catalysts using different structural concepts. Chem Ing Tech 92:603–607
40. Lee H, Kim J, Lee D (2020) A new design and synthesis approach of supported metal catalysts via interfacial hydrothermal-oxidation/reductive-exolution chemistry of Al metal substrate. Appl Catal A 594:117461
41. Cored J, García-Ortiz A, Iborra S, Climent MJ, Liu LC, Chuang CH, Chan TS, Escudero C, Concepción P, Corma A (2019) Hydrothermal synthesis of ruthenium nanoparticles with a metallic core and a ruthenium carbide shell for low-temperature activation of CO_2 to methane. J Am Chem Soc 141:19304–19311
42. Zhang B, Qin Y (2018) Interface tailoring of heterogeneous catalysts by atomic layer deposition. ACS Catal 8.10064–10081

Chapter 5
Multimetallic Catalysts and Electrocatalysts: Dynamic Core–Shell Nanostructures

Zhi-Peng Wu, Shiyao Shan, Shan Wang, Dominic Caracciolo, Aolin Lu, Zhijie Kong, Richard Robinson, Guojun Shang, and Chuan-Jian Zhong

5.1 Introduction

Heterogeneous catalysis plays a critical role in the global drive to sustainable energy and environment [1]. As nanostructured catalysts have found increasing applications in heterogeneous catalysis involving many sustainable energy and environmental reactions, the ability to control the surface composition, structure, and morphology of the nanocatalysts is challenging due to the highly dynamic nature of atoms on the surface or in the bulk phase of the metal or alloy nanomaterials [2–4]. Nanoalloy catalysts and electrocatalysts containing platinum group metals (PGMs) and 3d-transition metals (3d-TMs) have been proven to be an effective method to enhance the catalytic performance while reducing the cost by taking the advantages of the modulated charge distribution and the optimized strain effect [3–7]. The structure of catalysts, whether in core–shell or alloy states, exhibits a significant influence on their properties in terms of thermal stability and catalytic performance.

The study of core–shell and alloy structured catalysts dates back decades ago [8, 9]. With the rapid development of nanomaterials, there have been extensive studies of core–shell and alloy structures of different types (Fig. 5.1). In addition to composition-homogeneous alloy (CHA), which can also be considered as disordering (defect)–ordering core–shell structure (DOCS) with an ordered core and a disordered shell containing atomic defects, we can conceptually group core–shell nanomaterials in four major types. Phase-segregated core–shell (PSCS) features a core–shell nanoparticle which is traditionally defined as an inner core (e.g., metal M) conformally coated with a shell (e.g., metal N). Many core–shell structures often exhibit different elemental enrichment at the core and the shell, which are defined

Z.-P. Wu · S. Shan · S. Wang · D. Caracciolo · A. Lu · Z. Kong · R. Robinson · G. Shang ·
C.-J. Zhong (✉)
Department of Chemistry, State University of New York at Binghamton, Binghamton, NY 13902, USA
e-mail: cjzhong@binghamton.edu

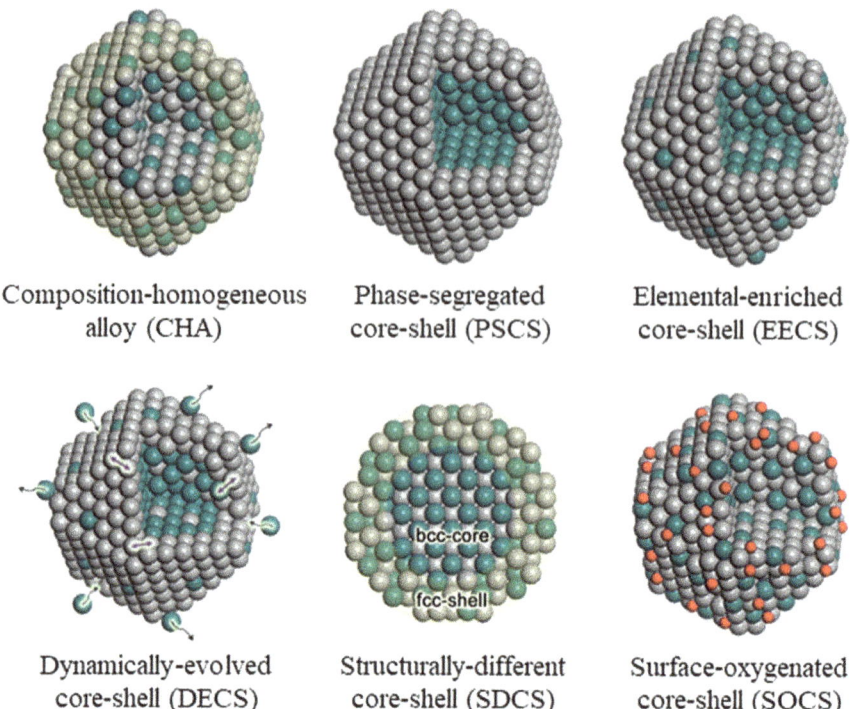

Composition-homogeneous   Phase-segregated   Elemental-enriched
alloy (CHA)   core-shell (PSCS)   core-shell (EECS)

Dynamically-evolved   Structurally-different   Surface-oxygenated
core-shell (DECS)   core-shell (SDCS)   core-shell (SOCS)

Fig. 5.1 Schematic illustrations of different types of core–shell and alloy structures using binary metal nanoparticles as examples. Reproduced with permission from ref. [3] Copyright 2020 American Chemical Society

as elemental-enriched core–shell (EECS) structures. There are a few emerging core–shell structures, such as dynamically evolved core–shell (DECS), structurally different core–shell (SDCS), and surface-oxygenated core–shell (SOCS) structures. Often there is no clear boundary between the different core–shell and alloy structures since the transformations among these structures can be highly dynamic, especially under different post-synthesis or catalytic reaction conditions [3].

A variety of methods for the synthesis and processing of core–shell structures have been successfully developed in the past few decades, as can be classified in (i) as-synthesis method such as seeded growth method [10]; (ii) post-synthesis processes including thermochemical treatment [5, 11], adsorbate-induced segregation [12], chemical dealloying [13], and underpotential deposition (UPD) followed by galvanic displacement [14]; (iii) dynamic core–shell evolution under electrochemical or fuel cell operating conditions [6, 15]. Despite significant advances that have been achieved on nanostructured catalysts with core–shell and alloy structures, a rarely asked question is, how the structure of nanomaterials dynamically evolve under reaction conditions between core–shell and alloy characteristics along with the impact on the catalytic synergy [3]? In this chapter, the dynamic structure evolution of nanostructured catalysts is highlighted with focal points on the impacts of the thermochemical treatment, electrochemical potential cycling, and fuel cell operating condition, as well as gas-phase reactions.

PGMs play a major role in catalysis. Core–shell and alloy structures enable not only efficient utilization of the precious PGM, but also produce remarkable catalytic synergies. In a core–shell structured catalyst, cost-prohibitive PGMs expose on the shell participate in the reaction with high intrinsic activity while non-noble metals in monometallic or alloy states form the core to influence the structural properties of the shell. One important area of catalysis in the global drive to sustainable energy and environment is the development of clean energy conversion devices such as fuel cells. Fuel cell technologies represent a crucial vector of clean, efficient, and sustainable energy sources when applied in portable electronic devices and automobiles [9, 16]. Electrocatalysts are the soul components in fuel cells both at the cathodic site for the oxygen reduction reaction (ORR) and the anodic site for alcohol oxidation reaction and hydrogen oxidation reaction. Another important area of catalysis in the global drive to sustainable energy and environment is the development of effective catalysts for oxidation reactions. There have been increasing demands for enhanced production of valuable chemicals, effective remediation of hydrocarbon pollutants, and sustainable energy conversion of various fuels. Oxidation reactions (total or partial oxidations) account for more than 60% of the chemicals and intermediates that are widely used in the chemical industry to produce pharmaceuticals, fine chemicals, agricultural chemicals, and various functionalized hydrocarbon molecules. Oxidation reactions also play a pivotal role in the remediation of hydrocarbon pollutants and the production of biomass fuels.

Indeed, PGMs have played and continue to play important roles in these catalytic reactions. PGM-based nanoalloy electrocatalysts exhibit excellent performance in both activity and durability in acidic fuel cells. Although alkaline fuel cells enable PGM-free electrocatalysts (such as non-noble metal-based and carbon materials), good performance in fuel cell reactions, their mass commercialization, however, is greatly hindered by the relatively low activities and the currently limited progress on the alkaline membrane. Therefore, the focus is limited on PGM-containing alloy or core–shell electrocatalysts in this chapter. In general, alloying PGMs with 3d-TMs has been proven as an effective method to enhance the activity while reducing the cost of electrocatalysts. However, the current state of PGM-based nanoalloy electrocatalyst is still suffering from the high cost with a high PGM usage greater than 75 atomic percent (PGM mass fraction >90% in nanoalloy) [17]. On the other hand, the development of low-temperature active catalysts with a reduced amount of PGMs will contribute to advanced emission control systems to meet the increasingly stringent emission standards, fuel-efficient technologies, and environmental sustainability. There is increasing evidence showing fuel economy advantage with advanced fuel injection systems over the conventional gasoline counterparts, which translates to a significant reduction in greenhouse gas emissions. However, the fuel efficiency improvement resulted in lower emission temperatures where conventional aftertreatment systems are not suitable, leading to the emission of significant amounts of HC, CO, particulate matter (PM), and nitrogen oxides (NO_x). PGMs are widely used in emission control systems to meet the ever-tightening emission standards, but they are expensive and fluctuating in the market, driving up the catalyst manufacturing cost [18]. With the reinforcement of federal regulations for the reduction of green house gases (GHG), advanced engines and powertrain systems with enhanced fuel economy are expected to be operated at much lower exhaust temperatures than the conventional emission control systems. Many design strategies focus on active PGM catalyst-support peripheral interfaces by exploring isolated single atoms, reactive surface intermediates, surface-oxidized species, metal carbonylation, meso/nano porous structure, and lattice oxygen [19, 20]. Despite extensive studies of conventional PGM or non-PGM catalysts, the understanding of the role of PGM components in catalyst activation and deactivation, especially in terms of simultaneous evolution of the surface sites and atomic-scale structures, remains elusive.

5.2 Core–Shell/Alloy Structures of Metal and Alloy Nanomaterials

Core–shell nanomaterials can be broadly defined as core and shell with differences in composition and structures [8]. Significant progress has been made in the synthesis of alloy and core–shell nanoparticles of different types using different approaches. With the as-synthesized nanoparticles, thermochemical treatment is often applied to remove the surface contaminations and reconstruct the alloy structure. The chemical structures obtained after the thermochemical treatment are highly dependent on the treatment temperatures and atmospheres [5]. Interestingly, the high temperature-induced strong dynamic evolution could alter the structure of nanomaterials from alloy to core–shell structure, or from one core–shell structure to another core–shell type, or even from core–shell structure to alloy state. A dynamically atomic reconstruction under thermochemical treatment enables a mutual transformation between alloy and core–shell structures.

For supported bimetallic or trimetallic nanoparticles, the thermochemically induced phase structure evolution could include alloying and partial or complete phase segregation (Fig. 5.2a). For example, carbon-supported AuPt (AuPt/C) CHA NPs are thermochemically treated under 20% O_2/N_2 at 280 °C and then under 15% H_2/N_2 at 300–800 °C as in sequence. Different core–shell and alloy structures are obtained after the thermochemical calcination (Fig. 5.2a) [11]. AuPt/C NPs annealed under H_2 at low temperatures (300–400 °C) exhibit CHA characters while a phase-segregated Pt-rich alloy core and Au shell EECS or PSCS structure is achieved at high temperature (800 °C)-treated AuPt/C NPs. This dynamic evolution of the core–shell structure is validated by XRD characterizations (Fig. 5.2b). A single alloy phase structure was shown at low temperature annealed catalysts (300 °C). With the thermochemical treatment temperature increasing, more obvious phase separations were exhibited by showing two diffraction peaks. The analyses of the lattice parameters further substantiated this finding (Fig. 5.2c). In contrast to Au, Pt exhibits

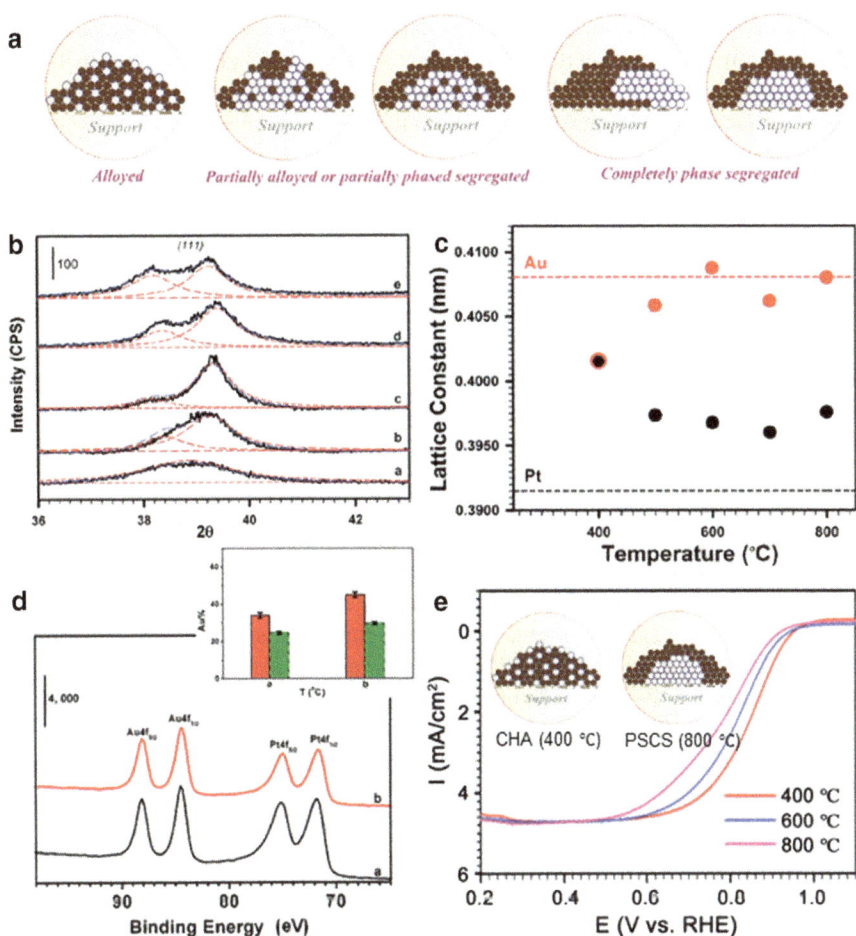

Fig. 5.2 a Idealized illustrations of the nanoscale alloyed, partially alloyed/partially phase segregated, or completely phase segregated bimetallic metals on a support. **b** XRD patterns of AuPt/C NP treated at different temperatures under H_2. **a**, 300 °C; **b**, 400 °C; **c**, 500 °C; **d**, 600 °C; and **e**, 800 °C. **c** The corresponding lattice parameters of AuPt/C NP catalysts treated at different temperatures. **d** XPS spectra of AuPt/C NP thermochemically treated at different temperatures. a, 400 °C; b 800 °C under H_2. Inset shows a comparison of XPS-determined Au% (red) and the Au% calculated (green) based on idealized Pt (core)/Au (shell) nanoparticle. **e** RDE curves of AuPt/C NP treated at different temperatures. Inset is the illustration of AuPt NPs thermochemically treated at different temperatures under H_2 in varying structures. Reproduced with permission from ref. [11] Copyright 2010 American Chemical Society

higher intrinsic catalytic activity toward ORR. This conclusion is confirmed by the significantly reduced Pt-specific active surface area and ORR mass activity on the Pt-rich alloy@Au EECS/PSCS NPs compared with its fully alloyed CHA counterpart (Fig. 5.2e). Density functional theory (DFT) studies were carried out to investigate the thermodynamic stability of AuPt nanoalloy catalysts with different phase structures. A phase-segregated cluster model with Au-enriched shell structure was identified as the most stable configuration compared with other alloy counterparts.

Multimetallic NPs have been widely explored for the design of efficient catalysts. One example involves the exploration of PtAuNi NPs as electrocatalysts for fuel cell reactions. By thermochemical annealing under H_2, the as-synthesized Au-rich core@PtNi-rich shell EECS structure is dynamically evolved to form a Ni-rich core@PtAu-rich shell EECS structure (Fig. 5.3). The surface enrichment of noble metal Pt and Au with optimized Pt–Pt bond distances are found to be responsible for the enhanced methanol oxidation activity by facilitating the dehydrogenation of methanol and effectively removing the surface poisonous carbonaceous species [21]. The as-synthesized ternary PtPdCu NPs feature a PdCu alloy core and a Pt-rich shell EECS structure. However, after a thermochemical annealing treatment under H_2, the structure of PtPdCu NPs dynamically transforms to a uniform CHA state, which is responsible for the enhanced high activity and durability for the methanol oxidation reaction (MOR). The dynamic structure reconstruction under thermochemical treatment is believed to be linked to the surface energy difference between atoms, phase state stability at different temperatures, and varying bonding strength in metal combinations. Mechanistically, the surface catalytic sites control the formation of surface species in an indirect pathway for the electrocatalytic MOR over the Ni-rich core@PtAu-rich shell NPs [17].

Beyond elemental distribution-specified EECS structures, core–shell structures with the same bimetallic combination but different nanophase types have also been demonstrated by thermochemical processing of as-synthesized NPs. One example involves the thermochemical treatment of $Pd_{50}Cu_{50}$ SDCS NPs, which is shown to exhibit a significant impact on the electrocatalytic properties of the electrocatalyst for

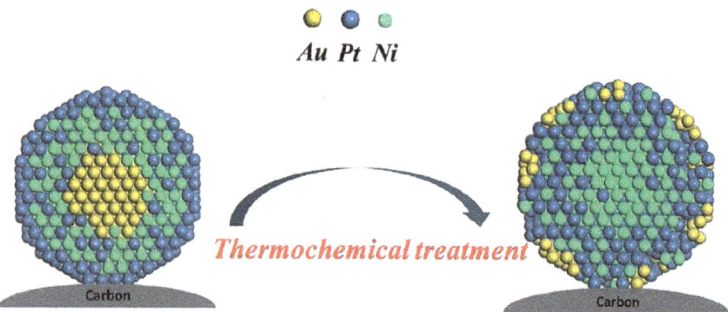

Fig. 5.3 Illustration of the atomic reconstruction of AuPtNi/C NP catalyst from Au-rich core@PtNi-rich shell to Ni-rich core@PtAu-rich shell structure under thermochemical treatment. Reproduced with permission from ref. [21] Copyright 2018 Royal Society of Chemistry

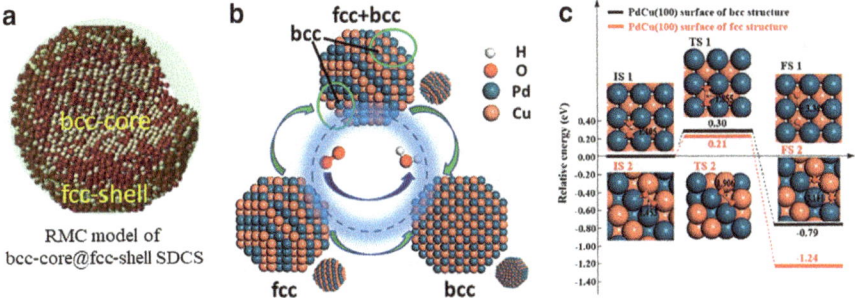

Fig. 5.4 **a** RMC model of the bcc-core@fcc-shell SDCS structure. **b** A schematic illustration of phase evolution in terms of fcc and bcc manipulated by thermochemical treatment. **c** DFT calculations on O–O bond cleavage in O_2 molecule on fcc- and bcc-structured PdCu alloy models. Reproduced with permission from ref. [5] Copyright 2018 American Chemical Society

oxygen reduction reaction [5]. For example, 100 °C/H_2-treated $Pd_{50}Cu_{50}$ NPs yield a pure fcc type CHA phase structure, while 400 °C/H_2-treated $Pd_{50}Cu_{50}$ NPs show a mixed SDCS structure of body-centered cubic (bcc) alloy core and face-centered cubic (fcc) alloy shell (Fig. 5.4a). The thermochemical treatment is shown to induce phase structure transformation between fcc, bcc, and fcc/bcc mixture depending on the calcination temperatures (Fig. 5.4b). The formation of bcc-core is induced by high thermotreatment temperatures ($\geq$200 °C). The ORR catalytic activity is highly dependent on the nanophase type of electrocatalysts. Surprisingly, $Pd_{50}Cu_{50}$ NPs with a pure fcc CHA structure show much better ORR catalytic activity but inferior durability than the structural type of SDCS with bcc-core@fcc-shell. This phenomenon demonstrates a general rule of thumb in catalysis, that is to say, high stability of the catalyst is sometimes achieved at the sacrifice of the activity [5, 22, 23].

To further obtain the reaction mechanism from the theoretical perspective, DFT calculations of the elementary step of O–O cleavage reaction in molecular oxygen based on PdCu models with fcc and bcc structures were performed (Fig. 5.4c). The results showed a lower reaction barrier on the fcc-phase model compared with that of the bcc-phase model, which coincides with the experimental results.

Besides using a single type of as-synthesized NPs as the precursor for the core–shell evolution by the thermochemical treatment of the supported NPs, the use of one type of as-synthesized NPs in the presence of another type of as-synthesized NPs or another metal precursors has also been demonstrated for the preparation of core–shell NPs. One approach involves thermal activation of two types of NPs as precursors in a solution for the formation of core–shell NPs. This dynamic core–shell structure evolution was achieved in the case of a Fe_2O_3@Au core@shell structure [24]. Molecularly capped Fe_2O_3 and Au NPs are homogeneously mixed together followed by a thermal activation process. Au NPs in a relatively small size adsorb on the surface of Fe_2O_3 NPs and then aggregate to form well-defined shells induced by thermal activation. The whole process is highly dynamic to reach a balanced state at every moment. The shell thickness can be readily controlled by varying the ratio between core and shell materials, which can be extended to diverse types of NPs. Another approach involves seeded growth for producing other types of core–shell nanoparticles with different chemical nature of the materials. The core materials can be in the form of metallic NPs to metal oxides. For example, $Fe_3O_4@Au$ core–shell NPs were synthesized via sequential deposition of Au atoms on Fe_3O_4 seeds by the wet chemical approach [25, 26]. Besides metal cations dissolved in chemical solutions, small Au NPs can also serve as the shell precursors. In this case, Au and Fe_2O_3 NPs are mixed in the solution and are subject to heating which thermally activates melting of the Au NPs on the surface of the core and hetero-interparticle aggregative growth and coalescence, forming a Au shell on the core, i.e., $Fe_2O_3@Au$ core–shell NPs [24]. The core–shell formation could also be a result of the reaction occurring in the surface layer during the seeded growth process, leading to a core and shell with different compositions and structures, e.g., core–shell MnZn ferrite nanocubes [27]. These core–shell nanoparticles have found intriguing applications where the magnetic core and the plasmonic gold or silver shell are exploited for biomolecular assays [28, 29].

5.3 Core–Shell/Alloy Structures and Evolution Under Electrolyte-Phase Reaction Conditions

It has been recognized that the core–shell/alloy structures can undergo dynamic evolution under catalytic reaction conditions. There are many examples demonstrating such evolution, especially under electrolyte-phase reaction conditions. By *in operando* HE-XRD and atomic PDF analysis of NP catalysts in an in situ performed PEMFC (Fig. 5.5), dynamic core–shell evolution has been observed under electrochemical or fuel cell operating conditions [30]. In this study, the high-energy synchrotron X-rays penetrate through a custom-designed PEMFC device, through

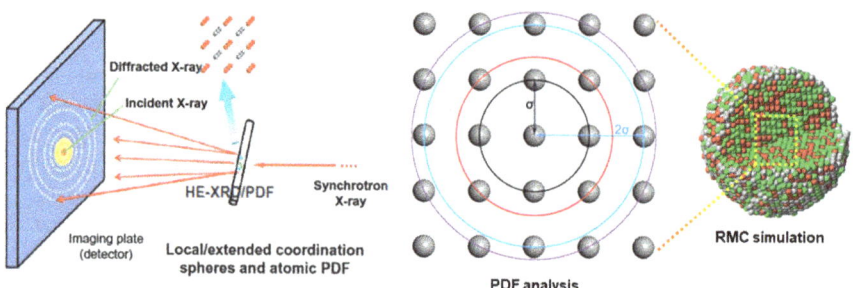

Fig. 5.5 Schematic illustrations of the HE-XRD measurement, PDF analysis, and RMC simulation for characterization of multimetallic NPs. Reproduced with permission from ref. [18] Copyright 2018 Royal Society of Chemistry

which the diffraction signals are collected by a panel detector and are transformed to HE-XRD patterns. The data are analyzed by the PDFs technique, in which the numbers of atom pairs in the first and the second coordination shells constitute the first (σ) and the second (2σ) strong peaks, and so on. Detailed structures of the catalysts are analyzed by 3D Reverse Monte Carlo (RMC) models, reflecting lattice parameters, atomic coordination numbers, nanophase contents, and atomic compositions and distributions, etc. PDFs converted from experimental HE-XRD patterns and 3D RMC simulations are depicted in black symbols and red lines, respectively. The better the fitting between the experimental and the simulative PDFs, the more accurate the information can be extracted from the best fitted 3D RMC models.

It is important to note that X-ray absorption fine structure (XAFS) techniques have been widely used to analyze the first-neighbor atomic coordination structures. As another powerful tool, HE-XRD/PDFs analysis allows the characterization of the nanophase structures, along with the assessment of the overall coordination numbers of metallic NPs at the atomic level with the aid of structural modeling [5]. In the following subsection, selected examples will be described to highlight the characterization of the dynamic core–shell nanostructures.

5.3.1 Pd/Pt-Based Nanoparticle Catalysts

Both Pt- and Pd-based NP catalysts have been widely studied for a variety of catalytic and electrocatalytic reactions. For commonly used metals in fuel cell electrocatalysts, the stabilities in an acidic environment fall in a sequence of Pt > Pd > 3d-TMs. Pd, which shows the most similar catalytic properties with Pt among all PGMs, serves as an important element of fuel cell electrocatalysts in pure metallic or alloy states [5]. Generally, the overall performance including both activity and durability of Pd-based fuel cell electrocatalysts in acidic media are inferior to Pt-based alloy counterparts. Sometimes, hence, Pd-based nanoalloys are selected as model catalysts to study the structure evolution process during electrochemical or fuel cell operating conditions due to their large degrees of 3d-TMs leaching and obvious dynamics during structure evolution. The composition–structure–activity synergy in PGM containing alloy electrocatalysts is one of the most crucial topics for a better design and fabrication of efficient catalysts. PdNi NP ORR electrocatalysts were studied under a combination of in situ HE-XRD/PDFs analyses and electrochemical/PEMFC tests to mechanically illustrate this synergistic relationship [15]. The in situ atomic PDFs are derived from in situ HE-XRD patterns for $Pd_{30}Ni_{70}$ NPs generated in a custom-designed PEMFC. The interatomic distances of the catalyst show an oscillatory behavior, reflecting a dynamic reconstruction of PdNi NP catalysts under fuel cell operation condition potential cycles. Interestingly, the evolution of the NP size of the catalyst determined by HE-XRD/PDFs shows an irregularly dynamic fluctuation favor along with the potential cycling line, which coincides with the trend of ex situ TEM-determined particle sizes. The lattice constant of $Pd_{30}Ni_{70}$ NPs derived from HE-XRD/PDFs patterns mirrors that of the Ni leaching rate during potential cycling and tends to reach a plateau after a certain number of potential cycles, indicative of an alloy phase.

As depicted in Fig. 5.6 for the dynamically structural evolution of PdNi NPs under the electrochemical or fuel cell operating condition, two possible scenarios are considered in the process of Ni leaching from the PdNi NP upon potential cycling. One scenario considers a dynamic reconstruction or realloying to reach another nanoalloy state with a Pd-enriched composition. The other envisages the formation of PdNi core with Pd-rich shell as well-documented in the literature for other bimetallic nanoalloys, such as PtCu alloy NPs [31]. PdNi NPs experience an oscillatory structural evolution process in regard to interatomic distances but always feature an alloy state during the reconstruction process. If PdNi NPs after Ni leaching transform into a complete "Pd-skin" structure, it is unlikely that such interatomic distance oscillation and NP size fluctuation would happen. The "Pd-skin" protects Ni species from leaching out and a steady plateau should be presented. Those oscillatory behaviors might be attributed to the strong dynamic process in which Ni species

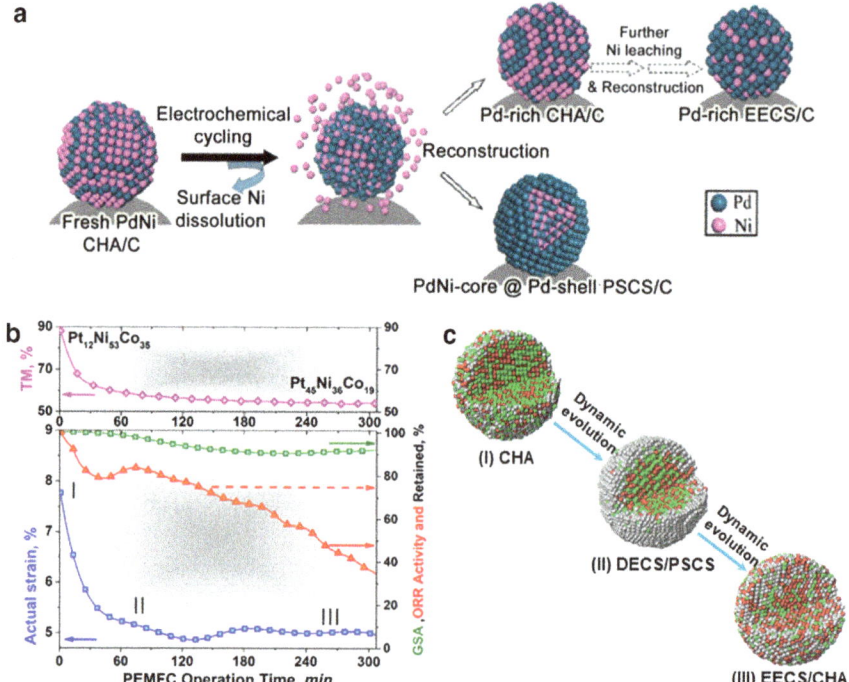

Fig. 5.6 **a** A schematic illustration of the dynamic evolution process for PdNi/C NPs under fuel cell operating condition. Reproduced with permission from ref. [15] Copyright 2015 American Chemical Society. **b** Evolution of atomic-level actual strain (blue), GSA (green), chemical composition of transition metal atomic percentage in NPs (magenta), and apparent ORR activity (red) for $Pt_{12}Ni_{53}Co_{35}$ nanoalloy catalysts operated inside a PEMFC. **c** The corresponding RMC models during structure evolution. Reproduced with permission from ref. [30] Copyright 2019 Royal Society of Chemistry

leach out and then redeposit on the NP surface in a realloying manner. This leaching–redepositing realloying phenomenon would repeat for several cycles along with the electrochemical potential cycling and overall Ni leaching going on. Such a surface alloying process is referred to as the kinetically controlled self-diffusion process [32].

The dynamic nature of the composition–activity relationship during potential cycling is supported by the correlation between ORR mass activity and bimetallic compositions before, during, and after potential cycles. It is evident that Pd% in the alloy NPs changes due to Ni leaching during potential cycling. However, activity–composition correlations reported in a large number of previous works were based on the composition of fresh catalyst, which is inappropriate in view of such strongly dynamic behavior. The maximum ORR mass activities of PdNi NPs are located at 30 at% Pd for fresh catalysts and 70 at% Pd for cycled catalysts after extensive potential cycles. The maximum ORR mass activity after a certain number of potential cycles is very likely to be linked to a PdNi alloy NP with composition near 50 at%. To clearly show the structural difference of PdNi NPs before and after potential cycling, along with a comparison between results obtained by *in operando* and ex situ approaches, RMC models were carried out. In general, similar evidences were exhibited by both *in operando* and ex situ studies, ensuring the high reliability of both *in operando* and ex situ data. $Pd_{30}Ni_{70}$ NPs with an initially low Pd content experience a significant Ni loss (80%) after 3,000 potential cycles and then finalize at a dynamically stable state with 70–80 at% Pd enrichment in the NPs, an increase in particle size about 2 nm, and a great expansion in lattice parameters. The *in operando* approach could provide very useful information of the transient states during the potential cycling, which is crucial for the understanding of the dynamic evolution process [33]. The dynamic process is indicated oscillations of the PDF peak in position and intensity for the nanoalloy catalysts under the fuel cell operating condition, reflecting the dealloying/realloying-induced fluctuations of the metal–metal coordination radii and numbers [33]. Taken together, a dynamically structural evolution dominates the whole process during the electrochemical cycling, in which PdNi NPs dynamically undergo a slow dealloying–realloying process and finalize at a PdNi core and Pd-enriched shell structure after extensive potential cycles.

In comparison with Pd-based NP catalysts, Pt-based NP catalysts have been extensively studied, demonstrating better performance for ORR and PEMFC. Examples include metal organic framework-derived Pt_3Co NPs [34], TM-doped Pt_3Ni octahedral [35], etc. Compared with Pd-based NPs, Pt-based NPs usually exhibit better anti-corrosive properties to harsh electrolytes hence better catalytic stability. Pt–TM alloy NP electrocatalysts also experience a dynamically structural evolution process under electrochemical and fuel cell operating conditions, which show less obvious dynamics than Pd-based counterparts and are discussed in this section. A set of PtNiCo ternary NPs with varying compositions, a binary PtCo, and pure Pt counterparts were recently studied by *in operando* HE-XRD and atomic PDF analysis at an in situ performed PEMFC [30]. The detailed phase structures and lattice parameters of fresh and post-cycled pure Pt- and Pt-based nanoalloy catalysts are studied. The as-prepared and 5 h-cycled pure Pt NPs are well simulated by uniform fcc-type structural models with lattice parameters of 3.919 Å and 3.922 Å, respectively. The lattice parameter of Pt NPs approaches the bulk value during potential cycling but is always slightly shorter than the bulk value of 3.924 Å. The fresh binary $Pt_{68}Co_{32}$ alloy NPs show a chemically ordered fcc-type structure with a compressed lattice constant of 3.855 Å. After about 1600 potential cycles (300 min), the structure of $Pt_{68}Co_{32}$ alloy NPs dynamically evolve to a chemically disordered phase with an expanded lattice parameter of 3.862 Å. Three ternary PtNiCo NPs with high ($Pt_{58}Ni_{17}Co_{25}$), middle ($Pt_{37}Ni_{39}Co_{24}$), and low ($Pt_{12}Ni_{53}Co_{35}$) Pt contents all feature chemically disordered fcc-type structures in the as-prepared states. Under operating conditions, the ternary PtNiCo NPs maintain the chemically disordered fcc-type structures but undergo a lattice parameter expansion to different degrees. Generally, Pt-based alloy NPs undergo a significant loss of base metals during potential cycling and a relaxation of compressive strain, and finally relatively stabilize in a structure of alloy core and Pt-rich shell.

The *in-situ/operando* results provide detailed information assessing the evolution of chemical composition, atomic compressive strain in correlation with the electrocatalytic activity of Pt-based NPs in the MEA and inside a PEMFC. The best fitted models reflecting the nanophase state of the NPs are also provided. In the case of pure Pt NPs, a low-level compressive strain relaxes at the beginning of the potential cycling. Meanwhile, the ORR catalytic activity of pure Pt NPs concurrently drops by a few percent. The further activity decay during potential cycling is most likely to be linked with the gradual decrease of the geometric surface area (GSA) or ECSA of the NPs. In contrast, the ORR activity losses of fresh Pt–TM alloy NPs at the beginning of electrochemical operating are much larger than that of pure Pt NPs. This significant initial activity decay is found to be proportional to the percentage of the base metal leached out from the NPs and the fast relaxation of compressive strain. For example, fresh $Pt_{68}Co_{32}$ NPs in a chemically ordered nanoalloy state undergo an instant 20% ORR activity decrease after the initial few potential cycles. Concurrently, about 25% base metal leaches out from the alloy NPs coupled with the compressive strain in the nanoalloy decreases from 1.9 to 1.75%. The structure of the NPs quickly and dynamically transforms to a base metal-rich ternary alloy core and Pt-rich shell structure. After further 1200 potential cycles, the NPs continue to leach base metal out at a relatively much slower pace and an almost negligible atomic compressive strain release. Interestingly, during the period of further potential cycling, the GSA of the NPs diminishes by about 10%, accompanied by a further ORR activity decay of 20%. The structure of the NPs dynamically changes slowly and preserves the characteristics of base metal-rich core and Pt-rich shell upon further operation. Taken together, two clear stages can be divided during the whole operating time to seek the reason for activity loss, i.e., (i) initial stage at the beginning of the potential cycling dominated by 3d-TM leaching and compressive strain relaxation, (ii) further cycling stage taken over by gradual GSA increase as a result of NP aggregation. This dynamically structural evolution process is validated by other ternary Pt-based nanoalloy electrocatalysts, such as $Pt_{58}Ni_{17}Co_{25}$ and $Pt_{37}Ni_{39}Co_{24}$.

The dynamically structural evolution of Pt-poor $Pt_{12}Ni_{53}Co_{35}$ NPs is interesting (Fig. 5.6b, c). The NPs initially experience a sharp 3d-TM loss of 65% and "decompress" by about 2.5%. In the meantime, the ORR activity loses by 30%. Surprisingly, the NPs undergo a drastically dynamic transformation into a full base metal core@thick Pt shell-type structure and maintain about 1 h. The ORR activity seems to rebound a bit at this transient state with a Pt-rich surface. This base metal@Pt core@shell structure is expected to be stable under operating conditions conventionally. However, further cycling induces a small amount of 3d-TM leaching out but the NPs reconstruct to a 3d-TM rich core@Pt rich shell state or another alloy state relatively enriched by Pt. The structural instability and the increased GSA contribute together to a further obvious ORR activity loss. This dynamically structural evolution process is not only limited to the conventional dealloying phenomena, in which base metals keep leaching out in harsh operating conditions, but other transient transition states may also be involved during this process. The evolution of core–shell structure and its dynamics is highly dependent on the initial state of nanomaterials, including metal combinations and compositions, elements distributions, and morphologies.

In addition to the above examples highlighting the evolution in chemical composition, compressive strain, and atomic distribution during the electrocatalytic reaction, the changes in phase states of the catalysts in some cases have also been identified by in situ HE-XRD/PDF characterizations. In a similar study of $Pt_{37}Ni_{29}Cu_{34}$ NPs with an initial single fcc phase, the dissolutions of Ni and Cu were observed due to their rather low reduction potentials in the acidic environment. After the first hour of the PEMFC operation, the NPs were shown to undergo phase segregation forming a mixed phase with less-densely packed tetragonal (44 vol%) and fcc (56 vol%) nanophases. A further five-hour operation results in more Ni species dissolution but the structure still remains phase segregated. The transformation of the phase states in alloy NPs is believed to be responsible for the activity change during the PEMFC operation [36].

5.3.2 Pt-Based Nanowire Catalysts

In comparison with zero-dimensional (0D) NPs, PGM-based NWs feature a one-dimensional (1D) nanostructure with unique anisotropic nature, effective mass and charge transfer, and lower tendencies of atom dissolution and Ostwald ripening/aggregation. Pt-based alloy NWs are a family of highly promising fuel cell electrocatalysts, which evolve to core–sheath structures under electrochemical potential cycling. Selected binary Pt–TM NWs are described here as examples to illustrate the dynamic structure evolution.

One recent example involves the exploration of PtFe NWs with different bimetallic compositions as fuel cell electrocatalysts [6]. The NWs were synthesized by a one-step surfactant-free hydrothermal method. The dynamic structure evolution process of PtFe NWs under electrochemical and PEMFC operating conditions was examined by a combination of *in operando* together with ex situ techniques. In terms of the phase structures of fresh PtFe NWs, both $Pt_{24}Fe_{76}$ and $Pt_{42}Fe_{58}$ NWs show a mixed-phase structure consisting of fcc and bcc types, while $Pt_{71}Fe_{29}$ NWs feature a long-range correlation/ordered fcc phase. In a set of systematic ex situ experiments, the chemical compositions, lattice parameters, and ORR mass activities of PtFe NWs were studied. $Pt_{24}Fe_{76}$ and $Pt_{42}Fe_{58}$ NWs both experience an instant Fe metal leaching to a final Pt content of about 70 at% within the first few potential cycles and then a slight composition fluctuation during further 40,000 cycles. $Pt_{71}Fe_{29}$ NWs with a high initial Pt content 70 at% show a steady composition throughout the entire potential cycling process. Interestingly, no matter how much Pt is fed in the fresh PtFe NWs with low (24 at%), middle (42 at%), and high (71 at%) Fe content, after the initial 50 potential cycles in the electrochemical environment, all of the PtFe NWs reach a chemically steady state containing 70 at% Pt. The trend for the increase of lattice constant is found to be very similar to the Pt content changing tendency. A maximized ORR mass activity is located at $Pt_{24}Fe_{76}$ NWs compared to PtFe NWs with other compositions. During the potential cycling of $Pt_{24}Fe_{76}$ and $Pt_{42}Fe_{58}$ NWs, the ORR mass activity undergoes three processes, (i) a slight increase after 50 cycles, (ii) keeps increasing and finally reaches a maximum mass activity after 20,000 cycles, and (iii) starts to decay from 20,000 to 40,000 and continuing potential cycles. This tendency coincides with that of the Pt content evolution under the operating condition.

As shown by the in situ HE-XRD results, the lattice parameters experience a dramatic increase in the first 6 potential cycles and then keep fluctuating slightly in the further extended 1,500 potential cycles (Fig. 5.7a, b). The EDS compositions of $Pt_{24}Fe_{76}$ NWs undergo a sharp Pt enrichment after the first 6 cycles and then keep leaching Fe at a relatively slow rate in the next 1,500 cycles.

The pristine Fe-rich PtFe NWs experience a quick TM (Fe) leaching during the initial 6 potential cycles (Fig. 5.7c). As a result of TM leaching, PtFe NWs quickly transform into a PtFe alloy core and Pt-rich sheath structure with a decreased NW diameter. This step is highly dynamic due to the super-fast TM leaching rate. The as-formed core–sheath structure undergoes a slowly dynamic reconstruction upon the further 1,500 potential cycles, where TM leach out very slowly and PtFe NW mainly maintains the core–sheath structure feature but transforms slightly back with a little bit alloy favor. Based on the modeling results, there is a great propensity of fcc and bcc mixed-phase structure to dynamically transform into a single fcc structure under electrochemical and PEMFC operations. Although PtFe NWs with varying compositions finalize in a chemical composition around $Pt_{70}Fe_{30}$, $Pt_{24}Fe_{76}$ NWs always show the highest ORR mass activity compared with others. The ORR activity shows an order of $Pt_{24}Fe_{76} > Pt_{42}Fe_{58} > Pt_{71}Fe_{29}$ throughout the whole potential cycling process, which is highly dependent on the initial Fe content or compressive strain instead of the final composition after sufficient cycles. Overall, $Pt_{24}Fe_{76}$ NWs dynamically undergo a structural evolution process from uniform alloy pristine state to a PtFe alloy core and Pt-rich sheath structure. This in situ-formed core–sheath structure originated from the initial quick dealloying, and the subsequent realloying processes enable super-high ORR activity and durability for PtFe NWs. The information provided here is very useful to guide the future fuel cell catalyst design for achieving high performance.

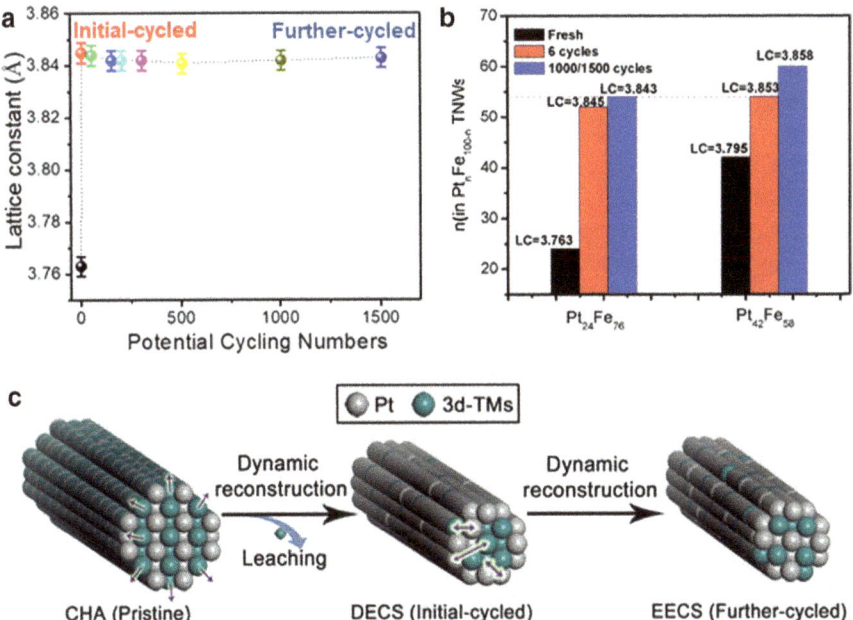

Fig. 5.7 **a** Plot of the lattice constant of Pt$_{24}$Fe$_{76}$ NWs versus potential cycling number. **b** Compositions and lattice parameters change during fuel cell operation. **c** A schematic illustration of TM-leaching and dynamic reconstruction of Pt–TM NWs during potential cycling. Reproduced with permission from ref. [6] Copyright 2020 American Chemical Society

Besides PtFe NWs, other Pt-based alloy NWs such as PtNi NWs and PtCu NWs are also studied as fuel cell electrocatalysts and experience similar dynamically structural evolution to yield Pt-based alloy core and Pt-rich sheath structures [37, 38]. For example, PtNi NWs with a Pt:Ni ratio of 3:2 exhibit a clear lattice expansion based on Vegard's law, which coincides with the maximum ORR mass activity among all PtNi NWs with other compositions. As Fe and Ni share similar chemical properties as 3d-TMs, Pt$_3$Ni$_2$ NWs also experience a similar dynamic evolution from a uniform alloy initial state to a PtNi alloy core and Pt-rich surface sheath after 5,000 potential cycles, which is evidenced by HR-TEM, ICP-OES, and XPS characterizations [37]. This dynamically structural evolution process on NW electrocatalyst is also validated in the EOR, as evidenced by a case of PtCu NWs. Ultrathin PtCu alloy NWs containing a Pt:Cu atomic ratio of 32:68 exhibit a maximum mass activity toward the EOR, which is twice as high as pure Pt NWs. Bimetallic Pt$_{32}$Cu$_{68}$ alloy NWs with uniform elemental distribution undergo a dealloying process during the EOR test. A portion of base metal Cu leaches out during the dealloying process as a result of potential cycling. Meanwhile, the structure of PtCu NWs dynamically transforms to a PtCu alloy core and Pt-enriched outmost surface sheath. This dynamic process is largely originated from the harsh operating conditions and the applied potential as two strong driving forces.

5.4 Core−Shell/Alloy Structures and Evolution Under Gas-Phase Reaction Conditions

In comparison with the core–shell/alloy structures and evolution under electrolyte-phase reaction conditions, the PGM-based NP catalyst under the gas-phase reaction conditions does not involve leaching of the base metals from the catalyst into the gas phase except in the case of sublimation under very high reaction temperatures. However, like those in electrolyte-phase reaction conditions, the nanophase and surface structures could undergo significant dynamic structural and compositional evolution under the gas-phase reaction condition. The structure evolution in the gas-phase reaction could be more pronounced than in the electrolyte phase given the fact that many gas-phase reactions, especially oxidation reactions, occur at high temperatures. Selected recent examples will be highlighted in this subsection to illustrate the core–shell/alloy structural evolution of NP catalysts in gas-phase oxidation reactions such as CO oxidation and hydrocarbon oxidation.

5.4.1 Carbon Monoxide Oxidation Reaction

The study of carbon monoxide oxidation over multimetallic NPs serves as an important probe to the detailed surface structure of the catalyst. One recent example involves PtAuNi NPs, as described earlier. The formation of a CHA-core@oxygenated-shell EECS structure occurs under reaction/thermochemical annealing conditions, which represents a structural evolution of the nanocatalysts with an Au-rich core/PtNi-rich shell nanostructure [21, 39]. The PtAuNi catalysts were activated under CO oxidation reaction (CO (1%) + O_2 (10%) in He) at 80 °C for 2 h) and successive oxygenation atmosphere (O_2 (20%) in He) and were evaluated by combined in situ HE-XRD/PDFs and DRIFTs techniques. The initial thermal expansion from 2.70 Å to 2.76 Å in experimental PDFs spectra under He atmosphere induced significant diminishing of the compressive atomic-level stresses at the NP surface (Fig. 5.8a). After exposure to CO oxidation conditions for 2 h, the cleaned-up (in He) surfaces of $Pt_{58}Au_{42}$, $Pt_{36}Au_9Ni_{55}$, and $Pt_{40}Au_{20}Ni_{40}$ alloy NPs appear very disordered at atomic-level evident by smearing of the oscillatory peaks in respective surface-specific atomic PDFs. $Pt_{36}Au_9Ni_{55}$ alloy NPs appear to display "hard" disordering characteristics of cubic-like NiO and/or tetragonal-like PtO in surface shell compared to relatively "soft" $Pt_{58}Au_{42}$ and $Pt_{40}Au_{20}Ni_{40}$ NPs (Fig. 5.8b). The degree of surface layer oxidation increases in the order $Pt_{36}Au_9Ni_{55}$ < $Pt_{58}Au_{42}$ ~ $Pt_{40}Au_{20}Ni_{40}$ alloy NPs.

Simultaneously, DRIFTS spectra for fresh $Pt_{36}Au_9Ni_{55}$ alloy NPs were collected, which did not show the presence of CO species adsorbed on the NP's surface, indicating a superb reactivity of CO on the NP's surface (Fig. 5.8c) which can be attributed to the fact that surface Pt atoms in the NPs largely remain under compressive stress (the average surface bonding distances remains $\leq$2.78 Å). The catalytic

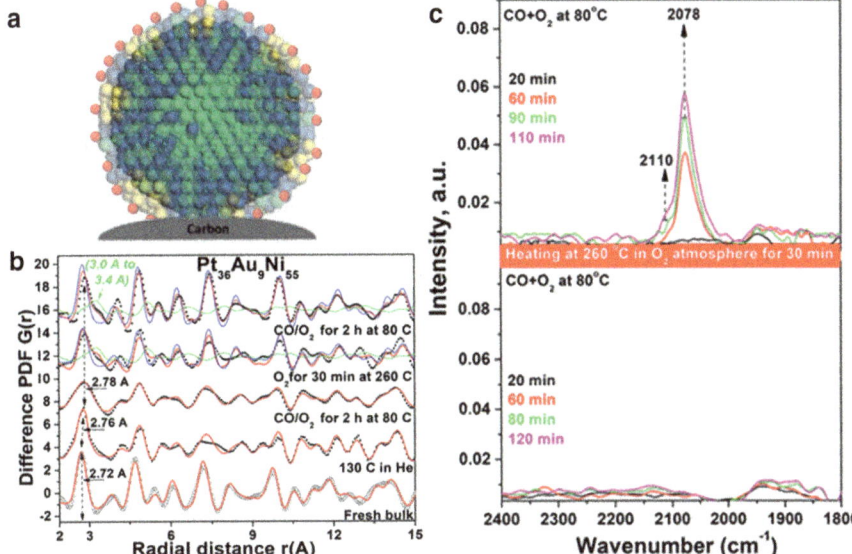

Fig. 5.8 **a** illustration of surface oxygenated PtAuNi catalysts featuring an Au core@PtNi shell structure; **b** Surface-specific atomic PDFs for $Pt_{36}Au_9Ni_{55}$ alloy NPs annealed in He atmosphere and then exposed to a sequence of He–CO oxidation–O_2 reactivation–CO oxidation reactions. Peaks in the surface-specific PDFs are seen to change both in position and intensity with changes in the reaction conditions, indicating a continuous reconstruction of the surface of probed $Pt_{36}Au_9Ni_{55}$ alloy NPs. **c** Selected DRIFTS spectra for adsorption of CO molecules on the surface of probed $Pt_{36}Au_9Ni_{55}$ alloy NPs. The CO oxidation exposure time is given for each data set in the respective color. Vertical broken lines mark characteristic C≡O stretching frequencies. Reproduced with permission from ref. [39] Copyright 2018 American Chemical Society

activity follows the order $Pt_{40}Au_{20}Ni_{40} < Pt_{58}Au_{42} < Pt_{36}Au_9Ni_{55}$ consistent with the "hard" disordering structure in $Pt_{36}Au_9Ni_{55}$ featuring a NiO/PtO surface shell. It is believed the driving force of the disordering may be a further softening of the surface metal-to-metal bonds arising from a transfer of charge from the nanoalloy's surface (back-donation) into the vacant (antibonding) $2\pi^*$-orbitals of the adsorbed CO molecules, and the presence of a significant surface structural disorder helps the formation of surface metal–oxygen species which, in turn, depends on NPs' chemical composition.

5.4.2 Propane Oxidation Reaction

As stated earlier, the need for active and stable oxidation catalysts is driven by the demands in the production of valuable chemicals, remediation of hydrocarbon pollutants, and energy sustainability. A new approach to oxidation catalysts for total oxidation of hydrocarbons (e.g., propane) was proposed based on surface oxygenation of platinum (Pt)-alloyed multicomponent nanoparticles (e.g., platinum–nickel cobalt (Pt–NiCo)) (Fig. 5.9) [40]. The as-synthesized ternary ($n = 42$, $m = 39$) alloy nanoparticles feature an average size of 4.9 (± 0.6) nm and a lattice spacing of 0.188 nm characteristic of (111) crystal plane (Fig. 5.9a, left). Upon thermochemical annealing, the catalysts feature an oxygenated Pt–NiCoO surface layer and disordered ternary alloy core, simply NA − SONA as evident by a lattice spacing of 0.204 nm at edges and 0.177 nm in the center (Fig. 5.9a), indicative of a surface layer of PtNiOCoO with 1.4 nm thickness which accounts for 6–7 atomic layers (Fig. 5.9a, right).

The ternary catalyst design features multifunctional surface active sites with an active Pt center and self-generated and self-perished surface oxygen-activating NiO/CoO sites, enabling not only composition-controllable but also dynamically tunable activity and stability since the degree of surface oxygenation depends on the alloying composition and phase structures. By analyzing the lattice parameters based on the HE-XRD/PDF data and the coordination number (CN) of metal components based on EXAFS spectral fitting (Fig. 5.9b), the catalysts were found to feature a long-range disordered alloy character in the core with different degrees of metal oxygenation. The detection of the high level of oxygenation for Ni and Co (N(Ni–O) −4.7, N(Co–O) −2.8) and the low level of Pt oxygenation (N(Pt–O) −1.1) reflect the ability to harnessing the surface oxygenation (Fig. 5.9b).

Remarkable catalytic activity and stability were obtained upon the formation of NA − SONA catalysts, which is evidenced by the fact that T_{50}−282 °C (Fig. 5.9c) was lower than that of commercial Pt/Al$_2$O$_3$ catalyst with the same total metal loading (T_{50}−301 °C). The NA − SONA features superior stability, showing no indication of deactivation after 800 °C hydrothermal aging (T_{50}−277 °C, 1.0 wt%) while a decay of catalytic activity over Pt catalysts is evidenced by an increase of T_{50} to 332 °C after aging (Fig. 5.9c). For further in-depth exploration of the catalytic synergy, in situ/operando time-resolved studies, including HE-XRD/PDF and DRIFTS, revealed largely irregular oscillatory kinetics associated with the dynamic lattice expansion/shrinking and ordering/disordering processes. The coupling of a partial positively charged Pt with the oxyphilic Ni–O and Co–O species was shown to play an important role in the active surface for the cleavage of carbon–carbon bonds of the adsorbed propane and the effective removal of reaction intermediates on the surface. The refinement of the surface oxygenation strategy may lead to a paradigm shift in the design of active and stable catalysts for catalytic oxidation reactions.

5.5 Summary

The explorations of multimetallic nanoparticle catalysts and electrocatalysts have generated significant insights into the dynamic nature of core–shell and alloy structures, which plays an important role in the catalytic and electrocatalytic synergies. The core–shell and alloy phase structures are shown to undergo dynamic evolution

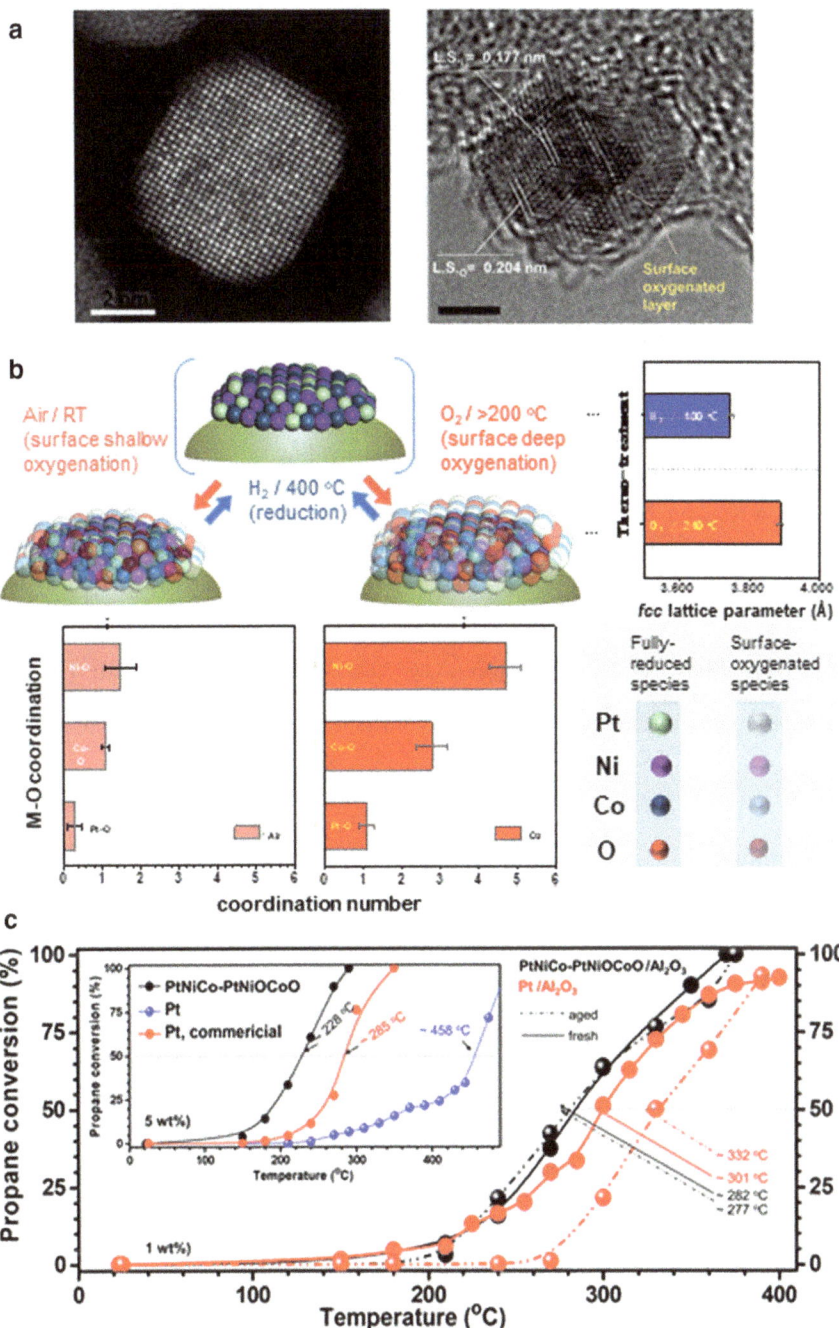

◀**Fig. 5.9** **a** Aberration-corrected HAADF-STEM images (left) for as-synthesized $Pt_nNi_mCo_{100-n-m}$ (n = 42, m = 39) and a representative HR-TEM of PtNiCo − $PtNiOCoO/Al_2O_3$ (1.0 wt%), scale bar = 2 nm. **b** Surface oxygenation of a ternary alloy catalyst: Top-left panel: illustrations of a fully deoxygenated state and the surface-oxygenated states of two different oxygenation degrees under the indicated conditions; Top-right panel: the corresponding fcc lattice parameter changes underdetermined by in situ HE-XRD/PDF analysis; Bottom-left panel: the corresponding M–O coordination number determined by ex situ EAXFS spectral analysis; **c** Plot of propane conversion over the ternary catalyst derived from PtNiCo − $PtNiOCoO/Al_2O_3$ (1.0 wt%, black) and Pt/Al_2O_3 (1.0 wt%, red): freshly prepared (solid curve), and hydrothermally aged under $10\% CO_2 + 10\% H_2O + N_2$ at 800 °C for 16 h (dash curve). Values of T_{50} (i.e., the temperature at which 50% conversion is achieved) are indicated in the plots. Reproduced with permission from ref. [40] Copyright 2020 Nature Publishing Group

under electrochemical potential cycling in electrolyte-phase reactions or under gas-phase oxidation conditions. The in-depth understanding of the atomic-scale details in the dynamic evolution is crucial for controlling the core–shell structures, which requires advanced in situ/*operando* investigations with the aid of computational modeling. Such investigations will provide new insights into the design of active, selective, robust, and low-cost PGM catalysts for a wide range of applications in the field of heterogeneous catalysis.

Acknowledgements This work was supported by National Science Foundation (CHE 1566283) and the Department of Energy – Basic Energy Sciences (DE-SC0006877). The authors also thank all collaborators for their contributions, especially Valeri Petkov, to the in-situ/operando studies.

References

1. Seh ZW, Kibsgaard J, Dickens CF, Chorkendorff I, Norskov JK, Jaramillo TF (2017) Combining theory and experiment in electrocatalysis: insights into materials design. Science 355:eaad4998
2. Li J, Yin HM, Li XB, Okunishi E, Shen YL, He J, Tang ZK, Wang WX, Yücelen E, Li C, Gong Y, Gu L, Miao S, Liu LM, Luo J, Ding Y (2017) Surface evolution of a Pt–Pd–Au electrocatalyst for stable oxygen reduction. Nat Energy 2:17111
3. Wu ZP, Shan S, Zang SQ, Zhong CJ (2020) Dynamic core–shell and alloy structures of multi-metallic nanomaterials and their catalytic synergies. Acc Chem Res 53:2913–2924. https://doi.org/10.1021/acs.accounts.1020c00564
4. Wu ZP, Caracciolo DT, Maswadeh Y, Wen J, Kong Z, Shan S, Vargas JA, Yan S, Hopkins E, Park K, Sharma A, Ren Y, Petkov V, Wang L, Zhong CJ (2021) Alloying–realloying enabled high durability for Pt–Pd–3d-transition metal nanoparticle fuel cell catalysts. Nat Commun 12:859. https://doi.org/10.21203/rs.21203.rs-54923/v21201
5. Wu ZP, Shan S, Xie ZH, Kang N, Park K, Hopkins E, Yan S, Sharma A, Luo J, Wang J, Petkov V, Wang L, Zhong CJ (2018) Revealing the role of phase structures of bimetallic nanocatalysts in the oxygen reduction reaction. ACS Catal 8:11302–11313
6. Kong Z, Maswadeh Y, Vargas JA, Shan S, Wu ZP, Kareem H, Leff AC, Tran DT, Chang F, Yan S, Nam S, Zhao X, Lee JM, Luo J, Shastri S, Yu G, Petkov V, Zhong CJ (2020) Origin of high activity and durability of twisty nanowire alloy catalysts under oxygen reduction and fuel cell operating conditions. J Am Chem Soc 142:1287–1299

7. Wu ZP, Miao B, Hopkins E, Park K, Jiang CY, H, Zhang M, Zhong CJ, Wang L (2019) Poisonous species in complete ethanol oxidation reaction on palladium catalysts. J Phys Chem C 123:20853–20868

8. Zhong CJ, Maye MM (2001) Core−shell assembled nanoparticles as catalysts. Adv Mater 13:1507–1511

9. Zhong CJ, Luo J, Njoki PN, Mott D, Wanjala BN, Loukrakpam R, Lim S, Wang L, Fang B, Xu Z (2008) Fuel cell technology: nano-engineered multimetallic catalysts. Energy Environ Sci 1:454–466

10. Xia Y, Gilroy KD, Peng HC, Xia X (2017) Seed-mediated growth of colloidal metal nanocrystals. Angew Chem Int Ed 56:60–95

11. Wanjala BN, Luo J, Loukrakpam R, Fang B, Mott D, Njoki PN, Engelhard M, Naslund HR, Wu JK, Wang L, Malis O, Zhong CJ (2010) Nanoscale alloying, phase-segregation, and core−shell evolution of gold−platinum nanoparticles and their electrocatalytic effect on oxygen reduction reaction. Chem Mater 22:4282–4294

12. Tao F, Grass ME, Zhang Y, Butcher DR, Renzas JR, Liu Z, Chung JY, Mun BS, Salmeron M, Somorjai GA (2008) Reaction-driven restructuring of Rh-Pd and Pt-Pd core-shell nanoparticles. Science 322:932–934

13. Bu L, Shao Q, E B, Guo J, Yao J, Huang X, (2017) PtPb/PtNi intermetallic core/atomic layer shell octahedra for efficient oxygen reduction electrocatalysis. J Am Chem Soc 139:9576–9582

14. Tian X, Luo J, Nan H, Zou H, Chen R, Shu T, Li X, Li Y, Song H, Liao S, Adzic RR (2016) Transition metal nitride coated with atomic layers of Pt as a low-cost, highly stable electrocatalyst for the oxygen reduction reaction. J Am Chem Soc 138:1575–1583

15. Wu J, Shan S, Petkov V, Prasai B, Cronk H, Joseph P, Luo J, Zhong CJ (2015) Composition-structure-activity relationships for palladiumalloyed nanocatalysts in oxygen reduction reaction: an Ex-Situ/In-Situ high energy X-ray diffraction study. ACS Catal 5:5317–5327

16. Wu Z, Zhang M, Jiang H, Zhong CJ, Wang CY, L, (2017) Competitive C-C and C–H bond scission in the ethanol oxidation reaction on Cu(100) and the effect of an alkaline environment. Phys Chem Chem Phys 19:15444–15453

17. Shan S, Luo J, Wu J, Kang N, Zhao W, Cronk H, Zhao Y, Joseph P, Petkov V, Zhong CJ (2014) Nanoalloy catalysts for electrochemical energy conversion and storage reactions. RSC Adv 4:42654–42669

18. Shan S, Yang L, Luo J, Zhong CJ (2014) Nanoalloy catalysts: structural and catalytic properties. Catal Sci Technol 4:3570–3588

19. Wu CH, Liu C, Su D, Xin HL, Fang HT, Eren B, Zhang S, Murray CB, Salmeron MB (2019) Bimetallic synergy in cobalt–palladium nanocatalysts for CO oxidation. Nat Catal 2:78–85

20. Huang ZF, Song J, Du Y, Xi S, Dou S, Nsanzimana JMV, Wang C, Xu ZJ, Wang X (2019) Chemical and structural origin of lattice oxygen oxidation in Co–Zn oxyhydroxide oxygen evolution electrocatalysts. Nat Energy 4:329–338

21. Lu A, Wu ZP, Chen B, Peng DL, Yan S, Shan S, Skeete Z, Chang F, Chen Y, Zheng H, Zeng D, Yang L, Sharma A, Luo J, Wang L, Petkov V, Zhong CJ (2018) From a Au-rich core/PtNi-rich shell to a Ni-rich core/PtAu-rich shell: an effective thermochemical pathway to nanoengineering catalysts for fuel cells. J Mater Chem A 6:5143–5155

22. Wu ZP, Lu XF, Zang SQ, Lou XW (2020) Non-noble-metal-based electrocatalysts toward the oxygen evolution reaction. Adv Funct Mater 30:1910274

23. Maswadeh Y, Shan S, Prasai B, Zhao Y, Xie ZH, Wu Z, Luo J, Ren Y, Zhong CJ, Petkov V (2017) Charting the relationship between phase type-surface area-interactions between the constituent atoms and oxygen reduction activity of Pd–Cu nanocatalysts inside fuel cells by in operando high-energy X-ray diffraction. J Mater Chem A 5:7355–7365

24. Park HY, Schadt MJ, Wang L, Lim IS, Njoki PN, Kim SH, Jang MY, Luo J, Zhong CJ (2007) Fabrication of magnetic core@shell Fe oxide@Au nanoparticles for interfacial bioactivity and bio-separation. Langmuir 23:9050–9056

25. Wang L, Luo J, Maye MM, Fan Q, Rendeng Q, Engelhard MH, Wang C, Lin Y, Zhong CJ (2005) Iron oxide–gold core–shell nanoparticles and thin film assembly. J Mater Chem 15:1821–1832

26. Wang L, Luo J, Fan Q, Suzuki M, Suzuki IS, Engelhard MH, Lin Y, Kim N, Wang JQ, Zhong CJ (2005) Monodispersed core−shell Fe_3O_4@Au nanoparticles. J Phys Chem B 109:21593–21601

27. Wang L, Wang X, Luo J, Wanjala BN, Wang C, Chernova NA, Engelhard MH, Liu Y, Bae IT, Zhong CJ (2010) Core−shell-structured magnetic ternary nanocubes. J Am Chem Soc 132:17686–17689

28. Li J, Skeete Z, Shan S, Yan S, Kurzatkowska K, Zhao W, Ngo QM, Holubovska P, Luo J, Hepel M, Zhong CJ (2015) Surface enhanced raman scattering detection of cancer biomarkers with bifunctional nanocomposite probes. Anal Chem 87:10698–10702

29. Lin L, Crew E, Yan H, Shan S, Skeete Z, Mott D, Krentsel T, Yin J, Chernova NA, Luo J, Engelhard MH, Wang C, Li Q, Zhong CJ (2013) Bifunctional nanoparticles for SERS monitoring and magnetic intervention of assembly and enzyme cutting of DNAs. J Mater Chem B 1:4320–4330

30. Petkov V, Maswadeh Y, Vargas JA, Shan S, Kareem H, Wu ZP, Luo J, Zhong CJ, Shastri S, Kenesei P (2019) Deviations from Vegard's law and evolution of the electrocatalytic activity and stability of Pt-alloys inside fuel cells by in operando X-ray spectroscopy and total scattering. Nanoscale 11:5512–5525

31. Strasser P, Koh S, Anniyev T, Greeley J, More K, Yu C, Liu Z, Kaya S, Nordlund D, Ogasawara H, Toney MF, Nilsson A (2010) Lattice-strain control of the activity in dealloyed core-shell fuel cell catalysts. Nat Chem 2:454–460

32. Rizzi M, Furlan S, Peressi M, Baldereschi A, Dri C, Peronio A, Africh C, Lacovig P, Vesselli E, Comelli G (2012) Tailoring bimetallic alloy surface properties by kinetic control of self-diffusion processes at the nanoscale. J Am Chem Soc 134:16827–16833

33. Petkov V, Prasai B, Shan S, Ren Y, Wu J, Cronk H, Luo J, Zhong CJ (2016) Structural dynamics and activity of nanocatalysts inside fuel cells by in operando atomic pair distribution studies. Nanoscale 8:10749–10767

34. Chong L, Wen J, Kubal J, Sen FG, Zou J, Greeley J, Chan M, Barkholtz H, Ding W, Liu DJ (2018) Ultralow-loading platinum-cobalt fuel cell catalysts derived from imidazolate frameworks. Science 362:1276–1281

35. Huang XQ, Zhao Z, Cao L, Chen Y, Zhu E, Lin Z, Li M, Yan A, Zettl A, Wang YM, Duan XF, Mueller T, Huang Y (2015) High-performance transition metal-doped Pt_3Ni octahedra for oxygen reduction reaction. Science 348:1230–1234

36. Petkov V, Maswadeh Y, Zhao Y, Lu A, Cronk H, Chang F, Shan S, Kareem H, Luo J, Zhong CJ, Shastri S, Kenesei P (2018) Nanoalloy catalysts inside fuel cells: an atomic-level perspective on the functionality by combined in operando x-ray spectroscopy and total scattering. Nano Energy 49:209–220

37. Chang F, Yu G, Shan S, Skeete Z, Wu J, Luo J, Ren Y, Petkov V, Zhong CJ (2017) Platinum–nickel nanowire catalysts with composition-tunable alloying and faceting for the oxygen reduction reaction. J Mater Chem A 5:12557–12568

38. Liao Y, Yu G, Zhang Y, Guo T, Chang F, Zhong CJ (2018) Composition-tunable PtCu alloy nanowires and electrocatalytic synergy for methanol oxidation reaction. J Phys Chem C 120:10476–10484

39. Petkov V, Maswadeh Y, Lu A, Shan S, Kareem H, Zhao Y, Luo J, Zhong CJ, Beyer K, Chapman K (2018) Evolution of active sites in Pt-based nanoalloy catalysts for the oxidation of carbonaceous species by combined in situ infrared spectroscopy and total X-ray scattering. ACS Appl Mater Interfaces 10:10870–10881

40. Shan S, Li J, Maswadeh Y, O'Brien C, Kareem H, Tran DT, Lee IC, Wu ZP, Wang S, Yan S, Cronk H, Mott D, Yang L, Luo J, Petkov V, Zhong CJ (2020) Surface oxygenation of multicomponent nanoparticles toward active and stable oxidation catalysts. Nat Commun 11:4201

Chapter 6
Fabrication of Core–Shell Structured Metal Nanoparticles@Metal–Organic Frameworks for Heterogeneous Thermal Catalysis

Guodong Li and Zhiyong Tang

6.1 Introduction

It is well known that catalysis is an old but evergreen field, now accounting for over 90% of chemical processes and 60% of chemical production in the world [1]. Typically, supported metal nanocatalysts have been widely studied in heterogeneous catalysis [2, 3], because they not only combine the function of individual metal nanoparticles (NPs), but also generate the unique synergetic properties with respect to single components. Traditionally, these composites are easily prepared via different methods such as impregnation, coprecipitation, and deposition–precipitation. The obtained products are characteristic of metal NPs dispersed on the surface of supports, in which metal NPs have high surface energy and tend to migrate and aggregate into larger particles, thus leading to loss of the unique properties of the original metal NPs [4]. To address above, metal NPs encapsulated by diverse supports with core-shell structures is recognized as one of the effective strategies, because they possess great potential for not only avoiding migration and aggregation of metal NPs but also generating the uniform and well-defined interfaces between metal cores and porous shells [5].

Among various porous supports such as carbon, metal oxides, and zeolites, metal-organic frameworks (MOFs) [6], also known as porous coordination polymers (PCPs), which are synthesized by self-assembly of metal ions or clusters with ditopic or polytopic organic linkers, have being attracted wide interest in heterogeneous catalysis, due to their intriguing features including extraordinarily large surface area, tunable pore dimensions, diverse metal nodes, and adjustable organic linkers [7]. Moreover, the uniform cavities with long-range order of MOFs could effectively

G. Li · Z. Tang (✉)
CAS Key Laboratory of Nanosystem and Hierarchical Fabrication, CAS Center for Excellence in Nanoscience, National Center for Nanoscience and Technology, Beijing 100190, P. R. China
e-mail: zytang@nanoctr.cn

© The Author(s), under exclusive license to Springer Nature Singapore Pte Ltd. 2021
H. Yamashita and H. Li (eds.), *Core-Shell and Yolk-Shell Nanocatalysts*,
Nanostructure Science and Technology,
https://doi.org/10.1007/978-981-16-0463-8_6

promote the mass transfer and diffusion of substrates and products with respect to traditional porous supports, and thus MOFs possess great potential to encapsulate metal NPs with core–shell typed structures as emerging catalysts. In addition, the encapsulated structure enables the maximized interfaces between metal NPs and MOFs, thus generating the strong electronic transfer between them and exhibiting the intriguing properties for catalysis, even for selective catalysis [8, 9].

6.2 Synthesis of Core–Shell Metal NPs@MOF Catalysts

The core-shell structured metal NPs@MOF catalysts have been widely studied and the developed synthesis strategies can be simply divided into the following two categories (Fig. 6.1) [8, 10]: (1) Metal NPs are confined into the cavities or channels of MOF matrix via ship-in-a-bottle approach; (2) Metal NPs are surrounded by MOFs via bottle-around-ship approach.

6.2.1 Metal NPs Confined in the Cavities or Channels of Host MOFs

To date, "ship-in-a-bottle" approaches including impregnation, coprecipitation, and deposition–precipitation are developed to encapsulate metal NPs by porous MOFs [11, 12], but it is difficult to realize the precise control of the position of metal NPs inside the host MOFs, and metal precursors often partially deposit on the external surface of MOFs to form aggregated counterparts. To effectively make metal NPs inside MOF matrix and prevent them from aggregation, the double-solvent approach has been developed as one of the effective "ship-in-a-bottle" ways [13], in which aqueous solution of equal or less than pore volume was absorbed into the hydrophilic

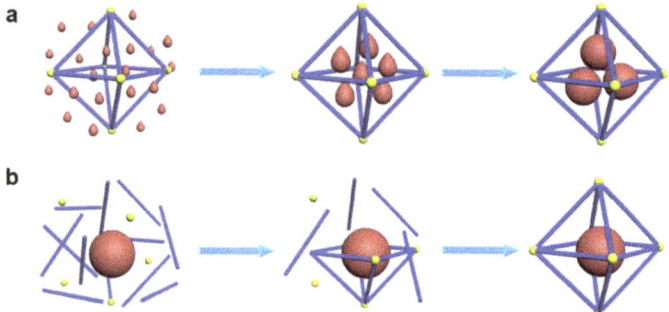

Fig. 6.1 Schematic route for the synthesis of metal NPs encapsulated by MOFs. **a** Ship-in-a-bottle method. **b** Bottle-around-ship method

pores of MOFs by capillary force, while an excess of organic solvent hexane was introduced to limit the metal precursors absorbed on the external surface of MOFs. After that, these samples were further dried, followed by reduction or heat treatment.

Xu et al. first developed a double-solvent approach to synthesize core–shell Pt@MIL-101(Cr) accompanied with hydrogen as reductant (Fig. 6.2) [13]. The obtained 1.8 nm Pt NPs exposed with (111) facets were small enough to be accommodated in the mesoporous cavities of MIL-101(Cr). Electron tomographic reconstruction clearly demonstrated the uniform distribution of Pt NPs throughout the interior cavities of MIL-101(Cr). Chen et al. synthesized mesoPd@NUS-6(Hf), in which a Pd^{2+} precursor at different concentrations was encapsulated in a microporous MOF modified with sulfonic acid group, NUS-6(Hf), via a double-solvent method, and then the Pd^{2+} was subjected to reduction by $NaBH_4$, yielding uniformly hierarchical porous mesoMOFs loaded with Pd NPs [14]. Moreover, the mesopore content could be controlled by modulating the Pd loading. 3.5 nm Pd NPs with exposed (111)

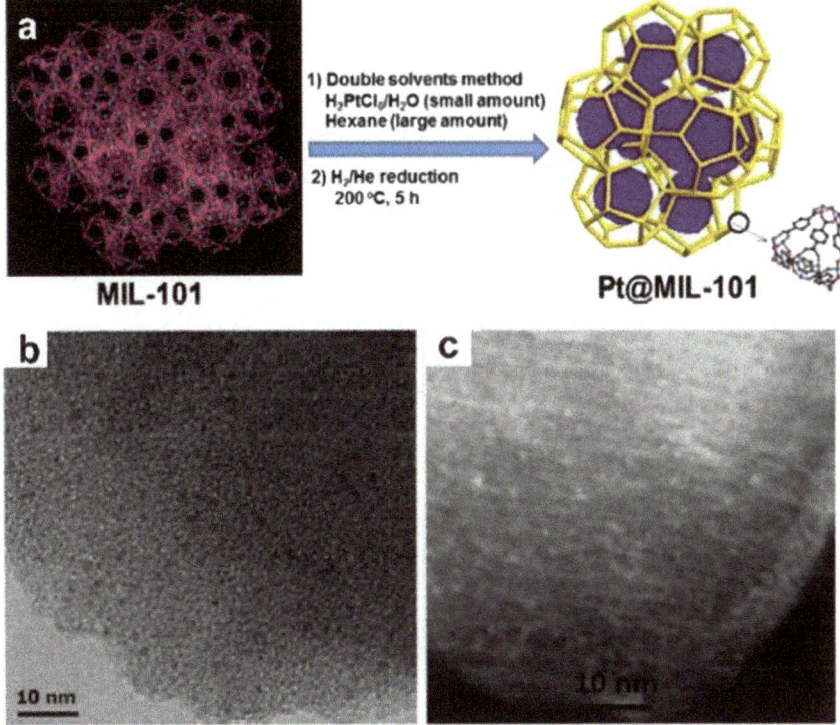

Fig. 6.2 a Scheme for the synthesis of Pt NPs inside MIL-101(Cr) matrix using double-solvent method. **b** TEM images and **c** reconstructed slice by tomography of Pt@MIL-101(Cr) [13]. Copyright 2012, American Chemical Society

facets were well dispersed within the NUS-6(Hf), indicating the successful encapsulation of Pd NPs within the mesopore. Jiang et al. employed UiO-66 with alterable linker groups to encapsulate Pd NPs via the ultrasound-assisted double-solvent approach (DSA), affording Pd@UiO-66-X (X = H, OMe, NH_2, 2OH, 2OH(Hf)) [15]. All the Pd@UiO-66-X samples well inherit the structure from UiO-66, but do not show any diffraction peak for Pd NPs, suggesting that Pd NPs could be small. The Pd contents in Pd@UiO-66-X are in the range of 2.13–2.84 wt%. Meanwhile, all the Pd@UiO-66-X catalysts show similar pore size distributions, making them ideal candidates to investigate the influence of chemical environment. In addition, the photoreduction is greener and milder, which is more favorable for preserving the MOF framework. Li et al. synthesized core–shell Pd@NH_2-UiO-66(Zr) via a double-solvent impregnation method coupled with the photoreduction process [16]. 1.2 nm Pd NPs smaller than the pore size of NH_2-UiO-66(Zr) were highly dispersed in Pd@NH_2-UiO-66(Zr), indicating that Pd NPs were confined in the cavities of NH_2-UiO-66(Zr).

Alloy NPs could be surrounded with MOFs by adding different metal precursors assisted with the coreduction method. Xu et al. developed the double-solvent method coupled with a liquid-phase concentration-controlled reduction strategy to immobilize AuNi NPs within MIL-101(Cr) [17]. When a high-concentration $NaBH_4$ solution (0.6 M) was used, 1.8 nm AuNi NPs were located within the pores of MIL-101(Cr). However, under moderate reduction condition, agglomeration of larger metal NPs on the external surface of MIL-101(Cr) was observed. Similarly, an aqueous solution of $Ni(NO_3)_2$ and $Cu(NO_3)_2$ was dropwise pumped into the hexane containing the suspended MIL-101(Cr), and subsequently, they were reduced by NH_3BH_3 to give CuNi@MIL-101(Cr), where ~3 nm CuNi NPs with a lattice fringe distance of 0.21 nm were observed [18]. Jiang et al. synthesized Pt_1Cu_2@MIL-101(Cr) via the double-solvent and H_2 reduction method. The obtained products are characteristic of 0.5 wt% Pt_1Cu_2 NPs of 1.7 nm in diameter encapsulated by MIL-101(Cr) [19]. No agglomerated large Pt_1Cu_2 NPs were observed, indicating that Pt_1Cu_2 NPs were highly dispersed and embedded inside the MIL-101(Cr) pores. In addition, Huang et al. prepared the ultrafine Pt nanoclusters of 1.4 nm encapsulated in NH_2-UiO-66(Zr), Pt@NH_2-UiO-66(Zr), through a 200 °C reduction in H_2 [20]. After that, an ethanolic solution of Sn^{2+} was impregnated into Pt@NH_2-UiO-66(Zr) based on the Pt content of 3.9wt%. After another gas-phase reduction at 200 °C, PtSn@NH_2-UiO-66(Zr) characteristic of 1.6 nm PtSn NPs was obtained. The Pt $4f$ XPS spectrum showed two peaks of metallic Pt^0 at 71.3 eV ($4f_{7/2}$) and 74.9 eV ($4f_{5/2}$) for Pt nanoclusters confined in NH_2-UiO-66(Zr). As for PtSn@NH_2-UiO-66(Zr), the peaks corresponded to Pt^0 binding energy shifted by less than 0.2 to 71.1 eV. This slight shift in the binding energy indicated the electron transfer from Sn to Pt in forming PtSn nanoclusters.

Altogether, using double-solvent approach effectively realizes the complete encapsulation of metal NPs inside the hydrophilic cavities or channels of MOFs. Nevertheless, it is still challengable to tune the morphology, composition, and spatial distribution of metal NPs inside MOFs.

6.2.2 Metal NPs Surrounded by MOFs

The "bottle-around-ship" approach is widely used to achieve the effective encapsulation of metal NPs by porous MOFs [18]. Generally, this strategy involves two steps. First, metal NPs with well-defined morphologies are prepared. Second, the metal NPs are mixed with the precursor solution of a MOF together to form the core-shell structure. The key issue is to develop the effective strategies to balance the heterogeneous nucleation and controllable growth of MOFs onto metal NP surface against self-aggregation and homogenous growth.

6.2.2.1 Surfactant/Ligand Molecules Induced Encapsulation

Ligands and/or surfactant molecules can be used as a bridge between MOFs and metal NPs to form core-shell nanostructures. Generally, ligand and/or surfactant molecules could reduce aggregation of metal NPs in solution via introducing additional charge and/or steric repulsion, because the precursors for synthesizing MOFs are usually the soluble forms of metal salts, which could shield the charge repulsion among metal NPs to promote their aggregation [5]. For example, Tang et al. dissolved the precursors $Zn(NO_3)_2$ and 2-amino terephthalic acid (NH_2-H_2BDC) in the double solvents containing polyvinylpyrrolidone (PVP), N, N-dimethylformamide (DMF) and ethanol, followed by addition of PVP-stabilized Pd NPs and subsequent solvothermal treatment to prepare the uniform core–shell Pd@IRMOF-3(Zn) nanostructures characteristic of a Pd NP core of about 35 nm in diameter and a uniform IRMOF-3(Zn) shell of about 145 nm in thickness (Fig. 6.3a) [21]. During the synthesis process, PVP not only functionalized as the stabilizer to make the Pd NPs well dispersed, but also provided the affinity to adsorb IRMOF-3(Zn) precursors onto the Pd NP surfaces. Similarly, Tang et al. synthesized core–shell Ag@ZIF-8(Zn) nanowires using a two-step synthesis method [22], in which the PVP-capped Ag nanowires were rather uniform diameters of 90–120 nm (Fig. 6.3b). Furthermore, Yaghi et al. synthesized core–shell Cu@UiO-66(Zr) composites, in which a single PVP-stabilized Cu NP of ~18 nm in diameter was uniformly surrounded by UiO-66(Zr) (Fig. 6.3c) [23]. Also, Tsung et al. introduced cetyltrimethylammonium bromide (CTAB) to bridge the metal NPs and ZIF-8(Zn) and to facilitate the controlled alignment (Fig. 6.3d) [24]. The obtained core–shell metal NP@MOF nanocomposites were composed of single shape-controlled metal NP encased in ZIF-8(Zn), and an alignment between the (100) planes of the Pd NPs and (110) planes of the ZIF-8(Zn) was discerned. Further, Fairen-Jimenez et al. first dispersed the isolated Zr-O clusters in a mixture of DMF and acetic acid, and then the PEG-stabilized AuNRs (AuNR@PEG-SH) were added into this suspension, resulting in a deep red mixture of DMF and acetic acid [25]. Under vigorous stirring, an H_4TBAPy-linker solution was added into DMF and stirred the solution overnight to get the AuNR@NU-901(Zr). The formation of core-shell structures was ascribed to the interactions between PEG and the MOF precursors. Nearly all AuNRs were in

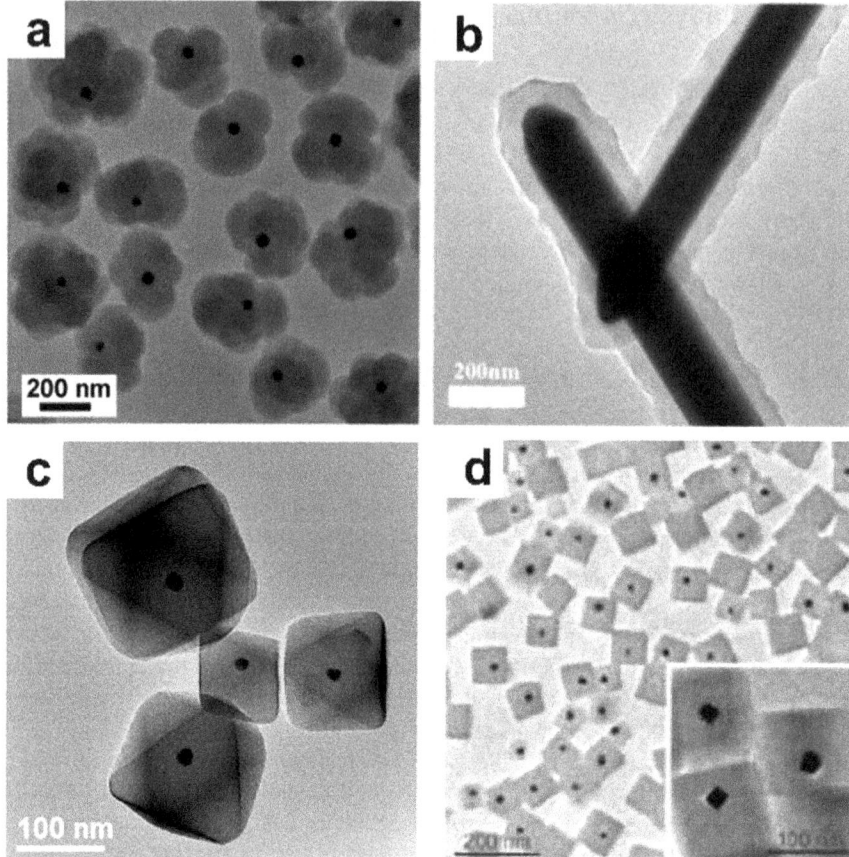

Fig. 6.3 TEM images of **a** Pd@IRMOF-3(Zn) [21], Copyright 2014, American Chemical Society, **b** Ag@ZIF-8(Zn) [22], Copyright 2015, WILEY-VCH Verlag GmbH & Co. KGaA, Weinheim, **c** Cu@UiO-66(Zr) [23], Copyright 2016, American Chemical Society and **d** Pd@ZIF-8(Zn) [24], Copyright 2014, American Chemical Society

the center of the crystallite and had a parallel orientation with respect to the principal prolate axis. Pérez-Juste et al. developed a general strategy for the encapsulation of individual quaternary ammonium surfactant-stabilized metal nanoparticles with the ZIF-8(Zn) [26]. The method can be readily applied to metal particles with arbitrary morphologies, such as gold nanorods, gold nanostars, and gold nanospheres. The hydrophobic hydrocarbon chains of CTA^+ molecules adsorbed on metal surfaces can readily complex with MOF crystals, thus inducing ZIF-8(Zn) nucleation and/or adsorption and further growth on the metal surface.

Different from single metal NP core, multiple metal NPs can be randomly encapsulated by porous MOFs via the time and sequence of adding the as-synthesized metal NPs into the precursors of MOFs. For example, Huo et al. developed PVP-capped nanostructures of various sizes, shapes, and compositions being enshrouded by ZIF-8(Zn) at room temperature, such as Au (13 nm), Pt (2.5, 3.3, and 4.1 nm), and Ag cubes (160 nm) [27]. Similarly, a series of MOF-based nanocomposites including Pt@UiO-66(Zr), Pt@MIL-53(Al), Pt@ZIF-8(Zn), Pt@ZIF-67(Zr), and Pd@UiO-66(Zr) were obtained [28–30]. Kitagawa et al. synthesized PVP-stabilized Pt NCs of 8 nm as the seeds and then mixed with $ZrCl_4$, H_2BDC, and acetic acid in N, N-dimethylformamide (DMF) to form Pt@UiO-66(Zr), which was characteristic of several Pt NCs as core and UiO-66(Zr) with different functional groups [31]. The loadings of Pt included in Pt@UiO-66-H, Pt@UiO-66-H(Hf), Pt@UiO-66-Br, and Pt@UiO-66-Me$_2$ were estimated to be 6.7, 4.6, 5.3, and 5.9 wt%, and the average particle sizes of Pt NCs for Pt@UiO-66-H, Pt@UiO-66-H(Hf), Pt@UiO-66-Br, and Pt@UiO-66-Me$_2$ were estimated to be 7.9, 8.0, 8.1, and 8.0 nm, respectively. Moreover, they also synthesized the Pd@UiO-66(Zr) characteristic with Pd NPs of 8 nm in diameter and 1 wt% loading [32]. This synthesis strategy realizes the effective encapsulation of metal NPs by porous MOFs; however, the local position of metal NPs is hard to be controlled.

To effectively control the local position of multiple metal NPs inside MOFs, sandwich structured composites were developed, in which the synthesis procedure involves synthesis of a MOF core, followed by adsorption of surfactant-stabilized metal NPs, and the subsequent overgrowth of a MOF shell on the MOF core. Tang et al. synthesized ~2.8 nm Pt NPs sandwiched by MIL-101 containing metal nodes of Fe^{3+}, Cr^{3+}, or both (Fig. 6.4) [33]. Octahedral MIL-101(Fe) of ~296 nm and MIL-101(Cr) of ~243 nm were synthesized under solvothermal conditions. Then, PVP-stabilized Pt NPs of ~ 2.8 nm in diameter were homogeneously absorbed onto MIL-101(Fe) or MIL-101(Cr). Finally, the sandwich structures were synthesized by coating another MIL-101(Fe) shell with different thicknesses on as-synthesized supported catalysts, such as MIL-101(Fe)@Pt@MIL-101(Fe) with the shell thickness of ~9.2 and ~22.0 nm, MIL 101(Cr)@Pt@MIL 101(Cr) with the shell thickness of ~5.1 nm, and MIL-101(Cr)@Pt@MIL-101(Fe) with the shell thickness of ~2.9 and ~8.8 nm. Furthermore, this synthesis strategy was applied to construct a series of sandwich structures by different MOFs such as UiO-66(Zr), UiO-67(Zr), MOF-74(Co), and MOF-525(Zr) [34]. Obviously, this synthesis strategy allows good control of the location of the metal NPs in MOF matrix, and the shell composition and thickness can be easily tuned by adjusting the metal precursors and reaction time.

6.2.2.2 Self-template Induced Encapsulation

Self-template direct method is another strategy to encapsulate metal NPs by porous MOFs, that is, the local parts of the metal or alloy NPs are dissolved into metal ions that can be used as metal source to coordinate with the organic linkers and produce MOFs around the left metal NPs. The key issue is how to effectively balance the etching rate of local position of metal NPs and the formation rate of MOF shells in order to avoid the self-nucleation and growth of MOFs.

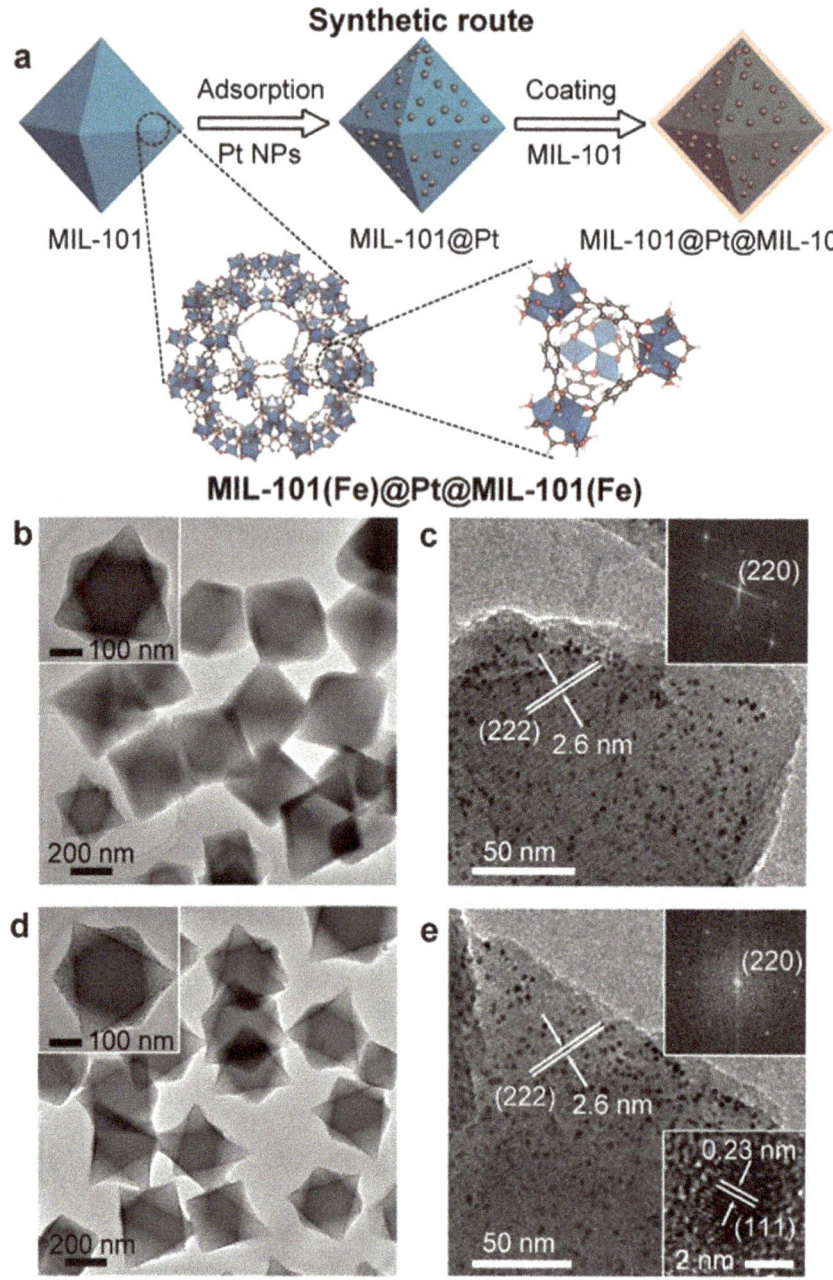

Fig. 6.4 Sandwich MIL-101@Pt@MIL-101 nanostructures. **a** Synthetic route for sandwich MIL-101@Pt@MIL-101. **b, d** TEM images of MIL-101(Fe)@Pt@MIL-101(Fe)$^{22.0}$ and MIL-101(Fe)@Pt@MIL-101(Fe)$^{9.2}$. **c, e** HRTEM images of MIL-101(Fe)@Pt@MIL-101(Fe)$^{22.0}$ and MIL-101(Fe)@Pt@MIL-101(Fe)$^{9.2}$. Insets: corresponding Fast Fourier transform (FFT) images of MIL-101(Fe)@Pt@MIL-101(Fe) and HRTEM image of Pt NPs [33]. Copyright 2016, Springer Nature

Zhang et al. synthesized the octahedron and flower Pt-Cu@HKUST-1(Cu) under a microwave irradiation condition within only 30 min [35]. During the synthesis process, as-prepared Pt-Cu alloys underwent oxidative dissolution at the presence of Fe^{3+} ions following the pathway of $Cu^0 + Fe^{3+} \rightarrow Cu^{2+} + Fe^{2+}$ and $4Fe^{3+} + 6Cl^- + Pt \rightarrow 4Fe^{2+} + PtCl_6^{2-}$ [36]. Pt is a relatively inert element compared to Cu, and thus, the former reaction occurs more easily than the latter, thus leading to many Cu^{2+} ions on the surface of Pt-Cu NPs. The formed Cu^{2+} ions reacted with trimesic acid for in situ growth of HKUST-1(Cu) around the alloy NPs. The obtained products were characteristic of octahedron and flower Pt-Cu core of 43.0 and 47.5 nm, respectively, and HKUST-1(Cu) shell of less than 25 nm. Similarly, Pt-Ni alloy NPs were adopted to construct core–shell Pt-Ni@MOF-74(Ni) nanostructures via in situ etching and coordination synthetic strategy [37]. Li et al. prepared Ni-rich Pt-Ni alloy with average size of ~ 20 nm and uniform truncated octahedral morphology. The PVP-capped Ni-rich Pt-Ni NPs were submersed in DMF solution to form a turbid solution, followed by adding a solution of 2,5-dioxidoterephthalate to form well-defined MOF-74(Ni) weaved on the surface of Pt-Ni nanoframes.

6.2.2.3 One-Pot Approach

One-pot synthesis method has been aroused great interest in constructing the core–shell metal NPs@MOFs, due to the reduced production procedures and ease of scaling up. However, it is more challengeable to effectively balance the simultaneous nucleation and growth of metal NP core and MOF shell together with the heterogeneous nucleation and growth of MOFs around metal NPs, because the specific experimental conditions are required to balance the heterogeneous nucleation and growth of MOFs around metal NPs.

Tang et al. prepared the core–shell Au@MOF-5(Zn) nanocomposites character istic of a single Au NP core coated with a uniform MOF-5(Zn) shell by directly mixing the precursors of $HAuCl_4$, $Zn(NO_3)_2 \cdot 6H_2O$ and H_2BDC in the reaction solution containing PVP, DMF, and ethanol [38]. Moreover, the formation rates of Au NPs and MOF-5(Zn) in the mixture solution could be tuned effectively. The shell thickness was controlled from 3.2 to 69 nm, while the diameter of Au NP cores varied from 54 to 30 nm, and correspondingly, the sizes of the core–shell Au@MOF-5(Zn) were tuned from 60 to 156 nm. In addition to single metal NP core, multiple metal NPs can be encapsulated by MOFs using one-pot method. Li et al. developed a facile one-pot method for synthesis of Pd@UiO-67(Zr) nanocomposites by directly dissolving the precursors of $Pd(NO_3)_2$, $ZrCl_4$, and 2,2′-bipyridine-5,5′-dicarboxylic acid with the reducing agent NH_3BH_3 in DMF solvent [39]. The obtained products were characteristic of Pd NPs of 3.2 nm homogeneously distributed inside UiO-67(Zr). Also, Li et al. reported a simple one-step protocol to encapsulate Pd NPs inside UiO-67(Zr) through a temperature control program [40]. This process involved in situ $PdCl_2(CH_3CN)_2$ incorporation in the mixture of $ZrCl_4$, 2,2′-bipyridine-5,5′-dicarboxylate, DMF, and HCl at 80 °C and on-site moderate reduction process to fabricate Pd NPs at an elevated temperature of 130 °C. The initial in situ incorporation of metal precursors, in which assembly of UiO-67(Zr) and encapsulation of metal precursor were achieved simultaneously, could allow the Pd precursors to be homogeneously distributed inside the pores of the UiO-67(Zr). As-reduced 1.5 nm Pd NPs with (111) facet exposed were highly dispersed in UiO-67(Zr).

6.3 Catalytic Performances of Core–Shell Metal NPs@MOF Catalysts

Generally, core–shell typed metal NPs@MOFs own the well-defined interfaces between core and shell, strong interfacial interaction as well as good thermal and chemical stability with respect to tranditional supported catalysts. These distinct properties have great potential to show the extraordinary catalytic performances, especially for selective catalysis. In the following, we mainly focus on discussion of various catalytic applications of core-shell structures as well as the relationships among the composition, structure, function synergy, and catalytic performances.

6.3.1 Molecular Sieving Effect of MOFs for Selective Catalysis

Regio- and stereo-selectivity is a key issue involved in many industrial heterogeneous catalysis. Due to the highly versatile and tunable pore structure of MOFs, the core–shell metal NP@MOF catalysts are particularly appealing to integrate metal NP functionality with the molecular sieving effect of MOFs to achieve size-selective catalysis [41].

Huo et al. reported the Pt@ZIF-8(Zn) as catalysts for liquid-phase hydrogenation of *n*-hexene versus *cis*-cyclooctene by considering the catalytic properties of Pt NPs and the molecular sieving capability of ZIF-8(Zn) matrix [27]. Pt@ZIF-8(Zn) catalyzed the hydrogenation of *n*-hexene with a conversion of 7.3%, but no propensity was found for hydrogenation of *cis*-cyclooctene, which was consistent with the small portal size for ZIF-8(Zn) (3.4 Å). Moreover, Pt@UiO-66(Zr) nanocomposites were used as catalysts for liquid-phase hydrogenation of different sized olefins including hex-1-ene (2.5 Å), cyclooctene (5.5 Å), trans-stilbene (5.6 Å), triphenylethylene (5.8 Å), and tetraphenyl ethylene (6.7 Å). The complete conversion of hex-1-ene molecules was obtained after 24 h reaction. With increasing the size of substrates, the conversion was decreased significantly and corresponding value was 66% for cyclooctene, 35% for trans-stilbene, 8% for triphenylethylene, and 0% for tetraphenyl ethylene. It is noted that the encapsulation of metal NPs inside MOFs provides an alternative way to achieve size-dependent selective catalysis. Also, selective hydrogenation of C = O groups versus C = C groups in citronellal was investigated by using Pt@MOFs with different pore environments including ZIF-8(Zn), ZIF-67(Co), and UiO-66(Zr) [29]. The selectivity to citronellol was > 99% for both Pt@ZIF-8(Zn) and Pt@ZIF-67(Co), which was much higher than 6.69% by Pt@UiO-66(Zr). These were ascribed to the small-size channels of Pt@ZIF-8(Zn) and Pt@ZIF-67(Co). Moreover, Huo et al. modified the as-prepared Pt@NH$_2$-UiO-66(Zr) with three anhydrides including acetic anhydride, butyric anhydride, and hexanoic anhydride, and the resulting composites were denoted as Pt@UiO-66(Zr)-AM1, Pt@UiO-66(Zr)-AM3 and Pt@UiO-66(Zr)-AM5, respectively, where the numbers stood for attached carbon chain length [42]. After modification, no obvious changes in morphology were observed. The catalytic selectivity of these four samples was tested in liquid-phase hydrogenation of triphenylethylene, trans-stilbene, and cyclooctene. All the four composites showed barely no catalytic activity for hydrogenation of triphenylethylene, indicating that the size of triphenylethylene molecules exceeds the pore apertures of NH$_2$-UiO-66(Zr) and the modified samples, namely, C = C double bonds could not contact with Pt NPs encapsulated inside MOFs. Being smaller in size than triphenylethylene (10.71 Å × 8.76 Å × 6.48 Å), trans-stilbene (11.65 Å × 5.06 Å) can be diffused into the MOF composites and then be hydrogenated. With lengths of the attached functional groups increased, conversions of trans-stilbene were reduced from 49 to 4% due to the retarded diffusions of trans-stilbene into micropores. Besides, hydrogenation of cyclooctene (5.29 Å × 5.21 Å × 3.55 Å) showed a similar trend.

Besides, MOFs with tunable pore structures might show the pore-dependent activity and selectivity for catalytic complex reactions including several possible reaction pathways. For example, Somorjai and Yaghi et al. encapsulated as-prepared multiple Pt NPs of 2.5 nm by UiO-66(Zr) for gas-phase hydrogenation of methylcyclopentane (MCP) compared with the contrast samples including pure UiO-66(Zr), Pt-on-UiO-66(Zr), and Pt-on-SiO$_2$ (Fig. 6.5) [43]. It should be pointed out that gas-phase hydrogenative conversion of MCP is relatively complex and it could carry out via five possible reaction pathways as follows. (I) MCP is converted into the dehydrogenated version of MCP. (II) MCP is ring-opened and isomerized into isomers. The C$_5$-cyclic ring of MCP is further enlarged to C$_6$-cyclic hydrocarbons through ring-enlargement reaction followed with (III) hydrogenation or (IV) dehydrogenation to produce cyclohexane or benzene, respectively. The last reaction pathway is cracking to produce C$_1$-C$_5$-based hydrocarbons that are undesired products in this reaction. (V) Catalytic results showed that no catalytic activity was found for pure UiO-66(Zr). Dehydrogenation and isomerization with selectivity of 18.2% and 81.8% were achieved by Pt-on-SiO$_2$, respectively, while as for Pt-on-UiO-66(Zr) catalyst, they catalyzed both dehydrogenation and isomerization with selectivity of 33.1% and 66.9%, respectively. Differently, Pt@UiO-66(Zr) showed C$_6$-cyclic hydrocarbon (> 60%) as main products with selectivity of 22.2% for cyclohexane and 41.2% for benzene. It should be pointed out that C$_6$-cyclic hydrocarbons were noteworthy for Pt@UiO-66(Zr) at 150 °C, while benzene might be produced over Pt-on-SiO$_2$ at the higher reaction temperature above 250 °C. These indicated that confinement of Pt NPs in UiO-66(Zr) contributed to decrease of the activation energy for formation of C$_6$-cyclic hydrocarbons.

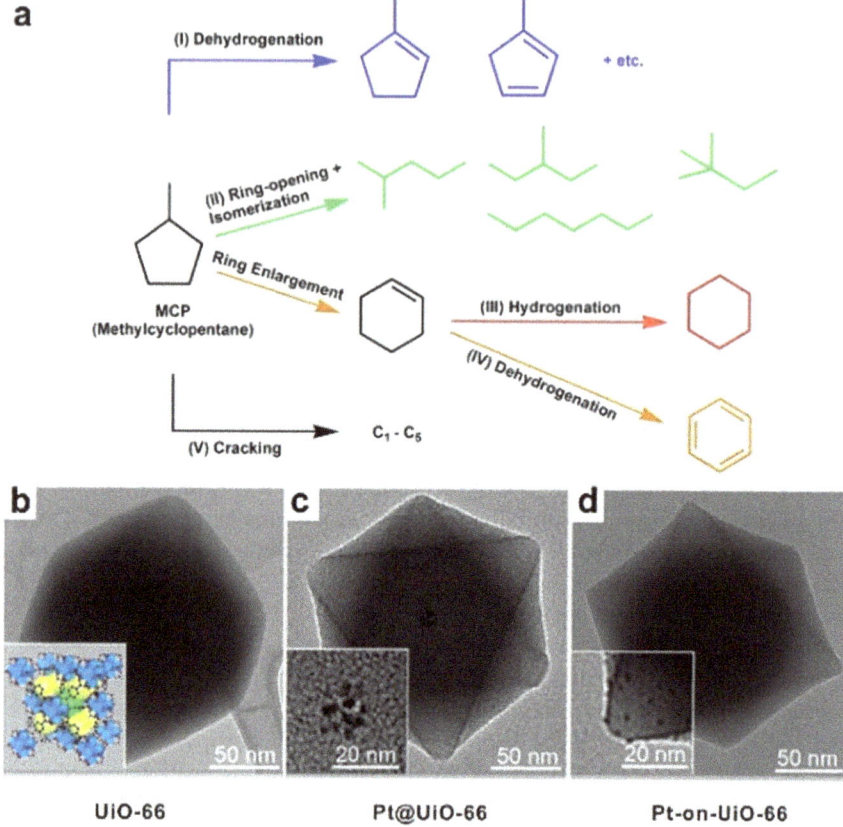

Fig. 6.5 a Schematic reaction diagram of hydrogenative conversion of MCP. TEM images of **b** UiO-66(Zr), **c** Pt@UiO-66(Zr) and (d) Pt-on-UiO-66(Zr) [43]. Copyright 2014, American Chemical Society

Altogether, the ordered and tunable pore structures of MOF provide the effective strategies for achieving regio-selective reactions, even for changing the reaction pathways.

6.3.2 Synergistic Effect Between Metal NPs and Unsaturated Metal Nodes in MOFs for Catalysis

Generally, MOFs could strongly affect the catalytic activity and selectivity of metal NPs via substrate adsorption or charge transfer. MOFs usually possess abundant coordinatively unsaturated metal nodes, which are used as Lewis acid sites and interact with metal NPs to achieve the intriguing and synergistic functions with respect to single counterparts.

Selective hydrogenation of the $C = O$ group against the conjugated $C = C$ group in α, β-unsaturated aldehydes (R-CH $=$ CH-CHO, (**A**) is an indispensable choice to produce the unsaturated alcohol (R-CH $=$ CH-CH$_2$OH, (**B**). However, hydrogenation of the $C = O$ group is thermodynamically unfavored when compared to the conjugated $C = C$ group, and the final products would be a mixture of **B**, hydrocinnamaldehyde (R-CH$_2$-CH$_2$-CHO, (**C**), and phenyl propanol (R-CH$_2$-CH$_2$-CH$_2$OH, (**D**). Until now it is still challengeable to realize high-performance catalysis. To address above, Tang et al. showed a new approach utilizing MOFs as selectivity regulators for hydrogenation of thermodynamic unfavorable $C = O$ bond in α, β-unsaturated aldehydes by sandwiched Pt NPs [33]. The enabling heterogeneous catalysts, where Pt NPs of 2.8 nm in diameter were sandwiched by MIL-101 containing metal nodes of Fe^{3+}, Cr^{3+}, or both. At the same conversion efficiency of cinnamaldehyde (~45%), no noticeable hydrogenation of cinnamaldehyde by MIL-101 was observed. Contrastingly, hydrogenation of cinnamaldehyde by Pt NPs was achieved with a turnover frequency (TOF) of 372.4 h^{-1}, but with only ~18.3% of the desired cinnamyl alcohol. Differing from bare Pt NPs and MIL-101, MIL-101@Pt nanostructures exhibited the dramatically increased selectivity of cinnamyl alcohol with 86.4% for MIL-101(Fe)@Pt and 44.0% for MIL-101(Cr)@Pt, indicating that MIL-101 remarkably promoted selective hydrogenation of the $C = O$ bond. More importantly, sandwich MIL-101@Pt@MIL-101 was considered as the ideal catalysts for selective hydrogenation of cinnamaldehyde to cinnamyl alcohol. As for coating MIL-101(Cr)@Pt or MIL-101(Fe)@Pt with MIL-101(Fe) shells, all the selectivities were higher than 94%. Furthermore, smaller-sized α, β-unsaturated aldehydes including acrolein (0.69 $\times$ 0.51 nm) with no substituents on $C = C$ bond, branched 3-methyl-2-butenal (0.79 $\times$ 0.60 nm), and furfural (0.81 $\times$ 0.64 nm) with a furan ring were hydrogenated by MIL 101@Pt@MIL 101 with the preferential selective hydrogenation of $C = O$ group compared to $C = C$ group. These were ascribed to the preferential interaction of coordinatively unsaturated metal sites in MOFs with the $C = O$ group, which could turn the thermodynamically unfavorable hydrogenation on $C = O$ group by the embedded Pt NPs into a favored reaction. To further enhance the catalytic activity along with excellent selectivity and good stability, MIL-101 shell was replaced by conjugated micro- and mesoporous polymers with iron(III) porphyrin (FeP-CMPs) to fabricate MIL-101(Cr)@Pt@FeP-CMP, the environment of Pt NPs was succeeded in modifying, and it is not only hydrophobic and porous for enriching reactants, but also possesses Fe sites to activate $C = O$ bonds, thereby regulating the selectivity for cinnamyl alcohol in the hydrogenation of cinnamaldehyde [44]. More importantly, MIL-101(Cr)@Pt@FeP-CMP sponge could achieve a high TOF value of 1516.1 h^{-1} with 97.3% selectivity for cinnamyl alcohol at 97.6% conversion.

Also, the synergistic effect between metal NPs and coordinatively unsaturated metal nodes in MOFs together with the pore confinement effect is of vital importance for catalysis. For example, Yaghi and Somorjai et al. synthesized the core–shell Cu@UiO-66(Zr) composites characteristic of a single PVP-stabilized ~ 18 nm Cu

NP as core and UiO-66(Zr) as the shell for CO_2 hydrogenation at 175 °C and 10 bar using CO_2 and H_2 in a 1:3 molar ratio along with Cu on UiO-66(Zr), Cu on ZrO_2, $Cu/ZnO/Al_2O_3$, Cu NPs on MIL-101(Cr) and Cu@ZIF-8(Zn) [23]. The initial TOF value of methanol formation was 3.7×10^{-3} s^{-1} for Cu@UiO-66(Zr), 1.7×10^{-3} s^{-1} for Cu on UiO-66(Zr), 0.42×10^{-3} s^{-1} for Cu on ZrO_2, and 0.45×10^{-3} s^{-1} for $Cu/ZnO/Al_2O_3$, but neither Cu NPs on MIL-101(Cr) nor Cu@ZIF-8(Zn) showed catalytic activity. It is noted that Zr oxide or Zn oxide was critical for CO_2 hydrogenation to methanol. Moreover, Cu@UiO-66(Zr) showed a twofold increase in activity compared with Cu on UiO-66(Zr), disclosing that the excellent catalytic performance of Cu@UiO-66(Zr) originated from more active Cu sites created by surrounding with Zr oxide SBU and confinement effect of Cu NPs in UiO-66(Zr). XPS measurements further indicated that the binding energy of Zr $3d_{5/2}$ for Cu on UiO-66(Zr) shifted from 182.8 to 182.2 eV compared with pure UiO-66(Zr) and the strong interfacial electron transfer between Cu NPs and Zr oxide nodes was present.

Altogether, the strong metal-support interactions between metal NPs and organic chelates/metal-oxo clusters together with the pore confinement effect of MOFs offer new opportunities in fine-tuning catalytic activity and selectivity of metal NPs@MOFs along with good stability.

6.3.3 Synergistic Effect Between Metal NPs and Functional Groups on MOFs

The chemical environment on the encapsulated metal NPs by MOFs could be studied by tuning the organic linkers of MOFs via utilizing different building units or post-modification of MOFs.

Jiang et al. developed the Pd@UiO-66-X with different functional groups for catalyzing the hydrogenation of benzoic acid [15]. Catalytic results showed that Pd@UiO-66-2OH possessed the highest activity followed by Pd@UiO-66-2OH(Hf), Pd@UiO-66-NH_2, Pd@UiO-66-OMe, and Pd@UiO-66, in a decreasing order of activity. Remarkably, the activity of Pd@UiO-66-2OH was around 14 times higher than that of Pd@UiO-66(Zr). The distinct activity was not only ascribed to the different electron transfer from Pd to the MOFs, but also due to the discriminated substrate adsorption energy of Pd@UiO-66-X. In addition, to clarify the impact of the functional groups on the reactivity of the H_2O absorbed in UiO-66(Zr) analogues, the water gas shift reaction (WGS, $H_2O + CO \rightarrow H_2 + CO_2$) was performed by the Pt@UiO-66(Zr) analogues with a fixed-bed reactor [31, 45]. Pt@UiO-66-Br exhibited a CO conversion of 21.1% at 340 °C, which was 1.7 times lower than that of Pt@UiO-66-H (35.6%), demonstrating that functionalization of the BDC ligand with the -Br group reduced the WGS reaction activity. On the other hand, Pt@UiO-66-Me_2 exhibited a CO conversion of 59.7% at 340 °C, which was 1.7 times higher than that of Pt@UiO-66-H. The functionalization with -Me_2 caused an enhancement of the reactivity. The trends of the reactivity (-Me_2 > -H > -Br) were observed in a wide range of flow rates of H_2O, demonstrating that the WGS reaction activity was systematically controllable through functionalization of the BDC ligands.

In brief, the aromatic linkers in MOFs can establish charge transfer interactions with metal NPs by coordination or π-π forces, thus generating the extraordinary capability toward selective catalysis.

6.3.4 Integrating Metal NPs with MOFs for Multistep Reactions

Multistep cascade catalysis has recently attracted significant attention in chemical industries because they are extremely important for sustainable synthesis with lower cost, fewer chemicals, and less energy consumptions.

The metal NPs and coordinatively unsaturated metal nodes in MOFs in the core-shell structures are easily combined as the bifunctional catalysts for cascade reactions. For example, Jiang et al. synthesized ~2.5 nm Pd@MIL-101(Cr) catalysts for the synthesis of 2-(4-aminophenyl)-1H-benzimidazole (**D**) via tandem transformation of acid catalysis and hydrogenation reaction (Fig. 6.6) [18]. In situ FT-IR spectra of CO adsorption and NH_3 temperature-programmed desorption (TPD) both confirmed the presence of rich coordinatively unsaturated Cr^{3+} ions inside MIL-101(Cr) and

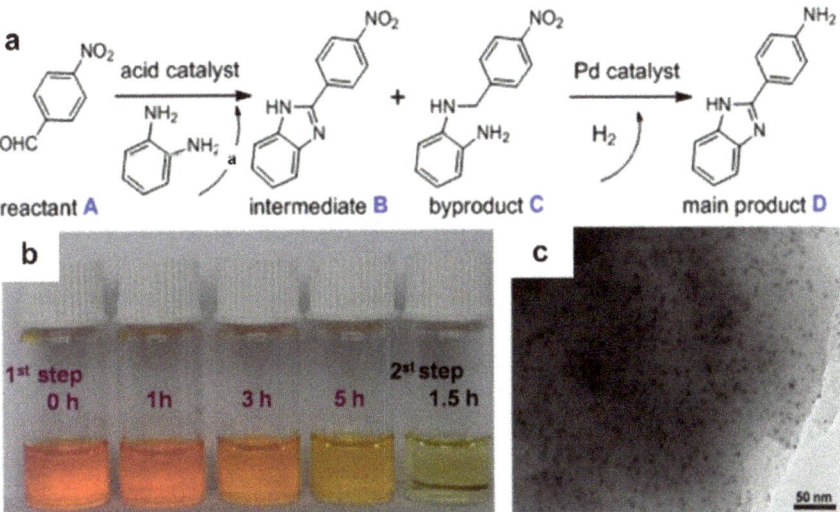

Fig. 6.6 (a) Illustration of the two-step tandem synthesis of 2-(4-aminophenyl)-1H-benzimidazole (D) involving a Lewis acid-based catalytic process and subsequent Pd based hydrogenation. (b) Solution color evolution during the reaction process. (c) Representative TEM image of Pd@MIL-101 catalyst [18]. Copyright 2015, American Chemical Society

Pd@MIL-101(Cr). When pure MIL-101(Cr) was used as catalyst, only product 2-(4-aminophenyl)-1H-benzimidazole (**B**) was obtained. As for Pd@MIL-101(Cr), the conversion rate of reactant 4-nitrobenzaldehyde (**A**) in the acid catalysis step was almost 100% and the corresponding yield of **D** was 99%. As contrast, about 50% of the reactants were converted to **C** in the first step due to lack of acidity when commercial Pd/C catalyst was used, and therefore the nitro group in both the residual reactant and **C** were coreduced to corresponding amino group in the second step. Furthermore, the core-shell structure showed that no deactivation occurred over three runs. Therefore, the obtained Pd@MIL-101 behaved as a bifunctional catalyst and presented excellent performance in tandem catalysis for the synthesis of target product **D**.

The metal NPs and functional groups on MOFs have already been integrated as bifunctional catalysts for tandem reactions. For example, the chemical, 2-(4-aminobenzylidene)malononitrile (**C**), is known as a key intermediate in synthesis of dyes and antihypertensive drugs. It is usually synthesized by Knoevenagel condensation of 4-nitrobenzaldehyde (**A**) and malononitrile into 2-(4-nitrobenzylidene)malononitrile (**B**) via the alkaline catalysts, followed by selective hydrogenation of -NO_2 group of the intermediate products **B** to **C** by metal NPs (Fig. 6.7). Based on the above, Tang et al. synthesized well-defined core–shell Pd@IRMOF-3(Zn) nanostructures, in which a single Pd NP of 35 nm in diameter was surrounded by uniform cubic-phase IRMOF-3(Zn) with average pore size of ~1.03 nm [21]. In the composites, the IRMOF-3(Zn) was used as the base catalyzing Knoevenagel condensation, while the Pd core was adopted for subsequent catalytic reaction. The performance of core–shell Pd@IRMOF-3(Zn) nanocomposites was evaluated using the cascade reactions along

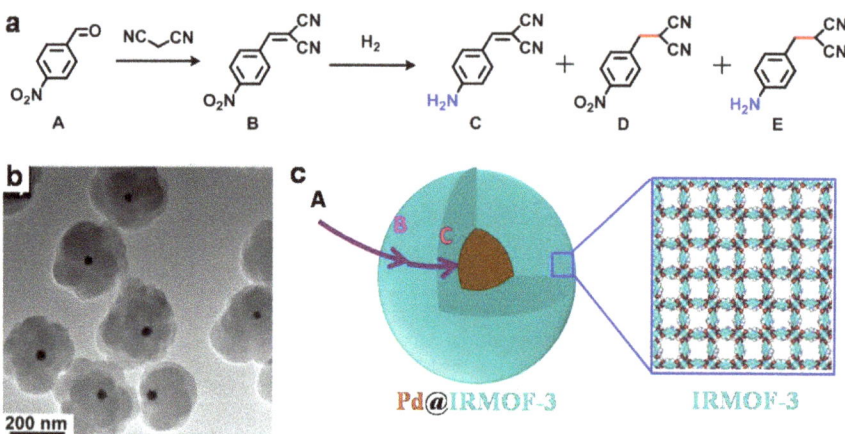

Fig. 6.7 **a** Cascade reactions involving Knoevenagel condensation of **A** and malononitrile via the IRMOF-3(Zn) shell, and subsequent selective hydrogenation of intermediate product **B** to **C** via Pd NP cores. **b** TEM image of core–shell Pd@IRMOF-3(Zn). **c** Scheme of cascade reaction process by core-shell structure [21]. Copyright 2014, American Chemical Society

with the contrast sample supported Pd/IRMOF-3(Zn). Both Pd@IRMOF-3(Zn) and Pd/IRMOF-3(Zn) could catalyze the cascade reactions, suggesting that integration of Pd NPs and IRMOF-3(Zn) was an effective strategy to achieve the multifunctional catalysis. As for Pd@IRMOF-3(Zn), the hydrogenation selectivity of **B** to **C**, **D**, and **E** was 86%, 8%, and 6%, respectively; as a comparison, the corresponding selectivity was 71%, 24%, and 5% for conventional Pd/IRMOF-3(Zn) catalysts. DFT calculations revealed that reactant **A** was a linear molecule (1.03×0.72 nm), and its selective conversion was substantially affected by the group-selective adsorption on the catalyst surfaces. The -NH$_2$ groups on the surface of IRMOF-3(Zn) showed an energetically preferable interaction with the -NO$_2$ groups of **A**, 7.83 kcal mol^{-1} lower than that of the C $=$ O groups of **A**. Accordingly, **A** preferred entering into the nanostructures with the -NO$_2$ group ahead when the core–shell Pd@IRMOF-3(Zn) hybrids were used as the catalysts, which would promote formation of target product **C**. Furthermore, the core-shell structure displayed excellent catalytic stability, while supported catalyst showed some activity decay during the recycle tests. To testify the pore effect, both a shorter molecule, 4-nitro-1-butene (0.87×0.53 nm) as an alternative for **B**, and a longer substrate, 4-nitrocinnamaldehyde (1.28×0.72 nm) as a replacement for **A**, are utilized in the cascade reaction under the same conditions. When the shorter is used, the selectivity of 4-amino-1-butane and 4-nitro-1-butane is 56% and 44%, respectively; whereas the selectivity reaches 96% for 2-(3-(4-aminophenyl)allylidene)malononitrile when the longer is adopted. These indicated that pore confinement effect also played an important role in achieving high selectivity of the target product.

Huang et al. developed the core–shell Pd NPs@UiO-66(Zr) for aerobic reaction between benzaldehyde and ethylene glycol, in which the functional groups on MOF linkers were changed from -H to -NH$_2$ or -OMe, denoted as Pd@UiO-66-X (X $=$ H, NH$_2$, OMe) (Fig. 6.8) [46]. The obtained UiO-66-X had two major pores at 7–10 and 12 Å, while Pd NPs with an average size of less than 1.2 nm were uniformly dispersed in UiO-66-X. It should be pointed out that aerobic reaction between benzaldehyde and ethylene glycol involves two steps: (i) condensation of benzaldehyde and ethylene glycol to yield hemiacetal, and (ii) further condensation to yield corresponding acetal or oxidation to yield ester. All of them could exhibit the high conversion rates of ~99%, ~91%, and ~98%, respectively, but the product distribution was distinctly different. Pd@UiO-66-NH$_2$ gave the corresponding acetal, benzaldehyde ethylene acetal (selectivity ~94%), as the major product, while Pd@UiO-66-H and -OMe showed high selectivity (~90% and ~97%, respectively) to the corresponding ester, 2-hydroxyethyl benzoate. -NH$_2$ groups coordinated to the Pd surface and donated electrons to the Pd NPs in Pd@UiO-66-NH$_2$ decreased the oxidation capability of encapsulated Pd NPs and resulted in the high acetal selectivity. DFT calculations showed that the NH$_2$-Pd NPs possessed higher chemical potential and weakened oxidation capability with respect to the Pd-OMe NPs, which agreed with the experimental results.

The integration of diverse metal nodes, functional organic linkers, and metal NPs along with pore confinement effect into single core-shell nanostructure is capable of more functions for multistep catalytic reactions. The key for achieving the excellent

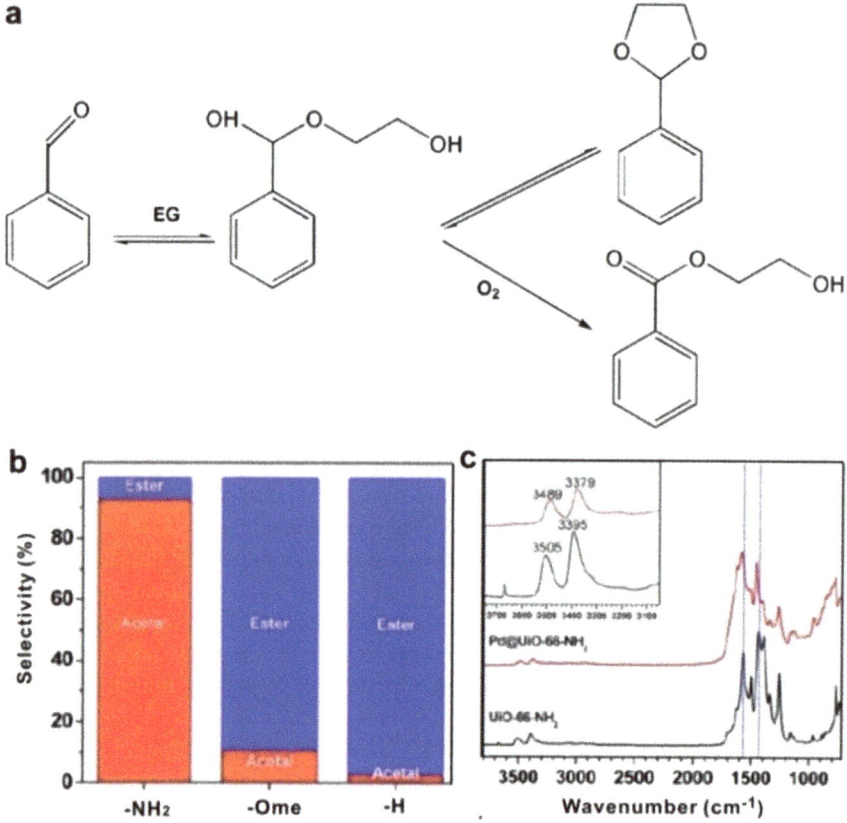

Fig. 6.8 a Aerobic reaction between benzaldehyde and ethylene glycol. **b** Product distribution using Pd@UiO-66-X (X = H, NH₂, OMe). Reaction conditions: benzaldehyde (0.1 mmol), ethylene glycol (1.5 mL), Pd/substrate 1/100, 1 atm O_2, 90 °C, 10 h. **c** DRIFTS spectra of UiO-66-NH₂ and 2.0wt%Pd@UiO-66-NH₂. The inset shows the enlarged N–H vibration region [46]. Copyright 2016, American Chemical Society

catalytic efficiency should be dependent on the strong synergistic interactions among them.

6.4 Summary and Outlook

This chapter summarizes the recent progresses in controllable encapsulation of metal NPs by porous MOFs with core-shell structures as well as their applications in heterogeneous catalysis. With the rapid development of synthesis chemistry, encapsulation of metal NPs by porous MOFs has been successfully constructed via different strategies including double-solvent method, surfactant induced assembly, template-assisted growth, and one-pot synthesis. Notably, when used as catalysts, they have exhibited many unique advantages as follows: (1) Molecular sieving effect of MOFs for selective catalysis. (2) Synergistic effect between active NPs and MOFs together with the pore confinement effect. (3) One-pot multifunctional catalysis. (4) Enhanced catalytic stability. The integration of inorganic and organic components together with function synergy between MOFs and NPs provides endless possibility for controllably constructing the desirable catalysts with atomically accurate structures.

Besides, it is still highly desirable for developing the novel and effective strategies to encapsulate metal NPs by porous MOFs, in which the sizes and shapes of metal NP cores as well as the thicknesses and structures of the MOF shells could be effectively tuned. Multifunctional properties must be imparted with the core-shell nanostructures in order to target different catalytic reactions. Both in situ and ex situ characterization techniques further need to be developed for revealing the catalytic mechanism. Meanwhile, theoretical calculations would provide the fundamental understanding on the catalytic reactions. All the above will contribute to rational design and construction of the core–shell typed metal NPs@MOFs at molecular level and realization of high catalytic activity, excellent selectivity, and good stability in the future.

References

1. Zhong CJ, Maye MM (2001) Core-shell assembled nanoparticles as catalysts. Adv Mater 13:1507–1511
2. Cargnello M, Doan-Nguyen VVT, Gordon TR, Diaz RE, Stach EA, Gorte RJ, Fornasiero P, Murray CB (2013) Control of metal nanocrystal size reveals metal-support interface role for ceria catalysts. Science 341:771–773
3. Corma A, Serna P (2006) Chemoselective hydrogenation of nitro compounds with supported gold catalysts. Science 313:332–334
4. Li WC, Comotti M, Schüth F (2006) Highly reproducible syntheses of active Au/TiO₂ catalysts for CO oxidation by deposition-precipitation or impregnation. J Catal 237:190–196
5. Li G, Tang Z (2014) Noble metal nanoparticle@metal oxide core/yolk-shell nanostructures as catalysts: recent progress and perspective. Nanoscale 6:3995–4011
6. Furukawa H, Cordova KE, O'Keeffe M, Yaghi OM (2013) The chemistry and applications of metal-organic frameworks. Science 341:1230444
7. Shimizu GKH, Vaidhyanathan R, Taylor JM (2009) Phosphonate and sulfonate metal organic frameworks. Chem Soc Rev 38:1430–1449
8. Yang Q, Xu Q, Jiang HL (2017) Metal-organic frameworks meet metal nanoparticles: synergistic effect for enhanced catalysis. Chem Soc Rev 46:4774–4808
9. Li G, Zhao S, Zhang Y, Tang Z (2018) Metal-organic frameworks encapsulating active nanoparticles as emerging composites for catalysis: recent progress and perspectives. Adv Mater 30:1800702
10. Hu P, Morabito JV, Tsung CK (2014) Core-shell catalysts of metal nanoparticle core and metal-organic framework shell. ACS Catal 4:4409–4419
11. Luan Y, Qi Y, Gao H, Zheng N, Wang G (2014) Synthesis of an amino-functionalized metal–organic framework at a nanoscale level for gold nanoparticle deposition and catalysis. J Mater Chem a 2:20588–20596

12. Volosskiy B, Niwa K, Chen Y, Zhao Z, Weiss NO, Zhong X, Ding M, Lee C, Huang Y, Duan X (2015) Metal-organic framework templated synthesis of ultrathin, well-aligned metallic nanowires. ACS Nano 9:3044–3049

13. Aijaz A, Karkamkar A, Choi YJ, Tsumori N, Rönnebro E, Autrey T, Shioyama H, Xu Q (2012) Immobilizing highly catalytically active Pt nanoparticles inside the pores of metal-organic framework: a double solvents approach. J Am Chem Soc 134:13926–13929

14. Xiao YY, Liu XL, Chang GG, Pu C, Tian G, Wang LY, Liu JW, Ma XC, Yang XY, Chen B (2020) Construction of a functionalized hierarchical pore metal-organic framework via a palladium-reduction induced strategy. Nanoscale 12:6250–6255

15. Chen DX, Yang WJ, Jiao L, Li LY, Yu SH, Jiang HL (2020) Boosting catalysis of Pd nanoparticles in MOFs by pore wall engineering: the roles of electron transfer and adsorption energy. Adv Mater 32:2000041

16. Sun D, Li Z (2016) Double-solvent method to Pd nanoclusters encapsulated inside the cavity of NH$_2$-UiO-66(Zr) for efficient visible-light-promoted suzuki coupling reaction. J Phys Chem C 120:19744–19750

17. Zhu QL, Li J, Xu Q (2013) Immobilizing metal nanoparticles to metal-organic frameworks with size and location control for optimizing catalytic performance. J Am Chem Soc 135:10210–10213

18. Chen YZ, Zhou YX, Wang HW, Lu JL, Uchida T, Xu Q, Yu SH, Jiang HL (2015) Multifunctional PdAg@MIL-101 for one-Pot cascade reactions: combination of host-guest cooperation and bimetallic synergy in catalysis. ACS Catal 5:2062–2069

19. Chen YZ, Gu BC, Uchida T, Liu JD, Liu XC, Ye BJ, Xu Q, Jiang HL (2019) Location determination of metal nanoparticles relative to a metal-organic framework. Nat Commun 10:3462

20. Goh TW, Tsung CK, Huang WY (2019) Spectroscopy identification of the bimetallic surface of metal-organic framework-confined Pt-Sn nanoclusters with enhanced chemoselectivity in furfural hydrogenation. ACS Appl Mater Interfaces 11:23254–23260

21. Zhao M, Deng K, He L, Liu Y, Li G, Zhao H, Tang Z (2014) Core-shell palladium nanoparticle@metal-organic frameworks as multifunctional catalysts for cascade reactions. J Am Chem Soc 136:1738–1741

22. Liu X, He L, Zheng J, Guo J, Bi F, Ma X, Zhao K, Liu Y, Song R, Tang Z (2015) Solar-light-driven renewable butanol separation by core-shell Ag@ZIF-8 nanowires. Adv Mater 27:3273–3277

23. Rungtaweevoranit B, Baek J, Araujo JR, Archanjo BS, Choi KM, Yaghi OM, Somorjai GA (2016) Copper nanocrystals encapsulated in Zr-based metal-organic frameworks for highly selective CO$_2$ hydrogenation to methanol. Nano Lett 16:7645–7649

24. Hu P, Zhuang J, Chou LY, Lee HK, Ling XY, Chuang YC, Tsung CK (2014) Surfactant-directed atomic to mesoscale alignment: metal nanocrystals encased individually in single-crystalline porous nanostructures. J Am Chem Soc 136:10561–10564

25. Osterrieth JWM, Wright D, Noh H, Kung CW, Vulpe D, Li A, Park JE, Van Duyne RP, Moghadam PZ, Baumberg JJ, Farha OK, Fairen-Jimenez D (2019) Core-shell gold nanorod@zirconium-based metal-organic framework composites as in situ size-selective Raman probes. J Am Chem Soc 141:3893–3900

26. Zheng G, de Marchi S, López-Puente V, Sentosun K, Polavarapu L, Pérez-Juste I, Hill EH, Bals S, Liz-Marzán LM, Pastoriza-Santos I, Pérez-Juste J (2016) Encapsulation of single plasmonic nanoparticles within ZIF-8 and SERS analysis of the MOF flexibility. Small 12:3935–3943

27. Lu G, Li SZ, Guo Z, Farha OK, Hauser BG, Qi XY, Wang Y, Wang X, Han SY, Liu XG, DuChene JS, Zhang H, Zhang QC, Chen XD, Ma J, Loo SCJ, Wei WD, Yang YH, Hupp JT, Huo FW (2012) Imparting functionality to a metal-organic framework material by controlled nanoparticle encapsulation. Nat Chem 4:310–316

28. Zhang W, Lu G, Cui C, Liu Y, Li S, Yan W, Xing C, Chi YR, Yang Y, Huo F (2014) A family of metal-organic frameworks exhibiting size-selective catalysis with encapsulated noble-metal nanoparticles. Adv Mater 26:4056–4060

29. Zhang WL, Shi WX, Ji WL, Wu HB, Gu ZD, Wang P, Li XH, Qin PS, Zhang J, Fan Y, Wu TY, Fu Y, Zhang WN, Huo FW (2020) Microenvironment of MOF channel coordination with Pt NPs for selective hydrogenation of unsaturated aldehydes. ACS Catal 10:5805–5813

30. Meng F, Zhang S, Ma L, Zhang W, Li M, Wu T, Li H, Zhang T, Lu X, Huo F, Lu J (2018) Construction of hierarchically porous nanoparticles@metal-organic frameworks composites by inherent defects for the enhancement of catalytic efficiency. Adv Mater 30:1803263

31. Ogiwara N, Kobayashi H, Inuka M, Nishiyama Y, Concepcion P, Rey F, Kitagawa H (2020) Ligand-functionalization-controlled activity of metal-organic framework-encapsulated Pt nanocatalyst toward activation of water. Nano Lett 20:426–432

32. Aoyama Y, Kobayashi H, Yamamoto T, Toriyama T, Matsumura S, Haneda M, Kitagawa H (2020) Significantly enhanced CO oxidation activity induced by a change in the CO adsorption site on Pd nanoparticles covered with metal-organic frameworks. Chem Commun 56:3839–3842

33. Zhao M, Yuan K, Wang Y, Li G, Guo J, Gu L, Hu W, Zhao H, Tang Z (2016) Metal–organic frameworks as selectivity regulators for hydrogenation reactions. Nature 539:76–80

34. Choe K, Zheng F, Wang H, Yuan Y, Zhao W, Xue G, Qiu X, Ri M, Shi X, Wang Y, Li G, Tang Z (2020) Fast and selective semihydrogenation of alkynes by palladium nanoparticles sandwiched in metal-organic frameworks. Angew Chem Int Ed 59:3650–3657

35. Jiang Y, Zhang X, Dai XP, Sheng Q, Zhuo HY, Yong JX, Wang Y, Yu KM, Yu L, Luan CL, Wang H, Zhu YC, Duan XN, Che PY (2017) In situ synthesis of core-shell Pt-Cu frame@metal-organic frameworks as multifunctional catalysts for hydrogenation reaction. Chem Mater 29:6336–6345

36. Guo Z, Dai X, Yang Y, Zhang Z, Zhang X, Mi S, Xu K, Li Y (2013) Highly stable and active PtNiFe dandelion-like alloys for methanol electrooxidation. J Mater Chem A 1:13252–13260

37. Li Z, Yu R, Huang J, Shi Y, Zhang D, Zhong X, Wang D, Wu Y, Li Y (2015) Platinum-nickel frame within metal-organic framework fabricated in situ for hydrogen enrichment and molecular sieving. Nat Commun 6:8248

38. He L, Liu Y, Liu J, Xiong Y, Zheng J, Liu Y, Tang Z (2013) Core-shell noble-metal@metal-organic-framework nanoparticles with highly selective sensing property. Angew Chem Int Ed 52:3741–3745

39. Chen LY, Chen HR, Li YW (2014) One-pot synthesis of Pd@MOF composites without the addition of stabilizing agents. Chem Commun 50:14752–14755

40. Chen LY, Chen XD, Liu HL, Bai CH, Li YW (2015) One-step encapsulation of Pd nanoparticles in MOFs via a temperature control program. J Mater Chem A 3:15259–15264

41. Li XC, Zhang ZH, Xiao WM, Deng SJ, Chen C, Zhang N (2019) Mechanochemistry-assisted encapsulation of metal nanoparticles in MOF matrices via a sacrificial strategy. J Mater Chem A 7:14504–14509

42. Liu Y, Shen Y, Zhang WN, Weng JN, Zhao MT, Zhu TS, Chi YGR, Yang YH, Zhang H, Huo FW (2019) Engineering channels of metal-organic frameworks to enhance catalytic selectivity. Chem Commun 55:11770–11773

43. Na K, Choi KM, Yaghi OM, Somorjai GA (2014) Metal nanocrystals embedded in single nanocrystals of MOFs give unusual selectivity as heterogeneous catalysts. Nano Lett 14:5979–5983

44. Yuan K, Song T, Wang D, Zhang X, Gao X, Zou Y, Dong H, Tang Z, Hu W (2018) Effective and selective catalysts for cinnamaldehyde hydrogenation: hydrophobic hybrids of metal-organic frameworks, metal nanoparticles, and micro- and mesoporous polymers. Angew Chem Int Ed 57:5708–5713

45. Ogiwara N, Kobayashi H, Concepción P, Rey F, Kitagawa H (2019) The first study on the reactivity of water vapor in metal-organic frameworks with platinum nanocrystals. Angew Chem Int Ed 58:11731–11736

46. Li X, Goh TW, Li L, Xiao C, Guo Z, Zeng XC, Huang W (2016) Controlling catalytic properties of Pd nanoclusters through their chemical environment at the atomic level using isoreticular metal-organic frameworks. ACS Catal 6:3461–3468

Chapter 7
Powerful and New Chemical Synthesis Reactions from CO_2 and C1 Chemistry Innovated by Tailor-Made Core–Shell Catalysts

Yang Wang and Noritatsu Tsubaki

7.1 Introduction

Catalytic conversion of single carbon molecules (C1 catalysis) such as syngas, CO_2, and CH_4, into valuable chemicals is a promising strategy to alleviate the pressure of sustainable development, which mainly stems from the depletion of traditional fossil fuels [1]. Syngas, as a mixture of CO and H_2, is mainly produced from the gasification of biomass, coal, shale gas, etc. The conversion of syngas has built a bridge between non-petroleum-based feedstocks and high-value-added chemicals synthesis [2]. With the diminishing of the crude oil, the oriented conversion of syngas to platform chemicals, such as light olefins (C_2–C_4) and aromatics, or liquid fuels, such as gasoline (C_5-C_{11} hydrocarbons), jet fuel (C_8–C_{16} hydrocarbons), and diesel fuel (C_{10}-C_{20} hydrocarbons), has attracted more and more attention [3]. CO_2, as a greenhouse gas, has aroused several environmental issues (ocean acidification, climate change, global warming, etc.) with the development of human society. Thermo-catalytic CO_2 hydrogenation with H2 supplied by water photo/electrolysis is a sustainable means of reducing CO_2 emissions [4]. Recently, the direct conversion of CO_2 to α-olefins has been realized by Fe-based catalyst, which seemed to be a promising strategy for high-value-added utilization of inert CO_2 molecule [5]. CH_4, as the major component of shale gas or natural gas, can be employed as feedstock for commodity chemicals synthesis via catalytic technology, such as aromatics and

Y. Wang · N. Tsubaki (✉)
Department of Applied Chemistry, Graduate School of Engineering, University of Toyama, Gofuku 3190, Toyama 930-8555, Japan
e-mail: tsubaki@eng.u-toyama.ac.jp

Y. Wang
e-mail: wangyang@upc.edu.cn

Y. Wang
College of New Energy, China University of Petroleum (East China), Qingdao 266580, PR China

© The Author(s), under exclusive license to Springer Nature Singapore Pte Ltd. 2021
H. Yamashita and H. Li (eds.), *Core-Shell and Yolk-Shell Nanocatalysts*,
Nanostructure Science and Technology,
https://doi.org/10.1007/978-981-16-0463-8_7

methanol synthesis from CH_4 catalyzed by Fe and Cu doped zeolites, respectively [6]. As a highly energy-intensive fuel source, CH_4 hydrate, also called flammable ice, is thought to bring a revolutionary of energy source due to its huge reserves, which are estimated to be larger than the total reserves of coal, crude oil, and shale gas. It is no doubt that the successful commercial exploitation of CH_4 hydrate will inspire the catalytic transformation of CH_4 to high-value-added chemicals. On the other hand, CH_4 is also a greenhouse gas, whose greenhouse effect is 25 times stronger than that of CO_2. The development of efficient CH_4 conversion strategy is also meaningful to eliminate the environmental concerns. Therefore, the above-mentioned C1 molecules are playing more and more important roles not only in reducing the over-dependence on non-renewable crude oil resources but also in eliminating the environmental impact caused by the excessive greenhouse gas emissions.

The rational design of efficient catalysts and the discovery of powerful synthesis reactions to boost the oriented conversion of C1 molecules to target products are hot topics in the field of C1 catalysis. However, the traditional catalytic process of syngas or CO_2 molecule usually produced linear hydrocarbons due to the carbon-chain growth mechanism. For example, the product distribution of Fischer–Tropsch synthesis (FTs) for syngas conversion obeys the Anderson-Schulz-Flory (ASF) law and subsequent hydrocracking or hydrorefining process is needed to upgrade the heavy waxes. α-Olefins are the main products from CO_2 hydrogenation process catalyzed by alkali metal modified Fe-based catalysts due to the similar carbene oligomerization mechanism with FTs reaction. To realize the direct conversion of syngas and CO_2 to target chemicals in a single-pass, the second catalytic component is usually combined with the main active site to regulate the traditional product distribution, which is the so-called tandem catalyst [7]. Even though several kinds of tandem catalysts with two or more functionalities have been proposed to break the selectivity limitation in syngas or CO_2 conversion, the final products are strongly dependent on the scales or intimacy modes of the different active sites. Therefore, the rational design of the tandem catalyst with smart architecture and active site arrangement is the key factor to realize the controlled synthesis from syngas or CO_2. Core–shell structured tandem catalyst has attracted great attention due to the spatial confinement effect of the wrapping shell and the perfect combination mode between different active components (Fig. 7.1) [8]. The distinctive functionalities of the core and shell components guaranteed extraordinary catalytic performance of the core-shell structured catalyst. In addition to the wide applications in syngas and CO_2 conversions, the core-shell structured catalyst also exhibited unique catalytic performance in CH_4 conversion. However, different from the tandem functionality endowed by the core-shell structured catalyst for syngas and CO_2 conversions, the application of core-shell structured catalyst in CH_4 conversion mainly depends on the confinement effect for unique active site fabrication (single-atom catalyst) and coke deposition suppression.

This chapter contributes to summarizing the innovations of core-shell structured catalysts for C1 molecules conversion. The catalytic performances of C1 catalysis processes are determined by the compositions, structures, and scales of the core-shell structured catalysts. A deeper understanding of the reaction mechanism will guide

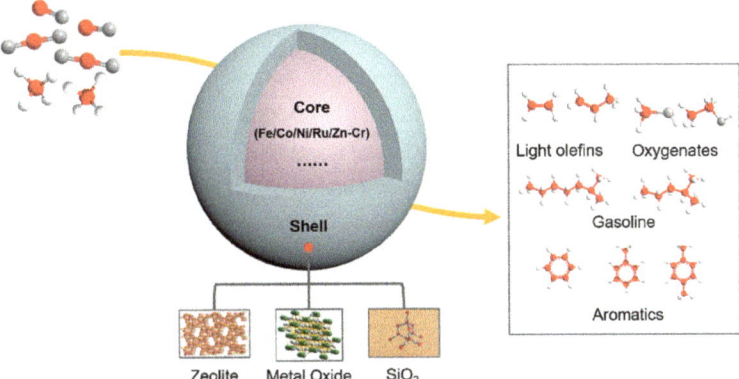

Fig. 7.1 Core-shell structured catalysts for C1 catalysis

the rational design of highly efficient core–shell structured C1 catalysts to realize the controlled synthesis.

7.2 Syngas Conversion by Core–Shell Structured Catalyst

7.2.1 Core–Shell Structured Capsule Catalyst

As mentioned in the introduction section, the traditional syngas conversion via FTs process produces hydrocarbons with broad distributions. Acidic zeolites with unique topologies and acidic properties were usually combined with the traditional FTs catalysts (Fe, Co, and Ru) to control the product distribution in a single-pass via the additional hydrocracking and isomerization reactions. The typical combination mode of these two active sites is the loading of metallic FTs active sites onto the acidic zeolite [9, 10]. Even though the supported method is the simplest approach to fabricate the bifunctional catalyst, the selectivity of desirable C_{5+} hydrocarbons is not high because of the over-hydrocracking reactions on the zeolite support that has close intimacy with the metallic active sites, also because the metallic nanoparticle inside the zeolite channel is too small to realize long-chain hydrocarbon growth. Another method to construct the bifunctional catalyst is the physical mixing strategy. However, the random arrangement of the two active sites leads to the incomplete transformation of the long-chain hydrocarbons [11]. Tsubaki and co-workers established a novel concept of capsule catalyst for the direct conversion of syngas to gasoline-range hydrocarbons, in which the millimeter-sized FTs active core was coated by a zeolite shell (Table 7.1) [12, 13]. In the early stage, the successful encapsulation of FTs catalyst was realized by an in-situ crystal growth strategy [14, 15]. However, the strong alkaline hydrothermal condition would destroy the FTs active core, especially when SiO_2 was employed as catalyst support. Therefore, researchers

Table 7.1 Typical core–shell structured capsule catalysts for syngas conversion

Reaction	Preparation method	Core	Shell	Products	Reference
Fischer–Tropsch synthesis	In-situ growth method	Co/SiO_2	H-ZSM-5	isoparaffins	[14, 15]
		Co/Al_2O_3	H-Beta	isoparaffins	[19]
		Fuse iron	H-MOR	isoparaffins	[20]
	Liquid-membrane crystallization	Co/SiO_2	H-Beta	isoparaffins	[17]
	Dual-membrane method	Fe/SiO_2	Silicalite-1 & H-ZSM-5	isoparaffins	[16]
	Physical adhesion	Co/SiO_2	H-ZSM-5	isoparaffins	[18]
Methanol-mediated pathway	In-situ growth method	CuZnAl	H-ZSM-5	DME	[21]
		CuZnAl	H-Beta	LPG	[22]
	Dual-membrane method	Pd/SiO_2	Silicalite-1 & H-ZSM-5	DME or LPG	[23, 24]
	Physical adhesion	CuZnAl	SAPO-11	DME	[25]
		Zn-Cr	SAPO-34	light olefins	[26]

try to synthesize zeolite shell under the neutral condition to avoid the corrosion of the core component. For example, one layer of Silicalite-1 was first synthesized on the alkali-sensitive SiO_2 supported Fe core catalyst under close-to-neutral conditions. The formed Silicalite-1 shell could be employed as an armor layer to protect the core component during the successive synthesis of the acidic H-ZSM-5 layer under strongly alkaline conditions. Even though the CO conversion (54.8%) of the dual-membrane capsule catalyst was slightly decreased compared with that of naked core catalyst (59.5%) due to the partial coverage of the active Fe sites, the selectivity of isoparaffins was dramatically enhanced from 12.9% to 29.8% after coating of dual zeolite membranes. This enhanced isoparaffins selectivity was attributed to the hydrogenation and isomerization of olefins (formed by the core catalyst Fe/SiO_2) in the confinement space constructed by the dual zeolite membranes [16]. Interestingly, an ingenious liquid-membrane crystallization method was developed to prepare the zeolite-encapsulated catalyst [17]. As shown in Fig. 7.2, the millimeter-sized SiO_2 pellet supported Co was used as the core component, and the H-Beta zeolite precursor was coated onto the surface of the core FTs catalyst and then the crystallization process was performed in a Teflon-lined autoclave containing the organic template and adjusted amount of H2O. The key factor for the successful synthesis of H-Beta zeolite on the FTs catalyst core lied in the separation of alkaline-sensitive Co and SiO_2 from the alkaline solution by a tailor-made Teflon meshy container, which was beneficial to avoid the corrosion of metallic active sites and catalyst supports that usually occurred in the in-situ crystallization process. This liquid-membrane crystallization method is a cost-effective and green technology because it consumes less substrate, produces less waste in the total process. A more convenient approach named physical adhesion method was proposed to increase the synthesis efficiency

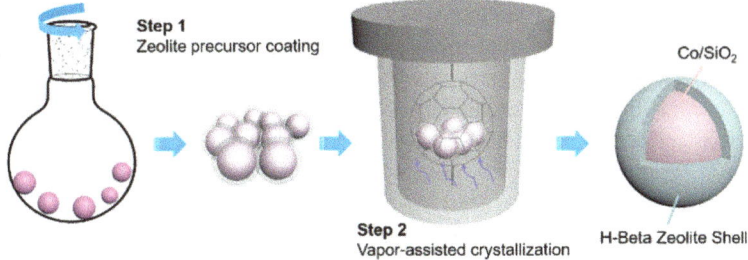

Step 1
Zeolite precursor coating

Co/SiO₂

Step 2
Vapor-assisted crystallization

H-Beta Zeolite Shell

Fig. 7.2 Liquid-membrane crystallization method for core–shell structured capsule catalyst synthesis

and industrial scalability of the zeolite capsule catalyst [18]. In this method, the prepared zeolite powder were directly and uniformly adhered to the surface of FTs catalyst using silica sol as adhesive. The adhesion process can be repeated several times, therefore desirable amount of zeolite can be coated on the FTs catalyst. Furthermore, the physical adhesion method is appropriate to more kinds of zeolites as the shells of capsule catalysts, if compared with the in-situ hydrothermal method and the liquid-membrane crystalization method.

In addition to the transformation of syngas via FTs process, syngas can also be converted to methanol by CuZnAl catalyst or oxygen vacancy-rich metal oxide directly. Therefore, the direct conversion of syngas to valuable chemicals can be realized by combining the methanol synthesis catalyst and the product regulators. Tsubaki and co-workers fabricated a capsule catalyst with H-ZSM-5 zeolite as shell and millimeter-sized methanol synthesis catalyst CuZnAl as core by a facile in-situ crystalization method [21]. This capsule catalyst can be employed as a tandem catalyst for direct conversion of syngas to dimethyl ether (DME) in a consecutive process, by which the methanol produced from CuZnAl core is inevitably dehydrated to DME when it leaves the capsule catalyst through the H-ZSM-5 shell with abundant acidic sites. To clarify the effect of the basicity of precursor solution on the catalytic activity, two types of capsule catalysts were synthesized under different conditions, one was synthesized under strong alkaline solution as the traditional acidic H-ZSM-5 zeolite synthesis, and another one was synthesized under close-to-neutral Silicalite-1 zeolite synthesis condition. Surprisingly, even though no additional Al sources were added into the close-to-neutral precursor solution, a complete and well-grown acidic zeolite shell was formed on the external surface of CuZnAl catalyst, which can be attributed to the migration of Al species from CuZnAl core to the zeolite shell during the close-to-neutral synthesis process. The capsule catalyst prepared under strong basic condition exhibited poor catalytic activity (CO conversion 5.6%) because of the severe corrosion of CuZnAl core component during the hydrothermal synthesis process. In contrast, the catalyst prepared under close-to-neutral conditions exhibited excellent catalytic performance (CO conversion 30.4%, DME selectivity 78.6%) due to the well preservation of CuZnAl core accompanied by the formation of acidic zeolite shell. The physical adhesion and dual-membrane methods were also applicable to fabricate millimeter-sized capsule catalysts for direct conversion of syngas to DME via the methanol-mediated pathway [25, 23]. It should be noted that the product distribution is also greatly influenced by the reaction conditions. A dual-membrane capsule catalyst prepared by coating Silicalite-1 and H-ZSM-5 orderly on the external surface of pellet SiO_2 supported Pd exhibited greatly different product distribution for the consecutive conversion of syngas at different reaction conditions, low reaction temperature (250 °C) only dehydrates two methanol molecules to DME, but high reaction temperature (350 °C) push the reaction forward to produce liquid petroleum gas (LPG) via the dehydration of methanol (olefins formation) and hydrogenation of olefins processes [23, 24]. It is no doubt that the hydrogenation process occurring on the metallic active sites is a vital step for LPG synthesis from syngas. However, it is inefficient that the olefins produced from the zeolite shell should diffuse back to the metallic core (Pd/SiO_2) for the hydrogenation reaction rather than leave the capsule catalyst directly. To increase the efficiency of LPG synthesis, a capsule catalyst composed of CuZnAl shell and H-Beta zeolite shell was rationally designed [22]. An intelligent catalytic interface with a controlled gradient of Cu between core and shell, where olefins could be hydrogenated to LPG directly without diffusing to the core component, guaranteed the highly efficient conversion of syngas to LPG (72.4%) in a single-pass. The formation of the catalytic interface was attributed to the leaching of Cu species from CuZnAl core to the zeolite shell during the hydrothermal process under strongly alkaline conditions. Recently, an elegant core-shell structured catalyst with high-temperature methanol synthesis catalyst Zn-Cr as the core and zeolite SAPO-34 as the shell was well designed to realize the coupling of methanol synthesis and methanol to olefins reactions in a single-pass [26]. The core-shell structured catalyst exhibited higher selectivity to light olefins compared with the bifunctional catalyst prepared by physical mixing, which was attributed to the regular arrangement of different active sites. Furthermore, the undesirable CO_2 was successfully suppressed due to the fast removal of H_2O from the Zn-Cr component to the SAPO-34 shell, thereby the water gas shift reaction (CO + $H_2O \rightarrow CO_2$ + H2) on the Zn-Cr catalyst was significantly weakened.

As discussed above, different types of zeolites can be coated on the external surface of FTs catalysts and methanol synthesis catalysts via different methods for the oriented conversion of syngas into value-added chemicals. The final product distributions obtained from the capsule catalysts are dependent on the reaction conditions, types of zeolites, the thickness of the zeolite shell, etc.

7.2.2 Other Types of Core-Shell Structured Catalysts for Syngas Conversion

In addition to the millimeter-sized capsule catalysts for syngas conversion, the metallic active sites can also be embedded into the nanosized SiO_2 or single-zeolite crystals to realize the controlled synthesis. Even though the wetness impregnation method can introduce the metal species into the framework of zeolites, most of the metal particles disperse on the external of zeolites, which is not the core-shell structured catalyst strictly speaking [27, 28]. Therefore, the traditional zeolite supported FTs catalysts will be not discussed in this chapter. Li et al. embedded Co nanoparticles of different crystallite sizes into an amorphous SiO_2 matrix to clarify the effect of spatial confinement on the product distributions [29]. This core-shell structured catalyst can effectively inhibit the aggregation of the Co particles during the FTs reactions, which avoids misunderstanding caused by size change. Benefiting from the spatial confinement effect, the trapped reaction intermediates can be enriched in the confined space, thereby boosting up the growth of long-chain hydrocarbons. On the other hand, the smaller Co particle with strong adsorption capability guarantees high CH_x coverage inside the confined space to stimulate the chain growth by the carbide mechanism. Khodakov et al. fabricated a metal and zeolite nanocomposite catalyst for direct gasoline-range hydrocarbons synthesis by three steps, the parent zeolite is first etched with an ammonium fluoride solution for mesopores creation, then the FTs active Ru nanoparticles are loaded into the mesopores of the etched zeolite, at last a zeolite shell of H-ZSM-5 is grown on the mesoporous zeolite, coating both the etched surface and metallic nanoparticles [30]. The comparative experiments indicate that the Brønsted acid sites in the shell are more important for iso-paraffins synthesis than that in the core component. The nanosized core-shell structured catalyst can also build a special coordination environment for the controlled synthesis from syngas. Xiao et al. fixed the RhMn nanoparticles into the Silicalite-1 nanocrystals [31]. This core-shell structured catalyst exhibited unexpected C_2-oxygenate selectivity with high CO conversion and excellent stability. Multiple studies demonstrated that the rigid zeolite framework efficiently stabilized the Mn–O–Rh structure, which has been widely known as the crucial site for C_2-oxygenate formation. Even though the nanosized core-shell structured catalysts have obtained tremendous progress in syngas conversion, the close intimacy between the core and shell components results in certain products because different intimacy modes give a totally different product distribution.

7.3 Oriented Hydrogenation of CO_2 into Valuable Chemicals

CO_2 hydrogenation is the easiest way for the direct conversion of CO_2 into valuable chemicals. Wide attention has been paid to CO_2 hydrogenation to basic chemicals, such as CO, CH_3OH, HCOOH, and CH_4 [32]. However, the synthesis of C_{2+} hydrocarbons from CO_2 is still a challenge due to the inertness of CO_2 and high kinetic barriers for C–C bond formation. Recently, two pathways for oriented hydrogenation of CO_2 into C_{2+} hydrocarbons have been developed, the modified FTs pathway and the methanol-mediated pathway [33].

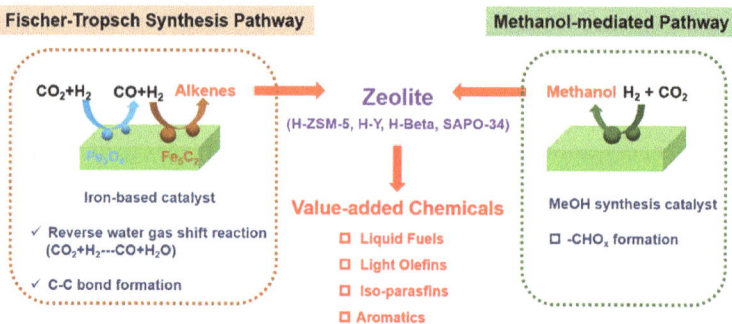

Fig. 7.3 CO_2 hydrogenation to value-added chemicals via the modified FTs and methanol-mediated pathways

During the modified FTs pathway, CO_2 is first converted to CO via the reverse water gas shift (RWGS, $CO_2 + H_2 \rightarrow CO + H_2O$) reaction, then the produced syngas can be transformed to hydrocarbons via the FTs process. Therefore, bifunctional or multifunctional catalysts are essential to realizing the direct conversion of CO_2 into valuable chemicals in a single-pass. Recently, the direct conversion of CO_2 to gasoline has been realized by combining the Fe-based catalyst and acidic zeolite H-ZSM-5, during which CO_2 was converted to alkenes via the RWGS and FTs reactions, then the alkenes were converted to gasoline on the acidic active sites through the hydro-isomerization reaction (Fig. 7.3). To enhance the efficiency of this tandem process, core-shell structured catalysts have been widely employed due to their unique spatial confinement effects. Yang and co-workers designed a nano-sized core-shell structured catalyst (CeO_2-Pt@$mSiO_2$-Co) with CeO_2 supported Pt as core (CeO_2-Pt) and mesoporous SiO_2 supported Co ($mSiO_2$-Co) as shell, in which the CeO_2-Pt interface converted CO_2 and H2 to CO via the RWGS reaction, and on the neighboring $mSiO_2$-Co interface yielded C_2-C_4 hydrocarbons through the subsequent FTs process [34]. Even though C_{2+} hydrocarbons can be produced from the modified FTs process catalyzed by core-shell structured catalysts, the product distributions are limited by the ASF distribution of FTs process with linear hydrocarbons as the main products. Therefore, the product regulator is necessary to realize the synthesis of target chemicals from CO_2 hydrogenation in a tandem process.

Similar to the syngas conversion strategy, CO_2 can also be converted to target products via the methanol-mediated pathway by the core-shell structured catalysts (Fig. 7.3). Tan and co-workers designed a millimeter-sized core-shell structured catalyst with methanol synthesis catalyst Fe-Zn-Zr as core and acidic zeolite as the shell (H-ZSM-5, H-Beta, and H-Y) for isoalkanes synthesis from CO_2 hydrogenation [35]. The double-zeolite shell composed of H-ZSM-5 and H-Beta exhibited an enhanced confinement effect to boost the isoalkanes synthesis from CO_2 (>80% of isoalkanes among all the hydrocarbons), if compared with the single-zeolite shell catalyst. Tsubaki and co-workers fabricated a dual-membrane capsule catalyst by hydrothermal crystalization method with Pd/SiO$_2$ as the core for CO$_2$ hydrogena-tion to methanol, Silicalite-1 as the protective shell and H-ZSM-5 as product regulator through the methanol to hydrocarbons reaction [36]. This capsule catalyst presented higher LPG selectivity compared with the physical mixing counterpart, which was attributed to the well-matched sub-steps work between Pd/SiO$_2$ core and dual-membrane shell.

Comparing the direct conversions of CO$_2$ and syngas pathways, the application of core-shell structured catalysts in syngas conversion has been studied earlier and it seemed to be more efficient and controllable. However, the core-shell structured catalysts with unique spatial confinement effects can provide more opportunities for CO$_2$ controllable conversion because the CO$_2$ hydrogenation process involves multiple steps, which can be integrated into a single-pass via the core-shell structured catalysts. Fe-based catalysts were usually employed to catalyze the modified FTs pathway for CO$_2$ hydrogenation into C$_{2+}$ hydrocarbons such as α-olefins, and Fe$_3$O$_4$ and Fe$_5$C$_2$ active sites have been clarified to be responsible for the RWGS and FTs, respectively. Recently, Fe-based catalysts have been combined with the acidic zeolite H-ZSM-5 to realize the direct conversion of CO$_2$ to gasoline and aromatics, which inspires the rational design of core-shell structured catalysts to expand the range of high-value-added chemicals synthesis [37, 38]. Furthermore, the successful transformations of CO$_2$ into light olefins and aromatics have been realized by physically mixing the methanol synthesis catalysts with SAPO-34 and H-ZSM-5 zeolites, respectively [39–40]. It is reasonable to believe that the extraordinary core-shell structured catalysts for oriented conversion of CO$_2$ into high-value-added chemicals, such as light olefins, aromatics, and gasoline, are coming soon in the near future.

7.4 Conversion of CH$_4$ Via Core–Shell Structured Catalyst

As shown in Fig. 7.4, there are several approaches for CH$_4$ conversion into valuable chemicals, for example, CH$_4$ dehydroaromatization for aromatics synthesis, oxidative coupling of CH$_4$ to light olefins, CH$_4$ dry/steam reforming to syngas, CH$_4$ partial oxidation to methanol or acetic acid, and so on [6, 41]. In brief, the different conversion routes can be classified into two kinds, CH$_4$ conversion process performed at high or low temperature. No matter which route is chosen for CH$_4$ conversion, a highly efficient catalyst is the pivotal factor to activate the stable CH$_4$ molecule. Core-shell structured catalysts have been widely investigated in the field of CH$_4$ conversion due to their spatial confinement effects and coke resistance properties.

7.4.1 CH$_4$ Dry Reforming Catalyzed by Core-Shell Structured Catalyst

$$CH_4 + CO_2 \rightarrow 2CO + 2H_2 \ \Delta H^0_{298} = 248 \, kJ \, mol^{-1}$$

Fig. 7.4 A variety of CH$_4$
conversion pathways

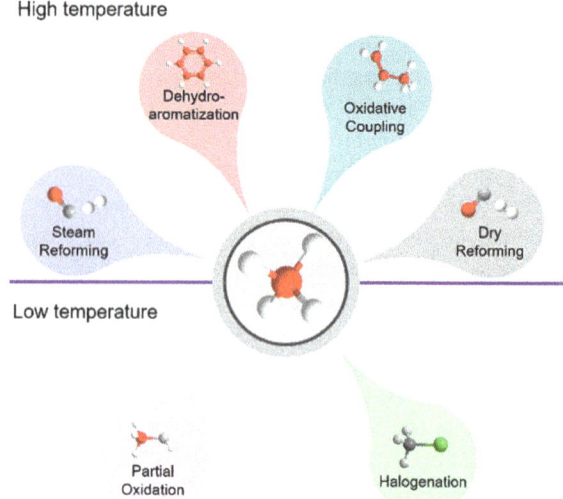

CH$_4$ dry reforming employing CO$_2$ as a soft oxidant has the potential to alleviate the environmental concerns stem from the emissions of greenhouse gases CO$_2$ and CH$_4$. The CH$_4$ dry reforming process is a highly endothermic reaction due to the high stability of CO$_2$ and CH$_4$, therefore high operation temperature is required to enhance the activation capability of the transition metal-based catalyst. However, the major challenge for the practical application of CH$_4$ dry reforming exists in the rapid deactivation of the catalysts, which results from metallic particles sintering and coke deposition at high reaction temperature. Even though noble metal catalysts are efficient to prolong the reaction lifetime by suppressing the undesirable coke deposition, the high-cost impedes their large-scale application. Therefore, core-shell structured catalysts with unique architecture have been rationally designed to overcome the challenges on the road to the industrial application of CH$_4$ dry reforming.

On the one hand, the CH$_4$ dry reforming is sensitive to the particle size of the active metal sites. A minimum Ni particle diameter of 7 nm is required for the growth of filamentous carbon, which covers the active sites for CH$_4$ and CO$_2$ activations [42]. For the traditional Ni-based supported catalysts, the aggregation of the Ni particles on the external of the support is inevitable, thus leading to serious coke deposition. Benefiting from the confinement effect, core-shell structured catalysts with solely metal particles (Ni, Pd, Co, alloys, *etc.*) or supported metals as core and SiO$_2$ or metallic oxides (TiO$_2$, CeO$_2$, Al$_2$O$_3$, *etc.*) as shell have been widely employed to suppress the unfavorable coke deposition phenomenon. SiO$_2$ shell has been widely reported to stabilize the metal nanoparticles and minimize metal sintering due to its high thermal stability, high specific surface area, tunable pore size distribution, and ease for synthesis. For the core-shell structured catalyst with supported metal as the core, the migration of metal particles on the external surface of the support usually leads to the formation of larger particles, which is inclined to be active sites for the growth of carbon nanotubes. However, as long as the SiO$_2$ shell remains thermally stable, the sintering of metal particles across different core-shell structured particles can be controlled. It should be noted that the thickness of SiO_2 shell renders an obvious effect on the catalytic performance, as thin shell leading to the collapse of the shell under high reaction temperature and thick shell slows down the mass-transfer efficiency of reactants [43]. Apart from the protective effect derived from the shell component, a few metal oxides, as shell, with high redox activity also endows CO_2 activation capability for the core-shell structured catalyst. For example, the CeO_2 shell can undergo substantial cation change between Ce^{4+} and Ce^{3+} accompanied by the consumption and formation of oxygen vacancies during the CH_4 dry reforming process. The transformation of Ce^{4+} and Ce^{3+} will release active lattice oxygen to oxidize the carbon species on the interface of the catalyst, during which oxygen vacancies can be formed and employed as catalytic centers for CO_2 activation. The replenished lattice oxygen will be reused to activate the carbon species from CH_4. The enhanced CO_2 activation ability is beneficial to reduce the coke deposition during the CH_4 dry reforming process. Therefore, the core-shell structured catalyst with reducible metal oxide as shell guarantees the bifunctional properties of enhancing CO_2 activation and suppressing metal sintering for the CH_4 dry reforming reaction [44].

On the other hand, the crystal structure of the metallic active sites is another vital factor for the CH_4 dry reforming. For example, Pd nanocubes with exposed [100] planes have been proven to have higher CH_4 activation capability. However, the aggregation of Pd nanocubes under high temperatures will break the original crystal structure of the metal catalyst, therefore leading to the decrease of CH_4 dry reforming activity. Through the confinement effect of the core-shell structured catalyst, the core component with certain active facets can be precisely controlled. Yue and co-workers investigated the protective effect of shell components by encapsulating the [100] planes exposed Pd nanocubes with mesoporous SiO_2. The morphology of the core-shell structured catalyst, especially the well defined core component, was well maintained after the long-term high-temperature reaction. In contrast, the traditional supported Pd-based catalyst exhibited poor reaction stability, which further confirmed the protective effect derived from the external shell [45].

7.4.2 Zeolites Encapsulating Metallic Species for Direct Conversion of CH₄ to Methanol

Another typical route for CH_4 conversion is the selective partial oxidation of CH_4 to methanol catalyzed by core-shell structured catalysts using zeolites encapsulating metallic species, which is an energy-saving strategy due to the mild operation temperatures. In nature, CH_4 monooxygenase enzymes selectively transform CH_4 into methanol at ambient temperature. Inspired by these enzymes, zeolites stabilized binuclear Cu cores have been rationally designed in a form analogous to that found in the CH_4 monooxygenase enzymes [46, 47]. Lercher and co-workers encapsulated the trinuclear copper oxygen clusters into the framework of mordenite zeolite by mimicking the nuclearity and reactivity of active sites in CH_4 monooxygenase enzymes [48]. Benefiting from the perfect confined environment for the highly selective stabilization of trinuclear copper oxygen clusters, the core-shell structured catalyst exhibited high reactivity toward C-H bond activation and the following methanol formation under gas-phase reaction conditions. However, this conversion process for CH_4 to methanol was realized by a chemical looping system that is composed of high-temperature catalyst activation for active sites formation and low-temperature CH_4 reaction and product extraction (Fig. 7.5a). Thus, the pseudo-catalytic process results in the low productivity of the target product methanol. Compared with the intermittent gas-phase reaction, CH_4 oxidation in a liquid solvent with H_2O_2 as an oxidant provides a facile method for high-value-added utilization of CH_4 under mild conditions. The partial oxidation of CH_4 using H_2O_2 as oxidant exhibited great potential for methanol synthesis. However, the high-cost of H_2O_2 limited its practical application. Recently, the partial oxidation of CH_4 using in-situ formed H_2O_2 from H_2 and O_2 has triggered much more attention, but the methanol productivity from this indirect strategy is still lower than that using H_2O_2 to oxidize CH_4 directly. Xiao and co-workers speculated that the slowly formed intermediate H_2O_2 diffused away from the metallic active site, which was unfavorable for the subsequent activation of CH_4. Therefore, they fabricated a hydrophobic core-shell structured catalyst, in which AuPd alloy was embedded into the framework of H-ZSM-5 zeolite by solvent-free method, followed by the transformation of the hydrophilic external surface of zeolite to the hydrophobic interface by long-chain organics (C_{16}) modification. The hydrophobic interface was employed as a molecular fence to prevent the diffusion of H_2O_2 away from the active sites without affecting the hydrophobic CH_4 molecules passing through the hydrophobic sheath to arrive the AuPd nanoparticles (Fig. 7.5b).

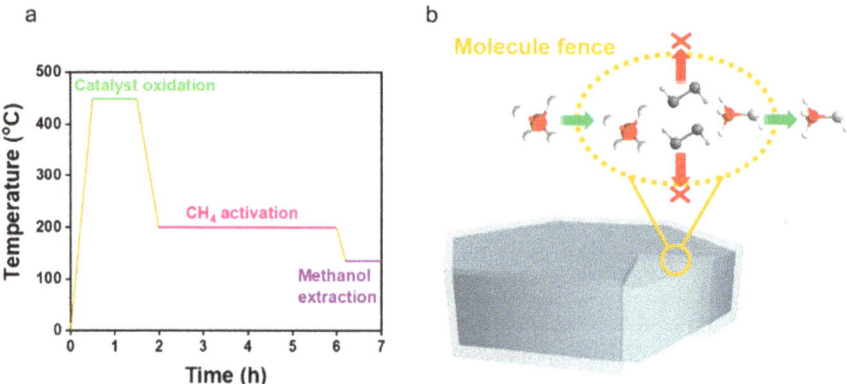

Fig. 7.5 Representative diagram for the stepped conversion of CH_4 **a**, and the concept of molecule fence for direct conversion of CH_4 to methanol proposed by Xiao et al. **b**

Through this rationally designed core-shell structured catalyst, a high local concentration of H2O2 around AuPd nanoparticles was kept, which was beneficial to address the challenge to accelerate CH_4 conversion to methanol [49].

7.4.3 Other Core-Shell Structured Catalysts for CH_4 Conversion

CH_4 can also be converted to high-value-added aromatics via the CH_4 dehydroaromatization (MDA) process in the absence of O_2. Since it was first reported in 1993, a variety of MDA catalysts based on metal ions (Fe, Mo, W, Zn, Cu, Ga, *etc.*) dispersed on various zeolites (ZSM-5, MCM-22, and MCM-49) have been emerged [6, 50]. However, due to the high operation temperature ($\geq 700\,°C$) of MDA reaction, the traditional supported catalysts suffered from severe coke deposition and active site aggregation, which hampered the industrial applications. Core-shell structured catalysts with unique confinement effects are promising candidates to prevent the sintering of metallic active sites. On the other hand, the protective effect derived from the shell can efficiently suppress the undesirable coke deposition. Tsubaki and co-workers designed a core-shell structured catalyst with Mo nanoparticles embedded into the hollow dual-layer zeolite capsule (Silicalite-1 and H-ZSM-5) for the MDA reaction [51]. In this capsule catalyst, the primary intermediates (CH_x and C_2H_y) produced from MoC_x that loaded on the inner shell were accumulated in the hollow cavity, where CH_4 and the primary intermediates were oligomerized and dehydrocyclized to benzene. It should be noted that the thin H-ZSM-5 membrane guaranteed the fast mass transfer of the formed benzene, which efficiently avoid the undesirable secondary reaction of benzene and suppress the coke deposition. Bao and co-workers embedded the single Fe sites into the SiO_2 matrix by fusing ferrous metasilicate with SiO_2 at 1700 °C in air, followed by leaching with aqueous HNO_3 [52]. Comprehensive characterizations confirmed that the single Fe sites enabled the activation of CH_4 to methyl radicals, which couple to ethylene and aromatics by a series of gas-phase reactions. The absence of adjacent Fe sites benefitted from the confinement effect of the SiO_2 matrix prevents the catalytic C–C coupling and the following oligomerization reactions on the catalysts surface, thereby suppressing the undesirable coke deposition.

Despite the challenges and difficulties of activating CH_4, there is significant progress in the development of direct conversion of CH_4 to hydrocarbons, particularly for the core-shell structured catalysts driving CH_4 conversion. Different functionalities from core-shell structured catalysts accommodate to different CH_4 conversion routes. CH_4 dry reforming and MDA need core-shell structured catalysts to prolong their lifetime by suppressing the coke deposition and metallic active sites sintering under high reaction temperatures. Because CH_4 selective partial oxidation is sensitive to the local concentration of oxidants, the core-shell structure is beneficial to create a special reaction environment, thereby boosting the CH_4 partial oxidation performance. No matter which route is chosen, further development of highly efficient catalysts is essential to meeting the high demand for the industrial application of CH_4 conversion to commodity chemicals.

7.5   Conclusion

Core-shell structured catalysts have been widely used in C1 molecules conversion due to their unique geometric architectures, ingenious confinement effects, and controllable active components for specific reactions. This chapter summarizes the typical examples of core-shell structured catalysts for C1 catalysis. Zeolites-based core-shell structured catalysts are essential to realizing the selective synthesis of target products from syngas conversion or CO_2 hydrogenation. The linear hydrocarbons or methanol synthesis from syngas or CO_2 can be transformed into value-added chemicals on the acidic sites of zeolites. Benefiting from the spatial confinement effect, the target product selectivity of the core–shell structured catalyst is higher than that obtained in the physical mixing counterpart. The widespread of nanosized core-shell structured catalyst in CH_4 conversion mainly relies on their excellent coke-resistance and metallic sites sintering suppression. Even though huge progress has been achieved in the rational design of core-shell structured catalysts for C1 catalysis, their industrial applications are still hampered by some limitations such as surface abrasion of zeolite shell, highly efficient separation engineering, and advanced reactor technology.

References

1. Zhou W, Cheng K, Kang J et al (2019) New horizon in C1 chemistry: breaking the selectivity limitation in transformation of syngas and hydrogenation of CO_2 into hydrocarbon chemicals and fuels. Chem Soc Rev 48:3193–3228
2. Zhang Q, Kang J, Wang Y (2010) Development of novel catalysts for Fischer-Tropsch synthesis: tuning the product selectivity. ChemCatChem 2:1030–1058
3. Li J, He Y, Tan L et al (2018) Integrated tuneable synthesis of liquid fuels via Fischer-Tropsch technology. Nat Catal 1:787–793
4. Álvarez A, Bansode A, Urakawa A et al (2017) Challenges in the greener production of formates/formic acid, methanol, and DME by heterogeneously catalyzed CO_2 hydrogenation processes. Chem Rev 117:9804–9838
5. Wei J, Sun J, Wen Z et al (2016) New insights into the effect of sodium on Fe_3O_4-based nanocatalysts for CO_2 hydrogenation to light olefins. Catal Sci Technol 6:4786–4793
6. Schwach P, Pan X, Bao X (2017) Direct conversion of methane to value-added chemicals over heterogeneous catalysts: challenges and prospects. Chem Rev 117:8497–8520
7. Zhang Q, Yu J, Corma A (2020) Applications of zeolites to C1 chemistry: recent advances, challenges, and opportunities. Adv Mater 30:1704439
8. Wang X, Feng J, Bai Y et al (2016) Synthesis, properties, and applications of hollow micro-/nanostructures. Chem Rev 116:10983–11060
9. Kang J, Cheng K, Zhang L et al (2011) Mesoporous zeolite-supported ruthenium nanoparticles as highly selective Fischer-Tropsch catalysts for the production of C_5–C_{11} isoparaffins. Angew Chemie-Int Ed 50:5200–5203

10. Peng X, Cheng K, Kang J et al (2015) Impact of hydrogenolysis on the selectivity of the Fischer-Tropsch synthesis: diesel fuel production over mesoporous zeolite-Y-supported cobalt nanoparticles. Angew Chemie-Int Ed 54:4553–4556

11. Khodakov AY, Chu W, Fongarland P (2007) Advances in the development of novel cobalt Fischer-Tropsch catalysts for synthesis of long-chain hydrocarbons and clean fuels. Chem Rev 107:1692–1744

12. Bao J, Tsubaki N (2018) Design and synthesis of powerful capsule catalysts aimed at applications in C1 chemistry and biomass conversion. Chem Rec 18:4–19

13. Bao J, Yang G, Yoneyama Y, Tsubaki N (2019) Significant advances in C1 catalysis: highly efficient catalysts and catalytic reactions. ACS Catal 9:3026–3053

14. He J, Liu Z, Yoneyama Y et al (2006) Multiple-functional capsule catalysts: a tailor-made confined reaction environment for the direct synthesis of middle isoparaffins from syngas. Chem-A Eur J 12:8296–8304

15. He J, Yoneyama Y, Xu B et al (2005) Designing a capsule catalyst and its application for direct synthesis of middle isoparaffins. Langmuir 21:1699–1702

16. Jin Y, Yang G, Chen Q et al (2015) Development of dual-membrane coated Fe/SiO$_2$ catalyst for efficient synthesis of isoparaffins directly from syngas. J Memb Sci 475:22–29

17. Li C, Xu H, Kido Y et al (2012) A capsule catalyst with a zeolite membrane prepared by direct liquid membrane crystallization. Chemsuschem 5:862–866

18. Yang G, Xing C, Hirohama W et al (2013) Tandem catalytic synthesis of light isoparaffin from syngas via Fischer-Tropsch synthesis by newly developed core-shell-like zeolite capsule catalysts. Catal Today 215:29–35

19. Bao J, He J, Zhang Y et al (2008) A core/shell catalyst produces a spatially confined effect and shape selectivity in a consecutive reaction. Angew Chemie-Int Ed 47:353–356

20. Lin Q, Yang G, Li X et al (2013) A catalyst for one-step isoparaffin production via Fischer-Tropsch synthesis: growth of a H-mordenite shell encapsulating a fused iron core. ChemCatChem 5:3101–3106

21. Yang G, Tsubaki N, Shamoto J, Yoneyama Y (2010) Confinement effect and synergistic function of H-ZSM-5/Cu-ZnO-Al$_2$O$_3$ capsule catalyst for one-step controlled synthesis. J Am Chem Soc 132:8129–8136

22. Lu P, Sun J, Shen D et al (2018) Direct syngas conversion to liquefied petroleum gas: Importance of a multifunctional metal-zeolite interface. Appl Energy 209:1–7

23. Yang G, Wang D, Yoneyama Y et al (2012) Facile synthesis of H-type zeolite shell on a silica substrate for tandem catalysis. Chem Commun 48:1263–1265

24. Zhang P, Yang G, Tan L et al (2018) Direct synthesis of liquefied petroleum gas from syngas over H-ZSM-5 enwrapped Pd-based zeolite capsule catalyst. Catal Today 303:77–85

25. Phienluphon R, Pinkaew K, Yang G et al (2015) Designing core (Cu/ZnO/Al$_2$O$_3$)-shell (SAPO-11) zeolite capsule catalyst with a facile physical way for dimethyl ether direct synthesis from syngas. Chem Eng J 270:605–611

26. Tan L, Wang F, Zhang P et al (2020) Design of a core-shell catalyst: an effective strategy for suppressing side reactions in syngas for direct selective conversion to light olefins. Chem Sci 11:4097–4105

27. Cheng K, Kang J, Huang S et al (2012) Mesoporous beta zeolite-supported ruthenium nanoparticles for selective conversion of synthesis gas to C$_5$–C$_{11}$ isoparaffins. ACS Catal 2:441–449

28. Kim JC, Lee S, Cho K et al (2014) Mesoporous MFI zeolite nanosponge supporting cobalt nanoparticles as a Fischer-Tropsch catalyst with high yield of branched hydrocarbons in the gasoline range. ACS Catal 4:3919–3927

29. Cheng Q, Tian Y, Lyu S et al (2018) Confined small-sized cobalt catalysts stimulate carbon-chain growth reversely by modifying ASF law of Fischer-Tropsch synthesis. Nat Commun 9:3250

30. Přech J, Strossi Pedrolo DR, Marcilio NR et al (2020) Core-shell metal zeolite composite catalysts for in situ processing of Fischer-Tropsch hydrocarbons to gasoline type fuels. ACS Catal 10:2544–2555

31. Wang C, Zhang J, Qin G et al (2020) Direct conversion of syngas to ethanol within zeolite crystals. Chem 6:646–657
32. Wang W, Wang S, Ma X, Gong J (2011) Recent advances in catalytic hydrogenation of carbon dioxide. Chem Soc Rev 40:3703
33. Guo L, Sun J (2018) Recent advances in direct catalytic hydrogenation of carbon dioxide to valuable C_{2+} hydrocarbons. J Mater Chem A 6:23244–23262
34. Xie C, Chen C, Yu Y et al (2017) Tandem catalysis for CO_2 hydrogenation to C_2–C_4 hydrocarbons. Nano Lett 17:3798–3802
35. Wang X, Yang G, Zhang J et al (2016) Synthesis of isoalkanes over a core (Fe-Zn-Zr)-shell (zeolite) catalyst by CO_2 hydrogenation. Chem Commun 52:7352–7355
36. Li H, Zhang P, Guo L et al (2020) A well-defined core-shell-structured capsule catalyst for direct conversion of CO_2 into liquefied petroleum gas. Chemsuschem 13:2060–2065
37. Wang Y, Kazumi S, Gao W et al (2020) Direct conversion of CO_2 to aromatics with high yield via a modified Fischer-Tropsch synthesis pathway. Appl Catal B Environ 269:118792
38. Wei J, Ge Q, Yao R et al (2017) Directly converting CO_2 into a gasoline fuel. Nat Commun 8:15174
39. Gao P, Dang S, Li S et al (2018) Direct production of lower olefins from CO_2 conversion via bifunctional catalysis. ACS Catal 8:571–578
40. Wang Y, Gao W, Kazumi S et al (2019) Direct and oriented conversion of CO_2 into value-added aromatics. Chem-A Eur J 25:5149–5153
41. Das S, Pérez-Ramírez J, Gong J et al (2020) Core-shell structured catalysts for thermocatalytic, photocatalytic, and electrocatalytic conversion of CO_2. Chem Soc Rev 49:2937–3004
42. Kim JH, Suh DJ, Park TJ, Kim KL (2000) Effect of metal particle size on coking during CO_2 reforming of CH_4 over Ni-alumina aerogel catalysts. Appl Catal A Gen 197:191–200
43. Li Z, Mo L, Kathiraser Y, Kawi S (2014) Yolk-satellite-shell structured Ni-Yolk@Ni@SiO_2 nanocomposite: Superb catalyst toward methane CO_2 reforming reaction. ACS Catal 4:1526–1536
44. Das S, Ashok J, Bian Z et al (2018) Silica-ceria sandwiched Ni core–shell catalyst for low temperature dry reforming of biogas: coke resistance and mechanistic insights. Appl Catal B Environ 230:220–236
45. Yue L, Li J, Chen C et al (2018) Thermal-stable Pd@mesoporous silica core-shell nanocatalysts for dry reforming of methane with good coke-resistant performance. Fuel 218:335–341
46. Tomkins P, Ranocchiari M, Van Bokhoven JA (2017) Direct conversion of methane to methanol under mild conditions over Cu-zeolites and beyond. Acc Chem Res 50:418–425
47. Ravi M, Ranocchiari M, van Bokhoven JA (2017) The direct catalytic oxidation of methane to methanol-a critical assessment. Angew Chemie-Int Ed 56:16464–16483
48. Grundner S, Markovits MAC, Li G et al (2015) Single-site trinuclear copper oxygen clusters in mordenite for selective conversion of methane to methanol. Nat Commun 6:7546
49. Jin Z, Wang L, Zuidema E et al (2020) Hydrophobic zeolite modification for in situ peroxide formation in methane oxidation to methanol. Science 367:193–197
50. Wang L, Tao L, Xie M et al (1993) Dehydrogenation and aromatization of methane under non-oxidizing conditions. Catal Letters 21:35–41
51. Zhu P, Yang G, Sun J et al (2017) A hollow Mo/HZSM-5 zeolite capsule catalyst: preparation and enhanced catalytic properties in methane dehydroaromatization. J Mater Chem a 5:8599–8607
52. Guo X, Fang G, Li G et al (2014) Direct, nonoxidative conversion of methane to ethylene, aromatics, and hydrogen. Science 344:616–619

Chapter 8
Core-Shell Structured Catalysts for Catalytic Conversion of CO_2 to Syngas

Sonali Das and Sibudjing Kawi

8.1 Introduction

Carbon dioxide management in terms of carbon capture and utilization (CCU) is an area of vigorous research because of an urgent need to limit the adverse environmental effects of excess CO_2 emissions [1, 2]. CO_2 or dry reforming of hydrocarbons is one of the potential routes of using CO_2 as a raw material into value chains for energy or chemical production on a large scale [3–5].

In dry reforming, CO_2 is used as a soft oxidant for the catalytic reforming of hydrocarbons to produce syngas, a mixture of CO and H_2. Due to wide availability of natural gas and shale gas, methane is the most commonly used hydrocarbon in CO_2 reforming. CO_2 (dry) reforming of methane (DRM) produces syngas with an equimolar ratio of H_2 and CO. The produced syngas may be used as a source of hydrogen for fuel cells or for other industrial processes. DRM can complement processes for hydrogen production from other, more established technologies such as steam reforming or autothermal reforming. Syngas components are among the key building blocks of the chemical industry and can be converted into hydrocarbons and oxygenates by Fischer–Tropsch synthesis or by methanol synthesis. Thus, dry reforming constitutes the first step of the indirect route of utilization of CO_2 and its conversion to liquid fuels or chemicals. In this indirect route, CO_2 is used to reform methane from natural gas, shale gas or biogas to produce syngas by DRM, which is subsequently converted to valuable fuels or chemicals by gas-to-liquid technologies. The syngas composition from DRM is ideal for the production of oxygenated compounds and Fischer–Tropsch synthesis of long-chain hydrocarbons and liquid fuel [6]. Use of CO_2 in dry reforming to form syngas and further conversion

S. Das · S. Kawi (✉)
Department of Chemical and Biomolecular Engineering, National University of Singapore, Singapore 119260, Republic of Singapore
e-mail: chekawis@nus.edu.sg

© The Author(s), under exclusive license to Springer Nature Singapore Pte Ltd. 2021
H. Yamashita and H. Li (eds.), *Core-Shell and Yolk-Shell Nanocatalysts*,
Nanostructure Science and Technology,
https://doi.org/10.1007/978-981-16-0463-8_8

to fuels does not lead to a net reduction in atmospheric CO_2 content because the ultimate consumption of the produced fuels releases the CO_2 to atmosphere again, but such CO_2 derived synthetic fuels may serve as a substitute of fossil-based fuels. This chapter covers the applications and benefits of core-shell structured catalysts in CO_2 reforming of methane to produce syngas.

8.2 Dry (CO_2) Reforming of Methane

CO_2 (dry) reforming of methane (Eq. 8.1) is conducted at elevated temperatures (600–1000 °C) over metal catalysts such as Ni, Pt, Ru, and Rh [6]. The DRM reaction is highly endothermic and requires high temperatures for significant conversion and product yield. Catalysts are required for the activation of CO_2 and CH_4, both of which are extremely stable compounds. Noble metal such as Pt, Pd, Ru, and Rh and non-noble metal nanoparticles such as Ni, Cu, and Co dispersed on suitable supports are effective in catalyzing this reaction. Side-reactions of DRM include the reverse water gas shift reaction (RWGS, Eq. 8.2), methane decomposition (Eq. 8.3) and CO disproportionation, commonly known as the Boudouard reaction (Eq. 8.4). RWGS is usually equilibrium limited and its occurrence lowers the hydrogen yield by converting it to water. CH_4 dissociation and CO disproportionation result in the formation of solid carbonaceous species or coke, which can deposit on the catalyst and cause catalyst deactivation.

$$CH_4 + CO_2 \rightarrow 2CO + 2H_2 \qquad \Delta H^0_{298} = 248 \, \frac{kJ}{mol} \qquad (8.1)$$

$$CO_2 + H_2 \rightarrow CO + H_2O \qquad \Delta H^0_{298} = 41.2 \, \frac{kJ}{mol} \qquad (8.2)$$

$$CH_4 \rightarrow C(s) + 2H_2 \qquad \Delta H^0_{298} = 75 \, \frac{kJ}{mol} \qquad (8.3)$$

$$2CO \rightarrow C(s) + CO_2 \qquad \Delta H^0_{298} = -172 \frac{kJ}{mol} \qquad (8.4)$$

The main challenge for the large-scale application of DRM is the rapid deactivation of the catalysts during operation. Catalyst deactivation is primarily caused by two factors: (1) sintering of the active metal nanoparticles at the high temperature operating conditions to form larger particles with low surface area, and (2) the deposition of solid coke on the catalyst that covers active sites and causes blockage of reactors. Because of cheaper cost, easy availability and high activity, Ni-based catalysts have the highest potential for commercial use as DRM catalysts, but the issues of metal sintering and coke formation are extremely severe for Ni [6, 7].

Metal sintering in catalysts occurs by the gradual agglomeration of the nanoparticles by coalescence or by Ostwald ripening [8]. High temperatures accelerate the motion of the metal nanoparticles on the surface of the support and facilitate their coalescence. The Tammann temperature of Ni (the temperature at which the onset of mobility of particles on a surface occurs) is lower than the DRM operating temperature of 600–1000 °C, leading to rapid particle growth and loss of active surface area. Nickel nanoparticles are also highly active in catalyzing coke formation during DRM. If the rate of formation of carbonaceous intermediates (from methane decomposition or CO disproportionation) on the metal surface is not balanced by the rate of oxidation of these intermediates, it results to the formation of solid coke. The formed carbon dissolves into the nickel crystallites to form a carbide phase followed by the nucleation of a separate coke phase upon exceeding the saturation concentration [9]. Crystalline coke may deposit as graphitic rings encapsulating the Ni nanoparticles or in the form of carbon nanotubes (Fig. 8.1). Carbon nanotube growth may occur with the Ni nanoparticle at the base or tip of the nanotube, and the growth continues to occur as long as the attached Ni nanoparticle is catalytically active. Thus, filamentous coke formation, especially with tip growth, may not immediately deactivate the catalyst; but it has severe implications on long-term operation. The uncontrolled growth of carbon nanotubes causes an expansion of the catalyst bed, ruptures catalyst

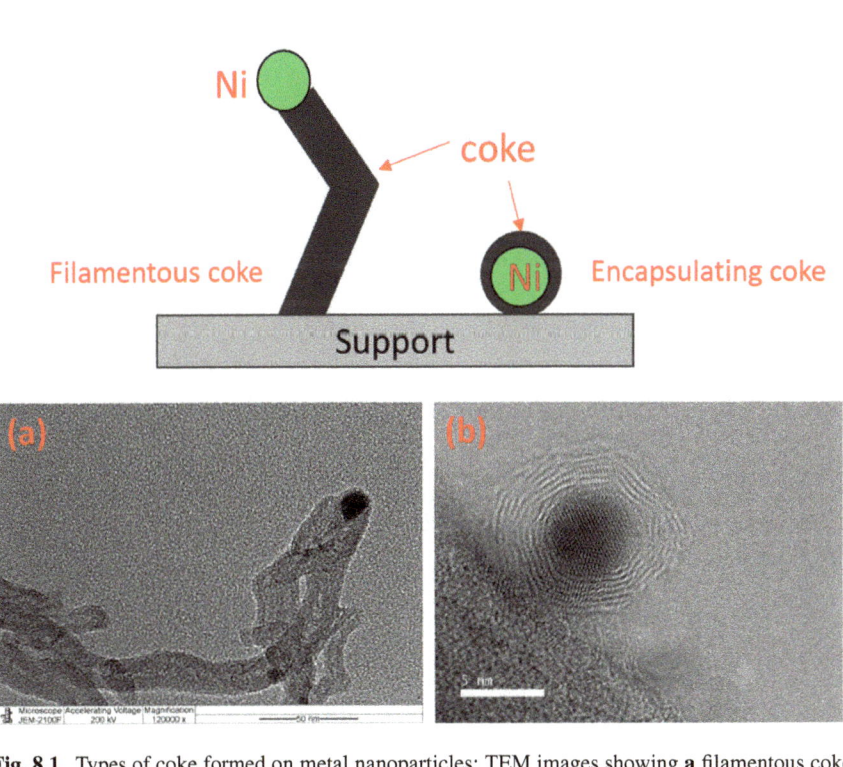

Fig. 8.1 Types of coke formed on metal nanoparticles; TEM images showing **a** filamentous coke with Ni nanoparticle at the tip (Reprinted from Ref. [10] Copyright (2018), with permission from Elsevier), **b** encapsulating coke surrounding a Ni nanoparticle (Reprinted from Ref. [11], Copyright (2019), with permission from The Royal Society of Chemistry)

pellets, and ultimately leads to blockage of the catalyst bed, preventing the reactant gases from going in. Tip growth of filamentous carbon dislodges the metal nanoparticles from the support, resulting in a complete loss of metal-support interaction and easier agglomeration of metal nanoparticles. More importantly, such wreckage of the original catalyst structure makes it extremely challenging to regenerate the catalyst because of irreversible sintering of the detached metal nanoparticles upon coke removal during regeneration.

Thermodynamically, coke formation by CH_4 dissociation is favored at high temperature, while CO disproportionation is favored at low temperature. Coke formed from the Boudouard reaction is often more inactive and difficult to oxidize than that produced by methane decomposition. Thus, coke deposition is more severe in relatively low temperature DRM operation. At very high temperatures (>900 °C for operation at 1 bar), coke formation in DRM may become negligible because of thermodynamic suppression of the Boudouard reaction and fast oxidation of coke [3]. But operation at lower temperature is more desirable for industrial applications. There is also a risk of formation of cold spots in the DRM reactor because of the endothermic nature of the reaction; and thus, even for high temperature operation, such local cold spots may initiate coke deposition.

Development of suitable catalysts that can resist metal sintering and coke formation during long-term DRM operation is, hence, the primary focus of research in this field. Nickel-containing catalysts, either monometallic or alloyed with other metals, are most investigated as potential candidates for future commercial applications [12].

8.3   Core-Shell Catalysts in CO_2 Reforming of Methane

In the past decade, core-shell structured materials have emerged as excellent candidates for coke resistant and thermally stable catalysts for DRM [5, 13]. Because of their unique morphology, core-shell materials offer high thermal stabilities, resistance to metal sintering, and inhibition of growth of carbon nanotubes during DRM, compared to conventional catalysts. At the same time, the flexibility of combining multiple materials with various desirable properties in a core-shell structure allows for bifunctional catalysis on such catalysts, which can help in coke removal during DRM and extend the catalyst life.

8.3.1   Benefits of Core-Shell Catalysts

In this section, the benefits of core-shell structured catalysts in increasing stability in dry reforming are discussed with relevant examples. The properties of core-shell structured catalysts that make them superior candidates for long-term application in CO_2 reforming reactions are illustrated in Fig. 8.2 and are discussed in detail in the subsequent sections. However, it is to be noted that notwithstanding these benefits

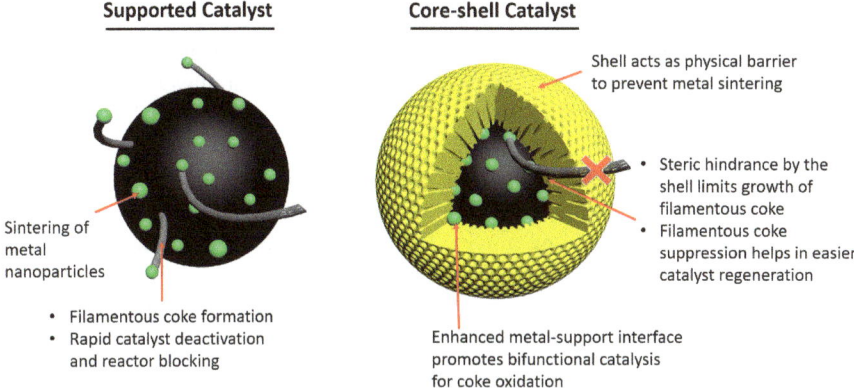

Fig. 8.2 Benefits of core-shell catalysts over conventional supported catalysts for CO_2 reforming of methane

and advantages of core-shell catalysts, they are still an emerging class of materials and there are several bottlenecks that need to be addressed before such catalysts can be used industrially. Some of the challenges of core-shell catalysts compared to conventional supported catalysts are their greater complexity and cost of synthesis and the possibility of lower overall activity associated with active site blockage or mass transfer limitations imposed by the shell. Readers may refer to a review article by Das et al. [5] for a detailed discussion on the advantages and disadvantages of core-shell catalysts versus conventional supported catalysts for various routes of catalytic conversion of CO_2.

8.3.1.1 Resistance to Sintering and Coking

As discussed in Sect. 8.2, the agglomeration of nanosized metal particles by coalescence and Ostwald ripening is accelerated at the high temperature of DRM operation. In conventional supported catalysts, the mobility of the metal nanoparticles over the support results in their rapid coalescence and formation of bigger nanoparticles with low catalytic surface area [8]. Not only that, metal sintering also leads to increased growth of filamentous carbon on the metal nanoparticles [7, 14, 15]. The rate of coke formation has been shown to be a strong function of the metal particle size, with coke formation being favored on bigger ensembles [14, 15]. It is generally observed that the carbon nanotube growth occurs with the metal particle at the tip or base, and the diameter of the nanotube is roughly equal to that of the metal nanoparticle. The thermodynamic properties and external tension energy of carbon nanofibers have been proposed to change as a function of the crystal size. A theoretical modeling study predicted that a small-sized Ni crystal results in a high saturation concentration of carbon nanofibers and, thus, a low driving force for carbon diffusion through the Ni crystals [9]. Theoretical calculations also indicate that a minimum cluster size of around 80 carbon atoms are required for the stability of a graphene island on Ni surface; thus, Ni nanoparticles with facets smaller than this critical size of carbon cluster may not be able to form coke. Some experimental studies suggest that Ni nanoparticles of size lower than 4–7 nm can inhibit the generation of filamentous carbon [15, 16]. Thus, limiting the metal particle size is crucial for maintaining good activity and coke resistance in DRM. However, even if very small metal particles are supported on conventional supported catalysts, the initial high metal dispersion cannot be maintained due to sintering in DRM reaction atmosphere, leading to coke formation and subsequent catalyst deactivation.

Core-shell structured catalysts provide one of the most effective ways to minimize catalyst sintering and to preserve the active structure of nanomaterials. In a core-shell structure, the active nanoparticles may be encapsulated or partially embedded in a layer of thermally stable material that acts as a physical barrier to hinder particle migration and agglomeration. Very small metal nanoparticles (e.g., Ni, Pt, Cu, Pd, etc. or their alloys) can be stabilized under extreme conditions by encapsulation in a thermally stable shell in metal@metal oxide type core-shell structures, where the metal oxide or silica shell physically prevent their coalescence. Examples of such structures abound in recent literature, like Ni@SiO$_2$ [17], NiCu@SiO$_2$ [18], Pd@SiO$_2$ [19], Au@SiO$_2$ [20], etc. Mesoporous silica is often employed as a suitable shell material because of its high thermal stability, high specific surface area, tunability of pore size, and ease of synthesis [21]. Other metal oxides such as alumina, cerium oxide, zirconium oxide, etc. have also been used to constitute the shells [10, 22–24]. Metal@SiO$_2$ or metal@metal oxide structures exhibit much higher resistance to sintering than conventional supported metal catalysts. If the initial metal particle size in the core-shell catalyst is small (lower than what is needed to promote the growth of carbon nanotubes) and metal sintering is inhibited by the core-shell structure, the catalyst may be able to completely resist formation of filamentous coke during the course of the DRM reaction.

Apart from preventing metal sintering, the encapsulation of the metal nanoparticles by the shell plays another important role in DRM. The growth of filamentous carbon on metal nanoparticles requires space for the carbon nanotube to grow. The presence of a shell on the metal surface limits this space and provides a steric hindrance to the growth of the carbon nanotubes, thereby largely suppressing filamentous carbon formation during DRM.

In the last few years, there has been vigorous research on sinter-resistant core-shell catalysts for DRM. Metal@SiO$_2$ catalysts are the most extensively investigated catalysts in this class. Metal nanoparticles of desired sizes are usually synthesized by colloidal routes, followed by coating the silica shell by sol–gel methods [13, 25]. The silica shell thickness and porosity can be easily tuned by adjusting the silica precursor concentration, hydrolysis time, and use of surfactants [21]. Several investigators have reported Ni@SiO$_2$ catalysts for DRM [13, 17, 26–28], and Ni@SiO$_2$ has been characterized by negligible coke formation and stable activity for more than 100 h in DRM [27]. Ni-containing alloys such as NiCu and NiCo have also been used as the core in similar structures [18, 29–31]. A NiCo@SiO$_2$ catalyst reported by Zhao et al. [29] demonstrated stable DRM performance for 1000 h at 800 °C.

Often, multiple metal cores may be encapsulated inside one porous silica sphere, forming multi core-shell structures. Multi core-shell structures are, in some cases, easier to synthesize (for example, by one-pot synthesis methods) than single core-shell structures because of lower precision requirement [27]. Multi core-shell structures also provide a higher specific surface area of the active core and a high interfacial area per unit volume between core and shell. However, during DRM reaction, these multiple metal cores confined inside the silica sphere can migrate inwards and agglomerate to form a single larger Ni core [26, 27]. Sintering of Ni nanoparticles across different silica spheres does not, however, occur; so the sintering is limited to every individual core-shell subunit.

Mesoporous silica shells have also been used to coat dispersed metal/support composites. For example, sandwich structured $SiO_2@Ni@SiO_2$ catalysts with a silica shell coated on a dispersed Ni/SiO_2 core have been reported [32, 33]. The purpose of the shell here is to compartmentalize the dispersed Ni nanoparticles on the inner support, keep them spatially dispersed, prevent their coalescence, and to sterically hinder carbon nanotube growth.

Attempts have been made to use core-shell structures to maintain not only the particle size but also the shape and exposed facets of the core. Depending on the properties of the material, certain planes of a crystal may be more catalytically active or selective than others for a given application. Developments in wet chemical synthesis techniques have now made it possible to synthesize nanomaterials with precise shapes such as nanocubes, nanorods, and nanopolyhedra that expose selective crystal planes that are desirable for the specific application. However, operation at high temperature may lead to structural modifications, and in extreme cases, a complete loss of the initial structural characteristics. For instance, CeO_2 nanorods exposing [100] and [110] planes are known to have catalytic activity superior to that of CeO_2 nanoparticles [34]. However, during reforming at elevated temperatures, the nanorod structure of Ni/CeO_2 collapses, leading to loss of active surface facets and surface area and metal sintering [35]. Coating a layer of mesoporous silica shell to form Ni/CeO_2 nanorod@SiO_2 could stabilize the ceria support structure during DRM, resulting in stable performance. Attempts have also been made to maintain the exposed facets of metal cores by coating with silica shell. Pd nanocubes with exposed [100] planes (that are expected to have higher activity for methane dissociation) were encapsulated in mesoporous silica shells [19]. However, the effectiveness of this approach in maintaining the metal nanoparticle surface facets needs further verification, since a slight "rounding" of the Pd nanoparticles was reported by the authors after 10 h of reaction.

The high thermal stability of core-shell structures also makes them ideal for chemical looping CO_2 reforming processes that subject the catalyst to continuous thermal and redox cycles. In chemical looping dry reforming process, the dry reforming reaction is separated into two spatially and temporally separate half-reactions in which the catalysts or oxygen storage materials (OSM) are reduced by reaction with the hydrocarbon in one step and regenerated by oxidation with CO_2 in the other step in a cyclic fashion. Since the oxidation and reduction half-reactions are spatially separated, pure hydrogen product stream can be obtained directly from chemical looping reforming process without the need of downstream separation [36]. Metal oxides incorporating Cu, Fe, Ni, Mn, etc. are promising oxygen carrier candidates, but severe sintering of these metal oxides in the repeated thermal and redox cycles impedes the practical application of the chemical looping process. Recently, some core-shell structured catalysts with chemically and thermally stable shell materials have been reported to possess long-term stability in chemical looping dry reforming of methane [37, 38]. For example, a $Fe_2O_3/ZrO_2@ZrO_2$ catalyst with Fe_2O_3 nanoparticles as the oxygen storage material and with a thin protective ZrO_2 shell exhibited excellent redox activity for the CO_2 reduction half-step for 100 redox cycles [37].

Core-shell catalysts have also been reported in recent years to inhibit coke formation in unconventional methods of CO_2 reforming such as plasma-assisted DRM [39–41].

8.3.1.2 Bifunctional Catalysis

A core-shell structure provides enormous flexibility in tuning the physicochemical and catalytic properties of the composite material by combining different materials with complementary properties at a desired spatial proximity with each other [5]. Thus, bifunctional catalysis is possible on core-shell catalysts, wherein different active sites in the same catalyst performs different catalytic functions, which work in tandem to complete the overall reaction process. It is to be noted that bifunctional catalysis is not a special property of core-shell structures—it is observed in traditional supported catalysts as well, but a core-shell structure allows for a higher degree of control over the spatial distribution of the various catalytic sites and maximizes the interface area between them, which often leads to more optimal reaction performance and higher reaction selectivity.

For CO_2 reforming of methane, the reaction mechanism involves (1) the dissociation of methane to CH_x* intermediates on the metal sites and further dehydrogenation to form $C*$ and adsorbed $H*$, (2) activation and dissociation of CO_2, (3) reaction of O-containing intermediate with $C*$ to form CO, and (4) desorption of H_2 and CO products [2, 42]. The activation and dissociation of CO_2 may occur on the metal sites or on the support sites, depending on the properties of the support. For inert supports like silica, both CH_4 and CO_2 are usually dissociated on the metal site (monofunctional mechanism), whereas for basic supports like MgO or redox supports like CeO_2, CO_2 can be activated on the support [2, 3, 43]. In the latter scenario, the oxidation of $C*$ intermediates on the metal sites to form CO also occurs through reaction with species involving the support, such as lattice oxygen species in redox supports or carbonate species in basic supports, thereby constituting a bifunctional mechanism. The involvement of the support in the bifunctional mechanism has been reported to facilitate CO_2 activation and oxidation of carbonaceous intermediates, thereby imparting better coke inhibition properties than the monofunctional mechanism.

To take advantage of the bifunctional mechanism to suppress coking, core-shell catalysts including materials like MgO [44], La_2O_3 [45], CeO_2 [10], ZrO_2 [23], etc. as constituents have been developed for DRM. The interface area between the metal nanoparticles and the oxide support is of significance in bifunctional catalysts, because only at this interface, the support facilitated coke oxidation may proceed. Compared to conventional supported catalysts, core-shell structures with encapsulated metal nanoparticles can increase this interface area and thus enhance the coke inhibition properties of the catalyst.

Some examples of bifunctional core-shell catalysts for DRM are discussed here. CeO_2 shells on Ni nanoparticles have been reported to significantly enhance coke resistance [10, 22]. CeO_2 is widely applied for oxidation-reduction reactions due to its redox properties and oxygen storage capacity. CeO_2 can undergo substantial stoichiometric changes between +4 and +3 oxidation states of Ce under different oxidizing/reducing environment, without a loss of its crystal structure. The partial reduction of Ce^{4+} to Ce^{3+} results in the release of lattice oxygen, which can oxidize carbon species at the metal/CeO_2 interface during DRM. The oxygen vacancies thereby formed in the CeO_2 lattice act as the primary catalytic centers for CO_2 activation and dissociation. Ni@CeO_2 was shown to exhibit better stability than Ni@SiO_2 in DRM [22]. The type of coke deposited on Ni@CeO_2 was also different (amorphous and more reactive coke) from that on Ni@SiO_2 (graphitic and less active coke). Another study reported sandwich structured Ni–SiO_2@CeO_2 multi core-shell catalyst that showed negligible coke formation for dry reforming of biogas with sub-stoichiometric CO_2 content [10]. In situ infrared spectra gave evidence of a bifunctional reaction mechanism involving the lattice oxygen of the CeO_2 shell. A direct compassion between the core-shell Ni–SiO_2@CeO_2 and a supported Ni/CeO_2 catalyst showed that although the bifunctional redox mechanism is followed on both the catalysts, coke resistance of Ni–SiO_2@CeO_2 is markedly higher than that of Ni/CeO_2. This improvement may be attributed to the confinement effect of the shell on metal sintering and also, possibly, to the higher Ni–CeO_2 interfacial area in the core-shell structure. Similarly, sandwiched core-shell SiO_2@Ni@ZrO_2 has also been reported, whereby the Ni/ZrO_2 interface was proposed to increase dry reforming activity by lowering the dissociation energy barriers for CH_4 and CO_2 [23]. In the examples mentioned above, the shell of the catalyst is the material adding the secondary function of coke oxidation, but such materials may also be incorporated in the core of the catalyst. For example, several sandwich structured catalysts have been reported where the core is constituted by metal nanoparticles dispersed on suitable active supports, and the whole structure is further coated by a mesoporous silica shell [44–46]. Here, the oxide support in the core is involved in the reaction, while the shell is responsible for protecting the core and preventing metal sintering or disintegration of the core structure at high temperatures. For example, Du et al. [46] designed a NiMgAl–LDH@m–SiO_2 catalyst for DRM based on the above rationale. Layered double hydroxides (LDH) are suitable supports for DRM catalysts because they have good CO_2– sorption capacity and ample basic sites and hydroxyl groups that impart good coke oxidation capability. However, high surface area LDH nanoplates easily aggregate at high temperatures and show poor coke- and sinter-resistance in DRM. The NiMgAl–LDH@m–SiO_2 catalyst integrated the conducive properties of LDH with the thermal stability of silica and was characterized by high stability under DRM conditions at 750 °C.

8.3.1.3 Catalyst Regeneration

Catalyst regenerability is an essential criterion for large-scale applications. Regeneration of a spent catalyst with accumulated coke after reforming reaction can be done by air calcination/reduction treatments. However, as discussed in Sect. 8.2, filamentous carbon can uproot metal nanoparticles from the support in supported catalyst during the DRM reaction. Removal of coke in the spent catalyst during regeneration results in free metal nanoparticles without any interaction or stabilization by the support. At the high temperature of regeneration, these free metal nanoparticles agglomerate to form larger particles with lower surface area. Thus, catalysts with high filamentous carbon deposits cannot be fully regenerated. More importantly, the regenerated catalysts, because of the larger metal particle size and weaker metal-support interaction, tend to form coke and deactivate faster than fresh catalysts.

By hindering filamentous coke formation and growth during DRM, core-shell structures have better potential for regenerations than supported catalysts. Li et al. [47] compared the behavior of Ni-phyllosilicate and Ni-phyllosilicate@SiO_2 catalysts in subsequent regenerations in DRM. On the supported Ni-phyllosilicate catalyst, the rate of coke accumulation accelerated after one regeneration cycle, and the regenerated catalyst deactivated by blocking the reactor within 3 h of regeneration. In contrast, the core-shell catalyst could be successfully regenerated multiple times. The effect of the shell in preventing metal detachment during regeneration is shown schematically in Fig. 8.3.

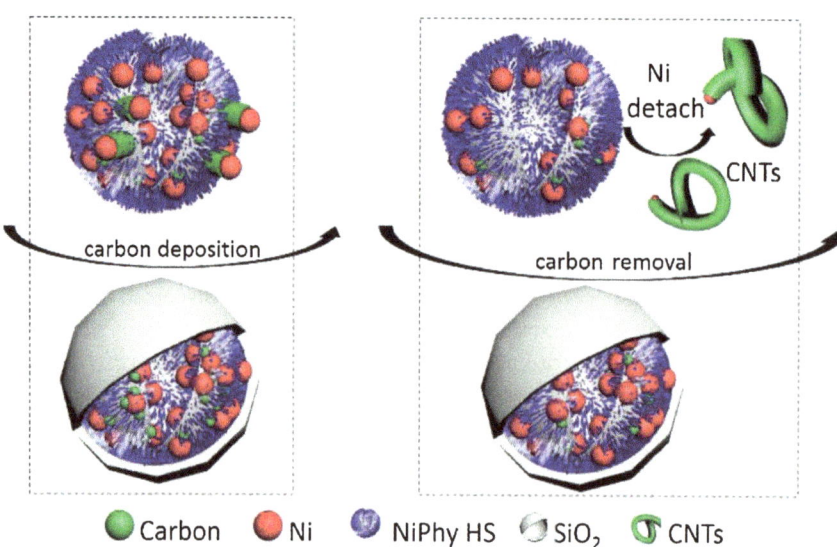

Fig. 8.3 Schematic of the role of silica shell in enhancing catalyst regeneration (carbon removal step) by inhibiting carbon nanotube (CNT) formation (Reprinted from Ref. [47], Copyright (2018), with permission from Elsevier)

8.3.2   Factors in Core-Shell Catalyst Design for DRM

8.3.2.1   Shell Thickness

The thickness of the shell, along with its porosity, plays a significant role in determining the activity and stability of a core-shell catalyst. One of the main challenges of core-shell catalysts is the potential mass transfer limitation imposed by the shell that can adversely affect the catalytic activity. While increased shell thickness may improve the thermal stability of the catalyst, it may also limit the accessibility of the reactants to the catalytically active metal sites in the core. A careful optimization of the shell thickness and porosity is hence required to achieve the desired thermal stability at minimum restrictions to the accessibility of active sites. Li et al. [48] compared the DRM activities of Ni@SiO_2 catalysts with shell thickness varying from 3.3 to 15.1 nm. They observed that a 11.2-nm-thick mesoporous silica shell on the ~12-nm-diameter Ni cores was optimum in increasing DRM activity while maintaining sinter-resistance of the catalyst structure. Too thin silica shells (3.3 nm) collapsed during DRM at 800 °C, resulting in metal sintering and deactivation. On the other hand, too thick shells (15.1 nm also) resulted in cross-linking of the silica shells and reduction in porosity, with a concurrent reduction in catalytic activity. Such a volcano shaped dependence of DRM activity on shell thickness was also reported for Co@SiO_2 catalysts [49].

Tuning shell thickness for mesoporous silica is quite straight-forward, wherein the hydrolysis time and concentration of precursors in sol–gel synthesis can be varied to control shell thickness. Similarly, atomic layer deposition (ALD) of alumina shells provide precise control on shell thickness by changing the number of ALD cycles. In a study by Baktash et al. [24] on Ni@Al_2O_3 catalysts, a clear trend was established between the number of ALD cycles and the DRM activity of the catalyst, presumably as a result of increased mass transfer resistance. However, all materials are not yet amenable to such precise control in synthesis, and further development of chemical synthesis techniques is required.

8.3.2.2   Shell Porosity

The porosity of the shell also has a similar effect as the shell thickness. Too small pores may cause diffusion limitations whereas, too big a pore size may be ineffective in preventing metal nanoparticle sintering. For example, in a multi core-shell catalyst, high porosity of the shell may result in the gradual agglomeration of all the multiple metal cores to form a single larger core [27]. Because of the relatively smaller size of the reactant and product molecules, DRM is less affected by internal diffusion limitations than reactions including bulkier molecules. Pore size of most reported catalysts are also in the mesoporous range, with very few studies reporting microporous shells. In one study, the effect of shell porosity on DRM activity was studied on RuCo@SiO_2 catalysts by varying the surfactants (CTAB, PVP or none) used during synthesis [50]. Although a higher shell porosity did improve the DRM activity, it was observed that at high reaction temperatures, the observed effect of shell porosity became less significant. Other methods have also been reported for tuning shell porosity, such as post-treatment of silica shells in alkaline media to create porous phyllosilicate structures [51]. Such treatments were shown to increase both the material porosity and activity in DRM; however, it is not clear from the studies whether the enhancement in activity can be related solely to the elimination of mass transfer limitations [51, 52].

8.3.2.3 Morphology

The catalytic activity of core-shell catalysts may be lowered not only by diffusion limitations but also by the blockage of the active sites by the deposition of the shell on its surface. Some variations of the core-shell morphology such as yolk-shell structures may be more advantageous to tackle such challenges. Yolk-shell or core@hollow structures contain an empty space between the active core/yolk and the shell, and thus minimize the blockage of active sites by the shell. At the same time, such hollow structures may also allow for thinner shells, reducing diffusion limitations [5].

Yolk-shell or hollow structures are usually synthesized using soft or hard templating methods, wherein a sacrificial layer of template such as silica or carbon is first coated in between the core and shell and subsequently removed to create a hollow space. Several yolk-shell catalysts such as Ni-yolk@SiO_2 [48], NiCe-yolk@SiO_2 [53], multi-Ni@hollow silica [54], and NiPt@hollow silicalite-1 [55] have been reported for DRM. A comparison of DRM activity at 800 °C between a core-shell Ni@SiO_2 and a yolk-shell Ni@SiO_2 showed a higher specific activity for the yolk-shell catalyst, while the turnover frequencies remained similar, indicating that the increase in DRM activity was a result of the higher exposed Ni surface area in the yolk-shell catalyst than the core-shell catalyst [48]. However, rigorous studies comparing the DRM performance of core-shell and yolk-shell structures, keeping all other parameters constant, are still largely missing.

However, because of the empty space in yolk-shell structures, there is also limited interaction between the core and the shell. So, the benefits of bifunctional catalysis and promotion of CO_2 activation by the support is very limited in yolk-shell structures compared to the core-shell morphology.

The coke suppression capacity of yolk-shell or hollow structures is also different from that of core-shell structures. The presence of the shell on the metal nanoparticles in core-shell structures provides steric hindrance to the growth of carbon nanotubes, thereby almost eliminating filamentous coke formation. Thus, coke on core-shell catalysts, if formed, are amorphous or encapsulating coke [56]. On the other hand, the void space in yolk-shell or hollow structures may allow for the growth of filamentous carbon inside the structure. Filamentous carbon formation may lead to the detachment of the metal nanoparticles from the support, and in certain cases, can even break the shells of the hollow structures, leading to the collapse of the structure. However, it should be noted that several factors decide whether growth of carbon nanotubes will occur inside the yolk-shell structure or cause the structure to rupture—such as the size of the metal nanoparticles, strength of metal-support interaction, thickness of the shell, and reaction conditions. Overall, both core-shell and yolk-shell structures have certain advantages and disadvantages, and careful structure optimization needs to be done, keeping in mind the trade-offs of each morphology.

While the term 'core-shell' or 'yolk-shell' were originally coined for concentric sphere-in-sphere architectures, recent developments have indicated certain benefits of using different shapes. For example, in Ni@SiO$_2$ yolk-shell structures with multiple Ni nanoparticle yolks encapsulated by a hollow SiO$_2$ shell, it was shown that a tubular shell with an elongated cavity was more effective in preventing coke in DRM than a spherical shell [57]. The Ni nanoparticle yolks were embedded along the wall of the shell. The elongated cavity in tubular shells provided a higher spatial separation of the multiple Ni yolks and prevented their agglomeration over time, while the more closely spaced yolks inside the spherical shell sintered into a bigger nanoparticle that supported coke formation. Thus, one-dimensional hollow structures appear to provide higher segregation and resistance to sintering in multi yolk-shell structures than the zero-dimensional spherical counterpart. Several studies have reported good DRM performance of multi-Ni@SiO$_2$ nanotube catalysts, wherein Ni nanoparticles are segregated and confined within the internal cavity of silica nanotubes [58, 59]. The nanotube structure with two open ends may also facilitate the diffusion of reactants and products to and from the confined metal nanoparticles, but the lower thermal stability of such structures at high temperatures needs to be taken into consideration.

Another recent development in core-shell morphology are sandwiched multi core-shell structures, some examples of which are mentioned in Sect. 8.3.1.2. In this structure, multiple active cores are embedded between two layers of the same or different materials [10, 33, 46, 59]. One of the benefits of such sandwich structures over typical core-shell morphology is that the metal nanoparticle cores can share interface with two materials with distinct and unique properties, which allows a higher flexibility in optimum catalyst design. Another possible benefit is easier and more facile synthesis than typical core-shell structures. The synthesis of core-shell structures often involves complex and elaborate recipes involving multiple steps. The synthesis of core nanoparticles by colloidal synthesis and the confinement of the individual colloidal nanoparticles inside uniform shells requires precise control and is characterized by low material yield and scalability. Sandwiched multi core-shell structure synthesis typically involves deposition of a shell material on a supported core/support composite by sol–gel, precipitation, or hydrothermal methods and is less complicated and possibly more scalable than the typical core-shell synthesis. Detailed techno-economic analyses are required to assess the scalability and practical potential of various types of core-shell catalysts.

8.3.2.4 Composition

Metal nanoparticles that are DRM-active constitute the cores of core-shell catalysts in DRM. Ni has been the most studied core material because its high DRM activity and low cost is challenged by its high tendency to form coke. However, there are also examples of other metals including noble metals like Pt [55], Pd [19], etc. and non-noble metals like Co [49] being used in core-shell catalysts for DRM. Bimetallic cores have also been used, e.g., $NiCo@SiO_2$ [29, 30], $NiFe@SiO_2$ [60], $NiCu@SiO_2$ [18], $RuCu@SiO_2$ [50], $NiPt@SiO_2$ [31], etc. Bimetallic core-shell structures can be synthesized by routes similar to those used for monometallic structures and can provide enhanced activity or selectivity in DRM by virtue of the synergy of the two metals. Transition metals such as Fe and Co are more oxophilic than Ni and can thus help in suppressing coke formation, when alloyed with the Ni core [29, 60]. However, the DRM activity of such alloys may be lower than that of Ni [43], and careful optimization of the alloy composition is needed to maximize both activity and stability of the resultant core-shell catalyst [7]. Alloying Ni with Cu in appropriate proportions has been reported to improve H_2 selectivity and yield in DRM in $NiCu@SiO_2$ catalysts [18]. Alloying Ni with noble metals such as Pt and Ru can increase the intrinsic DRM activity and also reduce coking by a dilution effect on the surface Ni ensemble size [7]. As per literature reports so far, the synergistic effects of alloying in conventional supported bimetallic catalysts may be expected to apply in a similar manner to core-shell systems.

As discussed in Sect. 8.3.1.2, the composition of the shell may determine the mechanism of CO_2 activation and coke oxidation during DRM. The literature of core-shell catalysts for DRM is dominated by materials with SiO_2 shells because of the ease of synthesis and high thermal stability, but silica is chemically quite inert and does not participate in the reaction. Some recent investigations have reported other shell materials, which can also activate the reactants and facilitate bifunctional reactions in addition to providing sinter-resistance.

Redox materials such as CeO_2 [10, 22] and ZrO_2 have been used as shells in core-shell structures, as discussed in Sect. 8.3.1.2. Improved CO_2 activation by these shells by virtue of the redox nature was shown to further reduce coke formation in DRM. Core-shell catalysts with alumina shells have also been reported for DRM [24, 61–63]. Baktash et al. [24] used ALD to synthesize $Ni@Al_2O_3$ and observed that five cycles of ALD yielded an Al_2O_3 layer a few nanometres thick that was sufficient to prevent Ni particle sintering at temperatures up to 800 °C. However, the $Ni@Al_2O_3$ catalyst formed coke and deactivated at lower reaction temperature of 525 °C, at which coke formation is thermodynamically more favored. Similarly, some coke formation was also reported in $Ni@Al_2O_3$ catalysts synthesized by other techniques [61]. It is likely that the acidic properties of alumina is responsible for the observed coke deposition, despite the core-shell structure and resistance to sintering. Alumina has been reported to facilitate the activation and decomposition of methane on metal sites. A high rate of methane decomposition, when not balanced by a fast carbon oxidation kinetics, can lead to deposition of coke. $Ni@TiO_2$ structures have also been applied for DRM [64, 65]; however, there are limited investigations and limited evidence of the effectiveness of a TiO_2 shell in inhibiting sintering or coke formation in DRM. While choosing the material for the shell, it is important to consider both the catalytic properties of the material (acidic/basic/redox, etc.) and its thermal stability. For example, even though TiO_2 is a reducible support capable of activating CO_2, it was observed that a TiO_2 shell in $Ni/SiO_2@TiO_2$ was unable to maintain its structure during DRM at 800 °C, whereas Al_2O_3, ZrO_2, MgO, and SiO_2 shells were stable and effective in preventing sintering [66]. Given the high stability and effectiveness of silica shells in preventing sintering, some researchers have reported silica coated core-shell catalysts with the doping of appropriate materials in the shell or core to promote the bifunctional mechanism. For example, $NiCe@SiO_2$ [53], $Ni@SiO_2–CeO_2$ [67], $NiZr@SiO_2$ [68] catalysts have been reported where small amounts of Ce and Zr has been added to promote CO_2 activation, while using silica shells to ensure thermal stability.

8.3.3   Summary of Recent Studies in Core-Shell Catalyst Development for DRM

Overall, the main advantages of core-shell catalyst in dry reforming reaction are higher catalyst stability and lower rate of coke deposition by virtue of reduced metal particle sintering and increased metal-support interfacial contact. A comprehensive summary of recent developments in core-shell catalyst synthesis and application in CO_2 reforming of methane is presented in Table 8.1.

Many of these studies listed in Table 8.1 have demonstrated long-term stability of the reported core-shell catalysts in DRM for 100–1000 h onstream, with relatively low amount of coke formation. Comparatively, conventional supported catalysts are characterized by lower stability and high coke formation. For instance, a supported commercial Ni-containing steam reforming catalyst (HiFUELTM R110, Alfa Aesar) was reported to accumulate more than 0.32 g_{coke}/g_{cat} within 6.4 h of DRM reaction at 850 °C [17]. This is not to say that stable supported catalysts have not been reported for DRM; several recent studies have reported various strategies to stabilize supported catalysts also for extended testing in DRM [69, 70]. However, for catalysts with similar composition, a catalyst with core-shell morphology is generally observed to be more stable and coke resistant than a supported one. Recent progress in core-shell catalyst development has also shown promising performance in dry reforming at coke-favoring operating conditions, such as lower reaction temperature and high CH_4/CO_2 content in feed, which are often too harsh for operation with supported catalysts [10, 33, 71]. Also, as discussed in Sect. 8.3.1.3, core-shell structures have also been shown to fare better in catalyst regeneration cycles than supported catalysts [47].

In terms of activity, some reports indicate lower reactant conversions on core-shell catalysts than on conventional catalysts. This may be a result of fewer exposed active sites and diffusion limitations imposed by the shell in core-shell structures.

Table 8.1 Recent progress in core-shell catalyst development for CO_2 (dry) reforming of methane

No.	Catalyst	Synthesis method	Reaction conditions			Catalyst performance: conversion, H_2/CO	Coke deposition at time onstream	Remarks	References
			Temperature (°C)	CO_2: CH_4: inert	WHSV				
Core-shell catalysts									
1.	Ni@SiO$_2$	Sol–gel method	750	1:1:2	48 L g$_{cat}^{-1}$ h^{-1}	$CH_4 = 58\%$ $CO_2 = 71\%$ $H_2/CO = 0.7$	1.2% (24 h)	Multiple Ni cores encapsulated by silica; mesoporous silica shell prevents coke formation	[17]
2.	Ni@SiO$_2$	Microemulsion method	800	1:1:0.67	18 L g$_{cat}^{-1}$ h^{-1}	$CH_4 = 85\%$ $CO_2 = 90\%$ $H_2/CO = 1$	0.7 wt% (50 h)	Small Ni core (~5 nm) encapsulated by silica; effect of TEOS hydrolysis time on texture and activity investigated	[26]
3.	Multi-Ni@SiO$_2$	Reverse micelle method	800	1:1:0	18 L g$_{cat}^{-1}$ h^{-1}	$CH_4 = 92\%$ $CO_2 = 90\%$ $H_2/CO = 1$	Negligible (100 h)	Confinement effect of shell inhibits sintering and coke; but, multiple Ni cores migrate inward during reaction to form single Ni particle	[27]
4.	Ni@SiO$_2$	Microemulsion method	700	3:3:4	60 L g$_{cat}^{-1}$ h^{-1}	CH_4 activity = 0.0028 mol CH_4 s^{-1} g$_{metal}^{-1}$	2.4% (40 h)	Effect of calcination temperature on Ni particle size, metal-support interaction and DRM performance investigated	[75]
5.	Ni@SiO$_2$	Reverse microemulsion method	550–750	3:3:4	18 L g$_{cat}^{-1}$ h^{-1}	$CH_4 =$ 22–80% $CO_2 =$ 32–81% $H_2/CO = 0.69$ – 0.87	17% (50 h)	Catalyst performance in both dry and steam reforming of methane was tested	[28]

(continued)

Table 8.1 (continued)

No.	Catalyst	Synthesis method	Reaction conditions			Catalyst performance: conversion, H_2/CO	Coke deposition at time onstream	Remarks	References
			Temperature (°C)	CO_2: CH_4: inert	WHSV				
6.	$Co@SiO_2$	Modified Stöber method	700	1.5:1.5:7	54 L g_{cat}^{-1} h^{-1}	$CH_4 = 72\%$ $CO_2 = 78\%$ $H_2/CO = 0.85$	33.9% (10 h)	Effect of core Co particle size and SiO_2 shell thickness on DRM performance has been studied in detail	[49]
7.	$NiCu@SiO_2$	Microemulsion method	700	2:2:6	13.3L g_{cat}^{-1} h^{-1}	$CH_4 = 76\%$	Not reported	NiCu alloy formation increases activity and H_2 selectivity by suppressing RWGS	[18]
8.	$NiCo@SiO_2$	Microemulsion method	800	1:1:1	300 Lg_{metal}^{-1} h^{-1}	$CH_4 = 87.2\%$ $CO_2 = 88.9\%$	Negligible (1000 h)	Long-term stability was demonstrated; NiCo alloy showed better performance than Ni as the core	[29]
9.	$Ni@SiO_2$-Co	Sol–gel microencapsulation method	750	1:1:1	36 L g_{cat}^{-1} h^{-1}	$CH_4 = 87\%$ $CO_2 = 93\%$ $H_2/CO = 0.84$	Negligible (12 h)	Effect of Co addition in the Ni core or the outer surface of SiO_2 shell on DRM performance was investigated	[30]
10.	$NiPt@SiO_2$	Stöber method	800	1:1:1	600 Lg_{metal}^{-1} h^{-1}	$CH_4 = 95\%$ $CO_2 = 95\%$	Negligible (200 h)	NiPt core has a hollow structure, which increases the specific active surface area and DRM activity	[31]

(continued)

Table 8.1 (continued)

No.	Catalyst	Synthesis method	Reaction conditions			Catalyst performance: conversion, H_2/CO	Coke deposition at time onstream	Remarks	References
			Temperature (°C)	CO_2: CH_4: inert	WHSV				
11.	RuCo@ SiO_2	Hydrothermal and Stöber method	700	1.5:1.5:7	54 L g_{cat}^{-1} h^{-1}	$CH_4 = 74.4\%$ $CO_2 = 84.7\%$ $H_2/CO = 0.98$	0.5 mg_{coke} g_{cat}^{-1} h^{-1}	Surface distribution of Ru and shell porosity are important factors in DRM performance	[50]
12.	Ni-Zr@ SiO_2	Microemulsion method	800	1:1:0	18 L g_{cat}^{-1} h^{-1}	$CH_4 = 90.5\%$ $CO_2 = 93.2\%$ $H_2/CO = 0.95$	Negligible (240 h)	ZrO_2 promoter was added in the core; ZrO_2 increased Ni reducibility and available oxygen species	[68]
13.	CeO_2 modified Ni@ SiO_2	Microemulsion method	600	3:3:4	150 L g_{cat}^{-1} h^{-1}	$CH_4 = 24\%$ $CO_2 = 27\%$ $H_2/CO = 0.55$	4% (30 h)	Ceria nanoparticles with smaller size activated more CO_2 and higher molecular oxygen generation by CeO_{-2} increased methane conversion	[67]
14.	Ni@ Al_2O_3	Atomic layer deposition	700	1:1:8	3600 L g_{cat}^{-1} h^{-1}	$CH_4 = 80\%$	Negligible (50 h)	Alumina coating inhibits metal sintering; effect of number of ALD cycles on DRM activity and stability investigated	[24]
15.	Ni@ Al_2O_3	Microemulsion method	800	1:1:0	36 L g_{cat}^{-1} h^{-1}	$CH_4 = 90\%$ $CO_2 = 87\%$ $H_2/CO = 1.1$	15% (50 h)	Active oxygen species was observed to be higher in Ni@ Al_2O_3 than Ni/Al_2O_3	[61]

(continued)

Table 8.1 (continued)

No.	Catalyst	Synthesis method	Reaction conditions		WHSV	Catalyst performance: conversion, H_2/CO	Coke deposition at time onstream	Remarks	References
			Temperature (°C)	CO_2:CH_4:inert					
16.	Ni@TiO_2	Atomic layer deposition	800	1:1:0	24 L g_{cat}^{-1} h^{-1}	$CH_4 = 38\%$ $CO_2 = 50\%$ $H_2/CO = 0.73$	80% (160 h)	Thin TiO_2 layer on Ni nanoparticles to prevent sintering; however, coke formation not prevented	[65]
17.	Ni@Hydrochar	Hydrothermal method	850	1:1:0	12 L g_{cat}^{-1} h^{-1}	$CH_4 = 80\%$ $CO_2 = 90\%$ $H_2/CO = 0.85$	Negligible (100 h)	Sugarcane bagasse was used to synthesize biochar; effect of Ni loading was studied; reaction mechanism on the core-shell catalyst was proposed	[76]
18.	Ni@CeO_2	Hydrothermal method	600	1:1:0	24 L g_{cat}^{-1} h^{-1}	$CH_4 = 88\%$ $CO_2 = 90\%$	12% (24 h)	CeO_2 shell can restrict Ni sintering and provide lattice oxygen for coke oxidation	[22]
Yolk-shell and hollow catalysts									
19.	Multi-Ni-yolk@SiO_2	Microemulsion method	700	1:1:0	60 L g_{cat}^{-1} h^{-1}	$CH_4 = 72\%$ $CO_2 = 80\%$ $H_2/CO = 0.95$	1.5% (30 h)	The effect of the length in inner cavity of the hollow tubular silica shell on DRM stability and coking has been investigated	[57]
20.	Ni-yolk@Ni@SiO_2	Microemulsion method	800	1:1:1	36 L g_{cat}^{-1} h^{-1}	$CH_4 = 90\%$ $CO_2 = 95\%$ $H_2/CO = 0.82$	Negligible (100 h)	Effect of core size and silica shell thickness has been investigated	[48]

(continued)

Table 8.1 (continued)

No.	Catalyst	Synthesis method	Reaction conditions			WHSV	Catalyst performance: conversion, H_2/CO	Coke deposition at time onstream	Remarks	References
			Temperature (°C)	CO_2: CH_4: inert						
21.	Ni@Ni-Mg phyllosilicate	Microemulsion and hydrothermal treatment	700	1:1:1		36 L g_{cat}^{-1} h^{-1}	$CH_4 = 79\%$ $CO_4 = 89\%$ $H_2/CO = 0.98$	4% (100 h)	Modification of shell porosity and catalyst basicity by post-treatment enhances catalyst activity	[51]
22.	Multi-Ni-yolk@SiO_2 nanorod	Hydrothermal method	700	1:1:1		36 L g_{cat}^{-1} h^{-1}	$CH_4 = 80\%$ $CO_4 = 83\%$ $H_2/CO = 0.8$	13.4% (70 h)	Ni nanoparticles were confined inside the channels of hollow silica nanotubes and showed good resistance to sintering	[58]
23.	ZnO/Ni@SiO_2	Stöber silica coating on Ni/ZIF-8	850	1:1:3		30 L g_{cat}^{-1} h^{-1}	$CH_4 = 90\%$ $CO_2 = 96\%$ $H_2/CO = 0.95$	Observed by XRD and Raman	Ni-ZnO yolk-shell structure showed good activity, but deactivated by 20% over 150 h	[77]
24.	NiCe-yolk@m-SiO_2	Modified Stöber method	750	1:1:0		6 L g_{cat}^{-1} h^{-1}	$CH_4 = 90\%$ $CO_2 = 95\%$	10.3% (40 h)	CeO_2 addition in the yolk suppressed carbon deposition, and yolk-shell structure increased activity	[53]
25.	Multi-Ni@ hollow silica sphere	Microemulsion method	800	1:1:2		144 L g_{cat}^{-1} h^{-1}	$CH_4 = 94.5\%$ $CO_2 = 95\%$	Negligible (55 h)	Multiple Ni nanoparticles anchored strongly inside silica hollow sphere; strong interaction with support lowers sintering	[54]

(continued)

Table 8.1 (continued)

No.	Catalyst	Synthesis method	Reaction conditions			Catalyst performance: conversion, H_2/CO	Coke deposition at time onstream	Remarks	References
			Temperature (°C)	CO_2: CH_4: inert	WHSV				
26.	Ni@NiPhy@SiO_2 Hollow Sphere	Hydrothermal and Stöber method	700	1:1:1	36 L g_{cat}^{-1} h^{-1}	CH_4 = 94.5% CO_2 = 95% H_2/CO = 0.8	5.5% (600 h)	Confinement effect improved metal-support interaction and prevented carbon nanotube growth	[47]
Sandwiched core-shell catalysts									
27.	Ni-SiO_2@CeO_2	Ammonia evaporation and precipitation	600	3:2:0	200 L g_{cat}^{-1} h^{-1}	CH_4 activity = 0.12 mol g_{metal}^{-1} min^{-1}	Negligible (72 h)	Multiple Ni nanoparticles sandwiched between SiO_2 and CeO_2 layers; CeO_2 provides oxygen species for coke removal by bifunctional mechanism	[10]
28.	NiMgAl-LDH@m-SiO_2	Hydrothermal and Stöber method	750	1:1:0	7.5 L g_{cat}^{-1} h^{-1}	CH_4 = 85% CO_2 = 87%	Negligible (8 h)	Combination of conducive catalytic properties of the LDH core and stability of silica shell leads to high activity and coke resistance	[46]
29.	SiO_2@Ni@ SiO_2	Ammonia evaporation and Stöber method	600	1:1:1	60 L g_{cat}^{-1} h^{-1}	CH_4 = 50% CO_2 = 60% H_2/CO = 0.78	5.1% (24 h)	Ni-phyllosilicate used to create highly dispersed Ni nanoparticles, that are further stabilized by silica shell; effect of shell thickness studied	[33]

(continued)

Table 8.1 (continued)

No.	Catalyst	Synthesis method	Reaction conditions			Catalyst performance: conversion, H_2/CO	Coke deposition at time onstream	Remarks	References
			Temperature (°C)	CO_2:CH_4: inert	WHSV				
30.	Ni@SiO_2	Microemulsion and Stöber method	800	9:9:2	$19\ Lg_{Ni}^{-1}\ min^{-1}$	$CH_4 = 42\%$ $CO_2 = 64\%$ $H_2/CO = 0.7$	Negligible (170 h)	5 nm Ni nanoparticles supported on silica spheres were coated with a silica overlayer, no deactivation observed for 170 h	[32]
31.	Ni@SiO_2/Al_2O_3/FeCrAl-fiber	Top-down macro–micro–nano organization	800	1:1:1	$5\ L\ g_{cat}^{-1}\ h^{-1}$	$CH_4 = 96.5\%$ $CO_2 = 91.9\%$	Negligible (500 h)	Confinement effect and strong metal-support interaction was responsible for better stability and activity of the catalyst	[78]
32.	Ni-Mg phyllosilicate nanotubes@ SiO_2	Hydrothermal and Stöber method	750	1:1:1	$60\ L\ g_{cat}^{-1}\ h^{-1}$	$CH_4 = 85\%$ $CO_2 = 90\%$ $H_2/CO = 0.8$	Negligible (72 h)	SiO_2 shell stabilized the nanotube structure of phyllosilicate at high reaction temperature	[59]
33.	Ni/$MgAl_2O_4$ @SiO_2	Sol–gel coating	750	5:5:1	$66\ L\ g_{cat}^{-1}\ h^{-1}$	$CH_4 = 70\%$ $CO_2 = 80\%$	Negligible (10 h)	DRM performance of core-shell catalyst with various methods of synthesis of the Ni/$MgAl_2O_4$ core were compared	[79]

(continued)

Table 8.1 (continued)

No.	Catalyst	Synthesis method	Reaction conditions		WHSV	Catalyst performance: conversion, H_2/CO	Coke deposition at time onstream	Remarks	References
			Temperature (°C)	CO_2: CH_4: inert					
34.	Ni/ZSM-5@SiO_2	Sol–gel coating	800	44:47:9	12 L g_{cat}^{-1} h^{-1}	CH_4 = 85% CO_2 = 85%	3% (50 h)	Silica shell prevented nickel sintering on ZSM-5 and increased coke resistance; effect of Ni loading investigated	[80]
35.	SiO_2@Ni@ZrO_2	Sol–gel coating	700	1:1:1	24 L g_{cat}^{-1} h^{-1}	CH_4 = 60% CO_2 = 60% H_2/CO = 0.75	Negligible (20 h)	ZrO_2 stabilized Ni clusters at the interface showed lower activation energy and higher activity	[23]
36.	Al_2O_3@Ni@Al_2O_3	Atomic layer deposition	800	1:1:1	300 L g_{Ni}^{-1} h^{-1}	CH_4 = 92% CO_2 = 95% H_2/CO = 0.75	Negligible (70 h)	Double interaction between Ni and γ-alumina support and alumina coating increased resistance to sintering and deactivation	[62]
37.	Ni@Al_2O_3/AlN	Self-assembly by degradation of AlN	800	1:1:0.5	50 L g_{cat}^{-1} h^{-1}	CH_4 = 65% CO_2 = 75%	Negligible (300 h)	Facile and room temperature method of synthesizing core-shell structure; dual confinement effects from Ni-AlN interaction and Al_2O_3 overlayers minimize coke	[63]

However, the intrinsic activities or turnover frequency (TOF) of core-shell materials have mostly been reported to be much higher (by almost one order of magnitude) than those of supported catalysts [5]. It is possible that the increased interface area between the metal core and the shell and the subsequent metal-support interaction in the encapsulated core-shell structure may modify the nature and electronic state of the metal sites for DRM, thereby enhancing their activity. The nano-reactor morphology and the effects of differences in local reactant and product concentrations inside the confined space of the core-shell sub-units compared to the bulk concentrations may also play a role in the observed catalytic activity. However, it is important to note that some of the TOF values reported may not be the true intrinsic activity because of interferences from equilibrium/transport limitations, or from the challenges in accurately measuring the density of active catalytic sites in core-shell catalysts under relevant conditions. Much of the fundamental catalytic phenomena occurring at the interface in confined core-shell structures are not yet understood fully and further investigations probing the effects of the core-shell structure on catalytic activity are required.

8.4   Outlook and Conclusion

In summary, core-shell catalysts exhibit good catalytic performance in CO_2 reforming of methane by suppressing particle sintering under reaction conditions, inhibiting the growth of filamentous coke, providing flexibility in integrating materials with bifunctional catalytic properties in coke removal, enhancing metal-support interface area and facilitating easier catalyst regeneration. However, the superior performance of core-shell catalysts come at a price. Core-shell structures are more complex than conventional supported catalysts, and the synthesis processes of these catalysts are complex, multi-step, and potentially much more expensive than those of conventional catalysts used industrially. It is, hence, crucial to develop cost-effective, scalable, and continuous synthesis techniques for core-shell structured catalysts. So far, there are limited studies focusing on improving the scalability and economics of core-shell catalyst synthesis. Some recent reports have proposed simpler techniques to create encapsulated structures (by self-assembly, gas treatments, flame synthesis, etc.) [63, 72, 73], which may lack the precision of typical core-shell catalysts but may still provide the necessary functionality. Given that the functional superiority of core-shell catalysts in DRM has been relatively well established by now, the time is probably ripe to attempt the scale-up and commercial utilization of such materials. Detailed techno-economic analyses and extensive benchmarking work is also required to assess whether the replacement of conventional materials by core-shell catalysts in CO_2 reforming reaction is justified by their potentially superior performance.

Another area for further development of core-shell catalysts involves the design of innovative core-shell structures that can circumvent some of the inherent challenges of such materials such as blockage of active sites and mass transfer limitations. Several factors that can be modified to address these challenges, such as shell thickness, porosity, and presence of cavities inside the shell have been discussed in Sect. 8.3.2. However, while the chemistry of silica synthesis is mature enough to allow such tunability of structure and morphology of silica-based core-shell catalysts, the same is not true for all other materials. Thus, there is significant scope for further development of versatile synthesis techniques applicable to a wide range of materials. New synthesis strategies may also be explored to minimize the adverse effects of the shell on catalytic activity; for example, one study attempted to selectively deposit the shell material around the metal nanoparticles to provide segregation while leaving a significant portion of the active metal surface exposed [74].

Overall, core-shell catalysts have higher stabilities and coke resistance in CO_2 reforming of hydrocarbons than conventional catalysts and thus have high potential for the long-term industrial application for these technologies in future. The research area is still growing, and efforts to scale-up synthesis at competitive costs, coupled with rigorous benchmarking studies are required for further translation of these materials to industrial use.

Acknowledgements The authors gratefully thank the National Environmental Agency of Singapore (NEA-ETRP Grant 1501 103), A*STAR (AME-IRG A1783c0016), National University of Singapore Flagship Green Energy Program, the Ministry of Education of Singapore (MOE2017-T2-2-130) and the National University of Singapore for supporting the research.

References

1. Artz J, Müller TE, Thenert K, Kleinekorte J, Meys R, Sternberg A, Bardow A, Leitner W (2018) Sustainable conversion of carbon dioxide: an integrated review of catalysis and life cycle assessment. Chem Rev 118(2):434–504
2. Jangam A, Das S, Dewangan N, Hongmanorom P, Hui WM, Kawi S (2019) Conversion of CO_2 to C1 chemicals: catalyst design, kinetics and mechanism aspects of the reactions. Catal Today
3. Pakhare D, Spivey J (2014) A review of dry (CO_2) reforming of methane over noble metal catalysts. Chem Soc Rev 43(22):7813–7837
4. Lavoie J-M (2014) Review on dry reforming of methane, a potentially more environmentally-friendly approach to the increasing natural gas exploitation. Front Chem 2(81)
5. Das S, Pérez-Ramírez J, Gong J, Dewangan N, Hidajat K, Gates BC, Kawi S (2020) Core–shell structured catalysts for thermocatalytic, photocatalytic, and electrocatalytic conversion of CO_2. Chem Soc Rev 49(10):2937–3004
6. Kawi S, Kathiraser Y, Ni J, Oemar U, Li Z, Saw ET (2015) Progress in synthesis of highly active and stable nickel-based catalysts for carbon dioxide reforming of methane. Chemsuschem 8(21):3556–3575
7. Bian Z, Das S, Wai MH, Hongmanorom P, Kawi S (2017) A review on bimetallic nickel-based catalysts for CO_2 reforming of methane. ChemPhysChem 18(22):3117–3134
8. Goodman ED, Schwalbe JA, Cargnello M (2017) Mechanistic understanding and the rational design of sinter-resistant heterogeneous catalysts. ACS Catal 7(10):7156–7173

9. Chen D, Christensen KO, Ochoa-Fernández E, Yu Z, Tøtdal B, Latorre N, Monzón A, Holmen A (2005) Synthesis of carbon nanofibers: effects of Ni crystal size during methane decomposition. J Catal 229(1):82–96

10. Das S, Ashok J, Bian Z, Dewangan N, Wai MH, Du Y, Borgna A, Hidajat K, Kawi S (2018) Silica-Ceria sandwiched Ni core–shell catalyst for low temperature dry reforming of biogas: coke resistance and mechanistic insights. Appl Catal B 230:220–236

11. Das S, Jangam A, Du Y, Hidajat K, Kawi S (2019) Highly dispersed nickel catalysts via a facile pyrolysis generated protective carbon layer. Chem Commun 55(43):6074–6077

12. Zhang G, Liu J, Xu Y, Sun Y (2018) A review of CH_4-CO_2 reforming to synthesis gas over Ni-based catalysts in recent years (2010–2017). Int J Hydrogen Energy 43(32):15030–15054

13. Li Z, Li M, Bian Z, Kathiraser Y, Kawi S (2016) Design of highly stable and selective core/yolk–shell nanocatalysts—a review. Appl Catal B 188:324–341

14. Abdel Karim Aramouni N, Zeaiter J, Kwapinski W, Ahmad MN (2017) Thermodynamic analysis of methane dry reforming: effect of the catalyst particle size on carbon formation. Energy Convers Manag 150:614–622

15. Das S, Jangam A, Xi S, Borgna A, Hidajat K, Kawi S (2020) Highly dispersed Ni/silica by carbonization-calcination of a chelated precursor for coke-free dry reforming of methane. ACS Appl Energy Mater 3(8):7719–7735

16. Kim J-H, Suh DJ, Park T-J, Kim K-L (2000) Effect of metal particle size on coking during CO_2 reforming of CH_4 over Ni–alumina aerogel catalysts. Appl Catal A 197(2):191–200

17. Zhang J, Li F (2015) Coke-resistant Ni@SiO_2 catalyst for dry reforming of methane. Appl Catal B 176–177:513–521

18. Wu T, Cai W, Zhang P, Song X, Gao L (2013) Cu–Ni@SiO_2 alloy nanocomposites for methane dry reforming catalysis. RSC Adv 3(46):23976–23979

19. Yue L, Li J, Chen C, Fu X, Gong Y, Xia X, Hou J, Xiao C, Chen X, Zhao L, Ran G, Wang H (2018) Thermal-stable Pd@mesoporous silica core-shell nanocatalysts for dry reforming of methane with good coke-resistant performance. Fuel 218:335–341

20. Montaño-Priede JL, Coelho JP, Guerrero-Martínez A, Peña-Rodríguez O, Pal U (2017) Fabrication of monodispersed Au@SiO_2 nanoparticles with highly stable silica layers by ultrasound-assisted Stöber method. J Phys Chem C 121(17):9543–9551

21. Li Z, Das S, Hongmanorom P, Dewangan N, Wai MH, Kawi S (2018) Silica-based micro- and mesoporous catalysts for dry reforming of methane. Catal Sci Technol 8(11):2763–2778

22. Tang C, Liping L, Zhang L, Tan L, Dong L (2017) High carbon-resistance Ni@CeO2 core-shell catalysts for dry reforming of methane. Kinet Catal 58(6):800–808

23. Dou J, Zhang R, Hao X, Bao Z, Wu T, Wang B, Yu F (2019) Sandwiched SiO_2@Ni@ZrO_2 as a coke resistant nanocatalyst for dry reforming of methane. Appl Catal B 254:612–623

24. Baktash E, Littlewood P, Schomäcker R, Thomas A, Stair PC (2015) Alumina coated nickel nanoparticles as a highly active catalyst for dry reforming of methane. Appl Catal B 179:122–127

25. Ghosh Chaudhuri R, Paria S (2012) Core/shell nanoparticles: classes, properties, synthesis mechanisms, characterization, and applications. Chem Rev 112(4):2373–2433

26. Wang F, Han B, Zhang L, Xu L, Yu H, Shi W (2018) CO_2 reforming with methane over small-sized Ni@SiO_2 catalysts with unique features of sintering-free and low carbon. Appl Catal B 235:26–35

27. Peng H, Zhang X, Zhang L, Rao C, Lian J, Liu W, Ying J, Zhang G, Wang Z, Zhang N, Wang X (2016) One-pot facile fabrication of multiple nickel nanoparticles confined in microporous silica giving a multiple-cores@shell structure as a highly efficient catalyst for methane dry reforming. ChemCatChem 9(1):127–136

28. Han B, Wang F, Zhang L, Wang Y, Fan W, Xu L, Yu H, Li Z (2020) Syngas production from methane steam reforming and dry reforming reactions over sintering-resistant Ni@SiO_2 catalyst. Res Chem Intermed 46(3):1735–1748

29. Zhao Y, Li H, Li H (2018) NiCo@SiO_2 core-shell catalyst with high activity and long lifetime for CO_2 conversion through DRM reaction. Nano Energy 45:101–108

30. Gunduz-Meric G, Kaytakoglu S, Degirmenci L (2020) Ni, Co/SiO2 and Ni/SiO2, Co bimetallic microsphere catalysts indicating high activity and stability in the dry reforming of methane. React Kinet Mech Catal 129(1):403–419
31. Wang G, Liang Y, Song J, Li H, Zhao Y (2020) Study on high activity and outstanding stability of hollow-NiPt@SiO$_2$ core–shell structure catalyst for DRM reaction. Front Chem 8(220)
32. Han JW, Kim C, Park JS, Lee H (2014) Highly coke-resistant Ni nanoparticle catalysts with minimal sintering in dry reforming of methane. Chemsuschem 7(2):451–456
33. Bian Z, Kawi S (2017) Sandwich-like silica@Ni@silica multicore-shell catalyst for the low-temperature dry reforming of methane: confinement effect against carbon formation. ChemCatChem 10(1):320–328
34. Du X, Zhang D, Shi L, Gao R, Zhang J (2012) Morphology dependence of catalytic properties of Ni/CeO$_2$ nanostructures for carbon dioxide reforming of methane. J Phys Chem C 116(18):10009–10016
35. Zhu S, Lian X, Fan T, Chen Z, Dong Y, Weng W, Yi X, Fang W (2018) Thermally stable core–shell Ni/nanorod-CeO$_2$@SiO$_2$ catalyst for partial oxidation of methane at high temperatures. Nanoscale 10(29):14031–14038
36. Hosseini D, Abdala PM, Donat F, Kim SM, Müller CR (2019) Bifunctional core-shell architecture allows stable H$_2$ production utilizing CH$_4$ and CO$_2$ in a catalytic chemical looping process. Appl Catal B 258:117946
37. Hu J, Galvita VV, Poelman H, Detavernier C, Marin GB (2017) A core-shell structured Fe$_2$O$_3$/ZrO$_2$@ZrO$_2$ nanomaterial with enhanced redox activity and stability for CO$_2$ conversion. J CO2 Utiliz 17:20–31
38. Shafiefarhood A, Galinsky N, Huang Y, Chen Y, Li F (2014) Fe2O3@LaxSr1 − xFeO3 core-shell redox catalyst for methane partial oxidation. ChemCatChem 6(3):790–799
39. Zheng X, Tan S, Dong L, Li S, Chen H (2015) Plasma-assisted catalytic dry reforming of methane: highly catalytic performance of nickel ferrite nanoparticles embedded in silica. J Power Sources 274:286–294
40. Zheng X, Tan S, Dong L, Li S, Chen H (2014) LaNiO$_3$@SiO$_2$ core–shell nano-particles for the dry reforming of CH$_4$ in the dielectric barrier discharge plasma. Int J Hydrogen Energy 39(22):11360–11367
41. Zheng X, Tan S, Dong L, Li S, Chen H (2015) Silica-coated LaNiO$_3$ nanoparticles for non-thermal plasma assisted dry reforming of methane: experimental and kinetic studies. Chem Eng J 265:147–156
42. Kathiraser Y, Oemar U, Saw ET, Li Z, Kawi S (2015) Kinetic and mechanistic aspects for CO$_2$ reforming of methane over Ni based catalysts. Chem Eng J 278:62–78
43. Das S, Bhattar S, Liu L, Wang Z, Xi S, Spivey JJ, Kawi S (2020) Effect of partial Fe substitution in La$_{0.9}$Sr$_{0.1}$NiO$_3$ Perovskite-derived catalysts on the reaction mechanism of methane dry reforming. ACS Catal 12466–12486
44. Xu J, Xiao Q, Zhang J, Sun Y, Zhu Y (2017) NiO-MgO nanoparticles confined inside SiO$_2$ frameworks to achieve highly catalytic performance for CO$_2$ reforming of methane. Mol Catal 432:31–36
45. Zhang L, Lian J, Li L, Peng C, Liu W, Xu X, Fang X, Wang Z, Wang X, Peng H (2018) LaNiO$_3$ nanocube embedded in mesoporous silica for dry reforming of methane with enhanced coking resistance. Microporous Mesoporous Mater 266:189–197
46. Du X, Zhang D, Gao R, Huang L, Shi L, Zhang J (2013) Design of modular catalysts derived from NiMgAl-LDH@m-SiO$_2$ with dual confinement effects for dry reforming of methane. Chem Commun 49(60):6770–6772
47. Li Z, Jiang B, Wang Z, Kawi S (2018) High carbon resistant Ni@Ni phyllosilicate@SiO$_2$ core shell hollow sphere catalysts for low temperature CH$_4$ dry reforming. J CO2 Utiliz 27:238–246
48. Li Z, Mo L, Kathiraser Y, Kawi S (2014) Yolk–satellite–shell structured Ni–Yolk@Ni@SiO$_2$ nanocomposite: superb catalyst toward methane CO$_2$ reforming reaction. ACS Catal 4(5):1526–1536
49. Pang Y, Zhong A, Xu Z, Jiang W, Gu L, Feng X, Ji W, Au C-T (2018) How do core-shell structure features impact on the activity/stability of the CO-based catalyst in dry reforming of methane? ChemCatChem 10(13):2845–2857

50. Pang Y, Dou Y, Zhong A, Jiang W, Gu L, Feng X, Ji W, Au C-T (2018) Nanostructured Ru-Co@SiO_2: highly efficient yet durable for CO_2 reforming of methane with a desirable H_2/CO ratio. Appl Catal A 555:27–35

51. Li Z, Kathiraser Y, Ashok J, Oemar U, Kawi S (2014) Simultaneous tuning porosity and basicity of Nickel@Nickel–magnesium phyllosilicate core-shell catalysts for CO2 reforming of CH4. Langmuir 30(48):14694–14705

52. Li Z, Kathiraser Y, Kawi S (2014) Facile synthesis of high surface area yolk-shell Ni@Ni embedded SiO2 via Ni phyllosilicate with enhanced performance for CO2 reforming of CH4. ChemCatChem 7(1):160–168

53. Zhao X, Li H, Zhang J, Shi L, Zhang D (2016) Design and synthesis of NiCe@m-SiO_2 yolk-shell framework catalysts with improved coke- and sintering-resistance in dry reforming of methane. Int J Hydrogen Energy 41(4):2447–2456

54. Lu Y, Guo D, Ruan Y, Zhao Y, Wang S, Ma X (2018) Facile one-pot synthesis of Ni@HSS as a novel yolk-shell structure catalyst for dry reforming of methane. J CO2 Utiliz 24:190–199

55. Dai C, Zhang S, Zhang A, Song C, Shi C, Guo X (2015) Hollow zeolite encapsulated Ni–Pt bimetals for sintering and coking resistant dry reforming of methane. J Mater Chem A 3(32):16461–16468

56. Yang W, Liu H, Li Y, Zhang J, Wu H, He D (2016) Properties of yolk–shell structured Ni@SiO_2 nanocatalyst and its catalytic performance in carbon dioxide reforming of methane to syngas. Catal Today 259:438–445

57. Wang C, Jie X, Qiu Y, Zhao Y, Al-Megren HA, Alshihri S, Edwards PP, Xiao T (2019) The importance of inner cavity space within Ni@SiO_2 nanocapsule catalysts for excellent coking resistance in the high-space-velocity dry reforming of methane. Appl Catal B 259:118019

58. Li Z, Wang Z, Jiang B, Kawi S (2018) Sintering resistant Ni nanoparticles exclusively confined within SiO_2 nanotubes for CH_4 dry reforming. Catal Sci Technol 8(13):3363–3371

59. Bian Z, Suryawinata IY, Kawi S (2016) Highly carbon resistant multicore-shell catalyst derived from Ni-Mg phyllosilicate nanotubes@silica for dry reforming of methane. Appl Catal B 195:1–8

60. Gunduz-Meric G, Kaytakoglu S, Degirmenci L (2020) Catalytic performance of silica covered bimetallic nickel-iron encapsulated core-shell microspheres for hydrogen production. Int J Hydrogen Energy

61. Huang Q, Fang X, Cheng Q, Li Q, Xu X, Xu L, Liu W, Gao Z, Zhou W, Wang X (2017) Synthesis of a highly active and stable nickel-embedded alumina catalyst for methane dry reforming: on the confinement effects of alumina shells for nickel nanoparticles. ChemCatChem 9(18):3563–3571

62. Zhao Y, Kang Y, Li H, Li H (2018) CO_2 conversion to synthesis gas via DRM on the durable Al_2O_3/Ni/Al_2O_3 sandwich catalyst with high activity and stability. Green Chem 20(12):2781–2787

63. Li S, Fu Y, Kong W, Pan B, Yuan C, Cai F, Zhu H, Zhang J, Sun Y (2020) Dually confined Ni nanoparticles by room-temperature degradation of AlN for dry reforming of methane. Appl Catal B 277:118921

64. Xu JY, Wu SF (2018) Stability of complex catalyst with NiO@TiO_2 core-shell structure for hydrogen production. Int J Hydrogen Energy 43(22):10294–10300

65. Kim DH, Kim SY, Han SW, Cho YK, Jeong M-G, Park EJ, Kim YD (2015) The catalytic stability of TiO_2-shell/Ni-core catalysts for CO_2 reforming of CH_4. Appl Catal A 495:184–191

66. Han JW, Park JS, Choi MS, Lee H (2017) Uncoupling the size and support effects of Ni catalysts for dry reforming of methane. Appl Catal B 203:625–632

67. Wang F, Han K, Yu W, Zhao L, Wang Y, Wang X, Yu H, Shi W (2020) Low temperature CO_2 reforming with methane reaction over CeO_2-modified Ni@SiO_2 catalysts. ACS Appl Mater Interfaces 12(31):35022–35034

68. Liu W, Li L, Zhang X, Wang Z, Wang X, Peng H (2018) Design of Ni-ZrO_2@SiO_2 catalyst with ultra-high sintering and coking resistance for dry reforming of methane to prepare syngas. J CO2 Utiliz 27:297–307

69. Dama S, Ghodke SR, Bobade R, Gurav HR, Chilukuri S (2018) Active and durable alkaline earth metal substituted perovskite catalysts for dry reforming of methane. Appl Catal B 224:146–158
70. Gurav HR, Dama S, Samuel V, Chilukuri S (2017) Influence of preparation method on activity and stability of Ni catalysts supported on Gd doped ceria in dry reforming of methane. J CO2 Utiliz 20:357–367
71. Song K, Lu M, Xu S, Chen C, Zhan Y, Li D, Au C, Jiang L, Tomishige K (2018) Effect of alloy composition on catalytic performance and coke-resistance property of Ni-Cu/Mg(Al)O catalysts for dry reforming of methane. Appl Catal B 239:324–333
72. Shim J-O, Hong YJ, Na H-S, Jang W-J, Kang YC, Roh H-S (2016) Highly active and stable Pt-loaded Ce$_{0.75}$Zr$_{0.25}$O$_2$ yolk–shell catalyst for water–gas shift reaction. ACS Appl Mater Interfaces 8(27):17239–17244
73. Cho JS, Ju HS, Kang YC (2016) Applying nanoscale Kirkendall diffusion for template-free, kilogram-scale production of SnO$_2$ hollow nanospheres via spray drying system. Sci Rep 6:23915
74. Lu P, Campbell CT, Xia Y (2013) A sinter-resistant catalytic system fabricated by maneuvering the selectivity of SiO$_2$ deposition onto the TiO$_2$ surface versus the Pt nanoparticle surface. Nano Lett 13(10):4957–4962
75. Zhang L, Wang F, Zhu J, Han B, Fan W, Zhao L, Cai W, Li Z, Xu L, Yu H, Shi W (2019) CO$_2$ reforming with methane reaction over Ni@SiO$_2$ catalysts coupled by size effect and metal-support interaction. Fuel 256:115954
76. Han J, Liang Y, Qin L, Zhao B, Wang H, Wang Y (2019) Ni@HC core-shell structured catalysts for dry reforming of methane and carbon dioxide. Catal Lett 149(11):3224–3237
77. Price C-AH, Pastor-Pérez L, Ramirez Reina T, Liu J (2018) Robust mesoporous bimetallic yolk–shell catalysts for chemical CO$_2$ upgrading via dry reforming of methane. React Chem Eng 3(4):433–436
78. Chai R, Zhao G, Zhang Z, Chen P, Liu Y, Lu Y (2017) High sintering-/coke-resistance Ni@SiO$_2$/Al$_2$O$_3$/FeCrAl-fiber catalyst for dry reforming of methane: one-step, macro-to-nano organization via cross-linking molecules. Catal Sci Technol 7(23):5500–5504
79. Wang Y, Fang Q, Shen W, Zhu Z, Fang Y (2018) (Ni/MgAl$_2$O$_4$)@SiO$_2$ core–shell catalyst with high coke-resistance for the dry reforming of methane. React Kinet Mech Catal 125(1):127–139
80. Han B, Amoo CC, Zhang G, Cheng S, Mazonde B, Javed M, Gai X, Lu C, Yang R, Xing C (2019) Spatial confinement effects of microcapsule catalyst for improved coking- and sintering-resistant behaviors toward CO$_2$ reforming of methane reaction. Energy Technol 7(4):1801033

Chapter 9
Supported Core–Shell Alloy Nanoparticle Catalysts for the Carbon Dioxide Hydrogenation to Formic Acid

Kohsuke Mori and Hiromi Yamashita

9.1 Introduction

The chemical transformation of carbon dioxide (CO_2) into synthetically valuable compounds is of great interest in industrial chemistry because this transformation not only reduces the emission of CO_2, but also supplies valuable chemical and fuel resources [1]. The hydrogenation of CO_2 to produce formic acid (FA; HCOOH), which is a liquid at room temperature and contains 4.4 wt% hydrogen, is a promising approach to establish FA as a renewable hydrogen storage material since the chemically stored H_2 in the FA can be liberated in controllable fashion in the presence of appropriate catalysts even at room temperature [2–6]. Thus, the design of novel catalysts for CO_2 hydrogenation to form FA is a crucial task in the realization of economical CO_2-mediated hydrogen energy cycles. The gas phase hydrogenation of CO_2 to produce formic acid has a positive free energy change [7].

$$CO_2(g) + H_2(g) \rightarrow HCOOH\ (l),\ \Delta G = +33\ \text{kJ mol}^{-1}$$

However, this reaction proceeds more readily in aqueous solution (CO_2 (aq) + H_2 (aq) → HCOOH (aq), $\Delta G = -4$ kJ mol^{-1}). The reaction is typically performed with the addition of a weak base, such as a tertiary amine or alkali/alkaline earth bicarbonate, which shifts the thermodynamic equilibrium to the product side.

$$(CO_2(aq) + H_2(aq) + B \rightarrow HCO_2^-(aq) + BH^+(B : \text{base}),$$

K. Mori (✉) · H. Yamashita
Division of Materials and Manufacturing Science, Graduate School of Engineering, Osaka University, 2-1 Yamada-oka, Suita, Osaka 565-0871, Japan
e-mail: mori@mat.eng.osaka-u.ac.jp

H. Yamashita
e-mail: yamashita@mat.eng.osaka-u.ac.jp

© The Author(s), under exclusive license to Springer Nature Singapore Pte Ltd. 2021 151
H. Yamashita and H. Li (eds.), *Core-Shell and Yolk-Shell Nanocatalysts*,
Nanostructure Science and Technology,
https://doi.org/10.1007/978-981-16-0463-8_9

$$\Delta G = -35.4\,\text{kJ mol}^{-1}).$$

Significant progress has been made utilizing homogeneous complexes in basic media [8, 9]. However, the development of heterogeneous catalysts lags significantly behind that of homogeneous catalysts, in spite of their practical utility [10–14], and, to make matters worse, their use frequently requires a high catalyst concentration, organic solvents, and extremely high pressures.

Metal nanoparticle-based catalysts are gaining increasing attention to bridge the gap between mononuclear metal complexes and heterogeneous bulk catalysts because of their existence on borderline molecular states with discrete quantum energy levels [15]. Their large surface area-to-volume ratio allows effective utilization of expensive metals. The variation in size, composition, morphology, and supports significantly influence the catalytic activities. Additionally, the accurate control of the geometric and electronic effect of bimetallic nanoparticles, in which the architectural configuration of two metals is as random alloys, segregated or a core–shell structure, is a key technology in attaining superior catalytic performances to the monometallic counterparts [16]. The interplay of the neighboring different metals creates specific new catalytically active sites, which frequently enables the fine-tuning of the geometric and electronic properties originating from synergic alloying effects [17–20]. Moreover, the replacement of the precious noble metal nanoparticles with inexpensive metals contributes to the atomic economy [21, 22]. Thus, the successful synthesis of bimetallic nanoparticles with controllable size, shape, and composition plays a crucial role in designing highly functionalized catalysts. However, the insights into the promising design strategy as well as the additional elucidation of the catalytically active species in the supported metal nanoparticles are required.

In this section, the state of the art of the nanostructured alloy metal nanoparticle catalysts, especially core–shell-type catalyst, reported for CO_2 hydrogenation to formic acid/formate is presented. The enhanced activity was demonstrated by the discussions based on kinetic and density functional theory (DFT) calculations.

9.2 Elucidating the Catalytically Active Species in Supported Pd@Ag Alloy Nanoparticles

The co-reduction of Pd and Ag precursors conventionally affords random PdAg alloy nanoparticles (PdAg/TiO$_2$) because of the complete solid solubility and similar reduction potentials of Pd and Ag ions. The precise tuning of the surface composition of PdAg nanoparticles was performed to investigate the effect of surface-exposed active Pd atoms in alloy NPs. By applying a surface engineering approach via the successive reduction of metal precursors, Pd@Ag/TiO$_2$ with a Pd$_{core}$Ag$_{shell}$ structure and Ag@Pd/TiO$_2$ with an Ag$_{core}$Pd$_{shell}$ structure were synthesized [23].

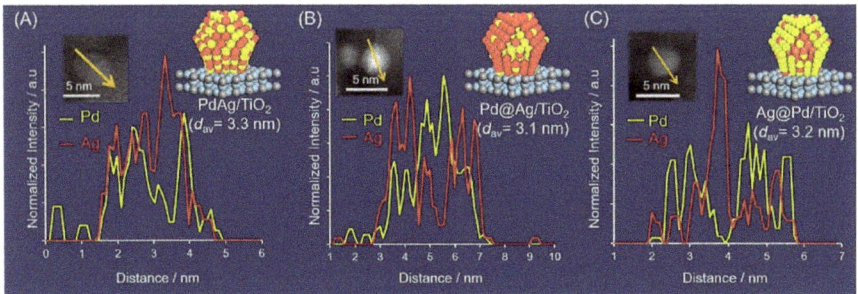

Fig. 9.1 TEM images of and elemental distributions along single NPs of **a** PdAg/TiO$_2$, **b** Pd@Ag/TiO$_2$, and **c** Ag@Pd/TiO$_2$ as determined by cross-sectional EDX line profiling (Reproduced with permission from [23]. Copyright © 2018 American Chemical Society.)

As shown in Fig. 9.1, high-angle annular dark-field scanning transmission electron microscopy (HAADF-STEM) images demonstrated highly dispersed PdAg alloy NPs with a narrow size distribution on the TiO$_2$ support for all samples (Pd:Ag = 30:70), with mean particle diameters of ca. 3 nm. This value is similar to that obtained for the Pd/TiO$_2$ (d_{ave} = 3.2 nm). Energy-dispersive X-ray spectroscopy (EDX) line analysis confirmed the successful surface engineering of the NPs. In the case of PdAg/TiO$_2$, both Pd and Ag were situated on the same particles, confirming the formation of a random PdAg alloy. By contrast, Pd atoms were preferentially located in the core region, and the Ag atoms were situated in the shell region in the Pd@Ag/TiO$_2$. Conversely, the Pd and Ag atoms had the opposite distribution for the Ag@Pd/TiO$_2$.

In the XPS spectra, the Pd 3d peaks of all PdAg samples were shifted to lower binding energies than those of the Pd/TiO$_2$, and this shift was decreased in the order of Pd@Ag/TiO$_2$ > PdAg/TiO$_2$ > Ag@Pd/TiO$_2$. Thus, the Pd atoms in the PdAg NPs were obviously electron enriched by the charge transfer from Ag atoms owing to the net difference in ionization potential between the two metals (Pd: 8.34 eV, Ag: 7.57 eV). A similar tendency in the electronic state of the Pd species was observed in the FT-IR experiments using CO as a probe molecule. Monometallic Pd/TiO$_2$ generated two distinct peaks assignable to the linear and bridging stretching vibrations of adsorbed CO at 2076 and 1943 cm^{-1}, respectively. The contribution of bridging-type CO decreased as the Pd/Ag ratio was lowered, and was completely absent in the case of the Pd@Ag/TiO$_2$, suggesting the isolation of Pd atoms. Additionally, linear-type CO was predominantly observed for all samples, and the peaks were gradually shifted to lower wavenumbers with decreases in the Pd/Ag ratio.

Figure 9.2 shows the comparison of catalytic activity in the CO$_2$ hydrogenation. The Pd@Ag/TiO$_2$ exhibited an elevated TON (2,496) based on the total quantities of Pd employed despite the low density of surface-exposed Pd atoms. As expected, a maximum TON value of 14839 was obtained from the Pd@Ag/TiO$_2$ based on the quantity of surface Pd atoms, which was determined by pulsed CO adsorption measurements. This TON value is more than ten times higher than that of the

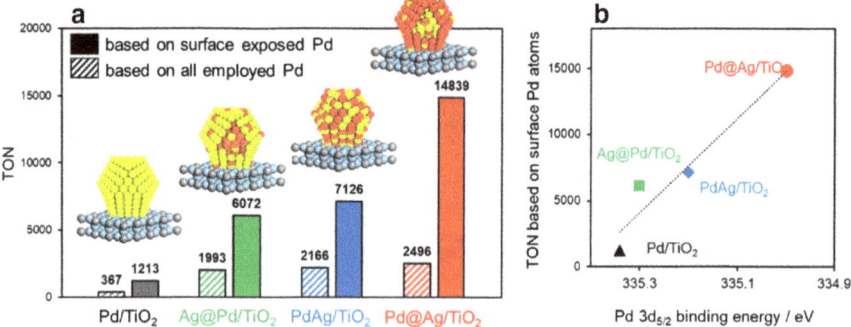

Fig. 9.2 **a** Comparison of the catalytic activities of a series of supported PdAg catalysts with different surface compositions and Pd/TiO$_2$ during CO$_2$ hydrogenation. **b** Relationship between the TON for CO$_2$ hydrogenation based on surface-exposed Pd atoms (as determined by CO pulse adsorption) and the Pd 3d binding energy (as determined by XPS) (Reproduced with permission from [23]. Copyright © 2018 American Chemical Society.)

monometallic Pd/TiO$_2$. Moreover, a good correlation between the TON based on surface Pd atoms and the Pd 3d$_{5/2}$ binding energy determined by XPS analysis is evidently observed (Fig. 9.2b).

In this study, the enhancement of activity by alloying was well evidenced based on the DFT calculations, employing Pd$_{22}$, Pd$_{11}$Ag$_{11}$, and Pd$_6$Ag$_{16}$ clusters as models for monometallic Pd and alloy nanoparticles (Fig. 9.3). The CO$_2$ hydrogenation over Pd$_{22}$ is initiated by the dissociation of H$_2$ to form a metal-hydride species via TS$_{I/II}$ with a barrier of 13.9 kcal/mol (*step 1*). Next is the adsorption of HCO$_3^-$ to produce intermediate III (*step 2*), followed by the attack of H atom to the C atom of HCO$_3^-$ via TS$_{III/IV}$, with a barrier of 77.4 kcal/mol (*step 3*). Finally, the formate that is produced accompanied by H$_2$O regenerates the initial active species (*step 4*). The activation energies for *step 1* using Pd$_{11}$Ag$_{11}$ and Pd$_6$Ag$_{16}$ clusters, were 11.9 and 11.0 kcal/mol, respectively, which were similar to that obtained with the Pd$_{22}$. On the other hand, the reduction of HCO$_3^-$ via TS$_{III/IV}$ occurs with a barrier of 58.7 and 46.2 kcal/mol for Pd$_{11}$Ag$_{11}$ and Pd$_6$Ag$_{16}$, respectively. These results show that the rate-determining step is *step 3*, and further demonstrate that the importance of low Pd/Ag ratio of the PdAg alloy nanoparticles in boosting the rate-determining step.

The kinetic analysis further supported the above results. In the case of a reaction under a flow of H$_2$ and D$_2$ through the catalyst, the TOF for HD formation was almost independent of the surface composition. On the contrary, the effect of the HCO$_3^-$ concentration was greatly dependent on the surface composition. The reaction rate decreased in the order Pd/TiO$_2$ (0.67) > Ag@Pd/TiO$_2$ (0.36) > PdAg/TiO$_2$ (0.33) > Pd@Ag/TiO$_2$ (0.22), which is consistent with the TON values based on the quantity of surface-exposed Pd atoms for the CO$_2$ hydrogenation.

This can be well explained by considering the electronic state in reaction intermediate III (Fig. 9.4). Mulliken atomic charges of Pd atoms decrease in the order of −0.115 (Pd$_{22}$) > −0.168 (Pd$_{11}$Ag$_{11}$) > −0.216 (Pd$_6$Ag$_{16}$), which consequently decreases

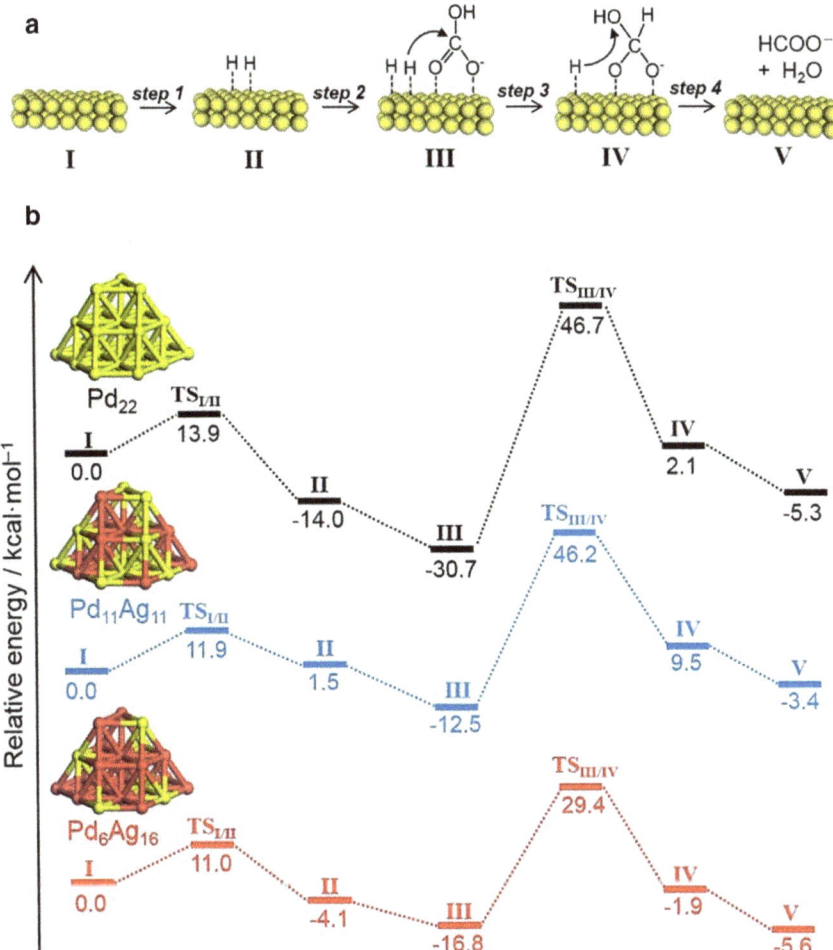

Fig. 9.3 **a** Possible reaction mechanism for CO$_2$ hydrogenation to formic acid. **b** Potential energy profiles as determined by DFT calculations for Pd$_{22}$, Pd$_{11}$Ag$_{11}$, and Pd$_6$Ag$_{11}$ cluster models (Reproduced with permission from [23]. Copyright © 2018 American Chemical Society.)

the electronegativity of the dissociated hydride species on the Pd atoms. Contrastingly, the electronic charges of the C atoms of the adsorbed HCO$_3^-$ are almost constant for all models, while keeping their positive charges. Thus, the hydride species on the Pd$_6$Ag$_{16}$ with more negative charge easily attack the C atoms of the adsorbed HCO$_3^-$, while a higher activation energy is necessary in the reaction between the less negative hydride species on the Pd$_{22}$ and the positively charged C atoms. It can be concluded that isolated and electron-rich Pd atoms created with the aid of neighboring Ag atoms explain the enhanced activity, which provides advanced

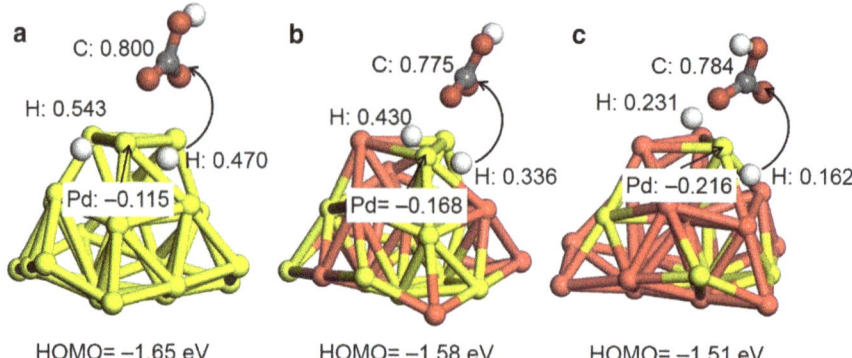

Fig. 9.4 Representative Mulliken atomic charges in the reaction intermediate III as determined by DFT calculations for **a** Pd_{22}, **b** $Pd_{11}Ag_{11}$, and **c** Pd_6Ag_{11} cluster models (Reproduced with permission from [23]. Copyright © 2018 American Chemical Society.)

insights into the architecture of catalytically active sites for CO_2 hydrogenation to FA.

9.3 Interfacial Engineering of PdAg/TiO₂ with a Metal–Organic Framework

In the drive to improve catalytic performances during CO_2 hydrogenation to produce FA, PdAg/TiO₂ was further modified with an MOF using a facile pretreatment method. A zeolitic imidazolate framework (ZIF-8), a product of the reaction between Zn^{2+} and 2-methylimidazole, was chosen as a modifying agent because of its high chemical and thermal inertness in aqueous solution, suitable mechanical stability even under high-pressure conditions, and ready synthesis at room temperature. The time allowed for the growth of the ZIF-8 was varied between 10 min and 3 h, affording a series of PdAg/TiO₂@ZIF-8. A schematic procedure is illustrated in Fig. 9.5 [24].

Because of the lower content of ZIF-8 (0.7 wt% from CHN elemental analysis) and its high dispersity on the support, the PdAg/TiO₂@ZIF-8 exhibited only peaks due to rutile TiO₂, with no characteristic peaks assignable to ZIF-8 or the PdAg NPs in the XRD pattern. No significant differences were observed in the surface area ($S_{BET} = 137.5$ m² g⁻¹ for PdAg/TiO₂ and 125.9 m² g⁻¹ PdAg/TiO₂@ZIF-8), suggesting that a very thin coating of ZIF-8 had been applied.

X-ray photoelectron spectroscopy (XPS) analysis showed the peak due to Zn $2p_{3/2}$ and Zn $2p_{1/2}$ at 1022.7 and 1045.8 eV, respectively. Peaks due to C–NH– and C=N– bonds in imidazole groups were also clearly seen in the N 1 s region at 399.7 eV. Scanning transmission electron microscopy (STEM) images confirmed that the PdAg/TiO₂ was covered with a thin shell of ZIF-8 having a thickness of approximately 1.6 nm. Elemental mapping further demonstrated that Zn, C, and N

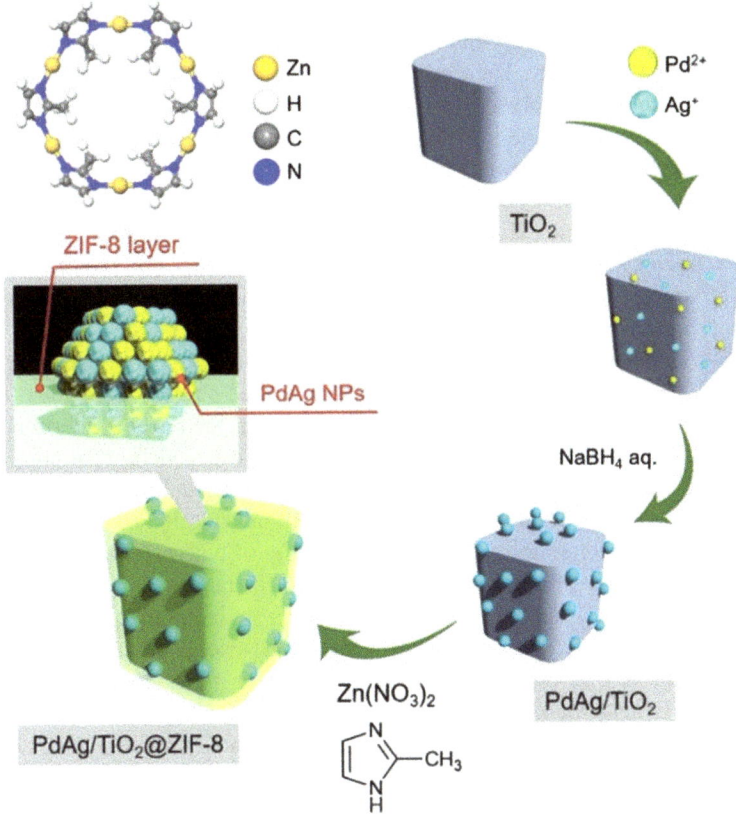

Fig. 9.5 A schematic illustration of the synthesis of PdAg/TiO$_2$@ZIF-8 (Reproduced with permission from [24]. Copyright © 2020 American Chemical Society.)

atoms, which were originated from ZIF-8 layer, were highly dispersed throughout the material. From the results of XAFS, a cross-sectional EDX line profile, and high-resolution TEM image, it is evident that the PdAg alloy NPs retained their original structure even after coating with the ZIF-8 layer.

The PdAg/TiO$_2$ without modification showed a TON of 488 at 6 h. In contrast, the PdAg/TiO$_2$@ZIF-8 specimen synthesized using a ZIF-8 growth time of 30 min exhibited the highest TON of 913 together with >99% selectivity, which is almost twice that of the unmodified material. Additionally, the catalytic activity was greatly affected by the growth time, such that the highest TON was obtained at 30 min, with decreases in the TON when longer times were applied. This volcano-type variation in activity suggests the formation of a uniform ZIF-8 layer on the surface of the PdAg/TiO$_2$ as well as a synergistic effect originating from integration with the ZIF-8.

Another important influence for the modification with ZIF-8 is the anchoring stabilization effect that inhibits the undesired agglomeration of NPs, which ultimately enhances the durability of the catalyst. In trials with the PdAg/TiO$_2$, the catalytic activity gradually decreased with continued use, such that the activity was reduced by half during the second recycling experiment. Conversely, the PdAg/TiO$_2$@ZIF-8 retained its original activity. Thus, each catalyst was isolated after catalytic reaction and subjected to TEM analysis for comparison of the average particle sizes with the initial particle sizes. The significant enlargement can be observed in the case of unmodified PdAg/TiO$_2$, where the average diameter of the isolated catalyst was determined to be 8.9 nm, which was almost twice that of the initial particle diameter. In contrast, PdAg/TiO$_2$@ZIF-8 catalyst suppressed the particle growth; the mean particle diameter after the reaction was determined to be 5.1 nm, which increased by only 6% compared with that of the initial catalyst particles, thus preserving their intriguing properties in catalysis applications.

To better understand the positive effect of ZIF-8 modification, adsorption energy (E_{ad}) of HCO$_3^-$ on PdAg surface and Mulliken atomic charges of selected atoms for the reaction intermediate in the rate-determining step were determined by DFT calculations (Table 9.1). Here, three representative models were considered for the calculation. These include PdAg(111) and PdAg(111) interacted with building unit of ZIF-8 framework (two 2-methylimidazole linked with Zn^{2+} ion) in two different configurations. As expected, the E_{ad} of both configurations for the PdAg(111) interacted with the building unit of ZIF-8 framework was estimated to be -133.9 and -185.2 kcal/mol, respectively, which are larger than that on the pristine PdAg(111) (Fig. 9.6a vs. b and c). In the lowest-energy adsorption structure, 2-methylimidazole ring vertically interacts with surface metal atoms. According to the Brønsted–Evans–Polanyi (BEP) relationship, the larger E_{ad} of the reaction intermediate on the metal catalyst corresponds to a lower reaction barrier. Such changes in the adsorption capacity accordingly alter the electronic charges of the C atoms of the adsorbed HCO$_3^-$, as shown in Table 9.1. In contrast, the electronic charges of the dissociated hydride species on the Pd atoms are almost constant for all calculation models. Therefore, the more positively charged C atoms of the adsorbed HCO$_3^-$ will tend to undergo the attack by the dissociated hydride species. It can be concluded that the electronic effect resulting from the interplay of the neighboring ZIF-8 unit explains the enhanced activity for CO$_2$ hydrogenation.

Table 9.1 Adsorption energy values (E_{ad}) for HCO$_3^-$ and representative Mulliken atomic charges as determined by DFT calculations involving the reaction intermediates in the rate-determining steps for bare PdAg (111) and PdAg (111) with the ZIF-8 framework (including two 2-methylimidazole molecules bonded to a Zn^{2+} ion) in two different configurations

Sample	E_{ad} of HCO$_3^-$ (kcal/mol)	Atomic charge		
		C atom of HCO$_3^-$	H atom	
			H$_1$	H$_2$
PdAg (111)	-121.5	0.633	0.044	0.032
PdAg (111) + ZIF-8 (1)	-133.9	0.665	0.041	0.020
PdAg (111) + ZIF-8 (2)	-185.2	0.730	0.043	0.019

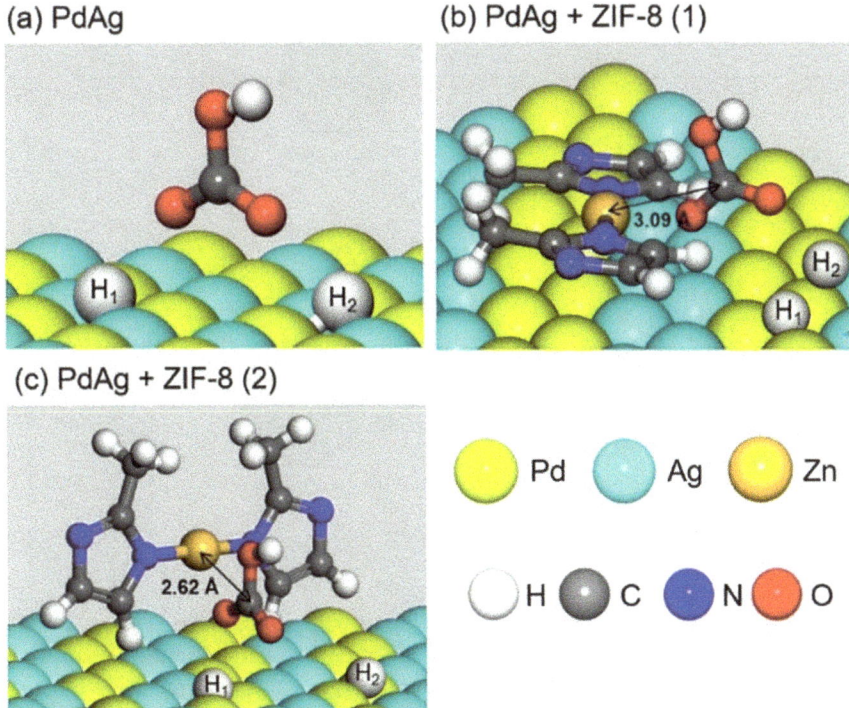

Fig. 9.6 DFT-optimized configurations for dissociated H atoms and HCO_3^- ions adsorbed on **a** PdAg (111) and **b** and **c** PdAg (111) with the ZIF-8 framework (showing two 2-methylimidazole molecules bonded to a Zn^{2+} ion) in two different configurations (Reproduced with permission from [24]. Copyright © 2020 American Chemical Society.)

9.4 Encapsulation of PdAg Nanoparticles Within ZIF-8 Framework with Core–Shell Structure

To overcome the general problems of the aggregation of supported metal NPs on the external surface of MOFs and damage of MOFs during the post-reduction process, a facile method to encapsulate metal NPs within MOFs has been developed. This "bottle around ship" approach involves the growth of ZIF-8 core at the initial stage by using 2-methylimindazole (Hmin) as an organic linker and Zn^{2+} as a connecting center, loading the PdAg NPs on the external surface of ZIF-8 core, and coating the PVP stabilized PdAg alloy NPs with ZIF-8. Figure 9.7 illustrates the synthetic route for the fabrication of ZIF-8@PdAg@ZIF-8 catalyst [25].

The diffraction pattern of ZIF-8@PdAg@ZIF-8 sample is similar to that of pure ZIF-8. This suggests that the encapsulation of PdAg NPs within ZIF-8 did not change the framework structure. However, the intensities of diffraction peaks are weaker than those of pure ZIF-8, which is because the encapsulated PdAg NPs cause the disorder in MOFs crystal. In addition, the diffraction peaks assigned to Pd and Ag cannot

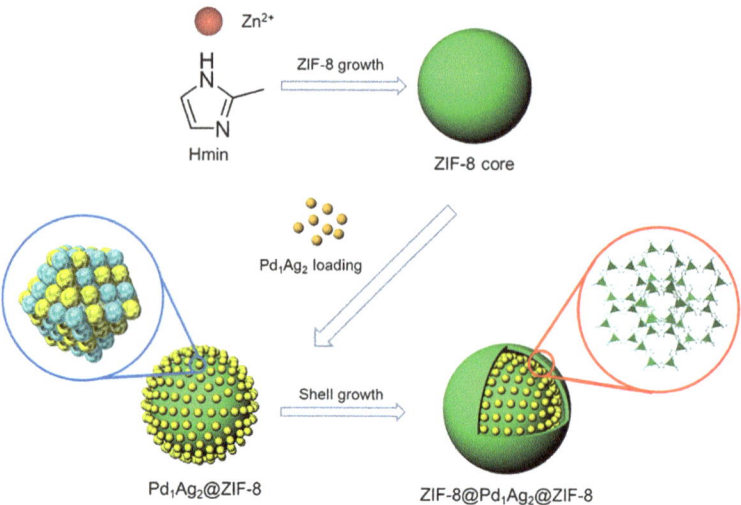

Fig. 9.7 Schematic illustration for the synthesis of ZIF-8@PdAg@ZIF-8

be observed, because of their low content and small particle size. The pure ZIF-8 shows a type-I isotherm with completely reversible isothermal in the N_2 sorption analysis, which is a typical feature of microporous materials. The Brunauer–Emmett–Teller (BET) surface area determined by N_2 adsorption-desorption is $1110 \, m^2 \, g^{-1}$. Meanwhile, ZIF-8@PdAg@ZIF-8 has a similar isotherm to pure ZIF-8 except for the slight decrease in the N_2 uptake, which suggests the decrease of some micropores after the encapsulation of PdAg NPs. This resulted in a slight decrease in their surface area to $926.3 \, m^2 \, g^{-1}$. Based on such results, it is concluded that the crystallinity and porosity of ZIF-8 are well preserved after PdAg encapsulation.

The morphologies of pure ZIF-8 and ZIF-8@Pd_1Ag_2@ZIF-8 are shown in Fig. 9.8. Pure ZIF-8 displays a rhombic dodecahedral morphology and the particle size is about 350 nm. Meanwhile, the morphology of ZIF-8@Pd_1Ag_2@ZIF-8 did not show obvious change. It should be noted that the external surface of ZIF-8@Pd_1Ag_2@ZIF-8 is very smooth, which indicates that the PdAg NPs are not loaded on the external surface of ZIF-8. From the TEM images, tiny PdAg NPs with a mean diameter of 2.8 nm can be observed, which was covered by a thin shell of ZIF-8, and the shell thickness was measured to be ca. 5 nm.

The ZIF-8@Pd_1Ag_2@ZIF-8 samples showed the highest catalytic activity among all samples, affording 16.68 mmol $g^{-1}_{(catal.)}$ after 24 h, almost two times the FA amount produced over ZIF-8@Pd_3@ZIF-8 under the identical reaction condition.

The poor catalytic performance was obtained on PdAg/ZIF-8 without core–shell structure. This is due to the large particle size of PdAg NPs, whose diameter was determined to be 10.5 nm. In comparison, ZIF-8@Pd_1Ag_2@ZIF-8 exhibited improved catalytic activity, which can be ascribed to the high dispersibility of PdAg NPs within ZIF-8 and the positive effect of the thin shell in protecting PdAg NPs during

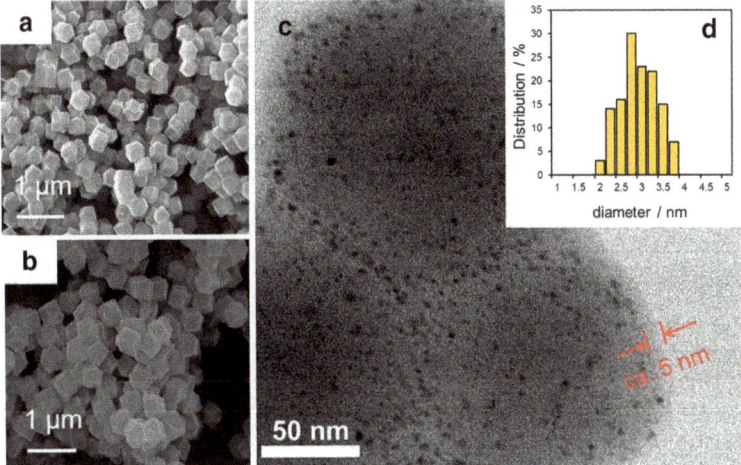

Fig. 9.8 SEM images of **a** pure ZIF-8 and **b** ZIF-8@Pd$_1$Ag$_2$@ZIF-8, and **c** TEM image and **d** size distribution diagram of ZIF-8@Pd$_1$Ag$_2$@ZIF-8

the reaction process. The catalyst was recovered from the reaction solution by using centrifugation and wash with water. TEM image of ZIF-8@PdAg@ZIF-8 after reaction showed that the PdAg NPs are still well dispersed within ZIF-8, and no significant aggregation occurred. Moreover, the recycled ZIF-8@PdAg@ZIF-8 could be re-used at least three times without significantly loss in activity. Based on the above result, it is clear that the presented synthetic approach has advantages for overcoming the general problems of the aggregation of metal nanoparticles on the external surface of MOFs and damage of MOFs during the post-reduction process, leading to the enhanced catalytic activity for CO$_2$ hydrogenation to produce FA.

9.5 Summary and Outlook

By tuning the surface-exposed Pd atoms in the alloy NPs, the isolated Pd atoms surrounded by large amount of Ag atoms in Pd@Ag/TiO$_2$ performed as an efficient catalyst for the CO$_2$ hydrogenation to FA even under low-pressure conditions. Kinetic and DFT calculations evidenced the enhanced electronegativity was found to facilitate the rate-determining reduction step of the adsorbed HCO$_3^-$ species. Moreover, the positive effects imparted by the interfacial modification of PdAg/TiO$_2$ with ZIF-8 and the encapsulation of PdAg NPs within the ZIF-8 Framework with core–shell structure were demonstrated. These studies not only provide advanced insights into the architecture of catalytically active centers for CO$_2$ hydrogenation to FA based on the surface engineering approach, but also emphasizes the importance of the interfacial surface engineering of catalysts for further improvement. Nevertheless, there are still some lacking aspects that should be tackled in future investigations. One of the principal issues is their insufficient stability and long durability under reaction conditions, which should be absolutely improved to meet the practical application criteria while preserving the unique surface characteristics of NPs. Further improvement is needed to develop reliable catalysts, which meet the practical application criteria in terms of efficiency, cost, and reusability, which will open a new avenue for environmentally benign CO_2-mediated hydrogen storage/release systems.

References

1. Wang W, Wang S, Ma X, Gong J (2011) Recent advances in catalytic hydrogenation of carbon dioxide. Chem Soc Rev 40:3703–3727
2. Singh AK, Singh S, Kumar A (2016) Hydrogen energy future with formic acid: a renewable chemical hydrogen storage system. Catal Sci Technol 6:12–40
3. Joó F (2008) Breakthroughs in hydrogen storage—formic acid as a sustainable storage material for hydrogen. Chemsuschem 1:805–808
4. Mori K, Tanaka H, Dojo M, Yoshizawa K, Yamashita H (2015) Synergic catalysis of PdCu alloy nanoparticles within a macroreticular basic resin for hydrogen production from formic acid. Chem Eur J 21:12085–12092
5. Mori K, Dojo M, Yamashita H (2013) Pd and Pd–Ag nanoparticles within a macroreticular basic resin: an efficient catalyst for hydrogen production from formic acid decomposition. ACS Catal 3:1114–1119
6. Masuda S, Mori K, Futamura Y, Yamashita H (2018) PdAg nanoparticles supported on functionalized mesoporous carbon: promotional effect of surface amine groups in reversible hydrogen delivery/storage mediated by formic acid/CO_2. ACS Catal 8:2277–2285
7. Moret S, Dyson PJ, Laurenczy G (2014) Direct synthesis of formic acid from carbon dioxide by hydrogenation in acidic media. Nat Commun 5:4017
8. Enthaler S, von Langermann J, Schmidt T (2010) Carbon dioxide and formic acid-the couple for environmental-friendly hydrogen storage? Energy Environ Sci 3:1207–1217
9. Mellmann D, Sponholz P, Junge H, Beller M (2016) Formic acid as a hydrogen storage material—development of homogeneous catalysts for selective hydrogen release. Chem Soc Rev 45:3954–3988
10. Preti D, Resta C, Squarcialupi S, Fachinetti G (2011) Carbon dioxide hydrogenation to formic acid by using a heterogeneous gold catalyst. Angew Chem Int Ed 50:12551–12554
11. Xu Z, McNamara ND, Neumann GT, Schneider WF, Hicks JC (2013) Catalytic hydrogenation of CO_2 to formic acid with silica-tethered iridium catalysts. ChemCatChem 5:1769–1771
12. Filonenko GA, Vrijburg WL, Hensen EJM, Pidko EA (2016) On the activity of supported Au catalysts in the liquid phase hydrogenation of CO_2 to formates. J Catal 343:97–105
13. Lee JH et al (2014) Carbon dioxide mediated, reversible chemical hydrogen storage using a Pd nanocatalyst supported on mesoporous graphitic carbon nitride. J Mater Chem A 2:9490–9495
14. Mori K, Taga T, Yamashita H (2017) Isolated single-atomic Ru catalyst bound on a layered double hydroxide for hydrogenation of CO_2 to formic acid. ACS Catal 7:3147–3151
15. Mori K, Yamashita H (2010) Progress in design and architecture of metal nanoparticles for catalytic applications. Phys Chem Chem Phys 12:14420–14432
16. Gao F, Goodman DW (2012) Pd-Au bimetallic catalysts: understanding alloy effects from planar models and (supported) nanoparticles. Chem Soc Rev 41:8009–8020
17. Tedsree K et al (2011) Hydrogen production from formic acid decomposition at room temperature using a Ag-Pd core-shell nanocatalyst. Nat Nanotechnol 6:302–307
18. Mori K, Naka K, Masuda S, Miyawaki K, Yamashita H (2017) Palladium copper chromium ternary nanoparticles constructed in situ within a basic resin: enhanced activity in the dehydrogenation of formic acid. ChemCatChem 9:3456–3462

19. Masuda S et al (2018) Simple route for the synthesis of highly active bimetallic nanoparticle catalysts with immiscible Ru and Ni combination by utilizing a TiO_2 support. ChemCatChem 10:3526–3531
20. Masuda S, Shun K, Mori K, Kuwahara Y, Yamashita H (2020) Synthesis of a binary alloy nanoparticle catalyst with an immiscible combination of Rh and Cu assisted by hydrogen spillover on a TiO_2 support. Chem Sci 11:4194–4203
21. Mori K, Kondo Y, Yamashita H (2009) Synthesis and characterization of FePd magnetic nanoparticles modified with chiral BINAP ligand as a recoverable catalyst vehicle for the asymmetric coupling reaction. Phys Chem Chem Phys 11:8949–8954
22. Mori K, Yoshioka N, Kondo Y, Takeuchi T, Yamashita H (2009) Catalytically active, magnetically separable, and water-soluble FePt nanoparticles modified with cyclodextrin for aqueous hydrogenation reactions. Green Chem 11:1337–1342
23. Mori K, Sano T, Kobayashi H, Yamashita H (2018) Surface engineering of a supported PdAg catalyst for hydrogenation of CO_2 to formic acid: elucidating the active pd atoms in alloy nanoparticles. J Am Chem Soc 140:8902–8909
24. Mori K, Konishi A, Yamashita H (2020) Interfacial engineering of $PdAg/TiO_2$ with a metal-organic framework to promote the hydrogenation of CO_2 to formic acid. J Phys Chem C 124:11499–11505
25. Wen M et al (2019) PdAg nanoparticles within core-shell structured zeolitic imidazolate framework as a dual catalyst for formic acid-based hydrogen storage/production. Sci Rep 9:15675

Chapter 10
Core–Shell Confinement MnCeO$_x$@ZSM-5 Catalyst for NO$_x$ Removal with Enhanced Performances to Water and SO$_2$ Resistance

Honggen Peng, Guiying Li, and Taicheng An

10.1 Introduction

Nitrogen oxides (NO$_x$, mainly existed as NO and NO$_2$) have become a global environmental problem during the last decades, because of their participation in the formation of acid rain, haze, and ozone [1–4]. The selective catalytic reduction of NO$_x$ by ammonia (NH$_3$-SCR of NO$_x$) is the most efficient technology for the elimination of NO$_x$ from various vehicle and stationary sources [5], and the standard NH$_3$-SCR reaction is expressed as follows (Eq. 10.1):

$$4NH_3 + 4NO + O_2 \rightarrow 4N_2 + 6H_2O \qquad (10.1)$$

To increase energy efficiency and satisfy the practical application, the low-deNO$_x$ technology through NH$_3$-SCR reaction (150–300 °C) is considered as one ideal method to control NO$_x$ emissions at low temperatures. Therefore, developing one kind of high-performance, low-temperature active deNO$_x$ catalyst is urgent. The commercialized V$_2$O$_5$/WO$_3$/TiO$_2$ and V$_2$O$_5$/MoO$_3$/TiO$_2$ catalysts have a limited application, owing to their active temperature window is too narrow and too high (300–400 °C). Importantly, the vanadium species involved in these catalysts are

H. Peng · G. Li · T. An (✉)
Guangdong Key Laboratory of Environmental Catalysis and Health Risk Control, School of Environmental Science and Engineering, Institute of Environmental Health and Pollution Control, Guangdong University of Technology, Guangzhou 510006, People's Republic of China
e-mail: antc99@gdut.edu.cn

H. Peng
e-mail: penghonggen@ncu.edu.cn

H. Peng
School of Resources, Environmental and Chemical Engineering, Nanchang University, Nanchang 330031, People's Republic of China

harmful to human health [6–8]. Thus, many researchers played much effort to design and synthesize new deNO$_x$ catalysts, including the traditional transition metal oxides or mixed metal oxides (non-vanadium catalysts) and the active metal (Cu, Fe, Co, etc.) exchanged zeolites [9, 10].

The Mn-based mixed metal oxides have received much attention because of their abundant active oxygen species and superior redox performance [11]. In addition, cerium oxide (CeO$_2$) is considered as a good catalyst promoter and plays a key role in NH$_3$-SCR of NO$_x$ reaction due to its superior oxygen storage/release capacity [8, 12–14]. However, the sulfur dioxide (SO$_2$) poisoning is still one critical hindrance for the commercial application of Mn-based catalysts [15–17]. Lots of researches have verified that the Mn-based mixed metal oxides are readily deactivated by the poisoning of SO$_2$, especially at the low reaction temperature [18, 19]. On the one hand, the formation of ammonium sulfates (NH$_4$HSO$_4$ and (NH$_4$)$_2$SO$_4$) can enclothe the active sites and block the pores in catalysts, thereby preventing the absorption and activation of NH$_3$. On the other hand, when the feed gases contain SO$_2$, the metal oxide is changed to metal sulfate, which evidently reduces their deNO$_x$ activities. Although the ammonium sulfates can be degraded by the simple thermal treatment process at relatively high temperatures, the process for the formation of metal sulfates is irreversible [12, 20, 21].

Silica-alumina zeolites, e.g., ZSM-5 with MFI crystal structure, have been widely applied as the support of metal or metal oxides, owing to their high specific surface areas and hydrothermal and thermal stability [22–25]. Our group has studied the encapsulating Pd-Ce mixed oxide nanowire into a silica sheath (Pd-Ce NW@SiO$_2$), which has been proved to be an effective method to protect active centers from SO$_2$ poisoning [26]. The silica sheath can efficiently retard the SO$_2$ poisoning to the catalyst. Ran and coauthors prepared a series of MnO$_x$-CeO$_x$ oxides supported on ordered mesoporous silica (SBA-15) and found that this kind of catalysts displayed improved resistance to SO$_2$ [27]. Dong and coauthors also found that the pore size in the catalyst could affect their sulfur resistance [28]. Using zeolite ZSM-5 as the carrier of manganese and cerium displayed excellent activity and stability for NH$_3$-SCR of NO$_x$ [11, 29]. However, the sulfur poisoning of the active sites is still a challenge. Thereby a new structure for separating the active sites from toxic species needs to be developed and constructed. Constructing a novel core-shell structure to limit or retard the adsorption of SO$_2$, the SO$_2$ poisoning, could be alleviated to some extent [13, 26, 27].

In this chapter, [30] a novel core-shell structured Mn-CeO$_x$ deNOx catalyst was designed and successfully prepared, which is the mesoporous ZSM-5 zeolite enveloping the Mn-CeOx mixed oxides (MnCeO$_x$@Z5) through a one-pot two-step method. MnCeO$_x$@Z5 catalyst exhibited superior deNOx activity and N$_2$ selectivity. Importantly, MnCeO$_x$@Z5 also exhibited enhanced water resistance, and the zeolite shell's shielding effect could also alleviate the SO$_2$ poisoning.

10.2   Results

10.2.1   Physicochemical Performance of Catalysts

The morphology of the mesoporous ZSM-5 zeolites confined MnCeO$_x$ (MnCeO$_x$@Z5) and related catalysts was confirmed by the transmission electron microscopy (TEM) technique. It is evident that the precursor (MnCeO$_x$@Al-SiO$_2$) had a worm-like core-shell morphology, in which the amorphous Al-SiO$_2$ shell grew along the MnCeO$_x$ mixed metal oxides nanowires [26, 30–32]. Subjected to a dry-gel crystallization process, the amorphous Al-SiO$_2$ was converted to a zeolite shell. The original TEM images of MnCeO$_x$@Z5 catalyst are presented in (Fig. 10.1a–c). One can observe that the sample has similar morphology of ZSM-5 zeolite [33–35]. The high-resolution TEM in Fig. 10.1d confirmed that the MnCeO$_x$ cores are enveloped in the zeolite. To more clearly visualize the interior structure of MnCeO$_x$@Z5, the Electron Tomography characterization for the MnCeO$_x$@Z5 sample was performed. Figure 10.1e and f shows the tomogram-section high-angle annular dark-field scanning transmission electron microscopy (HAADF-STEM) images of MnCeO$_x$@Z5. Evidently, these images verified that the MnCeO$_x$ cores are indeed enveloped in the inner of the zeolite shell with a watermelon-like morphology [36]. For the conventional supported MnCeO$_x$/Z5 catalyst, the MnCeO$_x$ species were supported on the outer surfaces of ZSM-5. Interestingly, a large amount of mesopores were observed,

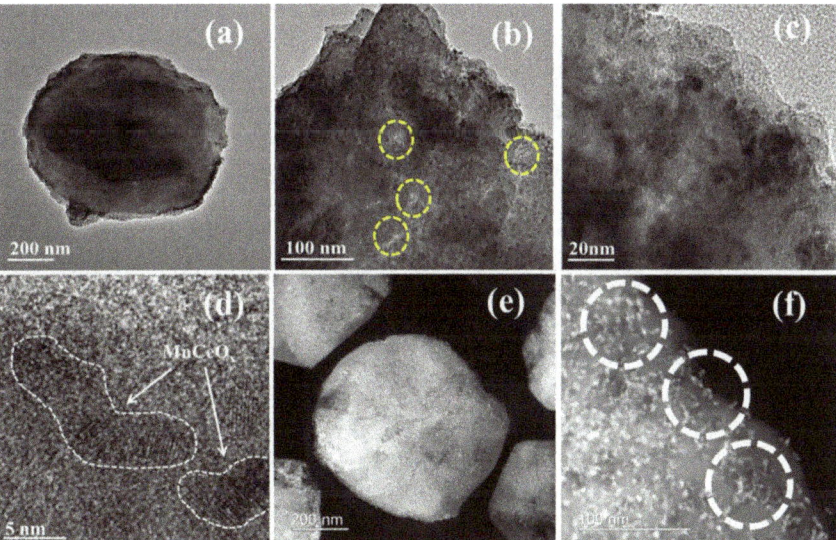

Fig. 10.1 The TEM **a–c**, and high-resolution (HR)TEM (d) and the tomogram-section AC-HAADF-STEM images **e–f** of MnCeO$_x$@Z5. The yellow circles highlight the mesopores and the white circles highlight the MnCeO$_x$ species. Reproduced with permission from ref. [30] Copyright 2020 Elsevier

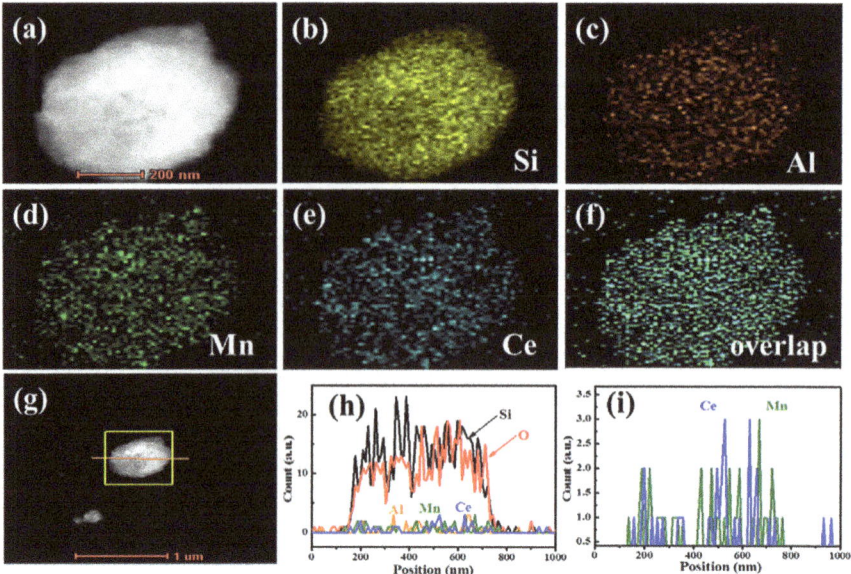

Fig. 10.2 The HAADF-STEM image (**a**), EDX elemental mapping images (**b–f**), and the line scan of MnCeO$_x$@Z5 (**g–i**). Reproduced with permission from ref. [30] Copyright 2020 Elsevier

which demonstrates that the as-prepared catalyst has micropores and mesopores together. The existence of additional mesopores might have a positive effect on the decomposition of NH$_4$HSO$_4$ [28, 37].

The HAADF-STEM and the energy-dispersive X-ray spectroscopy (EDX) elemental mapping technologies were adopted to confirm the component and dispersion of MnCeO$_x$ species in MnCeO$_x$@Z5. It is obvious that the Si, Al, Mn, and Ce species were homogeneously dispersed over one catalyst particle (Fig. 10.2b–e) and the Mn species were overlapped with the Ce species (Fig. 10.2f). To further confirm this confinement structure, the line scan measurement was performed, and the results are presented in Fig. 10.2h–i. The Mn and Ce signal was overlapped at the same positions, which indicates the Mn and Ce species were uniformly dispersed in MnCeO$_x$@Z5 at the same position.

The powder X-ray diffraction patterns (XRD) of confined MnCeO$_x$@Z5 catalyst and the related samples are displayed in Fig. 10.3. The diffract peaks about MnCeO$_x$ (Fig. 10.3a) are assigned to ceria with the cubic fluorite crystal structure. The pattern of MnCeOx@Al-SiO$_2$ precursor presented in Fig. 10.3b just has one broad diffraction peak between 15 and 35°, which should be attributed to the amorphous silica. After the dry-gel crystallization process, the as-prepared amorphous Al-SiO$_2$ shell was converted to the zeolite crystal shell. It is obvious that the typical diffraction peaks of the MFI crystal structure were appeared (Fig. 10.3c). The pattern of the hydro-type (H-Type) support (H-ZSM-5) and the supported catalyst (MnCeO$_x$/ZSM-5) was also presented in Fig. 10.3f and e, respectively, and the typical MFI crystal diffraction

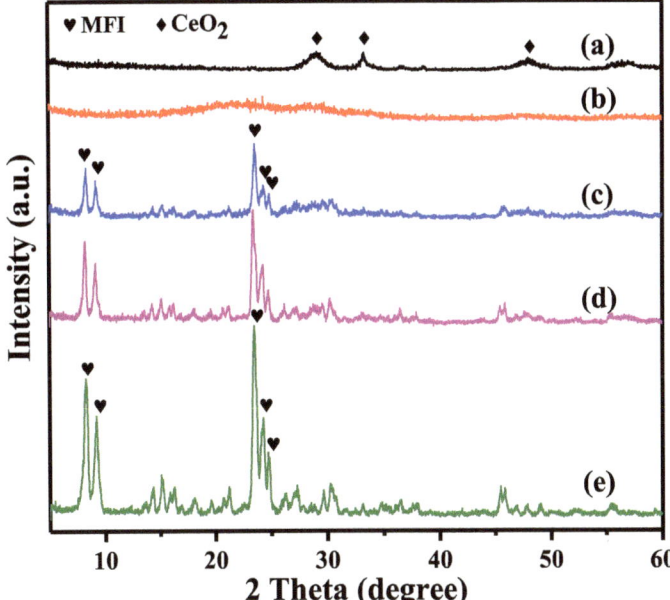

Fig. 10.3 XRD diffractograms from MnCeO$_x$@Z5 and the related samples: MnCeO$_x$ (**a**), MnCeO$_x$@Al-SiO$_2$ (**b**), MnCeO$_x$@Z5 (**c**), MnCeO$_x$/Z5 (**d**), and H-ZSM-5 (**e**). Reproduced with permission from ref. [30] Copyright 2020 Elsevier

peaks are evidently observed. In addition, the diffraction peaks about Mn-Ce mixed oxides species also existed (two theta degrees at ~ 28°), which demonstrates that Mn-Ce mixed oxide species are aggregated during the calcination process.

Scattered reflection Raman characterization was performed to investigate the surface metal-oxide structure of MnCeO$_x$@Z5, MnCeO$_x$/Z5, and MnCeO$_x$@Al-SiO$_2$ precursor, with the results presented in Fig. 10.4. For MnCeO$_x$/Z5, the scattering peak located at 454 cm^{-1} was the F$_{2g}$ vibration of the cubic fluorite structure of CeO$_2$, while the scattering peak located at -650 cm^{-1} is assigned to the oxygen defects [21]. While for MnCeO$_x$@Z5 and the MnCeO$_x$@Al-SiO$_2$ precursor, there was only one Raman scattering peak assigned to F$_{2g}$ mode. In addition, the F$_{2g}$ peak was very low; evidently, the signal about MnO$_x$ was even absent and should be attributed to most of the Mn species being enveloped by the zeolite shell. This result is well in line with the HRTEM analysis, which indirectly demonstrates that most of the MnCeO$_x$ species were enveloped in the inner of ZSM-5 zeolite.

The X-ray photoelectron spectroscopy (XPS) measurements were conducted to investigate the surface composition and the element valent state of MnCeO$_x$@Z5 and other related samples (Fig. 10.5). In Fig. 10.5a, two main peaks in the binding energy (BE) between 635 and 660 eV should be assigned to Mn 2p$_{3/2}$ and Mn 2p$_{1/2}$, respectively. The Mn 2p spectra over the enveloped MnCeO$_x$@Z5 and supported MnCeO$_x$/Z5 catalysts are assigned to the Mn^{2+} (BE = 641.3 eV), Mn^{3+} (BE =

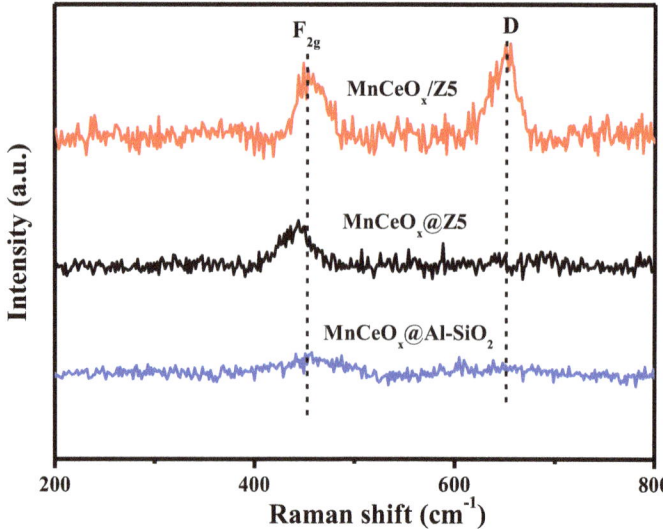

Fig. 10.4 Raman spectra of MnCeO$_x$@Z5, MnCeO$_x$/Z5, and MnCeO$_x$@Al-SiO$_2$. Reproduced with permission from ref. [30] Copyright 2020 Elsevier. Copyright 2020 Elsevier

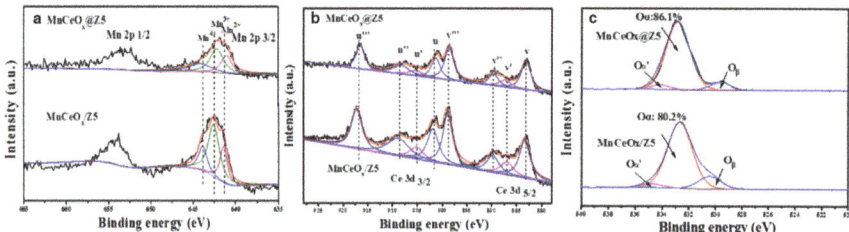

Fig. 10.5 XPS spectra of **a** Mn 2p, **b** Ce 3d, and **c** O1s orbitals of MnCeO$_x$@Z5 and MnCeO$_x$/Z5 catalysts. Reproduced with permission from ref. [30] Copyright 2020 Elsevier

642.6 eV), and the Mn^{4+} (BE = 643.9 eV) species. It is believed that the Mn^{4+} species can significantly impact the NO$_x$ through NH$_3$-SCR reaction and accelerate the oxidation of NO to NO$_2$ and increase the *de*NO$_x$ activity via the "fast SCR" reaction [20, 21, 38]. It should be noted that the Mn^{4+} content in the confined MnCeO$_x$@Z5 (26.8%) was evidently higher than that in the conventionally supported MnCeO$_x$/Z5 (20.2%) catalyst. Thereby, the higher molar ratio of Mn^{4+}/Mn^{3+} dispersed on MnCeO$_x$@Z5 might represent more active centers [39]. In Fig. 10.5b, there are eight peaks about Ce 3d over MnCeO$_x$/Z5. These peaks were labeled as V and U and were assigned to the spin–orbit of Ce 3d$_{5/2}$ and Ce 3d$_{3/2}$, respectively [40]. And the peaks labeled as u (BE = 901.1 eV), u″(BE = 908.2 eV), and u‴ (BE = 916.6 eV) were attributed to the Ce^{4+} 3d$_{3/2}$, and the peaks labeled as v (BE = 882.1 eV), v″(BE = 888.9 eV), and v‴ (BE = 898.2 eV) should be ascribed to the Ce^{4+} 3d$_{5/2}$. In addition, the existence of Ce³⁺ species was also verified by the peaks BE904.8 eV) and 886.2 eV [13, 40]. The pioneer researches have proved that the content of Ce³⁺ can be calculated by the formula of $Ce^{3+}/(Ce^{3+} + Ce^{4+})$ [21, 39]. The content of Ce³⁺ over MnCeOₓ@Z5 (13.4%) was also higher than that of MnCeOₓ/Z5 (12.1%). The higher content of Ce³⁺ could increase the amount of the surface active oxygen species and promote the adsorption and activation of reactants in the *de*NOₓ reaction [41–44]. Interestingly, the XPS results displayed a big difference between the enveloped MnCeOₓ@Z5 and the conventional supported MnCeOₓ/Z5 catalyst. In comparison with the supported MnCeOₓ/Z5 catalyst, the MnCeOₓ@Z5 catalyst displayed much weaker Mn 2p and Ce 3d signals. Therefore, it is verified that the Mn and Ce species were mainly located in the center of MnCeOₓ@Z5, and it is well in line with the aforementioned analysis [13, 26].

Figure 10.5c displays the O1s spectra of the three samples, and the fitted three peaks can be labeled as lattice oxygen (Oβ), chemisorbed oxygen (Oα), and adsorbed H2O (Oα'). It should be noted that the chemisorbed oxygen in MnCeOₓ@Z5 (86.1%) is relatively higher than that of MnCeOₓ/Z5 (80.2%) [8, 45]. Generally, the chemisorbed oxygen species are more active than that of the lattice oxygen due to their high reaction mobility. Therefore, the high Oα ratio can accelerate NO oxidized to NO₂, and increase the *de*NOₓ activities.

The surface acidity of catalysts was another main factor affecting the *de*NOₓ activity of NH₃-SCR reaction. The temperature-programmed desorption of ammonia (NH₃-TPD) was adopted to testify the acidity of the catalyst, and in the profile of MnCeOₓ@Z5 exists two ammonia desorption peaks between 100 and 350 °C(Fig. 10.6a). The desorption peaks at the low and middle temperatures were assigned to the weak and middle strength of acid sites. MnCeOₓ/Z5 exhibited similar desorption peaks and acid sites with MnCeOₓ@Z5. The acid sites over the precursor of MnCeOₓ@Al-SiO₂ were lower than that of enveloped MnCeOₓ@Z5 catalyst and the supported MnCeOₓ/Z5 catalyst, which can be applied to explain the poor *de*NOₓ

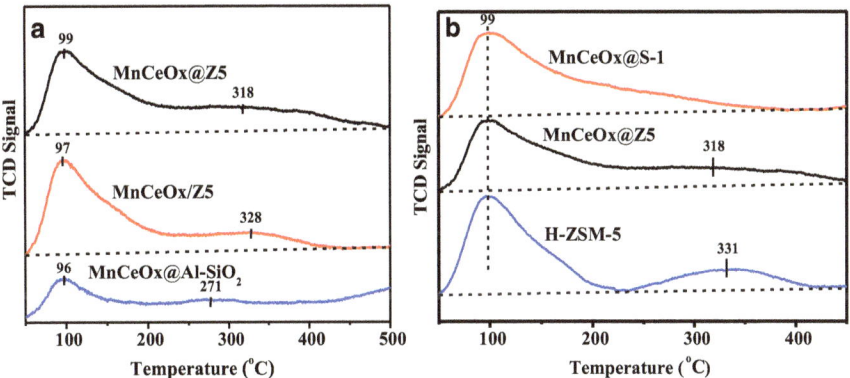

Fig. 10.6 NH₃-TPD profiles of MnCeOₓ@Z5 and related samples (**a, b**). Reproduced with permission from ref. [30] Copyright 2020 Elsevier

activity over $MnCeO_x@Al-SiO_2$ precursor. For comparison, the NH_3-TPD profiles of pure silica zeolite silicalite-1 ($MnCeO_x@S-1$), $MnCeO_x@Z5$, and H-ZSM-5 zeolite are displayed in (Fig. 10.6b). These three catalysts all have the same low-temperature ammonia desorption peak, which should be ascribed to the weak acid sites. Owing to the addition of Al species, $MnCeO_x@Z5$ has another desorption peak located at about 320 °C, which is absent over $MnCeO_x@S-1$, and it should be assigned to the middle acid sites. These results demonstrate that an aluminosilicate shell can induce the adsorption and activation of more NH_3 at a medium temperature [20, 46] and enhance its deNO$_x$ performance.

10.2.2 DeNOx Performances

The deNO$_x$ evaluations were performed between 150 and 400 °C, and the results are presented in Fig. 10.7a. The enveloped $MnCeO_x@Z5$ catalyst shows good activity at the temperature range of 200–400 °C, and the 80% NO_x conversion was obtained at about 200 °C. While the supported $MnCeO_x/Z5$ catalyst had relatively poor NO_x reduction performance. In contrast, the $MnCeO_x@Z5$ catalyst displayed higher deNOx activity at the low temperature, especially at a temperature range of 150–250 °C. The $MnCeO_x@Al-SiO_2$ precursor displayed nearly no activity during the whole measurement process, which is well in line with the previous characterization. The concentration of N_2O and the selectivity of N_2 over $MnCeO_x@Z5$ and $MnCeO_x/Z5$ catalysts are displayed in Fig. 10.7b. It should be noted that the N_2O concentration over $MnCeO_x@Z5$ catalyst was lower than that of $MnCeO_x/Z5$ catalyst in the reaction temperature range of 150–350 °C. Thus, $MnCeO_x@Z5$ had

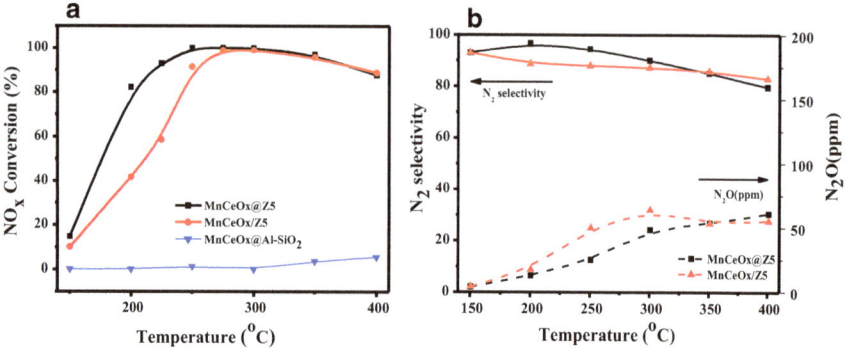

Fig. 10.7 DeNO$_x$ performance of $MnCeO_x@Z5$ and related catalysts: **a** NO_x conversion; **b** N_2 selectivity and N_2O concentration. Reaction conditions: $[NO_x] = [NH_3] = 500$ ppm, $[O_2] = 5$ vol.%, N_2 balance, total flow rate 100 mLmin^{-1}, and WHSV $= 60,000$ mLg$_{cat.}$$^{-1}$ h^{-1}. Reaction conditions: $[NO_x] = [NH_3] = 500$ ppm, $[O_2] = 5$ vol.%, N_2 balance, total flow rate 400 mL/min^{-1}, and WHSV $= 960,000$ mLh^{-1} g$_{cat.}$$^{-1}$, the NOx conversions were used below 15%. Reproduced with permission from ref. [30] Copyright 2020 Elsevier

better N$_2$ selectivity in the same temperature range. For comparison, silicalite-1 enveloped MnCeO$_x$ catalyst was also applied for the *de*NO$_x$ activity test. Unfortunately, for MnCeO$_x$@S-1, its deNOx activity was as low as 45% NO$_x$ at 200 °C due to its weak acid sites.

It has been established that a low amount of H$_2$O and SO$_2$ existed in real exhaust conditions, which can decrease the *de*NO$_x$ activity of the catalysts. Therefore, the resistance to H$_2$O and SO$_2$ over the MnCeO$_x$@Z5 and MnCeO$_x$/Z5 catalysts was conducted. Figure 10.8a displays the *de*NO$_x$ activity when H$_2$O and SO$_2$ were introduced. It can be observed that both catalysts showed stable *de*NO$_x$ performance in the absence of H$_2$O. For MnCeO$_x$/Z5, the introduction of H$_2$O led to a small but detectable decrease in *de*NO$_x$ activity; while without any decrease observed over MnCeO$_x$@Z5 catalyst, which could be attributed to the hydrophobic of the zeolite shell. It is evident that the *de*NO$_x$ activity over MnCeO$_x$@Z5 just decreased from 100 to 75% even after introducing 5% water and 100 ppm of SO$_2$ together. The zeolite shell seems to retard the deactivation of sulfur poisoning. In addition, when the SO$_2$ and H$_2$O are shut simultaneously, the removal rate can regenerate to about 85%. However, for MnCeO$_x$/Z5, its *de*NO$_x$ activity quickly decreased to about 40% after introduction of H$_2$O and SO$_2$. These results demonstrate that the zeolite shell can enhance the sulfur resistance during NH$_3$-SCR reaction. Then, the temperature of the stability test was further decreased to 200 °C to testify their low-temperature H$_2$O and SO$_2$ resistance (Fig. 10.8b). It is evident that the NO$_x$ conversion over MnCeO$_x$@Z5 was retained ~ 60% even when SO$_2$ and H2O were introduced together. However, under the same conditions, the *de*NO$_x$ activity over MnCeO$_x$/Z5 was decreased to 27%, which demonstrated that MnCeO$_x$@Z5 has superior SO$_2$ tolerance with the confinement structure.

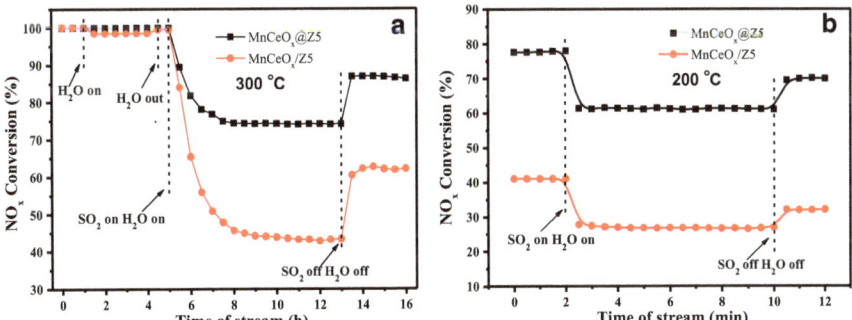

Fig. 10.8 The H$_2$O/SO$_2$ tolerance performances of MnCeO$_x$@Z5 and MnCeO$_x$/Z5 at 300 °C (**a**) and 200 °C (**b**). [SO$_2$] = 100 ppm (when used), [H2O] = 5 vol.% (when used), Reaction conditions: [NO$_x$] = [NH$_3$] = 500 ppm, [O$_2$] = 5 vol.%, N$_2$ balance, total flow rate 100 mLmin^{-1}, and WHSV = 60,000 mL h^{-1} g$_{cat.}$$^{-1}$ Reproduced with permission from ref. [30] Copyright 2020 Elsevier

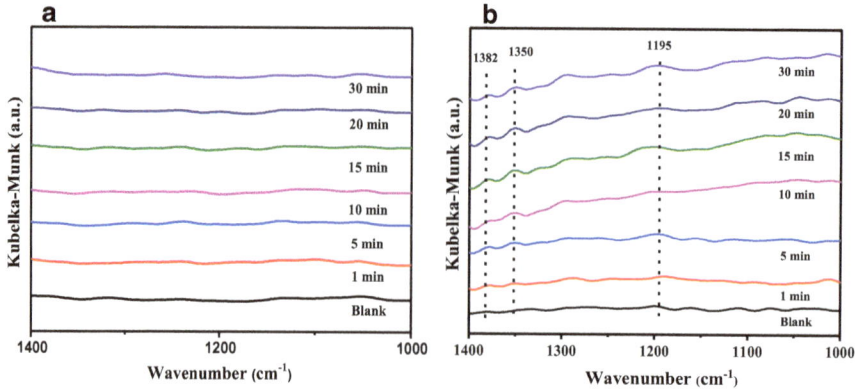

Fig. 10.9 The in situ*DRIFTS* results of 50 ppm SO_2 tolerance over $MnCeO_x$@Z5 (**a**) and $MnCeO_x$/Z5 (**b**) catalysts at 300 °C. Reproduced with permission from ref. [30] Copyright 2020 Elsevier

10.2.3 SO₂ Tolerance Mechanism Over MnCeOₓ@Z5

Figure 10.9a and bdisplays the in situ*DRIFTS* spectra of 50 ppm SO_2 adsorption over the enveloped $MnCeO_x$@Z5 and the supported $MnCeO_x$/Z5 catalysts at 300 °C. When SO_2 adsorbed over the $MnCeO_x$/Z5 catalyst, the peaks located at 1382 cm^{-1} and 1350 cm^{-1} should be attributed to the surface sulfate species (Fig. 10.9b) [47–50]. After 5 min of the introduction of SO_2, a broad absorption peak at 1195 cm^{-1} was observed, which can be assigned to metal sulfate species SO_4^{2-} species [51, 52]. For $MnCeO_x$@Z5 catalyst, it was also tested under the same adsorption conditions. Even after 30 min of introduction of SO_2, no sulfate species can be detected, including the surface sulfate species. These results evidently demonstrate that the confined $MnCeO_x$@Z5 catalyst had strong SO_2 tolerance, owing to the protective effect of ceria and the shielding effect of the zeolite shell. The opportunity for contact between the active components and SO_2 was reduced, which enabled the active components to avoid poisoning and sulfation. The conventional supported catalysts are prone to poisoning and sulfating the active sites because lots of active sites are exposed over the support surface. Thus, the enveloped catalysts exhibit superior sulfur tolerance and can be adopted to design and synthesize other high-performance deNO$_x$ or environment catalysts with good SO_2 tolerance.

10.2.4 Discussion

In this chapter, because the Mn-Ce oxides species were the main active components, the pathway of the NH$_3$-SCR of NOx reaction and the dominant factors over the enveloped $MnCeO_x$@Z5 catalyst should be similar to the conventional supported

MnCeO$_x$/Z5 catalyst. According to the aforementioned characterization results over the enveloped MnCeO$_x$@Z5 catalyst should be attributed to its stronger surface acidity and more surface active oxygen species. However, the deNO$_x$ activity over all these catalysts was obviously impacted by the existence of SO$_2$. The deposition of ammonia sulfates and the metal sulfates might be formed according to the following reaction steps (Eqs. 10.2–10.6):

$$NH_3(g) \rightarrow NH_3(a) \tag{10.2}$$

$$SO_2(g) \rightarrow SO_2(a) \tag{10.3}$$

$$4NH_3(a) + 2SO_2 + O_2 + 2H_2O \rightarrow 4(NH_4)_2SO_4 \tag{10.4}$$

$$MnO_2 + SO_2(a) \rightarrow MnSO_4 \tag{10.5}$$

$$2CeO_2 + 3SO_2(a) + O_2 \rightarrow Ce_2(SO_4)_3 \tag{10.6}$$

The MnCeO$_x$ active sites over MnCeO$_x$@Z5 were confined in the shell of ZSM-5 zeolite. Importantly, the ZSM-5 shell can provide the shielding effect to the active sites; in addition, the zeolite shell can also provide large amounts of acid sites, which can increase adsorption and activation of NH$_3$ species (Eq. 10.2). When SO$_2$ existed in the deNO$_x$ reaction, the activation of SO$_2$ occurred (Eq. 10.3). Thus, the active sites were exposed to the outer of the shell, which can react with the adsorbed SO$_2$ to form the inactive metal sulfates (Eqs. 10.5–10.6). However, The XPS and in situ DRIFTS results demonstrated that the adsorbed NH$_3$ species over the supported MnCeO$_x$/Z5 catalysts were deactivated by adsorbed SO$_2$; in addition, the formed ammonium sulfates can cause the pore blocking and cover the active sites (Eq. 10.4). These were the main reasons for catalyst deactivation over the conventional supported catalyst. Fortunately, with the shielding effect of the enveloped MnCeO$_x$@Z5 catalyst, the possibility for the contacting between the active sites and SO$_2$ decreased, which can inhibit the sulfating of the active sites (Eqs. 10.5–10.6). Additionally, the reported XPS and the in situ DRIFTS results [28, 37] demonstrated that the zeolite sheath can also prevent the reaction between the adsorbed SO$_2$ and NH$_3$ species (Eq. 10.4) to improve its SO$_2$ tolerance.

10.3 Conclusions

In conclusion, the zeolite enveloped MnCeO$_x$ mixed metal catalyst (MnCeO$_x$@Z5) with superior deNO$_x$ performance was discussed in this chapter [30]. Due to its high content of Mn^{4+} and Ce^{3+} species, and the synergy between acidic zeolite shell and the redox mixed oxides cores, MnCeO$_x$@Z5 exhibited good deNO$_x$ activity at low temperature. Furthermore, the shielding effect of zeolite sheath (ZSM-5) can effectively prevent the direct contact between the active site ($MnCeO_x$) and SO_2. It can evidently increase its SO_2/H_2O resistance. Therefore, the synergy strategy of the envelopment and the acid-redox adopted in this chapter can be applied to design other high-performance SO_2 resistance catalysts for deNO$_x$ and related air pollution control.

Though the deNO$_x$ catalysts have been intensively studied and commercialized for decades, the water and sulfur deactivation problems still exist. A large number of catalysts including various metal or metal oxides were developed for deNO$_x$. To further improve the water and sulfur tolerance of the deNO$_x$ catalysts, one of the most efficient strategies should be the re-construction of the conventional active catalysts, e.g., core-shell or yolk-shell confinement, nano-pore confinement, to inhibit or alleviate the water and SO_2 poisoning.

References

1. Peng C, Liang J, Peng H, Yan R, Liu W, Wang Z, Wu P, Wang X (2018) Design and synthesis of Cu/ZSM-5 catalyst via a facile one-pot dual-template strategy with controllable cu content for removal of NO$_x$. Ind Eng Chem Res 57:14967–14976
2. Damma D, Ettireddy PR, Reddy BM, Smirniotis PG (2019) A review of low temperature NH$_3$-SCR for removal of NOx. Catalysts 9:349
3. Li Y, Liu W, Yan R, Liang J, Dong T, Mi Y, Wu P, Wang Z, Peng H, An T (2020) Hierarchical three-dimensionally ordered macroporous Fe-V binary metal oxide catalyst for low temperature selective catalytic reduction of NOx from marine diesel engine exhaust. Appl Catal B: Environ 268, 118455
4. Tang C, Zhang H, Dong L (2016) Ceria-based catalysts for low-temperature selective catalytic reduction of NO with NH$_3$. Catal Sci Technol 6:1248–1264
5. Han L, Cai S, Gao M, Hasegawa JY, Wang P, Zhang J, Shi L, Zhang D (2019) Selective catalytic reduction of NOx with NH$_3$ by using novel catalysts: state of the art and future prospects. Chem Rev 119:10916–10976
6. Arfaoui J, Ghorbel A, Petitto C, Delahay G (2018) Novel V$_2$O$_5$-CeO$_2$-TiO$_2$-SO$_4{}^{2-}$ nanostructured aerogel catalyst for the low temperature selective catalytic reduction of NO by NH$_3$ in excess O$_2$. Appl Catal B 224:264–275
7. Cai S, Hu H, Li H, Shi L, Zhang D (2016) Design of multi-shell Fe$_2$O$_3$@MnO(x)@CNTs for the selective catalytic reduction of NO with NH3: improvement of catalytic activity and SO$_2$ tolerance. Nanoscale 8:3588–3598
8. Shan W, Liu F, He H, Shi X, Zhang C (2012) A superior Ce-W-Ti mixed oxide catalyst for the selective catalytic reduction of NO$_x$ with NH$_3$. Appl Catal B 115–116:100–106
9. Du T, Qu H, Liu Q, Zhong Q, Ma W (2015) Synthesis, activity and hydrophobicity of Fe-ZSM-5@silicalite-1 for NH$_3$-SCR. Chem Eng J 262:1199–1207
10. Wijayanti K, Andonova S, Kumar A, Li J, Kamasamudram K, Currier NW, Yezerets A, Olsson L (2015) Impact of sulfur oxide on NH$_3$-SCR over Cu-SAPO-34. Appl Catal B 166–167:568–579
11. Wang T, Liu H, Zhang X, Liu J, Zhang Y, Guo Y, Sun B (2018) Catalytic conversion of NO assisted by plasma over Mn-Ce/ZSM5-multi-walled carbon nanotubes composites: Investigation of acidity, activity and stability of catalyst in the synergic system. Appl Surf Sci 457:187–199
12. Chang H, Li J, Chen X, Ma L, Yang S, Schwank JW, Hao J (2012) Effect of Sn on MnOx–CeO$_2$ catalyst for SCR of NOx by ammonia: enhancement of activity and remarkable resistance to SO$_2$. Catal Commun 27:54–57

13. Chen X, Wang P, Fang P, Ren T, Liu Y, Cen C, Wang H, Wu Z (2017) Tuning the property of Mn-Ce composite oxides by titanate nanotubes to improve the activity, selectivity and SO2/H2O tolerance in middle temperature NH$_3$-SCR reaction. Fuel Process Technol 167:221–228

14. Qi G, Yang RT, Chang R (2004) MnO$_x$-CeO$_2$ mixed oxides prepared by co-precipitation for selective catalytic reduction of NO with NH$_3$ at low temperatures. Appl Catal B 51:93–106

15. Han L, Gao M, Feng C, Shi L, Zhang D (2019) Fe$_2$O$_3$-CeO$_2$@Al$_2$O$_3$ nanoarrays on Al-Mesh as SO$_2$-tolerant monolith catalysts for NO$_x$ reduction by NH3. Environ Sci Technol 53:5946–5956

16. Han L, Gao M, Hasegawa JY, Li S, Shen Y, Li H, Shi L, Zhang D (2019) SO$_2$-tolerant selective catalytic reduction of NO$_x$ over Meso-TiO$_2$@Fe$_2$O$_3$@Al$_2$O$_3$ metal-based monolith catalysts. Environ Sci Technol 53:6462–6473

17. Kang L, Han L, He J, Li H, Yan T, Chen G, Zhang J, Shi L, Zhang D (2019) Improved NO x reduction in the presence of SO2 by using Fe2O3-promoted halloysite-supported CeO$_2$-WO$_3$ catalysts. Environ Sci Technol 53:938–945

18. Kijlstra WS, Biervliet M, Poels EK, Bliek A (1998) Deactivation by SO$_2$ of MnO$_x$/Al$_2$O$_3$ catalysts used for the selective catalytic reduction of NO with NH$_3$ at low temperatures. Appl Catal B: Environ 16, 327–337

19. Xu L, Wang C, Chang H, Wu Q, Zhang T, Li J (2018) New insight into SO$_2$ poisoning and regeneration of CeO$_2$-WO$_3$/TiO$_2$ and V$_2$O$_5$-WO$_3$/TiO$_2$ catalysts for low-temperature NH$_3$-SCR. Environ Sci Technol 52:7064–7071

20. Jiang L, Liu Q, Ran G, Kong M, Ren S, Yang J, Li J (2019) V$_2$O$_5$-modified Mn-Ce/AC catalyst with high SO$_2$ tolerance for low-temperature NH$_3$-SCR of NO. Chem Eng J 370:810–821

21. Yao X, Chen L, Cao J, Chen Y, Tian M, Yang F, Sun J, Tang C, Dong L (2019) Enhancing the deNO performance of MnO /CeO$_2$-ZrO$_2$ nanorod catalyst for low-temperature NH$_3$-SCR by TiO$_2$ modification. Chem Eng J 369:46–56

22. Dedecek J, Balgová V, Pashkova V, Klein P, Wichterlová B (2012) Synthesis of ZSM-5 zeolites with defined distribution of Al atoms in the framework and multinuclear MAS NMR analysis of the control of Al distribution. Chem Mater 24:3231–3239

23. de Oliveira ML, Silva CM, Moreno-Tost R, Farias TL, Jiménez-López A, Rodríguez-Castellón E (2009) A study of copper-exchanged mordenite natural and ZSM-5 zeolites as SCR–NOx catalysts for diesel road vehicles: Simulation by neural networks approach. Appl Catal B: Environ 88, 420–429

24. Zhang Q, Chen G, Wang Y, Chen M, Guo G, Shi J, Luo J, Yu J (2018) High-quality single-crystalline MFI-type nanozeolites: a facile synthetic strategy and MTP catalytic studies. Chem Mater 30:2750–2758

25. Dai Q, Bai S, Lou Y, Wang X, Guo Y, Lu G (2016) Sandwich-like PdO/CeO$_2$ nanosheet@HZSM-5 membrane hybrid composite for methane combustion: self-redispersion, sintering-resistance and oxygen, water-tolerance. Nanoscale 8:9621–9628

26. Peng H, Rao C, Zhang N, Wang X, Liu W, Mao W, Han L, Zhang P, Dai S (2018) Confined ultra-thin Pd-Ce nanowires with outstanding moisture and SO$_2$ tolerance in methane combustion. Angew Chem Int Ed Engl 57:8953–8957

27. Ran X, Li M, Wang K, Qian X, Fan J, Sun Y, Luo W, Teng W, Zhang WX, Yang J (2019) Spatially confined tuning the interfacial synergistic catalysis in mesochannels toward selective catalytic reduction. ACS Appl Mater Interfaces 11:19242–19251

28. Guo K, Fan G, Gu D, Yu S, Ma K, Liu A, Tan W, Wang J, Du X, Zou W, Tang C, Dong L (2019) Pore size expansion accelerates ammonium bisulfate decomposition for improved sulfur resistance in low-temperature NH$_3$-SCR. ACS Appl Mater Interfaces 11:4900–4907

29. Carja G, Kameshima Y, Okada K, Madhusoodana CD (2007) Mn–Ce/ZSM5 as a new superior catalyst for NO reduction with NH$_3$. Appl Catal B 73:60–64

30. Yan R, Lin S, Li Y, Liu W, Mi Y, Tang C, Wang L, Wu P, Peng H (2020) Novel shielding and synergy effects of Mn-Ce oxides confined in mesoporous zeolite for low temperature selective catalytic reduction of NO$_x$ with enhanced SO$_2$/H$_2$O tolerance. J Hazard Mater 396:122592

31. Peng H, Dong T, Zhang L, Wang C, Liu W, Bao J, Wang X, Zhang N, Wang Z, Wu P, Zhang P, Dai S (2019) Active and stable Pt-Ceria nanowires@silica shell catalyst: design, formation mechanism and total oxidation of CO and toluene. Appl Catal B 256:117807

32. Liu W, Li L, Zhang X, Wang Z, Wang X, Peng H (2018) Design of Ni-ZrO_2@SiO_2 catalyst with ultra-high sintering and coking resistance for dry reforming of methane to prepare syngas. J CO_2 Util 27, 297–307

33. Wang L, Wang G, Zhang J, Bian C, Meng X, Xiao FS (2017) Controllable cyanation of carbon-hydrogen bonds by zeolite crystals over manganese oxide catalyst. Nat Commun 8:15240

34. Ma R, Wang L, Wang S, Wang C, Xiao F-S (2017) Eco-friendly photocatalysts achieved by zeolite fixing. Appl Catal B 212:193–200

35. Zhang J, Wang L, Shao Y, Wang Y, Gates BC, Xiao FS (2017) A Pd@Zeolite catalyst for nitroarene hydrogenation with high product selectivity by sterically controlled adsorption in the zeolite micropores. Angew Chem Int Ed Engl 56:9747–9751

36. Gu J, Zhang Z, Hu P, Ding L, Xue N, Peng L, Guo X, Lin M, Ding W (2015) Platinum nanoparticles encapsulated in MFI zeolite crystals by a two-step dry gel conversion method as a highly selective hydrogenation catalyst. ACS Catalysis 5:6893–6901

37. Peng C, Yan R, Peng H, Mi Y, Liang J, Liu W, Wang X, Song G, Wu P, Liu F (2019) One-pot synthesis of mesoporous ZSM-5 plus Cu ion-exchange: Enhanced NH_3-SCR performance on Cu-ZSM-5 with hierarchical pore structures, J Hazard Mater 121593

38. Boningari T, Pappas DK, Ettireddy PR, Kotrba A, Smirniotis PG (2015) Influence of SiO_2 on M/TiO_2 (M = Cu, Mn, and Ce) formulations for low-temperature selective catalytic reduction of NO_x with NH_3: surface properties and key components in relation to the activity of NOx reduction. Ind Eng Chem Res 54:2261–2273

39. Shen Q, Zhang L, Sun N, Wang H, Zhong L, He C, Wei W, Sun Y (2017) Hollow MnO_x-CeO_2 mixed oxides as highly efficient catalysts in NO oxidation. Chem Eng J 322:46–55

40. Liu H, Fan Z, Sun C, Yu S, Feng S, Chen W, Chen D, Tang C, Gao F, Dong L (2018) Improved activity and significant SO_2 tolerance of samarium modified CeO_2-TiO_2 catalyst for NO selective catalytic reduction with NH_3. Appl Catal B: Environ 244, 671–683

41. Devaiah D, Tsuzuki T, Boningari T, Smirniotis PG, Reddy BM (2015) $Ce_{0.80}M0.12Sn_{0.08}O_{2-\delta}$ (M = Hf, Zr, Pr, and La) ternary oxide solid solutions with superior properties for CO oxidation. RSC Adv 5, 30275–30285

42. Devaiah D, Jampaiah D, Saikia P, Reddy BM (2014) Structure dependent catalytic activity of $Ce_{0.8}Tb_{0.2}O_{2-\delta}$ and TiO_2 supported $Ce_{0.8}Tb_{0.2}O_{2-\delta}$ solid solutions for CO oxidation. J Ind Eng Chem 20, 444–453

43. Boningari T, Somogyvari A, Smirniotis PG (2017) Ce-based catalysts for the selective catalytic reduction of NO_x in the presence of excess oxygen and simulated diesel engine exhaust conditions. Ind Eng Chem Res 56:5483–5494

44. Chen L, Wang Q, Wang X, Cong Q, Ma H, Guo T, Li S, Li W (2020) High-performance CeO_2/halloysite hierarchical catalysts with promotional redox property and acidity for the selective catalytic reduction of NO with NH_3. Chem Eng J 390:124251

45. Wang P, Sun H, Quan X, Chen S (2016) Enhanced catalytic activity over MIL-100(Fe) loaded ceria catalysts for the selective catalytic reduction of NO_x with NH(3) at low temperature. J Hazard Mater 301:512–521

46. Han JW, Park JS, Choi MS, Lee H (2017) Uncoupling the size and support effects of Ni catalysts for dry reforming of methane. Appl Catal B 203:625–632

47. Luo T, Gorte RJ (2004) Characterization of SO_2-poisoned ceria-zirconia mixed oxides. Appl Catal B 53:77–85

48. Xie T, Zhao X, Zhang J, Shi L, Zhang D (2015) Ni nanoparticles immobilized Ce-modified mesoporous silica via a novel sublimation-deposition strategy for catalytic reforming of methane with carbon dioxide. Int J Hydrogen Energy 40:9685–9695

49. Kylhammar L, Carlsson P-A, Ingelsten HH, Grönbeck H, Skoglundh M (2008) Regenerable ceria-based SOx traps for sulfur removal in lean exhausts. Appl Catal B 84:268–276

50. Kylhammar L, Carlsson P-A, Skoglundh M (2011) Sulfur promoted low-temperature oxidation of methane over ceria supported platinum catalysts. J Catal 284:50–59
51. Zhuang K, Zhang Y-P, Huang T-J, Lu B, Shen K (2017) Sulfur-poisoning and thermal reduction regeneration of holmium-modified Fe-Mn/TiO$_2$ catalyst for low-temperature SCR. J Fuel Chem Technol 45:1356–1364
52. Yang S, Guo Y, Chang H, Ma L, Peng Y, Qu Z, Yan N, Wang C, Li J (2013) Novel effect of SO$_2$ on the SCR reaction over CeO$_2$: mechanism and significance. Appl Catal B 136–137:19–28

Chapter 11
Core–Shell Structured Zeolite Catalysts with Enhanced Shape Selectivity

Koji Miyake and Norikazu Nishiyama

11.1 Introduction

Zeolites are basically defined as crystalline microporous aluminosilicates while other types of zeolites, microporous crystalline materials formed by TO_4 (T = Si, Ge, Al, P, etc.) tetrahedral units, have been developed [1, 2]. Zeolites show unique shape selectivity due to their uniform micropore structure. Focusing on the chemical states of zeolites, zeolites are composed of SiO_4 and $[AlO_4]^-$ tetrahedrally, and cations are located near negative charged Al to balance the overall electronic charge [3–5]. These cations can be exchanged by other cations. The ion exchange capacity of zeolites depends on Al content in the zeolite framework. Zeolites work as solid Brønsted acid catalysts when the cations of zeolites are protons. In these functions, shape selectivity is the most unique property in zeolites and one of the most important factors determining their catalytic performance.

Shape selectivity is derived from ordered micropores of zeolites. Obviously, shape selectivity works well in micropores of zeolites. However, active sites such as acid sites are often located on the external surface of zeolites where shape selectivity does not work, which results in decreasing selectivity. Production of *p*-xylene over MFI-type zeolites is the most typical example. MFI-type zeolite has three-dimensional medium micropores (ca. 0.55 nm). The ratio of diffusivities of xylene isomers is 10: 1: 1000 for *o*: *m*: *p*, respectively, in MFI zeolitic micropores [6]. Therefore, xylene isomers are formed via reactions in MFI zeolitic micropores, and then *p*-xylene

K. Miyake (✉)
Department of Applied Chemistry and Biochemical Engineering, Faculty of Engineering, Shizuoka University, 3-5-1 Johoku Naka-ku, Hamamatsu 432-8561, Japan
e-mail: miyake.koji@shizuoka.ac.jp

N. Nishiyama
Division of Chemical Engineering, Graduate School of Engineering Science, Osaka University, 1-3 Machikaneyama, Toyonaka, Osaka 560-8531, Japan
e-mail: nisiyama@cheng.es.osaka-u.ac.jp

© The Author(s), under exclusive license to Springer Nature Singapore Pte Ltd. 2021 181
H. Yamashita and H. Li (eds.), *Core-Shell and Yolk-Shell Nanocatalysts*,
Nanostructure Science and Technology,
https://doi.org/10.1007/978-981-16-0463-8_11

selectively diffuses out of the zeolite crystal, leading to highly selective production of p-xylene. However, the isomerization of p-xylene occurs due to the existence of acid sites on the external surface. As a result, the selectivity of p-xylene decreases. The reaction on the external surface of zeolites is at high conversion or high contact time, which means that there is a trade-off relationship between conversion or yield and selectivity of p-xylene [7].

As mentioned above, it is obvious that active sites must be located in zeolite crystals to improve shape selectivity. In addition, shape selectivity should improve without decreasing porosity and hydrothermal stability. In our work, all silica MFI-type zeolite (silicalite-1) with no active sites is epitaxially overgrown on the core of MFI-type zeolite with various active sites. Silicalite-1 zeolite shell is continuously connected to the core MFI zeolite. We believe that silicalite-1-coated core-shell zeolites have accessible porosity and mechanical durability. In addition, we have introduced various active sites derived from protons, metal ions, and metal nanoparticles into core zeolite. The multi-functional core-shell zeolite catalysts have been developed for shape-selective one-pot production and shape-selective hydrogenation, respectively. In this chapter, we introduce three types of core-shell structured zeolite catalysts as shown below [7–9].

11.2   Core–Shell ZSM-5/Silicalite-1 Catalyst for Shape-Selective Production of Aromatics from C_{1-3} Feedstocks

Firstly, core-shell zeolite composite crystals consisting of an MFI structure were prepared by a zeolite overgrowth method. The core and shell zeolites are ZSM-5 with acid sites and silicalite-1 (Al-free MFI zeolite) without acid sites, respectively. The structured core-shell ZSM-5/silicalite-1 catalyst was used for the reactions to produce benzene, toluene, and p-xylene (BTpX) from five C_{1-3} feedstocks (methanol, dimethyl ether: DME, ethanol, ethylene, and acetone).

To examine the surface coverage of ZSM-5 crystals with silicalite-1, isomerization of o-xylene was performed. The o-xylene conversions on ZSM-5 and ZSM-5/slicalite-1 were 46% and <5%, respectively. The ZSM-5/silicalite-1 catalyst showed a much lower activity for o-xylene conversion compared with the uncoated ZSM-5. The diffusivity of o-xylene is estimated to be 1/100 of that of p-xylene because of its larger molecular size [6]. o-Xylene seems to react near the external surface of ZSM-5 in a diffusion control region. Thus, this result indicates that the surface of ZSM-5/silicalite-1 was almost covered with a silicalite-1 layer and there are few acid sites on the external surface of ZSM-5/silicalite-1. The results of reactions using various reactants over uncoated ZSM-5 and silicalite-1-coated ZSM-5 were listed in Table 11.1. The conversions of the reactants were the same (89–99 C-mol%) for both uncoated and coated catalysts, which is a different trend from the results of the o-xylene conversion. These C_{1-3} feedstocks can easily diffuse into the

Table 11.1 Results of reaction tests on C$_{1-3}$ feedstocks to hydrocarbons over uncoated ZSM-5 and core-shell ZSM-5/silicalite-1

Reactant	Conv (C-mol%)	BT*p*X Sel (C-mol%)[a]	C$_{9+}$ Sel (C-mol%)[b]	*para*-Sel (C-mol%)[c]
MeOH	>99 (>99)	57 (28)	2 (11)	94 (23)
DME	>99 (>99)	61 (34)	6 (13)	99 (42)
EtOH	>99 (>99)	53 (29)	4 (12)	99 (28)
Ethylene	89 (90)	47 (24)	2 (18)	99 (52)
Acetone	>99 (>99)	71 (42)	6 (16)	98 (32)

*The values in the parentheses represent the results of uncoated ZSM-5
[a]The values represent the selectivity to benzene, toluene, and *para*-xylene
[b]The values represent the selectivity to aromatics compounds that have 9 or more carbon atoms
[c]The values represent the selectivity to *para*-xylene in xylene isomers

MFI pores even after the silicalite-1 coating because of the smaller molecular sizes. The diffusion resistance for the C$_{1-3}$ feedstocks through the silicalite-1 layer seems to be negligible. The decrease of *o*-xylene conversion as mentioned above cannot be explained by an increase in diffusion resistance for *o*-xylene in the silicalite-1 layer. These results suggest that silicalite-1 pores are directly connected to the pores of ZSM-5. The accessibility into core ZSM-5 was not decreased by the silicalite-1 coating. The selectivity to total BT*p*X was significantly increased up to 47–71 C-mol% by the silicalite-1 coating. Instead, the selectivity to bulky molecules such as C$_{9+}$ aromatics was decreased. Furthermore, *para*-selectivity (*p*-xylene selectivity in xylenes) was dramatically improved to 94–99 C-mol% by the silicalite-1 coating. These values are much higher than the thermal equilibrium value (23 C-mol%). The difference of selectivity over ZSM-5 and ZSM-5/silicalite-1 lies on the existence of the external acid sites where shape selectivity cannot work. When the external acid sites exist, xylene isomerization and successive reactions to C$_{9+}$ aromatics occur there. By contrast, those reactions do not occur on the surface of ZSM-5/silicalite-1 crystals. As a result, ZSM-5/silicalite-1 showed higher BT*p*X selectivity and lower C$_{9+}$ aromatics selectivity even at high conversion conditions.

11.3   Zn Ion-Doped Core–Shell ZSM-5/Silicalite-1 Catalyst for Enhanced Production of *p*-Xylene from Methanol

In previous Sect. 11.2, silicalite-1-coated catalysts showed much higher catalytic performance on various reaction systems [8]. In this Sect. 11.3, to improve *p*-xylene yield, we designed Zn ion-doped core-shell zeolite (Zn/ZSM-5/silicalite-1), which consists of Zn ion-doped MFI zeolite with acid sites (Zn/ZSM-5) in the core and MFI zeolite without acid sites (silicalite-1) in the shell. Then, we evaluated the catalytic property of Zn/ZSM-5/silicalite-1 on methanol to *p*-xylene (MT*p*X).

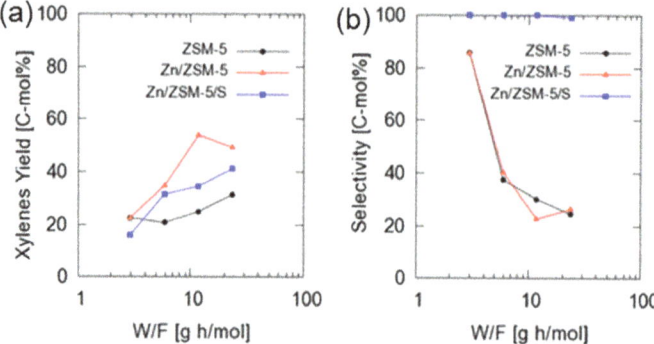

Fig. 11.1 The changes of **a** xylenes yield and **b** *para*-selectivity with contact time (*W/F*). Reprinted with permission from Ref. [7]. Copyright 2016 Elsevier

Figure 11.1 shows the results of the reaction test (MT*p*X). Xylenes yield increased with the increase of contact time (*W/F* values) on all samples. Then, xylenes yield increased by doping Zn, which was in accord with previous works [10–12] Uncoated samples (ZSM-5 and Zn/ZSM-5) showed a trade-off relationship regarding xylenes yield and *para*-xylene selectivity in xylenes (*para*-sel.). Meanwhile, as for Zn/ZSM-5/S, high *para*-sel. (>99%) maintained even in the case of high *W/F* values (high xylenes yield), which indicated that the undesired reactions on the external surface were inhibited even in the case of high *W/F* values (high xylenes yield). Moreover, in an optimum condition, Zn/ZSM-5/S showed *para*-xylene yield (40.7 C-mol%) with high *para*-sel. (>99%).

11.4 Pt Nanoparticle Catalyst Encapsulated in Silicalite-1 for Shape-Selective Hydrogenation

In previous Sects. 11.2 and 11.3, we have introduced core-shell structured MFI zeolite catalysts with Brønsted acid and metal ion. In this Sect. 11.4, we introduced metal nanoparticle catalyst into core-shell structured zeolite to expand the utilization of core-shell structured zeolite catalysts. In particular, we have fabricated Pt nanoparticles encapsulated in core-shell single crystal-like silicalite-1 zeolite (CS-Pt/s-1) by an epitaxial crystalline overgrowth method, and we discuss the shape selectivity of the Pt-encapsulated zeolite catalyst on hydrogenation.

Firstly, to confirm the encapsulation of Pt nanoparticles, we performed hydrogenation of mesitylene. The molecular size of mesitylene was larger than MFI zeolitic micropores [13]. Therefore, hydrogenation of mesitylene does not occur if Pt nanoparticles are completely covered with silicalite-1 layers. The conversion of mesitylene over Pt/s-1 was increased with a rise of reaction temperature and reached almost 100% at 373 K, while mesitylene did not react over CS-Pt/s-1 even

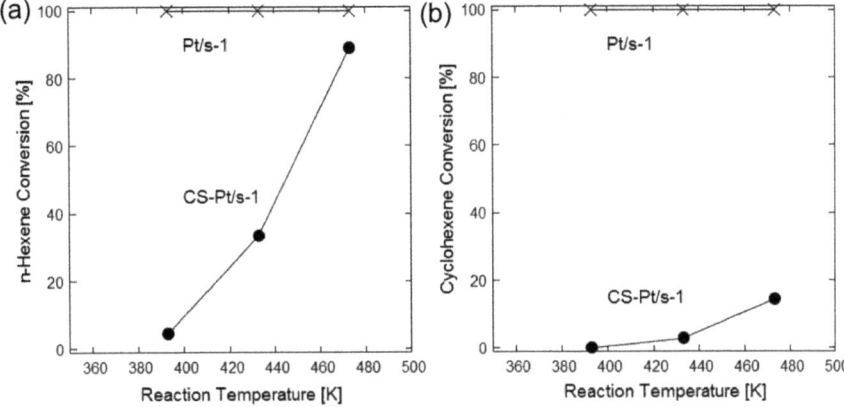

Fig. 11.2 Conversion of **a** *n*-hexene and **b** cyclohexene on hydrogenation over Pt/s-1 and CS-Pt/s-1. Reprinted with permission from Ref. [9]. Copyright 2018 Elsevier

at 373 K. This result exhibits that Pt nanoparticles were successfully encapsulated by silicalite-1 layers.

To investigate the shape selectivity of CS-Pt/s-1, we also performed hydrogenation of *n*-hexene and cyclohexene as shown in Fig. 11.2. The catalytic activity of CS-Pt/s-1 was lower than that of Pt/s-1. One possible reason is that Pt content of CS-Pt/s-1 was lower than that of Pt/s-1 due to mass gain of silicalite-1 layers. Another possible reason is the diffusion resistance of *n*-hexene through the silicalite-1 layers. On the Pt/s-1 catalyst, the reaction occurs without mass transfer resistance, but the reaction on CS-Pt/s-1 should be governed by diffusion control even for *n*-hexene. In the diffusion control region, we can expect the effect of shape selectivity of zeolite. CS-Pt/s-1 showed a higher catalytic activity on hydrogenation of *n*-hexene compared to cyclohexene. The molecular size of cyclohexene was larger than that of *n*-hexene. Silicalite-1 layers made the difference of diffusivity between *n*-hexene and cyclohexene, which results in selective hydrogenation.

11.5 Summary and Outlook

This chapter deals with three types of core-shell zeolite catalysts coated with inert silicalite-1, and the core-shell structured zeolite catalysts showed much high shape selectivity on various reactions. In other words, core-shell structure coated with inert silicalite-1 is an effective strategy to maximize shape selectivity. In addition, various active sites such as Brønsted acid, metal ion, and metal nanoparticle catalysts can be introduced into the core-shell structured zeolite catalysts, which enables the wide utilization of the core-shell structured zeolite catalysts for shape-selective transformation on various reaction systems. Furthermore, this concept can be applied to other types of zeolites as well as MFI-type zeolite. Shape selectivity can be controlled using different types of zeolites. In the future, many types of core-shell structured zeolite catalysts with various active sites and types of zeolites are expected to be developed, leading to more advanced shape-selective transformation.

References

1. Li Y, Yu J (2014) New stories of zeolite structures: their descriptions, determinations, predictions, and evaluations. Chem Rev 114:7268–7316
2. Roth WJ, Nachtigall P, Morris RE, Čejka J (2014) Two dimensional zeolites: current status and perspectives. Chem Rev 114:4807–4837
3. Uytterhoeven JB, Christner LG, Hall WK (1965) Studies of the hydrogen held by solids. VIII. The decationated zeolites. J Phys Chem 69:2117–2126
4. Haag WO, Lago RM, Weisz PB (1984) The active site of acidic aluminosilicate catalysts. Nature 309:589–591
5. Mortier WJ, Sauer J, Lercher JA, Noller H (1984) Bridging and terminal hydroxyls. A structural chemical and quantum chemical discussion. J Phys Chem 88:905–912
6. Mirth G, Cejka J, Lercher JA (1993) Transport and isomerization of xylenes over HZSM-5 zeolites. J Catal 139:24–33
7. Miyake K, Hirota Y, Ono K, Uchida Y, Tanaka S, Nishiyama N (2016) Direct and selective conversion of methanol to paraxylene over Zn ion doped ZSM-5/silicalite-1 core-shell zeolite catalyst. J Catal 342:63–66
8. Miyake K, Hirota Y, Ono K, Uchida Y, Nishiyama N (2016) Selective production of benzene, toluene and p-Xylene (BTpX) from various C$_{1-3}$ feedstocks over ZSM-5/silicalite-1 core-shell zeolite catalyst. Chem Select 1:967–969
9. Miyake K, Inoue R, Nakai M, Hirota Y, Uchida Y, Tanaka S, Miyamoto M, Nishiyama N (2018) Fabrication of Pt nanoparticles encapsulated in single crystal like silicalite-1 zeolite as a catalyst for shape-selective hydrogenation of C$_6$ olefins. Microporous Mesoporous Mater 271:156–159
10. Wang N, Qian W, Shen K, Su C, Wei F (2016) Bayberry-like ZnO/MFI zeolite as high performance methanol-to-aromatics catalyst. Chem Commun 52(10):2011–2014
11. Ni Y, Sun A, Wu X, Hai G, Hu J, Li T, Li G (2011) The preparation of nano-sized H[Zn, Al]ZSM-5 zeolite and its application in the aromatization of methanol. Microporous Mesoporous Mater 143:435–442
12. Niu XJ, Gao J, Miao Q, Dong M, Wang GF, Fan WB, Qin ZF, Wang JG (2014) Influence of preparation method on the performance of Zn-containing HZSM-5 catalysts in methanol-toaromatics. Microporous Mesoporous Mater 197:252–261
13. Csicsery SM (1984) Shape-selective catalysis in zeolites. Zeolites 4:202–213

Chapter 12
Core–Shell Structured Zeolite Catalysts with Minimal Defects for Improvement of Shape Selectivity

Masaki Okamoto

12.1 Introduction

Zeolite catalysts are widely used in the chemical industry because of their high catalytic activity and high product selectivity. Zeolites have pores with uniform pore size, which originated from their crystal structures, causing high surface area leading to high catalytic activity. Moreover, the uniform size of the pores brings about shape selectivity, i.e., reactant, transition state, and product selectivities [1], leading to obtaining the desirable products selectively. However, high product selectivity can be attained only when the reactions proceed inside the pores and/or at pore mouths. The catalysis of the active sites on the external surface of the zeolite particles results in nonselective (undesirable) reactions.

To prevent reactions to occur on the external surface, passivation of the active sites, which are mostly acid sites, on the external surface is generally used. Many efforts to passivate the external surface of zeolites have been reported, especially for MFI-type zeolite [2–20]. This chapter reviews the passivation of the external surface of MFI- and TON-type zeolites by crystal growth over aluminum-containing zeolites (core) with aluminum-free zeolites (shell) to form zeolites with a core–shell structure (core–shell zeolites). Generally, zeolites are synthesized using hydroxide ion as a mineralizer, and the obtained zeolites are small crystallites with intercrystalline voids. When fluoride ion is used as a mineralizer, large and single-like crystals with few defects are formed [21–23]. To completely passivate the active sites on the external surface, no defects in the shell and epitaxial growth of the core with the shell are necessary. To achieve them, the use of fluoride ion as a mineralizer is appropriate for the formation of core–shell zeolites (Fig. 12.1), and to evaluate the passivation

M. Okamoto (✉)
Research and Education Center for Natural Sciences, Keio University, 4-1-1 Hiyoshi, Kohoku-ku, Yokohama, Kanagawa 223-8521, Japan
e-mail: okamoto.m@keio.jp

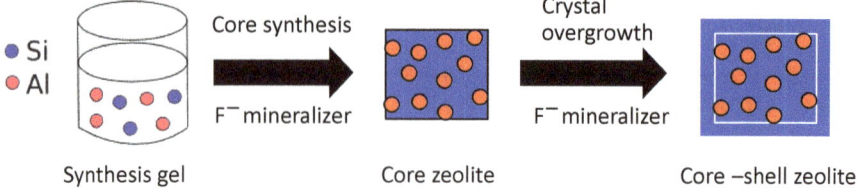

Fig. 12.1 Synthesis of the core–shell zeolite by crystal overgrowth in the presence of fluoride ion

effect, the catalytic activity test is suitable, especially on the analysis of the product selectivity.

12.2 *p*-Xylene Synthesis by Toluene Methylation with Methanol Catalyzed by MFI-Type Zeolite

Aluminum-containing MFI-type zeolite (ZSM-5) is industrially used as a catalyst with high product selectivity. To increase the selectivity, there are numerous reports on the passivation of the external surface of MFI-type zeolite catalysts [2–20]. Chemical vapor deposition of silicon alkoxide for covering the external surface of an MFI-type zeolite with amorphous silica is one of the effective passivation methods [2–8]. Another method is covering the external surface with an aluminum-free MFI-type zeolite (silicalite-1) shell to form the core–shell zeolite [9–20]. This was synthesized by hydrothermal crystallization of silicalite-1 gel in the presence of ZSM-5 crystals under conventional synthesis conditions, i.e., hydroxide ion was a mineralizer. However, pinholes, which decrease shape selectivity, exist in the silicalite-1 shell [12]. Acid treatment of the core zeolite (ZSM-5) is suitable for the growth of the core [13] and may decrease the formation of the pinholes. Nishiyama et al. reported the coated aluminum-containing core zeolite with small crystallites of aluminum-free zeolite [15, 16] and a single crystalline silicalite-1/ZSM-5 core–shell zeolite [17, 18]. They showed high catalytic performance for methylation of toluene with methanol to give the highly selective formation of *p*-xylene [15, 16, 18].

In these reports, the core and the shell zeolites were synthesized in the presence of hydroxide ion as a mineralizer. Fluoride ion was chosen as a mineralizer for the syntheses of the core and the shell zeolites to decrease the number of defects, which hinder passivation of the external surface [19]. The ideal method is crystal overgrowth of aluminum-free zeolite (the shell zeolite) on single crystals of zeolite (the core zeolite) with no defects. Lombard et al. reported the core–shell zeolite prepared in the presence of fluoride ion [20]. However, aluminum composition in the shell was gradient (Fig. 12.2c), and it was difficult to passivate the acid sites on the external surface because the core–shell zeolite was prepared in one step.

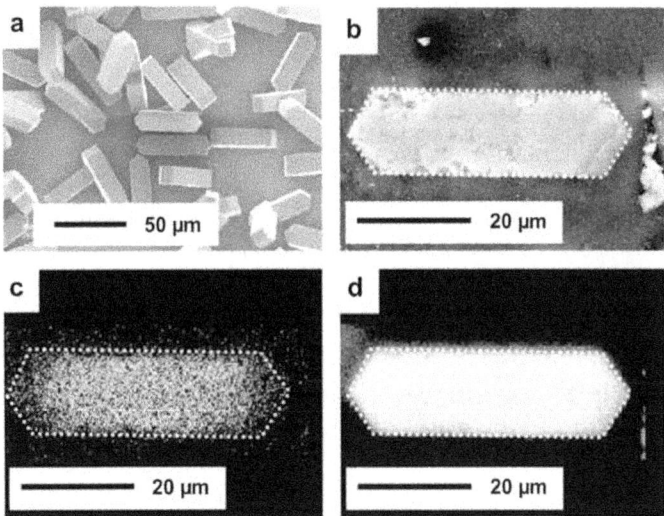

Fig. 12.2   EDX analysis of the core–shell zeolite was prepared in one step in the presence of fluoride ion: **a** scanning electron micrograph, **b** section of a crystal and its **c** Al Kα X-ray emission mapping, and **d** Si Kα X-ray emission mapping. Dotted lines denote the contour of the crystal. Reprinted from [20], Copyright 2010, with permission from Elsevier

12.2.1   Preparation of Core–Shell MFI-Type Zeolite with Minimal Defects

Core zeolite was synthesized using ammonium fluoride (NH₄F) as a fluoride ion source (the gel composition ratio was Si/tetrapropylammonium bromide (TPABr, structure-directing agent)/NH₄F/Al/H₂O = 1:0.125:0.9:0.05:33.) as well as using hydroxide ion as a conventional method. Crystal overgrowth was achieved by the use of fluoride ion as a mineralizer. The gel composition ratio was Si (fumed silica)/TPABr/NH₄F/H₂O/Si (core zeolite) = 1:0.125:0.9:33:1. Crystal overgrowth was also attempted under the synthesis conditions for the preparation of the conventional zeolite.

Table 12.1 shows terminology of the core and core–shell zeolites prepared under various conditions. For the core zeolites, CF(NC), CF(C), CA(NC), and CA(C) denoted the core prepared in the presence of fluoride ion but not calcined, the core prepared in the presence of fluoride ion and calcined, the core prepared in the absence of fluoride ion but not calcined, and the core prepared in the absence of fluoride ion and calcined, respectively. CF(NC)SF, CF(C)SF, and CA(NC)SF were prepared in the presence of fluoride ion by crystal overgrowth of CF(NC), CF(C), and CA(NC), respectively. CF(NC)SA was obtained from crystal overgrowth of CA(NC) in the absence of fluoride ion. To determine the particle size, the morphology of the core and core–shell zeolites was examined by scanning electron microscopy (SEM). The SEM images are shown in Fig. 12.3. The crystal shape of CF(C) was like a coffin. After

Table 12.1 Catalyst names and characterization of core and core–shell zeolites

Catalyst name	Core synthesis conditions		Shell synthesis conditions	Length of the particle's major axis [μm][a]	Si/Al	
					ICP[b]	NH₃ TPD[c]
CF(C)	With F⁻	Calcined	–	15	38	38
CF(C)SF	With F⁻	Calcined	With F⁻	18	60	67
CF(NC)SF	With F⁻	Not calcined	With F⁻	18	63	62
CF(NC)SA	With F⁻	Not calcined	Without F⁻	16	n.d	n.d
CA(C)	Without F⁻	Calcined	–	9	29	n.d
CA(NC)SF	Without F⁻	Not calcined	With F⁻	15	69	n.d

Reprinted from [19], Copyright 2011, with permission from Elsevier
n.d. not determined
[a]Calculated using SEM images
[b]Calculated using inductively coupled plasma
[c]Calculated using temperature-programmed desorption of ammonia

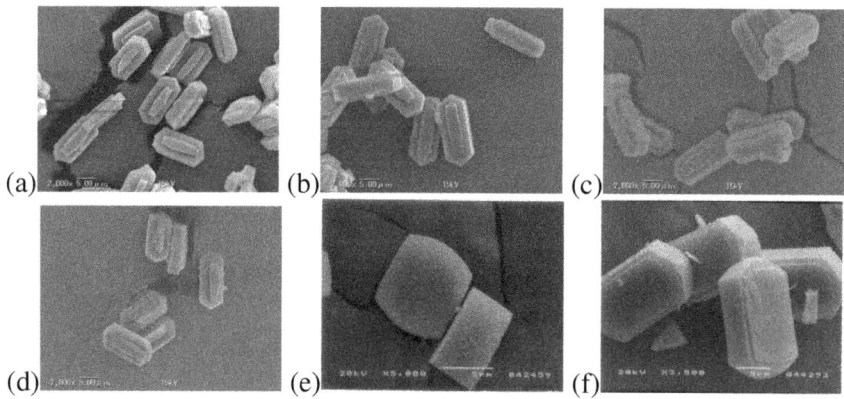

Fig. 12.3 SEM images of core and core–shell zeolites: **a** CF(C), **b** CF(C)SF, **c** CF(NC)SF, **d** CF(NC)SA, **e** CA(C), and **f** CA(NC)SF Reprinted from [19], Copyright 2011, with permission from Elsevier

the shell formation over CF(C) and CF(NC), the shapes of the core–shell zeolites were not changed, and the length of the particle's major size increased (Table 12.1). This indicated that the core zeolites were overgrown whether the core zeolites were calcined or not. When CA(NC) was used as the core zeolite, the size of CA(NC)SF increased. The amount of silicon in the synthesis gel used for the shell formation was equal to the amount of silicon in the core, so the shell volume should be the same as that of the core if perfect crystal overgrowth was to occur. However, the volume of each particle was more than double. This means the dissolution of part of the cores and formation of new zeolite particles during the step for the shell formation. When the shell was attempted to form in the absence of fluoride ion, the core zeolite was covered with the small particles. After removal of the small particles by sonication, the size of the obtained zeolite (CF(NC)SA) did not increase. Observations through SEM suggested that fluoride ion is necessary if the core zeolite is to grow while maintaining its shape.

To clarify the presence of small intercrystalline voids, namely, mesopores, nitrogen adsorption–desorption was measured. If defects form during the synthesis of the core and core–shell zeolites, the intercrystalline voids might also form. While the hysteresis loop was observed in the isotherm of the zeolite prepared by the conventional method (CA(C)), there was no loop for the zeolite prepared in the presence of fluoride ion CF(C), indicating the absence of mesopores. After the overgrowth of the shell zeolites, in the isotherm of CF(C)SF, the loop was observed. For the core–shell zeolite (CF(NC)SF) formed from the uncalcined core (CF(NC)), there were no mesopores. When the core zeolite prepared by the conventional method without calcination (CA(NC)) was used, the mesopores still remained.

To show the passivation of the acid sites on the external surface, the distribution of aluminum in a cross-section of CF(NC)SF was investigated by energy-dispersive X-ray spectroscopy. The linear analysis revealed that aluminum did not exist on the external surface, indicating that there are no acid sites on the external surface. Therefore, CF(NC)SF must be the catalyst showing high product selectivity.

Thus, CF(NC)SF is only the core–shell zeolite without mesopores, and that both core and shell must be synthesized in the presence of fluoride ion.

12.2.2 Toluene Methylation Catalyzed by Core–Shell MFI-Type Zeolite with Minimal Defects

Among xylenes, p-xylene is the most important, because it is a raw material for polyethylene terephthalate, which is industrially produced in a large scale. The mixture of xylenes, however, is formed by toluene methylation with methanol using acid catalysts. Selective synthesis of p-xylene is desirable in the industry. MFI-type zeolite catalyst shows high selectivity for p-xylene due to shape selectivity. However, the conventional MFI-type zeolite has acid sites on the external surface as well as on the pore walls. Acid sites on the external surface cause the formation of m- and o-xylenes.

Core–shell zeolites with minimal defects prepared by crystal overgrowth were catalysts with high performance for selective synthesis of p-xylene. The reaction results are summarized in Table 12.2. The fraction of p-xylene for CF(C) was higher than for CA(C). Because the particle size was higher for CF(C), the external surface area of CF(C) was lower. Because acid sites on the external surface lead to low p-xylene selectivity, CF(C) showed high selectivity for p-xylene.

Formation of shell zeolite over core zeolite in the presence of fluoride ion (CF(C)SF and CF(NC)SF) increased the fraction of p-xylene. The use of CF(NC)SF, which had no acid sites on the external surface and no mesopores as mentioned above,

Table 12.2 Toluene methylation with methanol catalyzed by the core and core–shell zeolites

Catalyst	CF(C)	CF(C)SF	CF(NC)SF	CA(C)	CA(NC)SF
Toluene conversion (%)	24.9	20.8	20.8	27.3	26.1
Yield (%)					
p-Xylene	14.4	17.0	17.8	11.5	11.3
m-Xylene	4.3	1.0	0.5	8.2	8.1
o-Xylene	1.9	0.5	0.3	3.3	3.0
Ethylbenzene	0.4	0.3	0.2	0.1	0.3
Trimethylbenzene and ethylmethylbenzene	3.9	2.0	2.0	4.2	3.4
Fraction of xylenes (%)					
p-Xylene	69.9	91.9	95.8	50.0	50.4
m-Xylene	20.9	5.4	2.9	35.7	36.2
o-Xylene	9.2	2.7	1.3	14.3	13.4

Reprinted from [19], Copyright 2011, with permission from Elsevier
Reaction temperature 400 °C, time of stream 2 h, W/F 0.15 kg(catalyst) h mol^{-1}, toluene 42.0 kPa, methanol 42.0 kPa

resulted in a higher fraction for *p*-xylene. This suggested that the shell successfully passivated the acid sites on the external surface of the core zeolite. When CF(C)SF was used, through the mesopores, the formed *o*- and *m*-xylenes, both of which were larger than the micropore size of the MFI-type structure, could penetrate the shell zeolite. The catalysis experiments indicated that only the shell of CF(NC)SF completely covered the core zeolite.

12.3 Skeletal Isomerization of *n*-Alkane Catalyzed by TON-Type Zeolite

Skeletal isomerization of *n*-alkanes is an important reaction for synthesizing high-octane gasoline and lubricants and preventing the solidification of gas oil at lower temperatures. Generally, the acid-catalyzed isomerization is attendant on the cracking of alkanes. Inhibition of the cracking is essential for selecting a suitable catalyst. TON- [24–39] and MTT-type [35, 40, 41] zeolites, which have one-dimensional 10-membered ring pores, are highly selective catalysts for skeletal isomerization of *n*-alkanes. Crystals of TON- and MTT-type zeolites are needle-shaped, and the one-dimensional micropores are parallel to the major axis of the crystal. The high selectivity for the skeletal isomerization using a TON-type zeolite catalyst was explained by pore mouth catalysis [24–30] or shape-selective catalysis inside the pores [32, 34, 35]. In any case, external acid sites on the side surface of the needle-shaped crystals may cause cracking of alkanes rather than isomerization. To increase selectivity for the isomerization, the formation of zeolite nanorods [31] and dealumination with

HCl [32] were reported. The crystal length was shortened by adding inhibitors of crystal growth along the [001] direction to the synthesis gel to increase the number of pore mouths for enhancement of the selectivity [37]. Passivation of the acid sites on the side surface of the needle-shaped crystals may also lead to highly selective catalysts for isomerization.

Niu et al. reported the core–shell TON-type zeolite prepared by epitaxial growth of aluminum-containing core zeolite with aluminum-free shell zeolite [38]. They used hydroxide ion as a mineralizer in the synthesis of the core zeolite and the crystal growth. The core–shell structure and filling the micropores with carbon resulted in high selectivity for skeletal isomerization. In this study, the method for passivation of the external surface of MFI-type zeolite was used to prepare the core–shell TON-type zeolite, i.e., an aluminum-containing TON-type core zeolite prepared in the presence of fluoride ion was epitaxially grown with aluminum-free TON-type shell zeolite in the presence of fluoride ion to give defect-free crystals [39]. TON-type zeolite can be also prepared in the presence of fluoride ion to form large and defect-free crystals [42]. Since the isomerization of n-alkanes occurs on the acid sites near the pore mouths, the passivation of the external surface except for the surface around the pore mouths may be effective. However, crystal overgrowth causes complete passivation of the external surface including the pore mouths. Therefore, the zeolite crystals were broken to form shorter pieces and consequently create acid sites near the new pore mouths formed at the cross-sections.

12.3.1   Preparation of Broken Core–Shell TON-Type Zeolite with Minimal Defects

Figure 12.4 shows the preparation scheme of the broken core–shell TON-type zeolite. After the core zeolite containing aluminum was prepared in the presence of fluoride ion, the core zeolite was epitaxially grown with the aluminum-free shell zeolite in the presence of fluoride ion to form the core–shell zeolite with minimal defects. Finally, to create the acid sites on the pore mouths, the needle-like core–shell zeolite was broken by a press at 10 and 20 MPa.

When the core zeolite was synthesized in the presence of fluoride ion, large needle-shaped crystals were obtained (Fig. 12.5a). The needle-like shape is typical of TON-type zeolite [42]. After crystal overgrowth, both the length and width of the particles increased. The aluminum distribution in the cross-section of the core–shell zeolite was examined using a wavelength-dispersive spectrometer. Linear analysis of aluminum concentration revealed that aluminum atoms did not exist in the shell zeolite. This indicated the success of the passivation of the external surface by crystal overgrowth.

The core–shell zeolite was pressed at 10 and 20 MPa. The needle-like particles were shortened by the press, and the size of the particles after the break at 20 MPa

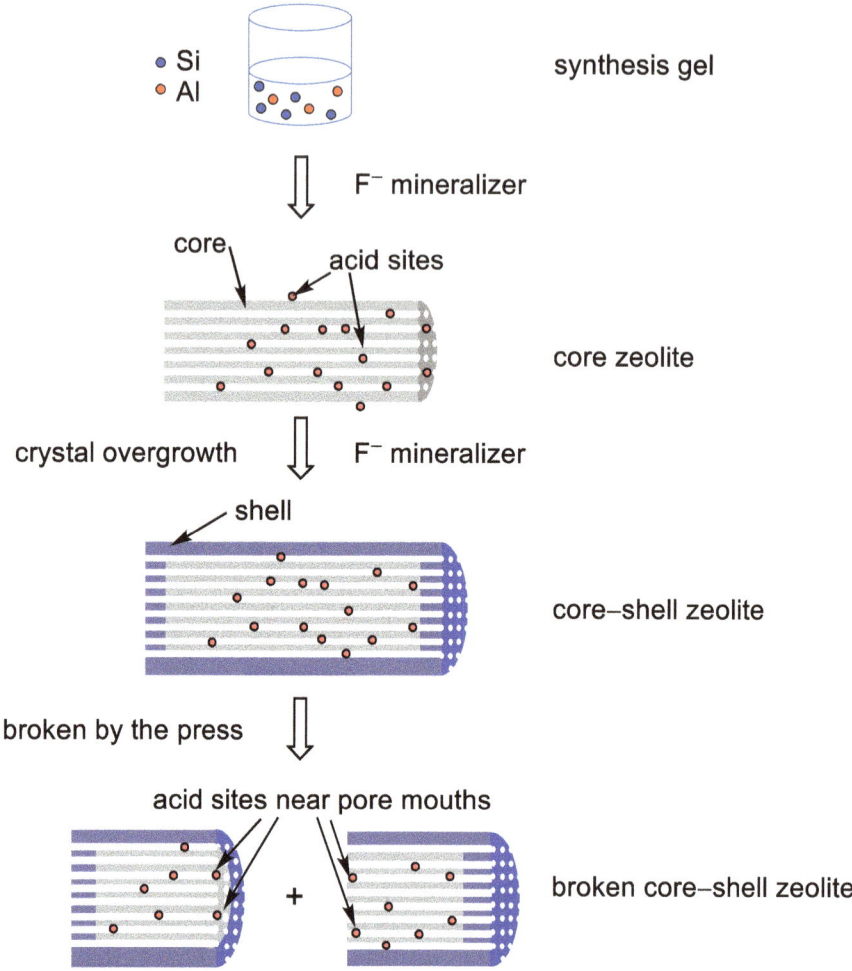

Fig. 12.4 Preparation scheme of the broken core–shell TON-type zeolite. Reprinted from [39], Copyright 2013, with permission from Elsevier

was shorter than that after the break at 10 MPa, i.e., the number of the pore mouths in the particles broken at 20 MPa may be larger.

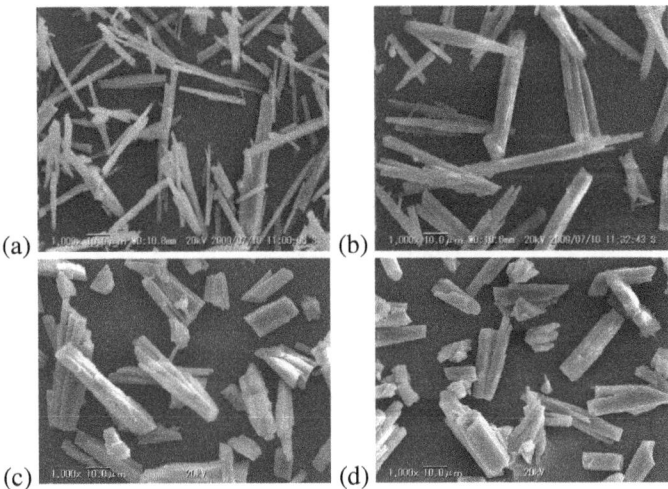

Fig. 12.5 SEM images of **a** the core zeolite, **b** the core–shell zeolite, and **c** the core–shell zeolites pressed at 10, and **d** 20 MPa. Reprinted from [39], Copyright 2013, with permission from Elsevier

12.3.2 Skeletal Isomerization of n-Tetradecane Catalyzed by Broken Core–Shell TON-Type Zeolite with Minimal Defects

For the skeletal isomerization, a platinum catalyst is necessary with the acid sites. Before the reaction, tetraammineplatinum(II) chloride monohydrate (Pt, 0.5% (w/w)) was loaded on the prepared zeolites, and the Pt-loaded zeolites were reduced in a hydrogen stream at 350 °C for 1 h.

The reaction results are summarized in Table 12.3. The overgrowth of the core zeolite with the shell zeolite decreased the tetradecane conversion because the number of the acid sites per weight of the catalyst decreased. On the other hand, the selectivity for the isomerization increased. The passivation of the external surface caused inhibition of the cracking. The break of the core–shell zeolite was effective for enhancement of the tetradecane conversion without a decrease of the selectivity, especially the break at 10 MPa. Higher pressure increased the acid sites on the pore mouths to increase the conversion but probably cracked the external surface to form new acid sites on the side surface of the needle-like particles. Compared to the catalytic activity of the conventional TON-type zeolite, both the conversion and the selectivity were enhanced. The broken core–shell zeolite was the most active and selective catalysts.

Table 12.3 Skeletal isomerization of *n*-tetradecane catalyzed by platinum-loaded TON-type zeolites

Catalyst	*n*-Tetradecane conversion (%)	Selectivity for isomerization (%)
Core zeolite	65	54
Core–shell zeolite	49	76
Core–shell zeolite broken at 10 MPa	73	77
Core–shell zeolite broken at 20 MPa	76	64
Conventional TON-type zeolite	67	47

Reprinted from [39], Copyright 2013, with permission from Elsevier

Reaction temperature 380 °C, time on stream 6 h, catalyst amount 1.5 g, Pt amount 0.5% (*w/w*), *n*-tetradecane 0.12 mmol min^{-1}, H$_2$ 0.58 mmol min^{-1}, total pressure 3 MPa-guage

12.4 Summary and Outlook

The core–shell zeolites with minimal defects can be prepared by crystal overgrowth of aluminum-containing core zeolite prepared in the presence of fluoride ion with aluminum-free shell in the presence of fluoride ion. The acid sites on the external surface of the core zeolite were completely passivated, and the core–shell zeolite showed high shape selectivity due to the zeolite crystal structure. In toluene methylation and skeletal isomerization of *n*-tetradecane, the MFI- and TON-type core–shell zeolite showed high shape selectivity, respectively.

Since the property of the core–shell zeolite as being almost a single crystal with minimal defects is unique, applications other than catalysts are also expected. For example, hollow *BEA-type zeolites with minimal defects were prepared by selective removal of the core zeolite from the core–shell zeolite [43]. Although many researchers have reported the hollow zeolites, almost all hollow zeolites were fabricated from polycrystalline zeolites [44–46], which must have mesopores due to intercrystalline voids. The hollow zeolite prepared from core–shell zeolite had only micropores originated from the crystal structure. This property is suitable for use as a drug vessel for controlled release and microreactor. The core–shell zeolite with minimal defects will be widely used in various fields.

References

1. Csicsery SM (1984) Shape-selective catalysis in zeolites. Zeolites 4:202–213
2. Niwa M, Kato M, Hattori T, Murakami Y (1986) Fine control of the pore-opening size of zeolite ZSM-5 by chemical vapor deposition of silicon methoxide. J Phys Chem 90:6233–6237
3. Hibino T, Niwa M, Murakami Y (1991) Shape-selectivity over HZSM-5 modified by chemical vapor deposition of silicon alkoxide. J Catal 128:551–558

4. Wang I, Ay CL, Lee BJ, Chen MH (1989) Para-selectivity of dialkylbenzenes over modified HZSM-5 by vapour phase deposition of silica. Appl Catal 54:257–266

5. Röger HP, Krämer M, Müller KP, O'Connor CT (1998) Effects of in-situ chemical vapour deposition using tetraethoxysilane on the catalytic and sorption properties of ZSM-5. Micropor Mesopor Mater 21:607–614

6. Manstein H, Möller KP, Böhringer W, O'Connor CT (2002) Effect of the deposition temperature on the chemical vapour deposition of tetraethoxysilane on ZSM-5. Micropor Mesopor Mater 51:35–42

7. O'Connor CT, Möller KP, Manstein H (2002) The effect of temperature and cyclic alkoxysilane deposition procedures on the silanisation and subsequent catalytic and sorption properties of zeolites. J Mol Catal A 181:15–24

8. O'Connor CT, Möller KP, Manstein H (2007) The effect of silanisation on the catalytic and sorption properties of zeolites. KONA 25:230–236

9. Rollmann LD (1980) ZSM-5 Containing aluminum-free shells on its surface. US Patent 4203869

10. Lee CS, Park TJ, Lee XY (1993) Alkylation of toluene over double structure ZSM-5 type catalysts covered with a silicalite shell. Appl Catal A 96:151–161

11. Weber RW, Fletcher JCQ, Möller KP, O'Connor CT (1996) The characterization and elimination of the external acidity of ZSM-5. Micropor Mater 7:15–25

12. Bouizi Y, Rouleau L, Valtchev VP (2006) Factors controlling the formation of core–shell zeolite–zeolite composites. Chem Mater 18:4959–4966

13. Li Q, Wang Z, Hedlund J, Creaser D, Zhang H, Zou X, Bons AJ (2005) Synthesis and characterization of colloidal zoned MFI crystals. Micropor Mesopor Mater 78:1–10

14. Ghorbanpour A, Gumidyala A, Grabow LC, Crossley SP, Rimer JD (2015) Epitaxial growth of ZSM-5@silicalite-1: a core–shell zeolite designed with passivated surface acidity. ACS Nano 9:4006–4016

15. Vu DV, Miyamoto M, Nishiyama N, Egashira Y, Ueyama K (2006) Selective formation of para-xylene over H-ZSM-5 coated with polycrystalline silicalite crystals. J Catal 243:389–394

16. Vu DV, Miyamoto M, Nishiyama N, Ichikawa S, Egashira Y, Ueyama K (2008) Catalytic activities and structures of silicalite-1/H-ZSM-5 zeolite composites. Micropor Mesopor Mater 115:106–112

17. Miyamoto M, Kamei T, Nishiyama N, Egashira Y, Ueyama K (2005) Single crystals of ZSM-5/silicate composites. Adv Mater 17:1985–1988

18. Vu DV, Miyamoto M, Nishiyama N, Egashira Y, Ueyama K (2009) Morphology control of silicalite/HZSM-5 composite catalysts for the formation of para-xylene. Catal Lett 127:233–238

19. Okamoto M, Osafune Y (2011) MFI-type zeolite with a core–shell structure with minimal defects synthesized by crystal overgrowth of aluminum-free MFI-type zeolite on aluminum-containing zeolite and its catalytic performance. Micropor Mesopor Mater 142:413–418

20. Lombard A, Simon-Masseron A, Rouleau L, Cabiac A, Patarin J (2010) Synthesis and characterization of core/shell Al-ZSM-5/silicalite-1 zeolite composites prepared in one step. Micropor Mesopor Mater 129:220–227

21. Guth JL, Kessler H, Wey R (1986) New route to pentasil-type zeolites using a non alkaline medium in the presence of fluoride ions. Stud Surf Sci Catal 28:121–128

22. Axon SA, Klinowski J (1992) Synthesis and characterization of defect-free crystals of MFI-type zeolites. Appl Catal A 81:27–34

23. Koller H, Wöllker A, Villaescusa LA, Díaz-Cabañas MJ, Valencia S, Camblor MA (1999) Five-coordinate silicon in high-silica zeolites. J Am Chem Soc 121:3368–3376

24. Ernst S, Weitkamp J, Martens JA, Jacobs PA (1989) Synthesis and shape-selective properties of ZSM-22. Appl Catal 48:137–148

25. Martens JA, Parton R, Uytterhoeven L, Jacobs PA, Froment GF (1991) Selective conversion decane into branched isomers- a comparison of platinum/ZSM-22, platinum/ZSM-5 and platinum/USY zeolite catalysts. Appl Catal 76:95–116

26. Martens JA, Souverijns W, Verrelst W, Parton R, Froment GF, Jacobs PA (1995) Selective Isomerization of hydrocarbon chains on external surfaces of zeolite crystals. Angew Chem Int Ed Engl 34:2528–2530
27. Claude MC, Martens JA (2000) Monomethyl-branching of long *n*-alkanes in the range from decane to tetracosane on Pt/H-ZSM-22 bifunctional catalyst. J Catal 190:39–48
28. Claude MC, Vanbutsele G, Martens JA (2001) Dimethyl branching of long *n*-alkanes in the range from decane to tetracosane on Pt/H–ZSM-22 bifunctional catalyst. J Catal 203:213–231
29. Laxmi Narasimhan CS, Thybaut JW, Marin GB, Jacobs PA, Martens JA, Denayer JF, Baron GV (2003) Kinetic modeling of pore mouth catalysis in the hydroconversion of *n*-octane on Pt-H-ZSM-22. J Catal 220:399–413
30. Denayer JF, Baron GV, Vanbutsele G, Jacobs PA, Martens JA (1999) Modeling of adsorption and bifunctional conversion of *n*-alkanes on Pt/H-ZSM-22 zeolite catalyst. Chem Eng Sci 54:3553–3561
31. Hayasaka K, Liang D, Huybrechts W, De Waele BR, Houthoofd KJ, Eloy P, Gaigneaux EM, van Tendeloo G, Thybaut JW, Marin GB, Denayer JFM, Baron GV, Jacobs PA, Kirschhock CEA, Martens JA (2007) Formation of ZSM-22 zeolite catalytic particles by fusion of elementary nanorods. Chem Eur J 13:10070–10077
32. Sastre G, Chica A, Corma A (2000) On the mechanism of alkane isomerisation (isodewaxing) with unidirectional 10-member ring zeolites-a molecular dynamics and catalytic study. J Catal 195:227–236
33. Wang G, Liu Q, Su W, Li X, Jiang Z, Fang X, Han C, Li C (2008) Hydroisomerization activity and selectivity of *n*-dodecane over modified Pt/ZSM-22 catalysts. Appl Catal A 335:20–27
34. Webb EB III, Grest GS (1998) Influence of intracrystalline diffusion in shape selective catalytic test reactions. Catal Lett 56:95–104
35. Maesen TLM, Schenk M, Vlugt TJH, de Jonge JP, Smit B (1999) The shape selectivity of paraffin hydroconversion on TON-, MTT-, and AEL-type sieves. J Catal 188:403–412
36. Wang X, Zhang X, Wang Q (2020) *N*-dodecane hydroisomerization over Pt/ZSM-22: Controllable microporous Brönsted acidity distribution and shape-selectivity. Appl Catal A 590:117335
37. Okamoto M, Nishimura Y, Takahashi M, Chen WH (2018) Synthesis of short, needle-shaped crystals of TON-type zeolite by addition of inhibitors of crystal growth along the [001] direction. Cyst Growth Des 18:6573–6580
38. Niu P, Xi H, Lin M, Wang Q, Chen X, Wang P, Jia L, Hou B, Li D (2018) Micropore blocked core–shell ZSM-22 designed via epitaxial growth with enhanced shape selectivity and high *n*-dodecane hydroisomerization performance. Catal Sci Technol 8:6407–6419
39. Okamoto M, Huang L, Yamano M, Sawayama S, Nishimura Y (2013) Skeletal isomerization of tetradecane catalyzed by TON-type zeolites with a fragmented core–shell structure. Appl Catal A 455:122–128
40. Huybrechts W, Thybaut JW, De Waele BR, Vanbutsele G, Houthoofd KJ, Bertinchamps F, Denayer JFM, Gaigneaux EM, Marin GB, Baron GV, Jacobs PA, Martens JA (2006) Bifunctional catalytic isomerization of decane over MTT-type aluminosilicate zeolite crystals with siliceous rim. J Catal 239:451–459
41. Ernst S, Kumar R, Weitkamp J (1988) Synthesis and catalytic properties of zeolite ZSM-23. Catal Today 3:1–10
42. Patarin J, Lamblin JM, Faust AC, Guth JL, Raatz F (1989) Nouvelles zéollithes de type structural ton, leur préparation et leur utilisation. European Patent 0,345,106
43. Okamoto M, You M, Iwamoto H (2009) Method of synthesizing hollow zeolite, hollow zeolite and drug support comprising hollow zeolite. Jpn Patent 2009269788
44. Wang XD, Yang WL, Tang Y, Wang YJ, Fu SK, Gao Z (2000) Fabrication of hollow zeolite spheres. Chem Commun 2161–2162
45. Dong A, Wang Y, Tang Y, Ren N, Zhang Y, Gao Z (2002) Hollow zeolite capsules: a novel approach for fabrication and guest encapsulation. Chem Mater 14:3217–3219
46. Valtchev V, Mintova S (2001) Layer-by-layer preparation of zeolite coatings of nanosized crystals. Micropor Mesopor Mater 43:41–49

Part II
Core-Shell for Photocatalysis and Electrocatalysis

Chapter 13
Core–Shell Materials for Photocatalytic CO_2 Reduction

Weixin Zou, Xiaoqian Wei, and Lin Dong

13.1 Introduction

With the growth of industry and population, various serious energy and environmental crises have attracted a great deal of attention. Carbon dioxide (CO_2), as one of the main greenhouse gases, recently have been focused on photoreduction into useful chemical fuels (hydrocarbons, alcohol, etc.) under solar light irradiation, with many advantages of friendly to environment, low energy consumption, solving the problems of both global warming and energy shortage, etc. [1–3] Since 1979, Inoue et al. investigated the CO_2 photoconversion to HCOOH, CH_4, and CH_3OH over a series of semiconductors [4]; a great number of efforts have been devoted to the highly efficient reaction performance to produce solar fuels [5–8].

As we know, the photocatalyst, with the abilities of light absorption, photogenerated charge transfer, and the reactant molecule adsorption/activation, is the key factor for the photocatalytic CO_2 reduction. Therefore, the design and development of the catalyst structures have already been recognized as a promising technology [9–11]. Engineering of the structural details of photocatalysts, including the particle size, shape, exposed crystal plane, and pore structure, could properly tune the reaction efficiency, and then the controllable manipulation with nanoscale structures would open a new horizon for the catalyst generation with the superior catalytic performance [12–14]. Core–shell structural catalysts have emerged as a significant class

W. Zou · X. Wei · L. Dong (✉)
Jiangsu Key Laboratory of Vehicle Emissions Control, School of the Environment, Center of Modern Analysis, Nanjing University, Nanjing 210093, People's Republic of China
e-mail: donglin@nju.edu.cn

W. Zou
e-mail: wxzou2016@nju.edu.cn

X. Wei
e-mail: weixiaoqian0920@163.com

© The Author(s), under exclusive license to Springer Nature Singapore Pte Ltd. 2021
H. Yamashita and H. Li (eds.), *Core-Shell and Yolk-Shell Nanocatalysts*,
Nanostructure Science and Technology,
https://doi.org/10.1007/978-981-16-0463-8_13

of catalysts with unique properties, including the following: (1) the stabilization of dispersed active phases; (2) maximization of the specific surface areas; (3) enhancing the synergistic interfacial interactions between multiple components; (4) the regulation of the electronic structures through strain engineering [15, 16]. On the basis of that, core–shell structured materials are regarded as promising photocatalysts for CO_2 reduction.

In this part, we review recent achievements of the core–shell materials in the application of CO_2 photoreduction. Firstly, an introduction of photocatalytic CO_2 reduction is briefly provided. Secondly, the synthesis, structures, and advantages of various core–shell nanomaterials, such as zero-dimensional (0D) nanoparticles, 1D nanocomposites, 2D nanosheets, and 3D hierarchical materials, are addressed in detail. Furthermore, the synergistic interaction mechanisms of multistep electrons transfer and CO_2 adsorption–photoreduction are discussed. Finally, a conclusion with a perspective on the future developments of this area is present.

13.2 CO_2 Photoreduction

For the photocatalytic CO_2 reduction, the process is mainly involved three steps, i.e., light absorption, photo-induced charge migration, and CO_2 molecule adsorption and activation. For the first step, effective light harvesting could thermodynamically promote the generation of electrons and holes. Generally, wavelengths of ultraviolet ($\lambda < 400$ nm), visible (400 nm $< \lambda < 800$ nm), and infrared ($\lambda > 800$ nm) account for 4%, 53%, and 43% of the sunlight, respectively [17]. Therefore, the photocatalyst design with a suitable bandgap energy is of great importance to effectively use the sunlight.

The migration from bulk to surface further kinetically affects the overall performance of the solar-to-fuels conversion, and thus the highly efficient separation of photo-induced electrons and holes is another key issue for enhancing the photocatalytic activity. On the basis of the fact that the charge recombination ($\sim 10^{-9}$ s) is considerably faster than the surface reaction (10^{-8} to 10^{-1} s) [18], the engineering of the internal electric field (IEF), i.e., built-in electric field, on the composite interface acts as an emerging and clearly viable method to efficiently facilitate charge separation and transfer [19, 20]. For example, the semiconductor with the positive and negative layered structure creates dipoles between the layers and then produces IEF (Fig. 13.1a); the photocatalyst with different crystal facets results in distinct potentials and the IEF (Fig. 13.1b); the various external environments (uneven doping, heterojunction with different work functions) generate an induced IEF (Fig. 13.1c).

What's more, the adsorption and activation of CO_2 molecule on photocatalyst surface are especially significant to the heterogeneous catalytic reaction because of the following factors: (1) the linear molecule of CO_2 with much higher bond energy of C=O (750 kJ mol^{-1}) is thermodynamically stable and demanded numerous energy to activation; (2) the surface one-electron CO_2 reduction is highly unfavorable with the high negative redox potential of CO_2/CO_2^- (-1.90 V vs. NHE, pH $= 7.00$); (3) the

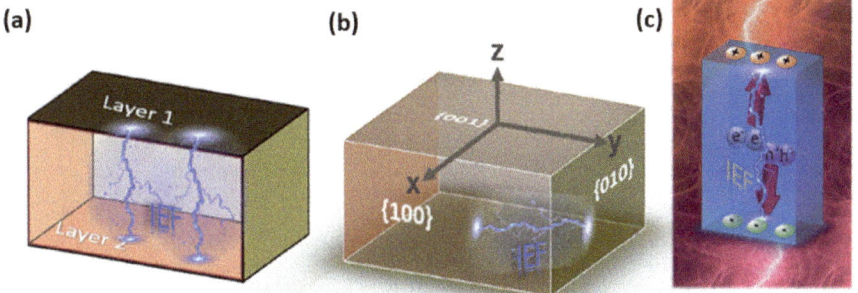

Fig. 13.1 The generated IEF on materials with **a** the layered structure; **b** different crystal faces; **c** external environment (Reproduced from Guo et al. 2019.)

products of CO_2 reduction are various (e.g., CO, CH_4, CH_3OH, etc.) determined by multiple electron/proton steps and redox potentials [21]. Fortunately, the structure design of photocatalyst is able to act as an effective way to modulate the activity and selectivity of CO_2 photocatalytic reduction. For example, photocatalysts with atomic-level reactive sites, i.e., vacancies, surface functional groups, single atoms, and frustrated Lewis pairs, show excellent affinity with CO_2 molecules, appropriate conduction band edges, lower energy barrier of CO_2 adsorption and activation, and then affect the redox reaction pathways [22].

13.3   Core–Shell Photocatalyst

The interaction of reactant CO_2 on surface, available light absorption, and fast photo-generated charge migration can be controlled by the overall architecture of core–shell of photocatalysts. Generally, the core–shell structures are divided into four categories, i.e., 0D nanoparticles, 1D nanocomposites, 2D nanosheets, and 3D hierarchical materials, each nanostructure has unique advantages and is worthy of being discussed. In this section, we summarize the merits, fabrications, structures, and CO_2 photoconversion efficiency of different core–shell photocatalysts, showed in Table 13.1.

13.3.1   0D Nanoparticles

Noble metal-based core–shell nanocomposites: Noble metals (NM) such as Ag, Au, Pt, and Pd are widely used as cocatalysts for CO_2 photoreduction, due to the following merits: (1) under light irradiation, the generation of a strong localized surface plasmon resonance (LSPR) leads to more available light absorption; (2) the formation of Schottky barriers drives the separation of photogenerated electrons and

Table 13.1 Summary of CO_2 photocatalytic reduction on core–shell nanocomposites

Core–shell photocatalyst		Preparation method	Light source	Catalytic performance	Refs.
0D nano particles	Au@CdS/TiO$_2$	Gas bubbling assisted reduction precipitation	300 W Xe lamp (320–780 nm)	CH$_4$: 41.6 μmol g^{-1} h^{-1} CO: 0.6 μmol g^{-1} h^{-1}	[23]
	Au@TiO$_2$	Turkevich method via reduction	200 W Hg/Xe lamp	CH$_4$: 1.0 μmol g^{-1} h^{-1}	[24]
	Pt/TiO$_2$/Au@SiO$_2$	Modified Stöber process	5 W LED lamp (365, 530 nm)	CH$_4$: 2.9 μmol g^{-1} h^{-1}	[26]
	TiO$_2$–Pd@Au	Ascorbic acid reduction	300 W Xe lamp (320–420 nm)	CO: 166.3 μmol g^{-1} h^{-1}	[29]
	TiO$_2$@TiO$_{2-x}$	Solvothermal method	300 W Xe lamp (400–780 nm)	CH$_4$: 16.2 μmol g^{-1} h^{-1}	[30]
	TiO$_2$@SiO$_2$	Sol–gel method	300 W Xe lamp (400–80 nm)	CO: 3.2 μmol g^{-1} h^{-1}	[31]
	Ag-MWCNT@TiO$_2$	Hydrolysis method	15 W energy saving light	CH$_4$: 0.85 μmol g^{-1} h^{-1} C$_2$H$_6$: 0.09 μmol g^{-1} h^{-1}	[34]
1D nano composites	Carbon nanofibers@TiO$_2$	Solvothermal method	350 W Xe lamp (UV and visible light)	CH$_4$: 13.5 μmol g^{-1} h^{-1}	[35]
	LaPO$_4$/g-C$_3$N$_4$ nanowires	Hydrothermal method	300 W Xe lamp	CO: 14.4 μmol g^{-1} h^{-1}	[36]
2D nanosheets	2D/2D/0D TiO$_2$/C$_3$N$_4$/Ti$_3$C$_2$	Hydrothermal induced self-assembly	350 W Xe lamp	CH$_4$: 1.20 μmol g^{-1} h^{-1} CO: 4.39 μmol g^{-1} h^{-1}	[39]
	CdS/rGO/TiO$_2$	Coating method	300 W Xe lamp	CH$_4$: 0.12 μmol g^{-1} h^{-1}	[41]

(continued)

Table 13.1 (continued)

Core–shell photocatalyst		Preparation method	Light source	Catalytic performance	Refs.
	Pt/TiO$_2$@rGO	A self-assembly method	300 W Xe lamp (320–780 nm)	CH$_4$: 41.3 μmol g^{-1} h^{-1}	[42]
3D hierarchical materials	Cu$_3$(BTC)$_2$@TiO$_2$	Hydrothermal method	300 W Xe lamp (<400 nm)	CH$_4$: 2.64 μmol g^{-1} h^{-1}	[43]
	AgNPs@Co-MOF-74	Solvothermal method	300 W Xe lamp (400–1000 nm)	CO: 5.5 μmol	[44]
	Cr$_2$O$_3$@TiO$_2$ microsphere	Hydrothermal method	UV lamp (365 nm)	CH$_4$: 168 μmol g^{-1} h^{-1}	[48]
	TiO$_2$–SiO$_2$ mesoporous	Sol–gel method	Xe lamp	CH$_4$: 0.70 μmol CO: 0.45 μmol	[50]
	CdS@CeO$_2$	Hydrothermal method	300 W Xe lamp (>400 nm)	CH$_4$: 0.94 μmol g^{-1} h^{-1} CH$_3$OH: 143.75 μmol g^{-1} h^{-1}	[52]

holes, and thus improves the quantum efficiency [23–27]. The LSPR intensity is strongly dependent on the nanoparticle (NP) size, and the larger size makes LSPR increased, leading to more plasmon absorption in the longer wavelength region. Bera et al. controlled the Au@SiO$_2$ core–shell with different Au NP sizes (4, 8, 18, and 26 nm), and suggested that as the NP size increased from 4 nm to 18 nm, the reaction activity in CH$_4$ generation gradually enhanced, but further enlarging size to 26 nm did not increase the catalytic performance, that is, Au core with a size of 18 nm had the best LSPR effect and activity [27].

Generally, the corrosion and dissolution of noble metals are hard to avoid during the photocatalytic process; core–shell nanostructures encapsulated by semiconductor shell acts as an efficient way to overcome the above drawback. In addition, the core–shell structured bi-noble metals are controllably synthesized, in which the differing work functions/electronegativities lead to the interfacial polarization, and then the generated driving force promotes the fast interfacial charge transfer [28, 29]. Cai et al. designed an ideal platform based on Pd@Au core–shell cocatalysts for CO_2 photocatalytic reduction (Fig. 13.2), which were provided with superior activity, selectivity, and stability.

Other NPs-based core–shell nanocomposites: Silica and vacancies species as shells to wrap the core of metal oxides have the excellent ability to entrap reactant

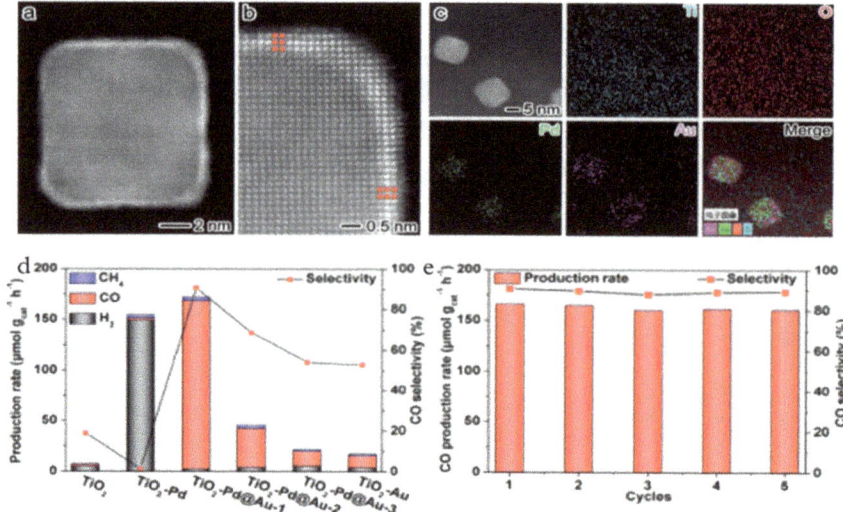

Fig. 13.2 High-magnification (**a**, **b**) and atomic-resolution (**c**) HAADF-STEM images of Pd@Au core–shell, and the activity (**d**), selectivity, and cycle test (**e**) of CO_2 photoreduction of TiO_2–Pd@Au (Reproduced from Cai et al. 2020.)

molecules [30–32]. Yin et al. used a facile low-temperature solvothermal method to obtain a uniform distinct crystalline core—amorphous shell structured blue H-TiO_{2-x}, and the numerous oxygen vacancies of shells were beneficial to a wide spectrum response and the formation of active intermediate CO_2^-, which promoted the enhanced CO_2 photoconversion [30].

13.3.2 1D Nanocomposites

Carbon-based core–shell nanocomposites: 1D carbon materials (i.e., nanofibers, nanotubes) have been widely considered as an important substrate to core–shell materials, due to the following advantages: (1) low cost, high specific surface area, good conductivity, and chemical stability; (2) photosensitive to the semiconductor photoactivation under visible light irradiation; (3) more electron sinking on carbon because of different Femi levels and the decreased electron diffusion length [33–35]. However, the hydrophobic nature of carbon materials leads to the weak interaction of core and shell components. Therefore, surface modification is employed to add oxygen-containing functional groups on the carbon surface. For example, the in situ growth TiO_2 nanoparticles onto acid-treated carbon nanofibers possessed excellent intimate contact, which reduced the self-agglomeration of TiO_2 and was beneficial for the charge transfer. Simultaneously, the carbon nanofibers act as core, leading to the enhanced conversion of light energy to heat energy and the fast diffusion of reactants and products [35].

Other 1D-based core–shell nanocomposites: LaPO$_4$ is a typical member of rare-earth phosphates and recognized as an effective photocatalyst for CO$_2$ reduction, due to the fact that lanthanum ions are active sites for CO$_2$ adsorption. Li et al. designed a novel 1D core–shell LaPO$_4$/g-C$_3$N$_4$ by a facile hydrothermal method (Fig. 13.3), and a synergic effect between LaPO$_4$ and g-C$_3$N$_4$ promoted the fast charge transfer [36]. WO$_3$ is another kind of polymorph semiconductor with many crystal phases including orthorhombic, tetragonal, monoclinic, hexagonal, and so on, used in CO$_2$ photoreduction. Li et al. constructed the vertical 1D/1D phase junction, in which

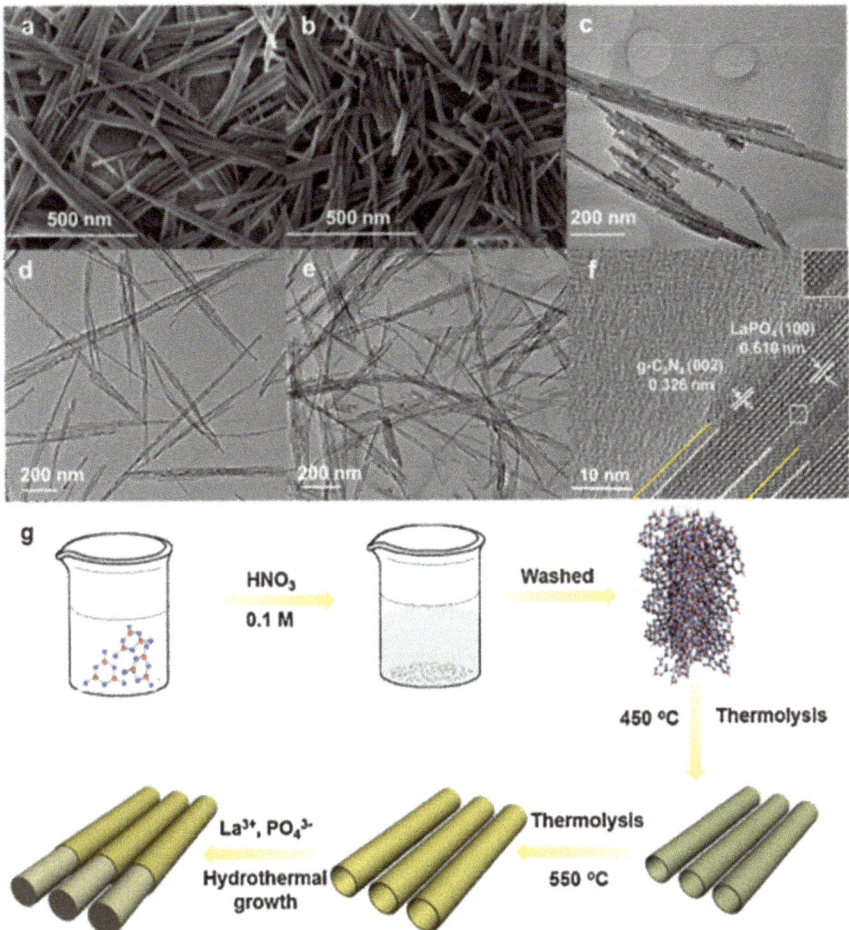

Fig. 13.3 FE-SEM, TEM, and HRTEM images; and schematic illustration of the synthesis of the core–shell LaPO$_4$/g-C$_3$N$_4$ nanocomposites (Reproduced from Li et al. 2017.)

exposed ends of the nanowires and nanorods were the spatial photogenerated charge separation [37].

13.3.3 2D Nanosheets

Carbon nitride-based core–shell nanocomposites: Graphitic carbon nitride (g-C_3N_4) with a special 2D layered structure has been widely considered in the application of photocatalytic CO_2 reduction, due to excellent physical/chemical stability, favorable negative conduction band energy, and narrow bandgap [38, 39]. It is known that the abilities of surface enrichments of photogenerated electrons and adsorbed CO_2 molecules are very critical for improving the activity and selectivity of photocatalytic CO_2 reduction. Wrapped g-C_3N_4 nanosheets can not only trap the photogenerated electrons, but also the surface π bond is able to improve adsorption capabilities for CO_2. Wang et al. proposed that the Au/TiO_2@g-C_3N_4 core–shell catalysts improved the surface density of adsorbed CO_2 and regional photoelectrons, leading to the high photocatalytic performance of visible-light-driven CO_2 conversion [38]. Furthermore, the g-C_3N_4 shell is able to inhabit the photo-corrosion, reaggregation, oxidation, or dissolution of nanocore [40].

Graphene-based core–shell nanocomposites: Graphene as another high-efficient acceptor and transporter of electrons is potential to improve the separation of photogenerated charges in artificial photo-synthesis reaction. Kuai et al. constructed CdS/reduced graphene oxide (rGO)/TiO_2 core–shell nanostructure, and rGO acted as the solid electron mediator for photogenerated electrons from TiO_2 to CdS, which promoted the faster electron transfer and better photoreduction activity [41]. Moreover, rGO materials with the surface hydroxyl species and extended π bond can increase the adsorption and activation capacities for both CO_2 and H_2O molecules, which are further helpful to the photocatalytic reaction [42].

13.3.4 3D Hierarchical Materials

MOFs-based core–shell nanocomposites: Metal–organic frameworks (MOFs), with the capability to expose adsorptive sites, are regarded as a class of potential porous materials for gas capture. Owing to the flexible tunability in composition and structure, MOFs are employed to combine with semiconductors in the hollow structure with a thin shell and cavity inside, which not only provide more active sites, but also offer confined space for photocatalytic CO_2 reduction [43–45]. Li et al. developed a method to synthesize $Cu_3(BTC)_2$@TiO_2 core–shell structures and suggested that CO_2 molecules easily penetrated the shells and then captured in the cores, dramatically improving the photocatalytic performance in terms of both activity and selectivity [43]. However, the applications of MOFs are limited, due to little visible and NIR light absorption. Many technologies have been developed. Li et al. designed core–shell structured upconversion nanoparticles (UCNPs)-Pt@MOF/Au photocatalysts, and the MOF and plasmonic Au were responsive to UV and visible light, respectively. Furthermore, the UCNPs absorbed NIR light to emit the visible and UV and light, which was harvested by the Au and MOF again [45].

Zeolitic imidazolate frameworks (ZIF), a subclass of MOFs, are provided with excellent CO_2 uptake and simple fabrication. For instance, ZIF-8 show high porosity, large surface area, and good thermal stability [46, 47]. Pipelzadeh et al. designed core–shell ZIF/TiO_2 nanocomposites, which not only made good use of the CO_2 adsorption, but also exhibited efficient photocatalytical CO_2 conversion [46].

Metal oxide-based core–shell nanocomposites: Metal oxide semiconductors like TiO_2 [48–50], CeO_2 [51, 52], etc. with 3D hierarchical mesoporous/microsphere structures have attracted much interest in the photocatalytic CO_2 reduction, due to the high surface-to-volume ratio, tunable contact area, and inter-connected porous network, which greatly improve the diffusion of reactants and the synergistic interaction. The pristine Cr_2O_3 microspheres coated by a thin layer of TiO_2 shell were synthesized by a five-step process involving hydrothermal and calcination treatments, and the formed novel core–shell nanostructure possessed sufficient interfacial contact, resulting in the enhanced CO_2 photocatalytic reduction efficiency [48].

13.4 Interfacial Effect on the Reaction Mechanism

13.4.1 Z-scheme

Due to the advantages of wide light absorption, high charge separation efficiency, and superior redox ability, the Z-scheme system has attracted considerable attention in the application of photocatalysis [53]. Photocatalytic thermodynamics in Z-scheme systems are worthy to be investigated, because, in the system, powerful holes and electrons with strong oxidation and reduction would be generated, respectively. The all-solid-state Z-scheme photocatalyst with the CdS (shell)–Pt (core)–TiO_2 (support) heterojunction was designed, and the vertical photogenerated electrons transfer of $TiO_2 \rightarrow Pt \rightarrow CdS$ is favorable for the charge separation. Simultaneously, the obtained electrons on the CB potential of CdS possessed more negative reduction potential, leading to the intensive driving force to CO_2 photoreduction, shown in Fig. 13.4a [54].

Except for the all-solid-state Z-scheme systems fabricated by noble metals (Au, Pt, Ag, etc.), the so-called "direct" Z-scheme systems with no redox mediators exhibit stronger reduction/oxidation power than all-solid-state Z-scheme systems, because electrons/holes in the reduction/oxidation catalysts don't react with the mediator. If the CB/valence band (VB) potential of the oxidation/reduction catalyst is not negative/positive enough to reduce and oxidize, the generated direct Z-scheme system would have enhanced reduction and oxidation to the CO_2, H_2, and •OH radical

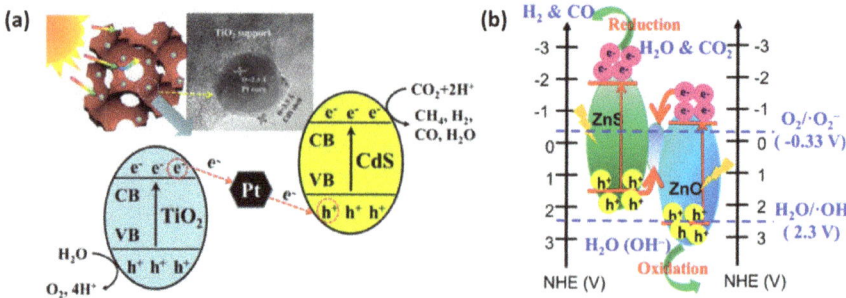

Fig. 13.4 a All-solid-state Z-scheme mechanism with the CdS (shell)–Pt (core)–TiO₂ (support) heterojunction, and **b** direct Z-scheme mechanism with ZnS@ZnO core–shell nanostructures for photocatalytic CO₂ reduction (Reproduced from Wei et al. 2015 and Li et al. 2018.)

production, respectively. The direct Z-scheme ZnS@ZnO photocatalysts with core–shell nanostructures were fabricated, which was in favor of CO_2 reduction with more driving force (Fig. 13.4b) [55].

13.4.2 Polarization Effect

Interfacial polarization, resulted from materials of different work functions/electronegativities, acts as a driving force to improve interfacial charge transfer and electrons accumulation on catalyst surface [56]. On the basis of that, the investigation of polarization effects on reaction mechanism is worth investigating. For example, Cai et al. designed an ideal platform based on Pd@Au core–shell cocatalysts for CO_2 photocatalytic reduction and demonstrated that the different electronegativities and lattice constants between Au and Pd resulted in a compressive strain on the Au surface, which shifted up the d-band center and enhanced the key intermediate adsorption of *COOH. Furthermore, the observed charge polarization at the interface between Au and Pd was beneficial to charge separation and photocatalytic activity [29]. He et al. adopted DFT calculation to investigate the polarization effect on the reaction mechanism. The Fermi level of g-C₃N₄ was higher than TiO₂ and Ti₃C₂ quantum dots (TCQD), and the different Fermi levels forced electrons to migrate from g-C₃N₄ to TCQD and TiO₂ once contact. The proposed charge transfer mechanism led to the formed interfacial built-in electric fields and the efficient separation of photogenerated electrons and holes [39].

13.4.3 Synergistic Interaction of Multiple Components

For CO$_2$ photocatalytic reaction, the synergistic interaction of active sites of photocatalysts on the adsorption/activation of reactant CO$_2$ molecules is worthy of being investigated, especially the reaction mechanism. Reactive sites on photocatalysts are mainly clarified as four categories: vacancies in the crystal lattice, anchoring functional groups, doping single heterogeneous atoms, and fabricating Lewis acid/base pairs [22].

Vacancies in the crystal lattice can change the surrounding electron density, which benefits the affinity of CO$_2$ molecules. In addition, vacancies not only favor light adsorption, but also act as photogenerated electrons capture sites and promote charge separation. Yin et al. proposed that the defects (Ti^{3+}) on TiO$_2$ could efficiently accelerate the adsorption and activation of the stable CO$_2$ molecules via the synergistic interaction of Ti^{3+} and Ti^{4+} redox through in situ diffuse reflectance infrared Fourier transform method [30]. Furthermore, the synergistic interaction of the adsorption sites and photo-electron generation sites plays important roles in CO$_2$ reaction mechanism. In hybrid carbon@TiO$_2$ hollow spheres, CO$_2$ is adsorbed on the carbon core via π–π conjugation interactions, and the photoexcited electrons are transferred from the CB of TiO$_2$ to neighboring carbon, in which the synergistic interaction greatly enhanced the photocatalytic activity [49].

13.5 Conclusions and Perspectives

In this chapter, we comprehensively summarize recent progress in the research area. The main conclusions of this review are as follows: (1) the core–shell structures of photocatalysts are of great significance to the performance of photocatalytic CO$_2$ reduction, in which the effective light harvesting and the migration from bulk to the surface could thermodynamically and kinetically affect the overall performance of the solar-to-fuels conversion, respectively. (2) The core–shell structure designs of photocatalysts, i.e., 0D nanoparticles, 1D nanocomposites, 2D nanosheets, and 3D hierarchical materials, are able to act as an effective way to modulate the activity and selectivity of CO$_2$ photocatalytic reduction. (3) The interfacial effect mechanism, such as Z-scheme, polarization effect, internal electric field, and multi-component synergy, of core–shell photocatalysts have a positive impact on available light absorption, photogenerated charge transfer, and CO$_2$ adsorption and activation.

However, it should be addressed that the current state-of-the-art for CO$_2$ photocatalytic conversion is still far from the practical realization. Fundamental studies are required to reveal the key challenges in the field. For example, the synthetic method of core–shell structures is still facing challenges in establishing controllable fabrication with desired morphology, shell thickness, specific crystal facet, surface chemistry, etc., and developing simple and fast methods for large-scale preparations of core–shell structured nanomaterials for practical applications is very important.

Furthermore, the insight into the absorption, activation, and subsequent conversion of CO_2 during the photocatalytic reduction is highly deficient.

Overall, the development of core–shell structured nanomaterials has been regarded as a new and potential research area and would explosively increase in the future.

References

1. Watson CS, White NJ, Church JA, King MA, Burgette RJ, Legresy B (2015) Unabated global mean sea-level rise over the satellite altimeter era. Nat Clim Change 5:565–568
2. Liu X, Iocozzia J, Wang Y, Cui X, Chen Y, Zhao S, Li Z, Lin Z (2017) Noble metal–metal oxide nanohybrids with tailored nanostructures for efficient solar energy conversion, photocatalysis and environmental remediation. Energy Environ Sci 10:402–434
3. Kim W, Edri E, Frei H (2016) Hierarchical inorganic assemblies for artificial photosynthesis. Acc Chem Res 49:1634–1645
4. Inoue T, Fujishima A, Konishi S, Honda K (1979) Photoelectrocatalytic reduction of carbon dioxide in aqueous suspensions of semiconductor powders. Nature 277:637–638
5. Tu W, Zhou Y, Zou Z (2014) Photocatalytic conversion of CO_2 into renewable hydrocarbon fuels: state-of-the-art accomplishment, challenges, and prospects. Adv Mater 26:4607–4626
6. Marszewski M, Cao S, Yu J, Jaroniec M (2015) Semiconductor-based photocatalytic CO_2 conversion. Mater Horiz 2:261–278
7. Li K, Peng B, Peng T (2016) Recent advances in heterogeneous photocatalytic CO_2 conversion to solar fuels. ACS Catal 6:7485–7527
8. Zhao G, Huang X, Wang X, Wang X (2017) Progress in catalyst exploration for heterogeneous CO_2 reduction and utilization: a critical review. J Mater Chem A 5:21625–21649
9. Zhang Q, Lee I, Joo JB, Zaera F, Yin Y (2013) Core-shell nanostructured catalysts. Acc Chem Res 46:1816–1824
10. Bell A (2003) The impact of nanoscience on heterogeneous catalysis. Science 299:1688–1691
11. Zou W, Xu L, Pu Y, Cai H, Wei X, Luo Y, Li L, Gao B, Wan H, Dong L (2019) Advantageous interfacial effects of AgPd/g-C$_3$N$_4$ for photocatalytic hydrogen evolution: electronic structure and H$_2$O dissociation. Chem Eur J 25:1–8
12. Xiong Y, Wiley BJ, Xia Y (2007) Nanocrystals with unconventional shapes—a class of promising catalysts. Angew Chem Int Ed 46:7157–7159
13. Zou W, Deng B, Hu X, Zhou Y, Pu Y, Yu S, Ma K, Sun J, Wan H, Dong L (2018) Crystal-plane-dependent metal oxide-support interaction in CeO$_2$/g-C$_3$N$_4$ for photocatalytic hydrogen evolution. Appl Catal B-Environ 238:111–118
14. Zhu C, Wang Y, Jiang Z, Xu F, Xian Q, Sun C, Tong Q, Zou W, Duan X, Wang S (2019) CeO$_2$ nanocrystal-modified layered MoS$_2$/g-C$_3$N$_4$ as 0D/2D ternary composite for visible-light photocatalytic hydrogen evolution: Interfacial consecutive multi-step electron transfer and enhanced H$_2$O reactant adsorption. Appl Catal B-Environ 259:118072–118081
15. Shao Q, Wang P, Liu S, Huang X (2019) Advanced engineering of core/shell nanostructures for electrochemical carbon dioxide reduction. J Mater Chem A 7:20478–20493
16. Li W, Elzatahry A, Aldhayan D, Zhao D (2018) Core-shell structured titanium dioxide nanomaterials for solar energy utilization. Chem Soc Rev 47:8203–8237
17. Zhang X, Peng T, Song S (2016) Recent advances in dye-sensitized semiconductor systems for photocatalytic hydrogen production. J Mater Chem A 4:2365–2402
18. Yuan Y, Ruan L, Barber J, Joachim Loo SC, Xue C (2014) Hetero-nanostructured suspended photocatalysts for solar-to-fuel conversion. Energy Environ Sci 7:3934–3951
19. Guo Y, Shi W, Zhu Y (2019) Internal electric field engineering for steering photogenerated charge separation and enhancing photoactivity. EcoMat 23:1–20

20. Zhu C, Wang Y, Jiang Z, Liu A, Pu Y, Xian Q, Zou W, Sun C (2019) Ultrafine Bi_3TaO_7 nanodot-decorated V, N codoped TiO_2 nanoblocks for visible-light photocatalytic activity: interfacial effect and mechanism insight. ACS Appl Mater Interfaces 11:13011–13021

21. Lee J, Sorescu DC, Deng X (2011) Electron-induced dissociation of CO_2 on $TiO_2(110)$. J Am Chem Soc 133:10066–10069

22. Zhang Y, Xia B, Ran J, Davey K, Qiao SZ (2020) Atomic-level reactive sites for semiconductor-based photocatalytic CO_2 reduction. Adv Energy Mater 10:1903879–1903901

23. Wei Y, Jiao J, Zhao Z, Liu J, Li J, Jiang G, Wang Y, Duan A (2015) Fabrication of inverse opal TiO_2-supported Au@CdS core–shell nanoparticles for efficient photocatalytic CO_2 conversion. Appl Catal B-Environ 179:422–432

24. Pougin A, Dodekatos G, Dilla M, Tuysuz H, Strunk J (2018) Au@TiO_2 core-shell composites for the photocatalytic reduction of CO_2. Chem Eur J 24:12416–12425

25. Zhai Q, Xie S, Fan W, Zhang Q, Wang Y, Deng W, Wang Y (2013) Photocatalytic conversion of carbon dioxide with water into methane: platinum and copper(I) oxide co-catalysts with a core-shell structure. Angew Chem Int Ed 52:5776–5779

26. Kumar D, Park CH, Kim CS (2018) Robust multimetallic plasmonic core-satellite nanodendrites: highly effective visible-light-induced colloidal CO_2 photoconversion system. ACS Sustain Chem Eng 6:8604–8614

27. Bera S, Lee JE, Rawal SB, Lee WI (2016) Size-dependent plasmonic effects of Au and Au@SiO_2 nanoparticles in photocatalytic CO_2 conversion reaction of Pt/TiO_2. Appl Catal B-Environ 199:55–63

28. Kamimura S, Yamashita S, Abe S, Tsubota T, Ohno T (2017) Effect of core@shell (Au@Ag) nanostructure on surface plasmon-induced photocatalytic activity under visible light irradiation. Appl Catal B-Environ 211:11–17

29. Cai X, Wang F, Wang R, Xi Y, Wang A, Wang J, Teng B, Bai S (2020) Synergism of surface strain and interfacial polarization on Pd@Au core–shell cocatalysts for highly efficient photocatalytic CO_2 reduction over TiO_2. J Mater Chem A 8:7350–7359

30. Yin G, Huang X, Chen T, Zhao W, Bi Q, Xu J, Han Y, Huang F (2018) Hydrogenated blue titania for efficient solar to chemical conversions: preparation, characterization, and reaction mechanism of CO_2 reduction. ACS Catal 8:1009–1017

31. Yuan L, Han C, Pagliaro M, Xu Y (2015) Origin of enhancing the photocatalytic performance of TiO_2 for artificial photoreduction of CO_2 through a SiO_2 coating strategy. J Phys Chem C 120:265–273

32. Yin G, Bi Q, Zhao W, Xu J, Lin T, Huang F (2017) Efficient conversion of CO_2 to methane photocatalyzed by conductive black titania. ChemCatChem 9:4389–4396

33. Gui M, Chai S, Xu B, Mohamed A (2014) Enhanced visible light responsive MWCNT/TiO_2 core-shell nanocomposites as the potential photocatalyst for reduction of CO_2 into methane. Sol Energy Mater Sol Cells 122:183–189

34. Gui MM, Wong WMP, Chai S-P, Mohamed AR (2015) One-pot synthesis of Ag-MWCNT@TiO_2 core–shell nanocomposites for photocatalytic reduction of CO_2 with water under visible light irradiation. Chem Eng J 278:272–278

35. Zhang J, Fu J, Chen S, Lv J, Dai K (2018) 1D carbon nanofibers@TiO_2 core-shell nanocomposites with enhanced photocatalytic activity toward CO_2 reduction. J Alloys Compd 746:168–176

36. Li M, Zhang L, Fan X, Wu M, Wang M, Cheng R, Zhang L, Yao H, Shi J (2017) Core-shell $LaPO_4$/g-C_3N_4 nanowires for highly active and selective CO_2 reduction. Appl Catal B-Environ 201:629–635

37. Li Y, Tang Z, Zhang J, Zhang Z (2017) Fabrication of vertical orthorhombic/hexagonal tungsten oxide phase junction with high photocatalytic performance. Appl Catal B-Environ 207:207–217

38. Wang C, Zhao Y, Xu H, Li Y, Wei Y, Liu J, Zhao Z (2020) Efficient Z-scheme photocatalysts of ultrathin g-C_3N_4-wrapped Au/TiO_2-nanocrystals for enhanced visible-light-driven conversion of CO_2 with H_2O. Appl Catal B-Environ 263:118314–118326

39. He F, Zhu B, Cheng B, Yu J, Ho W, Macyk W (2020) 2D/2D/0D TiO_2/C_3N_4/Ti_3C_2 MXene composite S-scheme photocatalyst with enhanced CO_2 reduction activity. Appl Catal B-Environ 272:119006–119017

40. Guo Q, Wan Y, Hu B, Wang X (2018) Carbon-nitride-based core–shell nanomaterials: synthesis and applications. J Mater Sci-Mater El 29:20280–20301
41. Kuai L, Zhou Y, Tu W, Li P, Li H, Xu Q, Tang L, Wang X, Xiao M, Zou Z (2015) Rational construction of a CdS/reduced graphene oxide/TiO_2 core–shell nanostructure as an all-solid-state Z-scheme system for CO_2 photoreduction into solar fuels. RSC Adv 5:88409–88413
42. Zhao Y, Wei Y, Wu X, Zheng H, Zhao Z, Liu J, Li J (2018) Graphene-wrapped Pt/TiO_2 photocatalysts with enhanced photogenerated charges separation and reactant adsorption for high selective photoreduction of CO_2 to CH_4. Appl Catal B-Environ 226:360–372
43. Li R, Hu J, Deng M, Wang H, Wang X, Hu Y, Jiang H, Jiang J, Zhang Q, Xie Y, Xiong Y (2014) Integration of an inorganic semiconductor with a metal-organic framework: a platform for enhanced gaseous photocatalytic reactions. Adv Mater 26:4783–4788
44. Deng X, Yang L, Huang H, Yang Y, Feng S, Zeng M, Li Q, Xu D (2019) Shape-defined hollow structural Co-MOF-74 and metal nanoparticles@Co-MOF-74 composite through a transformation strategy for enhanced photocatalysis performance. Small 15:1902287–1902293
45. Li D, Yu S, Jiang H (2018) From UV to near-infrared light-responsive metal-organic framework composites: plasmon and upconversion enhanced photocatalysis. Adv Mater 30:1707377–1707383
46. Pipelzadeh E, Rudolph V, Hanson G, Noble C, Wang L (2017) Photoreduction of CO_2 on ZIF-8/TiO_2 nanocomposites in a gaseous photoreactor under pressure swing. Appl Catal B-Environ 218:672–678
47. Su Y, Xu H, Wang J, Luo X, Xu Z, Wang K, Wang W (2018) Nanorattle Au@PtAg encapsulated in ZIF-8 for enhancing CO_2 photoreduction to CO. Nano Res 12:625–630
48. Tan JZY, Xia F, Maroto-Valer MM (2019) Raspberry-like microspheres of core-shell Cr_2O_3@TiO_2 nanoparticles for CO_2 photoreduction. Chemsuschem 12:5246–5252
49. Wang W, Xu D, Cheng B, Yu J, Jiang C (2017) Hybrid carbon@TiO_2 hollow spheres with enhanced photocatalytic CO_2 reduction activity. J Mater Chem A 5:5020–5029
50. Gong Y, Wang DP, Wu R, Gazi S, Soo HS, Sritharan T, Chen Z (2017) New insights into the photocatalytic activity of 3-D core–shell P25@silica nanocomposites: impact of mesoporous coating. Dalton Trans 46:4994–5002
51. Yang X, Zhang Y, Wang Y, Xin C, Zhang P, Liu D, Mamba BB, Kefeni KK, Kuvarega AT, Gui J (2020) Hollow β-Bi_2O_3@CeO_2 heterostructure microsphere with controllable crystal phase for efficient photocatalysis. Chem Eng J 387:124100–124108
52. Ijaz S, Ehsan MF, Ashiq MN, Karamat N, He T (2016) Preparation of CdS@CeO_2 core/shell composite for photocatalytic reduction of CO_2 under visible-light irradiation. Appl Surf Sci 390:550–559
53. Wang F, Li Q, Xu D (2017) Recent progress in semiconductor-based nanocomposite photocatalysts for solar-to-chemical energy conversion. Adv Energy Mater 7:1700529–1700547
54. Wei Y, Jiao J, Zhao Z, Zhong W, Li J, Liu J, Jiang G, Duan A (2015) 3D ordered macro-porous TiO_2-supported Pt@CdS core–shell nanoparticles: design, synthesis and efficient photocatalytic conversion of CO_2 with water to methane. J Mater Chem A 3:11074–11085
55. Li P, He T (2018) Common-cation based Z-scheme ZnS@ZnO core-shell nanostructure for efficient solar-fuel production. Appl Catal B-Environ 238:518–524
56. Liang Z, Yan C-F, Rtimi S, Bandara J (2019) Piezoelectric materials for catalytic/photocatalytic removal of pollutants: recent advances and outlook. Appl Catal B-Environ 241:256–269

Chapter 14
Self-supported CPs Materials for Photodegrading Toxic Organics in Water

Shu-Jin Bao, Ze-Ming Xu, Zheng Niu, and Jian-Ping Lang

14.1 Introduction

With the rapid development of industrialization and urbanization, the problem of water pollution has been becoming more and more serious, and the whole ecological environment has been destroyed by toxic, non-biodegradable, persistent dyes, and nitroaromatic compounds, which is a leading worldwide cause of death and diseases [1]. For treating industrial wastewater, traditional methods including adsorption, precipitation, and coagulation have been widely used in industrial laboratories. However, these approaches have many potential limitations because they are not efficient in dealing with organic pollutants with low concentrations in wastewater, which cannot be fully eliminated or decomposed via these traditional methods [2]. In this respect, it is imperative to develop greener and more efficient water treatment technologies to reduce pollution expansion in the world.

Heterogeneous photocatalysis has been proven to be one of the most adopted options for wastewater treatment [3]. This approach can be performed with low costs at ambient operating temperatures and pressures under sunlight or other light sources. A wide range of organic pollutants can be decomposed into easily biodegradable

---

S.-J. Bao · Z.-M. Xu · Z. Niu (✉) · J.-P. Lang (✉)
College of Chemistry, Chemical Engineering and Materials Science, Soochow University, Suzhou 215123, Jiangsu, People's Republic of China
e-mail: nkniuzheng@163.com

J.-P. Lang
e-mail: jplang@suda.edu.cn

S.-J. Bao
e-mail: 569571364@qq.com

Z.-M. Xu
e-mail: 2532468626@qq.com

compounds or less toxic molecules using this method, and even eventually mineralized into innocuous CO_2 and H_2O without leaving secondary pollution. To date, some semiconductor materials such as TiO_2, ZnO, ZnS, CdS have been widely used as photocatalysts for the degradation of organic pollutants due to their relatively high efficiency, corrosion resistance, and low-cost availability [4, 5]. Due to their poor visible light absorption, these materials can only be activated in the ultraviolet (UV) region. At the same time, they would be suffered from their intrinsic poor corrosion resistance. Besides, to avoid material agglomeration in the process of photocatalysis, some carrier materials, such as activated carbon, alumina, porous silica gel, have been employed but not with a remarkable performance [6]. Therefore, it is of great significance to develop other new catalytic materials which are carrier-free, stable, and have excellent visible light absorption.

In the past decades, coordination polymers (CPs), a class of crystalline inorganic–organic hybrid materials, have been widely applied to photocatalytic degradation of pollutants [7]. The CPs, in which metal ions or cluster cores are linked by organic ligands to form extended networks, are relatively stable in the photocatalytic process and cannot agglomerate without any carrier materials. For CPs, the conduction band (CB) is mainly composed of empty d orbitals with the LUMOs of organic linkers. Therefore, by changing their ligands, the absorption of light could be extended from UV to UV-visible to the visible region. These results break through the limitation that most traditional inorganic materials only work well under UV irradiation to some extent. Meanwhile, doping other metal ions into the framework of CPs is also a useful strategy to narrow the bandgap, thus improving the photocatalytic activity [8]. Besides, if some photocatalytically active species could be incorporated into CPs to generate so-called core-shell-like CP-based materials, it is possible to further tune the electronic structures of CPs and realize the synergic effects of CPs and the active species and finally enhance the photocatalytic activities of CPs. To our knowledge, some pristine CPs, metal-ion-doped CPs, and core-shell-like CP-based materials can exhibit excellent catalytic abilities in photodegrading various toxic organics in water. This chapter summarizes some recent progress in this respect.

14.2 Photodegradation of Dyes and Nitroaromatics Using Pristine CPs

Recently, much effort has been devoted to employing various CPs as photocatalysts to decompose organic dyes and organic pollutants in wastewater. A notable feature of CPs is that they are controllably synthesized by choosing appropriate organic ligands and central metals, hence enabling the tuning of their absorption bands and utilizing various light sources for degrading organic pollutants. Thus, CPs can be served as an alternative platform for developing efficient photocatalysts for decomposing organic pollutants in wastewater. Herein, some CPs containing different transition metal ions, such as Zn(II), Cd(II), Mn(II), Co(III), and Ni(II), Ag(I), and Au(I), were examined as self-supported photocatalysts to degrade organic pollutants such as methyl orange (MO), methyl blue (MB), rhodamine B (RhB), and so on under UV, visible, or UV-vis light (Table 14.1).

Three different CPs $\{[Zn_2(H_2O)-(1,4\text{-ndc})_2(\text{tpcb})]\}_n$ **(1)**, $\{[Zn(1,4\text{-ndc})(\text{tpcb})_{0.5}]\}_n$ **(2)**, and $\{[Zn_2(H_2O)(2,3\text{-ndc})_2(\text{tpcb})]\}_n$ **(3)** were synthesized by solvothermal reactions of $Zn(NO_3)_2$ with tetrakis(4-pyridyl)cyclobutane (tpcb) and 1,4-naphthalenedicarboxylic acid (1,4-H_2ndc) and 2,3-naphthalenedicarboxylic acid (2,3-H_2ndc) and their photocatalytic activities were assessed by the degradation of MO and MB in aqueous solution [9]. The absorption (α/S) data can be calculated from the reflectance using the Kubelka–Munk function: $\alpha/S = \frac{(1-R)^2}{2R}$, where α is the absorption, S is the scattering coefficient, and R is the reflectance at a given energy. If the energy gap (E_{onset}) obtained by extrapolation of the linear portion of the absorption edges of one material is in the range of 0–4 eV, it is usually considered as having semiconducting properties, which is critical in evaluating the photocatalysts. In these cases, the E_{onset} value of these three CPs was estimated to be 2.62 eV, 2.90 eV, and 3.34 eV, respectively, indicating they may behave as semiconductors if exposed to UV light and act as potential photocatalysts. Subsequently, these compounds were applied as photocatalysts to degrade different organic dyes. For **1**, the degradation of MO or MB was completed in about 24 h, displaying a relatively low activity for the photocatalytic reaction. In comparison, **2** showed a higher capacity than **1**, and the complete degradation time for MO and MB were 14 and 10 h, respectively (Fig. 14.1). The photocatalytic reaction for the decomposition of MO and MB initiated by **3** was faster than those catalyzed by **1** and **2**. The MO and MB can be totally decomposed by **3** upon UV light irradiation for 10 and 8 h. These results demonstrated that the photocatalytic efficiency is **3 > 2 > 1**, which may largely depend on the energy gap between the UV light ($\lambda = 365$ nm, $h\nu = 3.40$ eV) and these coordination polymers (for **1**, $E_g = 2.62$ eV; for **2**, $E_g = 2.90$ eV; for **3**, $E_g = 3.34$ eV), decreasing from **1** (0.78 eV) to **2** (0.50 eV) to **3** (0.06 eV). Such decreases may lead to the faster decomposition of MO or MB initiated by **3**.

Motivated by the synthesis of functional CPs, we have been interested in the preparation of some polydimensional CPs and would like to investigate whether they could be used to photocatalytically decompose organic dyes in polluted water. For this purpose, four different topological CPs formulated as $\{[Cd_3(\text{tpcb})_2)(\eta,\eta\text{-}\mu\text{-SO}_4)_2(H_2O)_6]SO_4\cdot16H_2O\}_n$ **(4)**, $\{[Cd(\text{tpcb})_{0.75}(OH)(H_2O)_2](NO_3)\}_n$ **(5)**, $\{[Cd_2(\text{tpcb})(SO_4)_2(H_2O)_6]\cdot2MeOH\cdot3H_2O\}_n$ **(6)** and $\{[Cd(\text{tpcb})(NO_3)(H_2O)_2](NO_3)\}_n$ **(7)** were prepared by reactions of $CdSO_4$ or $Cd(NO_3)_2$ with tetrakis(4-pyridyl)cyclobutane (tpcb) under solvothermal conditions or at ambient temperature [10]. Their corresponding band gap values were 2.96, 3.10, 2.68, and 3.24 eV, respectively. Thus these compounds were used to photocatalytically degrade three common organic dyes (MO, MB, and RhB) in water (Fig. 14.2). The degradation of MO in water was completed by **4–7** after 5 h, displaying a similar photocatalytic activity. For MB, its degradation in water was almost completed after UV irradiation for 2.5 h (**4**) or 1.5 h (**5–7**), indicating that **4** had a lower photocatalytic activity compared with **5–7**. For RhB, the degradation

Table 14.1 The photocatalytic performances of various CPs for the degradation of dyes and nitroaromatics in aqueous media

Coordination polymers[a]	Eg (eV)	Light source	Organic pollutants[b]	Degradation efficiency (%)
$\{[Zn_2(H_2O)(1,4\text{-ndc})_2(tpcb)]\}_n$ **(1)**	2.62	UV	MO, MB	33, 73
$\{[Zn(1,4\text{-ndc})(tpcb)_{0.5}]\}_n$ **(2)**	2.90	UV	MO, MB	87, 78
$\{[Zn_2(H_2O)(2,3\text{-ndc})_2(tpcb)]\}_n$ **(3)**	3.34	UV	MO, MB	90, 91
$\{[Cd_3(tpcb)_2)(\eta,\eta\text{-}\mu\text{-}SO_4)_2(H_2O)_6]SO_4 \cdot 16H_2O\}_n$ **(4)**	2.96	UV	MO, MB, RhB	100, 100, 100
$\{[Cd(tpcb)_{0.75}(OH)(H_2O)_2](NO_3)\}_n$ **(5)**	3.10	UV	MO, MB, RhB	100, 100, 100
$\{[Cd_2(tpcb)(SO_4)_2(H_2O)_6] \cdot 2MeOH \cdot 3H_2O\}_n$ **(6)**	2.68	UV	MO, MB, RhB	100, 100, 100
$\{[Cd(tpcb)(NO_3)(H_2O)_2](NO_3)\}_n$ **(7)**	3.24	UV	MO, MB, RhB	100, 100, 100
$[Mn_3(oba)_3(bbm)(H_2O)_2] \cdot DMF \cdot 2H_2O$ **(8)**	3.44	UV	MB	95
$[Co_5(bpb)_3(oba)_4(Hoba)_2(H_2O)_2]$ **(9)**	3.27	UV	MB	95
$[Zn(pbda)(bpp)]$ **(10)**	3.40	Sunlight	MB	87
$[Zn(pbta)_{0.5}(bpp)(H_2O)]$ **(11)**	3.60	Sunlight	MB	83
$[Zn(1,4\text{-BDC})(bmimb)]_n$ **(12)**	2.94	UV	RhB	90
$[Cd(1,3\text{-BDC})(bmimb)]_n$ **(13)**	3.00	UV	RhB	90
$\{[Zn(1,3\text{-BDC})(abimb)] \cdot H_2O\}_n$ **(14)**	2.77	UV	MB	40
$\{[Zn(5\text{-Br-}1,3\text{-BDC})(abimb)]_2 \cdot 0.5H_2O\}_n$ **(15)**	2.94	UV	MB	84.8
$[Zn(1,3\text{-BDC})(nbimb)]_n$ **(16)**	2.90	UV	MB	54.4
$[Zn(5\text{-Br-}1,3\text{-BDC})(nbimb)]_n$ **(17)**	2.71	UV	MB	59.4
$\{[Zn(1,3\text{-BDC})(bimpa)] \cdot 2H_2O\}_n$ **(18)**	3.09	UV	MB	75.7
$\{[Zn(5\text{-Br-}1,3\text{-BDC})(bimpa)] \cdot 2H_2O\}_n$ **(19)**	3.29	UV	MB	75.0
$[Zn(1,3\text{-BDC})(bimcz)]_n$ **(20)**	2.81	UV	MB	62.8
$[Zn(5\text{-Br-}1,3\text{-BDC})(bimcz)]_n$ **(21)**	3.00	UV	MB	75.9
$\{Ag(Sal)(3\text{-bdppmapy})\}_n$ **(22)**	3.31	UV	RhB	60
$\{[Ag_4(Sal)_2(\mu\text{-}\eta,\eta^2\text{-Sal})_2](3\text{-bdppmapy})_2\}_n$ **(23)**	3.31	UV	RhB	100
$[Ag_4I_4(3\text{-bdppmapy})_2]_n$ **(24)**	3.27	UV	RhB, MO, OI, OII, OIV, OG, AB, SY, ACBK, AR, or EBT	All 100
$[Ag(\mu\text{-}\eta^2\text{-Sal})(3\text{-dppmapy})]$ **(25)**	3.35	UV	EBT, AB, NO	All 100
$\{[Cd_2(H_2O)_2(tpeb)_2(1,2\text{-CHDC})_2] \cdot H_2O\}_n$ **(26)**	2.23	Vis	CR	90

(continued)

Table 14.1 (continued)

Coordination polymers[a]	E_g (eV)	Light source	Organic pollutants[b]	Degradation efficiency (%)
[Ag$_4$(NO$_3$)$_4$(dpppda)]$_n$ (**27**)	2.78	UV	NB, PNP, 2,4-DNP	100, 100, 100
[Au$_2$(dppatc)$_2$]Cl$_2$ (**28**)	2.85	UV	NB, PNP, 2,4-DNP	100, 100, 100
[Au$_2$(dppmt)$_2$]$_n$ (**29**)	2.98	UV	NB, PNP, 2,4-DNP	100, 100, 100

[a]1,4-H$_2$ndc = 1,4-naphthalenedicarboxylicacid; tpcb = tetrakis(4-pyridyl)cyclobutene; 2,3-H$_2$ndc = 2,3-naphthalenedicarboxylicacid; H$_2$oba = 4,4′-oxydibenzoic acid; bbm = 1,4-di(1H-imidazol-1-yl)benzene; bpb = 1,4-di(pyridin-4-yl)benzene; 1,3-BDC = 1,3-benzenedicarboxylate; 1,4-BDC = 1,4-benzenedicarboxylate; bmimb = 4,4′-bis(4-methyl-1-imidazolyl)biphenyl; H$_2$pbda = 4,4′-{[1,4-phenylenebis(methylene)]bis(oxy)}dibenzoic acid; bpp = 1,3-bis(4-pyridyl)-propane; H$_4$pbta = 5,5′-phenylenebis(methylene)-1,1′-3,3′-(benzene-tetracarboxylic acid); abimb = 2-amine-4,4′-bis(1-imidazolyl)-bibenzene; 5-Br-1,3-BDC = 5-bromo-1,3-benzenedicarboxylate; nbimb = 2-nitro-4,4′-bis(1-imidazolyl)-bibenzene; bimpa = bis-(4-imidazol-1-yl-phenyl)-amine; bimcz = 3,6-bis(1-imidazolyl)-carbazole; 3-bdppmapy = N,N-bis(diphenylphosphanylmethyl)-3-aminopyridine; Sal = salicylate; tpeb = 1,3,5-tri-4-pyridyl-1,2-ethenylbenzene; 1,2-H$_2$CHDC = 1,2-cyclohexanedicarboxylicacid; dpppda = 1,4-N,N,N′,N′-tetra(diphenylphosphanylmethyl)benzenediamine; dppatc = N,N-bis(diphenylphosphanylmethyl)-amino-thiocarbamide; dppmtH = (diphenylphosphino)methanethiol
[b]MO = methyl orange; MB = methyl blue; RhB = rhodamine B; OI = orange I; OII = orange II; OIV = orange IV; OG = orange G; AB = amino black; SY = sunset yellow; ACBK = acid chrome blue K; AR = amaranth red; EBT = ericochrome black T; NO = nigrosine; CR = congo red; NB = nitrobenzene; PNP = paranitrophenol; 2,4-DNP = 2,4-dinitropheno

in water was finished after UV irradiation for 4.5 h (**4** and **6**), and 3 h (**5** and **7**). Thus, these results revealed both **5** and **7** held the better catalytic efficiency for the degradation of MO, MB, or RhB. Such a good performance of **5** and **7** may also be attributed to the smaller energy gap between the UV light (3.40 eV) and themselves.

Two entangled CPs formulated as [Mn$_3$(oba)$_3$(bbm)(H$_2$O)$_2$]·DMF·2H$_2$O (**8**) (three-dimensional (3D) net with {$4^{21}·5^4·6^3$} topology) and [Co$_5$(bpb)$_3$(oba)$_4$(Hoba)$_2$(H$_2$O)$_2$] (**9**) (3D net with a $(6^3)(6^6)$ topology) were prepared (Fig. 14.3a, b) [11]. Their energy band gaps (E_{onset}) were estimated to be 3.44 eV (**8**) and 3.27 eV (**9**), respectively. Subsequently, under a UV lamp with a 375 nm wavelength, the photocatalytic activities of **8** and **9** were assessed by the degradation of MB. When the solution was exposed to UV light for 150 min (**8**) or 200 min (**9**), the degradation ratio of MB over **8** or **9** reached 95%. Hence, compound **8** showed a better photocatalytic performance for the decomposition of MB in water than **9** upon UV irradiation. Whereas, if no photocatalyst existed, the degradation of MB could not be discerned even after 200 min of UV light irradiation. In addition, two entangled networks of [Zn(pbda)(bpp)] (**10**) and [Zn(pbta)$_{0.5}$(bpp)(H$_2$O)] (**11**) (Fig. 14.3c, d) with the bandgap values (E_{onset}) of 3.4 and 3.6 eV, respectively, were also isolated and used to degrade MB in water using sunlight irradiation [12], which was quite rare choosing natural sunlight as a light source. The photocatalytic activity

Fig. 14.1 **a** Curves of absorbance of the MO solution degraded by **1**, **2**, and **3** under UV light. **b** Curves of absorbance of the MB solution degraded by **1**, **2**, and **3** under UV light. Reprinted with permission from Ref. [9]. Copyright 2013 American Chemical Society

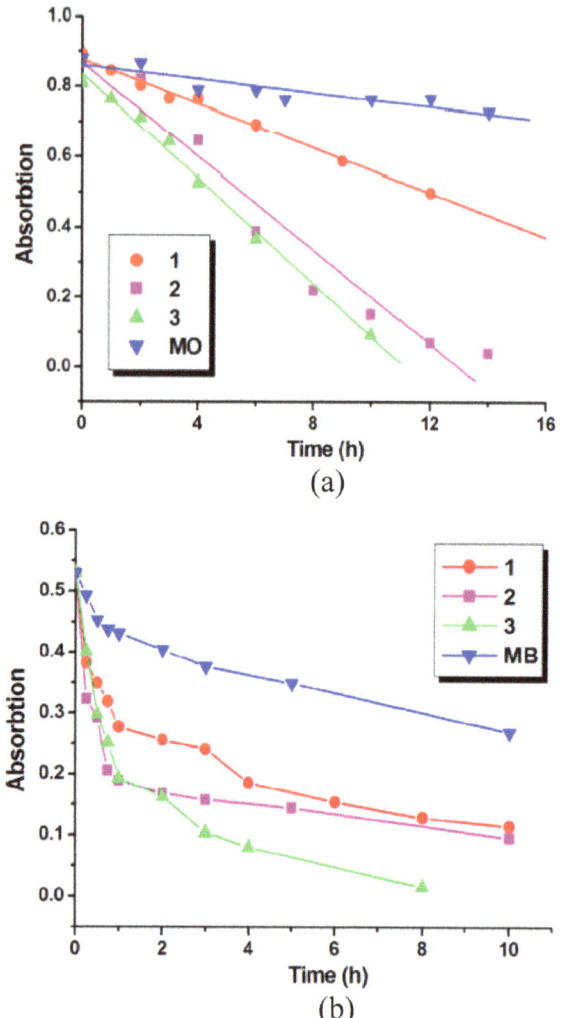

got increased from 25% (without any catalyst) to 87% for **10** and 83% for **11** after 180 min. Obviously, the degradation efficiency of **10** under sunlight irradiation was slightly higher than **11**, which was attributed to the smaller energy gap of **10**.

One CP, formulated as [Zn(1,4-BDC)(bmimb)]$_n$ (**12**) with an interpenetrating 3D framework (Fig. 14.4a), was successfully prepared [13]. Its photocatalytic activity was evaluated for the degradation of RhB in an aqueous solution under 400 W high-pressure mercury vapor lamp. About 90% RhB was decomposed within 4.0 h. In addition, the photocatalytic activity of another CP [Cd(1,3-BDC)(bmimb)]$_n$ (**13**) featuring a two-dimensional (2D) network without interpenetration (Fig. 14.4b) for the degradation of RhB was also investigated under the same conditions. About

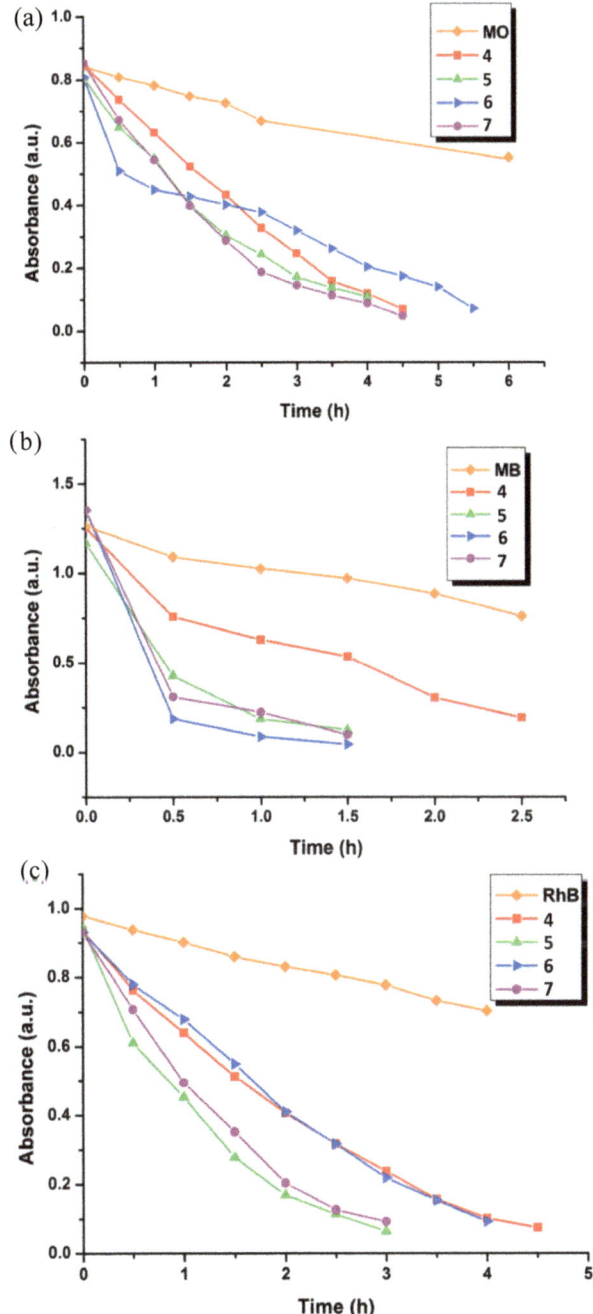

Fig. 14.2 a Curves of absorbance of the MO solution degraded by **4–7** under UV light. **b** Curves of absorbance of the MB solution degraded by **4–7** under UV light. **c** Curves of absorbance of the RhB solution degraded by **4–7** under UV light. Reprinted with permission from Ref. [10]. Copyright 2014 The Royal Society of Chemistry

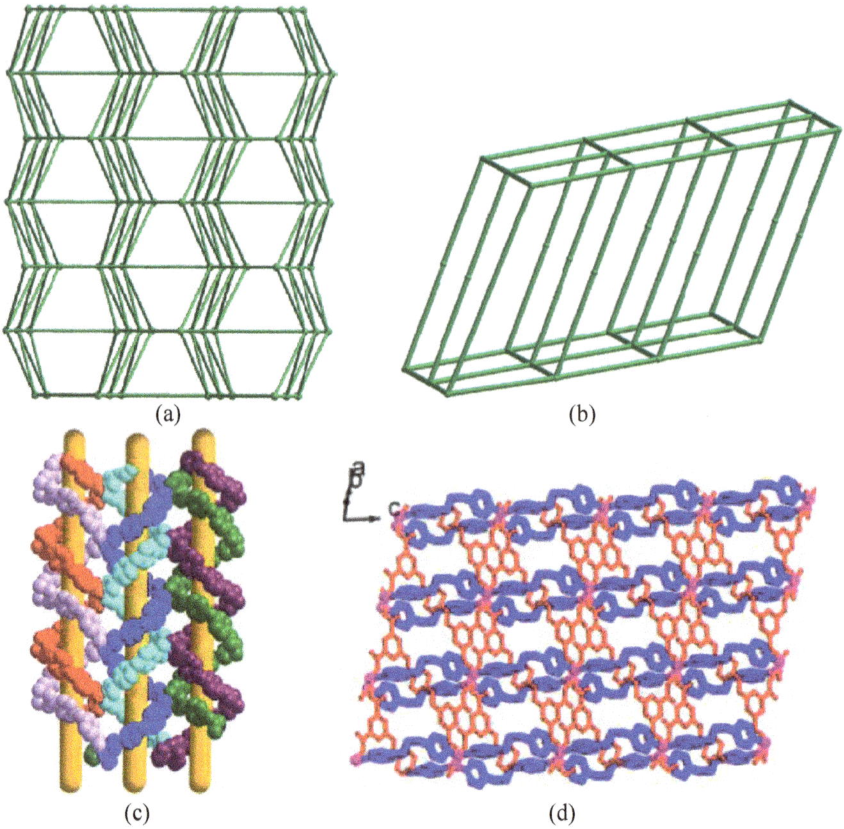

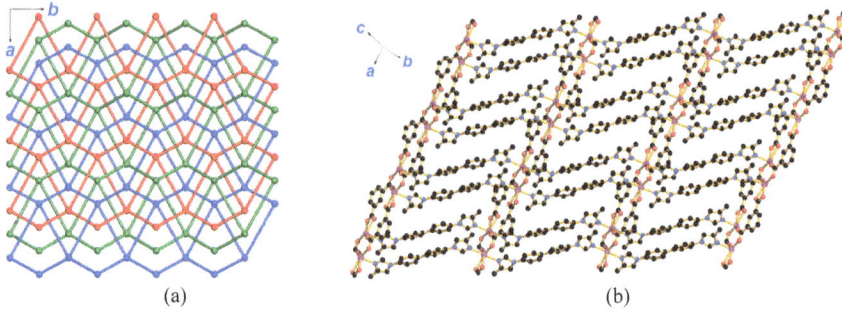

Fig. 14.3 **a** Schematic view of the $4^{21} \cdot 5^4 \cdot 6^3$ topological net of **8**. **b** Schematic view of the 3D net of **9** with a $(6^3)(6^6)$ topology. **c** Two identical layers interpenetrated into each other forming the 2D + 2D → 2D in **10**. **d** View of the 3D interpenetrating structure of **11**. Reprinted with permission from Ref. [11]. Copyright 2014 The Royal Society of Chemistry. Reprinted with permission from Ref. [12]. Copyright 2015 American Chemical Society

Fig. 14.4 **a** Threefold interpenetrating 3D structure of **12**. **b** 2D network of **13**. Reprinted with permission from Ref. [13]. Copyright 2015 The Royal Society of Chemistry

90% RhB was decomposed within 3.5 h. These results suggest that the degree of interpenetration has a slight influence on the degradation of RhB. Meanwhile, both **12** and **13** can be recovered from the catalytic systems and reused for five catalytic cycles without obvious loss of crystallinity as revealed by PXRD analysis.

Eight Zn(II) CPs with 5-R-1,3-benzenedicarboxylate (R = H, Br) ligands and bis-imidazolylligands, $\{[Zn(1,3\text{-BDC})(abimb)]\cdot H_2O\}_n$ (**14**) (fourfold interpenetrating 3D net), $\{[Zn(5\text{-Br-1,3-BDC})(abimb)]_2\cdot 0.5H_2O\}_n$ (**15**) (polythreaded 3D net), $[Zn(1,3\text{-BDC})(nbimb)]_n$(**16**) and $[Zn(5\text{-Br-1,3-BDC})(nbimb)]_n$ (**17**) (3D net), $\{[Zn(1,3\text{-BDC})(bimpa)]\cdot 2H_2O\}_n$ (**18**) (parallel (2D → 2D) twofold interwoven network), $\{[Zn(5\text{-Br-1,3-BDC})(bimpa)]\cdot 2H_2O\}_n$ (**19**) (3D net similar to that of **14**), $[Zn(1,3\text{-BDC})(bimcz)]_n$ (**20**) (3D polythreaded net), and $[Zn(5\text{-Br-1,3-BDC})(bimcz)]_n$ (**21**) (3D net) were prepared and subsequently evaluated for the degradation of MB in aqueous solutions [14]. Their energy band gaps (E_{onset}) were estimated to be 2.77 eV (**14**), 2.94 eV (**15**), 2.90 eV (**16**), 2.71 eV (**17**), 3.09 eV (**18**), 3.29 eV (**19**), 2.81 eV (**20**), and 3.00 eV (**21**), respectively. Surprisingly, the degradation of MB was slowed down by the presence of **14** rather than 'without catalysts'. One reason for this was that its solid surface might have a relatively low affinity to capture MB. The other was that the overall 3D entangled structure of **14** may not be preferable for the generation of electrons in the photo-induced charge separation process. For **16–21**, the degradation ratios after 5 h of UV light irradiation were 54.4%, 59.4%, 75.7%, 75.0%, 62.8%, and 75.9%, respectively. Compound **15** showed the highest photoactivity and the degradation ratio of 84.8% in 5 h under UV light irradiation. These complexes exhibited various photocatalytic performances, which may be consistent with their structural diversity and be related to the electronic charges on their solid surface.

Two Ag(I) complexes with a P–N hybrid ligand, $\{Ag(Sal)(3\text{-bdppmapy})\}_n$ (**22**) and $\{[Ag_4(Sal)_2(\mu\text{-}\eta,\eta^2\text{-Sal})_2](3\text{-bdppmapy})_2\}_n$ (**23**), displayed high catalytic activity toward the photodegradation of RhB in water [15]. The photodegradation of RhB followed zero-order kinetics, which can be written as: $c = c_0 - kt$, where c_0 and c are the initial and the remaining concentrations of RhB at regular time intervals, respectively, and k is the apparent zero-order rate constant. The k values were calculated to be 0.78×10^{-5} (without catalyst), 2.85×10^{-5} (3-bdppmapy), 4.68×10^{-5}(**22**) and 7.97×10^{-5} (**23**) mol L^{-1} h^{-1}, respectively. The results showed that the degradation of RhB without catalyst was quite slow, but greatly speeded up upon the addition of different catalysts.

One silver(I) complex $[Ag_4I_4(3\text{-bdppmapy})_2]_n$ (**24**) possesses a unique 2D network in which chairlike $[Ag_4I_4]$ units are interconnected by $\mu_3\text{-}3\text{-bdppmapy}$ bridges. It exhibited excellent catalytic activity toward the photodecomposition of a spectrum of 11 organic dyes in water under UV light irradiation and can be reused five times without noticeable decay of its catalytic efficiency [16]. Under dark conditions, the concentration of RhB was almost unchanged in the presence of **24**. UV light irradiation in the absence of **24** also did not result in the obvious photocatalytic decomposition of RhB. In the presence of the photocatalysts and UV irradiation, RhB was decomposed up to 99.6% within 75 min, which demonstrated that both UV light and photocatalyst were indispensable for the decomposition of RhB. The catalytic performance of **24** was also compared with that of TiO_2 under the same experimental conditions. In the case of TiO_2, only 69% of RhB was decomposed within 75 min, suggesting that the catalytic activity of **24** was better than that of TiO_2. Meanwhile, compound **24** was also examined its capacity in the degradation of other organic dyes such as RhB, MO, orange I (OI), orange II (OII), orange IV (OIV), orange G (OG), acid red 27 (AR 27), sunset yellow (SY), amido black 10B (AB 10B), acid chrome blue K (ACBK), eriochrome black T (EBT) in water. These dyes were found to be degraded within 150 − 360 min, implying that **24** possessed a big capacity for dye degradation under UV light irradiation.

$[Ag(\mu\text{-}\eta^2\text{-}Sal)(3\text{-}dppmapy)]$ (**25**) was synthesized by the reaction of a P–N hybrid ligand N-diphenylphosphanylmethyl-3-aminopyridine (3-dppmapy) and AgSal (Sal = salicylate), exhibiting good stability in water solution under UV irradiation [17]. Therefore, compound **25** was applied to degrade eriochrome black T (EBT), amido black (AB), and nigrosin (NO) in aqueous phase with high photocatalytic performance. In the presence of **25**, the dyes were completely degraded under UV irradiation after 1.5 h (for EBT), 6 h (for AB), and 6 h (for NO), respectively.

Solvothermal reactions of zinc and cadmium nitrates with 1,3,5-tri-4-pyridyl-1,2-ethenylbenzene (tpeb) in the presence of 1,2-cyclohexanedicarboxylic acid (1,2-H_2CHDC) gave rise to $\{[Cd_2(H_2O)_2(tpeb)_2(1,2\text{-}CHDC)_2]\cdot H_2O\}_n$ (**26**), which has a 2D wavelike layer structure. It has evident absorbance in the visible light region and can be employed to efficiently degrade Congo red (CR) in water without additional oxidizing or reducing reagents upon visible light irradiation [18]. The photobleaching of CR was negligible in the absence of **26**. However, in the presence of **26**, 90% of CR was decomposed upon visible light irradiation for 90 min. This photocatalyst could be reused five times without losing catalytic efficiency.

One Ag(I)/tetraphosphine 1D chain-like CP $[Ag_4(NO_3)_4(dpppda)]_n$ (**27**) was prepared by the reaction of $AgNO_3$ with one tetraphosphine ligand, 1,4-N,N,N′,N′-tetra(diphenylphosphanylmethyl)benzene diamine (dpppda), which exhibited good catalytic activity toward the photodegradation of nitrobenzene (NB), paranitrophenol (PNP), and 2,4-dinitrophenol (2,4-DNP) in aqueous solution under UV light irradiation [19]. These three nitroacromatics were all decomposed completely in the presence of **27** after UV light irradiation for 5 h (for NB and PNP) and 6 h (for 2,4-DNP) (Fig. 14.5a). The concentration of the remaining nitroaromatic was decreased regularly as the irradiation time was prolonged. This curve followed a pseudo-zero-order kinetics character, implying that such a catalytic photodegradation reaction might take place at the surface of **27**. Likewise, the zero-order kinetics model could be expressed by: $c = c_0 − kt$, where c_0 and c are the initial and the remaining concentrations of each nitroaromatic at regular time intervals, respectively, and k is the apparent zero-order rate constant. The k values were calculated to be 0.81×10^{-4} (NB), 0.79×10^{-4} (PNP) and 0.73×10^{-4} (2,4-DNP) mol L^{-1} h^{-1}, respectively. Compared with the photodegradation of NB catalyzed by TiO_2 nanoparticles coated on a quartz tube, the k value was *ca.* 3 times larger. Meanwhile, under the catalysis of TiO_2, three nitroacromatics were completely decomposed after taking longer irradiation time (7.5 h for NB, 5 h for PNP, and 7 h for 2,4-DNP), which indicated that **27** had a better catalytic activity than TiO_2 under the same experimental conditions.

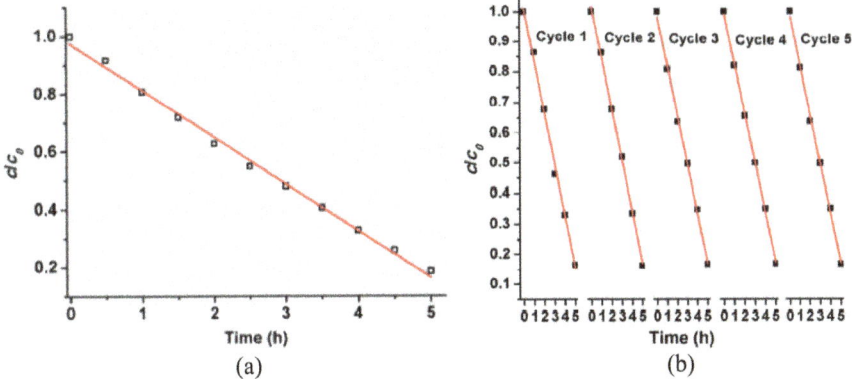

Fig. 14.5 **a** Pseudo-zero-order plot for the photodegradation of NB in the presence of **27** under UV light irradiation. **b** The repeated catalytic performance of **27** for the photodegradation of NB in water under UV light irradiation. The square dots and the red line are the experimental data and the fitted least-square line respectively. Reprinted with permission from Ref. [19]. Copyright 2014 Elsevier BV

In addition, catalyst **27** was quite stable and showed no obvious catalytic efficiency decay after the fifth cycle reaction (Fig. 14.5b).

Solvothermal reactions of $HAuCl_4 \cdot 4H_2O$ with a P–S hybrid ligand N,N-bis(diphenylphosphanylmethyl)-amino-thiocarbamide (dppatc) at 80 and 115 °C produced two Au–P–S complexes, $[Au_2(dppatc)_2]Cl_2$ (**28**) and $[Au_2(dppmt)_2]_n$ (**29**) (dppmtH = (diphenylphosphino)methanethiol) [20]. The E_g values were estimated to be 2.85 eV (**28**) and 2.98 eV (**29**), implying a semiconducting nature. Therefore, both **28** and **29** can be activated by UV light and have a potential capacity for photocatalytic reactions. When the reaction solution was mixed with the catalyst and irradiated by UV light, each substrate could be fully decomposed within 3.5 h to 6 h (for **28**: NB, 3.5 h; PNP, 3.5 h; DNP, 4 h; for **29**: NB, 4.5 h; PNP, 4 h; DNP, 6 h). The degradation reactions followed the zero-order kinetic model, in which three nitroaromatics could be converted into CO_2 and H_2O in 92–96% yields.

14.3   Photodegradation of Dyes Using Ion-Doped CPs

Although several CP-based photocatalysts for photodegrading dyes under UV light have been reported, to design and fabricate highly photocatalytic efficient and visible-light active CPs is still a big challenge because of the wide band gaps of the photocatalysts and fast recombination rate between photogenerated electrons and holes during the photocatalytic process. Therefore, to break through this challenge, it is necessary to modify CP-based photocatalysts with certain techniques. In the past few years, some synthetic approaches to modify CP-based materials have been developed. One is to dope different metal ions into the frameworks of CPs, which is a useful strategy to narrow the bandgap, thus improving the photocatalytic activity. Several ion-doped coordination polymers used to degrade dyes under UV light with high efficiency will be introduced below.

In order to contrast the catalytic photodegradation activities of CPs and ion-doped CPs, two isostructural CPs $[Zn_2(tipm)(1,3-BDC)_2]$ (**30**) and $[Co_2 (tipm)(1,3-BDC)_2]\cdot0.5CH_3CN$ (**31**$\cdot0.5CH_3CN$) were synthesized by the reactions of $Zn(NO_3)_2$ or $Co(NO_3)_2$ with tetrakis[4-(1-imidazolyl)phenyl]methane (tipm) and benzene-1,3-dicarboxylic acid (1,3-H_2BDC) under solvothermal conditions and two cobalt ion-doped CPs, $[Zn_{1.952}Co_{0.048}(tipm)(1,3-BDC)_2]$ (**32**) and $[Zn_{1.54}Co_{0.46}(tipm)(1,3-BDC)_2]\cdot H_2O$ (**33**) were also prepared (Table 14.2) [21]. The energy band gaps were estimated to be 2.25 (**30**), 1.38 (**31**), 1.80 V (**32**), and 1.77 eV (**33**), which were in the semiconductor range. Therefore, these four materials have been applied to degrade RhB in aqueous solution. When the reaction solutions were mixed with an equimolar catalyst and irradiated by UV light, the RhB was almost completely decomposed within 4 h (for **30** at 4 h,95%; for **31** at 3 h, 97%; for **32** at 2.5 h, 94%; and for **33** at 2 h, 90%). These results indicate that the Co-doped CPs **32** and **33** are superior photocatalysts than the undoped CPs **30** and **31** and further demonstrate the importance of metal-ion doping to the tuning of the properties of CP-based catalyst materials. Besides, the doped CP **33** showed no significant decrease in efficiency even after five cycles.

The solid-state reaction of $\{[Pb(Tab)_2]_2(PF_6)_4\}_n$ (**34**) (TabH $=$ 4-(trimethylammonio)benzenethiol) with 1,2-bis(4-pyridyl)ethylene (bpe) at room temperature afforded a unique 2D CP $\{[Pb(Tab)_2(bpe)]_2(PF_6)_4\}_n$ (**35**). The corresponding Ag(I) ion-doped CP $\{[Pb(Tab)_2(bpe)]_2(PF_6)_4\cdot1.64AgNO_3\}_n$ (**36**) was readily prepared by immersing **35** into $AgNO_3$ aqueous solution [22]. The energy band gaps of **34**-**36** are 2.51 eV, 2.52 eV, and 2.46 eV, respectively, implying a semiconductive nature, and **36** possessed the narrowest bandgap, suggesting that the 'doping strategy' may be a feasible approach to narrow the bandgap of the parent complex. These three CPs were used to photodegrade 12 different kinds of azo dyes including MO, AO 7, OI, OIV, OG, CR, AR 27, SY, AB 10B, NO, ACBK, and EBT in water under UV light irradiation. Under the same experimental conditions, **34**–**36** all exhibited good photocatalytic behaviors for the MO decomposition. Among them, **36** showed the highest photoactivity and 95% of MO was degraded in *ca.* 50 min under UV light irradiation, whereas for **34** or **35**, it took 120 min to decompose about 35% of MO. For **34** and **35**, AB 10B could be decomposed within 60 and 25 min, while it only took 12 min to degrade this dye by using **36** as a catalyst. In the case of CR, only 10% of the substrate was degraded in the absence of the catalysts in 2 h. Surprisingly, **36** showed significant enhancement in the photodegradation of CR compared with **34** and **35**. Almost 100% of CR was eliminated within 3 min (**36**), 18 min (**35**), and 30 min (**34**). Moreover, **36** also displayed better catalytic activity for the photodegradation of other azo dyes employed than that of either **34** or **35**. These results showed that the Ag(I)-doped complex **36** was more efficient than its precursors in the catalytic photodecomposition of azo dyes under UV light irradiation, which was ascribed to the fact that it holds the narrowest bandgap among the three complexes.

Table 14.2 The photocatalytic performances of various ions-doped CPs for the degradation of dyes and nitroaromatics in aqueous media

Coordination polymers[a]	Eg (eV)	Light sources	Organic pollutants[b]	Degradation efficiency (%)
$[Zn_2(tipm)(1,3\text{-}BDC)_2]$ (**30**)	2.25	UV	RhB	95
$[Co_2(tipm)(1,3\text{-}BDC)_2]\cdot0.5CH_3CN$ (**31**)	1.38	UV	RhB	97
$[Zn_{1.952}Co_{0.048}(tipm)(1,3\text{-}BDC)_2]$ (**32**)	1.80	UV	RhB	94
$[Zn_{1.54}Co_{0.46}(tipm)(1,3\text{-}BDC)_2]\cdot H_2O$ (**33**)	1.77	UV	RhB	90
$\{[Pb(Tab)_2]_2(PF_6)_4\}_n$ (**34**)	2.51	UV	MO, AO 7, OI, OIV, OG, CR, AR 27, SY, AB 10B, NO, ACBK, EBT	95, 100, 100, 30, 60, 100, 100, 100, 100, 100, 100, 100
$\{[Pb(Tab)_2(bpe)]_2(PF_6)_4\}_n$ (**35**)	2.52	UV	MO, AO 7, OI, OIV, OG, CR, AR 27, SY, AB 10B, NO, ACBK, EBT	95, 100, 100, 80, 60, 100, 100, 100, 100, 100, 100, 100
$\{[Pb(Tab)_2(bpe)]_2(PF_6)_4\cdot1.64AgNO_3\}_n$ (**36**)	2.46	UV	MO, AO 7, OI, OIV, OG, CR, AR 27, SY, AB 10B, NO, ACBK, EBT	95, 100, 100, 85, 95, 100, 100, 100, 100, 100, 100, 100
$[Cd(ppvppa)_2(1,3\text{-}bdc)]\cdot2H_2O$ (**37**)	2.62	UV	RhB, AR 27, theophylline	100, 100, 95
$[Co(ppvppa)_2(1,3\text{-}bdc)]\cdot1.5H_2O$ (**38**)	2.71	UV	RhB, AR 27, theophylline	100, 100, 95
$[Cd(ppvppa)_2(1,3\text{-}bdc)]\cdot AgNO_3$ (**39**)	2.58	UV	RhB, AR 27, theophylline	100, 100, 95
$[Co(ppvppa)_2(1,3\text{-}bdc)]\cdot0.75AgNO_3$ (**40**)	2.42	UV	RhB, AR 27, theophylline	100, 100, 95

[a]tipm = tetrakis[4-(1-imidazolyl)phenyl]methane; TabH = 4-(trimethylammonio)benzenethiol; bpe = 1,2-bis(4-pyridyl)ethylene; ppvppa = N-(pyridin-2-yl)-N-{4-[2-(pyridin-4-yl)vinyl]phenyl}pyridin-2-amine; 1,3-H$_2$bdc = isophthalic acid. [b]RhB = rhodamine B; MO = methyl orange; AO 7 = acid orange 7; OI = orange I; OIV = orange IV; OG = orange G; CR = congo red; AR 27 = acid red 27; SY = sunset yellow; AB 10B = amido black 10B; NO = nigrosin; ACBK = acid chrome blue K; EBT = eriochrome black T

In addition, the possible mechanism of photodegradation of various organic pollutants by Ag(I)-doped complex **36** was shown in Fig. 14.6. For **34** and **35**, electrons transfer to CB from VB and positively charged holes are formed in VB under UV light irradiation. After reaching the surface of the catalyst, the electrons and holes generate radical species such as $\cdot O_2^-$ and $\cdot OH$, which can destroy azo dyes. However,

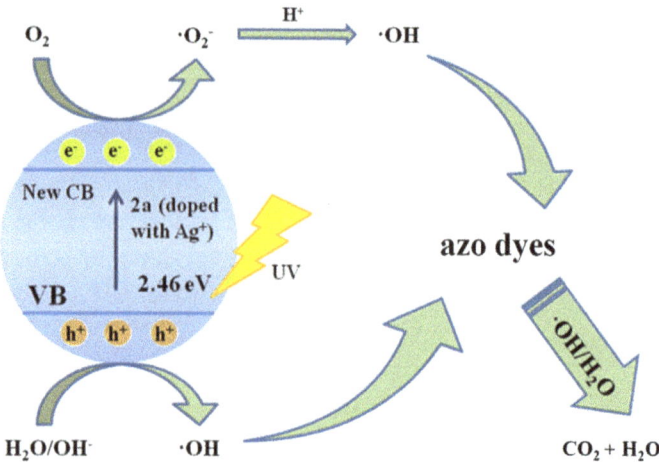

Fig. 14.6 Schematic illustration of the catalytic photodegradation process and the charge separation of **36** under UV light irradiation. Reprinted with permission from Ref. [22]. Copyright 2015 The Royal Society of Chemistry

with the doping of Ag$^+$ into the network of **35**, a new energy level can be generated between the 4d orbital of Ag$^+$ and the CB of **35**, which amounts to the CB of **36**. The VB energy level of **36** remains the same as that of **35**, while its CB energy level is lower compared with that of **35**, which leads to a decrease in the bandgap. Therefore, it is easier for **36** to excite electrons from VB to the newly formed CB than **34** and **35** and generate the corresponding radical species more quickly.

Two 2D CPs, [Cd(ppvppa)$_2$(1,3-bdc)]·2H$_2$O (**37**) and [Co(ppvppa)$_2$(1,3-bdc)]·1.5H$_2$O (**38**) were synthesized from solvothermal reactions of M(NO$_3$)$_2$ (M = Cd, Co) with isophthalic acid (1,3-H$_2$bdc) and N-(pyridin-2-yl)-N-{4-[2-(pyridin-4-yl)vinyl]phenyl}pyridin-2-amine (ppvppa) at 150 °C [23]. In both **37** and **38**, the uncoordinated di(pyridin-2-yl)amine of ppvppa can further bind additional metal ions. Therefore, two Ag(I) ion-doped CPs, [Cd(ppvppa)$_2$(1,3-bdc)]·AgNO$_3$ (**39**) and [Co(ppvppa)$_2$(1,3-bdc)]·0.75AgNO$_3$ (**40**), were successfully obtained via postmodification of **37** and **38**, respectively. The catalytic activities of **37–40** were evaluated by examining the photodegradation of RhB, AR 27, and theophylline. Compared with **37** and **38**, **39** and **40** exhibited greatly enhanced catalytic activities toward the photodegradation of RhB, AR 27, and theophylline in water due to their narrower bandgaps. For **37** and **38**, it took 110 min to destroy AR 27, whereas for **39** and **40**, only 70 and 45 min, respectively, were needed. Compounds **39** and **40** showed similar catalytic activity enhancement in the photodegradation of theophylline compared with **37** and **38**. For **40**, the apparent rate constants for the photodecomposition of the three pollutants are around 4.4, 2.3, or 1.76 times larger than those of **38**, respectively. The improved catalytic activities of **39** and **40** are likely due to the doping Ag(I) ions in the frameworks of the CPs which narrows the band gaps and speeds up the separation of photo-induced electrons and holes. The results provided an insight into a metal-ion doping strategy that can narrow the band gaps of the CPs and thus greatly improve their photocatalytic performance in degrading organic pollutants in water.

14.4 Photodegradation of Dyes Using Core-Shell-Like CP-Based Materials

Core-shell nanostructures containing a metal or metal oxide/sulfide species inside a void space surrounded by a semiconductor shell have been widely investigated, which can greatly combine the advantages of high catalytic activity of nanometal cores and the hollow structures of CPs [24]. Due to the ease of controlling the porosity and morphological properties, the large surface area, and a broad variety of metallic centers, CPs materials have also captured extensive attention in the field of photocatalysis in the past two decades. They can offer the broad platform to encapsulate photocatalytically active species as well as photo-sensitizers to form core-shell-like CP-based materials [25], which are similar to those of core-shell structures described in this book. Described below are two examples related to such structures and their performances of photodegrading dyes in water, though limited in number.

One example is that a nanomaterial Ag/AgCl@MIL-101 (MIL = Matériaux de l'Institut Lavoisier) possessing great visible-light response was prepared via vapor diffusion-photoreduction strategy, in which Ag/AgCl hybrid nanoparticles with 200–500 nm were uniformly distributed on the surface of MIL-101 submicrocubes (Fig. 14.7) [26]. Under visible light irradiation, this photocatalyst displayed improved photocatalytic activity to degrade RhB in water utilizing their synergistic effect, which is likely due to the large surface area of MIL-101 and good dispersion of Ag/AgCl on the surface of MIL-101.

The other example is that a nanocomposite AgBr@HPU-4 based on a 3D Cu-based CP, $\{[Cu(L)_{0.5} (H_2O)_2] \cdot 4H_2O\}_n$ (HPU-4) (L = 5,5′-(1H,1′H-[2,2′-biimidazole]-1,1′-diylbis(methylene))diisophthalic acid) was prepared through a facile reaction route [27]. This material has strong optical performances nearly in the whole range of visible light spectrum seen from its UV-Vis absorption spectrum. Therefore, the photocatalytic activity of AgBr@HPU-4 under visible-light irradiation was investigated toward the degradation of MB and MO dyes. The results revealed that 95% of MB was degraded in 60 min and 92% of MO could be decomposed in 120 min under visible-light irradiation. Compared with HPU-4, AgBr and the physical mixture of pure HPU-4 and AgBr, the photocatalytic activity of AgBr@HPU-4 is much higher. A possible mechanism is suggested and displayed in Fig. 14.8. Upon visible light irradiation, AgBr can absorb visible light, and many electron-hole pairs are produced. The electrons quickly transport to the HPU-4. Meanwhile, the photogenerated electron-hole pairs would also separate through the ligand to metal charge transfer process, keeping a photo-induced electron in the conduction band (CB) and a positive-charged

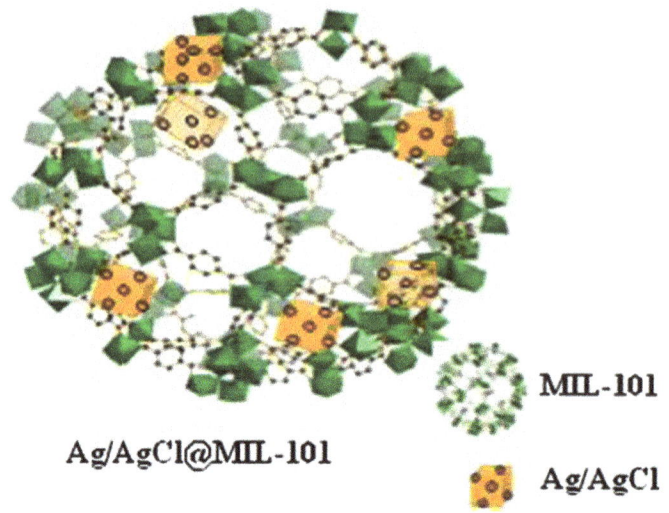

Fig. 14.7 Schematic view of the Ag/AgCl@MIL-101. Reprinted with permission from Ref. [26]. Copyright 2015 Elsevier Inc.

Fig. 14.8 The schematic illustration of degradation of dyes. Reprinted with permission from Ref. [27]. Copyright 2019 Elsevier Ltd.

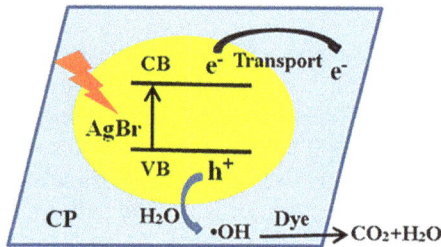

hole in the valence band (VB). The electrons may be adsorbed by dissolved O_2 and H_2O molecules on the photocatalyst surface to yield O_2^-, $\cdot O_2^-$, and other reactive oxygen species, which are further reacted with H_2O to afford $\cdot OH$ [28]. In addition, the holes in the valence band can directly oxidize H_2O to form $\cdot OH$. Then the highly reactive $\cdot OH$ can destroy the dye molecules.

14.5 Summary and Outlook

This chapter mainly deals with three types of self-supported photocatalysts, that is, pristine CPs, metal-ion-doped CPs, and core-shell-like CP-based materials, which show remarkable photocatalytic activities [24]. Under the carrier-free, low-cost, highly efficient, and ecofriendly conditions, organic dyes and nitrobenzene compounds can be readily degraded by using the above photocatalysts. It is anticipated that there is still much space to develop low-cost, highly efficient and eco-friendly CP-based or modified CP-based photocatalysts, which can be used not only for the decomposition of various organic pollutants, but also for other applications such as photocatalytic hydrogen evolution, photocatalytic oxygen absorption as well as photocatalytic carbon dioxide conversion. Meanwhile, the photoactivity of CPs may be further improved if some photosensitive nanomaterials such as CdS, CdSe, CdTe, and ZnS nanoparticles can be successfully incorporated or introduced into the frameworks of CPs to form core-shell like structures.

References

1. Wintgens T, Salehi F, Hochstrat R, Melin T (2008) Emerging contaminants and treatment options in water recycling for indirect potable use. Water Sci Technol 57:99–107
2. Wang M, Ioccozia J, Sun L, Lin C, Lin Z (2014) Inorganic-modified semiconductor TiO_2 nanotube arrays for photocatalysis. Energy Environ Sci 7:2182–2202
3. Dong SY, Feng JL, Fan MH, Pi YQ, Hu LM, Han X, Liu ML, Sun JY, Sun JH (2015) Recent developments in heterogeneous photocatalytic water treatment using visible light-responsive photocatalysts: a review. RSC Adv 5:14610–14630
4. Thompson TL, Yates JT (2006) Surface science studies of the photoactivation of TiO_2—new photochemical processes. Chem Rev 106:4428–4453
5. Kansala SK, Singh M, Sud D (2007) Studies on photodegradation of two commercial dyes in aqueousphase using different photocatalysts. J Hazard Mater 141:581–590
6. Li Y, Wang WN, Zhan Z, Woo MH, Wu CY, Biswas P (2010) Photocatalytic reduction of CO_2 with H_2O on mesoporous silica supported Cu/TiO_2 catalysts. Appl Cata B: Environ 100:386–392
7. Wang CC, Li JR, Lv XL, Zhang YQ, Guo G (2014) Photocatalytic organic pollutants degradation in metal–organic frameworks. Energy Environ Sci 7:2831–2867
8. Xu XX, Cui ZP, Gao X, Liu XX (2014) Photocatalytic activity of transition-metal-ion doped coordination polymer (CP): photoresponse region extension and quantum yields enhancement via doping of transition metal ions into the framework of CPs. Dalton Trans 43:8805–8813
9. Dai M, Su XR, Wang X, Wu B, Ren ZG, Zhou X, Lang JP (2014) Three zinc(II) coordination polymers based on tetrakis(4-pyridyl)cyclobutane and naphthalenedicarboxylate linkers: solvothermal syntheses, structures, and photocatalytic properties. Cryst Growth Des 14:240–248
10. Li DX, Ni CY, Chen MM, Dai M, Zhang WH, Yan WY, Qi HX, Ren ZG, Lang JP (2014) Construction of Cd(II) coordination polymers used as catalysts for the photodegradation of organic dyes in polluted water. CrystEngComm 16:2158–2167
11. Hu FL, Wang SL, Wu B, Yu H, Wang F, Lang JP (2014) Ligand geometry-directed assembly of seven entangled coordination polymers. CrystEngComm 16:6354–6363
12. Wang SL, Hu FL, Zhou JY, Zhou Y, Qin Huang, Lang JP (2015) Rigidity versus flexibility of ligands in the assembly of entangled coordination polymers based on bi- and tetra carboxylates and N-donor ligands. Cryst Growth Des 15:4087–4097
13. Lü CN, Chen MM, Zhang WH, Li DX, Dai M, Lang JP (2015) Construction of Zn(II) and Cd(II) metal-organic frameworks of diimidazole and dicarboxylate mixed ligands for the catalytic photodegradation of rhodamine B in water. CrystEngComm 17:1935–1943
14. Cheng HJ, Tang XY, Yuan RX, Lang JP (2016) Structural diversity of Zn(II) coordination polymers based on bis-imidazolyl ligands and 5-R-1,3-benzenedicarboxylate and their photocatalytic properties. CrystEngComm 18:4851–4862

15. Wang JF, Liu SY, Liu CY, Ren ZG, Lang JP (2016) Silver(I) complexes with a P-N hybrid ligand and oxyanions: synthesis, structures, photocatalysis and photocurrent responses. Dalton Trans 45:9294–9306

16. Liu CY, Xu LY, Ren ZG, Wang HF, Lang JP (2017) Assembly of silver(I)/N, N-Bis(diphenylphosphanylmethyl)-3-aminopyridine/halide or pseudohalide complexes for efficient photocatalytic degradation of organic dyes in water. Cryst Growth Des 17:4826–4834

17. Xu LY, Yang W, Liu CF, Ren ZG, Lang JP (2018) Assembly of silver-oxo complexes based on N-diphenylphosphanylmethyl-3-aminopyridineand their structures, photocatalysis and photocurrent responses. CrystEngComm 20:4049–4057

18. Zhang JG, Gong WJ, Guan YS, Li HX, Young DJ, Lang JP (2018) Carboxylate-assisted assembly of zinc and cadmium coordination complexes of 1,3,5-tri-4-pyridyl-1,2-ethenylbenzene: structures and visible-light-induced photocatalytic degradation of congo red in water. Cryst Growth Des 18:6172–6184

19. Wu XY, Qi HX, Ning JJ, Wang JF, Ren ZG, Lang JP (2015) One silver(I)/tetraphosphine coordination polymer showing good catalytic performance in the photodegradation of nitroaromatics in aqueous solution. Appl Cata B: Environ 168:98–104

20. Qi HX, Wang JF, Ren ZG, Ning JJ, Lang JP (2015) Syntheses and structures of two gold(I) coordination compounds derived from P-S hybrid ligands and their efficient catalytic performance in the photodegradation of nitroaromatics in water. Dalton Trans 44:5662–5671

21. Li DX, Ren ZG, Young DJ, Lang JP (2015) Synthesis of two coordination polymer photocatalysts and significant enhancement of their catalytic photodegradation activity by doping with Co^{2+} Ions. Eur J Inorg Chem 11:1981–1988

22. Wang F, Li FL, Xu MM, Yu H, Zhang JG, Xia HT, Lang JP (2015) Facile synthesis of a Ag(I)-doped coordination polymer with enhanced catalytic performance in the photodegradation of azo dyes in water. J Mater Chem A 3:5908–5916

23. Chen MM, Li HX, Lang JP (2016) Two coordination polymers and their silver(I)-doped species: synthesis, characterization, and high catalytic activity for the photodegradation of various organic pollutants in water. Eur J Inorg Chem 15–16:2508–2515

24. Sun H, Shen XS, Yao L, Xing SX, Wang H, Feng YH, Chen HY (2012) Measuring the unusually slow ionic diffusion in polyaniline via study of yolk-shell nanostructures. J Am Chem Soc 134:11243–11250

25. Abazari R, Mahjoub AR (2018) Amine-functionalized Al-MOF@ySm$_2$O$_3$-ZnO: a visible light-driven nanocomposite with excellent photocatalytic activity for the photodegradation of amoxicillin. Inorg Chem 57:2529–2545

26. Gao ST, Feng T, Feng C, Shang NZ, Wang C (2016) Novel visible-light-responsive Ag/AgCl@MIL-101 hybrid materials with synergistic photocatalytic activity. J Colloid Interface Sci 466:284–290

27. Li HJ, Li QQ, He XL, Xu ZQ, Wang Y, Jia L (2019) Synthesis of AgBr@MOFs nanocomposite and its photocatalytic activity for dye degradation. Polyhedron 165:31–37

28. Zhao HM, Xia QS, Xing HZ, Chen DS, Wang H (2017) Construction of pillared-layer MOF as efficient visible-light photocatalysts for aqueous Cr(VI) reduction and dye degradation. ACS Sustainable Chem Eng 5:4449–4456

Chapter 15
Synthesis of Plasmonic Catalyst with Core-Shell Structure for Visible Light Enhanced Catalytic Performance

Meicheng Wen, Kohsuke Mori, Yasutaka Kuwahara, Guiying Li, Taicheng An, and Hiromi Yamashita

15.1 Introduction

Pd, one of the most active catalysts, has been frequently employed in the synthesis of a large number of natural products and biologically active compounds [1–3] .Usually, the Pd-catalyzed chemical reactions usually require a high temperature to conquer the potential barrier to further drive chemical reactions, which leads to a huge energy consumption. In contrast, solar energy, as unlimited and sustainable energy sources [4, 5], is becoming a promising energy source. The conversion of solar energy into electric energy and thermal energy for driving Pd-catalyzed chemical reactions at ambient temperatures has been emerging as a most promising method for the synthesis of a large number of natural products and biologically active compounds [6–9]. Since the discovery of electrochemical photolysis of water on TiO_2 semiconductor-based electrode [10, 11], semiconductor has attracted widespread attention in energy exploitation and environmental decontamination [12, 13], which allows toxic pollutant molecule decomposition or water splitting into O_2 and H_2 [14]. However, the wide bandgap of TiO_2 is about 3.2 eV, which deeply limits its absorption wavelengths to less than 390 nm. In terms of energy proportion, sunlight contains less than 5% of UV light, whereas visible light (400–800 nm) accounts for the majority (ca. 43%). Therefore, it is desirable to develop visible light-driven photocatalysis,

M. Wen · G. Li · T. An (✉)
Guangdong Key Laboratory of Environmental Catalysis and Health Risk Control, School of Environmental Science and Engineering, Institute of Environmental Health and Pollution Control, Guangdong University of Technology, Guangzhou 510006, China
e-mail: antc99@gdut.edu.cn

K. Mori · Y. Kuwahara · H. Yamashita (✉)
Graduate School of Engineering, Osaka University, 2-1 Yamada-oka, Suita, Osaka 565-0871, Japan
e-mail: yamashita@mat.eng.osaka-u.ac.jp

© The Author(s), under exclusive license to Springer Nature Singapore Pte Ltd. 2021   233
H. Yamashita and H. Li (eds.), *Core-Shell and Yolk-Shell Nanocatalysts*,
Nanostructure Science and Technology,
https://doi.org/10.1007/978-981-16-0463-8_15

which entails broader visible light absorption and higher selectivity towards organic synthesis over Pd-based catalysts.

In recent years, plasmonic catalysts, such as nanostructured metals (Au, Ag, and Cu) [15–19], and highly-doped semiconductors (WO_{3-x}, and MoO_{3-x}) [20–23], represented as a class of promising visible light responsive materials, have been widely used to improve the efficiency of catalysis either by their intense visible/near-infrared response or efficient charge separation. A large amount of free charge carriers in plasmonic catalysts oscillate with incident light, leading to the localized surface plasmon resonances (LSPRs). LSPR is the free conduction electrons in solid materials that resonated with the electromagnetic field of the incident light [24]. Hot electrons with high energy, which are generated through the LSPR effect, have been widely used to improve the performances of many applications, especially for organic synthesis [25, 26]. The unique characteristic of plasmonic materials has given rise to a new approach to sensitize Pd-active sites and boost their intrinsic activity of Pd, which opens up new avenues for the efficient conversion of the abundant sunlight into chemical energy.

15.2 Plasmonic Au@Pd Nanoparticle Supported on Metal-Organic Framework for Boosting the Hydrogen Production from Formic Acid

It is well-known that nanostructured Au can strongly absorb visible light efficiently owing to its LSPR effect [27]. The free conduction electrons confined in Au nanoparticles resonate with the electromagnetic field of the incident light. As a result, energetic hot electrons induced by LSPR can either be transferred to the neighboring absorbent or come back to the ground state by releasing the excess energy to the surrounding environment. In the case of plasmonic Au@Pd nanoparticles with core-shell structure, a plasmonically excited hot electron will transfer across the Au and Pd interface due to the lower electronegativity of Pd (2.20) than that of Au (2.54) [28], resulting in an increase of the surface electron density of Pd [29], which is useful to enhance the catalytic activity for H_2 production by dehydrogenation of formic acid. Plasmonic Au@Pd nanoparticle with core-shell structures supported on titanium (Ti) doped metal-organic frameworks (MOFs) were prepared for boosting hydrogen production from formic acid [30].

The core-shell structure of Au@Pd nanoparticles was investigated by HAADF-STEM. The HAADF-STEM image with high spatial resolution offers direct information on the morphology and structure of Au@Pd nanoparticle (Fig. 15.1a), in which the dark shell and bright core suggested the formation of a core–shell structure. Two types of crystal lattice were observed on the metal nanoparticle, the shell interplanar fringes were determined to be 2.78 and 1.58 Å (Fig. 15.1b), which is ascribed to the crystal lattice of the (101) and (211) planes of Pd. The exposed facet of analyzed nanoparticles was measured to be Pd. The core interplanar fringes of the nanoparticle

Fig. 15.1 HAADF-STEM image of Au@Pd nanoparticle (**a**), **b**, **c** magnified cross sections of the image, and **d** the cross-sectional EDX line profile of a single Au@Pd core–shell nanoparticle, **e**, **f** high-magnification EDX spectra of Pd and Au for an individual Au@Pd nanoparticle. Reproduced with permission from Ref. [30]. Copyright@2017 American Chemical Society

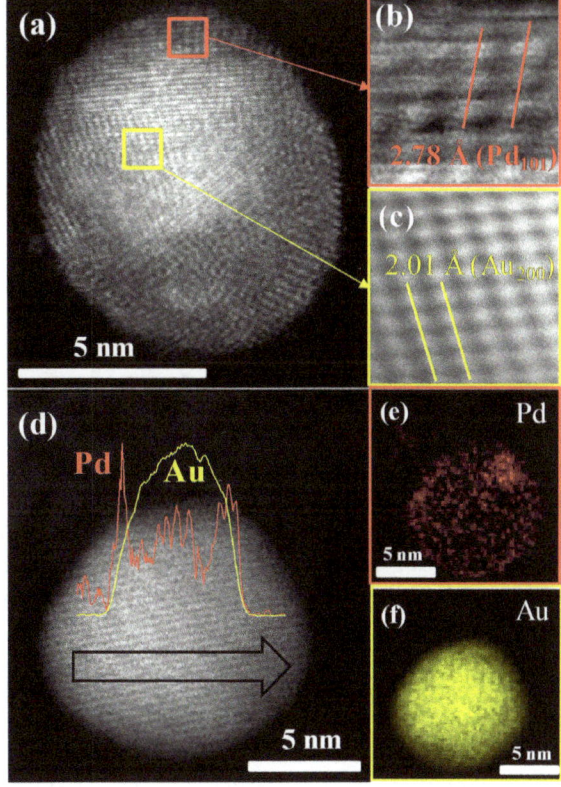

were measured to be 2.01 Å (Fig. 15.1c), which is ascribed to the crystal lattice of the (200) planes of Au. This result indicates that the Au core is covered by the Pd shell. In order to further investigate the distribution of Pd and Au elements in an individual Au@Pd nanoparticle, the HAADF image of the Au@Pd nanoparticle and the corresponding element mapping images are recorded and displayed as Fig. 15.1d. It can be observed that the Pd-L signals are weak in the core part and strong in the shell part, whereas the signal of Au-L in the center part is stronger than that in the periphery part, which clearly suggests that the Pd mainly existed in the shell part. The EDX line scan profile also suggests the core-shell structure of the metal nanoparticle, where the Pd signal is detected on the surface of the nanoparticle and the Au signal is detected in the core of nanoparticle. The presence of uniform Pd and Au in this individual nanoparticle was consistent with the formation of the proposed core-shell structure.

The catalytic activity of Au@Pd supported MOFs was evaluated by formic acid dehydrogenation. The reaction was carried out by stirring the catalyst (0.01 g) in 0.2 mL of formic acid and 4.8 mL of water under Ar atmosphere. As shown in Fig. 15.2, no H_2 was produced over Au/UiO-66($Zr_{85}Ti_{15}$) even under light irradiation. A small amount of H_2 was detected over Pd/UiO-66($Zr_{85}Ti_{15}$) under dark condition,

Fig. 15.2 Formic acid dehydrogenation with different catalysts in the dark (black bars) or under visible light irradiation (gray bars, λ > 420 nm, 320 mW cm^{-2})

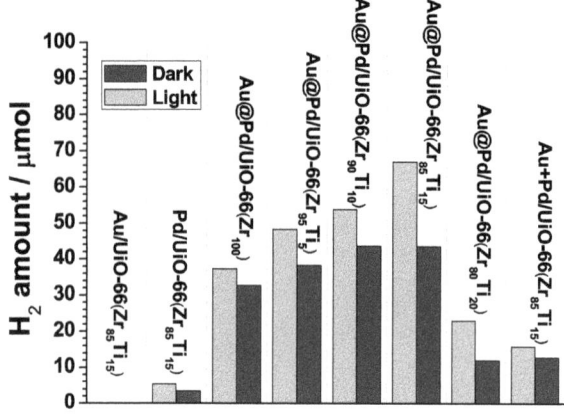

and enhancement of activity under light irradiation is slight, suggesting that the reaction is driven by Pd nanoparticles. However, it is interesting to observe that all Au@Pd supported MOFs exhibited much higher activities than monometallic nanoparticles supported MOFs owing to the synergistic effect of Au and Pd. The increase in the amount of Ti doping enhances the activity, and Au@Pd/UiO-66($Zr_{85}Ti_{15}$) produces the largest amount of H_2. Further increase of the doping amount of Ti caused a significant decrease in activity. The enhanced activity of Au@Pd nanoparticle supported sample can be ascribed to the electronically promoted Pd by the charge transfer arising from the difference in the work function of Au and Pd [31]. To elucidate this promoting effect, Au + Pd/UiO-66($Zr_{85}Ti_{15}$) was prepared by the physical mixture of Pd/UiO-66($Zr_{85}Ti_{15}$) and Au/UiO-66($Zr_{85}Ti_{15}$) was used as a reference sample. As shown in Fig. 15.2, considerable enhancement of Au + Pd/UiO-66($Zr_{85}Ti_{15}$) was observed, but still lower than that of Au@Pd/UiO-66($Zr_{85}Ti_{15}$). On the basis of the above results, it is evident that the presence of Au@Pd nanoparticles with core-shell structure is essentially important for promoting the catalytic activity.

Figure 15.3 proposes the possible reaction pathway for the visible light enhanced H_2 production from formic acid by using Au@Pd/UiO-66($Zr_{85}Ti_{15}$). It was reported that a large amount of weakly basic –NH_2 groups have a positive effect on O–H bond dissociation, in which the weakly basic –NH_2 groups act as a proton scavenger to form –$^+HNH_2$ and a Pd-formate intermediate. Pd-formate species then undergoes C–H bond dissociation to afford CO_2 and a Pd-hydride (Pd-[H]$^-$) species. Finally, the reaction of hydride species with –$^+HNH_2$ produces molecular hydrogen, following the regeneration of the metal species. After the visible light absorption by plasmonic Au@Pd nanoparticle and photoactive MOFs, the photo-generated electron migrates to Pd-active site. Such electron-rich Pd species significantly facilitates C-H bond cleavage from the Pd-formate intermediate. Meanwhile, formic acid serves as an electron donor, which donates electrons to the electron-deficient organic linkers and Au to continue the photo-assisted H_2 production from formic acid. It is reasonable to

Fig. 15.3 Possible reaction pathway for the visible light enhanced H_2 production from formic acid decomposition over Au@Pd/UiO-66($Zr_{85}Ti_{15}$) under visible light irradiation. Reproduced with permission from Ref. [30] Copyright @ 2017 American Chemical Society

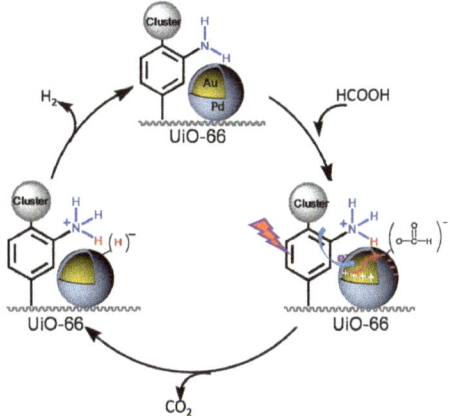

conclude that the synergistic effect of the weakly basic amine groups within metal-organic frameworks and the LSPR effect of Au@Pd nanoparticles and photoactive MOFs induced electronically promoted Pd species which accounts for its high catalytic activity. The dissociation of O–H and C–H bonds of formic acid is individually facilitated by the assist of amine groups within MOFs and active electron-rich Pd sites induced by plasmonic effect under visible light irradiation.

15.3 Pd-Decorated Au Nanorod for Enhancing Pd-Catalyzed Suzuki–Miyaura Coupling Reaction

In the section, Au nanorods are used as plasmonic converters to efficiently harvest light, Silica with high surface area and extremely low bulk density was selected as support, and Pd nanoparticle act as a catalyst to drive Suzuki–Miyaura reaction at ambient temperatures [32]. Au nanorods were obtained using the silver ion-assisted seed-mediated method and successfully deposited onto the surface of fumed SiO_2 by employing a simple ultrasonic-assisted method. Then, the Pd nanoparticles decorated on the surface of Au nanorod were realized by using sodium formate as a reducing agent. TEM was used to record the morphology of as-prepared sample. Au nanorods with ~40 nm in length and ~10 nm in diameter can be clearly observed. No significant agglomeration of Au nanorods occurred, indicating Au nanorods were well dispersed on SiO_2. After reducing Pd nanoparticles on the surface of Au nanorods, the smooth Au nanorods with dense core became rough. The morphological change suggests that Pd nanoparticles are successfully grown on the surface of the Au nanorods.

The optical properties of all prepared samples were investigated by UV-vis spectra. As shown in Fig. 15.4. The spectrum of Pd/SiO_2 is obviously different from Au/SiO_2. Pd/SiO_2 displayed a weak optical response in high-energy region ($\lambda < 500$ nm) assigned to LSPR of Pd nanoparticles [33]. In contrast, Au/SiO_2

Table 15.1 Yields of Pd-catalyzed Suzuki-Miyaura coupling reaction over Pd/SiO$_2$, Au/SiO$_2$, and Pd-Au/SiO$_2$ of various Pd/Au weight ratios under visible light irradiation at ambient temperature

Entry	Pd: Au	Yield/%			TON		
		Light (temp./°C)	Dark (temp./°C)	Heating[a] (temp./°C)	Light	Dark	Heating[a]
1	0: 1	0 (30.5)	0 (25)	0 (30.5)	0	0	0
2	0.1: 1	62 (32.2)	7 (25)	23 (32.2)	664	78	241
3	0.3: 1	73 (36.0)	25 (25)	49 (36.0)	259	89	173
4	0.5: 1	78 (36.8)	29 (25)	58 (36.8)	167	62	123
5	1: 1	71 (44.1)	25 (25)	66 (44.1)	76	27	71
6	0.5: 0	10 (27.3)	7 (25)	10 (27.3)	21	15	21

[a]Reaction conditions under conventional thermal heating using oil bath; iodobenzene (0.3 mmol), phenylboronic acid (0.2 mmol), catalyst (20 mg) 30 min, Ar atmosphere. The values in parentheses are the temperature of the reaction mixture. TON was calculated based on Pd

Fig. 15.4 UV-vis spectra of the Pd-Au/SiO$_2$ samples, Au/SiO$_2$, and Pd/SiO$_2$ catalysts. Reproduced with permission from Ref. [32] copyright 2014 Elsevier

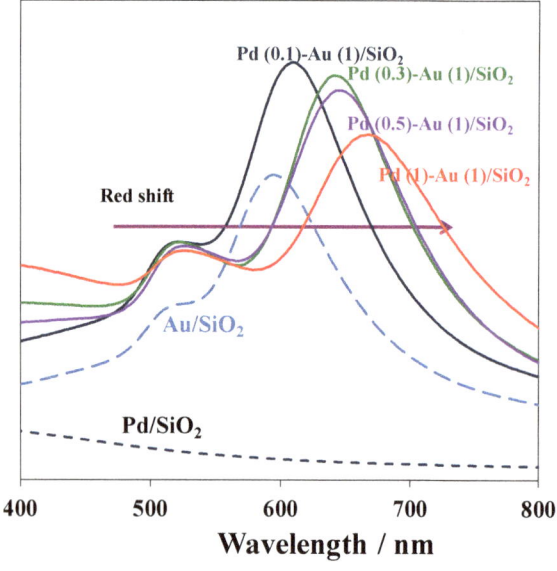

exhibits two intense absorption band in the visible region owing to the LSPR excitation. The two intense absorption peaks are corresponding to the transverse and longitudinal surface plasmon resonances of Au nanorods, respectively [27]. After Pd nanoparticle decorating, the absorption intensity in the visible region ($\lambda > 400$ nm) significantly increased due to the charge heterogeneity exited in the abrupt atomic interface between Au nanorods and Pd nanoparticles, leading to the energetic electrons transferred from Au nanorods to Pd nanoparticles [34]. Furthermore, the peak position of longitudinal surface plasmon resonance showed strong red-shift while transverse surface plasmon resonance exhibited slight red-shift with increasing Pd loading amount, suggesting the formation of Pd-Au hybrid catalyst with the increased length but a nearly unchanged diameter. However, the decrease in the absorption intensity for 1 wt% Pd loaded hybrid catalyst might be ascribed to the filter effect by Pd nanoparticles which prevent reaching of irradiated light on the surface of Au nanorods. In addition, with the increase of Pd loading from 0 to 1.0 wt%, the color of the catalyst changed from dark blue to gray, which visually demonstrated that the Pd nanoparticle was successfully loaded onto the surface of Au nanorods. This result provides the possibility of optically tuning gold-based plasmonic hybrid catalyst by simply changing the amount of Pd nanoparticles loading.

The catalytic activity of as-prepared sample was evaluated by Suzuki–Miyaura coupling reaction. As summarized in Table 15.1, no significant reaction occurred over Au/SiO_2 without Pd nanoparticles even under light irradiation, demonstrating the reaction was catalyzed by Pd. Little reaction occurred on Pd/SiO_2 under dark, and enhancement of activity under light irradiation or thermal heating is slight, suggesting that the coupling reaction is driven by Pd nanoparticles. After loading with Pd nanoparticles on the surface of Au nanorods, the $Pd-Au/SiO_2$ hybrid catalyst showed much higher activity than Pd/SiO_2, affording 62%, 7%, and 23% yields were obtained using 0.1 wt% Pd nanoparticles loaded Au/SiO_2 under light irradiation, dark, and thermal heating, respectively. This corresponds to turnover numbers 664, 78, and 241, respectively. The optimum Pd: Au mass ratio for Suzuki–Miyaura cross-coupling reaction was found to be around 0.5: 1, in which 78% yield was obtained under light irradiation while 58% yield was obtained under thermal heating. The enhanced activity under light irradiation can be attributed to the LSPR effect. With the increase of Pd loading from 0.1 to 0.5 wt%, the yields increased owing to the increased number of active sites at each condition. Further increase of Pd loading to 1.0 wt% caused a decrease in activity, which could be attributed to the filter effect by an excess amount of Pd nanoparticles and gathering of Pd into large nanoparticles. In addition, with the increase of Pd loading, the end temperatures of the reaction solutions were increased owing to the enhanced LSPR effect of Au nanorods in the near-infrared region.

In order to elucidate this promoting effect by the assistance of LSPR effect, a reference sample was prepared by the physical mixture of 0.5 wt% Pd/SiO_2 and 1 wt% Au/SiO_2 (sample c). Figure 15.5 displays the relative contributions of the LSPR effect and thermal heating processes to the yield of coupling products. Sample a displayed the higher activity as compared to sample b and sample c, which can be assigned to the LSPR effect of Au nanorods in the near-infrared region. Compared with thermal heating, sample a and sample c exhibit higher activity for the reaction under visible light irradiation due to the LSPR effect. In contrast, low yield was observed for sample b without Au nanorods even under light irradiation. The possible reason for the significantly enhanced activity under visible light irradiation might be attributed to the synergistic effect between Pd nanoparticles and Au nanorods. An energetic

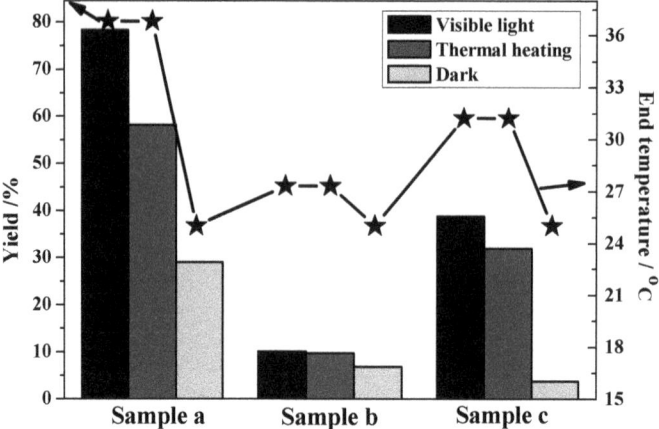

Fig. 15.5 Yield and the end temperature of sample a (0.5 wt% Pd-Au/SiO₂), sample b (0.5 wt% Pd/SiO₂), and sample c (physically mixture of 0.5 wt% Pd/SiO₂ and Au/SiO₂) for Suzuki–Miyaura coupling reaction. Reproduced with permission from Ref. [32], copyright 2014 Elsevier

conduction electron is formed on the surface of Au nanorod through the LSPR effect upon visible light irradiation. Subsequently, the energetic electron transfers to the Pd nanoparticle due to the charge heterogeneity that existed in the abrupt atomic interface between Au nanorods and Pd nanoparticles. The oxidative addition step, which is the rate-determining step in Suzuki–Miyaura coupling reaction, may be accelerated by the activated Pd species, leading to the enhancement of intrinsic catalytic activity of Pd-catalyzed reaction. On the other hand, the energetic electrons release the energy to the surrounding environment [34, 35], leading to an increase in temperature. The end temperatures of the reaction solutions were 36.8, 27.3, and 31.2 °C for sample a, sample b, and sample c, respectively. An increase in temperature indicated the effective conversion of solar light into thermal energy through photothermal heating processes achieved by the LSPR effect of Au nanorods and Pd nanoparticles. Sample c displayed low activity even under light irradiation, which means only photothermal heating play a role in the catalytic reaction. Therefore, Pd-Au/SiO₂ hybrid catalysts not only take advantage of generating active electron-rich Pd species, but also play an important role in increasing reaction temperature by photothermal heating, resulting in the enhancement of catalytic activity.

15.4 Summary and Outlook

This chapter deals with a new type of visible light responsive catalysts designed by the integration of plasmonic metal with Pd-active species. Which successfully showed strong visible light absorption capacity as well as an exceptional catalytic activity for various Pd-catalyzed chemical reaction (Suzuki–Miyaura coupling reaction, formic acid dehydrogenation) under visible light irradiation. The reason for the significantly enhanced activity under visible light irradiation might be attributed to the synergistic effect between Pd nanoparticles and plasmonic Au nanostructure in effective conversion of solar light, facilitating the energetic electron transfer, producing the electron-rich Pd species, increasing reaction temperature by photothermal heating. Although great achievements have been obtained, the plasmon-directed photocatalysis is still in its infancy. Many issues, such as the band alignment, interfacial structure, and contacting architectures between plasmonic metals and support materials, need more in-depth investigations. This chapter supplies a platform to design plasmonic material-metal nanoparticle hybrid catalyst for efficient utilization of sunlight for a wide range of chemical reactions.

References

1. Boukha Z, Choya A, Cortes-Reyes M, de Rivas B, Alemany LJ, Gonzalez-Velasco JR, Gutierrez-Ortiz JI, Lopez-Fonseca R (2020) Influence of the calcination temperature on the activity of hydroxyapatite-supported palladium catalyst in the methane oxidation reaction. Appl Catal B-Environ 277:119280
2. Tan C, Qasim M, Pang WM, Chen CL (2020) Ligand-metal secondary interactions in phosphine-sulfonate palladium and nickel catalyzed ethylene (co)polymerization. Polym Chem 11:411–416
3. Yin JJ, Zhan FK, Jiao TF, Deng HZ, Zou GD, Bai ZH, Zhang QR, Peng QM (2020) Highly efficient catalytic performances of nitro compounds via hierarchical PdNPs-loaded MXene/polymer nanocomposites synthesized through electrospinning strategy for wastewater treatment. Chin Chem Lett 31:992–995
4. Fu Y, Wang G, Mei T, Li JH, Wang JY, Wang XB (2017) Accessible graphene aerogel for efficiently harvesting solar energy. Acs Sustain Chem Eng 5:4665–4671
5. Hou Y, Vidu R, Stroeve P (2011) Solar energy storage methods. Ind Eng Chem Res 50:8954–8964
6. Mohapatra SK, Kondamudi N, Banerjee S, Misra M (2008) Functionalization of self-organized TiO_2 nanotubes with Pd nanoparticles for photocatalytic decomposition of dyes under solar light illumination. Langmuir 24:11276–11281
7. Xiang S, Zhang Z, Wu Z, Sun L, Radjenovic P, Ren H, Lin C, Tian Z, Li J (2018) 3D Heterostructured Ti-based Bi_2MoO_6/Pd/TiO_2 photocatalysts for high-efficiency solar light driven photoelectrocatalytic hydrogen generation. ACS Appl Energy Mater 2:558–568
8. Tudu B, Nalajala N, K PR, Saikia P, Gopinath CS (2019) Electronic integration and thin film aspects of Au-Pd/rGO/TiO_2 for improved solar hydrogen generation. ACS Appl Mater Interfaces 11:32869–32878
9. Cui J, Li Y, Liu L, Chen L, Xu J, Ma J, Fang G, Zhu E, Wu H, Zhao L, Wang L, Huang Y (2015) Near-infrared plasmonic-enhanced solar energy harvest for highly efficient photocatalytic reactions. Nano Lett 15:6295–6301
10. Kim HJ, Jackson DHK, Lee J, Guan YX, Kuech TF, Huber GW (2015) Enhanced activity and stability of TiO_2-coated cobalt/carbon catalysts for electrochemical water oxidation. Acs Cataly 5:3463–3469
11. Hoang S, Guo S, Hahn NT, Bard AJ, Mullins CB (2012) Visible light driven photoelectrochemical water oxidation on nitrogen-modified TiO_2 nanowires. Nano Lett 12:26–32
12. Kudo A, Miseki Y (2009) Heterogeneous photocatalyst materials for water splitting. Chem Soc Rev 38:253–278

13. Chen XB, Shen SH, Guo LJ, Mao SS (2010) Semiconductor-based photocatalytic hydrogen generation. Chem Rev 110:6503–6570
14. Shiraishi Y, Hirai T (2008) Selective organic transformations on titanium oxide-based photocatalysts. J Photoch Photobio C 9:157–170
15. Watanabe K, Menzel D, Nilius N, Freund HJ (2006) Photochemistry on metal nanoparticles. Chem Rev 106:4301–4320
16. Rycenga M, Cobley CM, Zeng J, Li W, Moran CH, Zhang Q, Qin D, Xia Y (2011) Controlling the synthesis and assembly of silver nanostructures for plasmonic applications. Chem Rev 111:3669–3712
17. Wang H, Tam F, Grady NK, Halas NJ (2005) Cu nanoshells: effects of interband transitions on the nanoparticle plasmon resonance. J Phys Chem B 109:18218–18222
18. Mulvaney P (1996) Surface plasmon spectroscopy of nanosized metal particles. Langmuir 12:788–800
19. Kelly KL, Coronado E, Zhao LL, Schatz GC (2003) The optical properties of metal nanoparticles: the influence of size, shape, and dielectric environment. J Phys Chem B 107:668–677
20. Manthiram K, Alivisatos AP (2012) Tunable localized surface plasmon resonances in tungsten oxide nanocrystals. J Am Chem Soc 134:3995–3998
21. Huang Q, Hu S, Zhuang J, Wang X (2012) MoO_{3-x}-based hybrids with tunable localized surface plasmon resonances: chemical oxidation driving transformation from ultrathin nanosheets to nanotubes. Chemistry 18:15283–15287
22. Cheng H, Kamegawa T, Mori K, Yamashita H (2014) Surfactant-free nonaqueous synthesis of plasmonic molybdenum oxide nanosheets with enhanced catalytic activity for hydrogen generation from ammonia borane under visible light. Angew Chem 53:2910–2914
23. Alsaif MM, Latham K, Field MR, Yao DD, Medhekar NV, Beane GA, Kaner RB, Russo SP, Ou JZ, Kalantar-zadeh K (2014) Tunable plasmon resonances in two-dimensional molybdenum oxide nanoflakes. Adv Mater 26:3931–3937
24. Huang X, Neretina S, El-Sayed MA (2009) Gold nanorods: from synthesis and properties to biological and biomedical applications. Adv Mater 21:4880–4910
25. Kong LN, Chen W, Ma DK, Yang Y, Liu SS, Huang SM (2012) Size control of $Au@Cu_2O$ octahedra for excellent photocatalytic performance. J Mater Chem 22:719–724
26. Cheng H, Meng X, He L, Lin W, Zhao F (2014) Supported polyethylene glycol stabilized platinum nanoparticles for chemoselective hydrogenation of halonitrobenzenes in $scCO_2$. J Colloid Interface Sci 415:1–6
27. Ni WH, Kou X, Yang Z, Wang JF (2008) Tailoring longitudinal surface plasmon wavelengths, scattering and absorption cross sections of gold nanorods. ACS Nano 2:677–686
28. Gao F, Wang YL, Goodman DW (2009) CO Oxidation over AuPd(100) from ultrahigh vacuum to near-atmospheric pressures: CO adsorption-induced surface segregation and reaction kinetics. J Phys Chem C 113:14993–15000
29. Sarina S, Zhu H, Jaatinen E, Xiao Q, Liu H, Jia J, Chen C, Zhao J (2013) Enhancing catalytic performance of palladium in gold and palladium alloy nanoparticles for organic synthesis reactions through visible light irradiation at ambient temperatures. J Am Chem Soc 135:5793–5801
30. Wen M, Mori K, Kuwahara Y, Yamashita H (2016) Plasmonic Au@Pd nanoparticles supported on a basic metal-organic framework: synergic boosting of H2 production from formic acid. ACS Energy Lett 2:1–7
31. Tedsree K, Li T, Jones S, Chan CW, Yu KM, Bagot PA, Marquis EA, Smith GD, Tsang SC (2011) Hydrogen production from formic acid decomposition at room temperature using a Ag-Pd core-shell nanocatalyst. Nat Nanotechnol 6:302–307
32. Wen MC, Takakura S, Fuku K, Mori K, Yamashita H (2015) Enhancement of Pd-catalyzed Suzuki-Miyaura coupling reaction assisted by localized surface plasmon resonance of Au nanorods. Catal Today 242:381–385
33. Jung S, Shuford KL, Park S (2011) Optical property of a colloidal solution of platinum and palladium nanorods: localized surface plasmon resonance. J Phys Chem C 115:19049–19053

34. Wang F, Li C, Chen H, Jiang R, Sun LD, Li Q, Wang J, Yu JC, Yan CH (2013) Plasmonic harvesting of light energy for Suzuki coupling reactions. J Am Chem Soc 135:5588–5601
35. Chen H, Shao L, Ming T, Sun Z, Zhao C, Yang B, Wang J (2010) Understanding the photothermal conversion efficiency of gold nanocrystals. Small 6:2272–2280

Chapter 16
Functionalization of Plasmonic Photocatalysts by the Introduction of Core–Shell Structure

Atsuhiro Tanaka and Hiroshi Kominami

16.1 Introduction

Unique optical properties of metallic nanoparticles have been applied in many different fields including biochemistry, sensing science, and catalysis. Nanoparticles of metals such as copper (Cu), silver (Ag), and gold (Au) show strong absorption of visible light due to surface plasmon resonance (SPR). Recently, supported Au nanoparticles have been applied to a visible-light-responding photocatalyst because Au nanoparticles are physically and chemically stable under irradiation of light. Now research on Au plasmonic photocatalysts are in the next stage of control of SPR and drastic increase in the reaction rate and the product selectivity. These challenges can be overcome through *functionalization* of Au plasmonic photocatalysts [1–3].

A combination of two or more kinds of metals has been widely applied in various materials to enhance the performance and reliability of the materials, and various strategies of combination such as physical mixing, doping, alloying and core–shell forming have been proposed. Among them, core–shell forming, especially in metal nanoparticles, has attracted much attention because of the unique optic [4, 5], magnetic [6], electronic [7], catalytic [8–14], and photocatalytic [15] properties. Formation of a core–shell structure is also a strong candidate for the functionalization of Au plasmonic photocatalysts. In this chapter, we introduce two examples of functionalization of Au plasmonic photocatalysts by introduction of core–shell structure, i.e., control of SPR and drastic improvement of the catalytic performance.

A. Tanaka · H. Kominami (✉)
Department of Applied Chemistry, Faculty of Science and Engineering, Kindai University, Kowakae, Higashiosaka, Osaka 577-8502, Japan
e-mail: hiro@apch.kindai.ac.jp

© The Author(s), under exclusive license to Springer Nature Singapore Pte Ltd. 2021
H. Yamashita and H. Li (eds.), *Core-Shell and Yolk-Shell Nanocatalysts*,
Nanostructure Science and Technology,
https://doi.org/10.1007/978-981-16-0463-8_16

16.2 Control of SPR Photoabsorption by the Introduction of Core–Shell Structure [16]

16.2.1 Preparation of Au@Ag/SnO₂ and Au@Cu/SnO₂

Multi-step photodeposition method (MSPD) [17] was used for the preparation of Au@Ag/SnO$_2$ and Au@Cu/SnO$_2$ by modification of Au(0.2 wt%)/SnO$_2$ with Ag and Cu. An aqueous solution of silver sulfate (Ag: 4.0 g dm^{-3}) or copper sulfate (Cu: 4.0 g dm^{-3}) was injected into an aqueous methanolic suspension of Au/SnO$_2$ and the mixture was photoirradiated by a mercury arc lamp under the same conditions as those for the preparation of Au(0.2)/SnO$_2$. The amount of Ag and Cu loading per photodeposition was fixed at 0.2 wt%, and this photodeposition of Ag and Cu was repeated for additional Ag and Cu loadings onto Au(0.2)@Ag/SnO$_2$ and Au@Cu/SnO$_2$. For example, the photodeposition of Ag was repeated four times for the preparation of Au(0.2)@Ag(0.8 wt%)/SnO$_2$ (0.8 wt% = 0.2 wt% × 4). Hereafter, this sample is designated as Au(0.2)@Ag(0.8)/SnO$_2$ and a sample modified with 0.8 wt%Cu is shown as Au(0.2)@Cu(0.8)/SnO$_2$. Analysis of the liquid phase after photodeposition revealed that the Ag and Cu sources had been almost completely (>99.9%) deposited as Ag and Cu metals on Au(0.2)/SnO$_2$. The resultant powder was washed repeatedly with distilled water and then dried in air at 310 K overnight.

16.2.2 Photoabsorption Properties

Figure 16.1 shows absorption spectra of Au(0.2)/SnO$_2$, Au(0.2)@Ag(0.8)/SnO$_2$, Au(1.0)/SnO$_2$, and Au(0.2)@Cu(0.8)/SnO$_2$. In the spectra of Au(0.2)/SnO$_2$ and Au(1.0)/SnO$_2$, photoabsorption was observed around $\lambda = 550$ nm (Fig. 16.1d, b), which was attributed to SPR of the supported Au particles, and more intense photoabsorption was achieved by increasing the Au contents of Au/SnO$_2$ and Au/CeO$_2$ [18].

Fig. 16.1 Absorption spectra of **a** Au(0.2)@Ag(0.8)/SnO$_2$, **b** Au(1.0)/SnO$_2$, **c** Au(0.2)@Cu(0.8)/SnO$_2$, and **d** Au(0.2)/SnO$_2$ Reprinted with permission from Ref. [16]. © 2016 Wiley-VCH Verlag GmbH & Co. KGaA, Weinheim

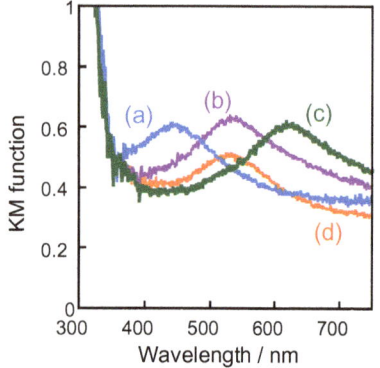

Both a shift of photoabsorption and an increase in intensity were achieved by the introduction of Ag and Cu into Au(0.2)/SnO$_2$ by MSPD (Fig. 16.1a, c). The photoabsorption at 550 nm of Au(0.2)@Ag(0.8)/SnO$_2$ and that of Au(0.2)@Cu(0.8)/SnO$_2$ at 550 nm were less intense than that of the Au(0.2)/SnO$_2$ mother material before modification. These results suggest that the SPR properties of Au(0.2)@Ag(0.8)/SnO$_2$ and Au(0.2)@Cu(0.8)/SnO$_2$ are not inherited from Au(0.2)/SnO$_2$ and originate from the properties of Ag and Cu themselves. In other experiments, however, we did not succeed in the preparation of Au-free Cu/SnO$_2$ exhibiting SPR by PD and MSPD. In addition, it is known that Cu nanoparticles are easily oxidized and lose their SPR gradually under ambient conditions [19]. Therefore, Au particles are clearly indispensable for the preparation of stable Cu-based particles supported on SnO$_2$ exhibiting intense photoabsorption at 630 nm due to SPR. The Au particles probably function as a kind of template and stabilizer for Cu and Ag metals.

Figure 16.2a, b shows the influence of Ag and Cu contents (X wt% and Y wt%, respectively) in the Au(0.2)@Ag(X)/SnO$_2$ and Au(0.2)@Cu(Y)/SnO$_2$ samples on the top of the wavelength due to SPR (λ_{top}) and photoabsorption (1-reflectance) at λ_{top}. The photoabsorption at λ_{top} increased with increases in Ag and Cu contents (X and Y). The maximum peaks gradually shifted to shorter and longer wavelengths with increases in X and Y and reached $\lambda = 450$ nm at X = 0.8 wt% and $\lambda = 630$ nm at Y = 0.8 wt%. These results indicate that the peak position of photoabsorption due to SPR can be controlled by modification of Au(0.2)/SnO$_2$ with Ag and Cu by using the MSPD.

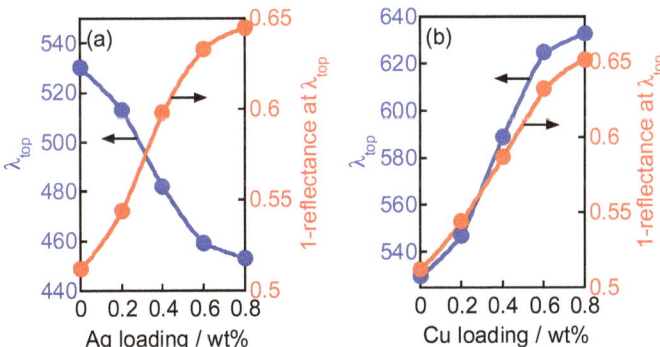

Fig. 16.2 Influence of **a** Ag loading (X) and **b** Cu loading (Y) to Au(0.2)/SnO$_2$ on the top of the wavelength due to SPR (λ_{top}) and 1-reflectance at λ_{top}. Reprinted with permission from Ref. [16]. © 2016 Wiley-VCH Verlag GmbH & Co. KGaA, Weinheim

16.2.3 Photocatalytic Mineralization of Formic Acid

Figure 16.3 shows time courses of the evolution of CO_2 from formic acid in aqueous suspensions of Au(1.0)/SnO_2, Au(0.2)@Ag(0.8)/SnO_2, and Au(0.2)@Cu(0.8)/SnO_2 samples under light irradiation from green, blue, and red LEDs. Visible light irradiated to the reaction system, and the maximum wavelength of each light was determined to be 475, 530, and 640 nm. In the presence of Au(1.0)/SnO_2 (shown as circles in Fig. 16.3), CO_2 evolved just after irradiation with green, blue, and red lights and CO_2 formation continued linearly with irradiation time, indicating zero-order kinetics over Au(1.0)/SnO_2. On the other hand, no gas was evolved in the dark, indicating that no thermocatalytic mineralization of formic acid occurred under the present conditions. From the slopes of time courses of CO_2 evolution, rates over Au(1.0)/SnO_2 under irradiation of green, blue, and red lights were determined to be 2.9, 0.55, and 0.29 μmol h^{-1}, respectively. Photoabsorption due to SPR of Au(1.0)/SnO_2 overlapped well with light from the green LED, and the higher utilization of light (photons) explains the largest rate under irradiation of green light. Similarly, rates of CO_2 formation over Au(0.2)@Cu(0.8)/SnO_2 (data plotted as diamonds in Fig. 16.3) under irradiation of green, blue, and red lights were determined to be 0.86, 0.93, and 2.1 μmol h^{-1}, respectively. As expected from the strong photoabsorption of Au(0.2)@Cu(0.8)/SnO_2 at 630 nm (Fig. 16.1c), the sample exhibited the largest rate under red light irradiation. In the case of Au(0.2)@Ag(0.8)/SnO_2 (data plotted as squares in Fig. 16.3), the CO_2 formation rates under irradiation of green, blue and red lights were determined to be 0.62, 2.4, and 0.11 μmol h^{-1}, respectively. The largest rate was obtained when blue light was irradiated to Au(0.2)@Ag(0.8)/SnO_2 as expected from the photoabsorption properties. These results for the three samples

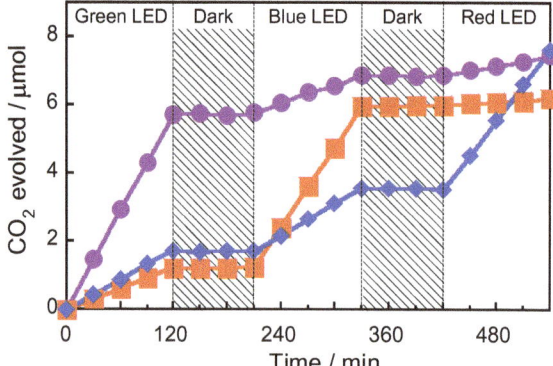

Fig. 16.3 Time courses of evolution of CO_2 from formic acid in aqueous suspensions of Au(1.0)/SnO_2 (circles), Au(0.2)@Ag(0.8)/SnO_2 (squares) and Au(0.2)@Cu(0.8)/SnO_2 (diamonds) under irradiation of visible lights from green, blue and red LEDs (1.7 mW cm^{-2}). Reprinted with permission from Ref. [16]. © 2016 Wiley-VCH Verlag GmbH & Co. KGaA, Weinheim

under three conditions of light irradiation indicate that the samples exhibited the best performance under irradiation of visible light overlapping with their SPR.

16.2.4 Action Spectrum

An action spectrum is a strong tool for determining whether an observed reaction occurs via a photoinduced process or a thermocatalytic process. To obtain an action spectrum in this reaction system, mineralization of formic acid in aqueous suspensions of Au(1.0)/SnO$_2$, Au(0.2)@Ag(0.8)/SnO$_2$, and Au(0.2)@Cu(0.8)/SnO$_2$ was carried out at 298 K under irradiation with monochromated visible light from a Xe lamp with light width of ±5 nm. The apparent quantum efficiency (AQE) at each centered wavelength of light was calculated from the ratio of the amount of CO$_2$ and the amount of photons irradiated using the following Eq. 16.1.

$$AQE = \frac{\text{amount of CO}_2}{\text{amount of incident photons}} \times 100 \tag{16.1}$$

The results are shown in Fig. 16.4. Wavelength-dependencies of AQEs over Au(1.0)/SnO$_2$, Au(0.2)@Ag(0.8)/SnO$_2$, and Au(0.2)@Cu(0.8)/SnO$_2$ were similar to those of their photoabsorption (1-reflectance), indicating that formation of CO$_2$ from formic acid in aqueous suspensions of the three samples was induced by photoabsorption due to their SPRs. AQE of more than 5% was obtained for each of the three samples when the samples were irradiated by lights matching their SPRs.

16.2.5 Selective Oxidation

Au(0.2)@Ag(0.8)/SnO$_2$, Au(1.0)/SnO$_2$, and Au(0.2)@Cu(0.8)/SnO$_2$ were used for oxidation of benzyl alcohol under irradiation of light from three LEDs in order to evaluate their performance in photocatalytic conversions other than mineralization under irradiation of visible light with different wavelengths. No oxidation of benzyl alcohol occurred over metal-free SnO$_2$, indicating that visible light coming from each LED did not cause the bandgap excitation of SnO$_2$. On the other hand, Au(0.2)@Ag(0.8)/SnO$_2$, Au(1.0)/SnO$_2$, and Au(0.2)@Cu(0.8)/SnO$_2$ were active in oxidation of benzyl alcohol and yielded benzaldehyde with a quite high selectivity (>99%) at >99% conversion of benzyl alcohol after 20 h when blue, green and red LEDs were used. We confirmed that a photocatalyst, irradiation of visible light, and O$_2$ were indispensable for the oxidation of benzyl alcohol. Since benzaldehyde increased linearly with photoirradiation time over the three samples under irradiation of visible lights from three LEDs, the formation rates were determined from slopes of the time courses of benzaldehyde formation and the rates are shown in Fig. 16.5.

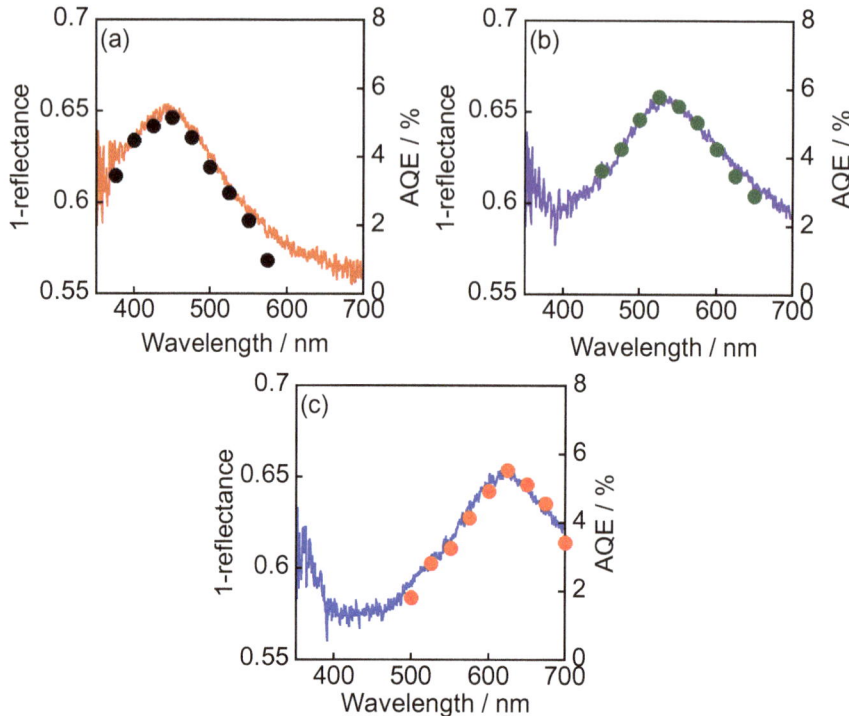

Fig. 16.4 Absorption spectra measured with barium sulfate as a reference (left axis) and action spectra (circles) in formic acid mineralization (right axis): **a** Au(0.2)@Ag(0.8)/SnO$_2$, **b** Au(1.0)/SnO$_2$, and **c** Au(0.2)@Cu(0.8)/SnO$_2$. Reprinted with permission from Ref. [16]. © 2016 Wiley-VCH Verlag GmbH & Co. KGaA, Weinheim

Fig. 16.5 Rates of formation of benzaldehyde from benzyl alcohol in aqueous suspensions of SnO$_2$, Au(0.2)@Ag(0.8)/SnO$_2$, Au(1.0)/SnO$_2$, and Au(0.2)@Cu(0.8)/SnO$_2$ under irradiation of visible lights from green, blue, and red LEDs (1.7 mW cm^{-2}). Reprinted with permission from Ref. [16]. © 2016 Wiley-VCH Verlag GmbH & Co. KGaA, Weinheim

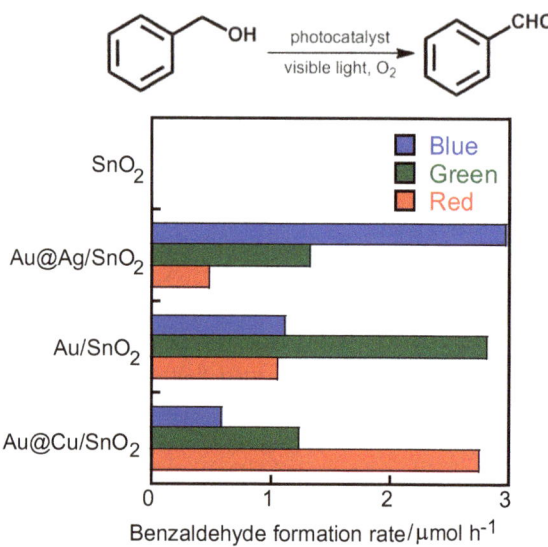

The largest reaction rates were obtained when irradiated light overlapped well with photoabsorption due to their SPR as in the case of formic acid mineralization.

16.3 Drastic Improvement of Catalytic Performance by Introduction of Core–Shell Structure [20]

16.3.1 Preparation of Au@Pd/TiO$_2$

Loading of 0.8 wt% Au on TiO$_2$ was performed by the photodeposition method. The bare TiO$_2$ powder (198 mg) was suspended in water (10 cm^3) in a test tube and the test tube was sealed with a rubber septum under argon (Ar). Aqueous solutions of citric acid (33 μmol) and tetrachloroauric acid (as 1.6 mg Au) were injected into the sealed test tube and then photoirradiated at $\lambda > 300$ nm by a 400-W high-pressure mercury arc under Ar with magnetic stirring in a water bath continuously kept at 298 K. The Au source was reduced by photogenerated electrons, and Au metal was deposited on TiO$_2$ particles, resulting in the formation of Au/TiO$_2$. For preparation of Au(0.8)@Pd(X)/TiO$_2$ by using an MSPD, a solution of palladium chloride (X = 0.1, 0.2, 0.5, and 1.0 wt%) was injected into aqueous suspensions of the Au(0.8)/TiO$_2$ sample and the mixture was photoirradiated by the same mercury arc under the same conditions as those for preparation of the Au(0.8)/TiO$_2$ sample. Analysis of the liquid phase after photodeposition revealed that the Au and Pd sources had been almost completely (>99.9%) deposited as Au and Pd metals on the TiO$_2$ particles. The resultant powder was washed repeatedly with distilled water and then dried at 310 K overnight under air.

16.3.2 TEM Observation

A transmission electron microscope (TEM) image and particle size distribution of Au(0.8)/TiO$_2$ are shown in Fig. 16.6a, b, respectively. Fine particles were observed and the average diameter of the Au particles (D$_{ave}$) was determined to be 9.6 nm, indicating that Au nanoparticles were deposited on TiO$_2$ by the photodeposition method. A TEM image and particle size distribution of Au(0.8)@Pd(0.2)/TiO$_2$ are shown in Fig. 16.6c, d, respectively. Slightly larger particles were observed and D$_{ave}$ of this sample was determined to be 11.5 nm. From the two values of D$_{ave}$ for Au(0.8)/TiO$_2$ and Au(0.8)@Pd(0.2)/TiO$_2$ samples, average shell thickness (T$_{ave}$) was calculated to be 1.0 nm (= (11.5 − 9.6)/2). Assuming that the shapes of Au and Au@Pd particles are regular icosahedrons and using the densities of Au and Pd metals of 19.3 and 12.0 g cm^{-3}, respectively, we estimated that Au and Au@Pd particles consist of 16 layers and 19 layers (16 layers for Au + 3 layers for Pd), respectively, at X = 0.2. The 3 Pd layers correspond to T$_{ave}$ = 1.0 nm. Au(0.8)@Pd(X)/TiO$_2$

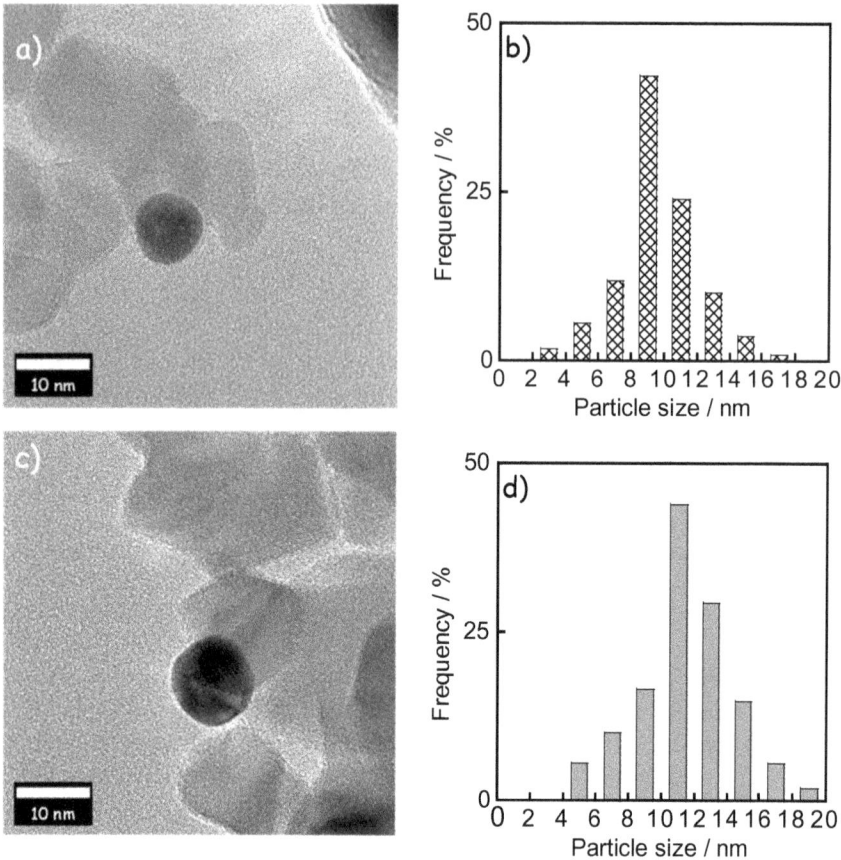

Fig. 16.6 TEM photographs and size distributions of Au(0.8 wt%)/TiO$_2$ (**a, b**) and Au(0.8 wt%)@Pd(0.2 wt%)/TiO$_2$ (**c, d**). Reprinted with permission from Ref. [20]. © 2013 American Chemical Society

samples (X = 0.1, 0.5 and 1.0) were also investigated, and D$_{ave}$ and T$_{ave}$ are plotted against X in Fig. 16.7. The values of D$_{ave}$ and T$_{ave}$ increased with an increase in X, indicating that the thickness of the Pd shell was controlled by using MSPD.

16.3.3 Photoabsorption

Figure 16.8 shows absorption spectra of Au(0.8)@Pd(X)/TiO$_2$ samples. In the spectra of Au(0.8)/TiO$_2$, Au(0.8)@Pd(0.2)/TiO$_2$ and Au(0.8)@Pd(0.5)/TiO$_2$ samples, strong absorption was observed at around 550 nm, which was attributed to SPR of the supported Au nanoparticles as reported previously. The introduction of Pd metal caused an increase in the baseline of absorption that was observed as a change in color

Fig. 16.7 Average diameter (D_{ave}, left, closed circles) and average shell thickness (T_{ave}, right, open circles) of Au@Pd particles loaded on Au(0.8 wt%)@Pd(X)/TiO$_2$ having different Pd loadings (X). Reprinted with permission from Ref. [20]. © 2013 American Chemical Society

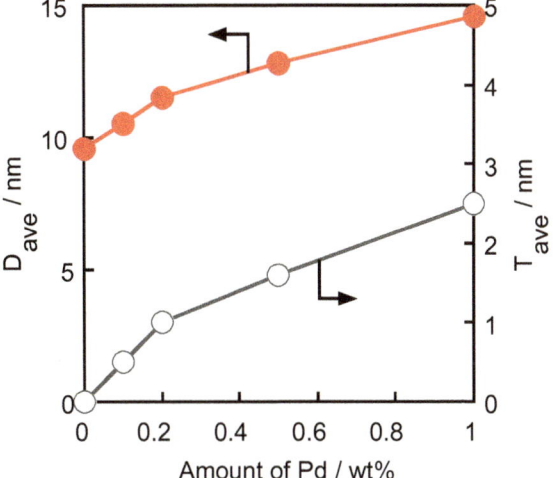

Fig. 16.8 Absorption spectra of Au(0.8)@Pd(X)/TiO$_2$ and visible light irradiated to reaction systems from a Xe lamp with a Y-48 cut filter. Reprinted with permission from Ref. [20]. © 2013 American Chemical Society

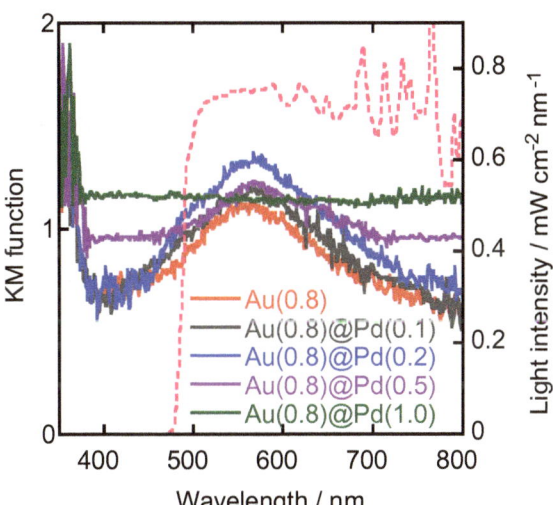

from purple to gray. We noted that absorption due to SPR of Au nanoparticles was slightly enhanced by the introduction of a thin Pd shell (T_{ave} = 0.5–1.0 nm), corresponding to 0.1 and 0.2 wt% Pd metal. On the other hand, the absorption was slightly weakened by the introduction of 0.5 wt% Pd and, finally, absorption disappeared in the Au(0.8)@Pd(1.0)/TiO$_2$ samples, indicating that SPR of Au nanoparticles was quenched by a Pd shell of T_{ave} = 2.5 nm.

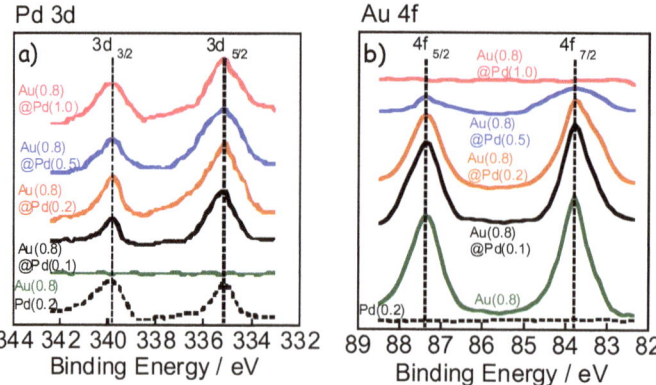

Fig. 16.9 XPS spectra of various photocatalysts around **a** Pd 3d and **b** Au 4f components. Reprinted with permission from Ref. [20]. © 2013 American Chemical Society

16.3.4 X-Ray Photoelectron Spectroscopy

X-ray photoelectron spectroscopy (XPS) was used to obtain information on surfaces of Au(0.8)@Pd(X)/TiO$_2$ samples, and Pd 3d and Au 4f XPS spectra of the samples are shown in Fig. 16.9a, b, respectively. For comparison, spectra of Au-free Pd(0.2)/TiO$_2$ and Pd-free Au(0.8)/TiO$_2$ samples are also shown. In spectra of all samples containing Pd metal, peaks due to Pd (3d$_{5/2, 3/2}$) assignable to Pd^0 were observed at 335.2 and 340.0 eV [21] and the peak intensities were almost the same (Fig. 16.9a). Intense Au (4f$_{7/2, 5/2}$) peaks assignable to Au^0 were observed at 83.8 and 87.4 eV [21] in the spectrum of a Pd-free Au(0.8)/TiO$_2$ sample and the intensity was gradually weakened with an increase in X (Fig. 16.9b). No peak of Au was detected in the spectrum of an Au(0.8)@Pd(1.0)/TiO$_2$ sample as or in the spectrum of an Au-free Pd(0.2)/TiO$_2$ sample. Results of TEM observation, absorption, and XPS revealed that Au nanoparticles were successfully covered with Pd by using MSPD and that optical and physical properties of the Au nanoparticles were completely shielded at X = 1.0 (T$_{ave}$ = 2.5 nm, Fig. 16.7). These results also indicate that functionalization of Au/TiO$_2$ with Pd, i.e., SPR plus Pd catalysis, is highly expected in Au(0.8)@Pd(X)/TiO$_2$ samples having X = 0.1–0.5 (T$_{ave}$ = 0.5–1.6 nm).

16.3.5 Photocatalytic Dechlorination

Au(0.8)@Pd(X)/TiO$_2$ samples were used for photocatalytic dechlorination of chlorobenzene (initially 50 µmol) in aqueous 2-propanol solutions under irradiation of visible light from a xenon (Xe) lamp with a Y-48 cut filter at 298 K. Visible light irradiated to the reaction system is shown in Fig. 16.8. We examined the reaction by using strictly limited visible light (460–800 nm in wavelength) in order to rule out the contribution of the original photocatalytic activity of TiO$_2$, which can be excited with UV light. Figure 16.10 shows the results for Au(0.8)@Pd(0.2)/TiO$_2$. The amount of chlorobenzene decreased linearly with photoirradiation, while benzene as the product of chlorobenzene dechlorination and acetone as the product of 2-propanol oxidation was formed. Chlorobenzene was completely consumed after irradiation for 20 h. The results indicate that benzene was formed with a quite high selectivity (>99%) at >99% conversion of chlorobenzene, i.e., almost quantitative conversion of chlorobenzene to benzene was achieved, in the present photocatalytic reaction system under irradiation of visible light. The amount of benzene (50 μmol) after photoirradiation for 20 h was larger than the amounts of Au (2.0 μmol) and Pd (0.9 μmol) loaded on TiO$_2$, indicating that the dechlorination of chlorobenzene observed in the present study was a (photo)catalytic reaction. To confirm the formation of chloride ions, the resultant liquid phase was analyzed using an ion chromatograph. The amount of chloride ions (50 μmol) was in good agreement with that of benzene determined by a gas chromatograph, indicating that chlorobenzene was completely dechlorinated, forming benzene and chloride ions. We noted that the plot of benzene overlapped with the plot of acetone almost completely in Fig. 16.10.

Totally, the photocatalytic reaction is expressed as Eq. (16.2).

$$Ph-Cl + (CH_3)_2CHOH \rightarrow Ph-H + (CH_3)_2CO + H^+ + Cl^- \qquad (16.2)$$

After photoirradiation for 20 h, 55.2 μmol of acetone was formed as the oxidized product and 4.5 μmol of hydrogen (H$_2$) was formed as the minor reduced product,

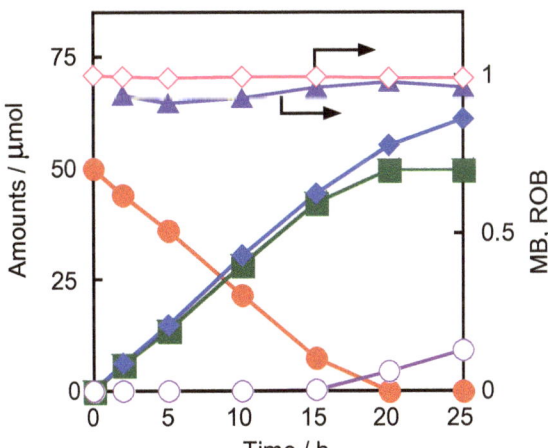

Fig. 16.10 Time courses of the amounts of chlorobenzene (closed circles), benzene (closed squares), acetone (closed diamonds), H$_2$ (open circles), MB (open diamonds), and ROB (closed triangles) in 2-propanol-water suspensions of Au(0.8 wt%)@Pd(0.2 wt%)/TiO$_2$ under irradiation of visible light from a Xe lamp with a Y-48 cut filter. Reprinted with permission from Ref. [20]. © 2013 American Chemical Society

while no CO_2 was detected during the photoirradiation. As clearly shown in Fig. 16.10, the formation of H_2 was observed after consumption of chlorobenzene, indicating that dechlorination occurred predominantly. Material balance (MB) and redox balance (ROB) were calculated from Eqs. (16.3) and (16.4), respectively:

$$MB = [n(Ph-Cl) + n(Ph-H)] / n_0(Ph-Cl) \qquad (16.3)$$

$$ROB = [n(Ph-Cl) + n(H_2)] / n(acetone) \qquad (16.4)$$

where n(Ph–Cl), n(Ph–H), n(H_2), and n(acetone) are the amounts of chlorobenzene, benzene, H_2, and acetone during the photocatalytic reaction, respectively, and n_0(Ph–Cl) is the amount of chlorobenzene before the photocatalytic reaction. As shown in Fig. 16.10, the values of ROB and MB were almost unity regardless of irradiation time. These results indicate that dechlorination of chlorobenzene to benzene and oxidation of 2-propanol to acetone occurred with high stoichiometry as shown in Eq. (16.2); in other words, 2-propanol was used for only reduction of chlorobenzene before its consumption. These results also indicate that, after consumption of chlorobenzene, hydrogenation of 2-propanol to acetone and H_2 occurred, maintaining its stoichiometry. We also confirmed that further irradiation to the reaction mixture did not alter the amount of benzene (Fig. 16.10).

Au(0.8)@Pd(X)/TiO$_2$ samples having different values of X, a Pd-free Au/TiO$_2$ sample, and an Au-free Pd-TiO$_2$ sample were used for photocatalytic dechlorination of chlorobenzene under the same conditions. Yields of benzene after photoirradiation for 10 h are shown in Fig. 16.11. For comparison, the results under a dark condition at 298 K are also shown in Fig. 16.11, indicating that thermocatalytic dechlorination of chlorobenzene is negligible at 298 K. No benzene was formed in the case of the Pd-free Au/TiO$_2$ sample under irradiation of visible light, although dehydrogenation of 2-propanol (formation of acetone and H_2) occurred. This result is consistent with the conclusion that Au/TiO$_2$ did not show any effects on photocatalytic dechlorination of chlorobenzene under irradiation of UV light [22–24] and indicates that functionalization of Au/TiO$_2$ with co-catalysts is essential to achieve photocatalytic dechlorination of chlorobenzene. Dechlorination activity under irradiation of visible light appeared in the Au(0.8)@Pd(0.1)/TiO$_2$ sample in which a thin Pd shell ($T_{ave} = 0.5$ nm) was introduced on the Au core. The yield of benzene over Au(0.8)@Pd(X)/TiO$_2$ after photoirradiation for 10 h increased until X = 0.2. Since Pd metal works as a reduction site for dechlorination of chlorobenzene, increase in the Pd content increases the reduction sites, resulting in an increase in benzene yield. Enhancement in SPR observed in the Au(0.8)@Pd(0.2)/TiO$_2$ sample (Fig. 16.8) also contributes to the larger yield. Further increase in Pd content (X ≥ 0.5) decreased the benzene yield. Absorption due to SPR was weakened when 0.5 wt% Pd was introduced as shown in Fig. 16.8, which probably accounts for the decrease. The benzene yield of the Au(0.8)@Pd(1.0)/TiO$_2$ sample under irradiation of visible light was similar to that under a dark condition. This means that photoinduced dechlorination of chlorobenzene negligibly occurred in the sample, which is explained by

Fig. 16.11 Benzene formation after irradiation for 10 h in aqueous suspensions of various photocatalysts under irradiation of visible light from a Xe lamp with a Y-48 cut filter. Reprinted with permission from Ref. [20]. © 2013 American Chemical Society

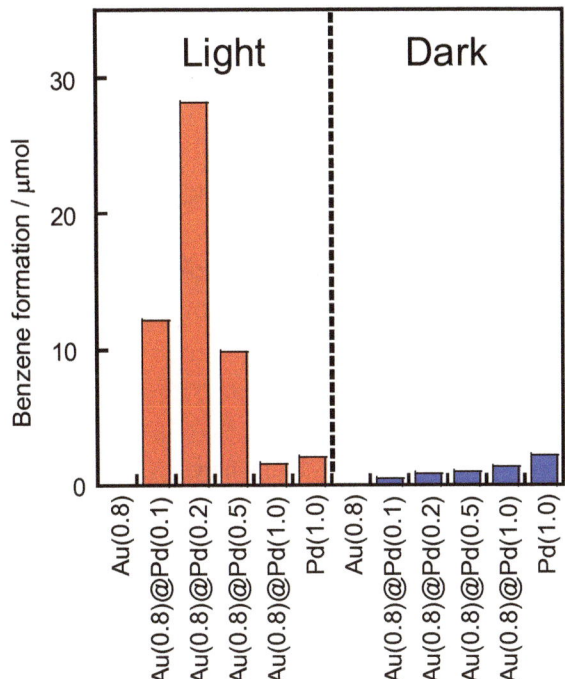

the disappearance of SPR absorption as shown in Fig. 16.8. These results indicate that control of the thickness of the Pd shell is very important for both a satisfactory co-catalyst effect and absorption due to SPR of Au nanoparticles. The findings are informative for the design of functionalization of plasmonic photocatalysts by the formation of a core–shell structure.

16.3.6   Expected Reaction Mechanism

Five blank reactions of chlorobenzene at 298 K, (1) photocatalytic reaction over Pd-free Au/TiO$_2$, (2) dark reaction in the presence of Au@Pd/TiO$_2$, (3) photocatalytic reaction over Au-free Pd-TiO$_2$, (4) photocatalytic reaction over TiO$_2$ and (5) photochemical reaction, gave only a trace or small amount of benzene. From the results of the five blank tests, it can be concluded that Au@Pd/TiO$_2$ and visible light are indispensable for the dechlorination of chlorobenzene to benzene. Rapid electron transfer from Au to the TiO$_2$ film under irradiation of visible light was observed using femtosecond transient absorption spectroscopy [25]. The following is an expected working mechanism for dechlorination of chlorobenzene to benzene in 2-propanol-water suspensions of Au@Pd/TiO$_2$ under irradiation of visible light as shown in Fig. 16.12. Four processes would occur: (1) the incident photons are

Fig. 16.12 Expected reaction mechanism for the production of benzene from chlorobenzene and acetone from 2-propanol over Au@Pd/TiO$_2$ under irradiation of visible light. Reprinted with permission from Ref. [20]. © 2013 American Chemical Society

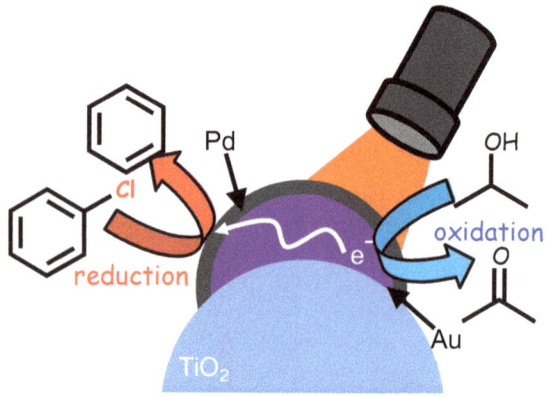

absorbed by Au particles through their SPR excitation, (2) electrons are transferred from the Au particles into the Pd particles, (3) chlorobenzene is reduced by electrons over Pd, resulting in the formation of benzene and elimination of chloride ions and (4) the resultant electron-deficient Au particles are reduced by electrons originated from 2-propanol and return to their original metallic state along with the formation of acetone. Since work function of Pd metal (5.00 eV from vacuum [26]) is larger than that of Au metal (4.78 eV from vacuum [26]) i.e., the Fermi level of Pd is lower than that of Au, electron transfer from Au to Pd in process (2) is reasonable. A direct or indirect route accounts for redox reaction between electron-deficient Au particles and 2-propanol, although we do not have clear evidence. In the direct route, some of the Au particles are exposed to the outer surface and accept electrons from 2-propanol. In the indirect route, this electron transfer occurs via redox of Pd-Pd^{2+} between electron-deficient Au particles and 2-propanol because $2Au^+ + Pd = 2Au + Pd^{2+}$ is electrochemically possible ($E^0 = +0.915$ V at 298 K) [27]. TiO$_2$ may not directly contribute to the photocatalytic redox reaction under the present condition. Formation of the Pd shell on the Au core was achieved only when we used the photodeposition method. In this sense, we think that TiO$_2$ is one of the essential components for the present photocatalytic reaction.

16.4 Conclusions

Using a simple multi-step photodeposition method, Au nanoparticles supported on semiconductor (TiO$_2$ and SnO$_2$) were successfully covered with metal (Ag, Cu, and Pd), forming core–shell Au@Ag/SnO$_2$, Au@Cu/SnO$_2$ and Au@Pd/TiO$_2$, indicating that the Au plasmonic photocatalyst was facilely functionalized with metal. In Sect. 16.1, absorption of the mother Au/SnO$_2$ at around $\lambda = 550$ nm gradually shifted to longer and shorter wavelengths with increases in the contents of Cu and Ag, respectively. Finally, Au@Cu/SnO$_2$ and Au@Ag/SnO$_2$ having 0.8 wt% Cu and Ag showed strong absorption at around $\lambda = 450$ and 620 nm, respectively. These samples were active for mineralization of formic acid and selective oxidation of alcohols to carbonyl compounds under visible light irradiation and exhibited the best performance when irradiated by light overlapping with their SPR. In Sect. 16.2, Pd-free Au/TiO_2 and Au-free Pd/TiO_2 were inactive, whereas the core–shell $Au@Pd/TiO_2$ samples were active in the photocatalytic dechlorination of chlorobenzene to benzene along with oxidation of 2-propanol to acetone under irradiation of visible light. In the reactions of the latter plasmonic photocatalysts, benzene was almost quantitatively formed with a stoichiometric amount of acetone. Thickness control of the Pd shell was very important for both a satisfactory co-catalyst effect and absorption due to the SPR of Au nanoparticles. The results of this study provide useful information for the design of functionalization of plasmonic photocatalysts by the formation of a core–shell structure.

References

1. Tanaka A, Sakaguchi S, Hashimoto K, Kominami H (2012) Catal. Sci Technol 2:907–909
2. Tanaka A, Sakaguchi S, Hashimoto K, Kominami H (2013) ACS Catal 3:79–85
3. Tanaka A, Nishino Y, Sakaguchi S, Yoshikawa T, Imamura K, Hashimoto K, Kominami H (2013) Chem Commun 49:2551–2553
4. Sobal NS, Hilgendorff M, Mҫhwald H, Giersig M, Spasova M, Radetic T, Farle M (2002) Nano Lett 2:621–624
5. Loo C, Lowery A, Halas N, West J, Drezek R (2005) Nano Lett 5:709–711
6. Sun S (2006) Adv Mater 18:393–403
7. Talapin DV, Lee J-S, Kovalenko MV, Shevchenko EV (2010) Chem Rev 110:389–458
8. Chen M, Kumar D, Yi C-W, Goodman DW (2005) Science 310:291–293
9. Enache DI, Edwards JK, Landon P, Solsona-Espriu B, Carley AF, Herzing AA, Watanabe M, Kiely CJ, Knight DW, Hutchings GJ (2006) Science 311:362–365
10. Alayoglu S, Nilekar AU, Mavrikakis M, Eichhorn B (2008) Nat Mater 7:333–338
11. Strasser P, Koh S, Anniyev T, Greeley J, More K, Yu CF, Liu ZC, Kaya S, Nordlund D, Ogasawara H, Toney MF, Nilsson A (2010) Nat Chem 2:454–460
12. Serpell CJ, Cookson J, Ozkaya D, Beer PD (2011) Nat Chem 3:478–483
13. Henning AM, Watt J, Miedziak PJ, Cheong S, Santonastaso M, Song M, Takeda Y, Kirkland AI, Taylor SH, Tilley RD (2013) Angew Chem Int Ed 52:1477–1480
14. Fang PP, Jutand A, Tian ZQ, Amatore C (2011) Angew Chem Int Ed 50:12184–12188
15. Tada H, Mitsui T, Kiyonaga T, Akita T, Tanaka K (2006) Nat Mater 5:782–786
16. Tanaka A, Hashimoto K, Kominami H (2016) Chem Eur J 22:4592–4599
17. Tanaka A, Hashimoto K, Kominami H (2011) ChemCatChem 3:1619–1623
18. Tanaka A, Hashimoto K, Kominami H (2012) J Am Chem Soc 134:14526–14533
19. Pastoriza-Santos I, Sánchez-Iglesias A, Rodríguez-González B, Marzán LM (2009) Small 5:440–443
20. Tanaka A, Fuku K, Nishi T, Hashimoto K, Kominami H (2013) J Phys Chem C 117:16983–16989
21. Moulder JF, Stickel WF, Sobol PE, Bomben KD (1992) Handbook of x-ray photoelectron spectroscopy. In: Chastain J (ed) Perkin-Elmer Co., Minnesota
22. Fuku K, Hashimoto K, Kominami H (2010) Chem Commun 46:5118–5120
23. Fuku K, Hashimoto K, Kominami H (2011) Catal. Sci Technol 1:586–592
24. Kominami H, Nishi T, Fuku K, Hashimoto K (2013) RSC Adv. 3:6058–6064

25. Furube A, Du L, Hara K, Katoh R, Tachiya M (2007) J Am Chem Soc 129:14852–14853
26. Trasatti S (1971) J Electroanal Chem 33:351–378
27. Bard AJ, Parsons R, Jordan J (eds) (1985) Standard potentials in aqueous solution, Marcel Dekker

Chapter 17
PdAu Core–Shell Nanostructures as Visible-Light Responsive Plasmonic Photocatalysts

Priyanka Verma, Robert Raja, and Hiromi Yamashita

17.1 Introduction

Noble metal nanostructures such as Ag [1], Au [2, 3] and Cu [4] can efficiently absorb visible light irradiation owing to the localized surface plasmon resonance (LSPR) characteristics, which occurs due to the collective oscillation of electrons in response to the incident light at resonant frequency [5, 6]. The non-radiative LSPR decay has recently been highlighted in the bond-breaking of adsorbate molecules on the semiconductor (TiO_2, CdS, C_3N_4) [7–9] or directly on the surface of plasmonic metal (Ag, Au, Cu) in combination with catalytically active metal nanoparticles (NPs) such as Pd, Ru and Pt [10–12]. Mechanistically, the LSPR excitation generates hot electrons and holes which get transferred to the lowest unoccupied molecular orbitals (LUMO) of the adsorbates, forming transient negative ion species, which assists in driving the chemical reaction by converting solar energy to chemical and thermal energy [2, 13]. The plasmon-mediated catalysis on bare Ag and Au NPs leads to lower reaction yields in catalytic applications because of their weak interactions with adsorbate molecules. It is strongly desired to explore the synthesis of highly active and selective multimetallic nanostructures for efficient plasmonic catalysis.

In the last decade, plasmonic bimetallic nanostructures have received significant research attention in order to prepare visible light-responsive multifunctional nanocatalysts for their improved catalytic, magnetic, and optical properties [13–16]. The inclusion of plasmonic metal led to the widespread potential applications in

P. Verma (✉) · R. Raja
School of Chemistry, University of Southampton, University Road, Highfield, Southampton SO17 1BJ, United Kingdom
e-mail: P.Verma@soton.ac.uk

H. Yamashita
Division of Materials and Manufacturing Science, Graduate School of Engineering, Osaka University, 2-1 Yamadaoka, Suita 565-0871, Osaka, Japan

© The Author(s), under exclusive license to Springer Nature Singapore Pte Ltd. 2021
H. Yamashita and H. Li (eds.), *Core-Shell and Yolk-Shell Nanocatalysts*,
Nanostructure Science and Technology,
https://doi.org/10.1007/978-981-16-0463-8_17

photobiology, photoelectricity, and photochemistry. Among all, photocatalysis has been under significant research attention exploring the use of bimetallic plasmonic nanostructures for various energy and environmental applications such as pollutant degradation, hydrogen generation, CO_2 reduction, and organic transformation reactions. The strategy for designing plasmonic nanomaterials is largely dependent on the size, morphology, aspect ratio of NPs, and the positioning of the LSPR in the absorption spectra. In an effort to design active catalysts, the LSPR excitation of noble metal NPs has recently been combined with the intrinsic catalytic activity of active metal species. Due to the chemical stability and strong optical absorption of Au in the visible regime, its combination with catalytically active species has been widely explored. AuPd core-shell nanostructures are one of the most researched bimetallic catalysts for plasmon-mediated catalysis [12, 17, 18]. It has also been shown that the SPR properties of AuPd in a core-shell arrangement are different and unique from their individual components. Pd is known to catalyze several important organic reactions and has become technologically important metal for various industrial applications and its incorporation with Au enhances the overall plasmonic properties. Moreover, an improvement in the yield of several reactions including cross-coupling, hydrogenation, and oxidation was also observed. The careful tuning of the composition and morphology of Au core and Pd shell is essential for preserving the plasmon behaviour of Au and catalytic performance of Pd NPs. Li et al. have proposed the existence of a special surface plasmon resonance (SPR) mode at the interface of core and shell NPs which determines the shape and shift observed in the absorption spectrum [19]. Especially, the LSPR band of Au@Pd NPs in the visible regime has higher tunability which is responsible for the enhanced catalytic activities under visible light irradiation.

To date, a myriad of chemical methods has been developed to synthesize AuPd nanostructures of different morphologies such as spheres, rods, dendrites, wheels, triangle, and bipyramid in order to increase the light absorption and improve the catalytic performances. Table 17.1 enlists the various morphology-controlled PdAu nanostructures along with their plasmonic-catalytic performances as discussed in this chapter.

We also encountered several published reports on plasmonic nanostructures in conjunction with semiconductor nanomaterials for heterogeneous catalytic reactions, mainly water splitting and selective oxidation reactions [27–32]. However, in this chapter, we will limit the discussion to AuPd plasmonic nanomaterials in colloidal form or supported on insulators (mesoporous silica) in order to study the direct catalysis on the surface of metal with higher photon efficiency [33]. The synthetic strategies to tailor the LSPR absorption and effect of different parameters such as surfactant, reaction time, temperature, and precursor concentration will be discussed briefly to control the morphology and hence the optical absorption. Various synthetic methods including seed-mediated growth, wet-chemical reduction, microwave synthesis, LSPR-assisted deposition and galvanic replacement methods have been introduced for the successful synthesis of core-shell nanostructures. The target catalytic reaction to study the plasmonic enhancements by AuPd NPs was the Suzuki–Miyaura coupling reaction including selective oxidation of alcohol and

Table 17.1 Morphology controlled AuPd core-shell nanostructures for plasmon-mediated catalysis under visible light irradiation

Plasmonic photocatalyst	Morphology	Synthesis method	Photocatalytic application	Catalytic performance		Ref.
				Yield dark (%)	Yield light (%)	
PdAu/SBA-15	Spherical	Microwave and LSPR assisted deposition	Suzuki–Miyaura coupling	45	70	[17]
PdAu/HPS	Spherical	Wet impregnation method	Suzuki–Miyaura coupling	55	72	[20]
Pd@Au nanowheels	Nanowheels	Wet-chemical reduction	Suzuki–Miyaura coupling	22	98	[21]
			Benzyl alcohol oxidation	18	97	
Au/SiO$_2$/Pd	Bipyramid-like	Seed-mediated growth	Suzuki–Miyaura coupling	54	–	[18]
Pd@Au nanorods	Rod-like	Seed-mediated growth	4-nitrothiophenol reduction to 4-aminothiophenol	–	$k = 5.3 \times 10^{-3}$ s^{-1}	[22]
PdAu nanotriangle	Triangular	CTAC and ascorbic acid	Suzuki–Miyaura coupling	71	85	[23]
Pd@Au concave	Cube-like	Galvanic replacement	Styrene oxidation	–	Conv. 44; selectivity-87	[24]
Pd@Au nanodendrites	Dendrites	CTAC and ascorbic acid	Suzuki–Miyaura coupling	100	–	[25]
PdAu nanostars	Nanostars	HAuCl$_4$ and ascorbic acid	Suzuki–Miyaura coupling	99	–	[26]
			4-nitrophenol reduction	$k = 8.14$ min^{-1}	–	

nitroaromatic reduction. We will describe the PdAu core-shell nanostructures in three different sections on the basis of their polarisation modes arising from the difference in their geometrical symmetry of spheres, rods, triangle, cubic and more complex morphologies.

17.2   Plasmonic PdAu Nanostructures with Spherical Geometry

The geometrical symmetry of metal NPs strongly affects the position and intensity of the LSPR band in the optical spectrum. The different morphologies of sphere, rod and cube increase the LSPR peaks from one, two and three, respectively due to their different polarization modes. The bimetallic nanostructures have been extensively studied because of their improved physicochemical properties, such as electronic, magnetic, catalytic and optical responses in comparison to the monometallic counterparts. Further, the synergistic bimetallic interactions between core and shell metal NPs displayed enhanced activity, selectivity and stability, when compared with monometallic species. In this section, some of the most prominent examples of PdAu core-shell nanostructures confined in a spherical symmetry will be discussed. For example; Huang et al. reinvented the plasmonic chemistry of novel AuPd nanowheels by a one-pot preparation method with Au core surrounded by Pd on its surface [21]. The nanowheels were prepared by a facile wet-chemical reduction method in which HNO_3 was used as the oxidant to control the sequence of reduction of Au and Pd precursors. The growth of Pd around the edges of Au core takes place by Ostwald ripening mechanism to form wheel-like nanostructure. An edge length of 290 nm along with a thickness of 6 nm was observed revealing a plate-like structure in the HAADF-STEM analysis. The LSPR absorption intensity and the size of nanostructure can be tailored by varying the molar ratio of metal precursors and reducing agent, ascorbic acid. The core-shell nanostructures were tested in benzyl alcohol oxidation and Suzuki coupling reaction for plasmon mediated catalysis under light irradiation conditions. The oxidation of benzyl alcohol gave 97% conversion with 98% selectivity for benzaldehyde formation under light irradiation conditions. The obtained yield under conventional heating conditions at 50 °C was merely 18% in 6 h. Similarly, the Suzuki coupling reaction displayed superior yields under light irradiation conditions than conventional heating. Further, a linear dependence in the enhancement of catalytic activity and intensity of light irradiation attributed to the significance of LSPR of AuPd nanostructures.

Minsker et al. reported the synthesis of Au (core)-Pd (shell) stabilised on hyper-cross-linked polystyrene (HPS) for visible light enhanced Suzuki–Miyaura coupling reaction [20]. Figure 17.1a, b shows the EDX mapping of Au–Pd NP with a very thin shell of Pd (less than 1 nm) on the surface of Au. The coupling reaction product shows significantly higher yields over Au–Pd/HPS under visible light irradiation (72%) in

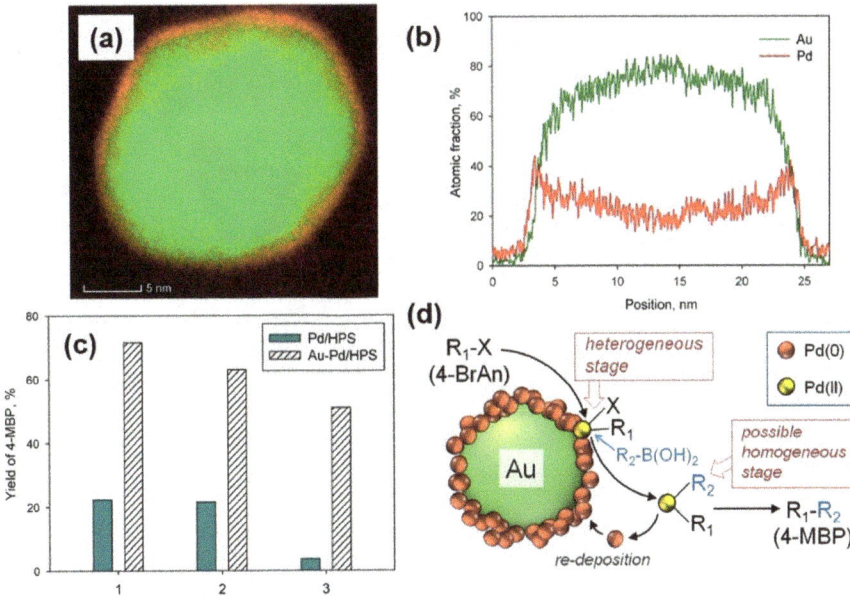

Fig. 17.1 **a** EDX mapping and **b** EDX line scan profile of an individual Au-Pd core-shell NP, **c** results of resue of Pd and Au–Pd/HPS catalyst in Suzuki–Miyaura coupling reaction, and **d** proposed reaction mechanism over Au–Pd NP. Reaction conditions: 60 °C, 5:1 EtOH/water mixture as the solvent, 0.75 mmol of NaOH, 0.5 mmol of 4-BrAn, 0.75 mmol of PBA. Adapted with permission from ref. [20] Copyright 2018 American Chemical Society

comparison to dark (55%) conditions owing to LSPR effect. The stability test indicated the bimetallic Au–Pd/HPS catalysts to be much more stable than monometallic catalysts (Fig. 17.1c). The Au core assisted in the redeposition of Pd from the reaction solution to the NP surface and hence improves the stability of Au–Pd/HPS catalyst. The proposed reaction pathway (Fig. 17.1d) explains the electron donation from Au to Pd under visible light irradiation to facilitate the oxidative addition step, and hence the rate of the reaction by forming more electron-rich Pd species.

In another report, PdAu bimetallic nanostructures were immobilised within the mesoporous silica framework by a two-step synthesis approach [17]. At first, plasmonic Au NPs were synthesised by a microwave-assisted deposition method in which 1-hexanol was used as the solvent and reducing agent. A clear suspension of mesoporous silica, SBA-15 in alcohol along with the Au precursor, $HAuCl_4$, was irradiated with microwave irradiation for a period of 3 min to form uniformly dispersed Au/SBA-15 catalyst. The second step of Pd deposition on the surface of Au NPs was carried out by LSPR-assisted deposition method in which an aqueous suspension of Au was irradiated with visible light followed by injection of fixed amounts of $Pd(OAc)_2$ to yield Pd/Au/SBA-15 catalyst. The average size of the spherical PdAu catalyst was 4.9 nm within the mesoporous channels of mesoporous silica, SBA-15. The Au L_{III}-edge EXAFS analysis of the PdAu catalyst exhibited the main peak at

Scheme 17.1 Schematic illustration of charge transfer pathway over PdAu plasmonic nanostructures under visible light irradiation. Reproduced with permission from ref. [17] Copyright 2016 The Royal Society of Chemistry

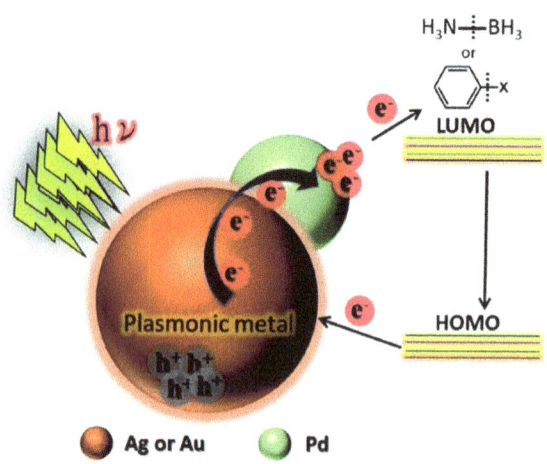

2.5 Å attributing to the presence of contiguous Au–Au bonding. The appearance of a shoulder peak in the Pd K-edge FT-EXAFS spectra confirmed the existence of heteroatomic bonding. The surface composition and chemical state of plasmonic nanostructures were confirmed by Au 4f and Pd 3d XPS analysis. The prepared catalyst was tested for the carbon–carbon bond formation reaction in Suzuki–Miyaura coupling reaction. PdAu/SBA-15 exhibited a significantly superior yield of biphenyl (70%) under visible light irradiation in comparison to 45% yield in dark conditions. The effect of substituents (electron-donating and electron-withdrawing) on the catalytic enhancements under visible light irradiation was also discussed in detail to understand the rate-determining step and predict the mechanistic pathway for the reaction. Scheme 17.1 illustrates the hot electron pathways from Au to Pd and ultimately to the LUMO of the adsorbed reactant molecules. This charge transfer is well supported due to the higher work function value of Pd (5.0 eV) than Au (4.7 eV). The accumulated electrons assist in the transformation of adsorbed species to transient ionic species which leads to bond weakening by elongation and hence faster reaction rates under light illumination conditions.

17.3 Plasmonic PdAu Nanostructures with Rod-Like Geometry

The rod-like morphology displays two LSPR absorption peaks in the visible and NIR region of the optical spectrum due to the transverse and longitudinal modes of polarisation. The increase in the aspect ratio of nanorods leads to an increase in the red-shift and broadening of longitudinal LSPR band of absorption. Wang et al. have recently reported the synthesis of Au nanorods and its site-selective deposition of Pd NPs on its surface. The anisotropic Pd/SiO$_2$/Au bimetallic nanostructures were prepared in which silica was selectively coated on the ends and sides of Au nanobipyramids (NBP) by varying the concentration of cetyltrimethylammonium bromide (CTAB) in the synthesis process [18]. The coating of silica provides the blocking for the deposition of second metal by enabling the overgrowth of Pd on the exposed surface of Au NBPs. The Au NBP/SiO$_2$ was used for site-selective Pd deposition on Au NBPs to form Au NBP/end-SiO$_2$/side-Pd and Au NBP/side-SiO$_2$/end-Pd nanocatalysts. The presence of silica can be considered as the hard template for guiding the deposition of Pd. The prepared nanostructures were tested for the Suzuki coupling reaction between bromobenzene and m-tolylboronic acid, as a function of laser intensity and the position of Pd was correlated with the plasmon-mediated photocatalytic activity. The rate-determining step in the Suzuki coupling reaction is the breaking of C–Br bond of bromobenzene adsorbed on the surface of Pd NPs. Under visible light irradiation, the plasmonic excitation of Au NBPs generates hot electrons which get injected into the LUMO of the adsorbed molecule via Pd and hence enhance the reaction rate. Under the laser excitation of 808 nm, Au NBP/end-Pd displayed the largest enhancement in the catalytic performance than Au NBP/side-Pd and Au NBP/all-Pd. The finite-difference time-domain (FDTD) simulations displayed the maximum electric field enhancements at the ends for the longitudinal plasmon excitation for three catalysts. Interestingly, the electric field enhancement effects were the largest for the Au NBP/end-Pd catalyst. This also correlates the superior catalytic performance of the Au NBP/end-Pd catalyst due to the presence of Pd at the end where the generation of plasmonic electric field enhancements is largest. Such selective coating for the synthesis of core-shell bimetallic nanostructures can be a promising approach for various plasmon-enabled applications.

The tunable LSPR absorption of Au NRs in the NIR region was studied for sensing and in situ monitoring of chemical reaction, as reported by Santos et al. [22]. Figure 17.2a, b shows the evolution of UV–vis spectra for the growth of Pd on Au in the presence of CTAB and cetyltrimethylammonium chloride (CTAC). The synthetic strategy for Au@Pd NRs involved the use of penta-twinned Au as seeds and Br$^-$ ions for the directional growth of NRs. The presence of CTAB allows the preferential reduction of Pd on the tips of Au NRs and CTAC led to the formation of dendritic Pd shell on the entire surface of Au NRs as depicted in Fig. 17.2c. The presence of Br$^-$ ions in CTAB facilitated the Pd deposition at the ends of Au NRs leading to an increase in the aspect ratio from 4.4 nm (in Au NRs) to 6.2 nm (Au@Pd NRs). The increase in the aspect ratio further enhanced and shifted the longitudinal plasmonic absorption from 900 to 1300 nm by LSPR peak broadening. The seeded growth strategy can tailor the optical absorption of Au@Pd hybrid through the careful tuning of parameters such as precursor concentration, type of surfactant, and reaction time. Figure 17.2d–f shows the representative TEM images of Au NRs before and after Pd deposition in the presence of CTAB or CTAC. Further, electron tomography, Energy-dispersive X-ray spectroscopy (XEDS) and electron diffraction techniques were used to clearly visualize the preferential deposition of Pd on penta-twinned (PTW) Au NRs. The designed heterostructures exhibited excellent catalytic performance in the 4-nitrothiophenol (4-NTP) reduction to 4-aminothiophenol (4-ATP) as monitored by in situ surface-enhanced Raman scattering (SERS) spectroscopy. The peak intensities

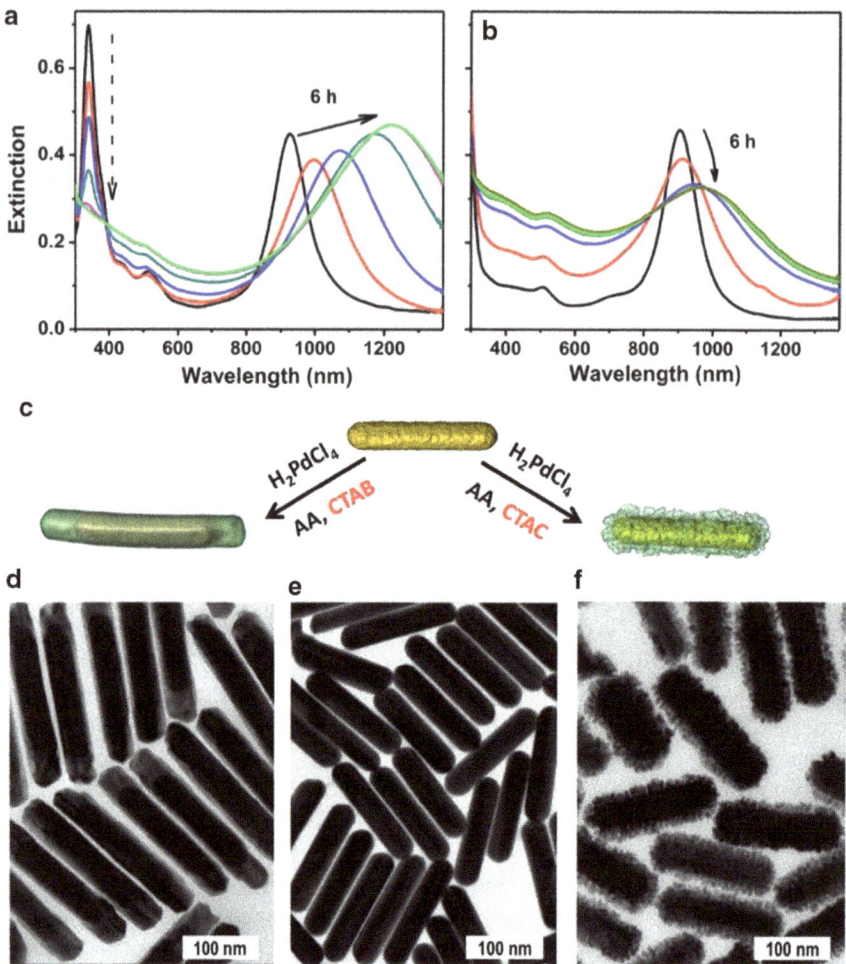

Fig. 17.2 a, b Time evolution of the U-visible-NIR extinction spectra during the overgrowth of PTW Au nanorods with Pd in the presence of CTAB (**a**) or CTAC (**b**), (**c**) Schematic illustration of Au@Pd NRs formation with CTAB and CTAC, **d–f** Representative TEM images of the original PTW Au nanorods before (**d**) and after the deposition of Pd in the presence of CTAB (**e**) and CTAC (**f**). Reproduced with permission from ref. [22] Copyright 2016 American Chemical Society

of vibrational bands due to the stretching modes in 4-NTP and 4-ATP in the SERS spectrum were compared to understand the kinetics of reaction. The observed rate constant value for the hydrogenation reaction on Pd@Au NRs (CTAB) was calculated to be $k = 5.3 \times 10^{-3}$ s^{-1}. The design and study of such hybrid nanostructures with intense plasmonic absorption will open a new avenue in the relatively new field of plasmon-mediated catalysis.

17.4 Plasmonic PdAu Nanostructures with Complex Geometries

The optical response of nanotriangles is expected to display strengthened plasmonic absorption due to its sharp features in comparison to other morphologies. Recently Scott et al. have reported the synthesis of Pd@Au core-shell nanotriangles in order to efficiently harvest visible light for enhancing the reaction rate of Suzuki–Miyaura coupling reaction [23]. The use of triangular morphology allows the strong plasmonic absorption and a larger number of active sites than conventional NPs. The Au nanotriangles can be prepared by using CTAC as the stabilising agent followed by the addition of ascorbic acid and potassium iodide as the reducing and oxidative etchant, respectively. Pd was introduced on Au nanotriangles by reducing the precursor with ascorbic acid. Figure 17.3a displays the plasmonic absorption band of Au and AuPd nanotriangles at 605 and 545 nm, respectively. The blue shift of 60 nm in the absorption of AuPd nanotriangles could be attributed to surface alloying between some Au and Pd NPs. The triangular morphology of Au and AuPd nanostructures was confirmed by TEM analysis Fig. 17.3b–d. The contrast difference in the TEM image confirmed the spatial distribution of Au core and polycrystalline

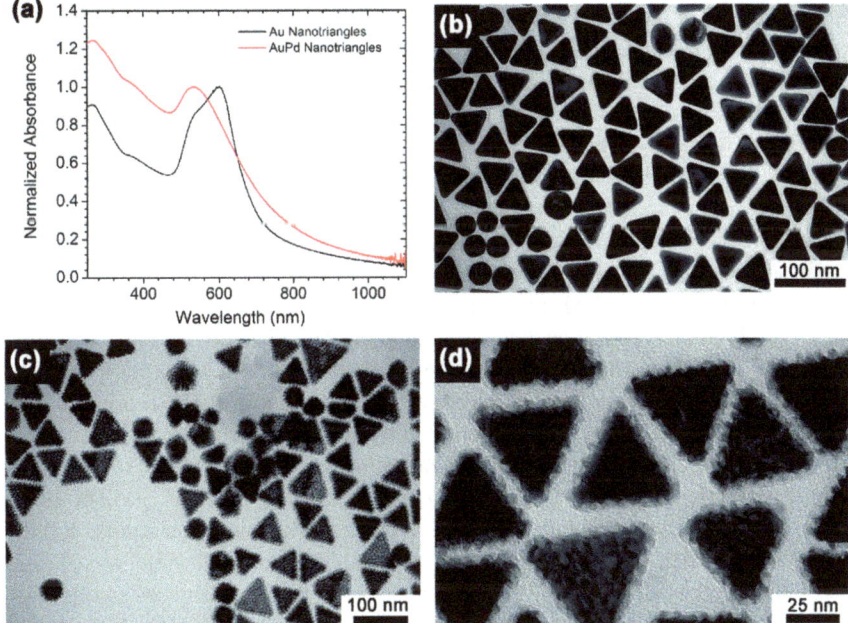

Fig. 17.3 **a** The UV–vis spectra of Au and AuPd bimetallic nanotriangles; TEM images of **b** as-synthesized Au nanotriangles, **c** AuPd bimetallic nanotriangles and **d** magnified image of AuPd nanotriangles. Reproduced with permission from ref. [23] Copyright 2017 The Royal Society of Chemistry

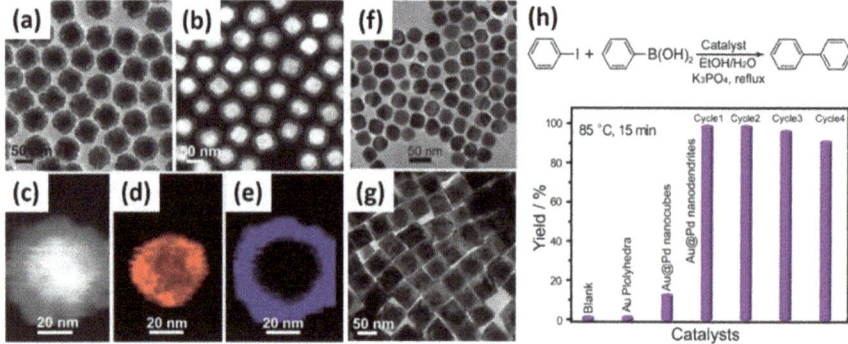

Fig. 17.4 Au@Pd core-shell nanodendrites. **a** TEM image, **b** and **c** HAADF-STEM images, **d** and **e** EDX elemental maps of Au and Pd, respectively, **f** TEM image of Au polyhedra, **g** TEM image of Au@Pd nanocubes obtained by using CTAB and **h** Yields of biphenyl in the Suzuki coupling reaction between phenylboronic acid and iodobenzene using different nanostructures as the catalyst and cycle performance of Au@Pd core-shell nanodendrites. Reproduced with permission from ref. [25] Copyright 2013 The Royal Society of Chemistry

Pd shell in the nanotriangular structure. The cross-coupling of *p*-iodobenzoic acid and phenyl boronic acid gave a significantly higher product yield (92%) under light irradiation than in dark conditions (73%) for 3 h reaction time. Further, controlled experiments were carried out to study the contribution of plasmonic effects by hot electron transfer and localised heating.

In another interesting work reported by Wang et al., Au@Pd core-shell nanodendrites were synthesised by growing non-compact Pd porous shell onto Au polyhedra at RT [25]. The bifunctional nanodendrites were found to display intense LSPR absorption and superior catalytic activities in the Suzuki coupling reaction when compared with Au@Pd nanocubes. The structure of Au core and Pd shell was studied by high-angle annular dark-field scanning transmission electron microscopy (HAADF-STEM) to confirm the dendritic nature of Au@Pd nanostructure (Fig. 17.4a–e). An average diameter of 3–7 nm of Pd was covered on the surface of Au. An increase in the thickness of the Pd shell and LSPR absorption intensity was observed on increasing the amounts of precursor solutions. The morphology of bimetallic plasmonic catalyst could be tuned by changing the molar ratio of surfactants, CTAB and CTAC. The controlled synthesis using pure CTAC led to the preparation of nanodendrites and CTAB formed nanocubes (Fig. 17.4f, g). The blank reaction in the absence of catalyst gave a 1.3% yield of biphenyl and similar activity was observed when Au was used as the catalyst. The biphenyl yield was significantly improved by 88% from Au@Pd nanocubes to nanodendrites as summarised in Fig. 17.4h along with their corresponding TEM micrographs. The Au core can act as an electron promotor for Pd shell to facilitate the organic reaction on the surface. The superior catalytic performance of bifunctional Au@Pd nanodendrites was ascribed to the large surface area due to the presence of porous Pd shell and a higher concentration of unsaturated surface Pd atoms.

Xiong et al. studied the AuPd concave nanostructures with Au/Pd ratio $= 0.08$ in the aerobic oxidation reaction of styrene under full-spectrum light illumination $(200 \, \text{mWcm}^{-2})$ [24]. The Pd nanostructure was prepared by using PVP polymer and the Au was introduced onto Pd by galvanic displacement reaction in the presence of ascorbic acid and NaOH. The NaOH was introduced to control the rate of reduction and enhance the reduction ability of ascorbic acid. In this way, Au was introduced in a precisely controlled manner without affecting the size and morphology of Pd concave nanostructure. An average diameter of prepared $AuPd_{0.08}$ concave nanostructures was measured to be 51 nm. The LSPR absorption was also observed at 400–900 nm which would assist in visible-light-driven catalytic oxidation of styrene. The styrene oxidation reaction to styrene oxide was carried out at 80 °C in the presence of molecular oxygen which also forms various unwanted by-products including benzaldehyde, acetophenone and benzoic acid. It is important to monitor the conversion and selectivity of reactant and main product, respectively, under visible light irradiation. A volcano-shaped relationship in the styrene oxide selectivity with the Au content in the catalyst was observed. The decrease in the selectivity with further increase in the Au content was due to reduced activation of adsorbed oxygen species. This was further studied by diffuse reflectance infrared Fourier-transform spectroscopy (DRIFTS) analysis which confirmed that the O_2 species on $AuPd_{0.08}$ prefers the peroxo-like position instead of superoxo-like configuration which leads to the selective oxidation of styrene. The rational design of nanocatalysts for visible light-sensitive oxidation reaction can further be developed by following the concept demonstrated here.

17.5   Conclusions and Outlook

The ability to manipulate and tailor the LSPR absorption in PdAu bimetallic nanostructures can be harnessed to develop multifunctional plasmonic materials with attractive properties that are difficult to achieve with a single component Au NPs. In this book chapter, we have introduced different types of morphology-controlled PdAu core-shell nanostructures for efficient catalysis under visible light irradiation. The sophisticated synthetic methodologies can develop materials with high tunability, controlled morphology and enhanced electric field enhancement effects in order to maximise the photon efficiency. The synthesis of bimetallic core-shell nanostructures with different morphologies, for example, spherical, nanowheel-type, rod, triangular, cube-like, bipyramidal, dendrite and nanostars, has been discussed for their enhanced photocatalytic activity in the Suzuki–Miyaura coupling and selective oxidation and reduction reactions. The most commonly employed synthetic methodology involved the use of seed-mediated growth process in which the careful modification of the concentration of surfactant and reductant could produce different morphologies of core-shell nanostructures as listed in Table 17.1. It was found that Pd@Au nanodendrites hybrid catalyst was found to exhibit superior catalytic performance with 100% yield in the coupling reaction attributing to its wide optical response and higher concentration of unsaturated Pd atoms on the surface. In contrast to spherical nanostructures, the presence of sharp features in more complex morphologies such as rod, triangle, dendrite, cube and star, are expected to display strong plasmonic absorption and hence much higher enhancement in the catalytic performances under visible light irradiation.

The core-shell bimetallic nanostructures offer a new avenue to merge the advantages of functionalities to engineer nanomaterials for more effective applications in optics, photocatalysis and photovoltaics. Although the research efforts on the fabrication of plasmonic nanocatalysts are still in their infancy, we expect that further developments in the design of well-defined hybrid nanomaterials will pave the way to enhance the catalytic performances and understanding of the mechanistic details with the help of various in situ spectroscopic tools.

Acknowledgements Dr. P. Verma would like to thank The Royal Society-Newton International Fellowship (NIF\R1\180185) for her postdoctoral research funding at the University of Southampton.

References

1. Mori K, Verma P, Hayashi R, Fuku K, Yamashita H (2015) Color-controlled Ag nanoparticles and nanorods within confined mesopores: microwave-assisted rapid synthesis and application in plasmonic catalysis under visible-light irradiation. Chem Eur J 21:11885–11893
2. Jo S, Verma P, Kuwahara Y, Mori K, Choi W, Yamashita H (2017) Enhanced hydrogen production from ammonia borane using controlled plasmonic performance of Au nanoparticles deposited on TiO_2. J Mater Chem A 5:21883–21892
3. Verma P, Mori K, Kuwahara Y, Cho SJ, Yamashita H (2020) Synthesis of plasmonic gold nanoparticles supported on morphology-controlled TiO_2 for aerobic alcohol oxidation. Catal Today 352:255–261
4. Fernández-Catalá J, Navlani-García M, Verma P, Berenguer-Murcia Á, Mori K, Kuwahara Y, Yamashita H, Cazorla-Amorós D (2020) Photocatalytically-driven H_2 production over Cu/TiO_2 catalysts decorated with multi-walled carbon nanotubes. Catal Today. https://doi.org/10.1016/j.cattod.2020.05.032
5. Yamashita H, Mori K, Kuwahara Y, Kamegawa T, Wen M, Verma P, Che M (2018) Single-site and nano-confined photocatalysts designed in porous materials for environmental uses and solar fuels. Chem Soc Rev 47:8072–8096
6. Verma P, Kuwahara Y, Mori K, Yamashita H (2019) Design of silver-based controlled nanostructures for plasmonic catalysis under visible light irradiation. Bull Chem Soc Jpn 92:19–29
7. Bai X, Zong R, Li C, Liu D, Liu Y, Zhu Y (2014) Enhancement of visible photocatalytic activity via $Ag@C_3N_4$ core-shell plasmonic composite. Appl Catal B Environ 147:82–91
8. Lee JE, Bera S, Choi YS, Lee WI (2017) Size-dependent plasmonic effects of M and $M@SiO_2$ (M = Au or Ag) deposited on TiO_2 in photocatalytic oxidation reactions. Appl Catal B Environ 214:15–22
9. Liu Z, Wang L, Li R, Huang M (2019) Synthesis of $Au@MoS_2$-CdS ternary composite structure with enhanced photocatalytic activity. NANO 14:1–8
10. Verma P, Yuan K, Kuwahara Y, Mori K, Yamashita H (2018) Enhancement of plasmonic activity by Pt/Ag bimetallic nanocatalyst supported on mesoporous silica in the hydrogen production from hydrogen storage material. Appl Catal B Environ 223:10–15

11. Verma P, Kuwahara Y, Mori K, Yamashita H (2015) Synthesis and characterization of a Pd/Ag bimetallic nanocatalyst on SBA-15 mesoporous silica as a plasmonic catalyst. J Mater Chem A 4:10142–10150

12. Xiao Q, Sarina S, Jaatinen E, Jia J, Arnold DP, Liu H, Zhu H (2014) Efficient photocatalytic suzuki cross-coupling reactions on Au-Pd alloy nanoparticles under visible light irradiation. Green Chem 16:4272–4285

13. Fan H, Li Y, Liu J, Cai R, Gao X, Zhang H, Ji Y, Nie G, Wu X (2019) Plasmon-enhanced oxidase-like activity and cellular effect of Pd-coated gold nanorods. ACS Appl Mater Interfaces 11:45416–45426

14. Bathla A, Pal B (2020) Superior co-catalytic activity of Pd(core)@Au(shell) nanocatalyst imparted to TiO_2 for the selective hydrogenation under solar radiations. Sol Energy 205:292–301

15. Verma P, Kuwahara Y, Mori K, Yamashita H (2017) Synthesis of mesoporous silica-supported Ag nanorod-based bimetallic catalysts and investigation of their plasmonic activity under visible light irradiation. Catal Sci Technol 7:2551–2558

16. Wadell C, Antosiewicz TJ, Langhammer C (2012) Optical absorption engineering in stacked plasmonic Au-SiO 2-Pd nanoantennas. Nano Lett 12:4784–4790

17. Verma P, Kuwahara Y, Mori K, Yamashita H (2016) Pd/Ag and Pd/Au bimetallic nanocatalysts on mesoporous silica for plasmon-mediated enhanced catalytic activity under visible light irradiation. J Mater Chem A 4:10142–10150

18. Zhu X, Jia H, Zhu XM, Cheng S, Zhuo X, Qin F, Yang Z, Wang J (2017) Selective Pd deposition on Au nanobipyramids and Pd site-dependent plasmonic photocatalytic activity. Adv Funct Mater 27:1–15

19. Zhang C, Chen BQ, Li ZY, Xia Y, Chen YG (2015) Surface plasmon resonance in bimetallic core-shell nanoparticles. J Phys Chem C 119:16836–16845

20. Nemygina NA, Nikoshvili LZ, Tiamina IY, Bykov AV, Smirnov IS, Lagrange T, Kaszkur Z, Matveeva VG, Sulman EM, Kiwi-Minsker L (2018) Au core-Pd shell bimetallic nanoparticles immobilized within hyper-cross-linked polystyrene for mechanistic study of suzuki cross-coupling: homogeneous or heterogeneous catalysis? Org Process Res Dev 22:1606–1613

21. Huang X, Li Y, Chen Y, Zhou H, Duan X, Huang Y (2013) Plasmonic and catalytic AuPd nanowheels for the efficient conversion of light into chemical energy. Angew Chem Int Ed 52:6063–6067

22. Rodal-Cedeira S, Montes García V, Polavarapu L, Solís DM, Heidari H, La Porta A, Angiola M, Martucci A, Taboada JM, Obelleiro F, Bals S, Pérez-Juste J, Pastoriza-Santos I (2016) Plasmonic Au@Pd nanorods with boosted refractive index susceptibility and SERS efficiency: a multifunctional platform for hydrogen sensing and monitoring of catalytic reactions. Chem Mater 28:9169–9180

23. Gangishetty MK, Fontes AM, Malta M, Kelly TL, Scott RWJ (2017) Improving the rates of Pd-catalyzed reactions by exciting the surface plasmons of AuPd bimetallic nanotriangles. RSC Adv 7:40218–40226

24. Hu C, Xia X, Jin J, Ju H, Wu D, Qi Z, Hu S, Long R, Zhu J, Song L, Xiong Y (2018) Surface modification on Pd nanostructures for selective styrene oxidation with molecular oxygen. ChemNanoMat 4:467–471

25. Wang H, Sun Z, Yang Y, Su D (2013) The growth and enhanced catalytic performance of Au@Pd core-shell nanodendrites. Nanoscale 5:139–142

26. Ma T, Liang F (2020) Au-Pd nanostars with low Pd content: controllable preparation and remarkable performance in catalysis. J Phys Chem C 124:7812–7822

27. Yu X, Liu F, Bi J, Wang B, Yang S (2017) Improving the plasmonic efficiency of the Au nanorod-semiconductor photocatalysis toward water reduction by constructing a unique hot-dog nanostructure. Nano Energy 33:469–475

28. Yoo SM, Rawal SB, Lee JE, Kim J, Ryu HY, Park DW, Lee WI (2015) Size-dependence of plasmonic Au nanoparticles in photocatalytic behavior of Au/TiO_2 and $Au@SiO_2/TiO_2$. Appl Catal A Gen 499:47–54

29. Rather RA, Singh S, Pal B (2016) Core-shell morphology of Au-TiO2@graphene oxide nanocomposite exhibiting enhanced hydrogen production from water. J Ind Eng Chem 37:288–294

30. Negishi R, Naya SI, Kobayashi H, Tada H (2017) Gold(Core)–Lead(Shell) nanoparticle-loaded Titanium(IV) oxide prepared by underpotential photodeposition: plasmonic water oxidation. Angew Chem Int Ed 56:10347–10351

31. Ma L, Chen YL, Yang DJ, Li HX, Ding SJ, Xiong L, Qin PL, Chen XB (2020) Multi-interfacial plasmon coupling in multigap (Au/AgAu)@CdS core-shell hybrids for efficient photocatalytic hydrogen generation. Nanoscale 12:4383–4392

32. Naya SI, Kume T, Akashi R, Fujishima M, Tada H (2018) Red-light-driven water splitting by Au(Core)-CdS(Shell) Half-Cut nanoegg with heteroepitaxial junction. J Am Chem Soc 140:1251–1254

33. Verma P, Kuwahara Y, Mori K, Raja R, Yamashita H (2020) Functionalized mesoporous SBA-15 silica: recent trends and catalytic applications. Nanoscale 12:11333–11363

Chapter 18
Core–Shell Nanoparticles as Cathode Catalysts for Polymer Electrolyte Fuel Cells

Hiroshi Inoue and Eiji Higuchi

18.1 Introduction

Polymer electrolyte fuel cells (PEFCs) using molecular hydrogen as a fuel are highly efficient and clean energy conversion devices, and utilize hydrogen fuel to generate electricity with water as the sole product, so they are currently commercialized for vehicles and home uses.

The cell reaction of a PEFC is represented by Eq. 18.1 is exothermic ($\Delta G^\circ = -237$ kJ mol^{-1}), and its standard electromotive force (U°) is 1.23 V.

$$H_2 + 1/2O_2 \rightarrow H_2O \quad U^\circ = 1.23 \text{ V} \tag{18.1}$$

$$H_2 \rightarrow 2H^+ + 2e^- \quad E^\circ = 0 \text{ V (vs. SHE)} \tag{18.2}$$

$$O_2 + 4H^+ + 4e^- \rightarrow 2H_2O \quad E^\circ = +1.23 \text{ V (vs. SHE)} \tag{18.3}$$

A PEFC typically consists of an anode and a cathode separated by a proton-exchange membrane (PEM) such as Nafion® [1, 2]. Both electrodes are gas diffusion electrodes, which have a laminated structure of a catalyst layer for electrochemical reactions and a gas diffusion layer for hydrogen and oxygen gas diffusion, to make a three-phase interface. At an anode of the PEFC, hydrogen is oxidized in a catalyst layer to produce two protons and two electrons, as shown in Eq. 18.2. The protons diffuse through a PEM, while the electrons go through an external circuit to generate electricity. At a cathode, oxygen is reduced to water, as shown in Eq. 18.3.

H. Inoue (✉) · E. Higuchi
Department of Applied Chemistry, Graduate School of Engineering, Osaka Prefecture University, Sakai, Osaka 599-8531, Japan
e-mail: inoue-h@chem.osakafu-u.ac.jp

© The Author(s), under exclusive license to Springer Nature Singapore Pte Ltd. 2021   275
H. Yamashita and H. Li (eds.), *Core-Shell and Yolk-Shell Nanocatalysts*,
Nanostructure Science and Technology,
https://doi.org/10.1007/978-981-16-0463-8_18

When a PEFC operates, its voltage typically drops from 1.23 V to 0.6–0.7 V. The hydrogen oxidation reaction at the anode is fast, while ORR at the cathode is sluggish, so overpotential is much larger for ORR [3–6], which is the main cause of the decrease in operating voltage. Thus, various electrocatalysts have been developed to decrease the overpotential for ORR.

Pt is known to be an element with the highest ORR activity. To further enhance ORR activity, size reduction of nanoparticles [7–12] and alloying of Pt with foreign transition metals [13–17] were effective due to the increase in specific surface area and electronic modification of Pt, respectively. The ORR activity is strongly influenced by the strength of the oxygen-metal bond interaction, which depends on the average energy of the metal d states (d-band center) relative to the Fermi level [18]. A downward shift of the d-band center results in a downward shift of the antibonding states formed by the coupling of oxygen $2p$ states and the metal d states, leading to a weaker oxygen-metal bond interaction. In this way, an approach for screening new catalysts with higher ORR activity by looking for surfaces that bind oxygen a little weaker than Pt or a downward shift of the d-band center of Pt has been established [18]. The Pt-M (M = Ni, Co, Fe) alloy catalysts prepared under such an idea had higher ORR activity than Pt, but deteriorated under fuel cell operation conditions [19, 20]. In addition, when the size of Pt NPs is reduced to 2–3 nm or less, the ORR activity per electrochemical surface area (ECSA), specific activity (SA_{Pt}), and thus that per mass, mass activity (MA_{Pt}), of the Pt NPs reduced [21]. These suggest that the conventional ideas to improve the ORR activity are reaching their limits.

The core–shell structure has recently attracted attention as a bimetal different from alloys [22–30]. To form a Pt shell with one or two monolayers on a foreign metal core NP is advantageous in the reduction in the Pt content and the increase in the utilization and specific surface area of Pt. Moreover, the foreign metal core will contribute to modify the electronic states of Pt, enhancing SA_{Pt} and MA_{Pt}. Zhang et al. have first reported the use of core–shell NPs to reduce Pt loading in fuel cell electrocatalysts [27, 28]. They formed a Cu monolayer on Pd core NPs using the underpotential deposition technique, and the Cu monolayer was replaced with a monolayer of Pt by galvanic replacement to form a Pt monolayer shell. Various noble and non-noble metals and alloys have been proposed as core material [20–29], and metal core-Pt shell NP-loaded carbon catalysts whose core material was Pd, Ir, Rh, Au, Ru, etc., demonstrated higher ORR activity than commercial Pt-loaded carbon (Pt/C) catalyst.

Adzić et al. found that a Pt monolayer deposited on M(111) (M = Au, Ag, Pd, Rh, Ir) and Ru(0001) exhibited a volcano-type ORR activity in alkaline solutions, and Pt/Pd(111) had the highest ORR activity because it had the optimum balance between the kinetics of O–O bonding breaking and the electroreduction of the oxygenated intermediates or O–H formation [24, 29]. The ORR activity, particularly SA_{Pt}, of core–shell NP catalysts with a Pt monolayer shell is influenced by strain effect as well as ligand (electronic) effect. The compressive strain leads to an extension of the metal d band due to the increased overlap of the wave functions. If the d band is more than half-filled, i.e., if the metal is a late transition metal, then the band extension leads to an increased population of the d band. Because of charge conservation, the *d* band moves down in order to preserve its degree of *d*-band filling. A downshift of the *d* band leads to a higher occupation of antibonding states in the molecule-surface interaction, thus lowering the interaction strength. For Pt, the *d*-band center moves downward or the oxygen-Pt bond strength is weakened by the compressive strain, which has experimentally been verified [31–34]. The effect of compressive strain on ORR activity is more important in Pt NP catalysts because the compressive strain on the NP surface is larger than that on the bulk surface. Shao et al. reported that an Au core-Pt shell NP catalyst whose core size was 3.0 nm improved the ORR activity due to the decrease in surface Pt–Pt distance by compressive strain [35]. We also found that the ORR activity of Au core-Pt shell NP catalysts whose size is 2.5–3.5 nm increased with decreasing the core size [36]. Wang et al. reported that the Pt–Pt distance of the Pt shell depended on the size of Pd core NPs [37].

So far, we have prepared carbon-supported Pd NP (Pd/C) with *ca.* 4.2 nm in size by using CO as a reducing agent in acetonitrile solution containing palladium acetate, and found that MA_{Pt} at 0.9 V versus RHE for the Pd core-Pt shell catalyst using Pd/C as a core (Pt/Pd/C) was about 5 times in as high as that for the commercial Pt/C [38]. To further enhance the MA_{Pt}, it is significant to realize the most appropriate Pt–Pt distance of the Pt shell by tuning the compressive strain by the composition and size of core NPs. In this paper, alloy NPs ($Pd_{100-x}Au_x$) of Pd with Au that has larger atomic radius than Pd were used as a core to tune the Pt–Pt distance of the Pt monolayer shell, and the ORR activity of the $Pd_{100-x}Au_x$ alloy core-Pt shell NP-loaded carbon catalysts was evaluated.

18.2 Characterization of $Pd_{100-x}Au_x$ Core Nanoparticles

18.2.1 Influence of Alloy Composition

$Pd_{100-x}Au_x$/C (x = 5, 10, 20) catalysts were prepared by the same method as Pd/C [38] except that potassium tetrachloroaurate (III) as well as palladium acetate were used as precursors. The bulk composition of $Pd_{100-x}Au_x$ NPs was evaluated by inductively coupled plasma mass spectrometry. The loading of $Pd_{100-x}Au_x$ on Ketjen black was 30 wt.%.

Figure 18.1 shows XRD patterns of the $Pd_{100-x}Au_x$/C (x = 5, 10, 20) and Pd/C. No peaks due to Au were observed for each sample and the diffraction peak for Pd(111) shifted to lower angles as the Au content increased, indicating that alloying of Pd with Au occurred.

The relationship between the nearest neighbor interatomic distance (R_{bulk}) and Au content is shown in Fig. 18.2. The R_{bulk} value increased with the Au content and deviated from the Vegard law, which suggests that the $Pd_{100-x}Au_x$ alloys were not uniform.

Figure 18.3a shows HAADF-STEM image of $Pd_{90}Au_{10}$ NPs. Lattice spacings of $Pd_{90}Au_{10}$ NPs were about 2.27 Å, which assigned to the (111) facet as shown in

Fig. 18.1 XRD patterns of
$Pd_{100-x}Au_x/C$ and Pd/C.
Scan rate: $1° \, min^{-1}$

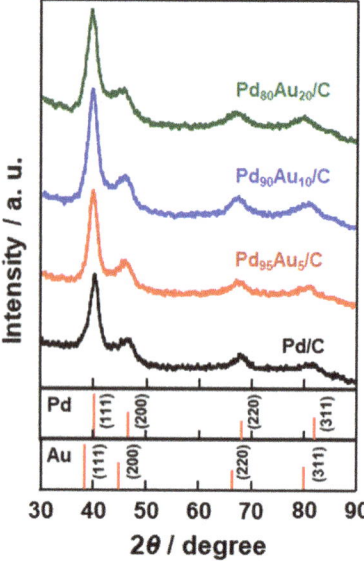

Fig. 18.2 Relationship
between the nearest neighbor
interatomic distance (R_{bulk})
and Au content

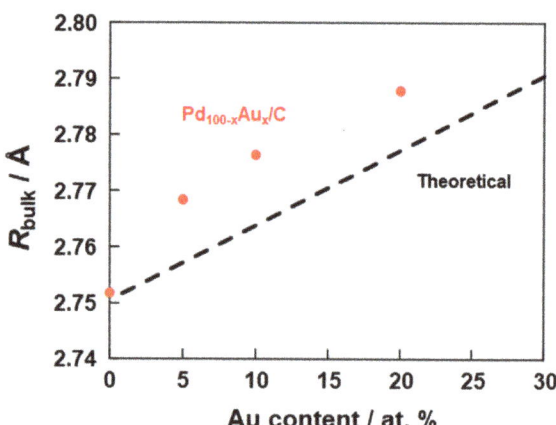

Fig. 18.3a. Also, the lattice constant (3.93 Å) calculated from the lattice spacing coincided with that estimated from XRD. EDX mapping images and EDX line profiles of Au and Pd for a $Pd_{90}Au_{10}$ NP are shown in Fig. 18.3b and c, respectively. Both figures suggest that Au atoms were concentrated in the interior of the NP, forming an inhomogeneous alloy with different Au contents near the surface and in the bulk, which coincides with the conclusion drawn from XRD data.

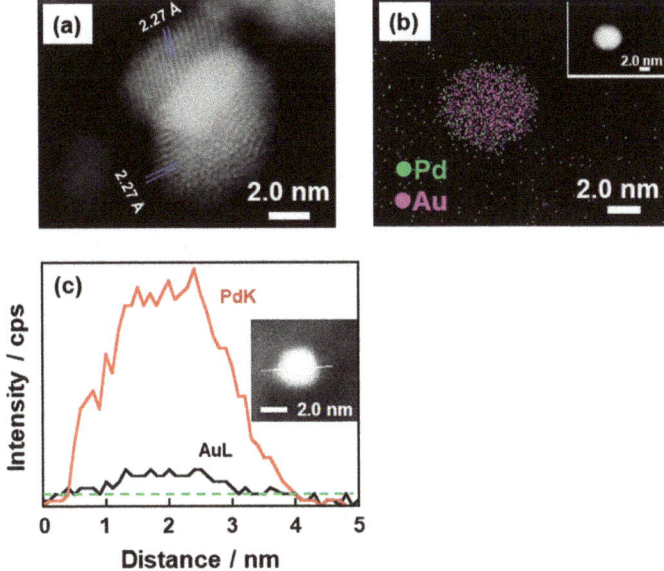

Fig. 18.3 a HAADF-STEM image, and **b** mapping images and **c** line profiles of Pd and Au for Pd$_{90}$Au$_{10}$/C

18.2.2   Influence of Particle Size

To change the size of alloy core NPs, the total concentration of the precursors was changed. The mean sizes of the prepared Pd$_{100-x}$Au$_x$/C (x = 5, 10, 20) were evaluated by applying the (111) peak of each XRD pattern to Scherrer's equation. The mean particle size as a function of the total concentration of the precursors is shown in Fig. 18.4a. Irrespective of the Au content, the mean particle size was 3.0, 3.5, 4.2, and 4.8 nm for the total concentrations of 0.25, 0.5, 1.0, and 2.0 mM, respectively,

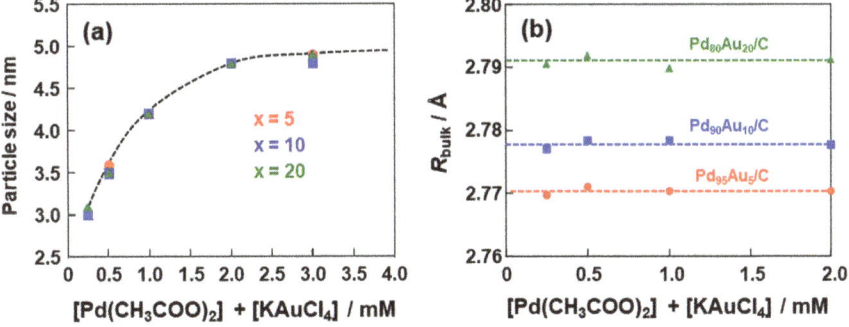

Fig. 18.4 a Mean particle size and **b** R_{bulk} value as a function of total concentration of precursors for the Pd$_{100-x}$Au$_x$/C (x = 5, 10, 20) catalysts

suggesting that the size of $Pd_{100-x}Au_x$ NPs can be controlled by the total concentrations of precursors in a range of 3–5 nm. In contrast, the R_{bulk} value was dependent on the Au content, and independent of the total concentration of precursors as shown in Fig. 18.4b, suggesting that the distribution of Au and Pd in the NPs did not change even if the particle size changed.

STEM images and size distribution profiles of $Pd_{100-x}Au_x$ ($x = 5, 10, 20$) NPs for $Pd_{100-x}Au_x$/C at the total concentration of precursors of 1 mM are shown in Fig. 18.5. The STEM images show that $Pd_{100-x}Au_x$ NPs were highly dispersed on carbon. The mean size and its deviation of $Pd_{100-x}Au_x$ NPs calculated from Fig. 18.5 were 4.2 ± 0.7, 4.2 ± 0.6, 4.2 ± 0.5 and 4.2 ± 0.6 for $x = 0, 5, 10,$ and 20, respectively. Each mean particle size evaluated from STEM was almost in agreement with that (4.2 nm) evaluated from XRD.

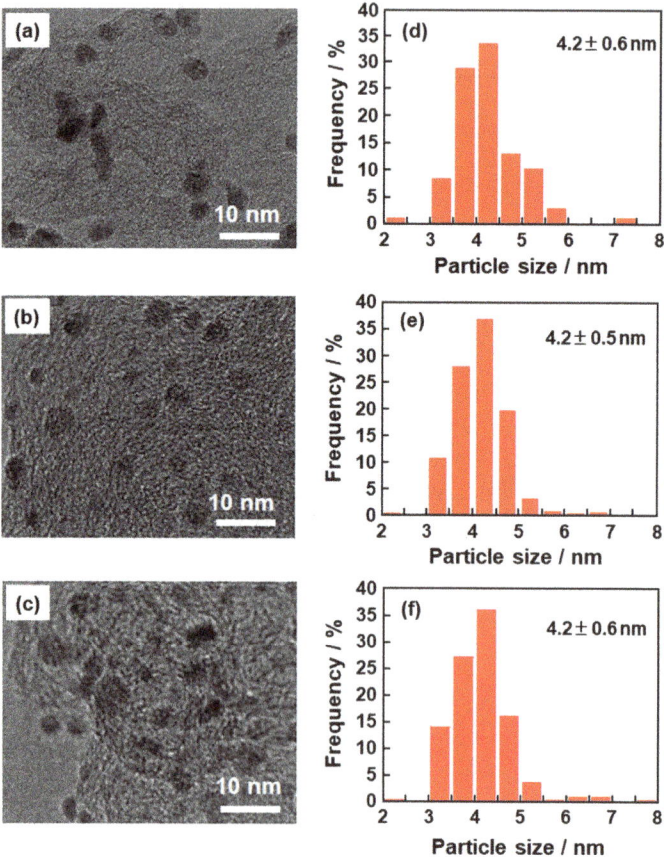

Fig. 18.5 **a, b, c** TEM image and **d, e, f** size distribution profile for $Pd_{100-x}Au_x$/C. **a, d** $x = 5$, **b, e** $x = 10$, **c, f** $x = 20$

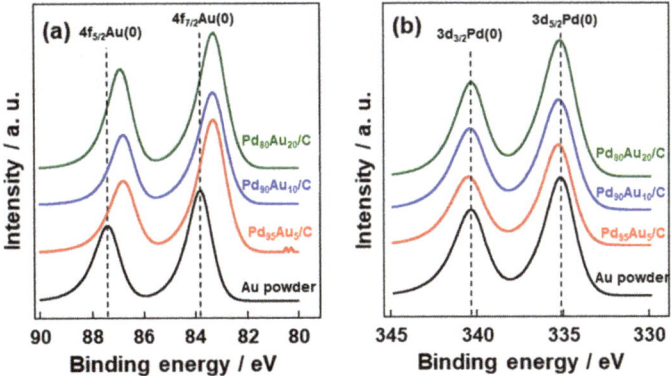

Fig. 18.6 XPS spectra of **a** Au 4f and **b** Pd 3d for the $Pd_{100-x}Au_x/C$ (x = 5, 10, 20), Au powder and Pd/C

XPS spectra of Pd 3d and Au 4f for $Pd_{100-x}Au_x/C$ (x = 5, 10, 20) are shown in Fig. 18.6. The shift of the Au 4f peak to the lower binding energy was distinctly observed, whereas the Pd 3d peak slightly shifted to the higher binding energy, suggesting the alloying of Pd with Au.

18.3 Electrochemical Properties of $Pd_{100-x}Au_x$ Core-Pt Shell Nanoparticle Catalysts

The preparation of $Pd_{100-x}Au_x/C$ and the formation of a Pt shell monolayer on the $Pd_{100-x}Au_x$ NP surface were performed according to the previous paper [38]. The metal loading was 6.5 $\mu g\ cm^{-2}$.

Figure 18.7 shows cyclic voltammograms (CVs) and CO-stripping voltammograms of the $Pd_{100-x}Au_x/C$- and Pd/C-loaded glassy carbon electrodes in an Ar-saturated 0.1 M $HClO_4$ aqueous solution. In Fig. 18.7a, typical oxide formation/reduction peaks of Pd were observed at each electrode. When the Au content

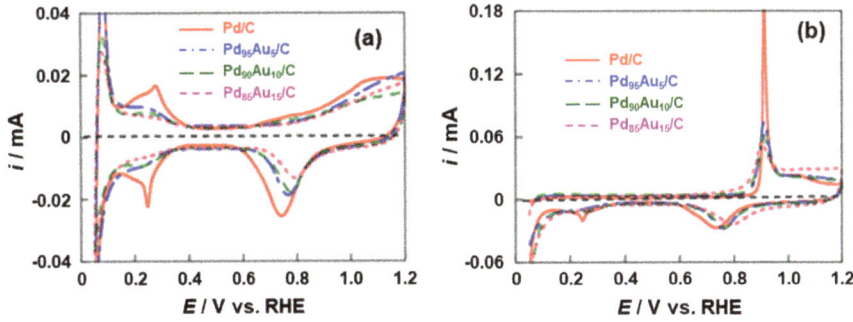

Fig. 18.7 **a** CVs and **b** CO-stripping voltammograms of $Pd_{100-x}Au_x/C$ electrodes

increased, the reduction peak shifted to higher potentials and hydrogen adsorption/desorption peaks decreased. Also, a CO-stripping peak decreased with an increase of the Au content. These results suggest that the Au content on the $Pd_{100-x}Au_x$ NP surface increased.

CVs and CO-stripping voltammograms of the $Pd_{90}Au_{10}/C$ and $Pt/Pd_{90}Au_{10}/C$ electrodes are shown in Fig. 18.8. In Fig. 18.8a, after forming a Pt shell, the oxide reduction peak shifted to higher potentials, and typical hydrogen absorption/desorption peaks of Pd around 0.1 V decreased. In Fig. 18.8b, the CO-stripping peak for the $Pd_{90}Au_{10}/C$ electrode was observed around 0.9 V, which is assigned to stripping of CO adsorbed at the Pd sites, while that for the $Pt/Pd_{90}Au_{10}/C$ electrode was observed around 0.8 V, which is assigned to the stripping of CO adsorbed on the Pt shell surface. The peak shift indicates that most of the $Pd_{90}Au_{10}$ NP surface was covered with a Pt monolayer.

Figure 18.9 shows hydrodynamic voltammograms at various rotation speeds in O_2-saturated 0.1 M $HClO_4$ aqueous solution for the $Pt/Pd_{90}Au_{10}/C$ electrodes. The onset potential of the ORR current was around 0.95 V irrespective of rotation speed,

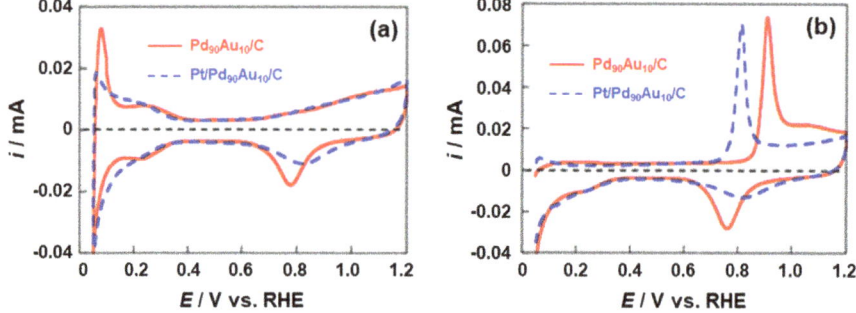

Fig. 18.8 **a** CVs and **b** CO-stripping voltammograms of $Pd_{90}Au_{10}/C$ and $Pt/Pd_{90}Au_{10}/C$ electrodes

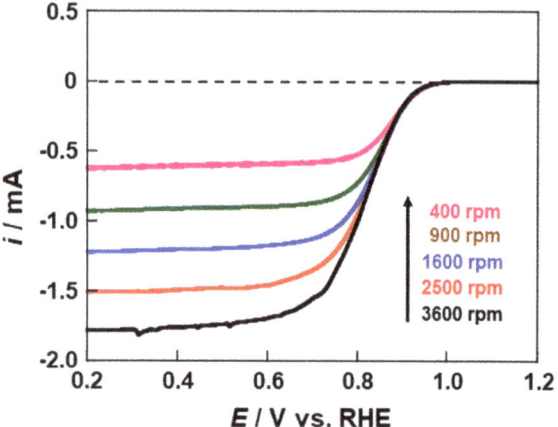

Fig. 18.9 Hydrodynamic voltammograms at various rotation speeds in an O_2-saturated 0.1 M $HClO_4$ aqueous solution for $Pt/Pd_{100-x}Au_x/C$ electrode. Sweep rate: 10 mV s^{-1}

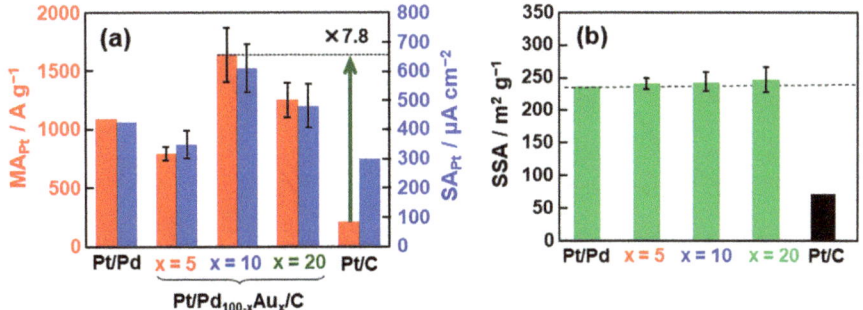

Fig. 18.10 SA_{Pt} and MA_{Pt} for ORR at 0.9 V and SSA_{Pt} for the $Pt/Pd_{100-x}Au_x/C$, $Pt/Pd/C$ and commercial Pt/C electrodes

which is similar to that of a commercial Pt/C electrode. The slope of Koutecky–Levich plots drawn from Fig. 18.9 showed that ORR proceeded in a 4-electron reduction mechanism, similar to the commercial Pt/C electrode, suggesting that the $Pd_{100-x}Au_x$ NPs were covered with Pt.

Figure 18.10 shows the MA_{Pt} and SA_{Pt} at 0.9 V and specific surface area of Pt (SSA_{Pt}) for the $Pt/Pd_{100-x}Au_x/C$, $Pt/Pd/C$ and commercial Pt/C electrodes when the particle size of core NPs was 4.2 nm. In Fig. 18.10a, SA_{Pt} and MA_{Pt} showed a volcano-type relationship with the Au content, which suggests that the optimal surface Pt–Pt distance was achieved by tuning the Au content or the R_{bulk} value of core NPs. In contrast, SSA_{Pt} was almost constant irrespective of the Au content, and larger than the theoretical value (180 m^2 g^{-1}) calculated from a Pt monolayer on Pd NPs with 4.2 nm in size, suggesting that the $Pd_{90}Au_{10}$ core NP surface was not completely covered with a Pt shell. As can be seen from Fig. 18.10a, the MA_{Pt} of the $Pt/Pd_{90}Au_{10}/C$ electrode was *ca.* 8 times and 1.5 times as high as that of the commercial Pt/C and Pt/Pd/C electrodes, respectively. Since MA_{Pt} is the product of SA_{Pt} and SSA_{Pt}, the improvement of MA_{Pt} is ascribed to that of SA_{Pt} which is influenced by compressive strain or surface Pt–Pt distance.

Figure 18.11a shows MA_{Pt} at 0.9 V as a function of size of core NPs for the $Pt/Pd_{100-x}Au_x/C$ (x = 5, 10, 20) electrodes. For the Au content of 10 and 20 at.%, a volcano-type relationship was found between MA_{Pt} and the size of core NPs. However, when the Au content was 5 at.%, MA_{Pt} increased with the core size. Since SSA_{Pt} did not change with the core size irrespective of the Au content, it is concluded that MA_{Pt} changed with SA_{Pt} which can be influenced by the change in compressive strain and Pt–Pt distance of Pt monolayer shell with the core size [33, 35].

Shao et al. clarified a relationship between surface compressive strain (S) and particle size for truncated octahedral Pd, Au, and Pt NPs with density functional theory (DFT) calculations [35]. The neighbor interatomic distances on the outermost surface of core NPs ($R_{surface}$) for $Pd_{100-x}Au_x$ core NPs with different sizes are estimated with the following equation.

$$R_{surface} = R_{bulk} \times (1-S) \tag{18.4}$$

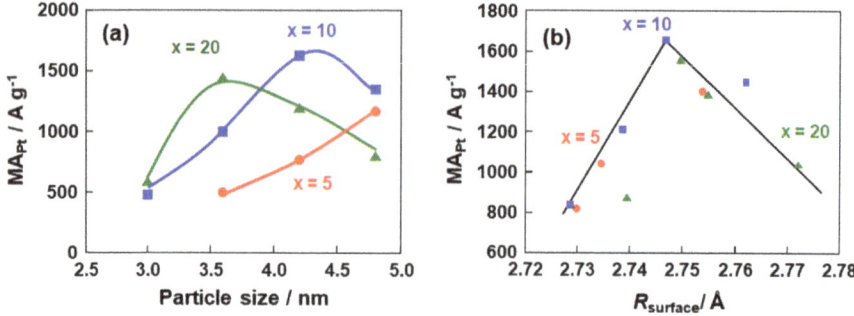

Fig. 18.11 MA_{Pt} at 0.9 V as a function of **a** size of core NPs and **b** $R_{surface}$ for the $Pt/Pd_{100-x}Au_x/C$ (x = 5, 10, 20) electrodes

Figure 18.11b shows a relationship between MA_{Pt} and $R_{surface}$ for the $Pt/Pd_{100-x}Au_x/C$ electrodes. MA_{Pt} exhibited a volcano-type dependence on $R_{surface}$, which clearly indicates that the size dependence of MA_{Pt} is due to the change in surface Pt–Pt distance caused by the increase in compressive strain with decreasing particle size.

The durability of $Pt/Pd_{100-x}Au_x/C$ electrodes was investigated by repeating square-wave potential cycling between 0.6 V for 3 s and 1.0 V for 3 s in an Ar-saturated 0.1 M $HClO_4$ aqueous solution at 60 °C [36, 38]. Figure 18.12 shows the change in normalized ECSA with cycle number for the $Pt/Pd_{100-x}Au_x/C$ (4.2 nm), Pt/Pd/C, and commercial Pt/C electrodes. The normalized ECSA was defined as the percentage of ECSA after potential cycling to that before potential cycling. After 10000 cycles, the normalized ECSAs of the $Pt/Pd_{95}Au_5/C$ and $Pt/Pd_{90}Au_{10}/C$ electrodes were similar to each other, and as large as those of the Pt/Pd/C and Pt/C-TKK electrodes. On the other hand, the normalized ECSA after 10,000 cycles for the $Pt/Pd_{80}Au_{20}/C/GC$ electrode was *ca.* 15% larger than the other electrodes, demonstrating better durability.

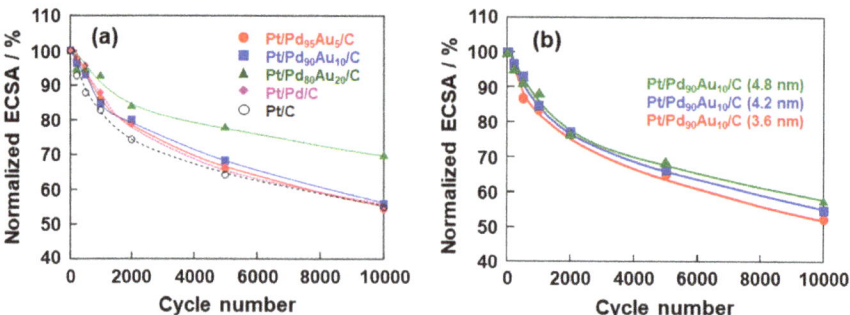

Fig. 18.12 Change in normalized ECSA of Pt with cycle number for **a** $Pt/Pd_{100-x}Au_x/C$, Pt/Pd/C and commercial Pt/C and **b** $Pt/Pd_{90}Au_{10}/C$ (3.6, 4.2 and 4.8 nm) electrodes in an Ar-saturated 0.1 M $HClO_4$ aqueous solution at 60 °C. Scan rate: 20 mV s^{-1}

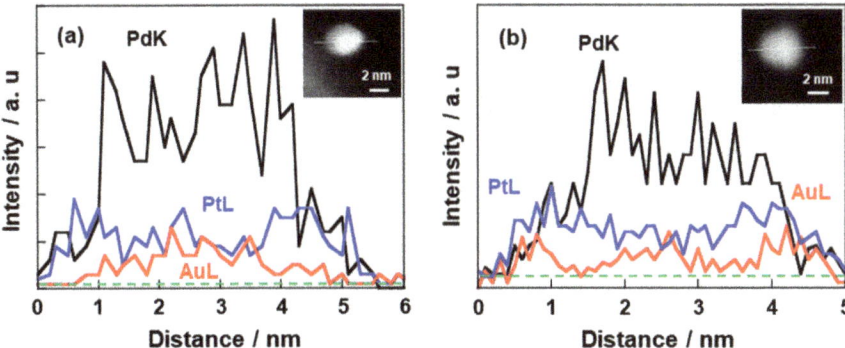

Fig. 18.13 Line profiles of Au, Pd and Pt components for Pt/Pd$_{80}$Au$_{20}$/C **a** before and **b** after 10,000 cycles

The analysis of Pt, Pd, and Au components dissolved in the electrolyte solution after potential cycling by ICP-MS showed that only Pt and Pd components were detected in the electrolyte solution. In addition, the amount of dissolved Pt and Pd components was smaller for Pt/Pd$_{100-x}$Au$_x$/C with higher Au contents, which suggests that the oxidative dissolution of Pt was suppressed. EDX line profiles of Au, Pd, and Pt components before and after 10,000 cycles for Pt/Pd$_{80}$Au$_{20}$/C were shown in Fig. 18.13. After 10,000 cycles, the Au atoms had moved from the inside of the core NP to the Pt monolayer shell surface, which was also observed for Pt/Pd$_{90}$Au$_{10}$/C. Sasaki et al. have reported that Au atoms in Pt/PdAu NPs segregated to the NP surfaces and protected defective sites on Pt monolayer shells [39], which can also be applied to the results for Pt/Pd$_{90}$Au$_{10}$/C. However, this was not the case for Pt/Pd$_{90}$Au$_{10}$/C electrodes probably because of the inhomogeneity of Pd$_{100-x}$Au$_x$ core NPs.

The change in normalized ECSA with cycle number for the Pt/Pd$_{90}$Au$_{10}$/C (3.6, 4.2, and 4.8 nm) electrodes was shown in Fig. 18.11b. The normalized ECSA slightly decreased with decreasing core size, suggesting that core size did not significantly affect durability. According to Ref. [35], the proportion of surface defective sites increases with decreasing particle size, and the increment is smaller for particle sizes of 3 nm and more. So, the difference in particle size in this study may not affect the protective effect of defective sites.

18.4   Summary and Outlook

Core–shell NP catalysts are attractive for enhancing MA$_{Pt}$ or decreasing Pt consumption in cathode catalysts for PEFCs. Pd$_{100-x}$Au$_x$ alloy core NPs supported on carbon (Pd$_{100-x}$Au$_x$/C, x = 5, 10, 20) were prepared by changing the composition of palladium acetate and potassium tetrachloroaurate (III) precursors, and the core size was controlled by the total concentration of precursors. The production of Pd$_{100-x}$Au$_x$

alloy NPs was confirmed by XRD, TEM, and XPS. The nearest neighbor interatomic distances estimated from XRD patterns increased with the Au content, but they are longer than the theoretical ones because of the inhomogeneity of $Pd_{100-x}Au_x$ alloys. The formation of a Pt monolayer shell on $Pd_{100-x}Au_x$ core NPs was confirmed by cyclic voltammetry and CO-stripping voltammetry. The MA_{Pt} for ORR at 0.9 V for the $Pt/Pd_{90}Au_{10}/C$ electrode was $ca.$ 8 times and $ca.$ 1.5 times as high as that for the commercial Pt/C and Pt/Pd/C electrodes, respectively. The increase in MA_{Pt} was ascribed to that of SA_{Pt} which was influenced by compressive strain or surface Pt–Pt distance. MA_{Pt} also showed a volcano-type relationship with the core size, although it increased for the Au content of 5 at %. Moreover, MA_{Pt} exhibited a volcano-type dependence on the $R_{surface}$ irrespective of the Au content, which clearly indicated that the size dependence of MA_{Pt} is due to the change in surface Pt–Pt distance caused by the increase in compressive strain with decreasing particle size. In terms of durability, the $Pt/Pd_{80}Au_{20}/C$ electrode was superior to the other electrodes because Au atoms segregated to the NP surfaces and protected defective sites on Pt monolayer shells. The durability was not significantly influenced by the core size.

In this way, the core–shell NP catalysts with a Pt monolayer shell are promising as cathode catalysts for PEFCs in terms of the activity and durability for ORR. For their practical use in the near future, research and development on the preparation of stable core NPs that consist of non-precious metals and easy formation technique of a Pt monolayer shell will be urgent tasks. If all these tasks are completed, the core–shell NP catalysts will be used not only for fuel cells but also as versatile electrocatalysts.

Acknowledgements This work was partly supported by the New Energy and Industrial Technology Development Organization (NEDO) through the Industrial Technology Research Grant Program (08002049-0). We appreciate Dr. Masanobu Chiku (Osaka Prefecture University) for his helpful discussion in this study, and Mr. Taiki Kuwahara (presently Toyota Motor Corporation) for his experimental assistance.

References

1. Kerres JA (2001) Development of ionomer membranes for fuel cells. J Membr Sci 185:3–27
2. Bruijn FA, Dam VA, Janssen GJM (2008) Review: durability and degradation issues of PEM fuel cell components. Fuel Cells 08:3–22
3. Chung HT, Cullen DA, Higgins D, Sneed BT, Holby EF, More KL, Zelenay P (2017) Direct atomic-level insight into the active sites of a high-performance PGM-free ORR catalyst. Science 357:479–484
4. Oezaslan M, Hasche F, Strasser P (2013) Pt-based core−shell catalyst architectures for oxygen fuel cell electrodes. Phys Chem Lett 4:3273–3291
5. Wang Y, Chen KS, Mishler J, Cho SC, Adroher XC (2011) A review of polymer electrolyte membrane fuel cells: technology, applications, and needs on fundamental research. Appl Energy 88:981–1007
6. Gasteiger HA, Kocha SS, Sompalli B, Wagner FT (2005) Activity benchmarks and requirements for Pt, Pt-alloy, and non-Pt oxygen reduction catalysts for PEMFCs. Appl Catal B 56:9–35

7. Yano H, Higuchi E, Uchida H, Watanabe M (2006) Temperature dependence of oxygen reduction activity at Nafion-coated bulk Pt and Pt/carbon black catalysts. J Phys Chem B 110:16544–16549

8. Yano H, Inukai J, Uchida H, Watanabe M, Babu PK, Kobayashi T, Chung JH, Oldfield E, Wieckowski A (2006) Particle-size effect of nanoscale platinum catalysts in oxygen reduction reaction: an electrochemical and [195]Pt EC-NMR study. Phys Chem Chem Phys 42:4932–4939

9. Chen S, Kucernak A (2004) Electrocatalysis under conditions of high mass transport rate: oxygen reduction on single submicrometer-sized Pt particles supported on carbon. J Phys Chem B 108:3262–3276

10. Giordano N, Passalacqua E, Pino L, Arico AS, Antonucci V, Vivaldi M, Kinoshita K (1991) Analysis of platinum particle size and oxygen reduction in phosphoric acid. Electrochim Acta 36:1979–1984

11. Rice C, Tong Y, Oldfield E, Wieckowski A (1998) Cyclic voltammetry and [195]Pt nuclear magnetic resonance characterization of graphite-supported commercial fuel cell grade platinum electrocatalysts. Electrochim Acta 43:2825–2830

12. Higuchi E, Taguchi A, Hayashi K, Inoue H (2011) Electrocatalytic activity for oxygen reduction reaction of Pt nanoparticle catalysts with narrow size distribution prepared from [Pt$_3$(CO)$_3$(μ-CO)$_3$]$_n{}^{2-}$ (n = 3–8) complexes. J Electroanal Chem 663:84–89

13. Toda T, Igarashi H, Uchida H, Watanabe M (1999) Enhancement of the electroreduction of oxygen on Pt alloys with Fe, Ni, and Co. J Electrochem Soc 146:3750–3756

14. Mukerjee S, Srinivasan S, Soriaga MP, McBreen J (1995) Role of structural and electronic properties of Pt and Pt alloys on electrocatalysis of oxygen reduction: an in situ XANES and EXAFS investigation. J Electrochem Soc 142:1409–1422

15. Greeley J, Stephens IEL, Bondarenko AS, Johansson TP, Hansen HA, Jaramillo TF, Rossmeisl J, Chorkendorff I, Norskov JK (2009) Alloys of platinum and early transition metals as oxygen reduction electrocatalysts. Nat Chem 1:552–556

16. Paulus UA, Wokaun A, Scherer GG, Schmidt TJ, Stamenkovic V, Radmilovic V, Markovic NM, Ross PN (2002) Oxygen reduction on carbon-supported Pt−Ni and Pt−Co alloy catalysts. J Phys Chem B 106:4181–4191

17. Yano H, Kataoka M, Yamashita H, Uchida H, Watanabe M (2007) Oxygen reduction activity of carbon-supported Pt-M (M = V, Ni, Cr Co, and Fe) alloys prepared by nanocapsule method. J Phys Chem C 112:6438–6445

18. Stamenkovic V, Mun BS, Mayrhofer KJJ, Ross PN, Markovic NM, Rossmeisl J, Greeley J, Nørskov JK (2006) Changing the activity of electrocatalysts for oxygen reduction by tuning the surface electronic structure. Angew Chem Int Ed 45:2897–2901

19. Hodnik N, Zorko M, Bele M, Hočevar S, Gaberšček M (2012) Identical location scanning electron microscopy: a case study of electrochemical degradation of PtNi nanoparticles using a new nondestructive method. J Phys Chem C 116:21326–21333

20. Hasché F, Oezaslan M, Strasser P (2011) structure and degradation of dealloyed PtNi$_3$ nanoparticle electrocatalyst for the oxygen reduction reaction in PEMFC. J Electrochem Soc 159:B24–B33

21. Shao M, Peles A, Shoemaker K (2011) Electrocatalysis on platinum nanoparticles: particle size effect on oxygen reduction reaction activity. Nano Lett 11:3714–3719

22. Kuttiyiel KA, Sasaki K, Su D, Vukmirovic MB, Marinkovic NS, Adzic RR (2013) Pt monolayer on Au-stabilized PdNi core–shell nanoparticles for oxygen reduction reaction. Electrochim Acta 110:267–272

23. Yang H (2011) Platinum-based electrocatalysts with core–shell nanostructures. Angew Chem Int Ed 50:2674–2676

24. Zhang J, Vukmirovic MB, Xu Y, Mavrikakis M, Adzic RR (2005) Controlling the catalytic activity of platinum-monolayer electrocatalysts for oxygen reduction with different substrates. Angew Chem Int Ed 44:2132–2135

25. Sasaki K, Naohara H, Cai Y, Choi YM, Liu P, Vukmirovic MB, Wang JX, Adzic RR (2010) Core-protected platinum monolayer shell high-stability electrocatalysts for fuel-cell cathodes. Angew Chem Int Ed 49:8602–8607

26. Adzic RR, Zhang J, Sasaki K, Vukmirovic MB, Shao M, Wang JX, Nilekar AU, Mavrikakis M, Valerio JA, Uribe F (2007) Platinum monolayer fuel cell electrocatalysts. Top Catal 46:249–262
27. Zhang J, Mo Y, Vukmirovic MB, Klie R, Sasaki K, Adzic RR (2004) Platinum monolayer electrocatalysts for O_2 reduction: Pt monolayer on Pd(111) and on carbon-supported Pd nanoparticles. J Phys Chem B 108:10955–10964
28. Zhang J, Lima FHB, Shao MH, Sasaki K, Wang JX, Hanson J, Adzic RR (2005) Platinum monolayer on nonnoble metal−noble metal core−shell nanoparticle electrocatalysts for O_2 reduction. J Phys Chem B 109:22701–22704
29. Lima FHB, Zhang J, Shao MH, Sasaki K, Vukmirovic MB, Ticianelli EA, Adzic RR (2007) Catalytic activity-d-band center correlation for the O_2 reduction reaction on platinum in alkaline solutions. J Phys Chem C 111:404–410
30. Shao M, Peles A, Odell J (2014) Enhanced oxygen reduction activity of platinum monolayer with a gold interlayer on palladium. J Phys Chem C 118:18505–18509
31. Xu Y, Ruban AV, Mavrikakis M (2004) Adsorption and dissociation of O_2 on Pt-Co and Pt-Fe alloys. J Am Chem Soc 126:4717–4725
32. Ruban A, Hammer B, Stoltze P, Skriver HL, Nørskov JK (1997) Surface electronic structure and reactivity of transition and noble metals. J Mol Catal a 115:421–429
33. Strasser P, Koh S, Anniyev T, Greeley J, More K, Yu C, Liu Z, Kaya S, Nordlund D, Ogasawara H, Toney MF, Nilsson A (2010) Lattice-strain control of the activity in dealloyed core-shell fuel cell catalysts. Nat Chem 2:454–460
34. Jia Q, Caldwell K, Strickland K, Ziegelbauer JM, Liu Z, Yu Z, Ramaker DE, Mukerjee S (2015) Improved oxygen reduction activity and durability of dealloyed $PtCo_x$ catalysts for proton exchange membrane fuel cells: strain, ligand, and particle size effects. ACS Catal 5:176–186
35. Shao M, Peles A, Shoemker K, Gummalla M, Njoki PN, Luo J, Zhong CJ (2011) Enhanced oxygen reduction activity of platinum monolayer on gold nanoparticle. J Phys Chem Lett 2:67–72
36. Higuchi E, Okada K, Chiku M, Inoue H (2015) Electrocatalytic activity for oxygen reduction reaction of Au core/Pt shell nanoparticle-loaded carbon black catalyst with different core sizes. Electrochim Acta 179:100–107
37. Wang X, Orikasa Y, Takesue Y, Inoue H, Nakamura M, Minato T, Hoshi N, Uchimoto Y (2013) Quantitating the lattice strain dependence of monolayer Pt shell activity toward oxygen reduction. J Am Chem Soc 135:5938–5941
38. Inoue H, Sakai R, Kuwahara T, Chiku M, Higuchi E (2015) Simple preparation of Pd core nanoparticles for Pd core/Pt shell catalyst and evaluation of activity and durability for oxygen reduction reaction. Catalysts 5:1375–1387
39. Sasaki K, Naohara H, Choi Y, Cai Y, Chen W, Liu P, Adzic RR (2012) Highly stable Pt monolayer on PdAu nanoparticle electrocatalysts for the oxygen reduction reaction. Nat Comm 3:1115

Chapter 19
Tailoring of Core Shell Like Structure in PdPt Bimetallic Catalyst for Catalytic Application

Shuai Chang, Soon Hee Park, Chang Hwan Kim, and Sung June Cho

19.1 Introduction

Novel nanoparticles have attracted much attention due to their unique catalytic performance in various reactions [1]. A nanoparticle can be classified as zero-dimensional entity. The electrochemical, magnetic, and optical properties of these nanoparticles are totally different from those of bulk metals. Preparation of such a nanoparticle has been investigated since 1970 [2]. There are many methods for the preparation of nanoparticles such as ion exchange, impregnation, chemical vapor deposition, etc. Traditional methods such as ion exchange and impregnation are still routines for the preparation of industrial catalysts. Microstructure of the obtained nanoparticles has been studied with gas adsorption, X-ray diffraction, and transmission electron microscopy [3–5].

Extended X-ray absorption fine structure (XAFS) has been used to characterize the local atomic structure in nanomaterials. XAFS has been applied to elucidate the structure of bimetallic nanoparticle catalyst used in catalytic conversion of hydrocarbon [6]. The bimetallic nanoparticles were prepared simply with the incipient impregnation method of both metal salts on alumina. In 1990, Iwasawa and coworker investigated the polymer supported nanoparticles with XAFS [7]. They showed that

S. Chang · S. J. Cho (✉)
Department of Chemical Engineering, Chonnam National University, Gwangju 61186, Republic of Korea
e-mail: sjcho@chonnam.ac.kr

S. H. Park
Super Ultra Low Energy & Emission Vehicle Center, Korea University, Seoul 02841, Republic of Korea

C. H. Kim
Institute of Fundamental and Advanced Technology, Hyundai Motor Group, Uiwang 16082, Republic of Korea

© The Author(s), under exclusive license to Springer Nature Singapore Pte Ltd. 2021   289
H. Yamashita and H. Li (eds.), *Core-Shell and Yolk-Shell Nanocatalysts*,
Nanostructure Science and Technology,
https://doi.org/10.1007/978-981-16-0463-8_19

the uniform-sized nanoparticles can be prepared upon reduction in the solution phase. However, there was a lack of information on the surface structure of nanoparticles.

The microstructure of the nanoparticles can depend on the sequence of the loading of metal salts and the choice of pretreatment methods. Tailoring of the surface structure of bimetallic nanoparticles is an essence in the current preparation technology [8]. The surface technique such as XPS is a possible characterization method for the surface structure of the nanoparticles. However, the supported nanoparticles are dispersed on high surface area substrate, resulting in much difficulty in measurement and analysis [9]. Thus, a more effective method such as ^{129}Xe NMR spectroscopy is necessary for the characterization of the surface structure. ^{129}Xe NMR spectroscopy was pioneered by Fraissard and coworker [10, 11]. It has been used extensively to probe the micropore and surface structure of zeolite related materials. The presence of the nanoparticles inside the cage of microporous materials gives a rise in ^{129}Xe NMR chemical shift due to the strong interaction between Xe atoms and metal surface. The degree of the interaction depended much on the concentration and type of metal presented at the corresponding surface.

This work was aimed to design the bimetallic nanoparticles based on the investigation of each monometallic nanoparticles. Thus, the electronic property and the catalytic activity can be controlled as a result of the rational design of the bimetallic nanoparticles. The design of the bimetallic nanoparticles has been achieved based on the investigation of the monometallic nanoparticles.

19.2 Design Strategy of Monometallic and Bimetallic Nanoparticles

Monometallic nanoparticles can be prepared with three different methods after the impregnation and ion exchange into suitable supports. The calcinations-reduction treatment is suitable for the generation of Pt [12], Pd [13], and Rh [14] nanoparticles. The evacuation or thermal decomposition-reduction is suitable for the Ir [15] and Ru [16] nanoparticles. The direct reduction is preferred to prepare the sponge like metal aggregates such as Pt and Pd. Depending on the pretreatments, the location of metal particle can be altered, and also the resulting adsorption and catalytic properties can be modified. If the Ir and Ru are subject to oxidation or calcination, the significant particle growth or sintering was resulted due to the volatile metal oxide complex such as $RuO_4(g)$ or $RuO_3(g)$ [17, 18]. Thereby, the gas adsorption and catalytic properties were deteriorated.

For the preparation of bimetallic nanoparticles, the pretreatment should be selected carefully since one metal can sinter and the other does not. Also, the order of the loading of metal salt is critical for the preparation of bimetallic nanoparticles. Figure 19.1 shows the imaginative bimetallic nanoparticle structures. The first one

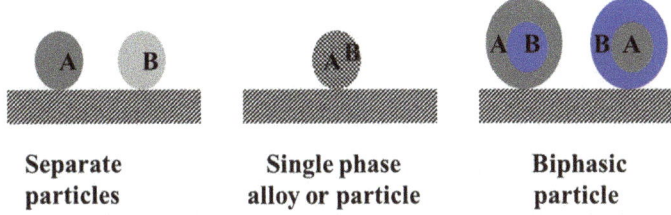

Separate **particles**   **Single phase** **alloy or particle**   **Biphasic** **particle**

Fig. 19.1 Schematics for the possible structure of the bimetallic nanoparticles

can be considered to be a just mixture of two independent monometallic nanoparticles. The second one is single phase alloy particles that can be obtained through simultaneous loading of the two types of metals. The last one is the bimetallic nanoparticle structure of cherry or core-shell like model. In this case, the other metal component can be supported subsequently on the substrate containing one metal component. The proper treatment can be selected to obtain such unique structure.

Figure 19.2 shows the transmission electron micrographs of the Pt nanoparticle supported on NaY zeolite. The Pt nanoparticles can be found in Fig. 19.2a. It was prepared with the calcination and the subsequent reduction by flowing hydrogen after the ion exchange of $Pt(NH_3)_4^{2+}$ into NaY zeolite. The size was estimated to be 1 nm, consistent with the result of TEM, hydrogen chemisorption, Xe adsorption, and XAFS. The loading was controlled to 2 wt%. Figure 19.2b showed the Pt

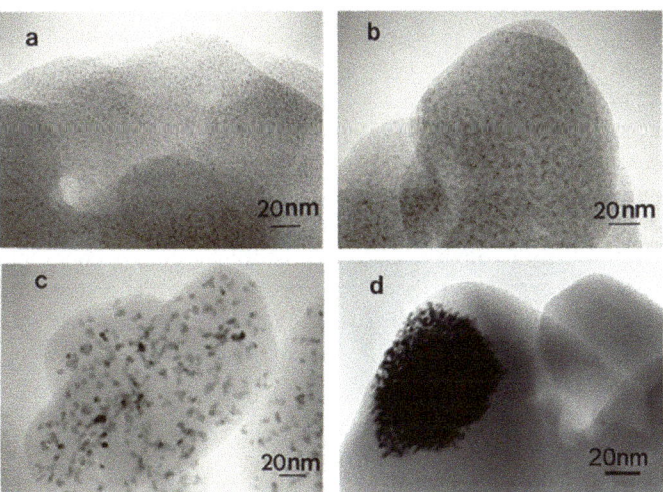

Fig. 19.2 Transmission electron micrograph of Pt nanoparticle supported on Y zeolite: **a** Pt/NaY obtained through the calcination and the subsequent reduction; **b** $Pt(NH_3)_4^{2+}$ was exchanged into Pt/NaY, treated in flowing oxygen at 593 K, and subsequently reduced at 573 K; **c** $Pt(NH_3)_4^{2+}$ was exchanged into Pt/NaY and subsequently reduced with hydrogen at 573 K; **d** $Pt(NH_3)_4^{2+}$ exchanged into NaY and subsequently reduced with hydrogen at 673 K

Table 19.1 Results of the characterization with xenon adsorption measurement and XAFS

Sample	Xe/Pt	n^a	CN^b	R (nm)^c
2 wt% Pt/NaY^e	0.069	58	5.0	0.275
2 + 8 wt% Pt/NaY(CR)^f	0.083	55	5.4	0.276
10 wt% Pt/NaY^e	0.066	61	5.1	0.276
2 + 8 wt% Pt/NaY(DR)^g	0.018	–	12.0	0.277

^{a}Average number of atoms per Pt particle calculated following the method presented in the reference [12]. ^{b}Pt-Pt coordination number (± 0.5). ^{c}Pt-Pt distance (± 0.001). ^{e}The samples were prepared through the calcinations-reduction. ^{f}The sample was prepared through the calcinations-reduction after the second ion exchange of additional Pt complex. ^{g}The sample was prepared through the direct reduction after the second ion exchange of additional Pt complex

nanoparticles after the ion exchange of additional $Pt(NH_3)_4{}^{2+}$ into NaY zeolite, 8 wt%, and subsequent calcination-reductions. The particle size in Fig. 19.2b was the same as that in Fig. 19.2a. It seems that the calcinations-reduction resulted in the formation of separate Pt nanoparticles after the additional loading of the Pt complex. Table 19.1 shows the results of the xenon adsorption measurement and atomic structural parameters from XAFS. The results on the sample in Fig. 19.2a, b indicated the same Pt nanoparticle size. Comparison with the results on 10 wt% Pt nanoparticle samples suggested that the calcination-reduction method led to the formation of separate nanoparticles after the second ion exchange [12].

Direct reduction of the second loaded Pt complex resulted in the formation of large Pt metal particles. The xenon adsorption decreased to 0.018 and the atomic structural parameter indicated the bulk Pt. The resulting particle size was limited to 3–4 nm. However, large agglomerates of Pt nanoparticles are obtained if there is no seed Pt nanoparticle as shown in Fig. 19.2d. The formation mechanism can be modeled as in Fig. 19.3. Initially, 2 wt% Pt sample contained 1 nm size Pt nanoparticles entrapped in NaY zeolite supercage. The additional loading of the Pt complex can result in either 1 nm Pt nanoparticle through the calcinations-reduction or 3–4 nm Pt nanoparticle through the direct reduction. It can be assumed that the preloaded Pt nanoparticle can be reduced readily through contact with hydrogen even at room temperature. Thus, the strong reducing agent, atomic hydrogen was present on the surface of the preloaded nanoparticle. The secondly loaded Pt complex can be reduced near or at the surface of the preloaded nanoparticle by the adsorbed atomic hydrogen [8]. This reduction mechanism is proposed in Fig. 19.4.

The reduction mechanism for the bimetallic nanoparticles has been tested for the PtPd nanoparticles supported on NaY zeolite [19]. Pt nanoparticle was prepared first and Pd was loaded secondly. Depending on the preparation methods, it was found that the same trend in the results of the xenon adsorption measurement as that of Pt nanoparticle above-mentioned. The Xe/M of the Pd/Pt/NaY(DR) was 0.076 when only the Pt content was considered. It meant that the preloaded Pt play a role as a seed for the reduction and formation of Pd nanoparticle over the Pt nanoparticle. This result suggested that the reduction mechanism outlined in Fig. 19.4 can be applied to the preparation of the well-defined nanoparticles.

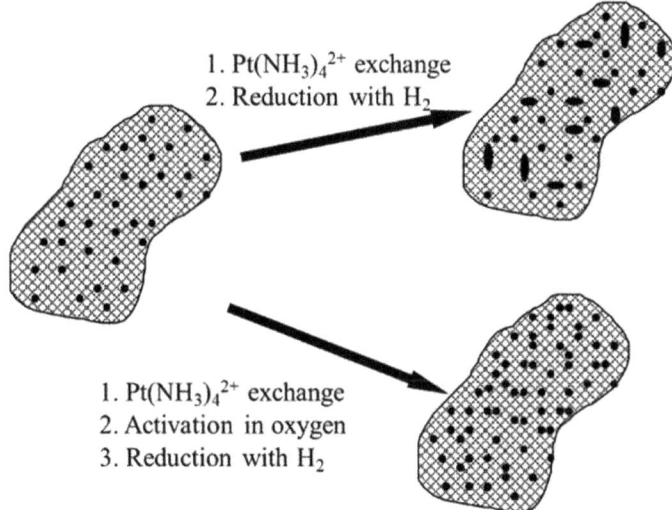

1. Pt(NH$_3$)$_4$$^{2+}$ exchange
2. Reduction with H$_2$

1. Pt(NH$_3$)$_4$$^{2+}$ exchange
2. Activation in oxygen
3. Reduction with H$_2$

Fig. 19.3 Schematics for the structural model for the formation of nanoparticles through the different preparation methods

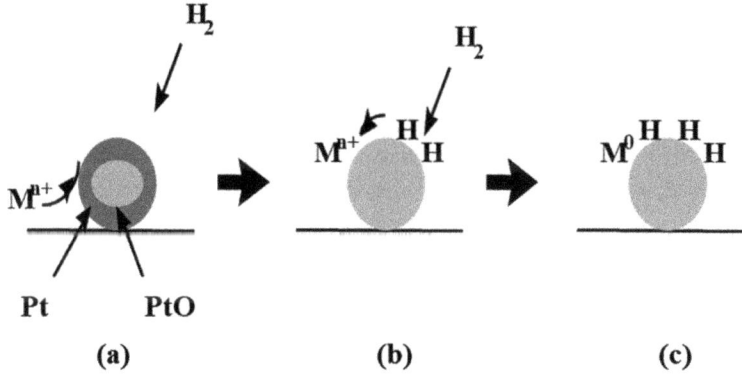

Fig. 19.4 Schematics for the reduction mechanism. The Pt nanoparticle was preloaded and subsequently M^{n+} is loaded secondly

In this work, the preloaded Pt nanoparticles or Pd nanoparticles were considered as a seed for the loading of other metals. The pretreatment condition adopted in this work was direct reduction since it aimed to cover the preloaded Pt nanoparticle with the secondly loaded metal, resulting in the cherry or core–shell like model structure as illustrated in Fig. 19.1.

19.3 Preparation, Characterization, and Catalysis of Bimetallic Nanoparticles

For the preparation of PtPd nanoparticles on alumina, $Pd(NO_3)_2$ (Engelhard, 19.9% Pd) was impregnated into La-doped Al_2O_3 (2 mol% La-Al_2O_3) supplied from Condea Inc. Its surface area was 90 m^2 g^{-1}. The activation procedure was the calcination in oxygen at 823 K for 6 h followed by the reduction with hydrogen at 823 K for 4 h. The Pd loading of the catalyst was 2 wt% on dry basis. The obtained Pd catalyst was denoted as Pd/La-Al_2O_3.

The bimetallization by platinum was performed as follows: the reduced Pd nanoparticle was impregnated with the aqueous solution containing the desired amount of H_2PtCl_6. The platinum complex impregnated Pd catalyst was dried in an oven at 373 K for overnight. The obtained sample was reduced up to 823 K by flowing hydrogen without the activation in oxygen flow. The platinum loading, the Pt/Pd ratio was controlled to 0, 0.5, 1.0, and 2.0, respectively. The bimetallic nanoparticle catalyst was also denoted as Pt/Pd/La-Al_2O_3.

Natural xenon gas (Matheson, 99.995%) was used for adsorption measurement. Xenon and hydrogen adsorption measurements were performed at 296 K with a conventional volumetric gas adsorption apparatus. The adsorption temperature was controlled to within 296 ± 0.1 K by a constant-temperature circulation bath.

Hydrogen chemisorption at 296 K was measured volumetrically after the pre-adsorbed hydrogen atoms were desorbed at 673 K in a vacuum (1×10^{-3} Pa) for 2 h and the sample cooled to 296 K. Extrapolation of this adsorption isotherm from 13–53 kPa to zero pressure was referred to as total hydrogen chemisorption value, H_{total}/Pt_{total}. The sample was then evacuated at 296 K for 2 h and a second isotherm was measured, which corresponded to reversible hydrogen chemisorption value, H_{rev}/Pt_{rev}.

X-ray absorption fine structure (XAFS) for the samples was obtained with self-supporting wafers of 10 mm in diameter pressed with 0.20 g powder sample. Since the sample wafers were exposed to air during the wafer pressing, each wafer was reduced again with H_2 flow at 573 K inside a Pyrex U-tube flow reactor and evacuated at 673 K to remove the chemisorbed hydrogen. The sample wafer was then moved to a joining XAFS cell. The XAFS cell was fabricated with Pyrex glass and Kapton windows (Du Pont, 125 μm thick). The Kapton windows were joined using Torr Seal (Varian). The XAFS cell containing sample wafer under He gas was sealed with flame and stored in a vacuum desiccator until XAFS measurement.

The XAFS in He gas was measured at the Pt L_{III} edge at room temperature using Beam Line 10B at the Photon Factory and at Pd K edge using Beam Line at Pohang Accelerator Laboratory. The ring current was maintained at 300–350 mA during the measurements. The X-ray energies for X-ray absorption near edge structure (XANES) and XAFS at the edge were increased by 0.5 eV and 1.9 eV, respectively. The X-ray intensity was measured by using a gas ionization chamber. The detector gases for I_0 and I were 100% Ar and 100% Kr, respectively.

XAFS for the sample was obtained above the Pd K and Pt L_{III} edges. The XAFS data in the wave vector (k) range between 30 and 140 nm^{-1} were analyzed using UWXAFS2 program package [20]. The EXAFS oscillation $(\chi(k))$ was multiplied by the wave vector cube (k^3) after background removal and normalization. Background was removed with a r-space technique in which low-r background components in the Fourier transform (FT) are optimized through the comparison with a standard XAFS generated using FEFF5 code [21]. Fourier transformations were performed from $30 \le k \le 140$ nm^{-1} to $0.18 \le r \le 0.30$ nm using Hanning window function. The curve fitting for the EXAFS was performed without Fourier filtering. The number of parameters used in the curve fitting was less than the allowed maximum number of parameters, $N_{free} = 2/\pi \cdot \delta r \cdot \delta k + 2$, where δk is the FT range in k-space, and δr is the fitting range in r-space.

The catalytic activity of methane combustion over the sample was measured using the microreactor system equipped with a gas chromatograph (HP5890, Hewlett Packard). The methane concentration was controlled to 1 vol% and 99% air. The space velocity, GHSV was 30,000 h^{-1}.

19.4 Pt/Pd/La-Al₂O₃ Catalyst

In the previous work, we suggested the reversible structural model of the Pd catalyst during the thermal cycling as shown in Fig. 19.5 [8, 22]. The PdO seems to convert to the surface attached PdO state at 950 K after the desorption of oxygen upon increasing the temperature. Rodriguez et al. proposed the PdAlOx as the surface attached PdO species [23]. The surface attached PdO species was further reduced to Pd metallic state at 1023–1073 K that is susceptible to particle growth upon excessive heating, thereby losing the catalytic activity.

The thermal stabilization of the Pd catalyst was investigated using the surface coating of titanium oxide that is known to have a strong metal support interaction [24]. The near edge spectra of PdO and Pd foil are presented in Fig. 19.6 to illustrate the difference of the near edge structure above the Pd K edge. The X-ray absorption at the Pd K edge in Pd foil was suppressed due to the dipole forbidden transition. The intensity of the $4s \to dp$ transition and the presence of $4s \to dsp, dp$ can be traced

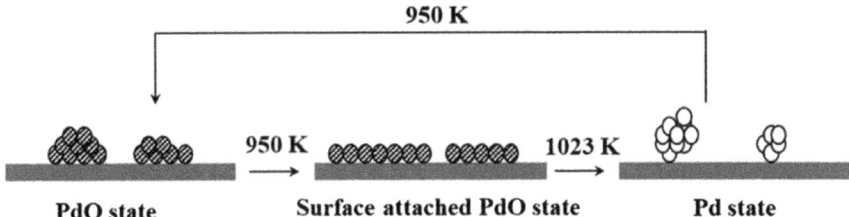

Fig. 19.5 The reversible structural transformation in Pd catalyst

Fig. 19.6 Near edge spectra
of the **a** Pd foil and **b** PdO.
The absorption inflection
corresponds to $4s \rightarrow d$ and
the second and third peaks
correspond to dp and
dsp transitions, respectively

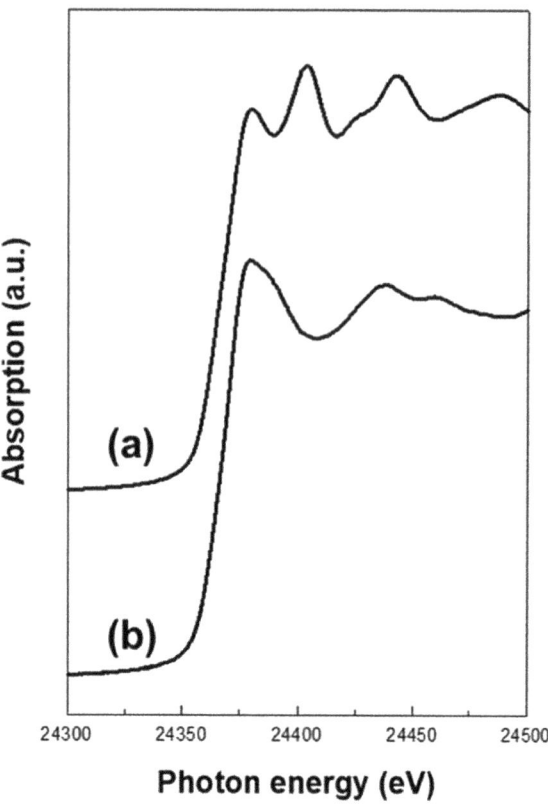

Fig. 19.6 Near edge spectra of the **a** Pd foil and **b** PdO. The absorption inflection corresponds to $4s \rightarrow d$ and the second and third peaks correspond to dp and dsp transitions, respectively

in order to clarify the atomic and electronic change upon the heating. In the XANES experiment, the transformation temperature from PdO to Pd was increased by the surface coating while the reverse transformation occurred at a similar temperature.

This characteristic change of the metal oxide surface coated Pd catalyst is also shown in Fig. 19.7. The sudden change of near edge spectra occurred at above 1123 K, which corresponded to the transformation of PdO to Pd. While, the reverse transformation from Pd to PdO occurred at 973 K. Such a transformation behavior is the same as that of Pd catalyst, except the high transformation temperature from PdO to Pd by 100 K. Thus, the thermal stability can be improved by the coating of metal oxides onto the Pd catalyst.

Further, the drastic change in the near edge spectra occurred in the bimetallized Pd nanoparticle by Pt as shown in Fig. 19.8. Upon the increase of the Pt content, the near edge structure changed progressively to resemble that of Pd foil. At the high loading of Pt, the XANES spectrum was almost the same as that of Pd foil. The model for the formation mechanism of bimetallic nanoparticles was proposed using the Pt nanoparticle in zeolite cage [12]. The preloaded Pd catalyst is easy to reduce upon heating in hydrogen flow at low temperature, ~373 K. During the reduction of the secondly loaded platinum complex without the activation in oxygen, the reduced

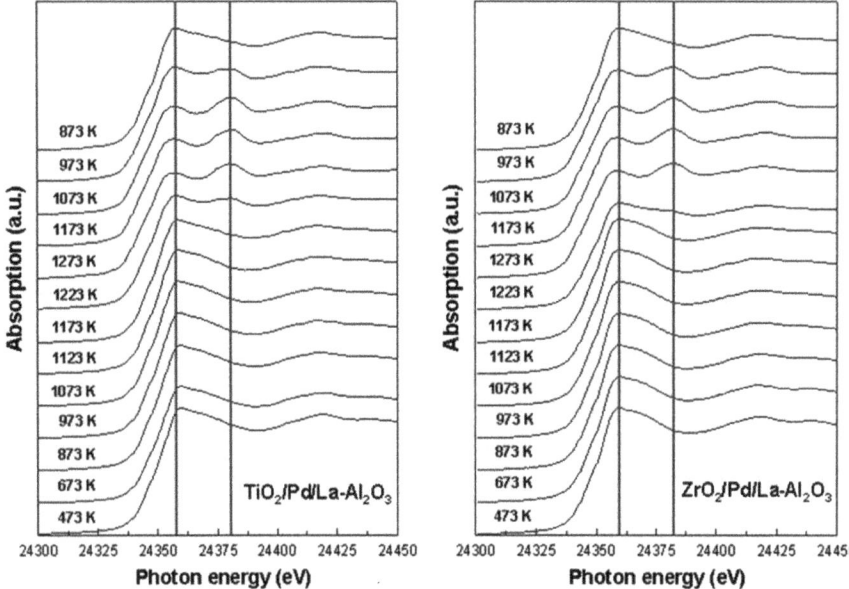

Fig. 19.7 Near edge spectra of the metal oxide surface coated Pd catalyst measured as a function of heating temperature (The inserted bars are the guides for the change of the edge features)

preloaded palladium metal nanoparticles adsorb atomic hydrogen and the platinum complex ion moves the site of palladium nanoparticles to be reduced readily. Such a mechanism also resulted in the formation of the M surface-enriched Pd particle as illustrated in Fig. 19.9. Thus, the palladium metal covered with platinum has no contact with oxygen or oxygen coordination when increasing Pt content [8]. The results of the data analysis of the XAFS spectrum of the bimetallized Pd catalyst are shown in Tables 19.2 and 19.3. At the Pd K edge, the oxygen coordination around the palladium atom disappeared with the increase of platinum content.

The overall coordination number was 9 ± 1 with 0.270 nm of metal distances, indicating the reduced metallic state in the bimetallic nanoparticles. The particle size estimated from the coordination number was 1–2 nm. The oxygen coordination, however, was detected at the Pt L_{III} edge in the series of the bimetallized catalysts. The coordination number of the bimetallic pair decreased progressively with the increase of the platinum content. The total metallic coordination number was 6 ± 1 with 0.271 nm of metal distances, consistent with the formation of the nanosized particles.

The bimetallic PtPd nanoparticle of Pt/Pd = 1.0 showed a little different change of XANES spectrum as a function of heating temperature in Fig. 19.10. The metallic XANES features were retained over the temperature range upon heating and cooling except for the slight change at around 973 K. The structure of the bimetallized Pd nanoparticle seems to remain the same. Thus, the inhibition of the reversible structural transformation of the bimetallic PtPd nanoparticles can be suggested based on the

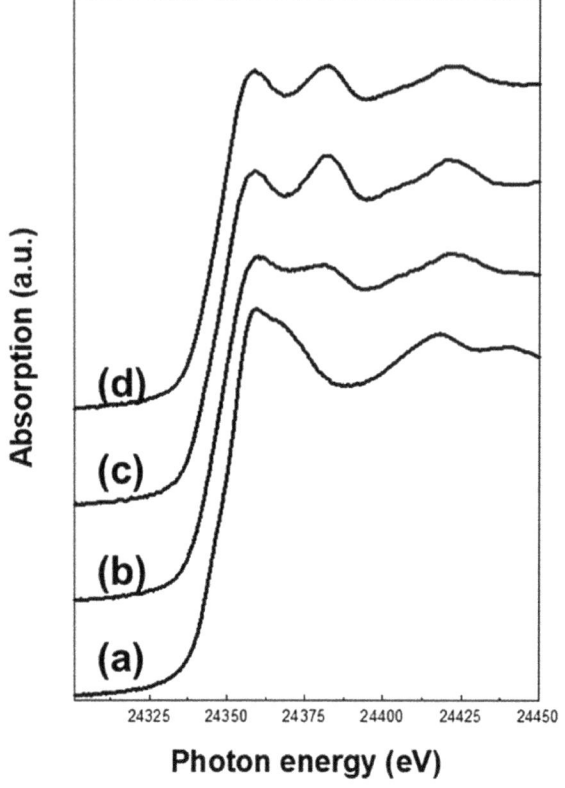

Fig. 19.8 Near edge spectra of the bimetallic PtPd nanoparticle supported on alumina measured as a function of Pt/Pd ratio

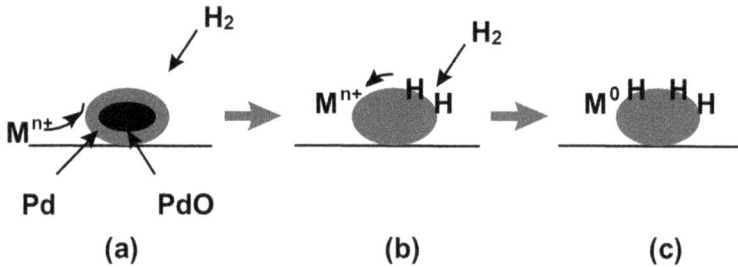

Fig. 19.9 Schematic of the reduction mechanism: **a** the reduction of the preloaded Pd nanoparticle, **b** the formation of adsorbed atomic hydrogen and the subsequent reduction of the secondly loaded metal ion, and **c** the formation of. M surface-enriched Pd particle

Table 19.2 Structural parameters obtained from the curve fitting of EXAFS data analysis of the EXAFS spectrum above the Pd K edge

Sample	Pair	N	R (nm)	σ^2 (pm^2)[a]
Pd/La-Al$_2$O$_3$	Pd–O	3.5	0.201	17
	Pd–Pd	4.8	0.305	58
	Pd–Pd	4.2	0.343	42
Pt/Pd/La-Al$_2$O$_3$(Pt/Pd = 0.5)	Pd–O	2.0	0.201	54
	Pd–Pd	2.7	0.271	56
	Pd–Pt	4.2	0.267	49
Pt/Pd/La-Al$_2$O$_3$(Pt/Pd = 1.0)	Pd–O	–	–	–
	Pd–Pd	5.2	0.273	57
	Pd–Pt	4.2	0.270	73
Pt/Pd/La-Al$_2$O$_3$(Pt/Pd = 2.0)	Pd–O	–	–	–
	Pd–Pd	2.3	0.273	63
	Pd–Pt	7.4	0.270	96

[a]The Debye–Waller factor

Table 19.3 Structural parameters obtained from the curve fitting of XAFS spectrum above the Pt L$_{III}$ edge

Sample	Pair	N	R (nm)	σ^2 (pm^2)[a]
Pt/Pd/La-Al$_2$O$_3$(Pt/Pd = 0.5)	Pt–O	3.4	0.193	142
	Pt–Pd	4.3	0.258	56
	Pt–Pt	4.7	0.259	49
Pt/Pd/La-Al$_2$O$_3$(Pt/Pd = 1.0)	Pt–O	1.3	0.201	82
	Pt–Pd	1.7	0.272	38
	Pt–Pt	3.4	0.271	70
Pt/Pd/La-Al$_2$O$_3$(Pt/Pd = 2.0)	Pt–O	1.3	0.199	58
	Pt–Pd	1.2	0.272	35
	Pt–Pt	5.1	0.271	70

[a]The Debye–Waller factor

results of the measurement of XANES as a function of temperature. The summary of the reversible transformation is presented in Table 19.4.

Figure 19.11 shows the catalytic activity of methane combustion as a function of reaction temperature. The catalytic activity of the Pd only catalyst had a strong catalytic hysteresis with 100 K temperature gap. The platinum catalyst also showed weak hysteresis, which the catalytic performance was inferior to that of the Pd catalyst. However, the bimetallized Pd catalyst with Pt showed no hysteresis in the methane combustion. It seems to be due to the formation of platinum-enriched palladium nanoparticles as suggested from the results of XANES.

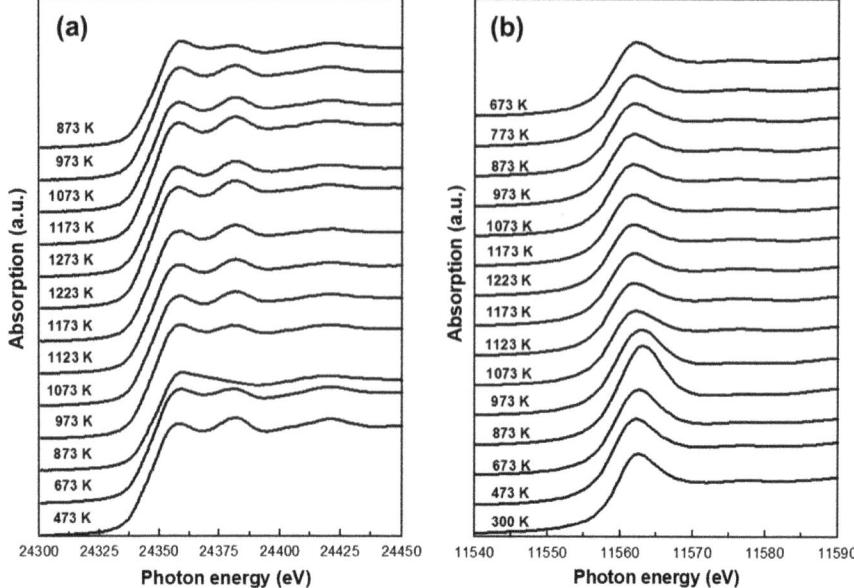

Fig. 19.10 Near edge spectra of the PtPd bimetallic nanoparticles measured as a function of heating temperatures at **a** the Pd K edge and **b** the Pt L_{III} edge

Table 19.4 Summary of the reversible transformation temperature obtained from the measurement of the XANES as a function of temperature

Sample	$T_{PdO \to Pd}$ (K)	$T_{Pd \to PdO}$ (K)
Pd/La-Al$_2$O$_3$	1123	950
TiO$_2$-Pd/La-Al$_2$O$_3$	1273	<973
ZrO$_2$-Pd/La-Al$_2$O$_3$	1273	<970
Pt/Pd/La-Al$_2$O$_3$	~973	<873

The bimetallization of the Pd catalyst with Pt changed completely the reversible structural transformation by forming the platinum enriched palladium nanoparticles. The effect of the bimetallization is the removal of catalytic hysteresis in methane combustion and the same catalytic entity over the temperature range, consistent with the results of XAFS/XANES.

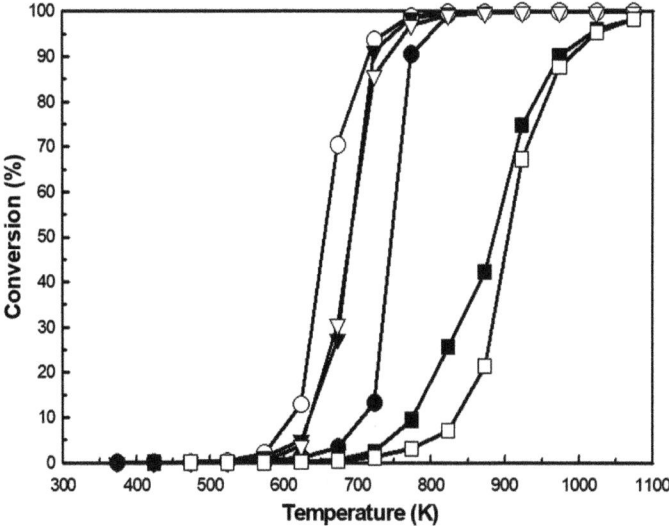

Fig. 19.11 Catalytic activity of (circle) 2 wt% Pd/La-Al$_2$O$_3$, (rectangular) 2 wt% Pt/La-Al$_2$O$_3$, and (triangle) Pt/Pd/La-Al$_2$O$_3$ (Pt/Pd = 0.5). The open and closed symbols indicate the increase and the decrease of the reaction temperature, respectively

19.5 Conclusion

Tailoring of the bimetallic nanoparticles comprising of core–shell model in which the one specific metal was deposited preferentially on the other metal was achieved with the proper pretreatments after the formation of the nanoparticle and subsequently loading of the second component metal. Such microstructural control can be possible based on the information on the formation and growth of monometallic nanoparticles. The obtained bimetallic nanoparticles showed the improved catalytic performance compared to that of monometallic nanoparticles. The reduction mechanism proposed in the present work provides a basic understanding of how the specific microstructure can be made using the conventional pretreatment method. Thereby the economic preparation technology for the bimetallic nanoparticles can be built employing the simple reduction mechanism.

Acknowledgements This research was supported by the National Research Foundation of Korea (grant numbers NRF-2016R1A5A1009592). The experiments at Pohang Accelerator Laboratory (PAL) were supported in part by MSIP and POSTECH.

References

1. Liu L et al (2018) Metal catalysts for heterogeneous catalysis: from single atoms to nanoclusters and nanoparticles. Chem Rev 118:4981–5079
2. Cuenya BR (2010) Synthesis and catalytic properties of metal nanoparticles: size, shape, support, composition, and oxidation state effects. Thin Solid Films 518:3127–3150
3. Liu T et al (2015) Synthesis, characterization and enhanced gas sensing performance of porous $ZnCo_2O_4$ nano/microspheres. Nanoscale 7:19714–19721
4. Li W et al (2013) CuTe nanocrystals: shape and size control, plasmonic properties, and use as SERS probes and photothermal agents. J Am Chem Soc 135:7098–7101
5. Berg R (2016) Revealing the formation of copper nanoparticles from a homogeneous solid precursor by electron microscopy. J Am Chem Soc 138:3433–3442
6. Johns TR (2013) Microstructure of bimetallic Pt-Pd catalysts under oxidizing conditions. ChemCatChem 5:2636–2645
7. Ichikuni N, Iwasawa Y (1993) In situ d electron density of Pt particles on supports by XANES. Catal Lett 20:87–95
8. Cho SJ, Kang SK (2004) Structural transformation of PdPt nanoparticles probed with X-ray absorption near edge structure. Catal Today 93–95:561–566
9. Haverkamp RG et al (2011) Energy resolved XPS depth profile of (IrO_2, RuO_2, Sb_2O_5, SnO_2) electrocatalyst powder to reveal core–shell nanoparticle structure. Surf Interface Anal 43:847–855
10. Chen QJ et al (1991) ^{129}Xe-n.m.r. study of rare earth-exchanged Y zeolites. Zeolites 11:239–243
11. Gedeon A et al (1993) ^{129}Xe NMR study of the Cu^+-Xe interaction in CuNaY zeolites. Generalization to the demonstration of Xe-nd10 interactions. J Phys Chem 97:4254–4255
12. Ryoo R et al (1993) Clustering of platinum atoms into nanoscale particle and network on NaY zeolite. Catal Lett 20:107–115
13. Pei X et al (2018) Size-controllable ultrafine palladium nanoparticles immobilized on calcined chitin microspheres as efficient and recyclable catalysts for hydrogenation. Nonoscale 10:14719–14725
14. Huang L et al (2018) Fabrication of rhodium nanoparticles with reduced sizes: an exploration of confined spaces. Ind Eng Chem Res 57:3561–3566
15. Celaje JJA et al (2016) A prolific catalyst for dehydrogenation of neat formic acid. Nat Commun 7:11308. https://doi.org/10.1038/ncomms11308
16. Aitbekova A et al (2018) Low-temperature restructuring of CeO_2-supported Ru nanoparticles determines selectivity in CO_2 catalytic reduction. J Am Chem Soc 140:13736–13745
17. Kleiman-Shwarsctein A et al (2012) A general route for RuO_2 deposition on metal oxides from RuO_4. Chem Commun 48:967–969
18. Liu K et al (2020) Strong metal-support interaction promoted scalable production of thermally stable singl-atom catalysts. Nat Commun 11:1263. https://doi.org/10.1038/s41467-020-14984-9
19. Cho SJ (1997) Structure and reactivity of Pt cluster and Pt-based bimetallic cluster supported on KL zeolite. PhD Thesis, Korea Advanced Institute of Science and Technology, p104
20. Frenkel A et al (1994) Solving the structure of disordered mixed salts. Phys Rev B 49:11662–11674
21. Rehr JJ et al (1992) High-order multiple-scattering calculations of x-ray absorption fine structure. Phys Rev Lett 69:3397–3400
22. Cho SJ, Kang SK (2000) Reversible structural transformation of palladium catalyst supported on La-Al_2O_3 probed with X-ray absorption fine structure. J Phys Chem B 104:8124–8128
23. Rodriguez NM et al (1995) In-situ electron microscopy studies of palladium supported on Al_2O_3, SiO_2, and ZrO_2 in oxygen. J Catal 157:676–686
24. Asakuara K et al (1992) Structure of one atomic layer titanium oxide on silica and its palladium-mediated restructuring. J Phys Chem 96:829–834

Chapter 20
Core–Shell Functional Materials for Electrocatalysis

Jinchen Fan, Qunjie Xu, Qiaoxia Li, and Juan Wang

20.1 Introduction

With the rapid development of industry, the large-scale development and utilisation of various resources has caused more and more serious environmental pollution problems, and at the same time threatened human life and health. Environmental governance methods are relatively limited, and the technological development and utilisation of new energy sources has broad prospects for development.

Electrochemical conversion and energy storage technologies, such as water splitting, metal-air batteries and fuel cells, play an important role in the development of new energy sources. The water-separating electrolysers that can convert electrical energy into storable hydrogen are a fascinating and scalable energy conversion technology for the use of renewable energy. In order to accelerate the slow hydrogen and oxygen evolution reaction (HER and OER), electrocatalysts are essential to reduce their kinetic energy barriers and ultimately improve the energy conversion efficiency. Metal-air batteries have gained new attention as potential energy storage/conversion solutions due to their high specific energy, low cost and safety. However, the development of metal-air batteries has been greatly hindered due to their relatively low rate capability and the lack of effective and stable air catalysts. The former is mainly due to the slow kinetics of the oxygen reduction reaction (ORR) and the high electrode potential To. Fuel cells have also attracted great attention as a promising clean energy conversion system. Amongst the different fuels that have been used in fuel

J. Fan · Q. Xu (✉) · Q. Li · J. Wang
Shanghai Key Laboratory of Materials Protection and Advanced Materials in Electric Power, Shanghai Engineering Research Center of Energy-Saving in Heat Exchange Systems, Shanghai University of Electric Power, Shanghai 200090, China
e-mail: xuqunjie@shiep.edu.cn

J. Fan · Q. Xu · Q. Li
Shanghai Institute of Pollution Control and Ecological Security, Shanghai 200092, China

© The Author(s), under exclusive license to Springer Nature Singapore Pte Ltd. 2021
H. Yamashita and H. Li (eds.), *Core-Shell and Yolk-Shell Nanocatalysts*,
Nanostructure Science and Technology,
https://doi.org/10.1007/978-981-16-0463-8_20

cells, hydrogen, methanol and ethanol have been the most extensively studied, and each has its advantages and disadvantages. The slow kinetics of the methanol/ethanol oxidation reaction (MOR/EOR) is the biggest obstacle to the development of fuel cells. Due to its environmental friendliness and cost-effectiveness, electrocatalytic nitrogen fixation (N_2) technology has emerged, which can achieve the green production of ammonia. However, the production of ammonia by electrocatalysis is still far from use in practical applications. In order to promote practical applications, a thorough understanding of nitrogen fixation is necessary for the future design of efficient catalysts. The electrochemical conversion of CO_2 into high-energy-density chemicals and fuels by electricity generated from renewable energy sources has been considered as a promising strategy for achieving sustainable energy. As a thermodynamically stable and kinetically inert molecule, the activation and reduction of CO_2 are extremely challenging. Although the participation of protons in the carbon dioxide electroreduction reaction (CO_2RR) helps lower the energy barrier, a high overpotential is still required to effectively promote the process. There is an urgent need for electrocatalysts with high activity and selectivity multi-carbon-based products to improve the energy efficiency of CO_2RR. The wide application of the above technologies is inseparable from high efficiency, low toxicity, low cost, green and sustainable electrocatalysts. At present, it is necessary to develop high-efficiency electrocatalysts with more reasonable control over the structure (such as particle size, morphology, surface structure and electronic structure) and chemical composition, and it is still a huge challenge. All or part of the problems can be solved by developing a catalytic system with a core–shell structure. On the one hand, using the core as a carrier can achieve the porosity, surface area, etc. of a specific surface (shell) nanostructure. On the other hand, the synergy between the shell and the core can achieve a higher efficiency/yield/selectivity combination. Third, the properties of the core and shell in catalytic applications are used for improvement/combination applications (for example, magnetically separable nanocatalysts that can be reused without loss of catalytic efficiency).

In this chapter, we will describe the recent progress of core–shell functional materials for electrocatalysis toward HER, OER, ORR, nitrogen reduction reaction (NRR), CO_2RR and MOR/EOR. The mechanisms of HER, OER, ORR, NRR, CO_2RR and MOR/EOR are discussed and the remaining challenges and future prospects of core–shell functional materials for electrocatalysis are presented.

20.2 Core–Shell Functional Materials for Electrocatalysis

20.2.1 Hydrogen Evolution Reaction (HER)

In the electrolysis water system, an external current is applied to overcome the energy barriers of the HER and OER, generating H_2 and O_2 at the cathode and anode, respectively. HER is a multi-step electrochemical process on the electrode surface where electrons are obtained to generate hydrogen. Although the electrolysis of water to produce hydrogen is simple and feasible in theory, there is still no large-scale application of water decomposition to produce H_2 [1]. At present, the electrode material with the best performance for water electrolysis is a Pt-based noble metal catalyst with disadvantages such as high price, limited reserves and poor electrocatalytic stability, which greatly limits the application of this material. [2] Therefore, when using Pt-based precious metals as HER electrocatalysts, it is necessary to develop highly efficient Pt-based catalysts with a reduced Pt amount, or non-precious metal catalysts to reduce the costs of the electrode materials. [3, 4].

20.2.1.1 Mechanism of HER

HER is a multi-step electrochemical process on the electrode surface to obtain electrons to generate H_2, which involves three major steps [5]. As shown in Fig. 20.1, the Volmer reaction is a necessary step for HER. It is generally assumed that the HER process must include an H*-generating process (Volmer reaction) and at least one desorption process (Heyrovsky or Tafel reaction). Therefore, the Volmer–Heyrovsky or Volmer–Tafel processes are the two main reaction mechanisms in the process of hydrogen production from water electrolysis. With minimal external potential, the proton or water molecules in the acidic medium are reduced to hydrogen molecules (H_2) on the electrode surface. This reaction includes the adsorption and desorption of intermediates and the diffusion of reactants and products at the interface of gas, solid and liquid.

i. Electrochemical hydrogen adsorption (Volmer reaction)

Proton and electron reactions generate adsorbed hydrogen atoms (H*) on the electrode material surface (M). Under acidic and alkaline conditions, the proton sources

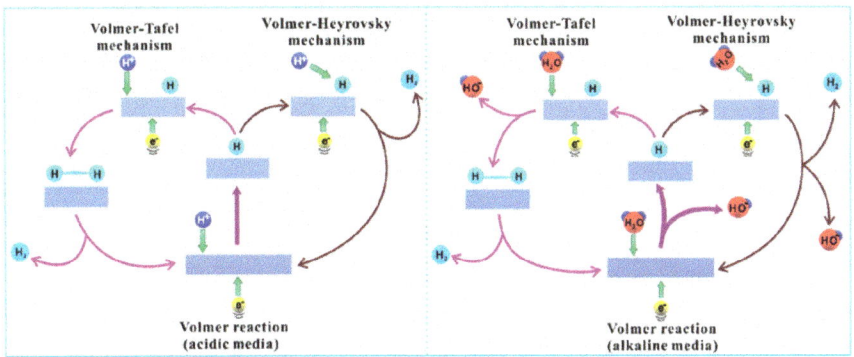

Fig. 20.1 Mechanism of hydrogen evolution on the surface of an electrode in acidic (left) and alkaline (right) solutions. Reproduced with permission from Ref. [5]. Copyright 2019 American Chemical Society

are H_3O^+ and H_2O, respectively. If the adsorption of H_3O^+ or H_2O on the catalyst surface is too weak, the Volmer step will be the rate-determining step (RDS), resulting in a Tafel slope of ≈ 120 mV dec^{-1} (25 °C).

$$H_3O^+ + M + e^- \rightleftharpoons M - H^* + H_2O (\text{Acidic medium})$$

$$H_2O + M + e^- \rightleftharpoons M - H^* + OH^- (\text{Alkaline medium})$$

ii.   Electrochemical desorption (Heyrovsky reaction)

It requires additional protons to diffuse to M–H* and then react with the second electron to form H_2. When the adsorption of H* is too strong on the catalyst surface, the overall reaction kinetics will be controlled by the desorption of hydrogen. If the concentration of the intermediate M–H* is very low at the surface, the Tafel slope is approximately 40 mV dec^{-1} (25 °C).

$$H^+ + M - H^* + e^- \rightleftharpoons M + H_2 (\text{Acidic medium})$$

$$H_2O + M - H^* + e^- \rightleftharpoons M + H_2 + OH^- (\text{Alkaline medium})$$

iii.   Chemical desorption (Tafel reaction)

A high concentration of H* atoms adsorbed on the catalyst surface allows them to combine directly, resulting in the generation of molecular hydrogen which is then released from the surface. The Tafel slope is approximately 30 mV dec^{-1} (25 °C).

$$2M - H^* \rightleftharpoons 2M + H_2 (\text{Both acidic and alkaline medium})$$

HER is due to the breaking of the O–H bond and the formation of the H–H bond. Therefore, determining which step is the rate-determining step is mainly based on the Tafel slope.

20.2.1.2   Core–Shell Functional Materials for HER

Noble Metal-Based Core–Shell HER Catalyst

Platinum group metals (PGM, including Pt, Pd, Ru, Ir and Rh) are located near the top of the volcanic plot and are good HER catalysts. Amongst them, Pt is often used as a benchmark to measure other catalysts, but its scarcity and high cost limits its application. Therefore, by reducing the amount of precious metal used, changing the thickness of its atomic layer, or combining with other non-precious metal materials to form a core–shell structure, the electrochemical activity can be improved.

For example, Li et al. [6] used a simple pyrolysis treatment to anchor ultrafine Ru nanoclusters on the surface of the electrocatalyst of tungsten oxynitride (WNO) nanowires, and coat them with nitrogen-doped carbon materials. The resulting core–shell catalyst, denoted as Ru/WNO@C, has an excellent HER performance with a Ru mass load of 3.37%, a current density of 10 mA cm^{-2} obtained at the overpotential of only 2 mV, and the Tafel slope is 33 mV dec^{-1}. When the overpotential is 50 mV, the mass-specific activity of Ru/WNO@C is 4095.6 mA mg^{-1}, and it also shows long-term stability for 100 h and nearly 100% faradaic efficiency. Wang et al. [7] synthesised octahedral nanocages (c-PtPd MTONs) with crystalline PtPd mesoporous truncated (c-PtPd MTONs) as the core and amorphous NiB (a-NiB), through three steps by wet chemical reduction, acid etching and epitaxial assembly, resulting in the construction of a new crystal PtPd@ amorphous NiB core–shell heterostructure (c-PtPd@a-NiB CSHs). This hybridisation leads to an increase of catalytically active sites, the adjustment of the surface electronic structure and the enhancement of hydrophilicity. It shows enhanced electrocatalytic activity for HER in 1.0 M KOH with a required overpotential of 31 mV (10 mA cm^{-2}) and a Tafel slope of 40.9 mV dec^{-1}. Choi et al. [8] first synthesised nano-scale Pd and β-PdH octahedrons as substrates, and prepared Pd@Pt and PdH@Pt core–shell octahedra by conformal deposition of Pt (111) shells. Due to the difference in lattice spacing between Pd and β-PdH and Pt (111), the thickness of the platinum shell can be adjusted from 1 to 5 atomic layers by changing the number of platinum precursors added during the synthesis process. In alkaline electrolytes, the HER activity is enhanced by the construction of the core–shell structure, or by increasing the number of Pt shells.

Transition Metal-Based Core–Shell HER Catalyst

The strain, the lattice distortion, electronic structure and the morphological changes induced by the core–shell structure could bring a significant improvement with the HER activity. Fu et al. [9] constructed a Ni$_3$N-NiMoN core–shell structure on a carbon cloth (CC) Ni-Mo-O precursor after two steps of hydrothermal and calcination. The Ni$_3$N-NiMoN core–shell catalyst achieved an overpotential of 31 mV of 10 mA cm^{-2} towards HER, which is similar to that of the Pt catalyst. Lee et al. [10] prepared the CoS$_x$@Cu$_2$MoS$_4$-MoS$_2$/NSG core–shell catalyst by ultrasonication, reflow and a two-step heat treatment. The reasonable combination of a highly active core–shell CoS$_x$@Cu$_2$MoS$_4$ and large-area, the high-porosity MoS$_2$/NSG produces unique physical and chemical properties with multiple integrated active centres and synergistic effects. At a current density of 10 mA cm^{-2}, the overpotential of CoS$_x$@Cu$_2$MoS$_4$-MoS$_2$/NSG is 118.1 mV at 0.1 M KOH, which is lower than those of NSG (321 mV), MoS$_2$/NSG (282.7 mV), Cu$_2$MoS$_4$-MoS$_2$/NSG (238.1 mV) and CoS$_x$-MoS$_2$/NSG (182.5 mV). Guo et al. [11] adjusted the tensile surface strain on the Co$_9$S$_8$/MoS$_2$ core/shell nanocrystals to increase the hydrogen release reaction (HER) activity by controlling the number of MoS$_2$ shells. It was found that the tensile surface strain of Co$_9$S$_8$/MoS$_2$ core–shell nanocrystals can be adjusted by changing the number of MoS$_2$ shell layers from 5 to 1. Amongst them, the strained

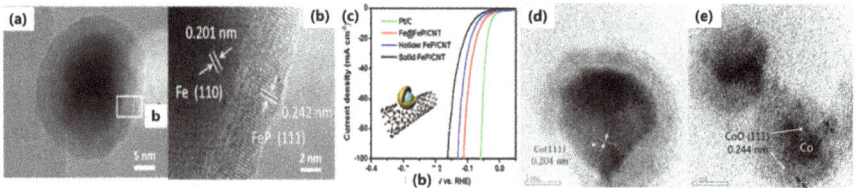

Fig. 20.2 **a** HRTEM images of Fe@FeP/CNT, **b** Enlarged image of the box area in panel a, **c** HER polarisation curves of the Fe@FeP/CNT catalyst compared with the hollow FeP/CNT, the solid FeP/CNT and a commercial Pt/C catalyst in 0.5 M H_2SO_4 at a scan rate of 5 mV s^{-1}. Reproduced with permission from Ref. [12]. Copyright 2017 American Chemical Society. HRTEM images (**d**) and (**e**) of an individual CoO@Co nanoparticle. Reproduced with permission from Ref. [13]. Copyright 2017 American Chemical Society

$Co_9S_8/1L\ MoS_2$ (3.5%) showed the best HER performance with an overpotential of only 97 mV (10 mA cm^{-2}) and a Tafel slope of 71 mV dec^{-1}. In addition, DFT calculations showed that the $Co_9S_8/1L\ MoS_2$ core/shell nanostructure produced the lowest hydrogen adsorption energy (ΔE_H) of 1.03 eV, with a transition state energy barrier (ΔE_{2H*}) of 0.29 eV (MoS_2, $\Delta E_H = 0.86$ eV, $\Delta E_{2H*} = 0.49$ eV), which is the key to enhancing the HER activity by stabilising HER intermediates and capturing H ions.

Carbon-Based Core–Shell HER Catalyst

Incorporating with carbon-based materials including carbon nanotubes, graphene, carbon nanofibres, etc. can provide a unique three-dimensional network structure which can maximise the exposure of the active sites. In the meantime, the carbon-based materials also can improve the electroconductivity of the HER catalyst system, even as direct current collectors. Wang et al. [12] designed and synthesised Fe-FeP core–shell nanoparticles on carbon nanotubes (CNT) (Fig. 20.2a, b). The Fe-FeP/CNT materials showed excellent catalytic activity for HER, and the catalytic current density of 10–100 mA cm^{-2} could be achieved at the overpotential of only 53–110 mV. Kalyan et al. [13] successfully synthesised $MoSe_2$-amorphous carbon nanotubes (*a*CNT) layered core–shell nanostructures, through a simple solvothermal method, which exhibited excellent electrocatalytic HER activity in acidic media, with a potential of −64 mV and a Tafel slope of 49 mV dec^{-1}. This is ascribed to the defects on the carbon network that promote the growth of $MoSe_2$ nanostructures on the *a*CNT's wall, thereby preventing the reuniting of $MoSe_2$ nanochips, while the active edges of the $MoSe_2$ nanostructures are exposed on the *a*CNT's amorphous walls, with low free energy barriers, which promotes the charge transfer to the Se atoms, making the Se atoms the active host of the HER reaction. Thus, the electrocatalytic performance of HER can be improved. Wang et al. [14] used a combination of the rapid microwave-polyol method and vacuum heat treatment to prepare new core–shell structure CoO@Co nanoparticles immobilised on N-doped reduced graphene oxide (CoO@Co/N-rGO) (Fig. 20.2d, e). After doping with N atoms, the hybrid had a large number of N donors and Co-N complexes as catalytically active sites. The catalyst exhibited significantly enhanced catalytic activity and excellent stability for HER. Under the conditions of 0.5 M H_2SO_4 and 0.1 M KOH, the overpotential at a current density of 10 mA cm^{-2} was 140 and 237 mV, respectively. Fang et al. [15] synthesised nano-Co nanoparticles by thermally induced reduction and carbonisation of the graphene oxide-coated core–shell bimetallic zeolite imidazole framework. These nanoparticles were wrapped in nitrogen-doped carbon nanotubes (N-CNTs) and grafted on both sides of the reduced graphene oxide (Co@N-CNTs@rGO). GO can be used as a matrix to adsorb CNTs, ensuring sufficient growth space for CNTs and avoiding particle and/or carbon accumulation at high temperature. Due to the uniform distribution of Co nanoparticles and the high graphitisation of N-CNTs@rGO with a large surface area and rich porosity, the Co@N-CNTs@rGO composite exhibited excellent electrocatalytic HER activity with an overpotential of ~108 and 87 mV in 1 M KOH and 0.5 M H_2SO_4, respectively, at a current density of 10 mA cm^{-2}.

20.2.2 Oxygen Evolution Reaction (OER)

The development of clean renewable energy is an inevitable choice for human society to achieve sustainable development [16]. OER plays a key role in many new energy technologies, especially in the field of alkaline water electrolysis (Fig. 20.3) [17]. However, OER is a four-electron-proton coupling reaction with slow kinetics, which is considered to be a major obstacle for high-efficiency water electrolysis. The use of high-efficiency catalysts can effectively reduce the overpotential of OER and improve the energy conversion efficiency in the electrolysis process [18–20].

Fig. 20.3 Schematic illustration of electrocatalytic water splitting

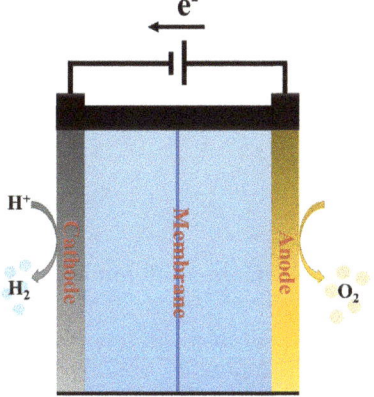

20.2.2.1 Mechanism of OER

In general, the water-splitting reaction shown in formula (1) is a non-spontaneous reaction and requires higher energy to overcome the reaction energy barrier. As a key half-reaction of water splitting, OER has different reaction processes under different pH values. Under acidic and alkaline conditions, the half-reaction equations are shown in formula (2) and formula (3), respectively. Due to the strong corrosiveness and high reaction potential under acidic conditions, OER is usually carried out under alkaline conditions. Therefore, alkaline water electrolysis technology is relatively mature and is favoured by the industry.

$$2H_2O \rightarrow 2H_2 + O_2 \tag{1}$$

$$2H_2O_{(l)} \rightarrow O_{2(g)} + 4H^+ + 4e^- \tag{2}$$

$$4OH^- \rightarrow 2H_2 + 2H_2O_{(l)} + 4e^- \tag{3}$$

At present, many scientific researchers have conducted a large number of related studies on the mechanism of OER under alkaline conditions. Although no unified understanding has been obtained, most of the mechanisms include transition metal element M and two important intermediate products M·OH and M·O. Possible reaction pathways are shown in formulas (4)–(8) [21].

$$M + OH^- \rightarrow M \cdot OH + e^- \tag{4}$$

$$M \cdot OH + OH^- \rightarrow M \cdot O + H_2O_{(l)} + e^- \tag{5}$$

$$2M \cdot O \rightarrow 2M + O_{2(g)} \tag{6}$$

$$M \cdot O + OH^- \rightarrow M \cdot OOH + e^- \tag{7}$$

$$M \cdot OOH + OH^- \rightarrow M + O_{2(g)} + H_2O_{(l)} + e^- \tag{8}$$

20.2.2.2 Core–Shell Functional Materials for OER

The OER electrocatalyst needs to maintain the activity and stability of the electrocatalyst. More active sites exposure and the fast diffusion of electrons/ions between the electrode/electrolyte interfaces are of crucial importance for OER catalysts [22].

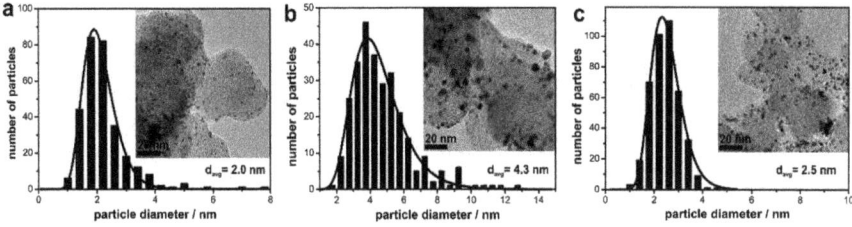

Fig. 20.4 TEM and particle size distribution histograms of Ir (**a**), Ru (**b**), and Pt (**c**) nanoparticle supported on Vulcan XC 72R. Reproduced with permission from Ref. [23] Copyright © 2012, American Chemical Society

Noble-Metal-Based Core–Shell OER Catalyst

Ru is the most active metal for OER catalysis, but Ru nanoparticles typically lose this high activity completely within 5 cycles [23]. By forming core–shell nanoparticles, the high activity of Ru is retained. The core–shell structure can maximise the exposure of the active Ru on the surface of the nanoparticles to achieve high activity, while the thin shell (<2 nm) allows the surface Ru atoms to interact with the stable core to improve stability, as shown in Fig. 20.4, Tilley et al. [24] presented a synthesis for Pd–Ru core–shell nanoparticles with tuneable shell thicknesses between 0.3 and 1.2 nm. The Pd core stabilises the Ru shell to increase the stability by up to 10×, while still maintaining the high current densities of pure Ru nanoparticles. The results show that the activity and stability of the nanoparticles are highly dependent on the nanoparticle shell thickness. Catalysts with thin Ru shells and full coverage of the Pd core are of vital importance to increase the activity and stability.

Non-precious Metal-Based Core–Shell OER Catalyst

At present, it is still a challenge to assemble low-dimensional nanomaterials on mesoscopic or microstructures to achieve high-performance hierarchical electrocatalysts. Shao et al. [25] reported a hierarchical electrocatalyst based on carbon microtube@nanotube core–shell nanostructure (CMT@CNT) with superior electrocatalytic activity towards ORR and OER, with a small potential gap of 0.678 V. Remarkably, when being employed as an air–cathode in a zinc-air battery (ZAB), the CMT@CNT presented an excellent performance with a high power density (160.6 mW cm^{-2}), specific capacity (781.7 mAh g_{Zn}^{-1}), as well as long cycle stability (117 h, 351 cycles). Wang et al. [26] prepared a new type of 3D self-supporting porous NiO@NiMoO$_4$ core–shell nanosheets by growing on nickel foam using a convenient stepwise hydrothermal method (Fig. 20.5). Ultrathin NiO nanosheets on the nickel foam, cross-linked to each other, are used as the core, and tiny NiMoO$_4$ nanosheets are further engineered to be immobilised uniformly on the NiO nanosheets to form the shell. This step-by-step construction of the architecture, composed of ultrathin primary and secondary nanosheets, efficiently avoids the agglomeration problems of

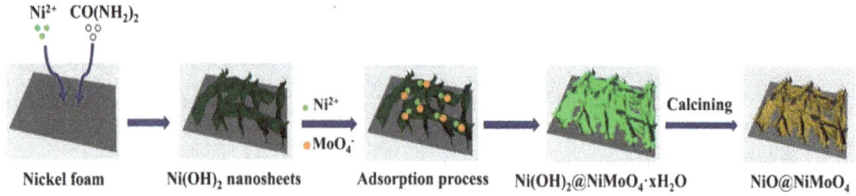

Fig. 20.5 Schematic diagram of the production for NiO@NiMoO₄ Nanosheets. Reproduced with permission from Ref. [26] Copyright © 2019, American Chemical Society

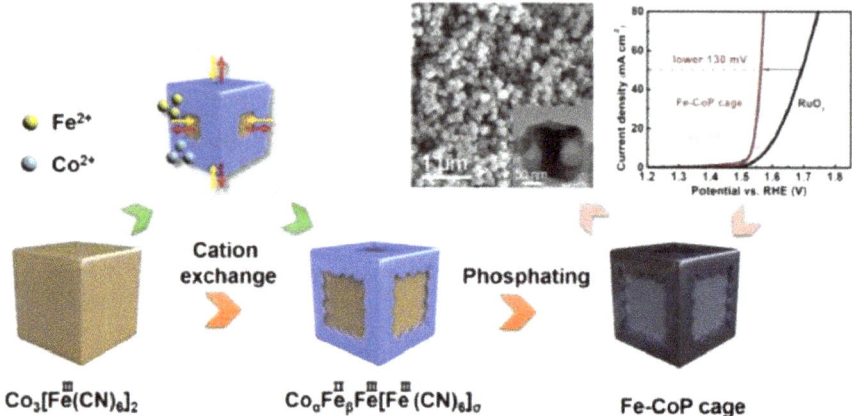

Fig. 20.6 Method for manufacturing core–shell OER catalyst [28] Copyright © 2020 Elsevier Ltd

individual ultrathin nanosheets. The ingenious architecture possesses the advantages of numerous diffusion channels for electrolyte ions, ideal pathways for electrons, and a large interfacial area for electrochemical reaction. The introduction of the NiMoO₄ secondary nanosheets on the NiO primary nanosheets not only endows the heterostructure with high electrical conductivity and a large active area, but also promotes an increase in the oxygen vacancy content which favours the improvement of the electrocatalytic properties for the oxygen evolution reaction. The Tafel plot for the NiO@NiMoO₄ core–shell architecture is as low as 32 mV dec^{-1}, and the overpotential needed to reach a current density of 10 mA·cm^{-2} for NiO@NiMoO₄ nanosheets, is only 0.28 V.

Yu et al. [27] induced the formation of MIL-88b nanorods from an amorphous 2-methylimidazole solution using the Fe-based crystalline organic framework with nanorods as the starting material. aMOF-NC was obtained by adding Co^{2+} to the reaction solution, including an iron-rich Fe Co-aMOF core, cobalt-rich Fe-Co-aMOFs nanorods, and amorphous Co(OH)₂ nanosheets as the outer layer. Benefiting from the structural and compositional heterogeneity, the aMOF-NC demonstrates an excellent OER activity with a low overpotential of 249 mV at a current density of 10.0 mA cm^{-2} and a Tafel slope of 39.5 mV dec^{-1}. Interestingly, as illustrated in Fig. 20.6,

Dong et al. [28] designed a Fe-CoP cage by a hydrothermal method through using the Fe-PBA cage as a precursor (Fig. 20.6). The centre of each face of the Fe-CoP cage has an opening, which exposes more active parts of the OER.

Wong et al. [29] prepared a dimensional open-cell multi-metal $Cu(OH)_2$@CoNiCH core–shell nanotubes, with foamed copper as the carrier, through a simple, easy-to-control, and highly controllable two-step method, and used them as an effective OER electrocatalyst. At room temperature, large-sized tubular $Cu(OH)_2$ arrays were directly grown on the copper foam, and high-porosity nano-spines were grown on the walls of these $Cu(OH)_2$ NTs by a low-temperature hydrothermal method, which simplifies the manufacturing process and reduces the inherent resistance of the electrode. This unique 3D layered core–shell structure with a hollow tubular $Cu(OH)_2$ core and porous CoNiCH nano-spine shell provides a large number of active sites, abundant defects, a large specific surface area, and fast electron transport, thus enhancing the electrocatalytic performance and the durability of its OER. Dai et al. [30] reported the nitride-core, oxide-shell-armour structured FeCoNi oxynitride as an efficient oxygen evolution electrocatalyst, with a homogeneous nitride ($Fe_{0.70}Co_{0.56}Ni_{0.92}N_{1.0}O_{0.06}$) core and oxide ($Fe_{0.48}Co_{0.1}Ni_{0.21}N_{0.05}O_{1.0}$) shell. The catalyst demonstrated an excellent activity for the oxygen evolution reaction with a current density of 10 mA cm^{-1} at a low overpotential of 0.291 V in alkaline media (1 M KOH), which is superior to the activities of commercial IrO_2, RuO_2, and Pt/C catalysts, and comparable to the state-of-the-art catalysts (e.g. NiFe-LDH, $NiCo_2O_4$, O-NiCoFe-LDH). Density functional theory (DFT) simulations suggested that the incorporation of multiple metal elements can indeed improve the reaction energetics with a synergistic effect from the core–shell structure. This unique structure of nitride-core with oxide-shell presents a new form of multimetallic oxynitride with an effective performance in electrolytic oxygen evolution.

Lu et al. [31] used a two step electrochemical strategy to produce core shell nanospheres with high activity OER electrocatalyst. $NiFe_xSn$ alloy nanospheres were prepared by a simple and rapid electrodeposition method. The surface was electro-oxidised to generate a NiFe (hydroxyl) hydroxide amorphous shell, thus forming a core–shell structure. The metal core of $NiFe_xSn$ helps electrons transfer to the shell of the amorphous NiFe (oxygen) hydroxide, which in turn prevents further oxidation of the metal core. The selective electrochemical etching of tin in alkaline solution produces a large surface area which can expose a large number of active sites and is beneficial to the diffusion and transport of substances.

20.2.3 Oxygen Reduction Reaction (ORR)

The energy and environmental crisis caused by the excessive consumption of fossil fuels and large amounts of pollution has stimulated scientists to seek new environmentally friendly energy, energy storage and conversion systems, including fuel cells and metal-air batteries, etc. ORR is an important part of energy conversion;

however, the slow kinetics greatly limits its development. Therefore, it is very important to develop ORR-based catalysts with high electrochemical activity and stability to accelerate the catalytic process. In recent years, the vigorous development of core–shell-structured catalysts initially meets the requirements. The core–shell structure not only ensures the maximum exposure of active sites, but also enhances its stability, thereby greatly improving the catalytic efficiency of ORR. Various core–shell electrocatalysts with advanced nanostructures have been developed and show excellent ORR activity and stability, and further exhibit great potential in proton exchange membrane (PEM) fuel cell applications. Herein, the latest research progress of core–shell electrocatalysts toward ORR are summarised, including the basic mechanism of catalytic reaction, different synthesis methods, and various strategies developed by researchers for the core–shell structure.

20.2.3.1   Mechanism of ORR

ORR is a multi-electron reaction involving a multi-step basic reaction and different intermediate products (Fig. 20.7). There are two main reaction pathways, direct reaction and indirect reaction. The direct reaction way is to produce H_2O directly through a 4-electron reaction. In the indirect reaction, H_2O_2 may continue to undergo 2-electron reactions to produce H_2O, which may be directly precipitated out of the solution to produce H_2O_2. The intermediate product is unstable and so can be decomposed into O_2 by a reversible reaction and participate in the reduction reaction again.

In an ideal fuel cell, oxygen is completely reduced, and the output voltage is high. This is a 4-electron reaction path, but the reduction potential of the 4-electron reaction is higher than that of the 2-electron reaction, and the dissociation energy of the O–O bond in O_2 is larger than that in H_2O_2. Therefore, when the catalytic activity

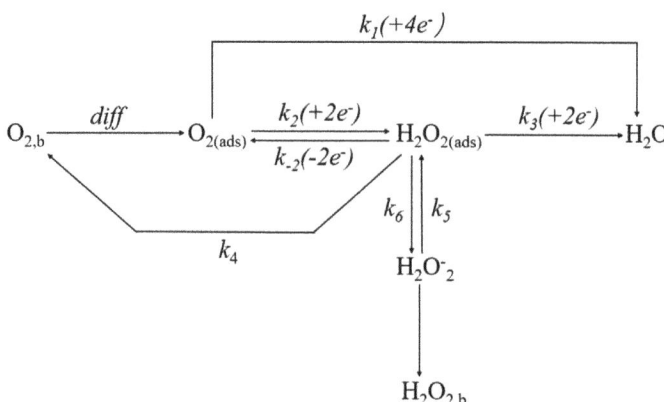

Fig. 20.7 The mechanism and reaction pathway of ORR

Fig. 20.8 Schematic illustration to show the synthesis of core–shell Au@Pd-I nanoparticle. Reproduced with permission from Ref. [33] Copyright © 2015, Springer Nature

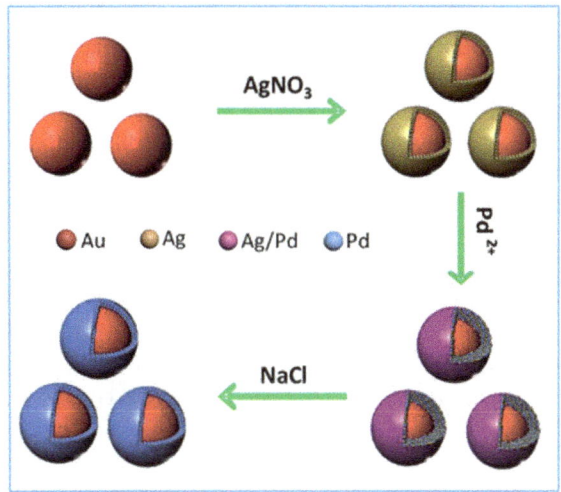

is not very strong, the 2-electron reaction or a mixed reaction of 2-electron and 4-electron may occur. H_2O_2 will damage the catalyst and proton exchange membrane and accelerate the ageing of the catalyst and proton exchange membrane. Therefore, the 4-electron reaction is an ideal way for oxygen reduction [32].

20.2.3.2   Core–Shell Catalyst for ORR

Noble Metal-Based Core–Shell ORR Electrocatalyst

The core–shell structure strategy has been widely used to balance the activity and stability while reducing the cost. There are two kinds of noble metal-based core–shell ORR electrocatalysts: noble metal@noble metal and non-noble metal@noble metal core-shell catalysts.

i.   Noble metal@noble metal core–shell catalyst

The lattice strain induced by the unique core–shell structure could bring an improvement in catalytic performances. Yang et al. [33] prepared bimetallic Au@Pd core–shell nanoparticles and investigated their electrocatalytic properties towards ORR. They used the core–shell Au@Ag/Pd nanoparticles with an Au core and an alloy Ag/Pd shell as an intermediate template. As shown in Fig. 20.8, the core–shell Au@Ag nanoparticles with an Au core and an Ag shell were firstly prepared by reducing the Ag⁺ precursors in the presence of pre-synthesised Au seed particles in oleylamine. Then, the pure Ag shells were converted into the shells made of Ag/Pd alloy by a galvanic replacement reaction between the Ag shells and Pd²⁺ precursors. Subsequently, the core–shell Au@Ag/Pd nanoparticles were agitated with a saturated aqueous NaCl solution for the removal of the Ag component from the Ag/Pd alloy shells, leading to the formation of Au@Pd core–shell nanoparticles. The as-prepared core–shell Au@Pd nanoparticles have a superior activity and durability towards ORR, which are higher than those of the core–shell Au@Pd nanoparticles produced by directly depositing a Pd shell on the Au seeds, and the commercial Pd/C catalysts from Johnson Matthey (JM). In addition to the Au core, the Ag removal from the alloy shell may induce additional lattice strain on the remaining Pd shell. The sufficient lattice strain imposed by the Au core and Ag removal, which could tailor the d-band centre of the Pd shell, may account for the enhanced ORR performance of the core–shell Au@Pd nanoparticles. Aoki et al. [34] synthesised a carbon-supported Pd core Pt shell structure catalyst (Pt/Pd/C) through the direct displacement reaction (DDR) in which the Pd core nanoparticles (NPS) and $[PtCl_4]^{2-}$ were directly replaced with Cu, at 70 °C in N_2 saturated H_2SO_4 aqueous solution, instead of using the modified copper under potential deposition (Cu-UPD)/Pt replacement method. In DDR, compared with the Cu shell and $[PtCl_4]^{2-}$ in the Cu-UPD/Pt replacement method, the potential difference between the PD core and $[PtCl_4]^{2-}$ was reduced, which inhibited the formation of the heterogeneous Pt shell, and increased the coverage of the Pt shell, and enhanced the activity of ORR. Cui et al. [35] designed and implemented a simple strategy to control the surface evolution of Pd@ Pt core–shell nanostructures by simply adjusting the amount of OH⁻ to control the reduction ability of ascorbic acid, and ultimately control the supersaturation in the reaction system. By increasing the pH value, the surface structure of the Pd@Pt bimetallic nanocrystals changes from an island shell with exposed Pt (111) facets to a conformal exposed Pt (100) facet. The well-aligned Pd@Pt core island-shell nanocubes exhibited a significantly enhanced electrocatalytic activity and good long-term stability for ORR in alkaline media.

ii.   Non-noble metal@noble metal core–shell catalysts

The core–shell structure catalyst constructed by the hybridisation of precious metals and non-precious metals has been the focus of research in recent years. For example, Pathak et al. [36] studied the potential applicability of cubic octahedral core–shell (Ti_{19}@Pt_{60}) nanoclusters (NC) to ORR activity, and compared it with pure Pt NC (Pt_{79}). Compared with the cubic octahedron Pt NC (Pt_{79}), the rate-determining step (*O_2 activation and *OH formation) of pure metals (Pt, Pd and Ag) and alloys (Pt_3M; M = Ni, Co, Ti) is much higher than that of the Ti_{19}@ Pt_{60} NC-based catalyst. The detailed study showed that the structural change caused by *O_2 is beneficial to the direct *O_2 dissociation on the Ti_{19}@Pt_{60} NC surface (Fig. 20.9a). Petkov et al. [37] reported the synthesis of an ORR catalyst with Mn core and Pt shell structure, and it was proved that the introduction of $3d$-metal core and $5d$-metal platinum shell as a flexible structural strategy was helpful for ORR reaction (Fig. 20.9b). Wang et al. [38] developed a composite ORR catalyst composed of ordered intermetallic Pt alloy nanoparticles attached to an N-doped carbon substrate, and atomically dispersed Fe–N–C positions. It exhibited enhanced catalytic activity and durability. The catalyst was prepared by depositing Pt nanoparticles on an N-doped carbon matrix with atom-dispersed Fe–N–C, followed by heat treatment. The latter leads to the formation

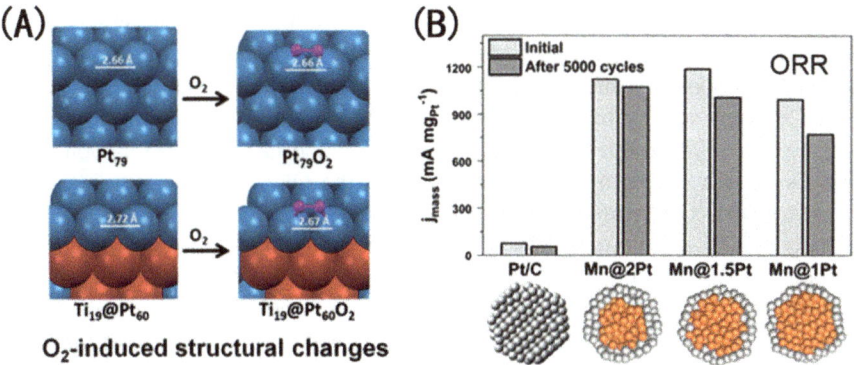

Fig. 20.9 a O_2-induced structural changes on the surfaces of Pt_{79} and $Ti_{19}@Pt_{60}$ NC surface. Reproduced with permission from ref 36 Copyright © 2016, American Chemical Society. **b** (First row) 3D model structures for pure fcc Pt and bcc-like Mn@fcc-like Pt NPs determined as described in the text. (Second row) Cross sections of the respective structures. Pt atoms are in grey and Mn atoms are in light brown. Reproduced with permission from Ref. [37] Copyright © 2020 Elsevier B.V

of core–shell structured Pt alloy nanoparticles in which the ordered intermetallic compound Pt_3M (M = Fe and Zn) is the core, and the Pt atoms are on the shell surface, which is beneficial to ORR activity and stability. The presence of Fe in the porous Fe–N–C matrix not only provides more active sites for ORR, but also effectively enhances the durability of the composite catalyst. Che et al. [39] developed $Fe/Fe_4N@Pd/C$ (FFPC) nanocomposites that successfully synthesised yolk–shells in two easy steps: interfacial polymerisation and an annealing treatment. The concentration of Pd^{2+} is a key factor in the density of Pd nanoparticles (Pd NPs) embedded in the carbon shell, and it plays a role in ORR and the surface enhanced Raman scattering (SERS) properties. The best ORR performance shows that the onset potential and Tafel slope can reach 0.937 V (relative to the reversible hydrogen electrode (RHE)) and 74 mV dec^{-1}, respectively. This is due to good electrical conductivity and greater electrochemical activity area, and strong interface charge polarisation. The interface charge polarisation can promote the ORR of the Pd NPs and defective carbon at the same time, and the shell with the low density Pd NPs is easier to form strong interface charge polarisation.

iii. Non-noble metal@noble metal core–shell catalysts

Non-noble metals used in OER catalysts include transition metal oxides, transition metal nitrides, transition metal carbides, transition metal sulphides and phosphides. On the one hand, the intrinsic activity can be improved through reasonable design to make it comparable to commercial Pt/C; on the other hand, it must be stable under acidic conditions. Based on the above two aspects, researchers have done a lot of meaningful work. For example, Lu et al. [40] demonstrated a new catalyst structure based on coupling non-precious metal Co_3O_4 nanocrystals to a nitrogen-doped core–shell carbon nanotube-graphene nanoribbon (N-csCNT-GNR) scaffold. The stent was prepared by the microwave-assisted control of multi-walled carbon nanotubes compressed upwards. Due to the synergistic interaction between Co_3O_4 and the N-csCNT-GNR substrate, the prepared Co_3O_4/N-csCNT-GNR composite catalyst has extremely high activity for both OER and ORR. Xiao et al. [41] synthesised core–shell Co@Co_3O_4 nanoparticles embedded in bamboo-like N-doped carbon tubes (BNCNT) by a simple method. The degree of the ORR catalytic activity of the Co@Co_3O_4/BNCNT composites and the oxidation of the Co nanoparticles is closely related to the N content in BNCNT. When oxidised at 300 °C, the N/C molar ratio is about 1.6%. The composite catalyst also shows higher ORR catalytic activity than the Co_3O_4/carbon nanotube (CNT) catalyst. The tolerance and cycle stability of the composite catalyst to methanol molecules are even better than that of the high-efficiency Pt/C catalyst. Ma et al. [42] used electrospun polyacrylonitrile (PAN), melamine and ferric chloride hexahydrate ($FeCl_3 \cdot 6H_2O$) composite nanofibres as templates, and prepared polythiophene (PT) on the surface of electrospun nanofibers by photo-polymerisation technology. Then, the core–shell nanofibres were pyrolysed and converted into Fe-S/NC nanofibres. This provides a convenient way due to its metal and S/N-doped structure and unique 1D structure, so it can be used as an ORR catalyst with efficient mass transportation and charge transfer. Oh et al. [43] reported on the process of producing highly active carbon-based ORR catalysts from a carefully designed core–shell mixed metal-organic framework (MOF). Cobalt and nitrogen-doped porous carbon leaves (Co, N-PCLs) were designed through the leaf-shaped core–shell hybrid MOF (ZIF-L @ ZIF-67). ZIF (zeolite imidazole framework) is a subclass of MOF. It contains two different metal ions (Zn^{2+} in the core and Co^{2+} in the shell) and sufficient nitrogen source, and has a thin flat morphology. They have the ideal structure and composition characteristics of ORR, such as many carbon nanotubes (CNT), a large amount of Co and N doping, large surface area and a high pore volume, while maintaining a favourable thin blade shape. Due to this unique structure and compositional characteristics, the ORR activity of Co, N-PCLs is better than the congeners prepared from the parent material (ZIF-L or ZIF-67). In addition, compared with the commercially available Pt/C materials, Co, N-PCL shows even better electrochemical stability and better methanol tolerance. Jiang et al. [44, 45] developed a new type of bifunctional oxygen catalyst by embedding ultra-fine NiFeO nanoparticles (NPs) in a porous amorphous MnO_x layer in which NiFeO nuclei contribute to the oxygen release reaction (OER). The active MnO_x shell layer is the active phase of the ORR, promoted by the synergy between the NiFeO core and the MnO_x shell layer. The synergistic effect is related to withdrawing the electrons of the NiFeO core from the MnO_x shell, which reduces the affinity and adsorption energy of oxygen on the MnO_x shell, and significantly improves the kinetics of ORR.

20.2.4 Nitrogen Reduction Reaction (NRR)

As one of the most important basic raw materials, ammonia is widely used in the industrial manufacturing of fertilizers, nitric acid, synthetic fibres and drugs. Owing to its high energy density, it is also used as a carrier of clean energy [45–47]. The raw material for synthetic ammonia is the abundant nitrogen in the atmosphere, and research on it has aroused the interest of scientists all over the world. Until now, industrial ammonia still mainly uses the Haber–Bosch (H-B) process, which must be carried out under high temperature and high-pressure conditions [48, 49]. The hydrogen energy consumed accounts for 1–2% of the world's energy, and the CO_2 produced annually also reaches 300–400 million metric tonnes [50, 51]. This consumes huge resources of the earth and also causes environmental pollution. In order to solve this problem, researchers are looking for a method of ammonia synthesis that consumes less energy and is environmentally friendly. Electrocatalytic NRR is of interest to researchers and has been widely studied because it can synthesise ammonia with low energy consumption and environmental friendliness under ambient conditions. At present, the catalysts in NRR are mainly divided into nanoparticles, quantum dots, nanorods, nanobelts, nanosheets, hollow structures and core–shell structures, etc. [52–54].

The core–shell structure of the catalyst has a highly exposed active surface, which can achieve synergy involving multiple sites on the core and shell and enhance the long-term durability of corrosive electrolytes by reasonably changing the electronic/chemical configuration of the interface sites [55, 56].

20.2.4.1 Mechanism of NRR

Regarding the NRR response mechanism, after years of debate, it has been constantly evolving. At present, it is mainly divided into two types: dissociation and association mechanisms (Table 20.1) [57, 58]. In the dissociation path, the $N \equiv N$ bond is directly broken and two independent N atoms are adsorbed on the surface of the catalyst, which are then hydrogenated. Since the bond energy of the nitrogen-nitrogen bond is very high (941 kJ mol^{-1}), this mechanism usually requires harsh conditions (high temperature, high pressure and high-efficiency catalyst), which only usually occur in the H-B process [59, 60]. As for the association mechanism, the step of adding the H atom to the N atom is completed before the $N \equiv N$ bond is broken. According to different hydrogenation sequences, the association mechanism can be further divided into remote pathways and alternative pathways. In the distal path, the distal N atoms away from the adsorption site first continuously acquire H to form NH_3. After releasing the first NH_3, another N atom begins the hydrogenation process. In an alternative path, the H atom is alternately bonded to two N atoms. At the end of the reaction, two NH_3 molecules are released. In all three ways, only one N atom in the nitrogen molecule is adsorbed on the active site, that is, the end-to-end configuration. The other path is enzymatic catalysis, which also involves alternating hydrogenation processes, but both N atoms are adsorbed on the active site to form a side configuration. The conversion from nitrogen to ammonia undergoes multi-step proton-electron transfer, and different intermediates are produced in each reaction path [61–63]. Considering that NRR involves a variety of intermediates, theoretical calculations are usually used to analyse the reaction process in detail. For example,

Table 20.1 Mechanisms of the electrocatalytic NRR

Mechanism	Elementary reaction steps
Dissociative pathway	$N_2 + 2^* \rightarrow 2^*N$
	$2^*N + 2e^- + 2H^+ \rightarrow 2^*NH$
	$2^*NH + 2e^- + 2H^+ \rightarrow 2^*NH_2$
	$2^*NH_2 + 2e^- + 2H^* \rightarrow 2NH_3 + 2^*$
Associative distal pathway	$N_2 + {}^* \rightarrow {}^*N_2$
	${}^*N_2 + e^- + H^+ \rightarrow {}^*NNH$
	${}^*NNH + e^- + H^+ \rightarrow {}^*NNH_2$
	${}^*NNH_2 + e^- + H^+ \rightarrow {}^*N + NH_3$
	${}^*N + e^- + H^+ \rightarrow {}^*NH$
	${}^*NH + e^- + H^+ \rightarrow {}^*NH_2$
	${}^*NH_2 + e^- + H^+ \rightarrow NH_3 + {}^*$
Associative alternating pathway	$N_2 + {}^* \rightarrow {}^*N_2$
	${}^*N_2 + e^- + H^+ \rightarrow {}^*NNH$
	${}^*NNH + e^- + H^+ \rightarrow {}^*NNNH$
	${}^*NHNH + e^- + H^+ \rightarrow {}^*NHNH_2$
	${}^*NHNH_2 + e^- + H^+ \rightarrow {}^*NH_2NH_2$
	${}^*NH_2NH_2 + e^- + H^+ \rightarrow {}^*NH_2 + NH_3$
	${}^*NH_2 + e^- + H^+ \rightarrow NH_3 + {}^*$

* Denotes an adsorption site on catalyst surface

Nørskov's group performed density functional theory (DFT) calculations to estimate the free energy of intermediates in NRR [64, 65]. They confirmed that the multiple reaction steps of NRR are not isolated from each other, but mutually restricted, and there is a linear proportional relationship between each intermediate step, so only one N* combination descriptor can be used to show the energy of catalytic performance [47].

20.2.4.2   Core–Shell-Structured Electrocatalysts for NRR

Au-Based Core–Shell-Structured NRR Electrocatalysts

As a noble metal, Au has the advantages of high activity and high stability, and is one of the most efficient NRR catalysts. By constructing a core–shell-structured catalyst, defects are generated and the active sites of the catalyst are increased, which can

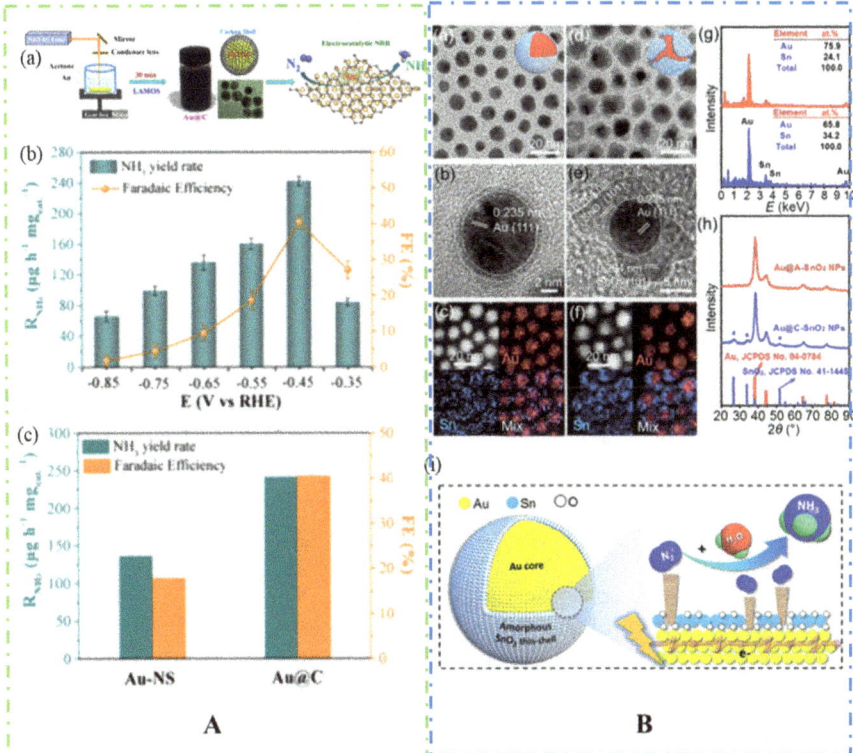

Fig. 20.10 (**A-a** Schematic illustration of the fabrication process of Au@C for the electrocatalytic NRR. **A-b** Dependence of NH₃ yield rates and faradaic efficiency of Au@C on applied potential in a N₂-saturated 0.1 M Na₂SO₄ electrolyte with an NRR measurement time of 1 h. **A-c** NRR performance comparison of Au@C and Au-NS at −0.45 V (vs RHE). Reproduced with permission from Ref. [66] Copyright 2019, American Chemical Society. Morphology and structure characterisation of Au@A-SnO₂NPs and Au@C-SnO₂NPs. TEM images and models (**B-a, B-d**), HRTEM images (**B-b, B-e**), HAADF-STEM images (**B-c, B-f**) and EDS elemental mappings of Au@A-SnO₂NPs (**B-a**)–(**B-c**) and Au@C-SnO₂NPs (**B-d**)–(**B-f**). EDS spectra (**B-g**) and XRD patterns (**h**) of Au@A-SnO₂NPs and Au@C-SnO₂NPs. (**B-i**) Schematic illustration for enhanced ENR on Au@A-SnO₂NPs/C catalyst. Reproduced with permission from Ref. [68]. Copyright 2019, Science China Press

improve the performance of NRR. Zhang et al. [66] reported the synthesis of a core–shell-structured Au@C composite through a simple one-step laser ablation technique (Fig. 20.10a). The results demonstrate that the Au@C with a mean nanosphere size of ~8.0 nm is composed of a spherical-shaped Au core and a 1–2 layered graphitic carbon shell with abundant defects. As a nitrogen reduction reaction (NRR) electrocatalyst, the Au@C gives a large NH₃ yield rate of 241.9 µg h⁻¹mg⁻¹cat. with a high faradaic efficiency of 40.5% at −0.45 V, versus a reversible hydrogen electrode (RHE) in a 0.1 M Na₂SO₄ electrolyte (pH = 6.3) under ambient conditions, surpassing the performances of most aqueous-based NRR electrocatalysts recently reported. The ^{15}N labelling experimental results demonstrate that the produced NH_3 is undoubtedly originated from the NRR process catalysed by Au@C. The superior NRR performance of Au@C can be ascribed to the ultrathin carbon layer, effectively inhibiting the aggregation of Au nanospheres during the NRR, and the abundant defects such as carbon vacancies, which exist in the ultrathin carbon layer, providing additional NRR catalytic active sites. Zhang et al. [67] reported a room-temperature spontaneous redox approach to fabricate a core–shell-structured Au@CeO$_2$ composite for the NRR, with Au nanoparticle sizes below about 10 nm, and a loading amount of 3.6 wt%. The results demonstrate that the as-synthesised Au@CeO$_2$ possesses a surface area of 40.7 m^2g^{-1} and a porous structure. As an electrocatalyst, it exhibits high NRR activity, with an NH_3 yield rate of 28.2 μgh^{-1} cm^{-2} (10.6 μgh^{-1} $mg_{cat.}^{-1}$, 293.8 μgh^{-1} mg_{Au}^{-1}) and a faradaic efficiency of 9.50% at -0.4 V versus RHE in a 0.01 M H_2SO_4 electrolyte. The characterisation results reveal the presence of rich oxygen vacancies in the CeO$_2$ nanoparticle shell of Au@CeO$_2$; these are favourable for N_2 adsorption and activation for the NRR. The abundant oxygen vacancies in the CeO$_2$ nanoparticle shell, combined with the Au nanoparticle core of Au@CeO$_2$, are electrocatalytically active sites for the NRR, and thus, synergistically enhance the conversion of N_2 into NH_3. Huang et al. [68] reported an efficient strategy to facilitate N_2 adsorption and activation for N_2 electroreduction into NH_3 by vacancy engineering of core@shell-structured Au@SnO$_2$ nanoparticles (NPs). They found that the ultrathin amorphous SnO$_2$ shell with enriched oxygen vacancies was conducive to adsorb N_2 as well as promote the N_2 activation, meanwhile the metallic Au core ensured the good electrical conductivity for accelerating electrons transport during the ENRR, synergistically boosting the N_2 electroreduction catalysis (Fig. 20.10b). As confirmed by the ^{15}N-labelling and controlled experiments, the core@shell Au@amorphous SnO$_2$ NPs with abundant oxygen vacancies show the best performance for N_2 electroreduction with an NH_3 yield rate of 21.9 μgh^{-1} $mg_{cat.}^{-1}$ and a faradaic efficiency of 15.2% at -0.2 V versus RHE, which surpasses the Au@crystalline SnO$_2$ NPs for NRR. The heterojunction between gold and the core–shell catalyst weakens the hydrogen adsorption of the catalyst and inhibits the HER reaction, thereby improving the NRR performance. Du et al. [69] reported the design and preparation of a core–shell nanostructured NPG@ZIF-8 composite for high-efficiency NRR electrocatalysis under ambient conditions. The high catalytic activity of the nanoporous gold (NPG) and the hydrophobicity and molecular concentrating effect of the zeolitic imidazolate framework-8 (ZIF-8) were incorporated in the NPG@ZIF-8 nanocomposite so that the ZIF-8 shell could weaken hydrogen evolution and retard reactant diffusion. A highest faradaic efficiency of 44% and an excellent rate of ammonia production of (28.7 $\pm$ 0.9) mgh^{-1} cm^{-2} were achieved, which are superior to traditional gold nanoparticles and NPG. Moreover, the composite catalyst shows high electrochemical stability and selectivity (98%). Ling et al. [70] reported the synthesis of Pt/Au@ZIF by a layer-by-layer overgrowth method of ZIF on the Pt/Au electrode. They electronically modified the Pt/Au electrocatalyst d-band structure using a zeolitic-imidazole framework (ZIF) to achieve a faradaic efficiency (FE) of >44% with a high ammonia yield rate of >161 μg $mg_{cat.}^{-1}$ h^{-1} at ambient conditions. Their strategy lowers the electrocatalyst d-band position to weaken H adsorption and concurrently creates electron deficient sites to kinetically drive NRR by promoting catalyst-N_2 interaction. The ZIF coating on the electrocatalyst doubles as a hydrophobic layer to suppress HER, further improving FE by >44-fold compared to without ZIF (~1%).

Fe-Based Core–Shell-Structured NRR Electrocatalysts

Non-precious metal materials have certain catalytic activity, are low in price and rich in content, so they are considered to be the best substitutes for precious metal catalysts and have broad prospects in catalytic applications. Inspired by iron-based nitrogenase, iron can be used as an NRR catalyst [71]. Through the catalytic activity trend of a series of metal surfaces, the volcano curve between the nitrogen adsorption energy of different metal catalysts and the ammonia synthesis activity is observed. It can be clearly seen that Fe is above the volcano map. This shows that an iron-based catalyst can be used as an effective NRR catalyst [65].

Wang et al. [72] synthesised sandwich-like reduced graphene oxide/yolk–shell-structured $Fe@Fe_3O_4$/carbonised paper. The electrocatalytic measurements show that the as-obtained freestanding electrode exhibits high electrocatalytic activity (NH_3 yield rate of 1.3×10^{-10} mol cm^{-2}s^{-1}), excellent selectivity (faradaic efficiency of 6.25%) and good stability, which are equivalent to (or even higher than) those of previously reported noble metal-based catalysts under comparable reaction conditions. The superior electrocatalytic performance of the rGO/$Fe@Fe_3O_4$/CP freestanding cathode for electrochemical synthesis of ammonia is mainly attributed to its unique sandwich-like nanoarchitecture with the middle yolk–shell-structured $Fe@Fe_3O_4$ nanoparticles, and the synergistic effect between rGO and $Fe@Fe_3O_4$. Inspiringly, Tan et al. [73] reported a bioinspired $Fe_3C@C$ composite as an efficient electrocatalyst for nitrogen reduction. The $Fe_3C@C$ core shell structure, as the real active centre, contributes to selective electrocatalytic synthesis of ammonia from nitrogen with a faradaic efficiency of 9.15% and production rate of 8.53 μg/(h mg$_{cat}$) or 12.80 μg/(h cm^2), at a low potential of -0.2 V versus RHE, which is comparable with that of noble-metal-based catalyst. Qiu et al. [74] reported the performance of electrocatalytic nitrogen reduction for the obtained α-Fe_2O_3 nanospindles coated with mesoporous TiO_2 with different crystallinity (denoted as α-$Fe_2O_3@mTiO_2$-X ($X = 300$, 400 and 500 °C)). The as-prepared α-$Fe_2O_3@mTiO_2$-400 composite exhibited a large NH_3 yield (27.2 μg h^{-1} mg$_{cat.}^{-1}$) at -0.5 V versus RHE, and a high faradaic efficiency (13.3%) in 0.1 M Na_2SO_4, with excellent electrochemical durability. Chen et al. [75] reported that NiFe–MoS_2 nanocubes (NiFe@MoS_2 NCs) were successfully synthesised from the corresponding Prussian blue analogue self-templating strategy. Owing to this four-pointed star face-dependent hollow structure and trimetallic synergistic interactions, it largely exposes abundant active sites, making it present superb electrocatalytic performance for N_2 conversion to NH_3. In 0.1 M Na_2SO_4 , this as-prepared Ni-Fe@MoS_2 NCs exhibited a significant NH_3 yield of 128.17 μg h^{-1} mg$_{cat.}^{-1}$ and a satisfactory faradaic efficiency of 11.34% at $-$

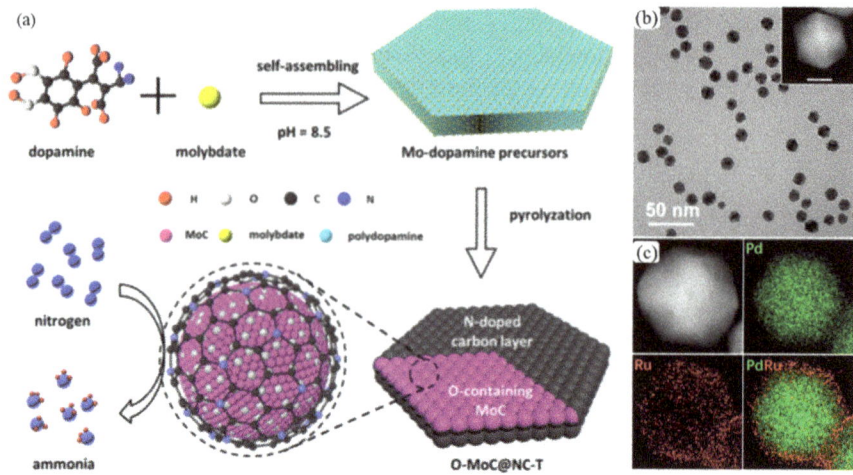

Fig. 20.11 **a** Schematic Illustration of the fabrication procedure of O-MoC@NC-T. Reproduced with permission from Ref. [76]. Copyright 2019, American Chemical Society. **b** TEM image of typical Pd@Ru core–shell icosahedra synthesised from 12 nm Pd icosahedra. The inset shows the dark-field STEM image of an individual Pd@Ru core–shell icosahedron (scale bar: 5 nm). Reproduced with permission from Ref [78]. **c** HAADF-STEM images and EDX mapping (green = Pd, red = Ru) of an individual core–shell icosahedron. Reproduced with permission from Ref. [78]. Copyright 2018, American Chemical Society

0.3 V versus RHE operation at 40 °C. The stability of the catalyst was determined by a 15-hour continuous N_2 reduction with a constant current density.

Other Metal-Based Core–Shell-Structured Electrocatalysts

In addition to Au and Fe, other metals also have certain NRR performance, such as Mo, Co, Bi, Ru and Rh; the catalytic activity of these metals has certain limitations, and the construction of a uniquecore–shell structure can improve its NRR performance. Sun et al. [76] prepared oxygen-containing molybdenum carbide (O-MoC) embedded in a nitrogen-doped carbon layer (N-doped carbon) by pyrolysing the chelate of dopamine and molybdate (Fig. 20.11a). The generation NH_3 yield rate is 22.5 $\mu g \cdot h^{-1} \cdot mg_{cat.}^{-1}$ at −0.35 V versus RHE, and the faradaic efficiency s 25.1% in 0.1 mM HCl + 0.5 M Li_2SO_4. It is worth noting that the synthesised O-MoC@NC-800 also shows high selectivity (no hydrazine formation) and electrochemical stability. The moderate electronic structure induced by the interaction between the O-MoC and the N-doped carbon shell can effectively weaken the activity of the hydrogen release reaction and increase the faradaic efficiency of NRR. Guo et al. [77] reported that multi-yolk–shell bismuth@porous carbon (MB@PC) composites were successfully synthesised via a facile simple hydrothermal reaction followed by a subsequent pyrolysation (Fig. 20.11b). The as-prepared MB@PC composite can act as an efficient NRR electrocatalyst under ambient conditions. Test results demonstrated that the MB@PC composite catalysts can deliver a high NH_3 yield of 28.63 $\mu g\ h^{-1}$ $mg_{cat.}^{-1}$, a faradaic efficiency of 10.58% at -0.5 V versus RHE, long-term electrochemical durability in N_2-saturated 0.1 M HCl solution, and an excellent selectivity for NH_3 formation, and are better than most reported bismuth-based electrocatalyst. Xia et al. [78] successfully synthesised Ru icosahedral nanocages with a face-centred cubic (*fcc*) structure by coating Pd icosahedral seeds with ultrathin Ru shells, followed by selective removal of the Pd cores via chemical etching (Fig. 20.11c, d). When benchmarked against the parental Pd@Ru core–shell nanocrystals, all the Ru nanocages displayed superior catalytic activities. First-principles density functional theory calculations also suggest that the *fcc*-Ru icosahedral nanocages containing residual Pd atoms are more promising than the conventional *hcp*-Ru solid nanoparticles in catalysing the nitrogen reduction for ammonia synthesis. With the subsurface impurities of Pd, the twin boundary regions of the icosahedral nanocages are able to stabilise the N_2 dissociation transition state, reducing the overall reaction barrier and promoting the competition with the N_2 desorption process. Huang et al. [79] reported an effective surface chalcogenation strategy to improve the NRR performance of pristine metal nanocrystals (NCs). Surprisingly, the NH_3 yield and faradaic efficiency (FE) (175.6 $\pm$ 23.6 mg h^{-1} g_{Rh}^{-1} and 13.3 $\pm$ 0.4%) of Rh-Se NCs is significantly enhanced by 16 and 15 times, respectively. Detailed investigations show that the superior activity and high FE are attributed to the effect of surface chalcogenation, which can not only decrease the apparent activation energy, but also inhibit the occurrence of the hydrogen evolution reaction (HER) process.

20.2.5 CO₂ Reduction Reaction (CO₂RR)

Due to the rapid development of modern industry, the use of coal-based fuels has increased sharply, leading to an increase in the CO_2 content in the atmosphere, resulting in the greenhouse effect. At the same time, CO_2 is also a rich and potentially low-cost carbon source and use it to produce fuels and organic chemicals. Electrochemical reduction of CO_2 is not only a potentially cost-effective chemical production process, but also uses this sustainable energy source. Therefore, the electrochemical reduction of CO_2 as a potential method for producing valuable chemicals with CO_2 as the sole carbon source has aroused great interest. The products obtained by CO_2 electroreduction generally include CO, CH_4, C_2H_4, HCOOH, HCHO, CH_3OH and C_2H_5OH. [80] The product of CO_2 electroreduction depends on the reaction conditions and the catalyst used. However, one of the main obstacles to reducing carbon dioxide emissions is low catalyst efficiency and fast deactivation. Core–shell nanoparticles are promising candidates for enhancing challenging reactions. Moreover, by using the synergy between the metal nanoparticles and the molecular sieve properties of the packaging material, the activity of the catalyst and the selectivity of the reaction can be improved.

20.2.5.1 Mechanism of CO₂RR

CO_2RR is a promising strategy to convert CO_2 into multifarious valuable chemicals and synthetic fuels. The entire reaction of carbon dioxide reduction is carried out in an "H" type device with a three-electrode system. The components of the system are a working electrode (WE, Working Electrode), a reference electrode (RE, Reference Electrode) and a counter electrode (CE, Counter Electrode). The electrolyte is generally potassium oxycarbonate, potassium hydroxide or an ionic liquid saturated with carbon dioxide, taking potassium bicarbonate solution as an example. The working electrode and the counter electrode are separated by Nafion film. The counter electrode is divided into two types: inert electrode and active electrode. The inert electrode mainly plays a role of conducting electricity, and it usually uses a pad or a carbon rod. The active electrode is usually an electrocatalyst that decomposes water to produce oxygen. The reference electrode plays a role in correcting the applied bias, usually an Ag/AgCl electrode or a saturated calomel electrode. The working electrode is our research object-the catalyst electrode. The gas-phase products produced are usually detected using gas chromatography. Liquid-phase products are usually detected by gas chromatograph mass spectrometer (GC/MS) or nuclear magnetic resonance (NMR). Figure 20.12 describes the process of reducing carbon dioxide [81].

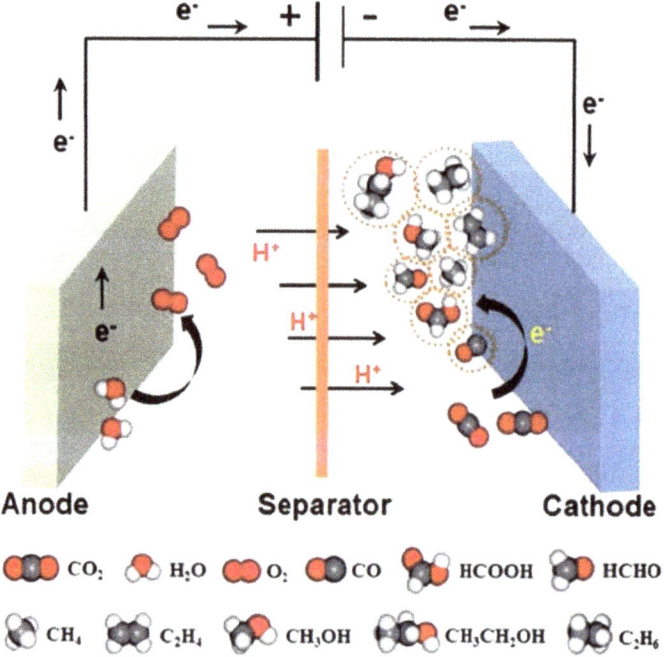

Fig. 20.12 Electrocatalytic CO_2 reduction process. Reproduced with permission from Ref. [81]. Copyright © 2011, American Chemical Society

Usually the reaction that occurs on the counter electrode is the oxidation of water, and the reduction of carbon dioxide occurs on the working electrode. The mechanism of electrocatalytic carbon dioxide reduction is as follows: the electrochemical workstation is used to apply a certain bias voltage to the electrode surface, and the high-energy electrons generated after the working electrode is in contact with the medium in the environment, combine with protons to break the carbon-oxygen bond in the carbon dioxide molecule, followed by reduction. Since the potential of carbon dioxide reduction is usually negative and exceeds the reduction potential of hydrogen ions in the electrolyte, carbon dioxide reduction and hydrogen production reactions usually exist on the working electrode.

The first step is the key to CO_2^-, because this is the rate-limiting step, and the coordination of this intermediate determines whether the $2e^-$ reduction product is CO or formic acid. The energy of CO_2^- intermediate is very high (-2.21 V vs. SCE) and easily reacts with water (formates or CO), or any other substances present in the solution (including another CO_2 molecule). Compared to the first step, the subsequent reduction step proceeds almost immediately. Therefore, the stabilisation of such high-energy intermediates is the key to achieving high-rate and energy-efficient carbon dioxide emission reduction processes. Group 1 consists of metals that do not bind to CO_2^- intermediates and cannot reduce CO. The second group of metals bind to CO_2^- intermediates but cannot reduce CO. The third group (copper) combines with CO_2^- intermediates and can reduce CO. The other group of metals strongly binds to hydrogen and therefore does not include the reduction of CO_2 in aqueous media. The first group includes metals such as Pb, Hg, In, Sn, Cd and Tl [82]. Their tendency to combine with CO_2^- intermediates is so low that it is believed that the reduction of CO_2 occurs through an outer layer mechanism, usually producing formic acid as a product. The second group consists of metals such as Au, Ag, Zn and Ga, which bind to intermediate CO_2^- to varying degrees, but cannot reduce CO, so they usually use CO as the main product of CO_2 reduction. Copper is the only metal that belongs to the third group that is used to reduce CO_2. It combines CO_2^- intermediates and reduces CO to higher reduction products, such as alcohols and hydrocarbons.

20.2.5.2 Core–Shell-Structured Electrocatalysts for CO_2RR

Cu-Based Core–Shell CO_2RR Electrocatalysts

There are many materials used to study this reaction process, and a copper-based catalyst has a unique advantage because it selectively produces hydrocarbons at a relatively low overpotential and has a higher efficiency than other metals. In heterogeneous catalysis, copper and copper oxide catalysts can selectively produce hydrocarbons from CO_2, but the efficiency is relatively low due to the high overpotential. It has always been a barrier to study the selective production of a single product, or a mixture mainly composed of the desired product, in a copper-based catalyst system. To boost CO generation while suppressing H_2 evolution on Cu during CO_2

electroreduction, it has been proposed that coupling Cu with an oxyphilic metal (e.g. Sn and In) would enhance the adsorption of *COOH at mild overpotentials. Previous studies demonstrate that binary Cu–Sn catalysts could favour CO formation. Li et al. [83] synthesised monodisperse core/shell Cu/SnO_2 NPs via a seed-mediated method, which showed a favourable catalytic performance for CO production in an H-type cell.

Wang et al. [84] utilised Cu/SnO_x hetero-structured nanoparticles supported on carbon nanotubes (CNTs) as a model catalyst system. By adjusting the Cu/Sn ratio in the catalyst material structure, one can tune the products of the CO_2 electrocatalytic reduction reaction from hydrocarbon-favourable to CO-selective to formic acid-dominant. In the Cu-rich regime, SnO_x dramatically alters the catalytic behaviour of Cu. An et al. [85] developed a facile oxidative etching of AuCu alloy for the synthesis of a monolithic nanoporous core–shell structured $AuCu_3$@Au electrode, which showed an FE of 97.27% with a partial current density of 5.3 mA cm^{-2} at -0.6 V versus RHE for CO production. The FE value is about 1.45 times higher than that of a Au nanocatalyst. Unlike single nanoporous Au, the $AuCu_3$@Au maintains an excellent performance at a broad potential window. Furthermore, a 23 cm long nanoporous $AuCu_3$@Au bulk electrode with good ductility has been prepared, over which the active current reaches up to 37.2 mA with a current density of 10.78 mA cm^{-2} at -0.7 V versus RHE, pushing the reduction of CO_2 to industrialisation levels..

Making use of strong metal/oxide interactions has recently been demonstrated to be effective in enhancing electrocatalysis in the liquid phase. Tang et al. [86] designed Cu–Sn core/shell nanowire arrays which were built on 3-dimensional macroporous Ni foams by a two-step deposition-annealing-electroreduction treatment. Cu was electroplated on the Ni foam substrates, and the sample was annealed at 500 °C followed by electroreduction, producing Cu nanowires of 150 nm diameter in arrays on the skeleton of Ni foams. Sn nanoparticles of 14–80 nm were then chemically deposited on the Cu nanowires in clusters, and a second annealing treatment at 200 °C followed by electroreduction re-organised the clusters into a Sn_xO/Sn shell of 8 nm thickness. Creating such a Sn shell on Cu nanowires suppressed the faradaic efficiencies for H_2 evolution from 55.7 to 10.1%, and for HCOOH formation from 13.2 to 2.0%, and enhanced CO generation from 32.0 to 90.0% at an applied potential of -0.8 V (vs. RHE). The faradaic efficiency for CO production remained almost constant at 90.0–91.4% with total current densities of -13.2 to -19.3 mA cm^{-2} between -0.8 and -1.2 V (vs. RHE). Huang et al. [87] developed monodisperse core/shell Cu/In_2O_3 nanoparticles (NPs) to boost efficient and tuneable syngas formation via electrochemical CO_2 reduction for the first time. The efficiency and composition of the syngas production on the developed carbon-supported Cu/In_2O_3 catalysts are highly dependent on the In_2O_3 shell thickness (0.4–1.5 nm). As a result, a wide H_2/CO ratio (4/1 to 0.4/1) was achieved on the Cu/In_2O_3 catalysts by controlling the shell thickness and the applied potential (from -0.4 to -0.9 V vs reversible hydrogen electrode), with a faradaic efficiency of syngas formation larger than 90%. Specifically, the best-performing Cu/In_2O_3 catalyst demonstrates remarkably large current densities under low overpotentials (4.6 and 12.7 mA/cm^2 at -0.6 and -0.9 V, respectively), which are competitive with most of the reported systems for syngas formation.

Metal–organic frameworks (MOFs), combining metal ions with organic ligands to form ordered networks, represent a class of emerging nanomaterials with ultra-high surface area and well-developed pore structure. Besides, the porous structure of the Cu-based MOF is helpful for CO_2 capture and adsorption, with the ability of CO_2 electro-reduction. The MOF catalyst can feature an enhanced ability to capture and adsorb CO_2, while the intrinsic catalytic activity of Cu_2O can be simultaneously well maintained. Qiu et al. [88] reported on a tailor-made multifunction-coupled Cu-metal−organic frameworks (Cu-MOF) electrocatalyst by time-resolved controllable restructuration from Cu_2O to $Cu_2O@Cu$-MOF. The restructured electrocatalyst features a time-responsive behaviour and is equipped with a high specific surface area for strong adsorption capacity of CO_2, and abundant active sites for high electrocatalyst activity based on the as-produced MOF on the surface of Cu_2O, as well as the accelerated charge transfer derived from the Cu_2O core in comparison with the Cu-MOF. These intriguing characteristics finally lead to a prominent performance towards hydrocarbons, with a high hydrocarbon of 79.4%, particularly, the CH_4 FE as high as 63.2% (at -1.71 V). This work presents a novel and efficient strategy to configure MOF-based materials in energy and catalysis fields, with the focus on a large surface area, high adsorption ability, and much more exposed active sites.

Transition Metal-Based Core–Shell CO_2RR Electrocatalysts

The incorporation of a transition metal in N-doped carbon was found to greatly strengthen the CO_2RR activity, especially for the Fe–N–C materials. Yang et al. [89] designed a Fe–N–C nanofibre catalyst featuring a core–shell structure consisting of iron nitride nano-particles encapsulated within Fe and N co-doped carbon layers that can efficiently catalyse CO_2 to CO with nearly 100% selectivity, high faradaic efficiency (~95%), and remarkable durability at -0.53 V versus reversible hydrogen electrode. Theoretical calculations reveal that the introduction of an iron nitride core can facilitate the CO intermediate desorption from the Fe and N co-doped shell, thus enhancing the catalytic performance of the CO_2 reduction. Moreover, Ye et al. [90] developed a low-temperature chemical vapour deposition strategy to prepare a sheet-like open nanostructure with Ni nanoparticles wrapped by Ni-N species dispersed on a carbon layer (Ni-NC@Ni). Such a Ni-NC@Ni catalyst exhibits a remarkable CO_2 electroreduction performance, including the high selectivity for the CO product (faradaic efficiency ~87%) and a high current density of 14.8 mA cm^{-2} at a moderate overpotential of 670 mV, as well as the long-term stability over 150 h. The synergistic interaction between cobalt sulphide and the cobalt oxide bicatalyst reduces the activation energy to convert CO_2 into adsorbed intermediates, and hereby enables CO_2RR to run at a low overpotential.

20.2.6   Methanol and Ethanol Oxidation Reaction (MOR/EOR)

With the growing demand for green energy technologies, fuel cells have attracted tremendous attention as promising clean energy-conversion systems. Amongst different fuels that have been used for fuel cells, hydrogen, methanol, and ethanol have been the most explored and each has its advantages and disadvantages [91–98]. The direct methanol fuel cells directly convert chemical energy into usable electrical energy through the redox reaction above the electrode, which contributes to the supplying of clean energy. At the same time, methanol has a higher theoretical energy density and has the potential value for commercialisation. Ethanol is one of the most promising renewable energy sources as a fuel because of its low toxicity, high availability of biomass production, and high energy density brought about by a 12-electron transfer after complete oxidation. Meanwhile, ethanol is also a green and renewable energy. However, the slow kinetics of MOR/EOR is the biggest obstacle to the development of direct alcohol fuel cells (DAFCs). A higher performance catalyst is needed to overcome this obstacle. The design of highly active catalysts is inseparable from the detailed understanding of the reaction mechanism in MOR/EOR, especially the rate-limiting steps [99, 100].

20.2.6.1   Mechanism of MOR/EOR

Methonal oxidation involves 6 electron transfer, and the process is complicated and slow. It is found that the main products of methanol oxidation on the Pt electrode are CO, COH, HCOH and H_2COH. In order to increase the rate of the anode reaction, it is necessary to study the methanol oxidation mechanism, especially the rate-determining step in the methanol oxidation process. There are many related studies, and it is generally considered to be carried out in two ways. The oxidation process is considered to be divided into two basic steps (Fig. 20.13): Methanol adsorbs to the surface of the catalyst and gradually dehydrogenates to form carbon-containing intermediate products. The dissociated water produces oxygen-containing species which react with carbon-containing intermediate products and release CO_2.

However, the research work on the EOR mechanism can be traced back to the 1950s. Now, the most recognised ethanol oxidation mechanism is the dual pathway theory, including the C1 oxidation pathway and the C2 oxidation pathway (Fig. 21) [96, 101, 102]. In the process of the C1 pathway, ethanol is oxidised to CO_2 (alkaline: carbonate CO^{3-}) by the adsorbed CO intermediate, which releases 12 electrons; the pathway of the C2 oxidation by ethanol is acetaldehyde, which transfers 2 electrons, or further oxidises to acetic acid and releases 4 electrons (alkaline: acetate). The C1 pathway is the complete oxidation of ethanol to CO_2, which involves the C–C bond breaking reaction which needs high energy. The C2 pathway is the partial oxidation of ethanol, which occurs relatively easily because it does not involve breaking of the C–C bond [103, 104]. Although higher electrical efficiency can be achieved through

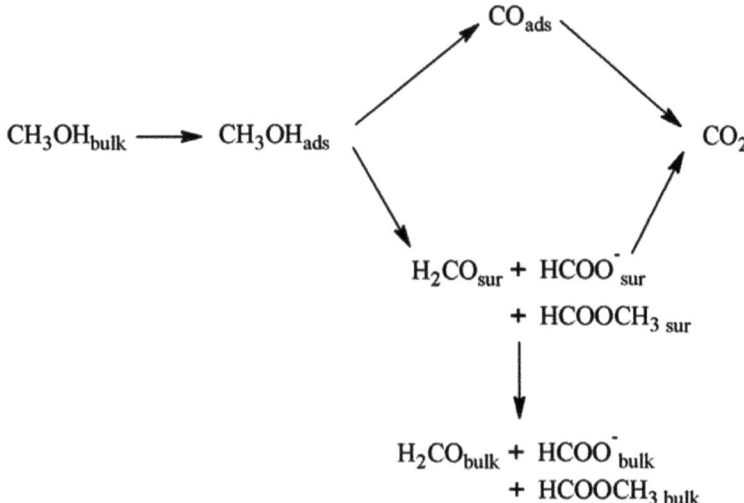

Fig. 20.13 Simplified reaction mechanism of the methanol oxidation on a platinum surface

the C1 pathway, the C2 pathway usually dominates the entire EOR. Therefore, it is an effective way to improve the efficiency of EOR to selectively enhance the C1 pathway through the rational design of high-performance catalysts. Meanwhile, during the EOR, strongly adsorbed intermediates (such as CO and CH_X) will be produced, which will be adsorbed on the surface of the catalyst (such as Pt) and greatly reduce the reaction kinetics. According to the mechanism of the ethanol oxidation reaction, the selection and design of catalysts are from three aspects: (1) there are C–C bond breaking active sites on the surface; (2) the surface composition can selectively improve the C1 pathway; (3) it can rapidly oxidise CO_{ads} and $-CH_X$. Pt and Pd-based nanomaterials are the best catalysts in the electrocatalytic oxidation of ethanol [92].

20.2.6.2   Core–Shell Electrocatalysts for MOR/EOR

Bimetallic Core/Yolk–Shell MOR/EOR Catalysts

Sun et al. [105] reported single noble metal Pt nanocubes assemblies (NCAMs) prepared by a one-pot hydrothermal method. After that, Pt NCAMs were used as the core, and a thin layer of PANI (or SiO_2) was coated on the outer surface to form a core–shell structure of Pt NCAMs@PANI (or Pt NCAMs@SiO_2) catalyst. This catalyst has good collective electronic properties due to its self-assembled core and conductive polymer PANI shell, so Pt NCAMs@PANI has a good proton conductivity, excellent catalytic activity (mass activity is 1.77 times that of the original Pt and 2.25 times that of commercial Pt/C) and outstanding stability (after 1000 cycles,

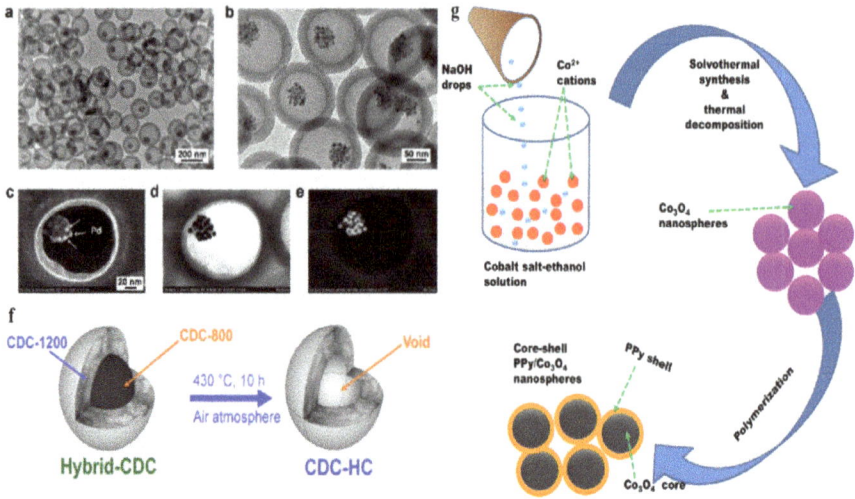

Fig. 20.14 **a, b** TEM images of Pd/C@HCS. **c** SEM, **d** bright field and **e** dark field STEM images of Pd/C@HCS after microtoming off the hollow structure [106]. **f** core–shell structured carbon materials. Reproduced with permission from Ref. [107]. Copyright © 2017 Elsevier B.V. All rights reserved. **g** Schematic illustration showing the assembly of the core–shell like PPy/Co$_3$O$_4$ nanospheres using a straightforward solvothermal method followed by a simple polymerisation process. Reproduced with permission from Ref. [108]. Copyright © 2017 Hydrogen Energy Publications LLC

Pt@ PANI catalyst retains 81.3% of the initial current density) under acidic conditions. Wang et al. [106] used polymer precursors to encapsulate the Pd^{2+} to form the yolk-shell polymer nanospheres with polymer-yolk supported Pd nanoparticles (Pd/P@HPS) by hydrothermal synthesis, then, formed Pd/C@HCS by annealing at 500 °C (Fig. 20.14a–e). In order to expose more metal active sites to the reaction liquid solution, Ariyanto et al. [107] proposed a new synthetic route for producing hollow structure mesopore/graphitic carbon. Due to the different thermal stability between carbon shell and carbon core, the hollow structure catalysts were synthesised by annealing at different temperatures (Fig. 20.14f). Then, through the methanol electro-oxidation test, the catalytic activity and stability of the hollow structure catalysts were greatly improved on the basis of the original materials.

Khalafallah et al. [108] synthesised Co$_3$O$_4$ nanospheres by a solvothermal method, and then PPy nanospheres were formed on the core structure of Co$_3$O$_4$ nanospheres by polymerisation of a pyrrole monomer (Fig. 20.14g). Due to the synergistic effect of the core–shell hierarchical structure, the high conductivity of PPy and the formation of highly active Co-Nx sites by the nitrogen functional groups in Co$_3$O$_4$ and PPy, the PPy/Co$_3$O$_4$ catalyst has an excellent MOR performance in the electrooxidation of methanol. Most core–shell structures take two different metals as the core and shell respectively, which puts forward higher requirements for the synthesis method. Sun et al. [109] reported a strategy of loading Pt$_3$Co and Pt skin bimetallic nanocatalysts directly on porous graphite carbon. By reducing the metal ions using hydrogen at high temperature, Pt precipitation occurs on the surface, and the high temperature provides sufficient energy for H_2 to preferentially combine with Pt on the surface. This process allows the preferential formation of Pt skins to obtain a core–shell structures. The thickness of platinum skin is only 1–2 atomic layers, about 0.5 nm. The bimetallic nanocatalyst composed of Pt_3Co and Pt skin was used as an electrocatalyst for ethanol. The mass activity of the catalyst was 0.79 mA μg_{Pt}^{-1}, which was 250% higher than that of commercial Pt/C (0.32 mA μg_{Pt}^{-1}). Guo et al. [110] reported on Cu@PdCu core–shell nanoparticles which were prepared by the galvanic replacement reactions (GRR) between Pd^{2+} ions and Cu particles. Compared with Pd/C (60 mA cm^{-2}), the peak current of ethanol oxidation on Cu@PdCu/C catalyst is very high and reaches 166.0 mA cm^{-2}. In addition to synthesising the core–shell structures by GRR, the template method is also a commonly used method. Wang et al. [111] explored the feasibility of utilising Te dendrites as a template and reducing agent to synthesise Te@Au core–shell hybrids. Compared with poly Au (2.24 mA cm^{-2}), the EOR performance of Te@Au is 11.58 mA cm^{-2}. Lei et al. [112] designed a self-contained shell Ni@PdNi nanowire array (Ni@PdNi NAs) without carbon carrier and binder for high-efficiency ethanol electrooxidation. Firstly, a simple template-assisted electrodeposition method was used to prepare Ni nanowire arrays (Ni NAs). Subsequently, Ni@PdNi NAs was formed by a one-step solution-based alloying reaction. The optimised Ni@PdNi NAs electrode has a high ECSA of 64.4 $m^2 g_{Pd}^{-1}$, excellent electrochemical performance (peak current density: 622 A g_{Pd}^{-1}) and cyclic stability of ethanol electrooxidation. Li et al. [113] reported a novel hybrid Pd/PANI/Pd sandwich nanotube array (SNTA) to oxidise small organic molecules in DEFCs by utilising the shape effect and synergistic effect of Pd-PANI composites (Fig. 20.15). Compared with Pd/C catalyst, the peak current density of Pd/PANI/Pd SNTAs is almost 3.6 and 1.7 times of that of Pd NTA and Pd/C catalysts.

Besides the GRR and template method, the seed-mediated method is also one of the methods to synthesise a core–shell structures. Ding et al. [114] synthesised gold nanoparticles by the seed-mediated method, and then reduced Pd on gold seeds Au@Pd Core–Shell Nano bricks (Au@Pd CNBs). The EOR activity of Au@Pd CNBs was 4.43 times higher than that of Pd black. Yan et al. [115] reported the successful synthesis of a highly efficient and high yield Au-M (M = Au, Pd and Pt) core–shell structures catalyst (Fig. 20.6). Taking Au-Pd core–shell nanostructures as an example, in 0.5 m KOH solution containing 0.5 M C_2H_5OH, the peak current density of C_2H_5OH oxidation is 151.9 mA mg^{-1}, which is 2.87 and 2.29 times of that of Pd NPs and Pt/C catalysts, respectively. The porous Pd shell greatly improves the quality activity.

Multi-metal Core/Yolk–Shell MOR/EOR Catalysts

Not only the bimetallic materials, but also the multi-metal materials core/yolk-shell structures have attracted the researcher's interest. Wang et al. [116] synthesised the Ni-C-Pt three-layer core–shell nanostructures with Ni-supported Pt nanoparticles by a simple method. Through this method, the loading content of Pt was largely improved

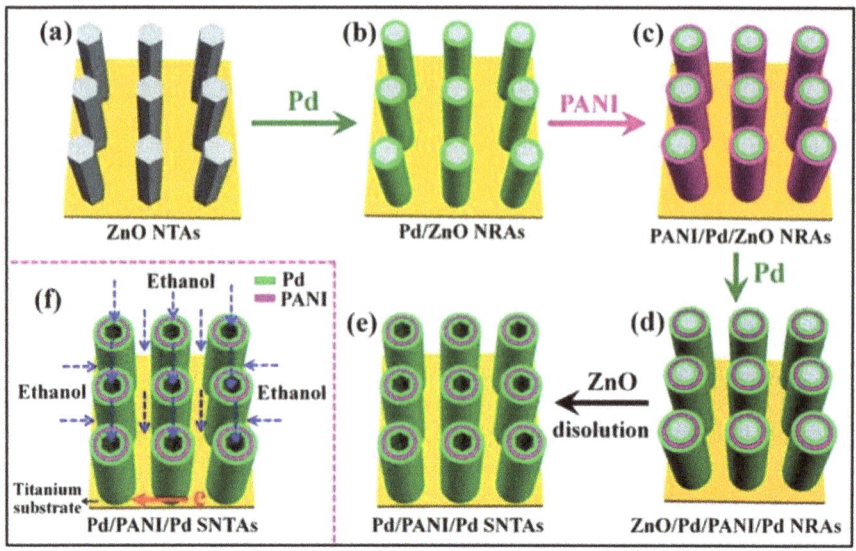

Fig. 20.15 a–e Schematic illustration for the fabrication of Pd/PANI/Pd SNTAs. **f** Schematic illustration for the advantages of Pd/PANI/Pd SNTAs as a catalyst, such as large surface area, large open space, fast transport of active species, and rapid electron transmission. Reproduced with permission from Ref. [113]. Copyright 2013 American Chemical Society

by the shell formed by many tiny Pt nanoparticles, which contributes to the higher MOR activity. Du et al. [117] prepared a new type of precious metal yolk–shell structure of precious metal microporous polymer nanoparticles. This microporous polymer shell could disperse the Au yolk and maintain activity. At the same time, this precious metal–polymer nanoparticles yolk–shell structure had a controllable thickness, high specific surface area and a certain rigidity, contributing to the good MOR activity. Yang et al. [118] prepared the core–shell nanoparticles containing a Ni octahedral shell on a cubic core saturated with Pt, by the stepwise co-deposition (SCD) method, and then the nanoparticles were transformed into yolk–shell nanostructures by the nickel coordination etching (NCE) process. Li et al. [119] used trisodium citrate as the reducing agent to synthesise an Au@PtAuAg yolk-shell nano-alloy by the wet chemical method. The methanol electrocatalytic oxidation test showed that the electrocatalytic performance and stability of Au@PtAuAg egg yolk–shell nano-alloys are higher than commercial Pt/C catalysts.

20.2.7   Others

As alternative fuels, formic acid, ethylene glycol and glycerine have attracted more and more attention. However, many huge obstacles related to the use of anode catalysts, such as slow reaction kinetics, soaring high costs and poor CO tolerance, greatly hinder the actual large-scale commercialisation of DFC. The most efficient electrocatalysts for this reaction are Pt and Pd. Feng et al. [120] reported the large-scale preparation of solid core-porous shell alloyed PtAg nanocrystals (PtAg NCs) by a simple one-pot co-reduction wet chemical method (Fig. 20.10). The prepared PtAg NCs coupled catalyst had a specific activity and mass activity of 77.91 mA cm^{-2} and 1303 mA mg$_{Pt}^{-1}$ towards the glycerine oxidation reaction (GOR), respectively. The doped Ag can stabilise Pt, decrease the CO-like poisoning of Pt and improve the catalytic activity of PtAg alloys. Feng et al. reported a simple, facile, and one-pot solvothermal method for the preparation of reduced graphene oxide (RGO) supported hollow Ag@Pt core–shell nanospheres (hAg@Pt). The hAg@Pt–RGO modified electrodes exhibited an enhanced electrocatalytic activity and a better stability for EGOR. Fan et al. [121] reported the facile synthesis of a core–shell structured Ru@Pd/multi-walled carbon nanotube (MWCNT) catalyst via a two-step chemical reduction process without any surfactant. The Ru@Pd/MWCNT also presented an enhanced electrocatalytic activity for formic acid oxidation (FAO) in an acidic medium.

20.3 Summary

Core–shell functional materials have shown great potential in the field of electrocatalysis. These core–shell functional materials can overcome the inherent defects of single core or shell materials, significantly improving the performance of electrocatalysis. In order to better design efficient catalysts, a basic understanding of the catalytic reaction process is very important.

So far, a large number of research works have reported that the core–shell functional materials exhibit excellent electrocatalytic activity in multiple reactions, but this is often simply attributed to "synergy" and there is a lack of in-detail discussion of synergy. The specific discussion of the role leads to an empirical result, which lacks considerable significance for further guidance on catalyst design. Therefore, in situ electrochemical correlation spectroscopy and electron microscopy will be powerful tools for understanding the electrocatalytic process. Density functional theory (DFT) calculation is a very effective way to design high-performance electrocatalysts.

The core–shell stands out mainly because of the following features: (i) More highly active catalytic sites are exposed to reduce the reaction overpotential; (ii) High structural stability and long service cycle; (iii) High conductivity and fast charge diffusion path; (iv) Effective mass transfer pathways and gas evolution; (v) Low-cost materials and scalable synthesis processes. In short, core–shell functional materials provide a new reference for the electrocatalysts to reduce metal loading, enhance catalytic activity and improve stability.

References

1. Tee SY, Win KY, Teo WS, Koh L-D, Liu S, Teng CP, Han M-Y (2017) Recent progress in energy-driven water splitting. Adv Sci 4(5):1600337
2. Yi J-D, Liu T-T, Huang Y-B, Cao R (2019) Solid-state synthesis of MoS2 nanorod from molybdenum-organic framework for efficient hydrogen evolution reaction. Sci China Mater 62(7):965–972
3. Zeng M, Li Y (2015) Recent advances in heterogeneous electrocatalysts for the hydrogen evolution reaction. J Mater Chem A 3(29):14942–14962
4. Zou X, Zhang Y (2015) Noble metal-free hydrogen evolution catalysts for water splitting. Chem Soc Rev 44(15):5148–5180
5. Zhu J, Hu L, Zhao P, Lee LYS, Wong K-Y (2020) Recent advances in electrocatalytic hydrogen evolution using nanoparticles. Chem Rev 120(2):851–918
6. Zhang L-N, Lang Z-L, Wang Y-H, Tan H-Q, Zang H-Y, Kang Z-H, Li Y-G (2019) Cable-like Ru/WNO@C nanowires for simultaneous high-efficiency hydrogen evolution and low-energy consumption chlor-alkali electrolysis. Energy Environ Sci 12(8):2569–2580
7. Deng K, Ren T, Xu Y, Liu S, Dai Z, Wang Z, Li X, Wang L, Wang H (2020) Crystalline core–amorphous shell heterostructures: epitaxial assembly of NiB nanosheets onto PtPd mesoporous hollow nanopolyhedra for enhanced hydrogen evolution electrocatalysis. J Mater Chem A 8(18):8927–8933
8. Kim J, Kim H, Lee W-J, Ruqia B, Baik H, Oh H-S, Paek S-M, Lim H-K, Choi CH, Choi S-I (2019) Theoretical and experimental understanding of hydrogen evolution reaction kinetics in alkaline electrolytes with Pt-based core-shell nanocrystals. J Am Chem Soc 141(45):18256–18263
9. Wu A, Xie Y, Ma H, Tian C, Gu Y, Yan H, Zhang X, Yang G, Fu H (2018) Integrating the active OER and HER components as the heterostructures for the efficient overall water splitting. Nano Energy 44:353–363
10. Nguyen DC, Tran DT, Doan TLL, Kim DH, Kim NH, Lee JH (2020) Rational design of Core@shell structured CoSx@Cu2MoS4 hybridized MoS2/N, S-codoped graphene as advanced electrocatalyst for water splitting and Zn-air battery. Adv Energy Mater 10(8):1903289
11. Zhu H, Gao G, Du M, Zhou J, Wang K, Wu W, Chen X, Li Y, Ma P, Dong W, Duan F, Chen M, Wu G, Wu J, Yang H, Guo S (2018) Atomic-scale core/shell structure engineering induces precise tensile strain to boost hydrogen evolution catalysis. Adv Mater 30(26):1707301
12. Li X, Liu W, Zhang M, Zhong Y, Weng Z, Mi Y, Zhou Y, Li M, Cha JJ, Tang Z, Jiang H, Li X, Wang H (2017) Strong metal-phosphide interactions in core-shell geometry for enhanced electrocatalysis. Nano Lett 17(3):2057–2063
13. Maity S, Das B, Samanta M, Das BK, Ghosh S, Chattopadhyay KK (2020) MoSe2-amorphous CNT hierarchical hybrid core-shell structure for efficient hydrogen evolution reaction. ACS Appl Energy Mater 3(5):5067–5076
14. Liu XX, Zang JB, Chen L, Chen LB, Chen X, Wu P, Zhou SY, Wang YH (2017) A microwave-assisted synthesis of CoO@Co core–shell structures coupled with N-doped reduced graphene oxide used as a superior multi-functional electrocatalyst for hydrogen evolution, oxygen reduction and oxygen evolution reactions. J Mater Chem A 5(12):5865–5872
15. Chen Z, Wu R, Liu Y, Ha Y, Guo Y, Sun D, Liu M, Fang F (2018) Ultrafine Co nanoparticles encapsulated in carbon-nanotubes-grafted graphene sheets as advanced electrocatalysts for the hydrogen evolution reaction. Adv Mater 30(30):1802011
16. Chu S, Majumdar A (2012) Opportunities and challenges for a sustainable energy future. Nature 488(7411):294–303
17. Kibsgaard J, Chorkendorff I (2019) Considerations for the scaling-up of water splitting catalysts. Nat Energy 4(6):430–433
18. Wang X, Li Q, Shi P, Fan J, Min Y, Xu Q (2019) Nickel nitride particles supported on 2D activated graphene–black phosphorus heterostructure: an efficient electrocatalyst for the oxygen evolution reaction. Small 15(48):1901530

19. Bai K, Fan J-C, Shi P-H, Min Y-L, Xu Q-J (2020) Directly ball milling red phosphorus and expended graphite for oxygen evolution reaction. J Power Sources 456:228003
20. Luo S, Gu R, Shi P, Fan J, Xu Q, Min Y (2020) π-π interaction boosts catalytic oxygen evolution by self-supporting metal-organic frameworks. J Power Sources 448:227406
21. Gonçalves JM, Martins PR, Angnes L, Araki K (2020) Recent advances in ternary layered double hydroxide electrocatalysts for the oxygen evolution reaction. New J Chem 44(24):9981–9997
22. Xie W, Song Y, Li S, Shao M, Wei M (2019) Integrated nanostructural electrodes based on layered double hydroxides. Energy Environ Mater 2(3):158–171
23. Reier T, Oezaslan M, Strasser P (2012) Electrocatalytic oxygen evolution reaction (OER) on Ru, Ir, and Pt catalysts: a comparative study of nanoparticles and bulk materials. ACS Cataly 2(8):1765–1772
24. Gloag L, Benedetti TM, Cheong S, Webster RF, Marjo CE, Gooding JJ, Tilley RD (2018) Pd–Ru core–shell nanoparticles with tunable shell thickness for active and stable oxygen evolution performance. Nanoscale 10(32):15173–15177
25. Xie W, Li J, Song Y, Li S, Li J, Shao M (2020) Hierarchical carbon microtube@nanotube core-shell structure for high-performance oxygen electrocatalysis and Zn–air battery. Nano-Micro Letters 12(1):97
26. Jia D, Gao H, Xing L, Chen X, Dong W, Huang X, Wang G (2019) 3D self-supported porous NiO@NiMoO4 core-shell nanosheets for highly efficient oxygen evolution reaction. Inorg Chem 58(10):6758–6764
27. Liu C, Wang J, Wan J, Cheng Y, Huang R, Zhang C, Hu W, Wei G, Yu C (2020) Amorphous metal-organic framework-dominated nanocomposites with both compositional and structural heterogeneity for oxygen evolution. Angew Chem Int Ed 59(9):3630–3637
28. Xie J-Y, Liu Z-Z, Li J, Feng L, Yang M, Ma Y, Liu D-P, Wang L, Chai Y-M, Dong B (2020) Fe-doped CoP core–shell structure with open cages as efficient electrocatalyst for oxygen evolution. J Energy Chem 48:328–333
29. Kang J, Sheng J, Xie J, Ye H, Chen J, Fu X-Z, Du G, Sun R, Wong C-P (2018) Tubular Cu(OH)2 arrays decorated with nanothorny Co–Ni bimetallic carbonate hydroxide supported on Cu foam: a 3D hierarchical core–shell efficient electrocatalyst for the oxygen evolution reaction. J Mater Chem A 6(21):10064–10073
30. Di J, Zhu H, Xia J, Bao J, Zhang P, Yang SZ, Li H, Dai S (2019) High-performance electrolytic oxygen evolution with a seamless armor core-shell FeCoNi oxynitride. Nanoscale 11(15):7239–7246
31. Chen M, Lu S, Fu X-Z, Luo J-L (2020) Core-shell structured NiFeSn@NiFe(Oxy)hydroxide nanospheres from an electrochemical strategy for electrocatalytic oxygen evolution reaction. Adv Sci 7(10):2070052
32. Li Y, Li Q, Wang H, Zhang L, Wilkinson DP, Zhang J (2019) Recent progresses in oxygen reduction reaction electrocatalysts for electrochemical energy applications. Electrochem Energy Rev 2(4):518–538
33. Chen D, Li C, Liu H, Ye F, Yang J (2015) Core-shell Au@Pd nanoparticles with enhanced catalytic activity for oxygen reduction reaction via core-shell Au@Ag/Pd constructions. Sci Rep 5(1):11949
34. Aoki N, Inoue H, Okawa T, Ikehata Y, Shirai A, Daimon H, Doi T, Orikasa Y, Uchimoto Y, Jinnai H, Inamoto S, Otsuka Y, Inaba M (2018) Enhancement of oxygen reduction reaction activity of pd core-Pt shell structured catalyst on a potential cycling accelerated durability test. Electrocatalysis 9(2):125–138
35. Qi K, Zheng W, Cui X (2016) Supersaturation-controlled surface structure evolution of Pd@Pt core–shell nanocrystals: enhancement of the ORR activity at a sub-10 nm scale. Nanoscale 8(3):1698–1703
36. Mahata A, Bhauriyal P, Rawat KS, Pathak B (2016) Pt3Ti (Ti19@Pt60)-based cuboctahedral core-shell nanocluster favors a direct over indirect oxygen reduction reaction. ACS Energy Lett 1(4):797–805

37. Matin MA, Lee J, Kim GW, Park H-U, Cha BJ, Shastri S, Kim G, Kim Y-D, Kwon Y-U, Petkov V (2020) Morphing Mncore@Ptshell nanoparticles: effects of core structure on the ORR performance of Pt shell. Appl Catal B 267:118727

38. Ao X, Zhang W, Zhao B, Ding Y, Nam G, Soule L, Abdelhafiz A, Wang C, Liu M (2020) Atomically dispersed Fe–N–C decorated with Pt-alloy core–shell nanoparticles for improved activity and durability towards oxygen reduction. Energy Environ Sci

39. Lu X, Chan HM, Sun C-L, Tseng C-M, Zhao C (2015) Interconnected core–shell carbon nanotube–graphene nanoribbon scaffolds for anchoring cobalt oxides as bifunctional electro-catalysts for oxygen evolution and reduction. J Mater ChemA 3(25):13371–13376

40. Xue H, Na Z, Wu Y, Wang X, Li Q, Liang F, Yin D, Wang L, Ming J (2018) Unique CO_3O_4/nitrogen-doped carbon nanospheres derived from metal–organic framework: insight into their superior lithium storage capabilities and electrochemical features in high-voltage batteries. J Mater Chem A 6(26):12466–12474

41. Xiao J, Chen C, Xi J, Xu Y, Xiao F, Wang S, Yang S (2015) Core–shell $CO@CO_3O_4$ nanoparticle-embedded bamboo-like nitrogen-doped carbon nanotubes (BNCNTs) as a highly active electrocatalyst for the oxygen reduction reaction. Nanoscale 7(16):7056–7064

42. Guo J, Niu Q, Yuan Y, Maitlo I, Nie J, Ma G (2017) Electrospun core–shell nanofibers derived Fe–S/N doped carbon material for oxygen reduction reaction. Appl Surf Sci 416:118–123

43. Park H, Oh S, Lee S, Choi S, Oh M (2019) Cobalt- and nitrogen-codoped porous carbon catalyst made from core–shell type hybrid metal–organic framework (ZIF-L@ZIF-67) and its efficient oxygen reduction reaction (ORR) activity. Appl Catal B 246:322–329

44. Cheng Y, Dou S, Veder J-P, Wang S, Saunders M, Jiang SP (2017) Efficient and durable bifunctional oxygen catalysts based on NiFeO@MnOx core-shell structures for rechargeable Zn–air batteries. ACS Appl Mater Interfaces 9(9):8121–8133

45. Klerke A, Christensen CH, Nørskov JK, Vegge T (2008) Ammonia for hydrogen storage: challenges and opportunities. J Mater Chem 18(20):2304–2310

46. Schlögl R (2003) Catalytic synthesis of ammonia—a "Never-Ending Story"? Angew Chem Int Ed 42(18):2004–2008

47. Zhu X, Mou S, Peng Q, Liu Q, Luo Y, Chen G, Gao S, Sun X (2020) Aqueous electrocat-alytic N2 reduction for ambient NH3 synthesis: recent advances in catalyst development and performance improvement. J Mater Chem A 8(4):1545–1556

48. Bao D, Zhang Q, Meng F-L, Zhong H-X, Shi M-M, Zhang Y, Yan J-M, Jiang Q, Zhang X-B (2017) Electrochemical reduction of N2 under ambient conditions for artificial N2 fixation and renewable energy storage using N2/NH3 cycle. Adv Mater 29(3):1604799

49. van der Ham CJM, Koper MTM, Hetterscheid DGH (2014) Challenges in reduction of dinitrogen by proton and electron transfer. Chem Soc Rev 43(15):5183–5191

50. Shipman MA, Symes MD (2017) Recent progress towards the electrosynthesis of ammonia from sustainable resources. Catal Today 286:57–68

51. Guo W, Zhang K, Liang Z, Zou R, Xu Q (2019) Electrochemical nitrogen fixation and utilization: theories, advanced catalyst materials and system design. Chem Soc Rev 48(24):5658–5716

52. Guo C, Ran J, Vasileff A, Qiao S-Z (2018) Rational design of electrocatalysts and photo(electro)catalysts for nitrogen reduction to ammonia (NH3) under ambient conditions. Energy Environ Sci 11(1):45–56

53. Manjunatha R, Karajić A, Liu M, Zhai Z, Dong L, Yan W, Wilkinson DP, Zhang J (2020) A review of composite/hybrid electrocatalysts and photocatalysts for nitrogen reduction reac-tions: advanced materials, mechanisms, challenges and perspectives. Electrochem Energy Rev

54. Yao R-Q, Lang X-Y, Jiang Q (2019) Recent advances of nanoporous metal-based catalyst: synthesis, application and perspectives. J Iron Steel Res Int 26(8):779–795

55. Jiang R, Tung SO, Tang Z, Li L, Ding L, Xi X, Liu Y, Zhang L, Zhang J (2018) A review of core-shell nanostructured electrocatalysts for oxygen reduction reaction. Energy Storage Mater 12:260–276

56. Yin X, Yang L, Gao Q (2020) Core–shell nanostructured electrocatalysts for water splitting. Nanoscale 12(30):15944–15969
57. Suryanto BHR, Du H-L, Wang D, Chen J, Simonov AN, MacFarlane DR (2019) Challenges and prospects in the catalysis of electroreduction of nitrogen to ammonia. Nat Cataly 2(4):290–296
58. Kong X, Peng H-Q, Bu S, Gao Q, Jiao T, Cheng J, Liu B, Hong G, Lee C-S, Zhang W (2020) Defect engineering of nanostructured electrocatalysts for enhancing nitrogen reduction. J Mater Chem A 8(16):7457–7473
59. Wang J, Chen S, Li Z, Li G, Liu X (2020) Recent advances in electrochemical synthesis of ammonia through nitrogen reduction under ambient conditions. Chem Electro Chem 7(5):1067–1079
60. Li Y, Wang H, Priest C, Li S, Xu P, Wu G (2020) Advanced electrocatalysis for energy and environmental sustainability via water and nitrogen reactions. Adv Mater 2000381
61. Cui X, Tang C, Zhang Q (2018) A review of electrocatalytic reduction of dinitrogen to ammonia under ambient conditions. Adv Energy Mater 8(22):1800369
62. Wan Y, Xu J, Lv R (2019) Heterogeneous electrocatalysts design for nitrogen reduction reaction under ambient conditions. Mater Today 27:69–90
63. Hou J, Yang M, Zhang J (2020) Recent advances in catalysts, electrolytes and electrode engineering for the nitrogen reduction reaction under ambient conditions. Nanoscale 12(13):6900–6920
64. Skúlason E, Bligaard T, Gudmundsdóttir S, Studt F, Rossmeisl J, Abild-Pedersen F, Vegge T, Jónsson H, Nørskov JK (2012) A theoretical evaluation of possible transition metal electrocatalysts for N2 reduction. Phys Chem Chem Phys 14(3):1235–1245
65. Montoya JH, Tsai C, Vojvodic A, Nørskov JK (2015) The challenge of electrochemical ammonia synthesis: a new perspective on the role of nitrogen scaling relations. Chemsuschem 8(13):2180–2186
66. Li W, Zhang C, Han M, Ye Y, Zhang S, Liu Y, Wang G, Liang C, Zhang H (2019) Ambient electrosynthesis of ammonia using core-shell structured Au@C catalyst fabricated by one-step laser ablation technique. ACS Appl Mater Interfaces 11(47):44186–44195
67. Liu G, Cui Z, Han M, Zhang S, Zhao C, Chen C, Wang G, Zhang H (2019) Ambient electrosynthesis of ammonia on a core–shell-structured Au@CeO2 catalyst: contribution of oxygen vacancies in CeO2. Chem Eur J 25(23):5904–5911
68. Wang P, Ji Y, Shao Q, Li Y, Huang X, Wang P, Ji Y, Shao Q, Li Y, Huang X (2020) Core@shell structured Au@SnO2 nanoparticles with improved N2 adsorption/activation and electrical conductivity for efficient N2 fixatio. Chin Sci Bull 65(5):350–358
69. Yang Y, Wang S-Q, Wen H, Ye T, Chen J, Li C-P, Du M (2019) Nanoporous gold embedded ZIF composite for enhanced electrochemical nitrogen fixation. Angew Chem 58(43):15362–15366
70. Sim HYF, Chen JRT, Koh CSL, Lee HK, Han X, Phan-Quang GC, Pang JY, Lay CL, Pedireddy S, Phang IY, Yeow EKL, Ling XY (2020) ZIF-induced d-band modification in a bimetallic nanocatalyst: achieving over 44% efficiency in the ambient nitrogen reduction reaction. Angew Chem Int Ed. https://doi.org/10.1002/anie.202006071
71. Harris DF, Lukoyanov DA, Shaw S, Compton P, Tokmina-Lukaszewska M, Bothner B, Kelleher N, Dean DR, Hoffman BM, Seefeldt LC (2018) Mechanism of N2 reduction catalyzed by Fe-nitrogenase involves reductive elimination of H2. Biochemistry 57(5):701–710
72. Li C, Fu Y, Wu Z, Xia J, Wang X (2019) Sandwich-like reduced graphene oxide/yolk–shell-structured Fe@Fe3O4/carbonized paper as an efficient freestanding electrode for electrochemical synthesis of ammonia directly from H2O and nitrogen. Nanoscale 11(27):12997–13006
73. Peng M, Qiao Y, Luo M, Wang M, Chu S, Zhao Y, Liu P, Liu J, Tan Y (2019) Bioinspired Fe3C@C as highly efficient electrocatalyst for nitrogen reduction reaction under ambient conditions. ACS Appl Mater Interfaces 11(43):40062–40068
74. Qiu W, Luo Y-X, Liang R-P, Qiu J-D (2020) Amorphous/crystalline hetero-phase TiO2-coated α-Fe2O3 core-shell nanospindles: a high-performance artificial nitrogen fixation electrocatalyst. Chem Eur J 26(45):10226–10229

75. Zeng L, Li X, Chen S, Wen J, Huang W, Chen A (2020) Unique hollow Ni–Fe@MoS2 nanocubes with boosted electrocatalytic activity for N2 reduction to NH3. J Mater Chem A 8(15):7339–7349

76. Qu X, Shen L, Mao Y, Lin J, Li Y, Li G, Zhang Y, Jiang Y, Sun S (2019) Facile preparation of carbon shells-coated O-doped molybdenum carbide nanoparticles as high selective electrocatalysts for nitrogen reduction reaction under ambient conditions. ACS Appl Mater Interfaces 11(35):31869–31877

77. Qiu Y, Zhao S, Qin M, Diao J, Liu S, Dai L, Zhang W, Guo X (2020) Multi-yolk–shell bismuth@porous carbon as a highly efficient electrocatalyst for artificial N2 fixation under ambient conditions. Inorganic Chem Front 7(10):2006–2016

78. Zhao M, Xu L, Vara M, Elnabawy AO, Gilroy KD, Hood ZD, Zhou S, Figueroa-Cosme L, Chi M, Mavrikakis M, Xia Y (2018) Synthesis of Ru icosahedral nanocages with a face-centered-cubic structure and evaluation of their catalytic properties. ACS Cataly 8(8):6948–6960

79. Yang C, Huang B, Bai S, Feng Y, Shao Q, Huang X (2020) A generalized surface chalcogenation strategy for boosting the electrochemical N2 fixation of metal nanocrystals. Adv Mater 32(24):2001267

80. Lan Y, Gai C, Kenis PJA, Lu J (2014) Electrochemical reduction of carbon dioxide on Cu/CuO core/shell catalysts. Chem Electro Chem 1(9):1577–1582

81. Thomann I, Pinaud BA, Chen Z, Clemens BM, Jaramillo TF, Brongersma ML (2011) Plasmon enhanced solar-to-fuel energy conversion. Nano Lett 11(8):3440–3446

82. Jones J-P, Prakash GKS, Olah GA (2014) Electrochemical CO_2 reduction: recent advances and current trends. Isr J Chem 54(10):1451–1466

83. Li Q, Fu J, Zhu W, Chen Z, Shen B, Wu L, Xi Z, Wang T, Lu G, Zhu J-J, Sun S (2017) Tuning Sn-catalysis for electrochemical reduction of CO_2 to CO via the core/shell Cu/SnO2 structure. J Am Chem Soc 139(12):4290–4293

84. Huo S, Weng Z, Wu Z, Zhong Y, Wu Y, Fang J, Wang H (2017) Coupled metal/oxide catalysts with tunable product selectivity for electrocatalytic CO_2 reduction. ACS Appl Mater Interfaces 9(34):28519–28526

85. Ma X, Shen Y, Yao S, An C, Zhang W, Zhu J, Si R, Guo C, An C (2020) Core–shell nanoporous AuCu3@Au monolithic electrode for efficient electrochemical CO2 reduction. J Mater Chem A 8(6):3344–3350

86. Hu H, Wang Y, Du N, Sun Y, Tang Y, Hu Q, Wan P, Dai L, Fisher AC, Yang XJ (2018) Thermal-treatment-induced Cu–Sn core/shell nanowire array catalysts for highly efficient CO_2 electroreduction. Chem Electro Chem 5(24):3854–3858

87. Xie H, Chen S, Ma F, Liang J, Miao Z, Wang T, Wang H-L, Huang Y, Li Q (2018) Boosting tunable syngas formation via electrochemical CO_2 reduction on Cu/In2O3 core/shell nanoparticles. ACS Appl Mater Interfaces 10(43):36996–37004

88. Tan X, Yu C, Zhao C, Huang H, Yao X, Han Y, Guo W, Cui S, Huang H, Qiu J (2019) Restructuring of Cu_2O to Cu_2O@Cu-metal–organic frameworks for selective electrochemical reduction of CO_2. ACS Appl Mater Interfaces 11(10):9904–9910

89. Cheng Q, Mao K, Ma L, Yang L, Zou L, Zou Z, Hu Z, Yang H (2018) Encapsulation of iron nitride by Fe–N–C shell enabling highly efficient electroreduction of CO_2 to CO. ACS Energy Letters 3(5):1205–1211

90. He Y, Li Y, Zhang J, Wang S, Huang D, Yang G, Yi X, Lin H, Han X, Hu W, Deng Y, Ye J (2020) Low-temperature strategy toward Ni-NC@Ni core-shell nanostructure with Single-Ni sites for efficient CO2 electroreduction. Nano Energy 77:105010

91. Lin Y, Liu Q, Fan J, Liao K, Xie J, Liu P, Chen Y, Min Y, Xu Q (2016) Highly dispersed palladium nanoparticles on poly (N 1, N 3-dimethylbenzimidazolium) iodide-functionalized multiwalled carbon nanotubes for ethanol oxidation in alkaline solution. RSC Adv 6(104):102582–102594

92. Liu Q, Fan J, Min Y, Wu T, Lin Y, Xu Q (2016) B, N-codoped graphene nanoribbons supported Pd nanoparticles for ethanol electrooxidation enhancement. J Mater Chem A 4(13):4929–4933

93. Liu Q, Jiang K, Fan J, Lin Y, Min Y, Xu Q, Cai W-B (2016) Manganese dioxide coated graphene nanoribbons supported palladium nanoparticles as an efficient catalyst for ethanol electrooxidation in alkaline media. Electrochim Acta 203:91–98

94. Liu Q, Lin Y, Fan J, Lv D, Min Y, Wu T, Xu Q (2016) Well-dispersed palladium nanoparticles on three-dimensional hollow N-doped graphene frameworks for enhancement of methanol electro-oxidation. Electrochem Commun 73:75–79

95. Wu T, Fan J, Li Q, Shi P, Xu Q, Min Y (2018) Palladium nanoparticles anchored on anatase titanium dioxide-black phosphorus hybrids with heterointerfaces: highly electroactive and durable catalysts for ethanol electrooxidation. Adv Energy Mater 8(1):1701799

96. Wu T, Ma Y, Qu Z, Fan J, Li Q, Shi P, Xu Q, Min Y (2019) Black phosphorus–graphene heterostructure-supported Pd nanoparticles with superior activity and stability for ethanol electro-oxidation. ACS Appl Mater Interfaces 11(5):5136–5145

97. Yu K, Lin Y, Fan J, Li Q, Shi P, Xu Q, Min Y (2019) Ternary N, S, and P-doped hollow carbon spheres derived from polyphosphazene as Pd supports for ethanol oxidation reaction. Catalysts 9(2):114

98. Geng D, Zhu S, Chai M, Zhang Z, Fan J, Xu Q, Min Y (2020) Pd x Fe y alloy nanoparticles decorated on carbon nanofibers with improved electrocatalytic activity for ethanol electrooxidation in alkaline media. New J Chem 44(13):5023–5032

99. Antolini E (2007) Catalysts for direct ethanol fuel cells. J Power Sources 170(1):1–12

100. Antolini E (2009) Palladium in fuel cell catalysis. Energy Environ Sci 2(9):915–931

101. Wang H-F, Liu Z-P (2007) Selectivity of direct ethanol fuel cell dictated by a unique partial oxidation channel. J Phys Chem C 111(33):12157–12160

102. Zhou W, Zhou Z, Song S, Li W, Sun G, Tsiakaras P, Xin Q (2003) Pt based anode catalysts for direct ethanol fuel cells. Appl Catal B 46(2):273–285

103. Wang Y, Zou S, Cai W-B (2015) Recent advances on electro-oxidation of ethanol on Pt- and Pd-based catalysts: from reaction mechanisms to catalytic materials. Catalysts 5(3):1507–1534

104. Zhou W, Li M, Zhang L, Chan SH (2014) Supported PtAu catalysts with different nanostructures for ethanol electrooxidation. Electrochim Acta 123:233–239

105. Sun X, Zhang N, Huang X (2016) Polyaniline-coated platinum nanocube assemblies as enhanced methanol oxidation electrocatalysts. Chem Cat Chem 8(22):3436–3440

106. Wang G-H, Chen K, Engelhardt J, Tüysüz H, Bongard H-J, Schmidt W, Schüth F (2018) Scalable one-pot synthesis of yolk-shell carbon nanospheres with yolk-supported pd nanoparticles for size-selective catalysis. Chem Mater 30(8):2483–2487

107. Ariyanto T, Kern AM, Etzold BJM, Zhang G-R (2017) Carbide-derived carbon with hollow core structure and its performance as catalyst support for methanol electro-oxidation. Electrochem Commun 82:12–15

108. Khalafallah D, Alothman OY, Fouad H, Abdelrazek Khalil K (2018) Hierarchical CO_3O_4 decorated PPy nanocasting core-shell nanospheres as a high performance electrocatalysts for methanol oxidation. Int J Hydrogen Energy 43(5):2742–2753

109. Zhang B-W, Sheng T, Wang Y-X, Qu X-M, Zhang J-M, Zhang Z-C, Liao H-G, Zhu F-C, Dou S-X, Jiang Y-X, Sun S-G (2017) Platinum-cobalt bimetallic nanoparticles with Pt skin for electro-oxidation of ethanol. ACS Cataly 7(1):892–895

110. Cai J, Zeng Y, Guo Y (2014) Copper@palladium–copper core–shell nanospheres as a highly effective electrocatalyst for ethanol electro-oxidation in alkaline media. J Power Sources 270:257–261

111. Jin H, Wang D, Zhao Y, Zhou H, Wang S, Wang J (2012) Fabrication of Te@Au core-shell hybrids for efficient ethanol oxidation. J Power Sources 215:227–232

112. Guo F, Li Y, Fan B, Liu Y, Lu L, Lei Y (2018) Carbon- and binder-free core-shell nanowire arrays for efficient ethanol electro-oxidation in alkaline medium. ACS Appl Mater Interfaces 10(5):4705–4714

113. Wang A-L, Xu H, Feng J-X, Ding L-X, Tong Y-X, Li G-R (2013) Design of Pd/PANI/Pd sandwich-structured nanotube array catalysts with special shape effects and synergistic effects for ethanol electrooxidation. J Am Chem Soc 135(29):10703–10709

114. Wang W, Zhang J, Yang S, Ding B, Song X (2013) Au@Pd core-shell nanobricks with concave structures and their catalysis of ethanol oxidation. Chemsuschem 6(10):1945–1951

115. Kuai L, Geng B, Wang S, Sang Y (2012) A general and high-yield galvanic displacement approach to Au-M (M = Au, Pd, and Pt) core-shell nanostructures with porous shells and enhanced electrocatalytic performances. Chem Eur J 18(30):9423–9429

116. Xiong S, Fan J, Wang Y, Zhu J, Yu J, Hu Z (2017) A facile template approach to nitrogen-doped hierarchical porous carbon nanospheres from polydopamine for high-performance supercapacitors. J Mater Chem A 5(34):18242–18252

117. Du Y, Huang Z, Wu S, Xiong K, Zhang X, Zheng B, Nadimicherla R, Fu R, Wu D (2018) Preparation of versatile yolk-shell nanoparticles with a precious metal yolk and a microporous polymer shell for high-performance catalysts and antibacterial agents. Polymer 137:195–200

118. Yang T, Wang Y, Wei W, Ding X, He M, Yu T, Zhao H, Zhang D (2019) Synthesis of octahedral Pt–Ni–Ir yolk–shell nanoparticles and their catalysis in oxygen reduction and methanol oxidization under both acidic and alkaline conditions. Nanoscale 11(48):23206–23216

119. Weng X, Liu Q, Wang A-J, Yuan J, Feng J-J (2017) Simple one-pot synthesis of solid-core@porous-shell alloyed PtAg nanocrystals for the superior catalytic activity toward hydrogen evolution and glycerol oxidation. J Colloid Interface Sci 494:15–21

120. Zheng J-N, Lv J-J, Li S-S, Xue M-W, Wang A-J, Feng J-J (2014) One-pot synthesis of reduced graphene oxide supported hollow Ag@Pt core–shell nanospheres with enhanced electrocatalytic activity for ethylene glycol oxidation. J Mater Chem A 2(10):3445–3451

121. Zhang X-J, Zhang J-M, Zhang P-Y, Li Y, Xiang S, Tang H-G, Fan Y-J (2017) Highly active carbon nanotube-supported Ru@Pd core-shell nanostructure as an efficient electrocatalyst toward ethanol and formic acid oxidation. Mol Cataly 436:138–144

Chapter 21
Pd-Core-Based Core–Shell Nanoparticles for Catalytic and Electrocatalytic Applications

Miriam Navlani-García, David Salinas-Torres, and Diego Cazorla-Amorós

21.1 Introduction

Catalysis has been crucial for humankind since the dawn of civilization and it is nowadays a field of research with a prominent interdisciplinary character. The term "Catalysis" (from the Greek words "*kata*" and "*lyein*", down and loosen, respectively) was coined in 1835 by the Swedish chemist Jöns Jakob Berzelius, who is considered one of the founders of modern chemistry [31], but its discovery date is difficult to establish [55]. Catalytic processes are traditionally divided into homogeneous and heterogeneous, the former referring to processes in which the catalyst and reactants are in the same phase, while heterogeneous catalytic processes take place when reactants and catalysts are in separated phases. Heterogeneous catalysts are preferred from the practical application standpoint and they are of crucial importance to the world's economy since they are involved in countless applications in the chemical, pharmaceutical, food, automotive, petrochemical, and energy sectors, among others [15].

Metal-based catalysts have sparked great interest and important efforts have been devoted to achieving maximum metal atom efficiency utilization, as well as to ascertain the relationship between the catalytic performance and the features of the metal

M. Navlani-García (✉) · D. Cazorla-Amorós
Department of Inorganic Chemistry and Materials Institute, University of Alicante, 03080 Alicante, Spain
e-mail: miriam.navlani@ua.es

D. Cazorla-Amorós
e-mail: cazorla@ua.es

D. Salinas-Torres
Department of Physical Chemistry and Materials Institute, University of Alicante, 03080 Alicante, Spain
e-mail: david.salinas@ua.es

© The Author(s), under exclusive license to Springer Nature Singapore Pte Ltd. 2021   343
H. Yamashita and H. Li (eds.), *Core-Shell and Yolk-Shell Nanocatalysts*,
Nanostructure Science and Technology,
https://doi.org/10.1007/978-981-16-0463-8_21

species [37]. Metal species in the form of metal nanoparticles and nanocrystals are preferred to bulk material due to their large surface area-to-volume ratio and the subsequent higher reactivities [7]. The consolidated yet burgeoning fields of nanoscience and nanotechnology have had a great significance in catalysis since their principles are crucial to tailor the active sites, as well as the site environments to attain active and selective catalysts for a certain application [30]. It was in the 1980s, with the advent of nanotechnology, that tools for the engineering of the structural details of the catalysts (i.e. size, shape, and surface properties) become available [64]. Nowadays, it is well-known that the properties of the metal nanoparticles and nanocrystals play vital roles in controlling the ultimate performance (activity, selectivity, and stability) of the catalytic systems, and the optimization of these aspects is still an essential cornerstone of most of the studies tacking with the development of metal-based catalysts [41].

Core–shell nanostructures have emerged as an interesting approach for the nano-engineering of the active sites [17]. They consist of an inner core material surrounded by a shell material, each having structural and/or compositional features with dimensions at the nanoscale [13]. Core–shell nanostructures are synthesized by following the procedures that have traditionally been utilized to prepare nanomaterials, which are divided into "top-down" and "bottom-up" methods (either with stepwise or one-pot fashion).

The significance of core–shell nanostructures in catalysis has been linked to the following advantages: (i) the core can serve as a support for the shell, thus resulting in the better catalytic performance of the material forming the shell; (ii) synergistic effect between the core and the shell, which lead to higher efficiency/yield/selectivity; (iii) combination of the properties of both core and shell [17]. Moreover, certain catalysts based on core–shell nanostructures can display better stability under reaction conditions than traditionally supported catalysts, which is ascribed to their resistance against metal leaching and sintering of the active phases. Such aspects are responsible for their unique structural, physical, and chemical properties, which provide them with great potential in the field of catalysis [13, 18, 63].

Broadly, core–shell structures can be classified by composition or morphology. As for the composition, core–shell structures can be comprised of inorganic materials (i.e. metal cores and metal or metal oxide shells, etc.), organic materials (i.e. polymers or other carbon-containing materials, such as carbon nanotubes, fullerenes, etc.), and inorganic–organic materials (i.e. metal nanoparticles covered by an organic polymer shell, etc.) [13]. Concerning the morphology of the core–shell structures, Pérez-Ramírez and Kawi used a general classification which divided the core–shell nanomaterials into core–shell, yolk–shell/hollow structures, and sandwiched core–shell structures. These different morphologies/structures are schematized in Fig. 21.1 [13].

Paria used a different classification of the core–shell nanostructures, dividing them into (a) concentric spherical core–shell nanoparticles; (b) different shaped core–shell nanoparticles; (c) multiple core nanoparticles formed by a single shell which is coated into various cores; (d) concentric nanoshells; and (e) moveable core particle within

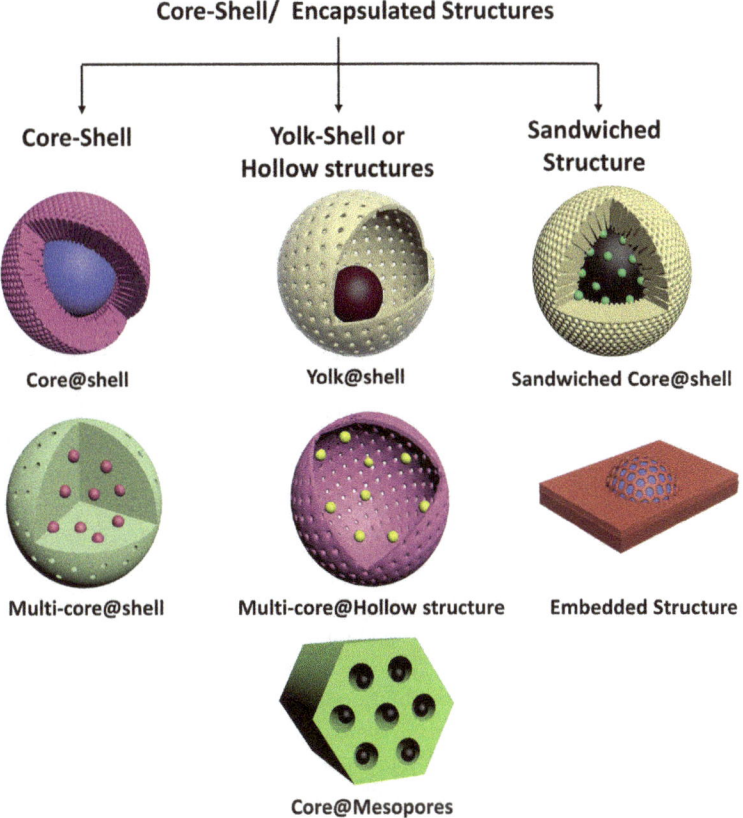

Fig. 21.1 Scheme of types of core–shell structures based on morphology. Reprinted with permission from Das et al. [13]. Copyright 2020, Royal Society of Chemistry

a uniformed hollow shell particle [19]. Varma et al. reported a more recent classification of core–shell nanoparticles, which was based on the properties of the shell. Attending to Varma´s classification, the following types of core–shell nanoparticles can be distinguished: (a) hollow-core–shell, (b) core-multishell, and (b) core-porous-shell [17]. There is a wide field of research dealing with core–shell nanostructures with different compositions and morphology, which have been documented in numerous studies [13, 18, 23, 59, 62]. Due to their importance, the present Chapter is addressed to review a selection of recent studies on core–shell-based catalysts and electrocatalysts comprised of Pd@oxide and Pd@Metal nanostructures.

21.2 Pd-Core Nanocatalysts Formed by Pd@oxide

Some of the most profusely investigated examples of Pd-core structures lie within the category of inorganic oxide shell-based nanomaterials, with SiO_2, TiO_2, and CeO_2 as the most investigated options. These structures attain enhanced performances which are often related to the stabilization of the Pd nanoparticle's active sites within the oxide shell or ligand effects resulting from the presence of additional functional groups in the oxide. Pd@oxide structures can be tuned in terms of both Pd characteristics (i.e. size, morphology, etc.) and oxide properties (i.e. porosity, shell thickness, etc.), which would ultimately affect the final catalytic performance for a certain application. Reactions of diverse nature (i.e. hydrogenations, oxidations, reductions, etc.) have been reported to be boosted by those Pd@oxide nanostructures, which usually attain better performance than those achieved by the oxide-supported counterparts.

In this line, Lee et al. evaluated the performance of a Pd core-porous silica shell catalysts (Pd@SiO_2) in the direct synthesis of H_2O_2 from hydrogen and oxygen [32]. That catalyst showed enhanced performance compared to the supported catalyst, which was related to the porosity of the shell, which allowed the reactant to reach the Pd cores, as well as the resistance of Pd cores against leaching under the acidic reaction conditions used. The same group further extended that investigation by assessing a more complex system comprised of Pd nanoparticles embedded within a shell oxide formed by SiO_2-Al_2O_3 (Pd@SiO_2–Al_2O_3), which displayed enhanced performance compared to that of Pd@SiO_2 [46]. Domínguez-Domínguez et al. [14] reported on the preparation of MCM-41 and Al-MCM-41 (with a SiO_2/Al_2O_3 ratio of 25) in presence of as-synthesized Pd nanoparticles. In that study, Pd nanoparticles were prepared using reduction by the solvent process (known as polyol method) with ethylene glycol acting as both solvent and reducing agent, and polyvinylpyrrolidone (PVP) as capping agent. The performance of those catalysts (denoted as Pd/MCM-41_s.s. and Pd/Al-MCM-41_s.s, respectively) in the semihydrogenation of phenylacetylene to styrene was compared to those of Pd/MCM-41 and Pd/Al-MCM-41, which were prepared by a traditional impregnation method using the same presynthesized Pd nanoparticles. Due to the synthetic steps, the average Pd nanoparticle size was much larger in Pd/Al-MCM-41_s.s and Pd/Al-MCM-41_s.s (11.25 and 12.86 nm, respectively) than in Pd/MCM-41 and Pd/Al-MCM-41 counterpart catalysts (2.81 and 3.10 nm, respectively), which turned out to be beneficial for the catalytic performance displayed in the application studied in that work. Yue et al. prepared Pd@mesoporous silica core–shell nanocatalysts with Pd nanocubes to be evaluated in the dry reforming of methane [60]. It was observed that the core–shell structure exhibited extraordinary thermal stability and coking-resistant ability, thus affording better performance than Pd cubes and Pd nanoparticles loaded onto SiO_2. It was also seen that the mesoporosity of the shell played a beneficial role by enhancing the contact between reactants and Pd cores.

Mitsudome et al. synthesized core–shell nanocomposites of Pd nanoparticles covered with a dimethylsulfoxide (DMSO)-like matrix on the surface of SiO_2 (Pd@MPSO/SiO_2) (MPSO $=$ methyl-3-trimethoxysilylpropylsulfoxide) [39]. In that case, the alkyl sulfoxide network contained in the shell served as a macroligand and allowed the selective access of alkynes to the Pd nanoparticles active sites, so that selective semihydrogenation of alkynes was achieved. Hu et al. prepared size-controlled Pd@SiO$_2$ core–shell nanocatalysts comprised of Pd cores with a nanoparticle size of 3.7 nm and mesoporous SiO$_2$ shells with a thickness from 10 to 30 nm [24]. Those nanostructures were achieved by using tetradecyl trimethyl ammonium bromide (TTAB) as the template for hydrolysis of the precursor of SiO$_2$ (TEOS). Since the pH of the synthesis solution affected the hydrolysis kinetics of TEOS, the structure of the Pd@SiO$_2$ nanocatalysts depended on the pH, and it was observed that increasing pH values led to the increasing thickness of the SiO$_2$ shells. The shell thickness was also depended on the ratio of TEOS/Pd used in the synthesis. The developed catalysts were used for the oxidation of CO at high temperatures, as well as for the hydrogenation of nitrobenzene. The results of the catalytic tests indicated that the sample with a shell thickness of 17.5 nm and pore size of 2.2 nm was very promising for both applications and exhibited good thermal stability, being suitable for long time use. Chen et al. reported a strategy to control the size of the Pd cores in Pd@SiO$_2$ catalysts, which was based on the utilization of different functional molecules, potassium bromide (KBr), cetyltrimethyl ammonium bromide (CTAB), and sodium citrate dihydrate (SC) [12]. The resulting catalysts had Pd size following the order $Pd_{CTAB} > Pd_{KBr} > Pd_{SC}$. It was also seen that the addition of such functional molecules had an important effect on the catalytic activity in the synthesis of H$_2$O$_2$ from hydrogen and oxygen, which was attributed to the different adsorption ability of the low coordinated Pd sites on the surface of the particles and the subsequent restrain side reactions.

Cai et al. prepared stable Pd@CeO$_2$ catalysts via a hydrothermal approach using H$_2$O$_2$ as an oxidant, citric acid as a complexing agent, and urea used to adjust the formation rate of hydrogen peroxide radicals [5]. The activity of those samples was assessed for the combustion of methane while analyzing the effect of the morphology of the catalysts and the chemical state of Pd. Highly dispersed and stabilized PdO species were detected within the CeO$_2$ shell. It was claimed that the porous CeO$_2$ shell favoured the full contact between reactants and PdO active sites, which preserved the catalytic activity for 50 h on-stream reaction at 500 °C. Wang et al. developed Pd@CeO$_2$ core–shell nanostructures with a tunable Pd core size, and shape, as well as with a controlled thickness of the CeO$_2$ sheath [53]. A biomolecule-assisted method was applied using L-arginine as a capping agent to promote the self-assembly of oxide nanosheets on Pd cores with various morphologies (i.e. Pd cubes@CeO$_2$, Pd octahedra@CeO$_2$, Pd cuboctahedron@CeO$_2$, and Pd nanowire@CeO$_2$). L-arginine had a crucial role in linking Pd and CeO$_2$ components, as well as serving as an adhesive to attain the assembly of the CeO$_2$ sheaths. Those nanostructures were supported on Al$_2$O$_3$ and the resulting heterogeneous catalysts were evaluated for the reduction of NO with CO so that the effect of the size and shape of the Pd core was analyzed. It was observed that Pd cube@CeO$_2$/Al$_2$O$_3$ with 6 nm Pd cubes displayed the best performance among prepared in that study.

Cargnello et al. designed catalysts based on Pd-core and CeO$_2$ shell homogeneously deposited on a modified hydrophobic Al$_2$O$_3$ (Pd@CeO$_2$/H-Al$_2$O$_3$) [9]. In that case, two active building blocks (i.e. Pd and CeO_2) were prepared separately and subsequently self-assembled and organized in solution to form supramolecular core–shell nanostructures held together by metal ion-ligand coordination interactions. After that, Al_2O_3 surface was modified to make it hydrophilic by reacting it with an organosilanetriethoxy(octyl)silane (TEOOS), which had greater adsorption of Pd@CeO_2 compared to the raw Al_2O_3. That catalytic system showed exceptionally high oxidation of methane, with complete conversion below 400 °C and excellent thermal stability under demanding conditions. Inspired by that study, Ali et al. designed catalysts formed by Pd@TiO_2 and Pd@CeO_2 core–shell supported over functionalized Al_2O_3 for their application in the oxidation of methane [2]. Better results were achieved for Pd@CeO_2/SiO_2.Al_2O_3 catalyst compared to the TiO_2-containing counterpart, reaching complete combustion of methane at around 400 °C, while ~550 °C was needed in the case Pd@TiO_2/SiO_2.Al_2O_3. The excellent performance of Pd@CeO_2/SiO_2.Al_2O_3 was attributed to the close Pd-CeO_2 contact and the efficient oxygen back-spillover at Pd and CeO_2 interface resulting from the core–shell structure.

Pi et al. compared the performance of Pd@SiO_2, Pd@CeO_2, and Pd@ZrO_2 supported on Si-modified Al_2O_3 towards the combustion of methane [42]. The enhancement observed for Pd@oxide/Si-Al_2O_3 as compared with Pd/Si-Al_2O_3 was attributed to the oxidation state of Pd species, and the PdO$_x$/Pd^0 mixed-phase was identified as the principal active phase for the combustion of methane. Moreover, the core–shell structure was claimed to be beneficial for oxidation activity. Among investigated, Pd@SiO_2 catalyst showed the highest activity in the presence of high concentrations of water vapour. Chen et al. checked the performance of Pd@ZrO_2/Si-Al_2O_3 for the oxidation of methane [10]. According to the observations of that study, Pd@ZrO_2-based catalysts displayed similar activity than Pd@CeO_2 counterparts in dry conditions, but Pd@ZrO_2 was more promising since it did not undergo deactivation in the presence of steam at high temperatures.

Zhang and Xu reported on the preparation of hollow core–shell nanocomposites formed by small Pd nanoparticle cores encapsulated within CeO_2 hollow shell (Pd@hCeO_2) [63]. A combination of sol–gel and hydrothermal processes was used for the preparation of the catalysts by following the experimental approach schematized in Fig. 21.2. Pd nanoparticles were loaded onto the surface of carbon spheres, which were previously synthesized from glucose. Ce^{3+} was loaded by interacting with the negatively charged carbon-Pd composites, and CeO_2 shell was subsequently formed after a hydrothermal treatment followed by a calcination step. The resulting Pd@hCeO_2 catalyst showed enhanced activity and great stability in the thermocatalytic and photocatalytic selective reduction of aromatic nitro compounds, which were related to its anti-aggregation and anti-leaching ability compared to the Pd/CeO_2 counterpart catalyst.

Adijanto et al. prepared Pd@CeO_2 core–shell structures (formed by 2 nm diameter Pd core surrounded by a 3 nm thick porous CeO_2 shell) grafted onto a yttria-stabilized zirconia (YSZ) support, which was functionalized with a deposited layer of triethoxy(octyl)silane to achieve a good dispersion of the Pd@CeO_2 on it [1]. It was observed that Pd cores showed outstanding stability against agglomeration.

Fig. 21.2 Schematic illustration for fabrication of Pd@hCeO$_2$ core–shell nanocomposite: (I) Synthesis of carbon sphere template; (II) Preparation of colloidal Pd nanoparticles; (III) Fabrication of carbon-Pd-Ce(III) nanocomposites via a hydrothermal treatment; (IV) Preparation of Pd@hCeO$_2$ core–shell nanocomposite with inner hollow space by a calcination process. Reprinted with permission from Zhang and Xu [63]. Copyright 2013, American Chemical Society Publications

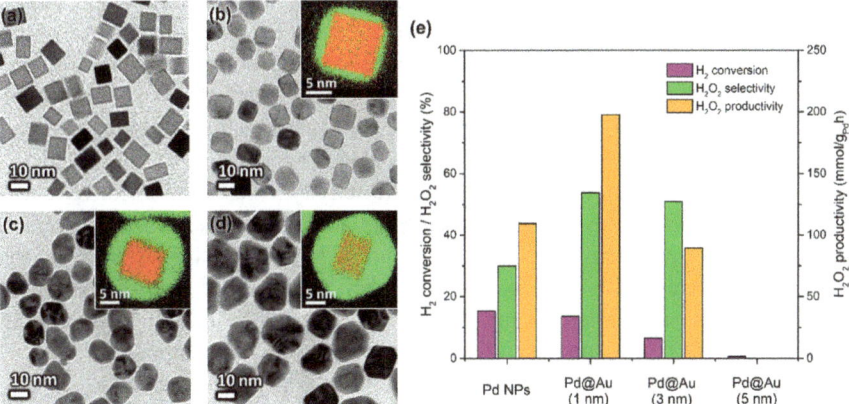

Fig. 21.3 Morphologies of Pd and Pd@Au NPs along with their catalytic properties. Bright-field TEM images of **a** Pd, **b** Pd@Au (1 nm), **c** Pd@Au (3 nm), and **d** Pd@Au (5 nm). The insets of **b**–**d** are high-resolution EDS images (green: Au, red: Pd) recorded from each core–shell nanoparticle. **e** Catalytic properties were evaluated from various NPs in terms of H$_2$ conversion, H$_2$O$_2$ selectivity, and H$_2$O$_2$ productivity. Reprinted with permission from Kim et al. [28]. Copyright 2019, American Chemical Society Publications

Furthermore, it was also claimed that the CeO$_2$ shells were more easily reduced than the bulk counterpart, which was related to the enhanced performance observed for Pd@CeO$_2$ core–shell structures as catalysts for the oxidation of CH$_4$. Li et al. checked the high-temperature hydrothermal stability of core–shell Pd@Ce$_{0.5}$Zr$_{0.5}$O$_2$/Al$_2$O$_3$ catalyst for its application in automotive three-way catalytic reactions (TWC) [33]. It was observed that Pd@Ce$_{0.5}$Zr$_{0.5}$O$_2$/Al$_2$O$_3$ system exhibited better performance and stability than those of Pd/Al$_2$O$_3$, Pd@CeO$_2$/Al$_2$O$_3$, and Pd@ZrO$_2$/Al$_2$O$_3$ catalysts. Pd species did not aggregate upon calcination at 1050 °C in the presence of 10% water, and Pd-core and Ce$_{0.5}$Zr$_{0.5}$O$_2$-shell interface in of Pd@Ce$_{0.5}$Zr$_{0.5}$O$_2$/Al$_2$O$_3$ catalyst had a key role in enhancing the catalytic performance for the TWC. The same research group evaluated the activity of moisture treated Pd@CeO$_2$/Al$_2$O$_3$ and

Pd/CeO$_2$/Al$_2$O$_3$ catalysts for the same application, confirming also the positive role of the core–shell structure [34].

The properties of Pd@oxide core–shell nanostructures can be also tuned by incorporating additional components which endow the final materials with interesting properties for the target application. For instance, Cargnello et al. designed a catalytic system formed by multiwalled carbon nanotubes (MWCNTs) embedded inside mesoporous layers of oxides (TiO$_2$, ZrO$_2$, or CeO$_2$), which contained dispersed metal nanoparticles (Pd or Pt) [8]. MWCNTs-free catalysts were also prepared as reference samples. The resulting materials were applied in different catalytic reactions based on the composition of the oxide: CeO$_2$-based catalyst was used for the water–gas shift reaction (WGS), TiO$_2$-based material was tested in the photocatalytic production of H$_2$, and ZrO$_2$-based catalyst was assessed in the Suzuki coupling reaction. The results of the catalytic tests revealed that MWCNTs played an important role in organizing the core–shell units forming accessible architectures. Additionally, it was also claimed that MWCNTs were also important for the photocatalytic application, in which they improved the lifetime of the electron–hole pairs by delocalizing the photogenerated electrons. Zhao et al. investigated the performance of core–shell structure formed by magnetic Fe$_3$O$_4$ microsphere loaded with Pd nanoparticles and coated by a porous anatase-TiO$_2$ shell (Fe$_3$O$_4$-Pd@TiO$_2$) [67]. It was observed that Fe$_3$O$_4$-Pd@TiO$_2$ displayed high photocatalytic activity towards the selective reduction of aromatic cyanides to aromatic primary amines, as well as excellent reusability and recovery of the catalyst from the reaction medium, which was attributed to the magnetic properties of Fe$_3$O$_4$. Also, the core–shell structure was responsible for the resistance of Pd nanoparticles against leaching during the reaction tests. Liu et al. prepared Pd-NiO@SiO$_2$ core–shell mesoporous nanocatalysts composed of Pd-NiO heteroaggregate nanoparticle cores of ~4 nm and mesoporous silica shells ~17 nm, and they were tested in the p-chloronitrobenzene hydrogenation with H2 [36]. That catalyst showed better performance than those of PdNi@SiO$_2$ and Pd@SiO$_2$, which was attributed to the role of the Pd-NiO interfaces in promoting the activity and selectivity of the materials.

Aside from the traditional Pd@oxide core–shell nanocatalysts described above, particles formed by Pd@PdO core–shell structures could be encompassed in this section. Examples of this are the studies of Jiang [27] and Guerrero-Ortega [20]. Jiang et al. reported the preparation of N-doped graphene-supported Pd@PdO core–shell clusters of 1–2 nm (Pd@PdO-NDG), which were assessed in the Suzuki–Miyaura reaction using phenylboronic acid and different substrates [27]. Experimental and theoretical calculations were combined in that study to check the binding strength between Pd or PdO clusters and N-doped graphene of N-free graphene. It was claimed that Pd or PdO clusters were physisorbed on graphene, while strong covalent chemical adsorption took place between PdO clusters and N-doped graphene. Such strong interaction was responsible for the control of the cluster size, with N atoms serving as local heterogeneous Pd nucleation sites. As for the catalytic results, it was observed that PdO layer located on the surface of the catalyst displayed high reactivity, which was related to the strong bonding interaction with the substrate molecules. Pd@PdO-NDG catalyst also showed a great stability, preserving the

Pd@PdO core–shell structure and its stability after five reaction runs. In the case of Guerrero-Ortega et al., Pd@PdO core–shell nanoparticles supported on Vulcan carbon XC-72R were prepared [20]. Those catalysts were synthesized from a one-pot synthesis using bis(dibenzylideneacetone) palladium (0) and subsequent heat treatment at 400 °C in air for 2 h to achieve the Pd@PdO structure. A PdO-free counterpart catalyst was also prepared for comparison by avoiding the heat-treatment step. The developed materials were assessed in the methanol electrooxidation in alkaline medium. It was observed that the heat-treated catalyst displayed better performance, which was attributed to the core–shell array and the interaction between Pd and PdO species, which promotes a major number of electrochemical active sites on the surface of the nanoparticles. It was also claimed that the core–shell structure had a positive effect in the redox process, showing good stability and resistance against poisoning by adsorption of species.

21.3 Pd-Core Nanocatalysts Formed by Pd@Metal

Core–shell nanostructures formed by Pd cores and different metal shells have received tremendous hope credited to their multiple possibilities and synergistic effect between the two metals arranged in the nanostructures. These core–shell bimetallic nanoparticles are one of the patterns in which bimetallic nanoalloys are divided into (i.e. core–shell alloys, sub-cluster segregated alloys, ordered and random homogeneous alloys, and multi-shell alloys) [45]. Such core–shell structures afford better control over the local atomic structure of the active sites than alloyed nanoparticles [54].

Examples of Pd@Metal nanostructures in which M can be either a precious metal or a non-precious metal are included in this section.

21.3.1 Pd@Precious Metal Core–Shell Catalysts

Precious-metal-based core–shell nanostructures have received significant attention in the past few decades because they are a platform for optimizing both catalytic performance and atomic utilization efficiency while retrenching the use of expensive noble metals. Important aspects, such as the coordination number, lattice strain, and electronic structure, which play a crucial role in controlling the catalytic performance of core–shell catalysts, depend to a great extent on the shell thickness. Broadly, noble-metal-based core–shell nanostructures can be classified into the following types: [18] (a) core–shell noble metal nanomaterials, which are formed by an ultrathin shell (2–6 atomic layers) which conformationally encapsulates an inner core of another metal; (b) Core-monolayer shell, formed by a nanocrystal core encapsulated with a monolayer of a different metal; and (c) Single-atom alloys, which are formed when isolated atoms of a metal are monodispersed on an array of a second metal.

One of the most widely reported compositions of Pd@Precious Metal is that formed by Pd@Au nanostructures with different morphologies, which have shown to display excellent performance in numerous catalytic and electrocatalytic applications.

Cai et al. reported on Pd@Au core–shell nanostructures with tunable Au thickness which were supported on TiO_2 and served as cocatalysts in the solar-driven photocatalytic reduction of CO_2 [6]. It was claimed that the core–shell configuration was optimum to induce the largest lattice strain, providing the maximal interfacial area for the polarization of the charge and maximizing the surface area for the reaction. It was observed that the catalysts with Au thickness of three atomic layers achieved a high average reaction rate and high selectivity for CO. Bathla et al. reported on the preparation of Pd-core Au shell nanocatalysts loaded to TiO_2, which were obtained by a galvanic replacement reaction, resulting in plasmonic Pd@Au-TiO_2 heterojunction [3]. Various Pd:Au weight ratios (i.e. 1:1, 1:2, and 1:3) were used, which resulted in different average particle sizes (i.e. 25 nm, 38 nm, and 57 nm, respectively), and the obtained photocatalysts were assessed in the cinnamaldehyde hydrogenation. Bimetallic core–shell structures displayed better performances than the monometallic catalysts. It was observed that the photocatalytic activity depended on the sell thickness (i.e. weight ratio of Au), being optimum for Pd_1@Au_2-TiO_2 with a Pd:Au weight ratio of 1:2. Upon irradiation of Pd@Au-TiO_2, the localized surface plasmon resonance (LSPR) of Au promotes the photoelectrons migration from the conduction band to the valence band of TiO_2. The carbonyl group of cinnamaldehyde is adsorbed selectively on the electron rich TiO_2 phase. The difference in electronegativity led to synergistic Pd-Au interaction and favoured the fragmentation of isopropyl alcohol to the alkoxide.

Raghavendra et al. fabricated Pd@Au nanoparticles on the surface of reduced graphene oxide (RGO) support by using a two-step protocol: (1) Synthesis of Pd_{core}-Ag_{shell} bimetallic nanoparticle on RGO by reducing the Pd^{2+} and Ag^+ with methyl ammonia borane; (2) Fabrication of Pd_{core}-Au_{shell} nanoparticles *via* galvanic replacement strategy involving sacrificial oxidation of metallic silver and reduction of gold ions [44]. The resulting materials were used as electrocatalysts for the oxygen reduction reaction (ORR), showing enhanced performance compared to a commercial Pt/C and reference Pd/RGO and Au/RGO catalysts, which was attributed to the optimized core–shell arrangement formed by small Pd core and thin Au shell, and the synergistic effects of both components.

Other studies evaluated the activity of unsupported Pd@Au nanostructures. This is the case of Zhao et al. [65] who prepared Pd@Au core–shell nanotetrapods and checked their activity towards the hydrogenation reduction of nitro functional groups, or Zhang et al. [61] who synthesized Pd@Au core–shell nanocrystals with concave cubic shapes, which were checked in the cathodic electrochemiluminescence reaction of luminol and H_2O_2. Kim et al. recently reported an interesting study on the use of strained ultrathin Au shell layers on Pd nanoparticles for the direct synthesis of H_2O_2 [28]. To carry out that research, Pd@Au nanostructures with different shell thicknesses were prepared to analyze the microstructural features of the Au shells and measure the lattice strains as a function of the thickness and crystallographic orientation of the shell. Moreover, Density-functional theory (DFT) calculations were used to determine the electronic structures of the shell to ascertain how both the effect of lattice strain and charge transfer made the Au shell catalytically active for the synthesis of H_2O_2. The nanostructures were comprised of Pd cubes of side lengths of ~10 nm and Au shell with a thickness of 1, 3, and 5 nm. The performance of the catalysts was evaluated in terms of conversion of H_2, H_2O_2 selectivity, and H_2O_2 productivity via the direct synthesis of H_2O_2. Figure 21.3 contains representative micrographs of the nanostructures together with their catalytic performance. It was observed that Pd@Au (1 nm) and Pd@Au (3 nm) catalysts, with strained Au layers, exhibited catalytic properties for the conversion of H_2 during the direct synthesis of H_2O_2, while bulk Au was not active. Besides, Pd@Au (1 nm) had better H_2O_2 selectivity than monometallic Pd catalyst, indicating that O_2 dissociation was hampered by the Au layer so that Pd@Au catalysts overcome the low selectivity linked to the dissociation of O_2.

Hu et al. succeeded in the development of Pd@Au catalysts formed by Au surface-modified Pd concave nanostructures for their application in the selective styrene oxidation with molecular oxygen [22]. It was claimed that the Pd crystals surface were the active sites for the activation of O_2 molecules, while the alloyed Pd-Au sites modified the activation of O_2 so that the activity and selectivity towards the oxidation of styrene could be controlled by tailoring the atomic arrangement of Pd and Au in the nanostructures. Catalysts with various compositions (i.e. Pd@Au$_{0.06}$, Pd@Au$_{0.08}$, Pd@Au$_{0.12}$, and Pd@Au$_{0.16}$) were prepared by galvanic replacement between Pd nanostructures and $HAuCl_4$, and using different concentrations of Au precursor. Nanocatalyst with a composition of Pd@Au$_{0.08}$ achieved a selectivity of 87% towards the main products (i.e. styrene oxide and benzaldehyde) and a conversion of 44% in the aerobic oxidation of styrene.

Luo et al. investigated the electrocatalytic performance of PdAu catalysts with variable atomic ratio (i.e. Pd/Au ratio of 2/1, 1/1, 1/2, and) for the oxidation of methanol [38]. It was observed that the morphology of the particles depended on their composition; Pd$_2$Au with a nanoflower core–shell structure, PdAu with nanochain and network structure, and PdAu^2 with solid and hollow nanosphere shape were synthesized. Nanoflower-like core–shell catalysts exhibited the best performance due to the high electrochemical active surface area (ECSA) and the presence of more active sites. Wang et al. also prepared Pd-based core–shell electrocatalysts for the oxidation of methanol [51]. They reported on the preparation of Pd@Au and Pd@Pt nanostructures, which were synthesized from (AcO)$_2$-Pd-(DMSO)$_2$ micro rods with well-defined morphology prepared with dimethyl sulfoxide (DMSO) as the organic ligand, and performing a subsequent heat treatment combined with galvanic replacement. The resulting materials showed better performance for the oxidation of methanol in alkaline media than a commercial Pd/C catalyst.

Pd@Pt nanostructures have also been explored for the oxidation of methanol. For instance, Kim et al. reported on star-shaped Pd@Pt core–shell catalysts supported RGO (Pd@Pt/RGO) [29]. That catalyst, as well as reference materials (i.e. PdPt/RGO, Pt/RGO, and Pt/C), were evaluated in the oxidation of methanol in an alkaline solution, achieving better performance, in terms of onset potential, current density, stability, and the charge transfer rate for Pd@Pt/RGO catalysts. Xu et al. also checked the activity of Pd@Pt core–shell structures in the oxidation of methanol, using, in this case, hydrangea-like core–shell structures supported on graphene [58]. Those catalysts exhibited better activity, stability, and tolerance of CO than the alloy PdPt counterpart catalysts and a commercial Pt/C sample.

Aside from that reaction, other oxidation reactions are catalyzed by Pd@Pt nanostructures. In this line, Xiao et al. checked the activity of carbon-supported Pd@Pt cubes with low Pt content for the oxidation of formic acid [56]. To achieve such core–shell structures, a direct seed-mediated growth was used by doing continuous injections of the Pt precursor to the solution which contained the Pd cores. Catalysts with various Pt contents (i.e. K_2PtCl_4 concentration used in the synthesis from 0.2 to 20 at.%) led to nanoparticles with very similar sizes (10–11 nm) and different morphologies (from cubic to concave cubic) originated from the anchoring of Pt to the corners of the cubes. It was observed that the Pd@Pt catalyst with a very low Pt content of 0.4 at.% displayed much higher activity than a commercial Pt/C catalyst.

Pd@Pt nanocatalysts with various morphologies have received interest for their application in various reactions. For example, Wang et al. prepared octahedral, rhombic dodecahedral, and cubic Pd@Pt nanoparticles for the oxidation of glucose. [52] Li et al. obtained Pd@Pt nanodots and reported their activity towards the ORR, [35] and Zhao et al. investigated the performance of Pd@Pt tetrapods in the ORR [66]. Some other interesting studies on the application of Pd@Pt catalysts for the ORR are also found in the recent literature. This is the case of the study addressed by Zheng et al., who prepared Pd@Pt/C catalysts by using a sequential sonochemical synthesis and short synthesis times, and the developed catalysts were evaluated in the ORR , showing promising activities [68].

Interesting works to get insight into the formation of Pd@Pt nanostructures have also been reported. For instance, Qi et al. designed a strategy to control the surface evolution of Pd@Pt core–shell structures by adjusting the pH of the synthesis solution to control the reducing ability of the reducing agent (i.e. ascorbic acid) and finally manipulating the supersaturation in the reaction system [43]. It was observed that by increasing the pH value, the surface structure of Pd@Pt nanocrystals changed from a Pt {111} facet-exposed island shell to a conformal Pt {100} facet-exposed shell. That study contains TEM imaging of the reaction-time-dependent morphology evolution to get insight into the growth mechanism of the Pd@Pt core–shell nanostructures. They fixed the initial reaction time when the ascorbic acid was added to the solution and monitored the changes in the morphology of the particles in two different conditions (i.e. low and high saturation). It was observed that for a low saturated solution, 30 min were needed for the deposition of small Pt nanoislands on the surface of Pd particles, and the Pt nanoislands were completed at 120 min. However, in the case of high saturated solutions, the Pd@Pt core–shell structure was formed after 5 min and the reaction was finished within 10 min. A detailed analysis by High Angle Annular Dark Field-Scanning Transmission Electron Microscopy (HAADF-STEM) and element mapping was performed to check the resulting surface structure of those Pd@Pt nanostructures in both low and high saturated conditions (See Fig. 21.4). It

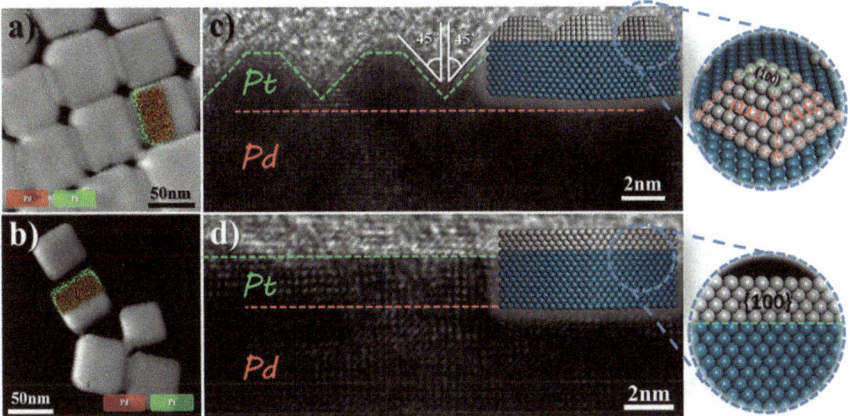

Fig. 21.4 STEM, elemental mapping, and HRTEM images of Pd@Pt core-island shell NCs (**a**, **c**) and Pd@Pt core-conformal shell nanocatalysts (**b**, **d**). Reprinted with permission from Qi et al. [43]. Copyright 2016, Royal Society of Chemistry

was observed that the Pt island achieved at low saturation had a truncated pyramidal shape, which exposed minor {100} facets at the upmost surface and mainly {111} facets on the four sides of the islands (Fig. 21.4c). However, Pd@Pt core-conformal shell NCs with Pt {100} facets exposed were achieved for a higher supersaturation (Fig. 21.4c). Both types of core–shell structures (i.e. Pd@Pt core-island shell and Pd@Pt core-conformal shell) were assessed in ORR and the electrocatalytic results indicated that Pd@Pt core-island shell nanocatalyst displayed better activity and favourable long-term stability toward ORR in alkaline media, which was attributed to both the enlargement in the proportion of the ORR active Pt {111}facets on the surface of the catalyst and the synergistic effect between Pd and Pt by the core–shell arrangement.

Zhou et al. performed theoretical calculations with DFT to understand the mechanism involved in the formation of the core–shell Pd@Pt structure, which is summarized in Fig. 21.5 [69]. According to the study, it was suggested that the as-synthesized Pd@Pt$_{1L}$ (1L: one atomic layer) core–shell nanoparticles would have a slightly mixed composition on the surface and that the seed terraces would be covered with a Pt overlayer in later stages of the synthesis.

Even though Pd@Au and Pd@Pt have possibly received the greatest interest among Pd@Precious-metal catalysts, some other compositions of the nanostructures have also sparked attention in the scientific community. Among them, Pd@Ag catalysts have merited numerous studies for their multiple applications.

One of the studies that deserve mention is that of Sun et al., who reported an extraordinary study in which catalysts based on highly dispersed and surfactant-free Pd@Ag particles were developed within the cell of Spirulina platensis (Sp.) [49]. The synthetic protocol used encompassed various steps: (1) Treatment of Sp. Cells with HCl to enhance their permeability; (2) Penetration of Pd in the cell

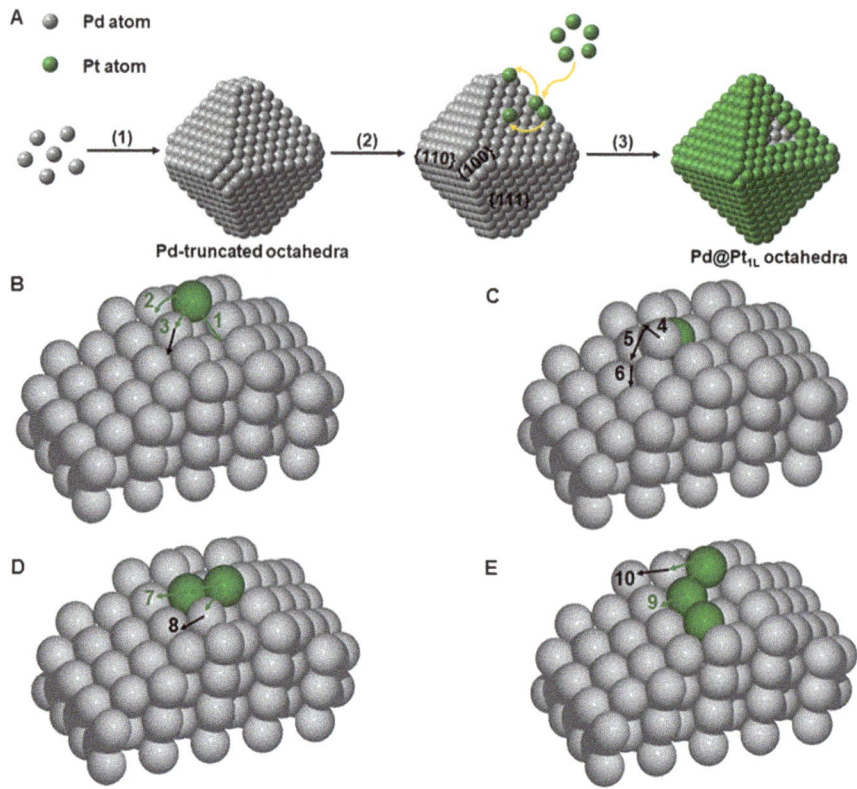

Fig. 21.5 **a** Schematic illustration of the growth mechanism of Pd@Pt$_{1L}$ octahedral nanocrystals: (1) nucleation and growth of Pd octahedra with truncation at the corners; (2) selective deposition of Pt atoms on the edges of the Pd truncated octahedra; (3) surface diffusion of the deposited Pt atoms to generate Pd@Pt$_{1L}$ octahedra with sharp corners. **b–e** DFT calculated pathways for Pt and Pd diffusion across the edge model: green and black arrows represent diffusion (via hopping or substitution) of a Pt and a Pd atom, respectively. Reprinted with permission from Zhou et al. [69]. Copyright 2019, American Chemical Society Publications. Copyright 2016, Royal Society of Chemistry

envelope, which served as catalytic seeds for the deposition of Ag; and (3) Electroless silver deposition. Various catalysts were prepared using different feeding times of Ag precursor (silver ammonia solution) to 1 g of activated Sp. cells (2.5, 5, and 7.5 to 15 mL, corresponding to (Pd@Ag)@Sp-2.5, (Pd@Ag)@Sp-5, (Pd@Ag)@Sp-7.5, and (Pd@Ag)@Sp-15 catalysts, respectively). The applicability of the resulting materials was evaluated in the reduction of 4-nitrophenol with NaBH$_4$, being (Pd@Ag)@Sp-5 the most promising catalyst among investigated in that study. Its performance was compared to that achieved by some reference materials and it was claimed that its superiority was due to several reasons, such as the synergistic effect of the Pd core and the Ag layer of preferable thickness, and the homogeneous cross-linked structure of cell matrix which ensured the good dispersion of small

Pd@Ag nanoparticles and protected them from aggregation during the reaction, among others.

An example of Pd@Ag catalysts used in hydrogenation reactions is also that reported by Mitsudome et al., who used core-Pd/shell-Ag nanocomposite with various Ag/Pd ratios (i.e. 0.10, 0.15, 0.20, 0.25, 0.50) for the selective semihydrogenation of alkynes [40]. The good activity and selectivity achieved by that catalysts were ascribed to the complementary Pd-Ag relationship in the core–shell structure. It was claimed that the Pd core improved the activity of Ag shell by acting as a hydrogen source, while the Ag shell improved the selectivity of the final catalytic system. Han et al. prepared TiO_2-supported Pd@Ag catalysts for the selective hydrogenation of acetylene in excess ethylene [21]. In that study, a sequential photodeposition method, with methanol as the electron donor, was used to synthesize catalysts with various Pd and Ag contents. Additionally, a reference sample was prepared by a conventional impregnation method. CO chemisorption experiments were used to confirm that, for the same composition of the particles, a more effective covering of the Pd sites was achieved by using the photodeposition method, which led to the blocking of the high coordination sites of the Pd cores and results in catalysts with improved ethylene selectivity for the catalytic acetylene hydrogenation.

21.3.2 Pd@Non-precious Metal Core–Shell Catalysts

Despite the great performance achieved by precious-metal-based catalysts, it is well-known that their main limitation is the scarcity of those metals so that introducing a different transition metal in the core–shell nanoparticles is highly desirable. From the economic viewpoint, the synthesis of core–shell nanocatalysts having a noble metal core and inexpensive metal shell is very interesting because it is an alternative to reduce the final use of noble metals, as well as to avoid their loss via leaching, and retaining, or even enhancing, the catalytic activity [17]. The synergistic effect in the core–shell depends on the metals and the way they are combined so that it is crucial to select two metals meticulously to obtain the desired properties of the final system. Due to the potential of PdCo and PdNi bimetallic nanostructures in catalysis, Co and Ni-based catalysts are among the most fruitfully investigated Pd@Non-Precious Metal core–shell catalysts and they are the focus of the present section.

As it is well-known, controlling the oxidation state of those transition metals under certain reaction conditions can be somehow challenging. However, it has been observed that the engineering of bimetallic Pd@Non-Precious Metal core–shell structures can be a promising approach to stabilize certain species. In this sense, Gopinath et al. recently reported an interesting study on the evolution of the electronic structure and morphology of Pd@Co nanocatalysts with different compositions (i.e. Pd:Co ratio of 2:1, 1:1, and 1:2) under oxidation and reduction conditions [25]. For that, the surface electronic, chemical, and structural changes during annealing under reducing and the oxidizing atmosphere was analyzed by

X-ray photoelectron spectroscopy (NAPXPS), while the morphological modifications were checked by high-resolution transmission electron microscopy (HRTEM). The results achieved indicate that the surface electronic structure of the catalysts could be tailored by performing variable-temperature reduction and oxidation treatments. Under reducing atmosphere, hydroxide and oxide species transformed into metallic cobalt at a temperature above 127 °C. However, oxidizing annealing did not modify the metallic percentage, but CoO and Co(OH)$_x$ transformed into spinel Co$_3$O$_4$ species. The control of the oxidation state of cobalt species was attributed to the role of Pd-Co interface in stabilizing the desirable oxidation state of cobalt. As for the morphology, it was demonstrated that the catalysts were thermally stable at least up to 300 °C and neither changes in size nor shape were observed. The same research group also reported on the preparation of Pd@Co nanocatalysts with tuned core and shell thickness [26]. The Co-shell thickness had an effect on the species or Co present on it (i.e. higher thickness led to the formation of oxidation/hydroxylation because of the vanishing influence of Pd-Co interface at the surface. In that study, they attempted at preparing dual-functional catalysts which were able to catalyze reactions based on selective alkyne hydrogenation to alkene and olefin oxidation to epoxide. It was observed that the catalyst with Pd/Co ratio of 1/1 was optimum among investigated.

Interesting studies in which classic supports have been used in the preparation of Pd@Co core–shell-based catalysts can be found in the literature. Xie et al. prepared Pd@Co catalysts with various compositions of the nanoparticles (i.e. Co/Pd molar ratio) = 2.4–13.6) and an average size of 3.5–4.5 nm, which were supported on three-dimensionally ordered macroporous CeO$_2$ (3DOM CeO$_2$), and applied in the oxidation of methane [57]. Among investigated, that catalyst with a Co/Pd ratio of 3.5 exhibited the best performance, which was related to its good oxygen and methane adsorption ability and the unique core–shell arrangement. Wang et al. reported on the synthesis of graphene-supported Pd@Co core–shell nanoparticles as efficient catalysts for hydrolytic dehydrogenation of ammonia borane [50]. The catalysts were *in-situ* synthesized by adding the reducing agent (ammonia borane) into a solution containing the metal precursors (i.e. Na$_2$PdCl$_4$, Co(NO$_3$)$_2$) and graphene oxide. The performance of that Pd@Co/graphene catalyst was compared to those of alloyed PdCo/graphene and graphene-free Pd@Co counterpart. core–shell catalyst displayed the best performance among those studied in that work, completing the hydrolytic dehydrogenation of ammonia borane in 3.5 min. That catalyst also displayed good stability and magnetically recyclability, which made it a promising candidate for the studied reaction. Chen et al. also studied the performance of Pd@Co core–shell nanocatalysts in the hydrolytic dehydrogenation of ammonia borane [11]. In that case, the bimetallic nanoparticles were stabilized in the mesopores of a Cr(III)-based MIL-101 Metal-Organic Framework (MOF). For that, the metal precursors were pre-incorporated in the MOF and they were in-situ reduced during the catalytic reaction, leading to the formation of Pd@Co nanoparticles with an average size of 2.5 nm which were mostly embedded within the MOF structure thanks to the diffusion of the metal precursors into the cavities of MIL-101 by a double solvent approach (DSA). Different structures (i.e. Pd@Co@MIL-101 with embedded Pd@Co nanoparticles,

Pd@Co/MIL-101 with Pd@Co nanoparticles deposited on the external surface of MOF, and PdCo@MIL-101 with PdCo alloy nanoparticles inside MOF pores) were achieved by changing the experimental conditions in terms of the reducing agent (i.e. $NaBH_4$ as a strong reducing agent or NH_3BH_3·as a mild reducing agent able to reduce only Pd precursor), and presence of a pre-incorporation step. The synthetic approach for the development of such nanocatalysts is schematized in Fig. 21.6. The catalytic test demonstrated the superiority of Pd@Co@MIL-101, especially in terms of stability.

Feng et al. reported on the preparation of TiO_2-supported Pd@Ni nanocatalysts for the synthesis of H_2O_2 [16]. It was observed that Pd@Ni nanocatalysts were not active towards the synthesis of H_2O_2, which was attributed to the blockage of Pd cores by Ni shell. In the light of that, Pd@NiO-x nanoparticles with porous NiO shell (x = 1, 2, 3, and 4) were prepared by direct thermal annealing of Pd@Ni-x nanoparticles with controlled annealing temperature and time. It was observed that the heat-treatment temperature affected the final structure of the catalysts, achieving "solid" and "void" structures at 250 and 300 °C, respectively. Those catalysts were denoted as Pd@NiO-3/TiO_2 and Pd@void@NiO-3/TiO_2, respectively, and displayed different activity towards the synthesis of H_2O_2, with Pd@NiO-3/TiO_2 exhibiting better performance due to the contact between Pd-core and NiO shell. The H_2O_2 productivity and selectivity were strongly depended on the NiO shell thickness. That study demonstrated that the catalytic performance of Pd@NiO-x towards the production of H_2O_2 could be controlled by modifying the composition of the catalysts. Among investigated, Pd@NiO-3/TiO_2 exhibiting the highest productivity and selectivity to H_2O_2. Pal et al. also reported on mesoporous TiO_2-supported Pd@Ni catalysts [4]. In that case, their performance in the photocatalytic hydrogenation of carbonyl compounds was

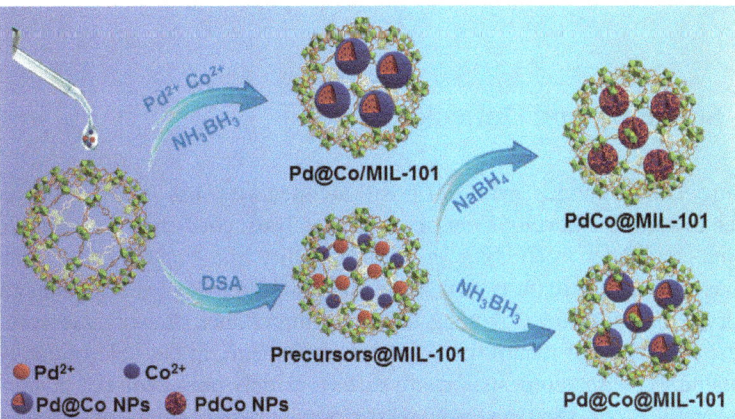

Fig. 21.6 Synthesis of Pd@Co@MIL-101, Pd@Co/MIL-101, and PdCo@MIL-101 catalysts by different procedures and reducing agents. Reprinted with permission from Chen et al. [11]. Copyright 2015, Wiley–VCH

explored and compared with that achieved by monometallic Pd and Ni samples. Catalysts with various compositions (i.e. $Pd_1@Ni_1$-TiO_2, $Pd_2@Ni_1$-TiO_2, and $Pd_3@Ni_1$-TiO_2) were prepared, and $Pd_3@Ni_1$-TiO_2 showed the most promising results among studied.

Aside from the composition of the nanostructures, in terms of Pd/metal molar ratio, the nanoarchitecture of the two metals is an important aspect to be considered. Shviro et al. developed PdNi alloy, and either Pd@Ni or Ni@Pd nanostructures by using different metal precursors [47]. They explored the use of both organometallic and metal-acetylacetonate complexes to get one-pot reactions of either alloyed or core–shell particles, depending on the kinetics of decomposition/reduction of the metal precursors. It was observed that the use of Palladium (II) acetylacetonate ($Pd(acac)_2$) and Nickel (II).

acetylacetonate ($Ni(acac)_2$) led to the formation of Ni surface-rich Pd@Ni nanostructures, while organometallic complex bis-(cyclooctadiene) nickel (0), $Ni(COD)_2$, and bis-(dibenzylideneacetone)-palladium (0), $Pd(dba)_2$, afforded the preparation of PdNi nanoalloys, and a mixed option, with $Ni(COD)_2$ and $Pd(acac)_2$ as metal precursors, formed Ni@Pd core–shell nanoparticles. It was found in that case that PdNi alloy resulted in better electrocatalytic performance in the hydrogen oxidation reaction (HOR) alkaline medium. A more complex system was prepared by Sneed et al. who developed shaped Pd-Ni-Pt core-sandwich-shell nanocrystals and investigated the influence of Ni sandwich layers on catalytic electrooxidation of formic acid and methanol [48]. In that case, Pd cubes substrated with two sizes (i.e. 10 and 30 nm) were prepared and subsequently mixed with Ni and Pt precursors in consecutive steps so that cubic Pd-Ni-Pt core-sandwich-shell nanostructures were formed. It was claimed that the combination of shaped Pd substrates and the mild reduction conditions used directed the overgrowth of Ni and Pt in an oriented layer-by-layer fashion.

21.4 Summary and Outlook

Designing nanostructures with different architectures has been a burgeoning approach towards the development of unique catalysts. core–shell nanocatalysts are promising candidates to combine multifunctionality in a single system. There are numerous core–shell structures, which are classified attending to either their composition or their arrangement. Among them, the importance of inorganic core–shell catalysts, comprised of metal@oxide and metal@metal, in the field of catalysis is indisputable. Pd-based core–shell catalysts have widely been explored in numerous catalytic applications, showing excellent performances in reactions of diverse nature (i.e. thermal catalysis, electrocatalysis, and photocatalysis). Pd@oxide-based catalysts usually show strong interactions between the Pd-core and the oxide shell which lead to improved catalytic performances. Such Pd@oxide-based catalysts are also an interesting alternative for those reactions in which harsh reaction conditions are required since the core–shell arrangement affords high stability against sintering and/or leaching of the metal active phase. Furthermore, additional functionalization and hence tailored catalytic properties can be also achieved by altering the oxide shell properties, which provide countless synthetic alternatives. In the case of Pd@metal catalysts, they have shown to exhibit superior performance than counterpart alloyed nanocatalysts, which has frequently been linked to the Pd-core and metal shell interaction in the interface of the core–shell structure. It has also been reported that such interaction can be also responsible for the control of the oxidation state of the metal which forms the shell, thus being a tool for tailoring the activity of the final catalytic system. core–shell nanocatalysts are, without a shadow of a doubt, one of the nanoarchitecture with the greatest potential; however, there is still room for improvement in the surface engineering of core–shell catalysts at the atomic level, which is crucial to tailor their catalytic performances.

Acknowledgements The authors thank Ministerio de Ciencia˙ Innovación y Universidades and FEDER (Project RTI2018-095291-B-I00) and the Generalitat Valenciana (PROMETEOII/2018/076) for financial support. MNG gratefully acknowledges the Plan GenT project (CDEIGENT/2018/027) for the research founding. DST thanks MICINN for the "Juan de la Cierva" contract (IJCI-2016-27636).

References

1. Adijanto L, Bennett DA, Chen C et al (2013) Exceptional thermal stability of Pd@CeO$_2$ core–shell catalyst nanostructures grafted onto an oxide surface. Nano Lett 13:2252–2257
2. Ali S, Al-Marri MJ, Al-Jaber AS et al (2018) Synthesis, characterization and performance of Pd-based core–shell methane oxidation nano-catalysts. J Nat Gas Sci Eng 55:625–633
3. Bathla A, Pal B (2020) Superior co-catalytic activity of Pd(core)@Au(shell) nanocatalyst imparted to TiO$_2$ for the selective hydrogenation under solar radiations. Sol Energy 205:292–301
4. Bathla A, Pal B (2018) Bimetallic Pd@Ni-mesoporous TiO$_2$ nanocatalyst for highly improved and selective hydrogenation of carbonyl compounds under UV light radiation. J Ind Eng Chem 67:486–496
5. Cai G, Luo W, Xiao Y et al (2018) Synthesis of a highly stable Pd@CeO$_2$ catalyst for methane combustion with the synergistic effect of urea and citric acid. ACS Omega 3:16769–16776
6. Cai X, Wang F, Wang R et al (2020) Synergism of surface strain and interfacial polarization on Pd@Au core–shell cocatalysts for highly efficient photocatalytic CO$_2$ reduction over TiO$_2$. J Mater Chem A 8:7350–7359
7. Campelo JM, Luna D, Luque R et al (2009) Sustainable preparation of supported metal nanoparticles and their applications in catalysis. Chemsuschem 2:18–45
8. Cargnello M, Grzelczak M, Rodríguez-González B, et al (2012) Multiwalled carbon nanotubes drive the activity of Metal@oxide core–shell catalysts in modular nanocomposites. J Am Chem Soc 134:11760–11766
9. Cargnello M, Jaén JJD, Garrido JCH et al (2012) Exceptional activity for methane combustion over modular Pd@CeO$_2$ subunits on functionalized Al$_2$O$_3$. Science (80-)337:713–717
10. Chen C, Yeh Y-H, Cargnello M et al (2014) Methane oxidation on Pd@ZrO$_2$/Si-Al$_2$O$_3$ is enhanced by surface reduction of ZrO$_2$. ACS Catal 4:3902–3909
11. Chen Y-Z, Xu Q, Yu S-H, Jiang H-L (2015) Tiny Pd@Co core–shell nanoparticles confined inside a metal-organic framework for highly efficient catalysis. Small 11:71–76

12. Chen Z, Pan H, Lin Q et al (2017) The modification of Pd core–silica shell catalysts by functional molecules (KBr, CTAB, SC) and their application to the direct synthesis of hydrogen peroxide from hydrogen and oxygen. Catal Sci Technol 7:1415–1422

13. Das S, Pérez-Ramírez J, Gong J et al (2020) Core–shell structured catalysts for thermocatalytic, photocatalytic, and electrocatalytic conversion of CO_2. Chem Soc Rev 49:2937–3004

14. Domínguez-Domínguez S, Berenguer-Murcia Á, Linares-Solano Á, Cazorla-Amorós D (2008) Inorganic materials as supports for palladium nanoparticles: application in the semi-hydrogenation of phenylacetylene. J Catal 257:87–95

15. Dumesic JA, Huber GW, Boudart M (2008) Principles of heterogeneous catalysis. Handb Heterog Catal

16. Feng Y, Shao Q, Huang B et al (2018) Surface engineering at the interface of core/shell nanoparticles promotes hydrogen peroxide generation. Natl Sci Rev 5:895–906

17. Gawande MB, Goswami A, Asefa T et al (2015) Core–shell nanoparticles: synthesis and applications in catalysis and electrocatalysis. Chem Soc Rev 44:7540–7590

18. Ge J, Li Z, Hong X, Li Y (2019) Surface atomic regulation of core–shell noble metal catalysts. Chem A Eur J 25:5113–5127

19. Ghosh Chaudhuri R, Paria S (2012) Core/shell nanoparticles: classes, properties, synthesis mechanisms, characterization, and applications. Chem Rev 112:2373–2433

20. Guerrero-Ortega LPA, Ramírez-Meneses E, Cabrera-Sierra R et al (2019) Pd and Pd@PdO core–shell nanoparticles supported on Vulcan carbon XC-72R: comparison of electroactivity for methanol electro-oxidation reaction. J Mater Sci 54:13694–13714

21. Han Y, Peng D, Xu Z et al (2013) TiO_2 supported Pd@Ag as highly selective catalysts for hydrogenation of acetylene in excess ethylene. Chem Commun 49:8350–8352

22. Hu C, Xia X, Jin J et al (2018) Surface modification on Pd nanostructures for selective styrene oxidation with molecular oxygen. ChemNanoMat 4:467–471

23. Hu P, Morabito JV, Tsung C-K (2014) Core–shell catalysts of metal nanoparticle core and metal-organic framework shell. ACS Catal 4:4409–4419

24. Hu Y, Tao K, Wu C et al (2013) Size-controlled synthesis of highly stable and active Pd@SiO_2 core–shell nanocatalysts for hydrogenation of nitrobenzene. J Phys Chem C 117:8974–8982

25. Jain R, Gopinath CS (2020) Electronic structure evolution of Pd@Co nanocatalysts under oxidation and reduction conditions and preferential CO oxidation. ChemCatChem n/a

26. Jain R, Gopinath CS (2018) New strategy toward a dual functional nanocatalyst at ambient conditions: influence of the Pd-Co interface in the catalytic activity of Pd@Co core–shell nanoparticles. ACS Appl Mater Interfaces 10:41268–41278

27. Jiang B, Song S, Wang J et al (2014) Nitrogen-doped graphene supported Pd@PdO core–shell clusters for C-C coupling reactions. Nano Res 7:1280–1290

28. Kim J-S, Kim H-K, Kim S-H et al (2019) Catalytically active Au layers grown on Pd nanoparticles for direct synthesis of H_2O_2: lattice strain and charge-transfer perspective analyses. ACS Nano 13:4761–4770

29. Kim Y, Noh Y, Lim EJ et al (2014) Star-shaped Pd@Pt core–shell catalysts supported on reduced graphene oxide with superior electrocatalytic performance. J Mater Chem A 2:6976–6986

30. Kung HH, Kung MC (2007) Nanotechnology and heterogeneous catalysis BT—nanotechnology in catalysis: volume 3. In: Zhou B, Han S, Raja R, Somorjai GA (eds) Springer. New York, NY, New York, pp 1–11

31. Kyle RA, Steensma DP (2018) Jöns Jacob Berzelius—a father of chemistry. Mayo Clin Proc 93:e53–e54

32. Lee H, Kim S, Lee D-W, Lee K-Y (2011) Direct synthesis of hydrogen peroxide from hydrogen and oxygen over a Pd core-silica shell catalyst. Catal Commun 12:968–971

33. Li L, Zhang N, Huang X et al (2018) Hydrothermal stability of core–shell Pd@$Ce_{0.5}Zr_{0.5}O_2$/Al_2O_3 catalyst for automobile three-way reaction. ACS Catal 8:3222–3231

34. Li L, Zhang N, Wu R et al (2020) Comparative study of moisture-treated Pd@CeO_2/Al_2O_3 and Pd/CeO_2/Al_2O_3 catalysts for automobile exhaust emission reactions: effect of core–shell interface. ACS Appl Mater Interfaces 12:10350–10358

35. Li S, Liu J, Zhu G, Han H (2019) Pd@Pt core–shell nanodots arrays for efficient electrocatalytic oxygen reduction. ACS Appl Nano Mater 2:3695–3700
36. Liu H, Tao K, Xiong C, Zhou S (2015) Controlled synthesis of Pd-NiO@SiO$_2$ mesoporous core–shell nanoparticles and their enhanced catalytic performance for p-chloronitrobenzene hydrogenation with H$_2$. Catal Sci Technol 5:405–414
37. Liu L, Corma A (2018) Metal catalysts for heterogeneous catalysis: from single atoms to nanoclusters and nanoparticles. Chem Rev 118:4981–5079
38. Luo L-M, Zhang R-H, Chen D et al (2018) Hydrothermal synthesis of PdAu nanocatalysts with variable atom ratio for methanol oxidation. Electrochim Acta 259:284–292
39. Mitsudome T, Takahashi Y, Ichikawa S et al (2013) Metal-ligand core–shell nanocomposite catalysts for the selective semihydrogenation of alkynes. Angew Chemie Int Ed 52:1481–1485
40. Mitsudome T, Urayama T, Yamazaki K et al (2016) Design of core-Pd/Shell-Ag nanocomposite catalyst for selective semihydrogenation of alkynes. ACS Catal 6:666–670
41. Navlani-García M, Salinas-Torres D, Mori K et al (2019) Tailoring the size and shape of colloidal noble metal nanocrystals as a valuable tool in catalysis. Catal Surv from Asia 23:127–148
42. Pi D, Li WZ, Lin QZ et al (2016) Highly active and thermally stable supported Pd@SiO$_2$ core–shell catalyst for catalytic methane combustion. Energy Technol 4:943–949
43. Qi K, Zheng W, Cui X (2016) Supersaturation-controlled surface structure evolution of Pd@Pt core–shell nanocrystals: enhancement of the ORR activity at a sub-10 nm scale. Nanoscale 8:1698–1703
44. Raghavendra P, Vishwakshan Reddy G, Sivasubramanian R et al (2018) Reduced graphene oxide-supported Pd@Au bimetallic nano electrocatalyst for enhanced oxygen reduction reaction in alkaline media. Int J Hydrogen Energy 43:4125–4135
45. Sankar M, Dimitratos N, Miedziak PJ et al (2012) Designing bimetallic catalysts for a green and sustainable future. Chem Soc Rev 41:8099–8139
46. Seo M, Kim S, Lee D-W et al (2016) Core–shell structured, nano-Pd-embedded SiO$_2$-Al$_2$O$_3$ catalyst (Pd@SiO$_2$-Al$_2$O$_3$) for direct hydrogen peroxide synthesis from hydrogen and oxygen. Appl Catal a Gen 511:87–94
47. Shviro M, Polani S, Dunin-Borkowski RE, Zitoun D (2018) Bifunctional electrocatalysis on Pd-Ni core–shell nanoparticles for hydrogen oxidation reaction in alkaline medium. Adv Mater Interfaces 5:1701666
48. Sneed BT, Young AP, Jalalpoor D et al (2014) Shaped Pd-Ni-Pt core-sandwich-shell nanoparticles: influence of ni sandwich layers on catalytic electrooxidations. ACS Nano 8:7239–7250
49. Sun L, Zhang D, Sun Y et al (2018) Facile fabrication of highly dispersed Pd@Ag core–shell nanoparticles embedded in spirulina platensis by electroless deposition and their catalytic properties. Adv Funct Mater 28:1707231
50. Wang J, Qin Y-L, Liu X, Zhang X-B (2012) In situ synthesis of magnetically recyclable graphene-supported Pd@Co core–shell nanoparticles as efficient catalysts for hydrolytic dehydrogenation of ammonia borane. J Mater Chem 22:12468–12470
51. Wang N, Xu Y, Han Y et al (2015) Mesoporous Pd@M (M=Pt, Au) microrods as excellent electrocatalysts for methanol oxidation. Nano Energy 17:111–119
52. Wang T-P, Hong B-D, Lin Y-M, Lee C-L (2020) Catalysis of the D-glucose oxidation reaction using octahedral, rhombic dodecahedral, and cubic Pd@Pt core–shell nanoparticles. Appl Catal B Environ 260:118140
53. Wang X, Zhang Y, Song S et al (2016) l-arginine-triggered self-assembly of ceo$_2$ nanosheaths on palladium nanoparticles in water. Angew Chemie Int Ed 55:4542–4546
54. Wang Y, Cao L, Libretto NJ et al (2019) Ensemble effect in bimetallic electrocatalysts for CO$_2$ reduction. J Am Chem Soc 141:16635–16642
55. Wisniak J (2010) The history of catalysis. from the beginning to nobel prizes. Educ Química 21:60–69
56. Xiao X, Jeong H, Song J et al (2019) Facile synthesis of Pd@Pt core–shell nanocubes with low Pt content via direct seed-mediated growth and their enhanced activity for formic acid oxidation. Chem Commun 55:11952–11955

57. Xie S, Liu Y, Deng J et al (2016) Three-dimensionally ordered macroporous CeO_2-supported Pd@Co nanoparticles: Highly active catalysts for methane oxidation. J Catal 342:17–26

58. Xu S, Li Z, Lei F et al (2017) Facile synthesis of hydrangea-like core–shell Pd@Pt/graphene composite as an efficient electrocatalyst for methanol oxidation. Appl Surf Sci 426:351–359

59. Ye R-P, Wang X, Price C-AH, et al (2020) Engineering of yolk/core–shell structured nanoreactors for thermal hydrogenations. Small n/a:1906250

60. Yue L, Li J, Chen C et al (2018) Thermal-stable Pd@mesoporous silica core–shell nanocatalysts for dry reforming of methane with good coke-resistant performance. Fuel 218:335–341

61. Zhang L, Niu W, Zhao J et al (2013) Pd@Au core–shell nanocrystals with concave cubic shapes: kinetically controlled synthesis and electrocatalytic properties. Faraday Discuss 164:175–188

62. Zhang L, Xie Z, Gong J (2016) Shape-controlled synthesis of Au-Pd bimetallic nanocrystals for catalytic applications. Chem Soc Rev 45:3916–3934

63. Zhang N, Xu Y-J (2013) Aggregation- and leaching-resistant, reusable, and multifunctional Pd@CeO_2 as a robust nanocatalyst achieved by a hollow core–shell strategy. Chem Mater 25:1979–1988

64. Zhang Q, Lee I, Joo JB et al (2013) Core–shell Nanostructured Catalysts. Acc Chem Res 46:1816–1824

65. Zhao R, Gong M, Zhu H et al (2014) Seed-assisted synthesis of Pd@Au core–shell nanotetrapods and their optical and catalytic properties. Nanoscale 6:9273–9278

66. Zhao R, Liu Y, Liu C et al (2014) Pd@Pt core–shell tetrapods as highly active and stable electrocatalysts for the oxygen reduction reaction. J Mater Chem A 2:20855–20860

67. Zhao Z, Long Y, Luo S et al (2019) Preparation of a magnetic mesoporous Fe_3O_4-Pd@TiO_2 photocatalyst for the efficient selective reduction of aromatic cyanides. New J Chem 43:6294–6302

68. Zheng H, Matseke MS, Munonde TS (2019) The unique Pd@Pt/C core–shell nanoparticles as methanol-tolerant catalysts using sonochemical synthesis. Ultrason Sonochem 57:166–171

69. Zhou M, Wang H, Elnabawy AO et al (2019) Facile one-pot synthesis of Pd@Pt1L octahedra with enhanced activity and durability toward oxygen reduction. Chem Mater 31:1370–1380

Part III
Yolk-Shell for Catalysis

Chapter 22
Structural and Catalytic Features of Metal Nanoparticles Encapsulated in a Hollow Carbon Sphere

Shigeru Ikeda and Takashi Harada

22.1 Introduction

Metal nanoparticles (MNPs) have attracted considerable attention for catalysts because of their large surface areas and size-dependent properties different from bulk metals [1–3]. The basic catalytic approach for these MNPs is to use them finely dispersed in an organic or aqueous solution in the presence of a protective agent to prevent aggregation and/or coalescence [4, 5]. Due to their difficulties of recovery and instabilities upon exposure in diverse conditions, e.g., high temperature and high pressure, supported heterogeneous systems using several host materials such as zeolites and activated carbons are also studied extensively [6–9]. Apart from these supported systems, creation of a core-shell structure is another approach to stabilize an MNP core in a solid shell [10–12]. Especially in catalytic applications, an MNP encapsulated by a hollow porous shell whose size is larger than that of the core MNP is a promising structure because the "yolk-shell" nanostructure enables stabilization of the core MNP without any protective agent even under severe reaction conditions. This chapter reviews our first discoveries on successful fabrications and demonstrations of heterogeneous catalytic properties of noble MNPs (i.e., Pt, Rh or Pd) encapsulated in a hollow porous carbon shell [13–16]. The carbon shell not only acts as a barrier to prevent coalescence between the core MNPs but also provides a

S. Ikeda (✉)
Faculty of Science and Engineering, Department of Chemistry, Konan University, 8-9-1 Okamoto, Higashinada-ku, Kobe 658-8501, Hyogo, Japan
e-mail: s-ikeda@konan-u.ac.jp

T. Harada
Research Center for Solar Energy Chemistry, Osaka University, 1-3 Machikaneyama, Toyonaka 560-8531, Osaka, Japan
e-mail: harada@chem.es.osaka-u.ac.jp

reaction space where organic transformation occurs on the surface of metal nanoparticles. As a result, these nanocomposites were proved to work as an effective and
robust heterogeneous catalyst with sufficient stabilities.

22.2 Fabrication of a "Yolk-Shell"-Structured Composite

As a typical system of an MNP core-hollow carbon shell composite, we firstly
prepared a Pt MNP encapsulated in a hollow mesoporous carbon shell (Pt@hmC)
following the procedure for fabrication of a hollow core/mesoporous carbon shell
containing an Au particle of ca. 20 nm in diameter [17]. Schematic synthesis procedure of the Pt@hmC composite is shown in Fig. 22.1. The procedure is based on
successive coatings of a silica layer and a mesoporous silica layer on the surface of
Pt MNPs and growth of a carbon shell inside the mesoporous silica shell followed by
removal of siliceous components by wet etching using an aqueous HF solution [13].
As a starting material, we used a Pt MNP stabilized by poly(N-vinyl-2-pyrrolidone)
(Pt-PVP) having an average particle size of 1.8 nm in diameter (Fig. 22.2a). After the
surface coverage with a double layer of silica and porous silica (Pt@m-SiO$_2$), particle
size of Pt MNP kept the same as that of the original Pt-PVP (Fig. 22.2b), while some
Pt@m-SiO$_2$ composites include a few Pt particles in each silica shell. Figure 22.2c
shows TEM images of Pt@hmC obtained from Pt@m-SiO$_2$ and the size distribution
of Pt MNPs in this material. The Pt@hmC thus obtained consists of a hollow carbon
cage (50–60 nm in diameter and c.a. 10 nm in shell thickness) and a central Pt MNP
with an average particle size of 2.2 nm. A slight increment (0.3 nm) of the Pt particle
size compared to that in Pt@m-SiO$_2$ is due to the coalescence of some Pt MNPs
included in the same silica shell. The content of Pt in Pt@hmC measured by ICP
analysis indicated the presence of 2.1 wt% of Pt in the present preparation condition.
In addition, the mechanical strength of the carbon shell of Pt@hmC was investigated
by giving high pressure. Figure 22.2d shows the TEM image of Pt@hmC after giving

Fig. 22.1 Schematic
synthesis procedure of
Pt@hmC

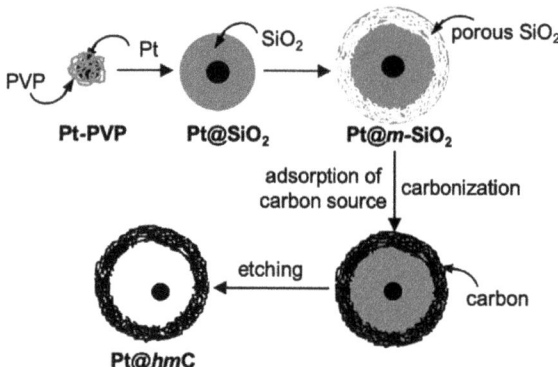

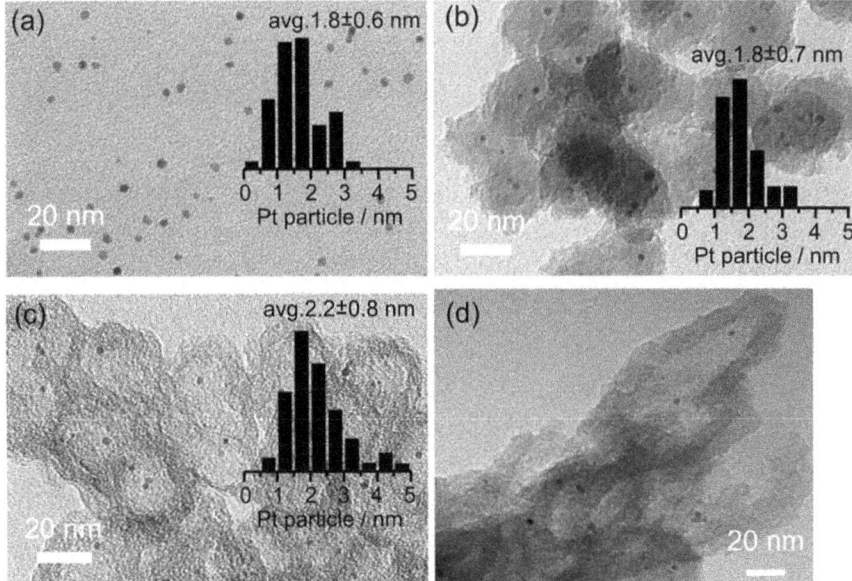

Fig. 22.2 TEM images and Pt size distributions of **a** Pt-PVP, **b** Pt@*m*-SiO$_2$, and **c** Pt@*hm*C. **d** TEM image of Pt@*hm*C pressurized at 80 MPa

a pressure of 80 MPa for three times. Although the carbon shell transmuted a spherical shape to an elongated form, almost no destruction of the hollow shell structure was obtained; the core-hollow shell structure was maintained. Thus, this core-shell material has a flexible and robust structure.

The powder XRD pattern of Pt@*hm*C exhibits typical (111), (200), and (220) reflections corresponding to Pt metal of an fcc structure (Fig. 22.3a). The crystallite size of the Pt nanoparticle determined by the Scherrer equation and the half-height width of the Pt(111) reflection was 2.5 nm. This value is consistent with the average particle size of Pt nanoparticles determined by TEM observation shown in Fig. 22.2c. Thus, each Pt nanoparticle in Pt@*hm*C is likely to be a single crystal. In addition to these reflections, two broad reflections at 2θ values of 10–30° and 35–50° were also observed in the XRD pattern. These broad reflections are assigned to C(002) and C(101) reflections of amorphous carbon composed of polycyclic aromatic carbon sheets oriented in a random manner [18, 19]. The Pt 4f core-level XP spectrum of Pt@*hm*C shown in Fig. 22.3b exhibited Pt^0 peaks at binding energies of 71.5 eV (4f$_{7/2}$) and 74.7 eV (4f$_{5/2}$) with slight contributions of Pt^{2+} components at binding energies of 73.6 eV (4f$_{7/2}$) and 76.8 eV (4f$_{5/2}$) [20], indicating surface oxidation of the Pt core nanoparticle. The fact that there is no component of Pt^{4+}, chlorine (Cl2p), and fluorine (F 1s) derived from the Pt precursor (H$_2$PtCl$_6$) and HF solution indicates that the surface of the Pt nanoparticle core was free from any impurity component.

The N$_2$ adsorption–desorption isotherm of Pt@*hm*C measured at 77 K is shown in Fig. 22.4. The characteristic of the isotherm is the presence of a significant hysteresis

Fig. 22.3 a XRD pattern and **b** Pt4f, Cl2p, and Fls XP spectra of Pt@*hm*C

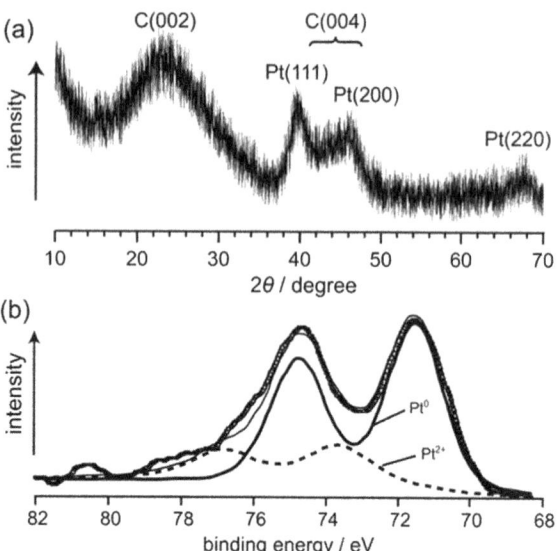

Fig. 22.4 N$_2$ adsorption–desorption isotherm of Pt@*hm*C

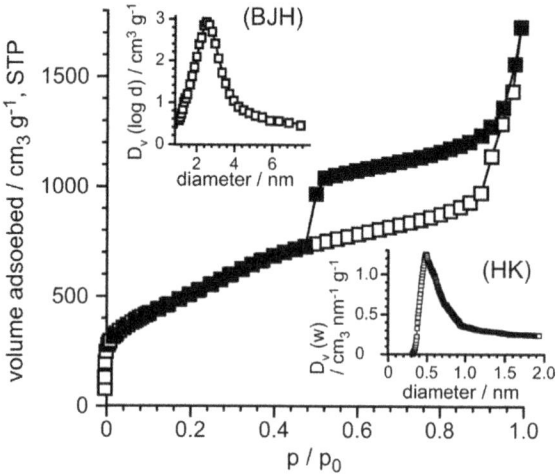

loop enclosed by a sudden drop in the volume adsorbed of the desorption isotherm in the p/p$_0$ range of 0.42–0.46. Because this phenomenon is observed in a material having a large mesopore encapsulated by a pore system of relatively small pore size [21], the carbon shell of Pt@*hm*C should consist of a hierarchical porous wall structure, i.e., large pores of 15–20 nm in diameter (hollow spaces) with a porous wall having relatively small pores. Actually, applications of the BJH model and the Horvath-Kawazoe (HK) model to the adsorption branch results in specification of two kinds of small pore systems centered at 2.2 nm and 0.6 nm, respectively, both of which are ascribed to pore systems in the carbon shell (insets of Fig. 22.4). It is noted that the BET surface area calculated from the isotherm at p/p_0 of 0.05–0.09 reaches 1830 m^2 g^{-1}.

22.3 Catalytic Activities of Pt@hmC

Figure 22.5a shows catalytic activities of Pt@hmC, colloidal Pt-PVP (the source of Pt nanoparticles in Pt@hmC), a commercial Pt catalyst supported on activated carbon (Pt/AC: purchased from N.E. Chemcat), and Pt@m-SiO$_2$ for room temperature hydrogenation of nitrobenzene. Remarkable activity was achieved on the Pt@hmC sample: Pt@hmC catalyzed complete conversion of nitrobenzene into aniline, whereas the reaction was not completed on Pt-PVP and Pt/AC. No catalytic activity obtained for the Pt@m-SiO$_2$ sample indicate that the presence of void spaces in Pt@hmC is indispensable for inducing the reaction. The original Pt-PVP exhibits a higher level of activity than that of Pt/AC. However, it cannot be recovered easily; thus, is difficult to be reused. On the other hand, the Pt@hmC catalyst was recovered simply by centrifugation and recycled for further reaction. TEM observation of Pt@hmC after reaction revealed that there is almost no change in the structure: Pt particle size distribution was well-corresponded to that before the catalytic reaction, as shown in Fig. 22.5b.

The scope for hydrogenation using Pt@hmC is summarized in Table 22.1. The Pt@hmC catalyst showed higher levels of catalytic activity for hydrogenation of primary, secondary, and cyclic olefins than those for Pt-PVP and Pt/AC catalysts. The turnover frequency (TOF) for Pt for Pt@hmC was calculated to be more than 20,000 h^{-1}, the value of which is much larger than that obtained on the Pt-nanoparticle-based system reported [22]. The Pt@hmC catalyst also gave relatively high conversion for the hydrogenation of trans-stilbene into 1,2-diphenylethane

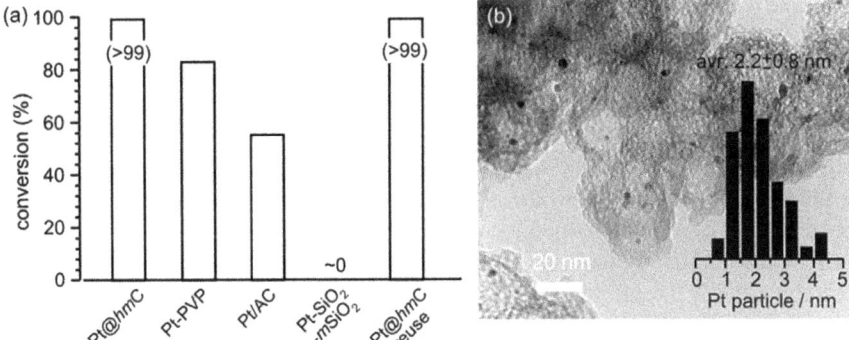

Fig. 22.5 **a** Hydrogenation of nitrobenzene into aniline on various Pt catalysts under H$_2$ bubbling at 303 K. Catalyst (Pt): 0.1 μmol; substrate (nitrobenzene): 5 mol; reaction time: 1.5 h. **b** TEM image and Pt size distribution of Pt@hmC after the reaction

Table 22.1 Hydrogenation of various olefins catalyzed by Pt@hmC, Pt/AC, and Pt-PVP.[a]

Substrate	Catalyst	Product	Time/h	Conv. (%)
1-hexene	Pt@hmC	1-hexane	2	>99
	Pt-PVP		2	91
	Pt/AC		2	7
2-hexene	Pt@hmC	2-hexane	2	96
	Pt-PVP		2	70
	Pt/AC		2	16
Cyclohexene	Pt@hmC	Cyclohexane	1	91
	Pt-PVP		1	42
	Pt/AC		1	3
Trans-stilbene	Pt@hmC	1,2-diphenylethane	15	72
	Pt-PVP		15	46
	Pt/AC		15	16

[a]All reactions were carried out with 0.1 μmol of catalysts (Pt) and of substrates (5 mmol) under H_2 (0.2 MPa in absolute pressure) at 348 K

compared to Pt-PVP or Pt/AC, suggesting that there is no significant effect of the porous carbon shell on mass transfer of such a bulky substrate.

Reductive alkylation of the aromatic amine with carbonyl compounds (i.e., aldehyde or ketone) is one of the useful reactions for producing amines which are important intermediates used for various industries. The reaction is considered to be started by addition of a carbonyl compound to aromatic amine followed by subsequent dehydration of thus formed hemiaminal to produce an imine intermediate. Finally, the corresponding amine is produced by hydrogenation of the C=N bond of the imine intermediate with molecular hydrogen (H_2) or hydride compounds. For the evaluation of further catalytic function of Pt@hmC, room temperature reductive alkylation of aniline with acetone under atmospheric pressure of H_2 was performed as a model reaction [14]. Table 22.2 summarizes the conversion of aniline and yield of the corresponding product, N-isopropylaniline, using various Pt catalysts. Among the catalysts, Pt@hmC showed the highest level of catalytic activity: aniline conversion of 93% and N-isopropylaniline yield of 75% were achieved for 1-h reaction. The estimated TOF for Pt for Pt@hmC ($3,120 h^{-1}$) reached remarkably high compared to several reported catalysts [23, 24]. When the reaction was continued for 2 h, the remainder of imine intermediate was solely hydrogenated to N-isopropylaniline without generating any by-products. Probably due to the surface coverage of PVP, the source Pt-PVP exhibited a lower level of activity than that of Pt@hmC. All of the supported Pt catalysts were also found to be less effective. Moreover, Pt@hmC was easily separated from the reaction solution by centrifugation. After drying the recovered Pt@hmC at 523 K for 8 h to remove the reaction residue, the sample was shown to retain a high level of catalytic activity for at least 2 successive runs.

Table 22.2 Reductive alkylation of aniline with acetone catalyzed by various Pt catalysts.[a] Reproduced with permission from Ref. [15] Copyright 2008 The Chemical Society of Japan

Catalyst	Conv. (%)	Yield (%)	Pt size/nm[b]
Pt@hmC	93	75	3.1
Pt@hmC[c]	99	99	3.1
Pt-PVP	52	23	2.2
Pt/AC	33	3	4.5
Pt/Al$_2$O$_3$	30	18	3.1
Pt/AC (Wako)[d]	31	7	4.2
Pt/AC (N. E. Chem)[d]	40	23	3.4
Pt@hmC[e]	92	72	3.1
Pt@hmC[f]	96	77	3.3[g]

[a]Reaction conditions: Pt (0.25 μmol), aniline (1 mmol), acetone (1 cm^3), r.t., H$_2$ (balloon), 1 h. [b]Average Pt particles size. [c]2-h reaction time. [d]Commercial Pt/AC (Pt: 5 wt %) purchased from Wako pure chemical and N. E. Chemcat. [e]2nd run of Entry 1. [f]3rd run of Entry 1. [g]After reaction

In separate experiments, the same reaction mixture as that used in the reductive alkylation of aniline with acetone was stirred under N$_2$ (i.e., instead of H$_2$) at room temperature. The imine intermediate should be accumulated because of the suppression of hydrogenation in this reaction condition. Therefore, the amount of decrease in aniline, i.e., aniline conversion, was used as the quantitative measure of imine production. As a result, carbon-supported samples gave ca. 30% conversion but that Pt/Al$_2$O$_3$ gave only 8% for 1-h reaction duration. Thus, the low rate of conversion of aniline on Pt/Al$_2$O$_3$ is explained by its poor activity for imine production. On the other hand, the formation of imine occurred in a similar manner on various carbon-supported catalysts independent of their structural differences. The high yield of N-isopropylaniline achieved over the Pt@hmC catalysts is therefore attributable to its high catalytic function for hydrogenation of the imine intermediate. In the Pt@hmC structure, the particle is physically separated by the porous carbon shell, and the Pt MNP inside the hollow porous carbon shell is free from any surface-covered agents. In this situation, the carbon shell efficiently supplies the imine intermediate into the hydrophobic void space where an active "ligand-free" Pt nanoparticle is present. Due to the synergetic effect, Pt@hmC exhibits a high level of catalytic activity.

22.4 Catalytic Performances of "Yolk-Shell"-Structured Composites Based on Other Metal Nanoparticles

Hydrogenation of aromatic ring is important for the synthesis of cyclohexane derivatives including the intermediates for the chemicals and pharmaceuticals [25–28]. Recently, Sajiki et al. reported a Rh/C catalyst showed the higher catalytic activity of hydrogenation of aromatic rings under a mild condition (353 K, 0.5 MPa) than other metals (Ir, Pt, Pd, and Ru) supported on the active carbon [29]. Interestingly, their process indicated higher conversion was achieved in water than in hexane which is known as a good solvent in general hydrogenation. Water is important as a typical solvent in physiological processes and is also a favored solvent in green chemistry compared with organic solvents. In essence, water is an interesting and useful solvent in organic chemistry in which hydrophilic organic substrates dissolved readily in water. However, when hydrophobic substrates are incorporated, the reaction rate is too slow to use in a commercial process due to limited interaction between the organic substrates and immobilized catalyst in water. In this section, the structural concept of Pt@hmC was expanded to Rh MNPs: Rh MNP encapsulated in porous hollow carbon (Rh@hmC) was fabricated and was applied to catalytic hydrogenation of aromatic and heterocyclic rings in aqueous media. Table 22.3 summarizes the catalytic activity of Rh@hmC for hydrogenation of various aromatic and heterocyclic compounds at 353 K under H$_2$ (0.5 MPa). The Rh@hmC catalyst was found to be active for the hydrogenation of benzoic acid, an electron-poor aromatic compound, and a heterocyclic compound of 3-hydroxypyridine to form cyclohexane carboxylic acid and 3-hydroxypiperidine, indicating applicability for relatively inactive unsaturated compounds. Moreover, biphenyl hydrogenation also progressed on the catalyst: bicyclohexane was obtained in 97% yield after 14-h reaction through the formation of benzylcyclohexane as an intermediate. As confirmed by the above Pt@hmC system, these results also indicate the achievement of efficient mass transfer of the lateral porous carbon shell even for such a bulky substrate. Besides, when phenol hydrogenation was performed, cyclohexanone, which is an important intermediate for the manufacture of adipic acid and ε-caprolactam, was directly obtained with high selectivity under these mild conditions.

In organic synthesis, liquid-phase oxidation of alcohols into the corresponding carbonyl compounds and carboxylic acids is one of the fundamental and important reactions. Among a variety of heterogeneous catalysts, Pd-supported catalysts have been studied widely since they are available not only for oxidations but also various other organic transformation reactions [30–33], though their catalytic performances

Table 22.3 Hydrogenation of aromatic and heterocyclic compounds catalyzed by Rh@hmC.[a]

Substrate	Product	Rh/μmol	Time/h	Conv. (%)	Yield (%)
t-butylbenzene	t-butylcyclohexane	0.23	2	>99	>99
Benzoic acid	Cyclohexane carboxylic acid	0.23	6	89	85
3-hydroxypyridine	3-hydroxypiperidine	0.23	6	98	98
Biphenyl	Bicyclohexane	0.46	6	92	59[b]
		0.46	14	99	97
Phenol	Cyclohexanol	0.12	6	99	20[c]

[a]Reaction conditions: substrate (0.25 mmol), water (5 cm^3), 0.5 MPa H$_2$, 353 K.
[b]Phenylcyclohexane (partially hydrogenated intermediate product) was obtained with 33% yield.
[c]Cyclohexanone was obtained with 79% yield

Table 22.4 Oxidation of benzyl alcohol into benzaldehyde by various Pd catalysts[a]. Reproduced with permission from Ref. [16] Copyright 2010 American Chemical Society

Catalyst	Conv. (%)	Yield (%)	TOF/h^{-1}
Pd@hmC	48	37	2940
Pd/AC[b]	57	50	2090
Pd/hmC[c]	26[d]	22[d]	1528
Pd@SiO$_2$	<1[d]	– [d]	–

[a]Reaction conditions: benzyl alcohol (0.25 mmol), 50 mM K$_2$CO$_3$ aqueous solution (5 cm^3), O$_2$ atmosphere, 353 K, 1 h. [b]Commercial Pd/AC purchased from N. E. Chemcat. [c]Pd MNPs supported on hollow mesoporous carbon (Pd/hmC) prepared by a conventional impregnation. [d]Reaction was carried out for 2 h

are usually lower than those of homogeneous catalysts. In order to achieve further improvement of catalytic activity for the oxidation of alcohols, we attempted to fabricate Pd-based core shell composite similar to the above-mentioned Pt and Rh composites, i.e., Pd@hmC; catalytic activity of thus obtained Pd@hmC was evaluated by oxidation of various alcohols. As a result, Pd@hmC showed a high level of catalytic activity for oxidation of benzyl alcohol into benzaldehyde using atmospheric pressure of O$_2$ as an oxidant (Table 22.4). Compared to the commercial Pd/AC sample, the Pd@hmC sample apparently shows lower conversion and yield than those of the Pd/AC sample. Superior catalytic ability of Pd@hmC compared to that of Pd/AC can be revealed by using TOF, indicating efficient utilization of the surface reaction sites of Pd nanoparticles in the Pd@hmC composite. The Pd@hmC catalysts also exhibited a high level of catalytic activity for aerobic oxidations of other primary benzylic and allylic alcohols into corresponding aldehydes [14].

Catalytic ability for benzyl alcohol oxidation over the sample recovered after 1 h reaction was examined. For this purpose, the Pd@hmC sample was reused after heat treatment at 623 K in air. Figure 22.6a shows the relative yield of benzaldehyde formed over regenerated Pd@hmC samples normalized by the yield of benzaldehyde over a fresh Pd@hmC sample. Results over Pd/AC are also shown for comparison. It is clear that a decrease in catalytic activity was significantly suppressed on Pd@hmC when compared to that on Pd/AC. As confirmed by CO chemisorption analyses, dispersions of Pd, defined as the amount of CO molecules irreversibly held (CO$_{irr}$) on the surface of the Pd nanoparticle (CO$_{irr}$/Pd), did not decline significantly for the Pd@hmC sample after regeneration at 623 K in air, whereas that for Pd/AC provided appreciable decrease (Fig. 22.6b, c). Thus, observed regeneration of catalytic activity of the Pd@hmC sample is attributed to effective suppression of Pd aggregation due to the physical separation of Pd nanoparticles by a porous carbon shell in Pd@hmC.

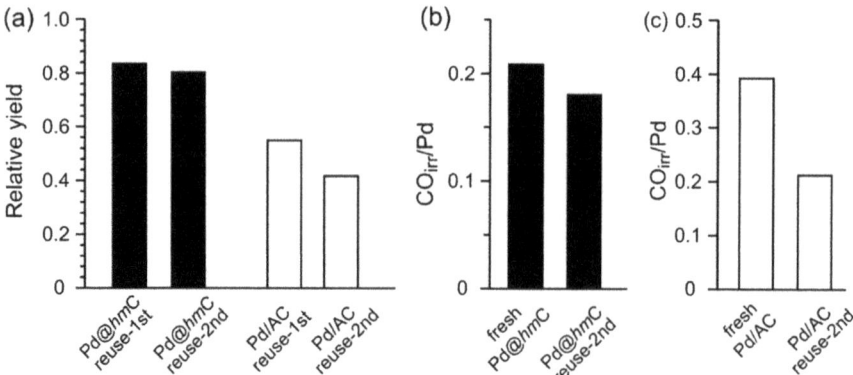

Fig. 22.6 **a** Relative yield of benzaldehyde over Pd@*hm*C and Pd/AC for oxidation of benzyl alcohol after heat treatment normalized by the yield of benzaldehyde over fresh catalyst. CO$_{irr}$/Pd of **b** Pd@*hm*C and **c** Pd/AC samples before and after the catalytic run followed by the regeneration treatment at 623 K in air

22.5 Summary and Outlook

In this chapter, we discussed the fabrication and characterization of MNPs encapsulated in a hollow porous carbon shell. The MNP core was proved to be free from any surface-covered stabilizing agents; the particle is physically separated by the carbon shell with a well-developed pore structure. The porous wall structure of the lateral carbon shell provides channels for efficient mass transfer of species into the void space where active catalytically MNPs are present. Thus, these composites exhibit excellent catalytic activity for various kinds of organic transformations depending on the kinds of core MNPs. Because of the controllability of porosity inside the shell, the use of the present core-shell structure is a promising strategy for designing catalysts with molecular selective properties toward the conversion of organics. Moreover, since the catalytic activity is not influenced by any stabilizing agents commonly present on the surface of metal nanoparticles, this heterogeneous catalytic system should also be useful for understanding essential catalytic functions and reaction mechanisms on the bare surface of various types of MNPs when appropriate MNPs with defined shapes, sizes, or crystalline planes are employed.

References

1. Haruta M, Daté M (2001) Advances in the catalysis of Au nanoparticles. Appl Catal A Gen 222:427–437
2. Kelly KL, Coronado E, Zhao LL, Schatz GC (2003) The optical properties of metal nanoparticles: the influence of size, shape, and dielectric environment. J Phys Chem B 107:668–677

3. Panigrahi S, Basu S, Praharaj S, Pande S, Jana S, Pal A, Ghosh SK, Pal T (2007) Synthesis and size-selective catalysis by supported gold nanoparticles: study on heterogeneous and homogeneous catalytic process. J Phys Chem C 111:4596–4605
4. Schmid G (1992) Large clusters and colloids. Metals in the embryonic state. Chem Rev 92:1709–1727
5. Toshima N, Yonezawa T (1998) Bimetallic nanoparticles—novel materials for chemical and physical applications. New J Chem 22:1179–1201
6. Su F, Zeng J, Yu Y, Lv L, Lee J Y, Zhao XS (2005) Template synthesis of microporous carbon for direct methanol fuel cell application. Carbon 43:2366–2373
7. Yang Z, Xia Y, Sun X, Mokaya R (2006) Preparation and hydrogen storage properties of zeolite-templated carbon materials nanocast via chemical vapor deposition: effect of the zeolite template and nitrogen doping. J Phys Chem B 110:18424–18431
8. Joo SH, Choi SJ, Oh I, Kwak J, Liu Z, Terasaki O, Ryoo R (2001) Ordered nanoporous arrays of carbon supporting high dispersions of platinum nanoparticles. Nature 412:169–172
9. Harada T, Ikeda S, Miyazaki M, Sakata T, Mori H, Matsumura M (2007) A simple preparation method of highly active palladium catalysts loaded on various carbon supports for liquid-phase oxidation and hydrogenation reactions. J Mol Catal A Chem 268:59–64
10. Schärtl W (2000) Crosslinked spherical nanoparticles with core-shell topology. Adv Mater 12:1899–1908
11. Graf C, Vossen DLJ, Imhof A, van Blaaderen A (2003) A general method to coat colloidal particles with silica. Langmuir 19:6693–6700
12. Sun X, Li Y (2004) Colloidal carbon spheres and their core/shell structures with noble-metal nanoparticles. Angew Chem Int Ed 43:597–601
13. Ikeda S, Ishino S, Harada T, Okamoto N, Sakata T, Mori H, Kuwabata S, Torimoto T, Matsumura M (2006) Ligand-free platinum nanoparticles encapsulated in a hollow porous carbon shell as a highly active heterogeneous hydrogenation catalyst. Angew Chem Int Ed 45:7063–7066
14. Harada T, Ikeda S, Ng YH, Sakata T, Mori H, Torimoto T, Matsumura M (2008) Rhodium nanoparticle encapsulated in porous carbon shell as active heterogeneous catalyst for aromatic hydrogenation. Adv Funct Mater 18:2190–2196
15. Harada T, Ikeda S, Okamoto N, Ng YH, Higashida S, Torimoto T, Matsumura M (2008) Efficient reductive alkylation of aniline with acetone over Pt nanoparticles encapsulated in hollow porous carbon. Chem Lett 37:948–949
16. Harada T, Ikeda S, Hashimoto F, Sakata T, Ikeue K, Torimoto T, Matsumura M (2010) Catalytic activity and regeneration property of a Pd nanoparticle encapsulated in a hollow porous carbon sphere for aerobic alcohol oxidation. Langmuir 26:17720–17725
17. Kim M, Sohn K, Na HB, Hyeon T (2002) Synthesis of nanorattles composed of gold nanoparticles encapsulated in mesoporous carbon and polymer shells. Nano Lett 2:1383–1387
18. Tsubouchi N, Xu C, Ohtsuka Y (2003) Carbon crystallization during high-temperature pyrolysis of coals and the enhancement by calcium. Energy Fuels 17:1119–1125
19. Hara M, Yoshida T, Takagaki A, Takata T, Kondo JN, Hayashi S, Domen K (2004) A carbon material as a strong protonic acid. Angew Chem Int Ed 43:2955–2958
20. Kim KS, Winograd N, Davis RE (1971) Electron spectroscopy of platinum-oxygen surfaces and application to electrochemical studies. J Am Chem Soc 93:6296–6297
21. Groen JC, Peffer LAA, Perez-Ramirez J (2003) Pore size determination in modified micro- and mesoporous materials. Pitfalls and limitations in gas adsorption data. Microporous Mesoporous Mater 60:1–17
22. Scheeren CW, Machado G, Dupont J, Fichtner PFP, Texeira SR (2003) Nanoscale Pt(0) particles prepared in imidazolium room temperature ionic liquids: synthesis from an organometallic precursor, characterization, and catalytic properties in hydrogenation reactions. Inorg Chem 42:4738–4742
23. Greenfield H (1994) Side reactions in reductive alkylation of aromatic amines with aldehydes and with ketones. Chem Ind (Catalysis of Organic Reactions) 53:265–277
24. Roy D, Jaganathan R, Chaudhari RV (2005) Kinetic modeling of reductive alkylation of aniline with acetone using Pd/Al2O3 catalyst in a batch slurry reactor. Ind Eng Chem Res 44:5388–5396

25. Chen B, Dingerdissen U, Krauter JGE, Lansink Rotgerink HGJ, Möbus K, Ostgard DJ, Panster P, Riermeier TH, Seebald S, Tacke T, Trauthwein H (2005) New developments in hydrogenation catalysis particularly in synthesis of fine and intermediate chemicals. Appl Catal A Gen 280:17–46

26. Odenbrand CUI, Lundin ST (1980) Hydrogenation of benzene to cyclohexene on a ruthenium catalyst: influence of some reaction parameters. J Chem Technol Biotechnol 30:677–687

27. Van Der Steen PJ, Scholten JJF (1990) Selectivity to cyclohexene in the gas phase hydrogenation of benzene over ruthenium, as influenced by reaction modifiers: II. Catalytic hydrogenation of benzene to cyclohexene and cyclohexane. Appl Catal 58:291–304

28. L'Argentière PC, Cagnola EA, Liprandi DA, Román-Martnez MC, Salinas-Martínez De Lecea C (1998) Carbon-supported Pd complex as catalyst for cyclohexene hydrogenation. Appl Catal A Gen 172:41–48

29. Maegawa T, Akashi A, Sajiki H (2006) A mild and facile method for complete hydrogenation of aromatic nuclei in water. Synlett 9:1440–1442

30. Milne JE, Buchwald SL (2004) An extremely active catalyst for the Negishi cross-coupling reaction. J Am Chem Soc 126:13028–13032

31. Shaughnessy KH, DeVasher RB (2005) Palladium-catalyzed cross-coupling in aqueous media: recent progress and current applications. Curr Org Chem 9:585–604

32. Felpin FX, Ayad T, Mitra S (2006) Pd/C: an old catalyst for new applications. Its use for the Suzuki-Miyaura reaction. Eur J Org Chem 12:2679–2690

33. Yin L, Liebscher J (2006) Carbon-carbon coupling reactions catalyzed by heterogeneous palladium catalysts. Chem Rev 107:133–173

Chapter 23
Yolk-Shell Structured Functional Nanoreactors for Organic Transformations

Fangfang Chang, Lingyan Jing, Yash Boyjoo, Jian Liu, and Qihua Yang

23.1 Introduction

The introduction of the term "nanotechnology" dates back to December 1959, when Richard Feynman gave his visionary talk entitled "There's Plenty of Room at the Bottom". Although at that time, nanotechnology was not being widely used, scientists took decades to realize the power behind it [1]. The development of electron microscopy techniques has provided technical support for researchers to explore the micro and nano world. With time, more and more nanomaterials with fancy structures have been constructed, some of them termed as "nanoreactor", which is like a cell mimicking the biocatalysis process at the nanometer/micrometer scale [2, 3]. Nanoreactor features are beneficial to performing parallel chemical reactions, eliminating undesirable products, and improving the catalytic performance as the result of special structure over convention one. The morphologies of nanoreactors are various, including hollow spheres [4, 5], core-shell particles [6, 7], Janus particles [8, 9], etc. Particularly, yolk-shell structured nanoreactors (YSNs) with unique yolk@void@shell morphologies have been designed, synthesized, and further developed due to their intriguing structure-induced properties and widespread applications in catalysis, drug delivery, and energy conversion and storage.

Organic transformations are one of the key processes for pharmaceutical and fine chemical production. However, organic transformations usually occur under harsh conditions, which means higher requirements are putting forward for catalysts. YSNs bring many benefits to organic transformations, such as hydrogenation reactions and

F. Chang · L. Jing · Y. Boyjoo · J. Liu (✉) · Q. Yang (✉)
State Key Laboratory of Catalysis, Dalian Institute of Chemical Physics, Chinese Academy of Sciences, 457 Zhongshan Road, Dalian 116023, China
e-mail: jianliu@dicp.ac.cn

Q. Yang
e-mail: yangqh@dicp.ac.cn

© The Author(s), under exclusive license to Springer Nature Singapore Pte Ltd. 2021
H. Yamashita and H. Li (eds.), *Core-Shell and Yolk-Shell Nanocatalysts*,
Nanostructure Science and Technology,
https://doi.org/10.1007/978-981-16-0463-8_23

cascade reactions, in virtue of the unique structure with void space between yolk and shell, controllable selectivity, diffusion enrichment, and confinement effect [10, 11]. For example, metal nanoparticles (NPs) are easily deactivated under high temperature or long reaction time even when these NPs are immobilized on a support, owing to sintering, leaching, and aggregation. If the active species (Au, Pd, etc.) are encapsulated inside a shell (carbon, silica, metal-oxide, etc.), the unique yolk-shell structure provides protection from these adverse effects, even under severe reaction conditions. Further, the active core of YSNs can move freely, exposing more active sites. In addition, energy consumption and high cost are inevitable during the subsequent separation and purification processes because of various products in organic transformation. The confinement effect and the molecular sieving capability of the shell are essential for regulating product selectivity [12, 13]. For cascade reaction, the integration and precise location of different active sites are especially important. The yolk, the inner and outer layers of the shell can each be functionalized differently via synthetic strategies, which make YSNs a promising candidate in cascade reactions.

So far, many review articles summarized the extraordinary advancements of YSNs in the past few years [14–17]. This review will mainly focus on the synthesis strategies of functional YSNs and their application in organic transformation processes.

23.2   Synthetic Methods

YSNs usually require multi-step synthetic steps due to the special yolk@void@shell structures. Recently, a series of synthesis strategies have been developed by researchers for YSNs with various compositions, properties, and applications. These methods can be classified as hard templating, soft templating, ship-in-bottle, and self templating [16, 18]. Here we will briefly describe these methods.

23.2.1   Hard-Templating Method

The hard-templating method is the most common and straightforward method to prepare YSNs. In this approach, a core material (metal, oxide or polymer) is coated with one or a few layers (oxide or polymer). Then the core or middle layers are removed selectively, leading to the yolk-shell structure. For example, Zhang et al. exploited the layer by layer method by coating SiO_2 and subsequently TiO_2 layers on the surface of Fe_3O_4 nanoparticles. This was followed by the removal of the SiO_2 layer, resulting in $Fe_3O_4@TiO_2$ yolk-shell nanoreactors [19] (Fig. 23.1a). In another work, Liu et al. used a similar protocol to prepare $CaCO_3@C$ yolk-shell nanoreactors [20]. Calcination under air can also be used to create void spaces. Jiang et al. fabricated carbon@TiO_2 yolk-shell spheres via changing the calcination duration of carbon@TiO_2 core-shell spheres in air. With the extension of calcination duration, the carbon core was shrinkage due to the combustion. In this process, air plays a role

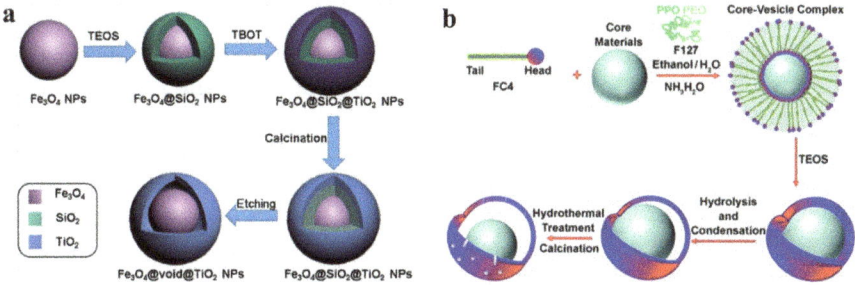

Fig. 23.1 Representative example of templating method. **a** Schematic illustration of the formation procedure of the yolk-shell structured Fe_3O_4@void@TiO_2NPs through hard templating method. Reproduced with permission from Ref. [19] Copyright 2017 Elsevier. **b** Procedure for the preparation of yolk–shell structured nanoreactor through soft templating method. FC4 = fluorocarbon surfactant, PPO = poly(propylene oxide), PEO = poly(ethylene oxide), TEOS = tetraethoxysilane. Reproduced with permission from Ref. [26] Copyright 2010 Wiley–VCH

of an "etchant" by partially removing the carbon core [21]. In recent years, metal–organic frameworks (MOFs) have regularly aroused attention due to their appealing properties, such as diverse structures, uniform porosity, and high surface area [22]. Additionally, MOF materials can act as a hard template and its metal can serve as the active site [23]. Zeolitic imidazolate frameworks (ZIFs), consisting of transition metal ions (such as Zn^{2+} and Co^{2+}) and imidazolate linkers, are a sub-family of MOFs, that exhibit exceptional thermal and chemical stability. Liu et al. successfully prepared various types of yolk-shell nanoreactors by adopting ZIFs or their derivative materials as hard template. ZIF materials were coated with either silica or polymer layer, followed by an etching process through hydrothermal or carbonization step [24, 25]. An unquestionable advantage of the hard template method is its easy tunability, in terms of the compartments' sizes. However, the tedious and complex process of removing the sacrificial template is unwelcomed.

23.2.2   Soft Templating Method

Soft templating methods have much similarity to hard-templating methods, but the preparation of fine homogenous nanocomposites by this means is simpler and more convenient. The soft templates include microemulsions, micelle/vesicle-based soft templates, as well as polymers, which can be easily removed by washing or calcination, without using harsh chemicals [16]. Usually, microemulsions are ternary systems which consist of an aqueous phase, an oil phase, and a surfactant. According to which phase is present in majority, microemulsions can be divided into direct (oil-in-water) and reverse (water-in-oil) microemulsions. Schüth et al. adopted an emulsion as soft-template to generate Pd/C@Hollow Crabon Sphere (HCS) yolk-shell structured nanoreactors. First, the emulsion solution was formed by mixing a

P123/sodium oleate/PdCl$_4{}^{2-}$ solution with an acidic solution of the polymer precursors (2,4-dihydroxybenzoic acid (DA) and hexamethylenetetramine (HMT)). In the process of hydrothermal treatment, formaldehyde produced by HMT could not only reduce the Pd^{2+} to Pd nanoparticles, but also polymerized with DA on the emulsion surface. Finally, the yolk-shell polymer nanospheres with polymer-yolk supported Pd nanoparticles (Pd/P@HPS) were generated. After carbonization in an inert atmosphere, Pd/C@HCS were obtained [26]. Micelles and vesicles are self-assembled by amphiphilic molecules in a single-phase solvent. Amphiphilic molecules are both hydrophilic and lipophilic. Liu et al. reported a facile strategy to generate yolk-shell structured materials by using fluorocarbon surfactant (FC4) as the vesicle-core complex template (Fig. 23.1b). With this method, SiO$_2$@SiO$_2$, Au@SiO$_2$, Fe$_3$O$_4$@SiO$_2$, and even enzyme@SiO$_2$ yolk-shell nanostructures with tunable shell thickness were prepared [27, 28]. Moreover, long-chained polymers are also common soft templates. For example, poly (vinyl pyrrolidone) (PVP) can serve as a soft-template to prepare Poly(3,4-thylenedioxythiophene) (PEDOT) hollow spheres via electrochemical polymerization. Subsequently, Ag$^+$ was introduced and reduced to form Ag @ PEDOT yolk-shell structure [29]. Polymers are convenient soft templates for the preparation of YSNs with organic shells. Currently, the main challenges of soft-templating methods are the low uniformity of the nanoparticle size, as well as difficulty in choosing appropriate surfactants. In addition, some soft templates are expensive, which limits their wide application.

23.2.3 Ship-in-Bottle

Ship-in-bottle offers an efficient, economical, and controllable route for the synthesis of metal@shell nanoparticles (NPs). Firstly, a highly robust hollow shell with pores (zeolites, silica, MOFs, etc.) is prepared. The metal precursors are then induced into the hollow space, followed by reduction during which metal NPs grow into the yolk. A series of metal@shell NPs (Au@SiO$_2$, Pt@zeolites, etc.) have been prepared this way [11, 30]. Xu and co-workers encapsulated Pt NPs into MFI zeolites via a cationic polymer-assisted synthetic strategy. In this case, the cationic polydiallydimethylammonium chloride (PDDA) polymer could electrostatically interact with the Pt precursor and zeolite [31, 32]. Xu et al. developed a "double solvent" method to introduce Pt NPs into the pores of MIL-101 through the power of capillary force [33] (Fig. 23.2a). Besides metal NPs, MOFs can also be introduced into the confined space. Very recently, Cao et al. encapsulated MOF into a hollow mesoporous carbon sphere (HMCS). In their study, they prepared the hollow carbon spheres using the SiO$_2$ nanospheres as the hard template and dopamine as the carbon source. The Co^{2+} ions were anchored onto abundant N sites in the HCMS, initiating the growth of ZIF within the confined space of the HMCS. Finally, ZIF@HMCS yolk-shell structured nanoreactors were obtained (Fig. 23.2b). Similarly, a series of ZIF@HMCS-m (m:

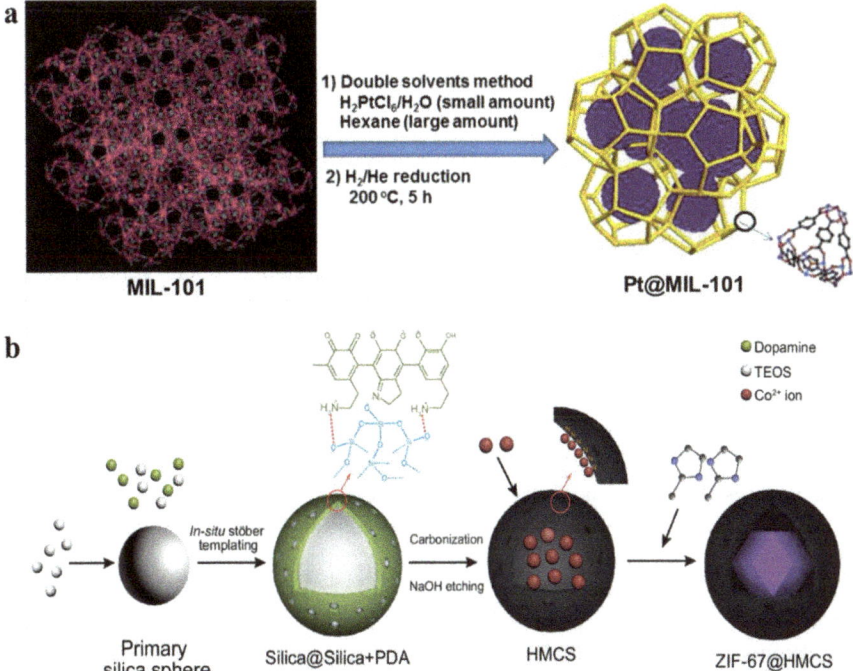

Fig. 23.2 Representative example of ship-in-bottle method. **a** Schematic representation of the formation of Pt@MIL-101. Reproduced with permission from Ref. [33] Copyright 2012 American Chemical Society. **b** Schematic illustration of synthetic procedure for ZIF@HMCS. Reproduced with permission from Ref. [34] Copyright 2020 Springer

weight ratio of HMCS to $Co(NO_3)_2 \cdot 6H_2O$) could be obtained [34]. The ship-in-bottle method offers a feasible path for the development of self-assembling fine nanoparticles in a confined space.

23.2.4 Self-Templating Method

Although the templating methods are straightforward in concept, difficulties arise in forming robust shells with desirable composition and in the efficient removal of the templates. In recent years, self-templating methods referred to as "smarter strategies" are of great interest due to the merit of facile synthesis of yolk-shell structures with desirable compositions and complex architectural features. The self-templating methods to form yolk-shell structures include thermally driven contraction process, the Ostwald ripening process, the Kirkendall effect, and galvanic replacement, among others. In addition, these self-templating methods hold great promise owing to their special advantages including reduced production cost, simple synthesis procedures, and potential in scaling up.

23.2.4.1 Thermally Driven Contraction Process

The occurrence of yolk-shell structures originating from the thermally driven contraction process is an interesting phenomenon. During thermal decomposition, significant mass/volume loss, especially under nonequilibrium conduction, can generate voids in solid particles. Metal carbonates/glycerates/glycolates, MOFs, and polymers can be employed as the decomposable precursors. For example, Liu et al. reported yolk-shell structured carbonaceous nanospheres prepared by using the confined pyrolysis method. In the first step, the Stöber method was used to coat silica shell on the surface of resorcinol–formaldehyde resin (RF) spheres, followed by calcination in an inert atmosphere, and etching of the silica shell to form yolk-shell structured carbonaceous spheres. When the polymer material was pyrolyzed in a confined space, the polymer shrank inward due to the oxidative degradation of the organic species, while the release of gas from the decomposition of organic species counteract the effect of inward shrinkage. Since the degree of inward shrinkage was larger, a void was formed. In a similar way, yolk-shell structured carbonaceous spheres doped with heteroatoms (N, S, etc.) could be prepared easily [35]. Recently, Lou et al. reported a simple annealing treatment of preformed metal acetate hydroxide precursors to form yolk-shell structured Ni–Co oxide nanoprisms, of which the composition can be easily tuned by changing the Ni/Co molar ratio [36] (Fig. 23.3a). In this case, there were two opposing forces which were the contraction force (Fc) and the adhesive force (Fa). If Fc is bigger than Fa, a void will form, leading to yolk-shell structured materials. Thermally driven contraction process is a facile method to obtain yolk-shell structured products with controllable composition.

23.2.4.2 The Ostwald Ripening Process

Ostwald ripening is a well-known physical phenomenon which is defined as the "dissolution of small crystals or sol particles and the redeposition of the dissolved species on the surfaces of larger crystals or sol particles", according to the International Union of Pure and Applied Chemistry in 2007. This effect is widely taken advantage of for fabricating metal oxides ($Pt@CeO_2$, etc.) [41] or metal sulfides (ZnS, MoS, etc.) [42] with complex structures. For instance, Zhu et al. demonstrated that a hierarchical ball-in-ball NiO/Ni/Graphene nanomaterial was derived from Ni-MOF via a solvothermal reaction (Fig. 23.3b). The hydrothermal environment was conducive to a dissolution and recrystallization process, whereby the core inside the sphere dissolved and diffused to the surface, forming a void inside the sphere [37]. Recently, Low et al. used $Fe(NO_3)_3 \cdot 6H_2O$ as the iron source, and glycerol, iso-propyl alcohol (IPA) together with a small amount of water as a solvent to prepare uniform Fe_3O_4 spheres via a one-pot solvothermal method. The amount of water played a

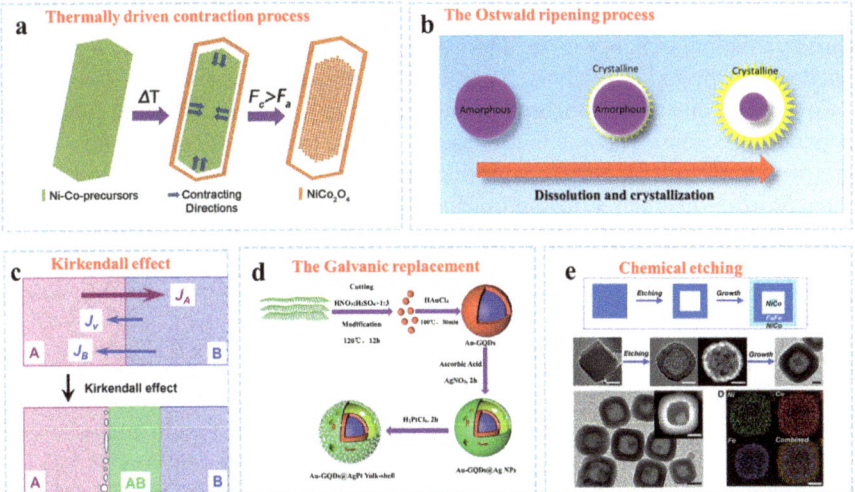

Fig. 23.3 Representative example of self-templating method. **a** Schematic illustration of the formation process of mesoporous yolk-shelled Ni–Co mixed oxide prisms. Reproduced with permission from Ref. [36] Copyright 2015 Wiley–VCH. **b** Schematic representation of the mechanism of structure evolution via Ostwald ripening process. Reproduced with permission from Ref. [37] Copyright 2016 American Chemical Society. **c** Schematic illustration of a nonequilibrium lattice diffusion at the interface in bulk phase by Kirkendall effect. Reproduced with permission from Ref. [38] Copyright 2013 American Chemical Society. **d** The synthesis process of Au-GQDs@AgPt Yolk-shell nanostructures. Reproduced with permission from Ref. [39] Copyright 2020 Elsevier. **e** Schematic illustration of chemical etching process of PBA. Reproduced with permission from Ref. [40] Copyright 2013 American Chemical Society

vital role in the process, regulating the formation of different Fe_3O_4 structures [43]. For example, when increasing the water amount, hollow particles with broken shells would be harvested. On the other hand, hierarchical Fe–glycerate solid spheres were obtained without water addition. Ostwald ripening is a thermodynamically favorable and spontaneous process, in which the void size greatly depends on the hydrothermal reaction time and temperature. In addition, the solvents can be used to modulate the ripening process, hence regulating the void size.

23.2.4.3   Kirkendall Effect

The Kirkendall effect is a well-known phenomenon in metallurgy. In this process, two opposites diffuse actions occur simultaneously at the interface between coupled materials A and B. If the diffusion rate of metal A to B (J_A) exceeds the opposite action of metal B to A (J_B), the unequal diffusion rates would result in the formation of lattice vacancies within the material's matrix (Fig. 23.3c). Thus, the condensation of these lattice vacancies leads to the formation of porous structures such as yolk-shell particles [38, 44]. Recently, Zhang et al. prepared hollow $CuInS_2$ nanododecahedrons by a cation exchange method based on the Kirkendall effect. Adopting $Cu_{2-x}S$ nanododecahedrons with elaborately designed shapes as templates, the different rates of Cu^+ extraction and In^{3+} incorporation formed $Cu_{2-x}S@CuInS_2$ hybrid core–shell nanododecahedrons. The continued reaction led to the formation of hollow structures with uniform shape and composition [45]. The Kirkendall effect can be used to synthesize yolk-shell structured nanocrystals with high uniformity in sizes and shapes. However, it is still a challenge to regulate the morphology of YSNs, because of the uncertainty and uncontrollability of the Kirkendall diffusion process.

23.2.4.4 Galvanic Replacement

Galvanic replacement is a sacrificial template strategy to obtain metal@metal (Au@Pt, Au@Au), metal@alloy (i.e., Au@AuAg, Au@PdAg, and Au@PtAg) and alloy@alloy (i.e., Au/Ag@Au/Ag) yolk-shell nanostructures, on the basis of different metals exhibiting different electrochemical potential [16]. Ag, Ni, Al, and Cu are the most common sacrificial templates in galvanic replacement processes. Recently, Zhou et al. developed a green wet chemical reduction method to synthesis Au-graphene quantum dots (GQDs)@AgPt yolk-shell nanostructures (Fig. 23.3d). The Au-GQDs@AgPt yolk-shell nanostructures were prepared by displacing metallic silver using $PtCl_6^{2-}$, driven by the galvanic replacement reaction [39]. Yu and co-workers introduced a double-template method to prepare Au@AgPt yolk-shell nanoparticles, in which Au NPs served as seeds, Ag shell grew on the surface of Au NPs to obtain Au@Ag NPs, then galvanic replacement reaction occurred between $PtCl_6^{2-}$ and Ag. In this process, Brij-58 surfactant was used as the soft-template to induce dendritic and porous shell [46]. Galvanic replacement is a facile method to obtain YSNs with controllable morphologies, compositions, and porosities and is particularly efficient with metallic particles. The shape and size can be easily tuned by the template nanoparticles and the reaction conditions. However, the main drawbacks of this method are the long reaction time and the instability of structure in the latter stage of replacing process.

23.2.4.5 Chemical Etching

Chemical etching method is deemed as an efficient strategy to construct yolk-shell structures, especially for preparing silica- and carbon-based yolk-shell structured nanoparticles. To date, a wide variety of yolk-shell nanoreactors ($Fe_3O_4@SiO_2$ [47], $Au@SiO_2$ [48], $SiO_2@SiO_2^3$...) have been reported via this method. Yamauchi et al. established a new concept of step-by-step coordination polymer crystal growth and subsequent etching process to prepare sophisticated types of shell-in-shell, yolk-shell, and yolk-double-shell structures (Fig. 23.3e). The research team chose Prussian Blue (PB) and its analogues as model materials having "soft" inside core and "hard" outside shell. This is because the inside core has more structural defects resulting in weaker stability. When the sample was dispersed in acid solution, the H^+ dissolved the material, with a faster etching rate at the core relative to the shell, creating a void [40]. Yu et al. developed a deflation–inflation asymmetric growth (DIAG) strategy to form tailored concave hollow spheres. The low crosslinked aminophenol-formaldehyde resin (APF) polymers could be etched by acetone. Through controlling the APF etching and deposition kinetics, the unique concave hollow nanostructures were obtained easily [49]. In addition, a surface-protected etching strategy was developed by Yin's group to prepare hollow silica spheres or $Au@SiO_2$ yolk-shell nanoparticles. During the process, the inner core was etched while the outer shell was protected by a layer of stabilizing agent (PVP) on the surface of SiO_2. At the same time, pores on the outer shell were created, which favored the diffusion of molecules to the inner active sites [50]. The nanoparticle size and shell thickness could be controlled by fine-tuning the amounts of etchants and surfactants. Chemical etching is attractive for fine-tuning the size and shell thickness of nanoparticles by the amount of etchants and the duration time of etching. However, the selection of appropriate types of etchants and surfactants is still an issue.

23.3 Application in Organic Transformations

Organic transformations are varied and some reactions are difficult to carry out under moderate conditions. Yolk-shell structured nanoreactors exhibit potential application for organic transformations due to their unique advantages compared to conventional catalysts embedded in bulk supports, such as high catalytic stability, controllable reaction microenvironment, and the advantages of compartmentalization or accumulation of various types of active sites for cascade reactions.

23.3.1 Single Site Nanoreactor

The hydrogenation reaction is one of the most important chemical reactions in pharmaceuticals, chemical engineering, energy conversion, and so on [51]. Liu et al. prepared ZnO@Carbon yolk-shell structured submicroreactors via solvothermal and carbonization processes. ZIF-8 as the sacrificial template would transform into the ZnO nanoribbons following coating a polymer layer on the surface of ZIF-8 and solvothermal treatment. Pd nanoparticles were then introduced by a wet chemical reduction process. The resulting Pd&ZnO@carbon submicroreactors exhibited a high selectivity (>99%) for hydrogenation of phenylacetylene to phenylethylene and excellent stability (>25 h) (Fig. 23.4). All these attractive performances were attributed to the special yolk-shell structure. The void could hold a large amount of phenylacetylene molecules, improving the ambient reactant concentration around the Pd nanoparticles, while the shell played a shielding role to inhibit the aggregation of Pd nanoparticles [25]. Lu et al. fabricated bifunctional yolk-shell structured

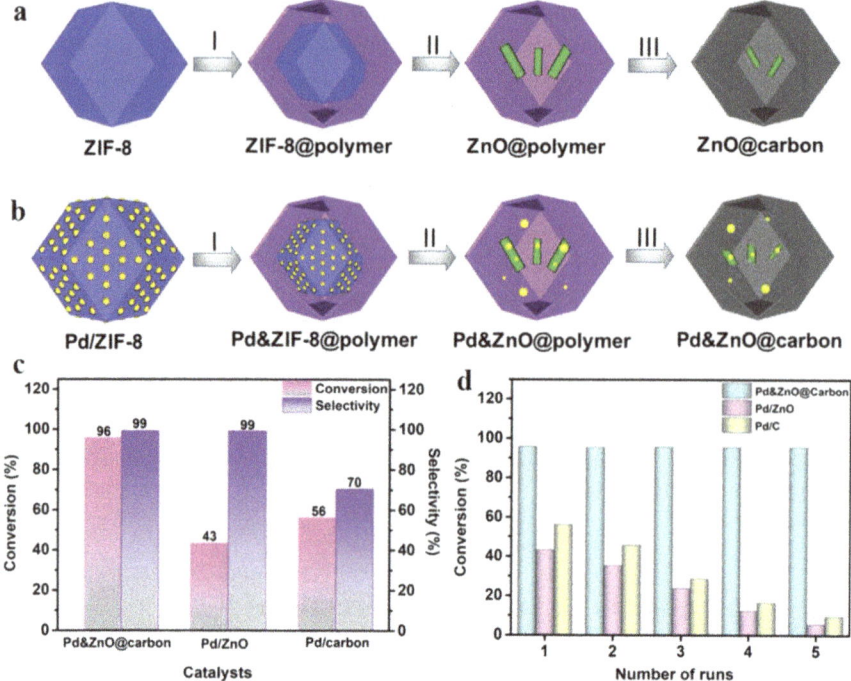

Fig. 23.4 Pd&ZnO@carbon nanoreactor for hydrogenation of phenylacetylene to phenylethylene. **a, b** Schematic illustration for the preparation of ZnO@carbon and Pd&ZnO@carbon, respectively. **c, d** Catalytic performance for Pd&ZnO@carbon, Pd/ZnO and Pd/C particles. Reproduced with permission from Ref. [25] Copyright 2018 Wiley–VCH

Fe_3O_4@h-C/Pt nanoreactors by a novel and facile synthesis method and their performance was evaluated for nitrobenzene hydrogenation. A mere 9% conversion was initially observed due to species covering the surface of Pt nanoparticles. However, when the used catalysts were calcined at 250 °C in air for 1 h, the catalysts demonstrated high conversion (>99%), even after seven cycles [52]. Yin et al. reported that Pt@CoO yolk-shell nanoreactors prepared through nanoscale Kirkendall effect showed high activity for the hydrogenation of ethylene [53].

Coupling reactions, such as Suzuki coupling, Sonogashira-, Heck-, and Ullmann-type reactions, etc., account for a large fraction of modern organic chemistry [54]. Noble metals and transition metals are extensively employed in these reactions. Song et al. fabricated a Pd@SiO_2 yolk-shell catalyst by using a hard-templating method. The yolk-shell structure was realized by coating silica shell on the surface of carbon spheres loaded with Pt NPs, followed by removal of the carbon. The Pd@SiO_2 catalyst showed superior activity in Suzuki coupling reactions with 99.5% yield after 3 min, which was ascribed to the high surface of the nanoreactor that was able to gather a large concentration of reactants around the Pd nanoparticles relative to the bulk solution. Furthermore, the thin layer of mesoporous silica allowed for easy diffusion of both reactants and products to and from the active sites [55].

23.3.2 Multi Sites Nanoreactor

Cascade reactions, which allow a series of chemical reactions to occur sequentially by integrating different active sites into one catalyst, have garnered intensive interests due to the advantage of high atom economy, step-saving, and biomimetic nature [56, 57]. From this point of view, yolk-shell structured catalysts exhibit promising potential in cascade reactions.

Yang et al. reported a yolk-shell nanoreactor with a basic core and an acidic shell through an organosilane-assisted selective etching method (Fig. 23.5). These YSNs showed excellent catalytic performance for the deacetalization-Henry cascade reaction [58]. Zhang et al. fabricated a three-dimensionally integrated yolk–shell (3D-IYS) nanoreactor via a protection–deprotection approach and core-shell colloidal crystal templates (CS-CCTs). Then, the team investigated their catalytic performance in the deacetalization-Henry cascade reaction and deacetalization-Knoevenagel cascade reaction. The authors prepared sulfonated polystyrene (CLPS) @SiO$_2$ by coating the silica layer on the surface of CLPS through sol–gel method. In situ polymerization of the functional monomer BOC-p-aminostyrene at the CLPS@SiO$_2$ CS-CCTs surface and then selectively removing the SiO$_2$ layer of CS-CCTs was employed to synthesize acid–base bifunctional integrated yolk-SO$_3$H@Shell-NH$_2$ (IY-SO$_3$H@S-NH$_2$). In the deacetalization-Henry cascade reaction, the acid–base bifunctional IY-SO$_3$H@S-NH$_2$ catalyst exhibited excellent activity (conversion =

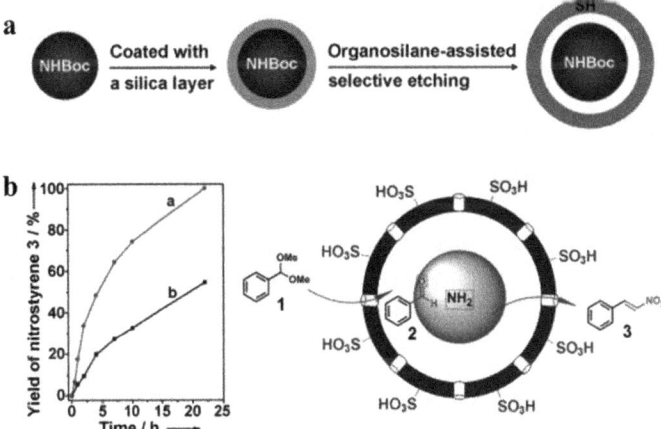

Fig. 23.5 Catalytic performance of the YSN nanoreactor. **a** Schematic illustration for the preparation of YS-NH$_2$@SO$_3$H. **b** Schematic illustration of the cascade reaction in a YS-NH$_2$@SO$_3$H acid–base nanoreactor. Reproduced with permission from Ref. [58] Copyright 2012 Wiley–VCH

100%, yield >99%) [59]. In order to extend and develop the core-shell colloidal crystal templating strategy, the authors prepared a novel bifunctional integrated yolk-shelled nanoreactor composed of monolithically interconnected ZIF-8 shell and sulfonated polystyrene yolks decorated with rhodium nanoparticles as IY-SO₃H/Rh@S-ZIF-8. The Knoevenagel condensation−hydrogenation cascade reaction was utilized to evaluate the catalytic activity of the obtained nanoreactors, which exhibited high activity and stability. This was attributed to the following advantages: (1) interconnected macropores improved the mass transfer of reactant and product; (2) the hierarchical pore structure exposed more active sites and enhanced catalytic performance; (3) the integrated yolk-shelled structures not only provided a confined space but also improved the mechanical stability; (4) 3D-ordered yolk-shelled nanoreactor with spatially isolated functionalities was able to realize cascade reaction [60]. In addition, enzymes can also be introduced into the nanoreactors. Yang et al. constructed a hierarchical yolk-shell@shell nanoreactor that spatially positioned Pd nanoparticles and Candida antarctic lipase B (CALB) enzyme in separated domains for tandem catalysis. In the one-pot dynamic kinetic resolution reaction of 1-phenylethylamine, the bifunctional nanoreactors showed excellent activity and selectivity [61]. Yang et al. reported that well-defined metal−organic cages were introduced into amino-functionalized mesoporous carbon (Cage@FDU-ED) through confinement self-assembly. The resultant catalyst demonstrated high activity, selectivity, and recyclability in the sequential oxidation-Knoevenagel condensation reaction from alcohols to α,β-unsaturated dinitriles, which was attributed to the yolk-shell structured nanoreactor combining the attractive features of metallosupramolecular complexes and mesoporous heterogeneous catalysts [62]. The metal−organic cages @shell structured nanoreactors as a promising platform will pave a new way for heterogeneous sequential reactions.

23.4 Summary and Outlook

Over the recent years, yolk-shell structured nanoreactors have attracted increasing research interests due to their appealing properties and widespread potential applications. The rational design and synthesis of yolk-shelled structured functional nanoreactors are of great significance as both fundamental challenges in materials science and practical solutions for organic transformation and other reactions in modern society. In this chapter, we have surveyed both the main synthesis methods and applications of yolk-shell structured nanoreactors in organic transformations. Special focus has been paid to the synthesis strategies of yolk-shell structured nanoreactors. However, these yolk-shell structured nanoreactors are still extremely basic, compared to biological cells with highly optimized spatial and functional control. It remains a challenge to develop novel yolk-shell structured nanoreactors with several, spatially compartmentalized catalytic functions. Existing examples of multifunctional nanoreactors are limited to few types of (acid, basic, and enzyme) organic moieties and metal nanoparticles as active sites. Expanding and integrating more types of active species, and meeting the need for more demanding sequential transformations are still difficult to realize. It should be noted that the simultaneous application of multiple synthesis strategies is the focus of synthetic functional catalysts. Therefore, it is essential to have an in-depth understanding of the current synthetic approaches. Better interpretation of reaction mechanism will not only help to design and produce advanced materials with complex structures, but also allow to fabricate novel functional materials through using yolk-shell structures as building blocks. We believe that via optimizing synthesis strategies, yolk-shell structured functional nanoreactors will show immense advantages in organic transformations and meet the requirements for industrial applications.

References

1. Toumey C (2009) Plenty of room, plenty of history. Nat Nanotechnol 4(12):783–784
2. Chi YG, Scroggins ST, Frechet JMJ (2008) One-pot multi-component asymmetric cascade reactions catalyzed by soluble star polymers with highly branched non-interpenetrating catalytic cores. J Am Chem Soc 130(20):6322–6323
3. Bai S, Yang H, Wang P, Gao J, Li B, Yang Q, Li C (2010) Enhancement of catalytic performance in asymmetric transfer hydrogenation by microenvironment engineering of the nanocage. Chem Commun 46(43):8145–8147
4. Yin YD, Erdonmez C, Aloni S, Alivisatos AP (2006) Faceting of nanocrystals during chemical transformation: from solid silver spheres to hollow gold octahedra. J Am Chem Soc 128(39):12671–12673
5. Chen D, Li L, Tang F, Qi S (2009) Facile and scalable synthesis of tailored silica "Nanorattle" structures. Adv Mater 21(37):3804–3807
6. Huang L, Gurav DD, Wu S, Xu W, Vedarethinam V, Yang J, Su H, Wan X, Fang Y, Shen B, Price C-AH, Velliou E, Liu J, Qian K (2019) A multifunctional platinum nanoreactor for point-of-care metabolic analysis. Matter 1(6):1669–1680
7. Arnal PM, Comotti M, Schuth F (2006) High-temperature-stable catalysts by hollow sphere encapsulation. Angew Chem Int Ed 45(48):8224–8227
8. Zhao T, Zhu X, Hung CT, Wang P, Elzatahry A, Al-Khalaf AA, Hozzein WN, Zhang F, Li X, Zhao D (2018) Spatial isolation of carbon and silica in a single janus mesoporous nanoparticle with tunable amphiphilicity. J Am Chem Soc 140(31):10009–10015
9. Yang T, Wei L, Jing L, Liang J, Zhang X, Tang M, Monteiro MJ, Chen YI, Wang Y, Gu S, Zhao D, Yang H, Liu J, Lu GQM (2017) Dumbbell-shaped bi-component mesoporous janus solid nanoparticles for biphasic interface catalysis. Angew Chem Int Ed 56(29):8459–8463
10. Guo M, Li C, Yang Q (2017) Accelerated catalytic activity of Pd NPs supported on amine-rich silica hollow nanospheres for quinoline hydrogenation. Catal Sci Technol 7(11):2221–2227
11. Ye RP, Wang X, Price CH, Liu X, Yang Q, Jaroniec M, Liu J (2020) Engineering of yolk/core-shell structured nanoreactors for thermal hydrogenations. Small e1906250
12. Dong C, Yu Q, Ye RP, Su P, Liu J, Wang GH (2020) Hollow carbon sphere nanoreactors loaded with PdCu nanoparticles: void-confinement effects in liquid-phase hydrogenations. Angew Chem Int Ed
13. Zhang X, Jing L, Wei L, Zhang F, Yang H (2017) Semipermeable organic-inorganic hybrid microreactors for highly efficient and size-selective asymmetric catalysis. ACS Catal 7(10):6711–6718
14. Wang MW, Boyjoo Y, Pan J, Wang SB, Liu J (2017) Advanced yolk-shell nanoparticles as nanoreactors for energy conversion. Chin J Catal 38(6):970–990

15. Wang X, Feng J, Bai Y, Zhang Q, Yin Y (2016) Synthesis, properties, and applications of hollow micro-/nanostructures. Chem Rev 116(18):10983–11060
16. Priebe M, Fromm KM (2015) Nanorattles or yolk-shell nanoparticles–what are they, how are they made, and what are they good for? Chemistry 21(10):3854–3874
17. Liu J, Qiao SZ, Chen JS, Lou XW, Xing X, Lu GQ (2011) Yolk/shell nanoparticles: new platforms for nanoreactors, drug delivery and lithium-ion batteries. Chem Commun 47(47):12578–12591
18. Purbia R, Paria S (2015) Yolk/shell nanoparticles: classifications, synthesis, properties, and applications. Nanoscale 7(47):19789–19873
19. Du D, Shi W, Wang L, Zhang J (2017) Yolk-shell structured Fe3O4@void@TiO2 as a photo-Fenton-like catalyst for the extremely efficient elimination of tetracycline. Appl Catal B Environ 200:484–492
20. Boyjoo Y, Merigot K, Lamonier J-F, Pareek VK, Tade MO, Liu J (2015) Synthesis of CaCO3@C yolk–shell particles for CO2 adsorption. RSC Adv 5(32):24872–24876
21. Wang W, Xu D, Cheng B, Yu J, Jiang C (2017) Hybrid carbon@TiO2 hollow spheres with enhanced photocatalytic CO2 reduction activity. J Mater Chem A 5(10):5020–5029
22. Chaikittisilp W, Ariga K, Yamauchi Y (2013) A new family of carbon materials: synthesis of MOF-derived nanoporous carbons and their promising applications. J Mater Chem A 1(1):14–19
23. Lan X, Ali B, Wang Y, Wang T (2020) Hollow and yolk-shell Co-N-C@SiO$_2$ nanoreactors: controllable synthesis with high selectivity and activity for nitroarene hydrogenation. ACS Appl Mater Interfaces 12(3):3624–3630
24. Tian H, Liu X, Dong L, Ren X, Liu H, Price CAH, Li Y, Wang G, Yang Q, Liu J (2019) Enhanced hydrogenation performance over hollow structured Co-CoOx@N-C capsules. Adv Sci 6(22):1900807
25. Tian H, Huang F, Zhu Y, Liu S, Han Y, Jaroniec M, Yang Q, Liu H, Lu GQM, Liu J (2018) The development of yolk-shell-structured Pd&ZnO@carbon submicroreactors with high selectivity and stability. Adv Funct Mater 28(32):1801737
26. Wang GH, Hilgert J, Richter FH, Wang F, Bongard HJ, Spliethoff B, Weidenthaler C, Schuth F (2014) Platinum-cobalt bimetallic nanoparticles in hollow carbon nanospheres for hydrogenolysis of 5-hydroxymethylfurfural. Nat Mater 13(3):293–300
27. Liu J, Qiao SZ, Budi Hartono S, Lu GQ (2010) Monodisperse yolk-shell nanoparticles with a hierarchical porous structure for delivery vehicles and nanoreactors. Angew Chem Int Ed 49(29):4981–4985
28. Zhao ZY, Liu J, Hahn M, Qiao S, Middelberg APJ, He L (2013) Encapsulation of lipase in mesoporous silica yolk–shell spheres with enhanced enzyme stability. RSC Adv 3(44):22008
29. Xia YY, Xu L (2010) Fabrication and catalytic property of an Ag@poly(3,4-ethylenedioxythiophene) yolk/shell structure. Synthetic Met 160(7–8):545–548
30. Goebl J, Yin Y (2013) Ship in a bottle: in situ confined growth of complex yolk-shell catalysts. ChemCatChem 5(6):1287–1288
31. Cho HJ, Kim D, Li S, Su D, Ma D, Xu BJ (2020) Molecular-level proximity of metal and acid sites in zeolite-encapsulated Pt nanoparticles for selective multistep tandem catalysis. ACS Catal 10(5):3340–3348
32. Cho HJ, Kim D, Li J, Su D, Xu BJ (2018) Zeolite-encapsulated Pt nanoparticles for tandem catalysis. J Am Chem Soc 140(41):13514–13520
33. Aijaz A, Karkamkar A, Choi YJ, Tsumori N, Ronnebro E, Autrey T, Shioyama H, Xu Q (2012) Immobilizing highly catalytically active Pt nanoparticles inside the pores of metal-organic framework: a double solvents approach. J Am Chem Soc 134(34):13926–13929
34. Xiong WF, Li HF, You HH, Cao MN, Cao R (2020) Encapsulating metal organic framework into hollow mesoporous carbon sphere as efficient oxygen bifunctional electrocatalyst. Natl Sci Rev 7(3):609–619
35. Yang T, Zhou R, Wang DW, Jiang SP, Yamauchi Y, Qiao SZ, Monteiro MJ, Liu J (2015) Hierarchical mesoporous yolk-shell structured carbonaceous nanospheres for high performance electrochemical capacitive energy storage. Chem Commun 51(13):2518–2521

36. Yu L, Guan BY, Xiao W, Lou XW (2015) Formation of yolk-shelled Ni-Co mixed oxide nanoprisms with enhanced electrochemical performance for hybrid supercapacitors and lithium ion batteries. Adv Energy Mater 5(21)
37. Zou F, Chen YM, Liu KW, Yu ZT, Liang WF, Bhaway SM, Gao M, Zhu Y (2016) Metal organic frameworks derived hierarchical hollow NiO/Ni/graphene composites for lithium and sodium storage. ACS Nano 10(1):377–386
38. Wang WS, Dahl M, Yin YD (2013) Hollow nanocrystals through the nanoscale Kirkendall effect. Chem Mater 25(8):1179–1189
39. Yang J, Shao T, Luo C, Li J, He S, Meng B, Zhang Q, Zhang D, Xue Z, Zhou X (2020) Simple synthesis of the Au-GQDs@AgPt yolk-shell nanostructures electrocatalyst for enhancing the methanol oxidation. J Alloy Compd 834:155056
40. Hu M, Belik AA, Imura M, Yamauchi Y (2013) Tailored design of multiple nanoarchitectures in metal-cyanide hybrid coordination polymers. J Am Chem Soc 135(1):384–391
41. Zhang N, Liu SQ, Xu YJ (2012) Recent progress on metal core@semiconductor shell nanocomposites as a promising type of photocatalyst. Nanoscale 4(7):2227–2238
42. Liu B, Zeng HC (2005) Symmetric and asymmetric Ostwald ripening in the fabrication of homogeneous core-shell semiconductors. Small 1(5):566–571
43. Ma FX, Hu H, Wu HB, Xu CY, Xu ZC, Zhen L, Lou XW (2015) Formation of uniform Fe3O4 hollow spheres organized by ultrathin nanosheets and their excellent lithium storage properties. Adv Mater 27(27):4097–4101
44. Fan HJ, Gosele U, Zacharias M (2007) Formation of nanotubes and hollow nanoparticles based on Kirkendall and diffusion processes: a review. Small 3(10):1660–1671
45. Li YM, Liu J, Li XY, Wan XD, Pan RR, Rong HP, Liu JJ, Chen WX, Zhang JT (2019) Evolution of hollow CuInS2 nanododecahedrons via Kirkendall effect driven by cation exchange for efficient solar water splitting. ACS Appl Mater Interfaces 11(30):27170–27177
46. Sui N, Yue RP, Wang YK, Bai Q, An RH, Xiao HL, Wang LN, Liu MH, Yu WW (2019) Boosting methanol oxidation reaction with Au@AgPt yolk-shell nanoparticles. J Alloy Compd 790:792–798
47. Chen Y, Chen HR, Guo LM, He QJ, Chen F, Zhou J, Feng JW, Shi JL (2010) Hollow/rattle-type mesoporous nanostructures by a structural difference-based selective etching strategy. ACS Nano 4(1):529–539
48. Wang C, Chen JC, Zhou XR, Li W, Liu Y, Yue Q, Xue ZT, Li YH, Elzatahry AA, Deng YH, Zhao DY (2015) Magnetic yolk-shell structured anatase-based microspheres loaded with Au nanoparticles for heterogeneous catalysis. Nano Res 8(1):238–245
49. Yu RT, Huang XD, Liu Y, Kong YQ, Gu ZY, Yang Y, Wang Y, Ban WH, Song H, Yu CZ (2020) Shaping nanoparticles for interface catalysis: concave hollow spheres via deflation-inflation asymmetric growth. Adv Sci 7(13)
50. Zhang Q, Lee I, Ge J, Zaera F, Yin Y (2010) Chemical etching surface-protected etching of mesoporous oxide shells for the stabilization of metal nanocatalysts. Adv Funct Mater 20(14):2201–2214
51. Zang WT, Li GZ, Wang L, Zhang XW (2015) Catalytic hydrogenation by noble-metal nanocrystals with well-defined facets: a review. Catal Sci Technol 5(5):2532–2553
52. Sun Q, Guo CZ, Wang GH, Li WC, Bongard HJ, Lu AH (2013) Fabrication of magnetic yolk-shell nanocatalysts with spatially resolved functionalities and high activity for nitrobenzene hydrogenation. Chem Eur J 19(20):6217–6220
53. Yin YD, Rioux RM, Erdonmez CK, Hughes S, Somorjai GA, Alivisatos AP (2004) Formation of hollow nanocrystals through the nanoscale Kirkendall effect. Science 304(5671):711–714
54. Chinchilla R, Najera C (2011) Recent advances in Sonogashira reactions. Chem Soc Rev 40(10):5084–5121
55. Chen Z, Cui ZM, Niu F, Jiang L, Song WG (2010) Pd nanoparticles in silica hollow spheres with mesoporous walls: a nanoreactor with extremely high activity. Chem Commun 46(35):6524–6526
56. Corma A (2016) Heterogeneous catalysis: understanding for designing, and designing for applications. Angew Chem Int Ed 55(21):6112–6113

57. Jia Z, Wang K, Tan B, Gu Y (2017) Hollow hyper-cross-linked nanospheres with acid and base sites as efficient and water-stable catalysts for one-pot tandem reactions. ACS Catal 7(5):3693–3702

58. Yang Y, Liu X, Li X, Zhao J, Bai S, Liu J, Yang Q (2012) A yolk-shell nanoreactor with a basic core and an acidic shell for cascade reactions. Angew Chem Int Ed 51(36):9164–9168

59. Guo Y, Feng L, Wang X, Zhang X (2019) Integration of yolk-shell units into a robust and highly reactive nanoreactor: a platform for cascade reactions. Chem Commun 55(21):3093–3096

60. Guo Y, Feng L, Wu C, Wang X, Zhang X (2019) Synthesis of 3D-ordered macro/microporous yolk-shelled nanoreactor with spatially separated functionalities for cascade reaction. ACS Appl Mater Interfaces 11(37):33978–33986

61. Zhang X, Jing L, Chang F, Chen S, Yang H, Yang Q (2017) Positional immobilization of Pd nanoparticles and enzymes in hierarchical yolk-shell@shell nanoreactors for tandem catalysis. Chem Commun 53(55):7780–7783

62. Zhu F-F, Chen L-J, Chen S, Wu G-Y, Jiang W-L, Shen J-C, Qin Y, Xu L, Yang H-B (2020) Confinement self-assembly of metal-organic cages within mesoporous carbon for one-pot sequential reactions. Chem

Chapter 24
Design and Synthesis of Yolk–Shell Nanostructured Silica Encapsulating Metal Nanoparticles and Aminopolymers for Selective Hydrogenation Reactions

Yasutaka Kuwahara and Hiromi Yamashita

24.1 Introduction

Yolk–shell nanostructured materials consisting of catalytically active core particles encapsulated by hollow silica materials are an emerging class of nanomaterials. Compared to the conventional periodic silica materials with narrow micro/mesoporous channels (e.g., zeolite and mesoporous silica), hollow silica materials provide enclosed large cavity spaces that can accommodate a variety of catalytically active components (e.g., metal nanoparticles (NPs), metal oxides, and metal complexes, etc.), and the outer silica shell serves as a physical barrier to protect the encapsulated components. Continuous pores created in the silica shell region allow the mass transfer of target reactant molecules and sometimes endow the molecular sieving effect. Furthermore, the tunability and functionality in the core and the shell regions can offer new catalytic properties, rendering them attractive platform materials for the design of multifunctionalized heterogeneous catalysts [1–7].

In most studies on Yolk–shell nanostructured catalysts, the enclosed inner void spaces are used for encapsulation of active catalytic components (such as metal NPs, metal oxides, and metal complexes). In general, the hollow cavity spaces provide a confined nanospace where reactant molecules are likely to be adsorbed and concentrated to efficiently react on catalytically active core particles, although the cause for boosting targeted chemical reactions has not fully been understood. On the other hand, decoration of interior space of the hollow cavity by incorporating additional catalytic components other than the core particles to add further functionalities has rarely been examined. It is expected that creation of further elaborated catalytic nano-environments is possible by introducing co-catalysts or modifier materials

Y. Kuwahara (✉) · H. Yamashita
Division of Materials and Manufacturing Science, Graduate School of Engineering, Osaka University, Suita, Osaka 565-0871, Japan
e-mail: kuwahara@mat.eng.osaka-u.ac.jp

within the hollow cavity spaces. Herein, we envisioned that hollow silica spheres having enclosed large cavity spaces could be a reasonable option to encapsulate functional polymers together with catalytically active metal NPs, which would lead to the design and synthesis of new types of Yolk–shell nanostructured catalysts with unprecedented unique catalytic performances. This is because the surrounding silica shell is expected to act as a protective barrier to circumvent the leaching of the polymer and sintering of metal NPs during the catalytic reactions, also leading to improved stability and reusability of catalyst. However, encapsulation of bulky functional polymers within hollow silica spheres to add functionalities has rarely been examined so far [8–10]. Although there are many reports for the synthesis of Yolk–shell nanostructured catalysts using costly polymers as organic templates, they are typically removed unprofitably by thermal decomposition to create hollow cavities.

In this context, we have developed a new method to synthesize hollow silica spheres encapsulating metal NPs and aminopolymer using aminopolymer itself as an organic template [11], in which aminopolymers can strongly coordinate on metal NPs to form aminopolymer-metal NP aggregate cores due to the presence of lone pairs on N atoms of amines [12]. Recently, poly(ethylenimine) (PEI), a type of aminopolymers, has attracted increasing attention as a promising organic-ligand for immobilizing Pd NP catalysts, as well as a promising adsorbent for capturing CO_2. PEI is a physically sticky polymer (the degree of stickiness is dependent on the molecular weight) and is soluble in water and alcohols. Two types of PEI are commercially available; linear PEI is almost exclusively composed of secondary amines, whereas branched PEI contains primary, secondary, and tertiary amines. Abundant amine groups in their polymer chains efficiently coordinate and immobilize Pd NPs, enabling chemoselective semihydrogenation of both internal and terminal alkynes [13]. In addition, these amine sites have an ability to reversibly adsorb/desorb CO_2 in temperature swing process under both dry and humid conditions, thereby being used as an adsorbent (typically combined with support solids) in Carbon Capture and Storage (CCS) technology [14–16].

In this chapter, the design and development of a new type of Yolk–shell nanostructured silica composites encapsulating metal nanoparticles (NPs) and aminopolymer, PEI, inside the hollow silicas are described. The synthesis of Yolk–shell structured Pd−PEI−silica nanocomposites by a self-assembly approach using PEI as an organic template is described. Such a Yolk–shell nanostructured catalyst shows high selectivity and reusability in the semihydrogenation of both internal and terminal alkynes to produce the corresponding alkenes, owing to the poisoning effect of PEI on Pd NP surface, as well as the ability of silica shell to prevent leaching/aggregation of the encapsulated components. Furthermore, a Yolk–shell nanostructured catalyst encapsulating PdAg NPs together with PEI shows an excellent catalytic activity under moderate reaction conditions and reusability in the CO_2 hydrogenation to produce formate, owing to the CO_2 capturing ability of PEI and the protective effect of the silica shell. The synergistic interaction mechanisms of metal NPs and PEI within the confined nanospace of hollow silicas are also addressed.

24.2 Design of Yolk–Shell Nanostructured Catalysts for Semihydrogenation of Alkynes

Semihydrogenation of alkynes to selectively produce alkenes is an important and fundamental reaction for the synthesis of commodity chemicals and fine chemicals [17]. Pd-based catalysts have been used as the overwhelming majority for liquid-phase semihydrogenation of alkynes, in which inorganic or organic modifiers are typically added to improve the selectivity toward alkene [18, 19]. For example, the Lindlar catalyst (Pd/CaCO$_3$ treated by Pb salts) has long been utilized as a benchmark heterogeneous catalyst, but it suffers from some critical drawbacks such as high toxicity of Pb and the low alkene selectivity toward terminal alkynes [20]. Pd NPs modified with alkylthiol surfactants are known to show improved alkene selectivities in the liquid-phase hydrogenation of alkynes [21–23]. Recently, poly(ethylenimine) (PEI) emerged as promising supports for Pd NP catalysts. Abundant amine groups in its polymer chain efficiently coordinate and immobilize Pd NPs, which provide high chemoselectivity in the semihydrogenation of both internal and terminal alkynes in the presence of 1 atm H$_2$ [13]. However, the inherent difficulties in separation and leaching of organic modifiers, as well as sintering of Pd species during catalytic use limit their practical applications. Jones et al. synthesized Pd−PEI−silica composite catalyst by immobilizing Pd NPs into a SBA-15 mesoporous silica functionalized with PEI polymer, which showed excellent activity and selectivity in the liquid-phase hydrogenation of diphenylacetylene [24]. Although porous silica appears to be effective for immobilizing PEI and Pd NPs with good dispersion and stability, leaching of amine-functional groups and Pd species during catalytic use through open-ended pores and the associated activity reduction are still matters of technical concern.

Herein, we envisioned that hollow silica spheres having enclosed cavity spaces could be used as an ideal support to immobilize Pd NPs together with PEI, since the hollow silicas are expected to act as rigid nanocages to circumvent the sintering of Pd NPs and leaching of PEI during the catalysis, which would endow improved stability and reusability of catalyst.

24.2.1 Synthesis of Hollow Silica Spheres Encapsulating Pd Nanoparticles and Poly(ethyleneimine)

Selective introduction of functional polymers with large molecular weights within the confined nanospace of hollow silica via a postsynthetic approach is quite difficult. Therefore, we have developed a new method to synthesize hollow silica spheres encapsulating Pd NPs and PEI (Pd + PEI@HSS) using PEI itself as an organic template [11]. Figure 24.1 schematically depicts the synthetic procedure of Pd + PEI@HSS composites using linear or branched PEI as organic templates. In the first step, NaBH$_4$ as a reducing agent was added to an EtOH–water–NH$_3$ mixed

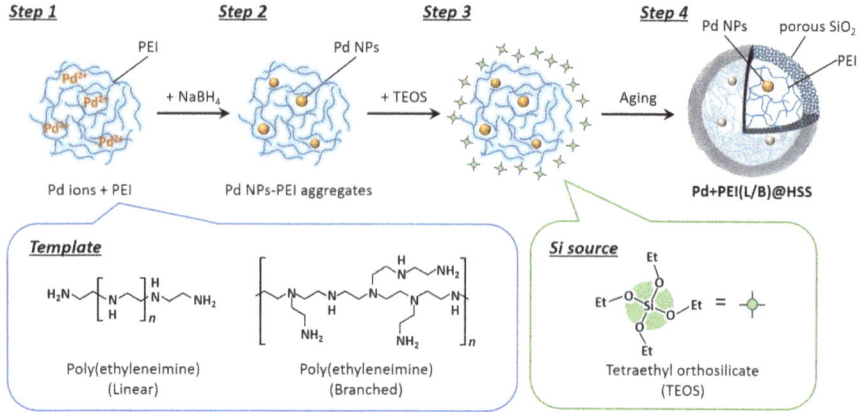

Fig. 24.1 Schematic representation of the synthetic procedure of hollow silica spheres encapsulating Pd NPs and poly(ethyleneimine) (Pd + PEI@HSS). Reprinted with permission from ACS Catalysis, 2019, 9, 1993 Ref. [11]. Copyright 2019 American Chemical Society

solution containing designated amounts of PEI and Pd precursor (Na_2PdCl_4) (steps 1 and 2), in which $PdCl_4^{2-}$ ions entrapped by the nitrogen ligands on the polymer chain are reduced to form Pd NP–PEI aggregates. In the following step, tetraethyl orthosilicate (TEOS) as a Si precursor was added to the above solution (step 3). In the presence of a base (NH_3), condensation of TEOS occurs to form a silica network, and silica shell is self-assembled around the Pd–PEI aggregates as nuclei through an electrostatic interaction between amine groups on the polymer chain and hydrolyzed TEOS. Following aging at moderate temperature leads to the formation of Pd + PEI@HSS nanocomposites (step 4).

Pd + PEI(L)@HSS synthesized with linear PEI ($M_w = 2,500$) was composed of spherical silicas with an average particle size of ca. 155 nm. TEM images clearly showed that the silica spheres have hollow nanostructure with a silica shell thickness of ca. 49 nm, and Pd NPs with an average diameter of 8.9 nm were observed in their hollow spaces. The morphology of the final solid was strongly dependent on the type and the molecular weight of PEI employed; branched-type PEI with a similar molecular weight ($M_w = 1,800$) afforded smaller silica particles (average particle size of ca. 107 nm) with thinner silica shells (ca. 36 nm). On the other hand, branched PEI with a smaller molecular weight ($M_w = 600$) or with a larger molecular weight ($M_w = 10,000$) resulted in the formation of solid silicas without any defined nanostructure. Thus, an appropriate choice of PEI is required to synthesize Yolk–shell nanostructured Pd–PEI–silica composite, since the size of Pd–PEI aggregates can be varied depending on the molecular weight of PEI.

In scanning transmission electron microscope (STEM) and high-angle annular dark-field STEM (HAADF-STEM) images of Pd + PEI@HSS, Pd NPs located inside the hollow cavities were clearly observed (Fig. 24.2a, b). Mapping analysis showed that the distributions of Si atoms and O atoms overlapped with the shell region, and Pd atoms were seen in the hollow region (Fig. 24.2c–e), evidencing the

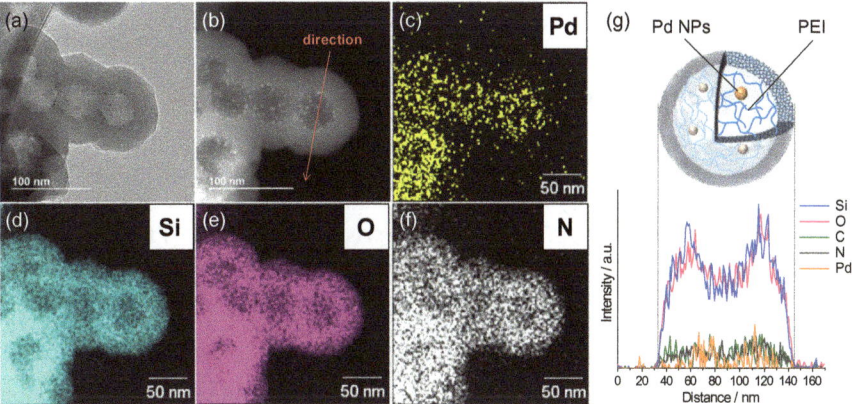

Fig. 24.2 a STEM image, **b** HAADF-STEM image, **c–f** the corresponding STEM elemental maps of **c** Pd, **d** Si, **e** O, and **f** N of Pd + PEI(L)@HSS. **g** STEM elemental line scan across the Pd + PEI(L)@HSS particle in **b**. Reprinted with permission from ACS Catalysis, 2019, 9, 1993 Ref. [11]. Copyright 2019 American Chemical Society

formation of Yolk–shell nanostructure consisting of Pd NPs as a core and silica as a shell. N atoms are uniformly distributed throughout the particles (Fig. 24.2f), indicating that the PEI is present in both shell and hollow regions. The line mapping of a selected silica particle confirmed the existence of N, C, and Pd atoms in the same axial region (Fig. 24.2g), suggesting a close proximity between Pd NPs and PEI. The presence of PEI (ca. 24.0 wt%) was confirmed by FTIR and thermogravimetric analyses, as well. Pd + PEI@HSS exhibited a clear hysteresis typical of hollow-structured solids in the range of $0.4 < p/p_0 < 1.0$ in N_2 adsorption–desorption isotherm. The Brunauer–Emmett–Teller (BET) surface area (S_{BET}) and total pore volume (V_{total}) were determined to be 34 m^2/g and 0.12 cm^3/g, respectively. The pore size distributions calculated by the Barrett–Joyner–Halenda (BJH) method confirmed the existence of broadly distributed mesopores with an average pore size of ca. 2.7 nm.

The coordination interaction between Pd NPs and PEI was characterized using X-ray absorption fine structure (XAFS) measurement. The X-ray absorption edges of the Pd + PEI@HSS were positioned between the absorption edges of Pd(0) foil and Pd(II)O, indicating the coexistence of Pd(0) and Pd(II) species. In the radial distribution functions (RDFs) obtained from the k^3-weighted Pd K-edge extended XAFS oscillations, Pd + PEI@HSS exhibited two distinct peaks ascribed to the Pd(II) species ligated to the nitrogen moieties of the PEI (at around $r = 1.6$ Å) and the Pd –Pd bond (at around $r = 2.5$ Å). These results, combined with elemental mapping analysis, indicate that the encapsulated Pd species are mainly present as Pd(0) NPs of which the surface is closely surrounded by PEI polymers.

24.2.2 Semihydrogenation of Alkynes by Hollow Silica Spheres Encapsulating Pd Nanoparticles and Poly(ethyleneimine)

The synthesized Pd + PEI@HSS catalysts were initially assessed in the semihy-drogenation of diphenylacetylene (**1**) with 0.5 mol % Pd in a flow of atmospheric pressure of H_2 at 30 °C (Fig. 24.3 (**A**)). Surprisingly, the Pd + PEI@HSS cata-lyst synthesized with linear PEI afforded *cis*-stilbene (**2**) with 90% selectivity, and *trans*-stilbene (**3**) and bibenzyl (**4**) were hardly produced even after an extended time of reaction. On the other hand, Pd + PEI@HSS catalyst calcined in air (a PEI-free analogue) gave **4** as the main product after an extended time of reaction. This result clearly demonstrates that PEI can efficiently suppress the overhydrogenation of stilbene owing to the productive poisoning effect on Pd NP. The Lindlar catalyst (used with 5 mol% quinoline) as a benchmark catalyst provided 94.1% selectivity toward **2** in 70 min of reaction but afterward showed a gradual decrease of selec-tivity due to the overhydrogenation into **4**. Pd/PEI as a silica-free analogue provided an excellent alkene selectivity (96% (*cis:trans* = 97:3)), but the reaction rate was apparently lower than that of Pd + PEI@HSS. Thus, the yolk –shell nanostructured Pd–PEI–silica composites provided a superior activity and alkene selectivity in the hydrogenation of internal alkyne.

The Yolk–shell nanostructured composite catalysts were also applicable to the selective hydrogenation of phenylacetylene (**5**) as a terminal alkyne (Fig. 24.3 (**B**)).

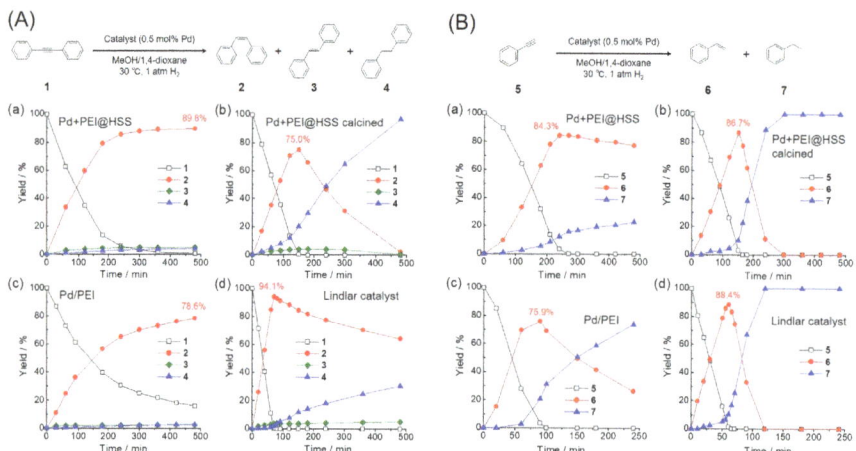

Fig. 24.3 Reaction kinetics in the semihydrogenation of (**A**) diphenylacetylene and (**B**) pheny-lacetylene over **a** Pd + PEI@HSS, **b** Pd + PEI@HSS calcined in air, **c** Pd/PEI, and **d** Lindlar catalyst (+5 mol% quinoline). *Reaction conditions:* catalyst (Pd 0.5 mol%), reactant (1 mmol), MeOH:1,4-dioxane = 1:1 (10 mL), 30 °C, under a flow of 1 atm H_2 (10 mL/min). Adapted with permission from ACS Catalysis, 2019, 9, 1993 Ref. [11]. Copyright 2019 American Chemical Society

The Pd + PEI@HSS catalyst synthesized with linear PEI afforded up to 84.3% yield of styrene (**6**) with an excellent alkene selectivity (87%), and subsequent hydrogenation into ethylbenzene (**7**) was significantly suppressed even after the complete consumption of **5**. Such a retention of high styrene selectivity was not observed for the PEI-removed catalyst, demonstrating that the surface poisoning of Pd NPs with PEI is the main cause for the improved alkene selectivity. A silica-free analogue, Pd/PEI, showed a similar reaction trend but showed a decreased alkene selectivity after an extended time of reaction. Lindlar catalyst was ineffective for the selective production of **6** under identical conditions, giving 100% yield of **7** after an extended time of reaction. Interestingly, Pd + PEI@HSS synthesized with branched PEI showed a significant decrease in styrene selectivity after the complete conversion of **5**, affording **7** as the main product. This result indicates that linear PEI more strongly coordinates on Pd NPs than branched PEI does. It is conceived that the different coordination ability of PEI is derived from the molecular structure of PEI; linear PEI, almost exclusively composed of secondary amines with less steric hindrance, can coordinate on Pd NPs more effectively, whereas branched PEI, containing primary, secondary, and tertiary amines, has a reduced ability to coordinate on Pd NPs due to the considerable steric hindrance around the tertiary amine sites. Thereby linear PEI allowed discriminative adsorption of alkynes over alkenes and prevented the overhydrogenation, thus lead to an increased alkene selectivity.

In the semihydrogenation of **1**, Pd + PEI@PEI exhibited excellent recyclability without any loss of activity and selectivity as well as reaction rate during five repeated runs. No significant structural changes were identified in TEM images and N_2 adsorption isotherms, and the leaching of PEI polymer and Pd NPs during the catalytic reaction was negligible as well. The Pd K-edge FT-EXAFS analysis indicated that the coordination between Pd(0) NPs and nitrogen ligands on PEI remained intact. Furthermore, Pd + PEI@HSS catalyst provided good selectivity for the semihydrogenation of other alkynes, including several kinds of aromatic and aliphatic internal alkynes, under the presence of atmospheric pressure of H_2. These results conclusively indicate that Pd + PEI@HSS catalyst can act as a stable, reusable, and efficient heterogeneous catalyst in the liquid-phase semihydrogenation of alkynes.

24.3 Design of Yolk–Shell Nanostructured Catalyst for Hydrogenation of CO_2

The effort toward the reduction of anthropogenic carbon dioxide (CO_2) and catalytic conversion of CO_2 into useful chemicals and fuels has been desirable in the past few decades [25, 26]. Catalytic transformation of CO_2 to formic acid (FA: HCOOH) has especially been regarded as a key reaction for realizing sustainable hydrogen energy cycles, because (i) FA can be used as a hydrogen storage compound with high stability, nontoxicity, and high H_2 storage capacity (4.4 wt.%) and (ii) CO_2 can be simultaneously stored in a chemically stable form [27, 28]. For the establishment of an efficient and economical hydrogen storage/utilization system utilizing CO_2 as an intermediate, the development of robust and reusable catalysts that can efficiently and selectively convert CO_2 to FA/formate is an urgent yet challenging task.

Recent studies demonstrated that the combination of Pd NPs (or PdAg NPs) with N-containing basic supports, such as amine-functionalized silica [29–31], amine-functionalized carbon [32, 33], and g-C_3N_4 [34–36], which promote CO_2 adsorption in the vicinity of the active Pd center, is the key for achieving high catalytic efficiency. Aminopolymers/polyamines containing a high density of amines are alternative compounds to those N-containing supports. Owing to their abilities to capture CO_2 under dry or wet environments at ambient temperature conditions, they have been employed as CO_2 adsorbents for Carbon Capture and Storage (CCS) technologies [14–16]. Besides, with their strong coordination ability with metals, they have been sometimes exploited as scaffolds to support metal complexes and metal NPs to demonstrate direct CO_2 transformation reactions [12, 37]. For example, Hicks et al. immobilized an Ir complex on a PEI-tethered iminophosphine ligand, which afforded formate from CO_2 at 120 °C and at a pressure of 40 bar (CO_2:H_2 = 1:1) because of a high CO_2-capturing ability of PEI [38]. Olah and Prakash et al. reported direct hydrogenation of CO_2 to methanol at 125–155 °C and at a pressure of 75 bar (CO_2:H_2 = 1:3) using a pincer-type Ru complex catalyst containing electron-donating ligands in the presence of pentaethylenehexamine, which can stabilize formate intermediate and boost the subsequent hydrogenation to produce methanol [39]. Thus, integration of a catalytically active center with high CO_2 hydrogenation ability and aminopolymer/polyamine with CO_2 capturing ability is expected to offer a promising strategy to promote efficient CO_2 hydrogenation; however, leaching of polymers and aggregation of active metals under such severe reaction conditions, and the resulting deactivation of catalyst still remain as major concerns.

Herein, we envisioned that hollow silica spheres having enclosed cavity spaces could be used as a promising platform to encapsulate PdAg NPs as a catalytic active center and PEI as a CO_2 adsorbent, since the silica shell is expected to act as a protective barrier to prevent the sintering of PdAg NPs and leaching of PEI under severe reaction conditions required for CO_2 hydrogenation.

24.3.1 Synthesis of Hollow Silica Spheres Encapsulating PdAg Alloy Nanoparticles and Poly(ethyleneimine)

The hollow mesoporous organosilica spheres encapsulating PdAg NPs and PEI (PdAg + PEI@HMOS) were synthesized on the basis of the method mentioned in the previous section [40]. Figure 24.4 schematically depicts the synthetic procedure of PdAg + PEI@HMOS composites using branched PEI (M_w = 1800) as an organic template. First, $NaBH_4$ as a reductant was added to an EtOH–water–NH_3 mixed solution containing designated amounts of PEI, Pd precursor (Na_2PdCl_4), and Ag precursor ($AgNO_3$) to form PdAg NP–PEI aggregates as nuclei (steps

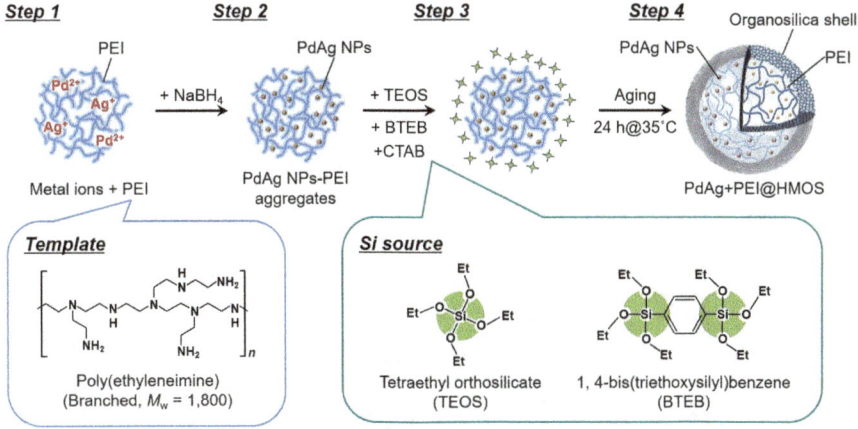

Fig. 24.4 Schematic illustration of the synthetic procedure for hollow mesoporous organosilica spheres encapsulating PdAg NPs and PEI (PdAg + PEI@HMOS). Reprinted with permission from ACS Catalysis, 2020, 10, 6356 Ref. [40]. Copyright 2020 American Chemical Society

1 and 2). Subsequently, two kinds of Si sources, tetraethyl orthosilicate (TEOS) and 1,4-bis(triethoxysilyl)benzene (BTEB), together with a pore-directing agent (cetyltrimethylammonium bromide (CTAB)) were added to create mesoporous organosilica network around the PdAg–PEI aggregates (step 3). BTEB was added to form organosilica shell, because organosilica network is known to be more tolerant to alkaline environment compared with pure siliceous network [41, 42]. The solution was aged at 35 °C for 24 h under gentle stirring, which induced a gradual transformation of solid spheres into hollow spheres via a continuous etching and recondensation of the silica species, thus resulted in a formation of PdAg + PEI@HMOS composite (step 4).

PdAg + PEI@HMOS was composed of spherical silicas with an average particle size of ca. 300 nm some of which are interconnected with each other (Fig. 24.5a). A hollow structure with an average shell thickness of ca. 50 nm was clearly observed in TEM images, and fine NPs (ave. diameter = 3.0 nm) with a uniform distribution were observed inside their hollow cavity spaces (Fig. 24.5b, c). The distributions of Si atoms and O atoms were well coincided with the shell region (Fig. 24.5d–g), verifying the formation of hollow silica structure. N, C, Pd, and Ag atoms were homogeneously distributed throughout the particle (Fig. 24.5h–k), indicating that PEI and PdAg NPs are located in both shell and hollow cavity regions. In the line mapping of a selected silica particle, N, C, Pd, and Ag atoms were uniformly distributed in the same axial region (Fig. 24.5m), suggesting a spatial proximity between PdAg NPs and PEI in the confined nanospace.

PdAg + PEI@HMOS exhibited an N_2 adsorption–desorption isotherm with a sharp increase in low relative pressure region and a large hysteresis in the range of $0.5 < p/p_0 < 1.0$, indicating the existence of hollow cavity spaces surrounded by

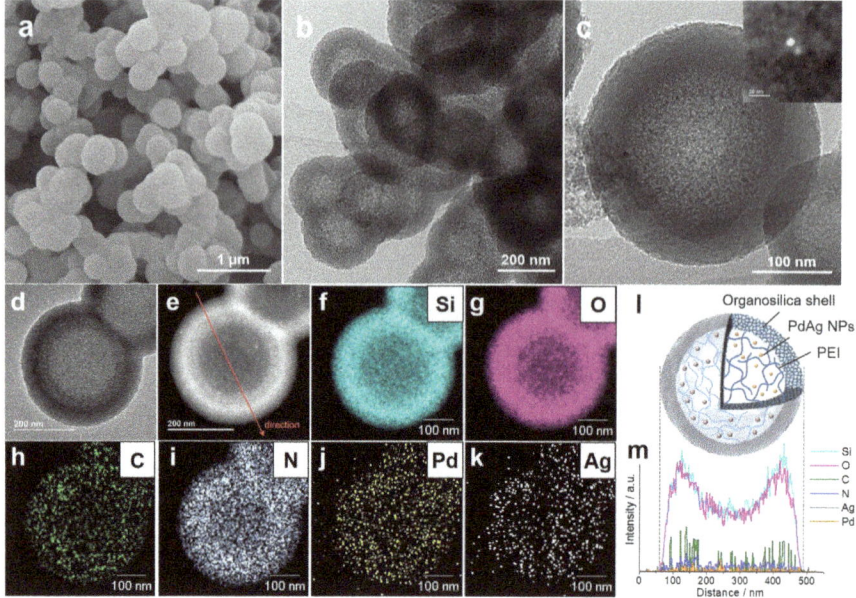

Fig. 24.5 **a** FE-SEM image, **b, c** TEM images (inset in **c** shows HAADF-SETM image of aggregated metal particle), **d** STEM image, **e** HAADF-STEM image, **f–k** and the corresponding STEM elemental maps of **f** Si, **g** O, **h** C, **i** N, **j** Pd and **k** Ag of PdAg + PEI@HMOS. **l** Illustration and **m** elemental line scan across the PdAg + PEI@HMOS particle along with the direction shown in **e**. Reprinted with permission from ACS Catalysis, 2020, 10, 6356 Ref. [40]. Copyright 2020 American Chemical Society

narrow mesopore channels. Based on the N_2 physisorption data, the peak pore diameter was estimated to be 1.8 nm by BJH method, and the BET surface area (S_{BET}) and total pore volume (V_{total}) were calculated to be 401 m^2/g and 0.50 cm^3/g, respectively, showing a highly porous nature sufficient for the reactants to access PdAg NPs. Thermogravimetric (TG) analysis showed two major weight losses, which were ascribed to the decomposition of PEI and benzene moieties in the shell. Solid-state ^{13}C cross-polarization magic angle spinning (CP/MAS) nuclear magnetic resonance (NMR) spectrum showed an intense signal at $\delta = 133.8$ ppm, which was assignable to the aromatic carbons connected to the siloxane networks, and several signals in the range of $\delta = 20$–70 ppm, which were ascribed to the carbon atoms contained in branched PEI. These results corroborate that the chemical structure of PEI was retained and Si-connected benzene units were successfully incorporated in the HMOS matrix upon the synthesis of PdAg + PEI@HMOS.

The chemical states of Pd and Ag atoms were analyzed with XPS and XAFS measurements. XPS spectra for Pd 3d and Ag 3d core levels showed obvious peaks after Ar etching, whereas no peaks were detected without etching treatment, verifying the selective encapsulation of Pd and Ag species inside the HMOS particles. The Pd component of PdAg + PEI@HMOS was found to be more electronegative in comparison with that of Pd + PEI@HMOS (synthesized without Ag). This result indicates the formation of more electron-enriched Pd species by alloying with Ag, due to a net charge transfer from Ag to contiguous Pd atoms possessing different ionization potentials [43]. In the Pd K-edge FT-EXAFS spectra, an intense peak for Pd–Pd bonds and a small peak for Pd–N bonds were observed due to the strong ligation of Pd atoms with amine sites of PEI [11, 32]. Furthermore, the Pd–Pd distance in the PdAg + PEI@HMOS catalyst was slightly longer compared with that for Pd foil and Ag-free analogue, which might be due to the creation of heteroatomic Pd–Ag bonds [32, 43]. These facts conclusively indicate that metallic PdAg NPs are encaged within HMOS particles together with PEI in close proximity.

24.3.2 Hydrogenation of CO_2 by Hollow Silica spheres Encapsulating PdAg alloy nanoparticles and Poly(ethyleneimine)

Gibbs free energy change for the direct CO_2 hydrogenation to FA in aqueous solution is slightly negative (CO_2(aq.) + H_2(aq.) → HCOOH(aq.), $\Delta G°_{298}$ = −4 kJ/mol). An alkaline aqueous solution containing homogeneous bases (such as NEt$_3$, NaHCO$_3$, and NaOH) can decrease Gibbs free energy change by transforming CO_2 to formate salts as products (CO_2(aq.) + H_2(aq.) + B → HCO_2^-(aq.) + BH^+ (B: base), $\Delta G°_{298}$ = − 35.4 kJ/mol), hence accelerates the reaction rate [44]. To eliminate the possibility of adventitious formate generation from initially added homogeneous bases (e.g. NaHCO$_3$), CO_2 hydrogenation in this study was examined using 0.1 M NaOH aqueous solution.

PdAg + PEI@HMOS catalyst exhibited the highest activity under a total pressure of 2.0 MPa (CO_2:H$_2$ = 1:1) at 100 °C, affording a substantial turnover frequency (TON) of 2754 in 22 h based on the number of Pd atoms contained in the catalysts, which corresponds to a turnover frequency (TOF) of 125 h^{-1} (entry 1 in Table 24.1). It is worth noting that the TOF achieved by the PdAg + PEI@HMOS is higher than the values of other Pd-based heterogeneous catalysts reported in the literature under similar conditions; Pd/mpg-C$_3$N$_4$ (1.6 h^{-1} at 100 °C, 4 MPa in NEt$_3$ aq.) [34, 35], PdAg/SBA-15-phenylamine (36 h^{-1} at 100 °C, 2 MPa in NaHCO$_3$ aq.) [29], PdAg/amine-modified mesoporous carbon (35 h^{-1} at 100 °C, 2 MPa in NaHCO$_3$ aq.) [32], PdAg/amine-modified resorcinol–formaldehyde polymers (36 h^{-1} at 100 °C, 2 MPa in NaHCO$_3$ aq.) [33], and PdAg@mesoporous hollow carbon sphere (112 h^{-1} at 100 °C, 2 MPa in NaHCO$_3$ aq.) [45], and it was even higher than that of Pd@Ag/TiO$_2$ (104 h^{-1} at 100 °C, 2 MPa in NaHCO$_3$ aq.) [43]. Unsupported PdAg/PEI in the form of colloidal solution gave a significantly lower activity than PdAg + PEI@HMOS (TON = 430, entry 8), because of the aggregation of PdAg NPs during the reaction. Ag-free analogue (Pd + PEI@HMOS) synthesized without the addition of Ag was found to be less active (TON = 455, entry 2) than

PdAg + PEI@HMOS. This result can be interpreted by the creation of more electronegative Pd atoms caused by charge transfer from the neighboring Ag(0) atoms. A PEI-free analogue (PdAg@HMOS), synthesized via calcination in air and a subsequent hydrogenation with H_2 at 200 °C, afforded TON of 279 under the identical conditions (entry 3), which was one tenth lower than that of the PdAg + PEI@HMOS. External addition of branched PEI together with PdAg@HMOS showed a substantial activity improvement compared with PdAg@HMOS alone, giving TON of 550 (entry 4), but it was far lower than that of the PdAg + PEI@HMOS. These results clearly demonstrate that PEI confined in the hollow cavity of HMOS plays a vital role for boosting the hydrogenation activity of PdAg NPs in the present catalytic system.

Kinetic analyses revealed that the CO_2 hydrogenation rate became less dependent on CO_2 pressure when PEI was encapsulated in HMOS (Fig. 24.6a), while the reaction order against H_2 pressure remained unchanged irrespective of the presence or absence of PEI (Fig. 24.6b). This result indicates that the CO_2 adsorption/insertion step is significantly facilitated by PEI, while the H_2 dissociation step to form active Pd-hydride is hardly influenced by PEI. In a gas-phase CO_2 adsorption measurement under humid CO_2 condition, a higher uptake of CO_2 was observed for PdAg + PEI@HMOS (0.64 mmol-CO_2/g), which was 7.7-fold higher than that of PEI-free analogue (0.08 mmol-CO_2/g) (Fig. 24.6c). CO_2 adsorption behavior over these two samples was further monitored by in-situ FTIR spectroscopy (Fig. 24.6d). The PdAg + PEI@HMOS catalyst exhibited several absorption bands originating from CO_2 upon exposure to water-saturated gaseous CO_2, while such bands were hardly

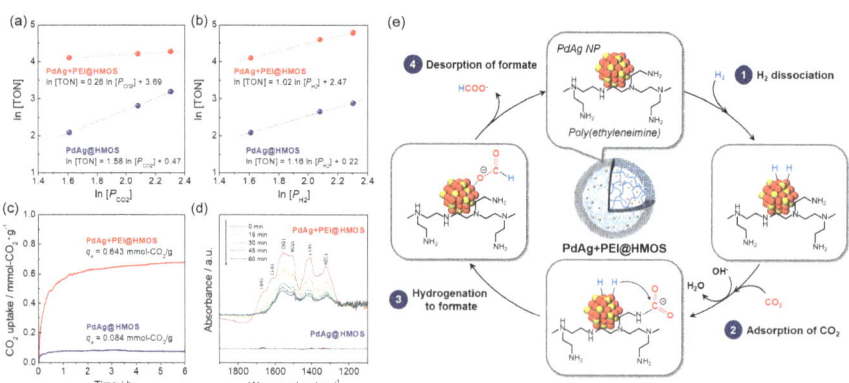

Fig. 24.6 **a** Double logarithm plots of TON and the partial pressure of CO_2 (P_{CO2}) and **b** double logarithm plots of TON and the partial pressure of H_2 (P_{H2}) in the CO_2 hydrogenation over PdAg + PEI@HMOS and PdAg@HMOS catalysts (*Reaction conditions*: catalyst (10 mg, Pd 0.38 μmol), solvent (0.1 M NaOH aq., 15 mL), Temp. = 100 °C, t = 2 h). **c** Kinetics in CO_2 adsorption at 40 °C under a flow of 10% CO_2/N_2. **d** in-situ FTIR difference spectra of CO_2 desorbed from CO_2-saturated samples at 100 °C as a function of time under a flow of N_2. **e** Plausible reaction mechanism for the CO_2 hydrogenation to produce formate over PdAg + PEI@HMOS catalyst. Reprinted with permission from ACS Catalysis, 2020, 10, 6356 Ref. [40]. Copyright 2020 American Chemical Society

observed on PEI-removed catalyst. In addition, the low-intensity absorption band assignable to the vibration of C=O bond derived from the weakly adsorbed carbamate species was detected, which was easily desorbed under a flow of N_2. These results conclusively indicate that the amine sites of PEI serve as CO_2-capturing sites to adsorb/enrich CO_2 molecules, most likely forming ionic carbamate species, and facilitate the access of CO_2 to the neighboring active PdAg NPs, thus affording a higher catalytic activity.

Figure 24.6e depicts a possible reaction mechanism for CO_2 hydrogenation over PdAg + PEI@HMOS catalyst. A gaseous H_2 is activated and dissociated on PdAg NPs to form active Pd-hydride species (step 1). Simultaneously, a gaseous CO_2 is captured by the amine sites on PEI through an ionic carbamate formation and is concentrated nearby the PdAg NPs, in which an initially formed zwitterionic intermediates (*i.e.*, $RNH_2 + CO_2 \rightarrow RN^+H_2...COO^-$) undergo deprotonation by a free strong base in aqueous solution (OH^-), thereby resulting in a carbamate formation (i.e., $RN^+H_2...COO^- + OH^- \rightarrow RNHCOO^- + H_2O$) (step 2). Then, nucleophilic attack by the hydride species onto C atoms of the carbamate intermediate takes place to give a formate intermediate (step 3). In this step, the electron-rich Pd species, caused by the electron donation from the surrounding Ag atoms, provides a more negative hydride species, which is more reactive for the nucleophilic attack to the C atoms than that formed on monometallic Pd NPs. Finally, formate is produced as a result of acid-base neutralization in an alkaline environment, which regenerates the initial active species (step 4). Thus, PEI plays an important role in this catalytic cycle primarily by promoting the adsorption/enrichment of CO_2 in the vicinity of PdAg NPs and possibly by entrapping the produced formate anions. Such cooperative action between PEI and PdAg NPs explains the high catalytic activity for CO_2 hydrogenation. Besides the above direct CO_2 hydrogenation mechanism, some fractions of CO_2 molecules may be hydrogenated through a typical bicarbonate hydrogenation route ($HCO_3^- + H_2 \rightarrow HCOO^- + H_2O$), in which PEI also serves as productive sites to promote the adsorption/enrichment of bicarbonate ions in the vicinity of PdAg NPs, thus accelerating the hydrogenation rate.

The PdAg + PEI@HMOS catalyst was easily recoverable and reusable at least five cycles without any appreciable loss of CO_2 hydrogenation ability, reaching a total TON of 13,700 over 110 h of reaction, demonstrating its long-term stability. STEM image and N_2 physisorption measurement revealed that the original hollow structure remained unchanged even after the repeated catalytic cycles, and the agglomeration of PdAg NPs was hardly observed. Furthermore, leaching of Pd, Ag, and PEI in the reaction solution was negligible, and the coordination structures of Pd and Ag atoms in PdAg NPs were unchanged upon the reusability test. Such excellent catalyst reusability and stability are likely to be endowed by the ability of HMOS to protect the encapsulated components and its high alkali-tolerant property.

24.4 Summary and Outlook

In this chapter, we showed the design and synthesis of new types of Yolk–shell nanostructured catalysts encapsulating Pd NPs together with PEI, which showed unprecedented unique catalytic performances in (i) semihydrogenation of alkynes to alkenes and (ii) CO_2 hydrogenation to formate. In the former case, the Yolk–shell nanostructured composite could act as an efficient and reusable heterogeneous catalyst in the semihydrogenation of alkynes in the presence of 1 atm H_2, giving a markedly improved alkene selectivity. This was attributed to the poisoning effect of PEI on the Pd NP surface to allow discriminative adsorption of alkynes over alkenes. In the latter case, the Yolk–shell nanostructured composite displayed a significantly higher catalytic activity for CO_2 hydrogenation to produce formate, under relatively mild conditions, compared with other supported Pd catalysts reported so far, owing to the ability of PEI to capture CO_2. Thus, it was demonstrated that integration of a catalytically active center with hydrogenation ability and aminopolymer with unique chemical properties within a confined hollow silica nanospace can offer new catalytic nano-environments. In both cases, hollow silica spheres having enclosed large cavity spaces could act as a rigid nano-cage to prevent leaching and aggregation of Pd NPs and PEI during the catalytic reactions, which led to an improved stability and reusability of the catalysts.

As demonstrated in the latter case, further tuning of core particles (alloying with a secondary metal) and controlling the porosity in the silica shell (creating mesopores), as well as adding further functionalities in the silica shell (incorporating organic moieties) is possible, while retaining the Yolk–shell nanostructure, which would allow us to design and synthesize further elaborated heterogeneous catalysts effective for targeted chemical reactions. Hollow carbon spheres and hollow-structured metal oxides have also been exploited recently as host materials alternative to hollow silica spheres in order to construct similar multifunctionalized heterogeneous catalysts. Furthermore, it is highly expected that encapsulation of other functional polymers, other than PEI, could offer undiscovered new catalytic opportunities and synergism for Yolk–shell nanostructured catalysts. However, there are still some technical challenges for the synthesis of Yolk–shell nanostructured catalysts; for example, establishing controllable fabrication with desired morphology, composition, shell thickness, and catalytic performances. Developing simple and low-cost methods for large-scale preparations is also important for their practical applications. Fundamental studies to understand the structure–activity relationship and the cooperative action between the encapsulated catalytic components are also important to optimize the nanostructures.

In summary, the development of Yolk–shell nanostructured catalysts with multi-functionalities can be regarded as a potential research area and is worthy of further investigation. We expect that our catalyst design strategy combined with the advantages of hollow silicas and aminopolymers opens up a new opportunity for the development of a wide range of multifunctionalized heterogeneous catalysts in the near future.

Table 24.1 CO_2 Hydrogenation to produce formate using various Pd catalysts[a]

Entry	Catalyst	Temp (°C)	Time (h)	TON[b]	TOF[b] (h^{-1})
1	PdAg + PEI@HMOS	100	22	2754	125
2	Pd + PEI@HMOS	100	22	455	20.7
3	PdAg@HMOS (without PEI)[c]	100	22	279	12.7
4	PdAg@HMOS[c] + PEI (physical mixture)[d]	100	22	550	25
5	PdAg/fumed SiO_2	100	22	63	2.9
6	PdAg/AC[e]	100	22	71	3.2
7	PdAg/TiO$_2$[f]	100	22	519	23.6
8	PdAg/PEI	100	22	430	19.5

[a]*Reaction conditions*: catalyst (10 mg, Pd 0.38 μmol), solvent (0.1 M NaOH aq., 15 mL), H_2 (1.0 MPa), CO_2 (1.0 MPa). [b]Calculated based on the number of Pd atoms contained in the catalysts. [c]Prepared from PdAg + PEI@HMOS by calcination in air at 500 °C, followed by H_2 reduction at 200 °C. [d]8.0 mg of PdAg@HMOS (without PEI) and 2.0 mg of branched PEI (M_w = 1,800) was used. [e]Active carbon (Osaka Gas Chemicals, Shirasagi®) was used as a support. [f]Titanium dioxide (Evonik, P25®) was used as a support

References

1. Perez-Lorenzo M, Vaz B, Salgueirino V, Correa-Duarte MA (2013) Hollow–shelled nanoreactors endowed with high catalytic activity. Chem Eur J 19(37):12196–12211
2. Li Y, Shi J (2014) Hollow-structured mesoporous materials: chemical synthesis, functionalization and applications. Adv Mater 26(20):3176–3205
3. Lee J, Kim SM, Lee IS (2014) Functionalization of hollow nanoparticles for nanoreactor applications. Nano Today 9(5):631–667
4. Purbia R, Paria S (2015) Yolk/shell nanoparticles: classifications, synthesis, properties, and applications. Nanoscale 7(47):19789–19873
5. El-Toni AM, Habila MA, Labis JP, ALOthman ZA, Alhoshan M, Elzatahry AA, Zhang F (2016) Design, synthesis and applications of core–shell, hollow core, and nanorattle multifunctional nanostructures. Nanoscale 8(5):2510–2531
6. Kuwahara Y, Ando T, Kango H, Yamashita H (2017) Palladium nanoparticles encapsulated in hollow titanosilicate spheres as an ideal nanoreactor for one-pot oxidation. Chem Eur J 23(2):380–389
7. Kuwahara Y, Matsumura R, Yamashita H (2019) Hollow titanosilicate nanospheres encapsulating PdAu alloy nanoparticles as reusable high-performance catalysts for a H_2O_2-mediated one-pot oxidation reaction. J Mater Chem A 7(12):7221–7231
8. Qiao ZA, Huo Q, Chi M, Veith GM, Binder AJ, Dai S (2012) A "ship-in-a-bottle" approach to synthesis of polymer dots@silica or polymer dots@carbon core–shell nanospheres. Adv Mater 24(45):6017–6021
9. Du X, Yao L, He J (2012) One-pot fabrication of noble-metal nanoparticles that are encapsulated in hollow silica nanospheres: Dual roles of poly(acrylic acid). Chem Eur J 18(25):7878–7885
10. Qiao ZA, Zhang P, Chai SH, Chi M, Veith GM, Gallego NC, Kidder M, Dai S (2014) Lab-in-a-shell: encapsulating metal clusters for size sieving catalysis. J Am Chem Soc 136(32):11260–11263

11. Kuwahara Y, Kango H, Yamashita H (2019) Pd nanoparticles and aminopolymers confined in hollow silica spheres as efficient and reusable heterogeneous catalysts for semihydrogenation of alkynes. ACS Catal 9(3):1993–2006
12. Kobayashi S, Hiroishi K, Tokunoh M, Saegusa T (1987) Chelating properties of linear and branched poly(ethyleneimines). Macromolecules 20(7):1496–1500
13. Sajiki H, Mori S, Ohkubo T, Ikawa T, Kume A, Maegawa T, Monguchi Y (2008) Partial hydrogenation of alkynes to *cis*-olefins by using a novel Pd(0)-polyethyleneimine catalyst. Chem Eur J 14(17):5109–5111
14. Sanz-Perez ES, Murdock CR, Didas SA, Jones CW (2016) Direct capture of CO_2 from ambient air. Chem Rev 116(19):11840–11876
15. Shen X, Du H, Mullins RH, Kommalapati RR (2017) Polyethylenimine applications in carbon dioxide capture and separation: From theoretical study to experimental work. Energy Technol 5(6):822–833
16. Kuwahara Y, Kang DY, Copeland JR, Brunelli NA, Didas SA, Bollini P, Sievers C, Kamegawa T, Yamashita H, Jones CW (2012) Dramatic enhancement of CO_2 uptake by poly(ethyleneimine) using zirconosilicate supports. J Am Chem Soc 134(26):10757–10760
17. Mitsudome T, Takahashi Y, Ichikawa S, Mizugaki T, Jitsukawa K, Kaneda K (2013) Metal-ligand core–shell nanocomposite catalysts for the selective semihydrogenation of alkynes. Angew Chem Int Ed 52(5):1481–1485
18. Crespo-Quesada M, Cárdenas-Lizana F, Dessimoz A-L, Kiwi-Minsker L (2012) Modern trends in catalyst and process design for alkyne hydrogenations. ACS Catal 2(8):1773–1786
19. Niu W, Gao Y, Zhang W, Yan N, Lu X (2015) Pd-Pb alloy nanocrystals with tailored composition for semihydrogenation: taking advantage of catalyst poisoning. Angew Chem Int Ed 54(28):8271–8274
20. Lindlar H (1952) Ein neuer katalysator für selektive hydrierungen. Helv Chim Acta 35:446–450
21. Moreno M, Kissell LN, Jasinski JB, Zamborini FP (2012) Selectivity and reactivity of alkylamine- and alkanethiolate-stabilized Pd and PdAg nanoparticles for hydrogenation and isomerization of allyl alcohol. ACS Catal 2(12):2602–2613
22. San KA, Chen V, Shon YS (2017) Preparation of partially poisoned alkanethiolate-capped platinum nanoparticles for hydrogenation of activated terminal alkynes. ACS Appl Mater Interfaces 9(11):9823–9832
23. Yoshii T, Umemoto D, Kuwahara Y, Mori K, Yamashita H (2019) Engineering of surface environment of Pd nanoparticle catalysts on carbon support with pyrene-thiol ligands for semihydrogenation of alkynes. ACS Appl Mater Interfaces 11(41):37708–37719
24. Long W, Brunelli NA, Didas SA, Ping EW, Jones CW (2013) Aminopolymer–silica composite-supported Pd catalysts for selective hydrogenation of alkynes. ACS Catal 3(8):1700–1708
25. Quadrelli EA, Centi G, Duplan JL, Perathoner S (2011) Carbon dioxide recycling: emerging large-scale technologies with industrial potential. ChemSusChem 4(9):1194–1215
26. Alvarez A, Bansode A, Urakawa A, Bavykina AV, Wezendonk TA, Makkee M, Gascon J, Kapteijn F (2017) Challenges in the greener production of formates/formic acid, methanol, and dme by heterogeneously catalyzed CO_2 hydrogenation processes. Chem Rev 117(14):9804–9838
27. Grasemann M, Laurenczy G (2012) Formic acid as a hydrogen source—recent developments and future trends. Energy Environ Sci 5(8):8171–8181
28. Wang WH, Himeda Y, Muckerman JT, Manbeck GF, Fujita E (2015) CO_2 hydrogenation to formate and methanol as an alternative to photo- and electrochemical CO_2 reduction. Chem Rev 115(23):12936–12973
29. Mori K, Masuda S, Tanaka H, Yoshizawa K, Che M, Yamashita H (2017) Phenylamine-functionalized mesoporous silica supported PdAg nanoparticles: A dual heterogeneous catalyst for formic acid/CO_2-mediated chemical hydrogen delivery/storage. Chem Commun 53(34):4677–4680
30. Liu Q, Yang X, Li L, Miao S, Li Y, Li Y, Wang X, Huang Y, Zhang T (2017) Direct catalytic hydrogenation of CO_2 to formate over a schiff-base-mediated gold nanocatalyst. Nat Commun 8(1):1407

31. Zhong H, Iguchi M, Chatterjee M, Ishizaka T, Kitta M, Xu Q, Kawanami H (2018) Inter-conversion between CO_2 and HCOOH under basic conditions catalyzed by PdAu nanoparticles supported by amine-functionalized reduced graphene oxide as a dual catalyst. ACS Catal 8(6):5355–5362

32. Masuda S, Mori K, Futamura Y, Yamashita H (2018) PdAg nanoparticles supported on functionalized mesoporous carbon: promotional effect of surface amine groups in reversible hydrogen delivery/storage mediated by formic acid/CO_2. ACS Catal 8(3):2277–2285

33. Masuda S, Mori K, Kuwahara Y, Yamashita H (2019) PdAg nanoparticles supported on resorcinol-formaldehyde polymers containing amine groups: the promotional effect of phenylamine moieties on CO_2 transformation to formic acid. J Mater Chem A 7(27):16356–16363

34. Lee JH, Ryu J, Kim JY, Nam S-W, Han JH, Lim T-H, Gautam S, Chae KH, Yoon CW (2014) Carbon dioxide mediated, reversible chemical hydrogen storage using a Pd nanocatalyst supported on mesoporous graphitic carbon nitride. J Mater Chem A 2(25):9490–9495

35. Park H, Lee JH, Kim EH, Kim KY, Choi YH, Youn DH, Lee JS (2016) A highly active and stable palladium catalyst on a g–C_3N_4 support for direct formic acid synthesis under neutral conditions. Chem Commun 52(99):14302–14305

36. Mondelli C, Puertolas B, Ackermann M, Chen Z, Perez-Ramirez J (2018) Enhanced base-free formic acid production from CO_2 on Pd/g-C_3N_4 by tuning of the carrier defects. Chemsuschem 11(17):2859–2869

37. Kretschmer F, Mansfeld U, Hoeppener S, Hager MD, Schubert US (2014) Tunable synthesis of poly(ethylene imine)-gold nanoparticle clusters. Chem Commun 50(1):88–90

38. McNamara ND, Hicks JC (2014) CO_2 capture and conversion with a multifunctional polyethyleneimine-tethered iminophosphine iridium catalyst/adsorbent. ChemSusChem 7(4):1114–1124

39. Kothandaraman J, Goeppert A, Czaun M, Olah GA, Prakash GK (2016) Conversion of CO_2 from air into methanol using a polyamine and a homogeneous ruthenium catalyst. J Am Chem Soc 138(3):778–781

40. Kuwahara Y, Fujie Y, Mihogi T, Yamashita H (2020) Hollow mesoporous organosilica spheres encapsulating PdAg nanoparticles and poly(ethyleneimine) as reusable catalysts for CO_2 hydrogenation to formate. ACS Catal 10(11):6356–6366

41. Ma N, Deng Y, Liu W, Li S, Xu J, Qu Y, Gan K, Sun X, Yang J (2016) A one-step synthesis of hollow periodic mesoporous organosilica spheres with radially oriented mesochannels. Chem Commun 52(17):3544–3547

42. Zou H, Wang R, Dai J, Wang Y, Wang X, Zhang Z, Qiu S (2015) Amphiphilic hollow porous shell encapsulated Au@Pd bimetal nanoparticles for aerobic oxidation of alcohols in water. Chem Commun 51(78):14601–14604

43. Mori K, Sano T, Kobayashi H, Yamashita H (2018) Surface engineering of a supported PdAg catalyst for hydrogenation of CO_2 to formic acid: elucidating the active Pd atoms in alloy nanoparticles. J Am Chem Soc 140(28):8902–8909

44. Laurenczy G, Dyson PJ (2014) Homogeneous catalytic dehydrogenation of formic acid: progress towards a hydrogen-based economy. J Braz Chem Soc 25(12):2157–2163

45. Yang G, Kuwahara Y, Masuda S, Mori K, Louis C, Yamashita H (2020) PdAg nanoparticles and aminopolymer confined within mesoporous hollow carbon spheres as an efficient catalyst for hydrogenation of CO_2 to formate. J Mater Chem A 8(8):4437–4446

Chapter 25
Inspiration of Yolk-Shell Nanostructures Toward Completely Adjustable Heterogeneous Catalysts

Hyunjoon Song

25.1 Introduction—Importance of Model Catalysts for Heterogeneous Catalytic Reactions

Today, more than 80% of industrial products are being made using catalysts, mostly in heterogeneous forms. There have been many catalytic reactions developed in a state-of-the-art manner, including methanol synthesis, Fischer-Tropsch reactions, CO insertion, and so on. Although these catalysts have intensively been studied for several decades, their fundamental aspects were still hardly understood mainly due to numerous factors essential to reaction activity, extreme sensitivity on reaction environments, and complex intermediate structures in multiple phases. Hence, heterogeneous catalysts are still in the state of 'black box', and their development depends on works-by-accident or trial-in-error in the lack of deep understanding.

To overcome these difficulties, many researchers tried to simplify the catalyst morphology and devise characterization tools to match the catalyst dimensions. They began to use a single crystalline metal surface as a two-dimensional 'model catalyst', and put gas-phase reactants to be adsorbed on the surface. Then, surface-sensitive spectroscopic techniques provided critical information on surface states and intermediate structures, which could successfully unveil reaction mechanisms on the surface. From these observations, the surface heterogeneity, including steps, kinks, and holes, was more significant in the reaction activity and selectivity than the flat surface was. Therefore, actual reaction sites and species became the most critical parts to be identified for understanding catalytic properties [1].

Although two-dimensional model catalysts made us understand the intermediate structures and mechanistic details in many heterogeneous systems, actual catalysts

H. Song (✉)
Department of Chemistry, Korea Advanced Institute of Science and Technology (KAIST), 291 Daehak-ro, Yuseong-gu, Daejeon 34141, Republic of Korea
e-mail: hsong@kaist.ac.kr

© The Author(s), under exclusive license to Springer Nature Singapore Pte Ltd. 2021   413
H. Yamashita and H. Li (eds.), *Core-Shell and Yolk-Shell Nanocatalysts*,
Nanostructure Science and Technology,
https://doi.org/10.1007/978-981-16-0463-8_25

are in three dimensions. Moreover, the most common forms of the heterogeneous catalysts are bifunctional with metal nanoparticles on metal oxide supports [2]. To approach these real systems, better catalyst models in three dimensions are required.

25.2 Bifunctional Catalysts in Discrete Structures

Since nanochemistry and nanotechnology have emerged, researchers could synthesize nanoparticles with various sizes, shapes, and compositions. It opened a new possibility to revisit structure-property relationships in various reaction systems. Yang and Somorjai et al. firstly designed new three-dimensional model catalysts. Bifunctional catalysts involve two separable parts—active nanoparticles and high surface area supports. The size and shape of the nanoparticles are modulated by synthetic techniques in nanoscale. The pore size and arrangement of the supports are also precisely controllable. If we simply mix together and induce the chemical interaction between two parts, all factors would be separately adjustable [3]. In the experiment, Pt nanoparticles were synthesized in the presence of the identical surfactant—poly(vinyl pyrrolidone) (PVP). As the reaction temperature increased, the reduction rate increased, and the particle size decreased. Consequently, the portfolio of the Pt nanoparticle size was obtained in the range of 1.7–7.1 nm. Then, the particles were embedded in the representative mesoporous silica, SBA-15, which had straight channels with a very regular pore size. This method was the so-called 'capillary inclusion' (Fig. 25.1a), which had the only variant of particle size, otherwise identical catalyst structure [4]. The activities of two distinct reactions showed the exact particle-size dependencies as expected in two-dimensional model systems. The catalysts showed constant activities in ethylene hydrogenation, but exhibited linear dependence on particle size in ethane hydrogenolysis. This system was the first example of

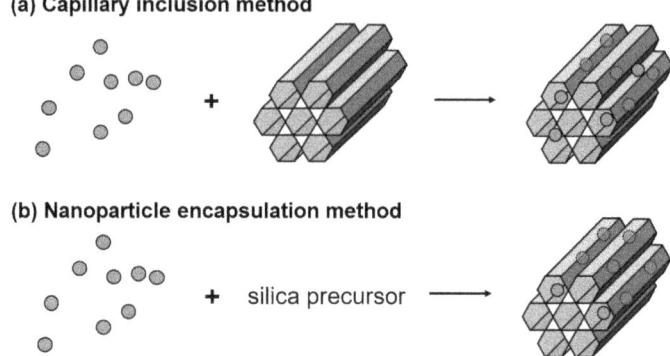

(a) Capillary inclusion method

(b) Nanoparticle encapsulation method

+ silica precursor

Fig. 25.1 Three-dimensional bifunctional model catalysts synthesized through (**a**) a capillary incipient method, and (**b**) a nanoparticle encapsulation method

a 'three-dimensional' model catalyst showing a direct particle size-catalytic activity relationship with only one control factor.

Although the capillary inclusion method was successful in showing the size-dependent reaction properties, the particles might behave as independent catalysts without interaction with supports. To make the real bifunctional catalysts, the reaction mixture was prepared with the pre-defined nanoparticles. It formed mesoporous silica walls bearing the nanoparticles under the synthetic conditions, which was the so-called 'nanoparticle encapsulation' (Fig. 25.1b) [5]. The resulting bifunctional catalysts showed no dependency for ethylene hydrogenation but a strong correlation on particle size for ethane hydrogenolysis. By these three-dimensional model catalysts, they confirmed the literal particle size dependencies of the heterogeneous catalytic reactions without ambiguous interference of other factors. Based on the results, the term 'nanostructured catalyst', briefly 'nanocatalyst', could be defined as the catalyst having a well-defined nanostructure where its size, morphology, composition, and surface structure are precisely adjusted by chemical and physical synthetic techniques [6].

25.3   Inspiration of Yolk-Shell Structure Design

Using discrete nanoparticles, constructing bifunctional catalysts were able to show the size-dependent reaction properties directly. However, the other part of the bifunctional catalysts, the support, still had a continuous morphology more than micrometers in size. Toward the perfect control of the bifunctional catalysts, the void space (or pores) as well as the active particle surface is better to be in the same dimension to be precisely controllable in the same way. In this respect, the support size would be reduced to nanometers, comparable to the size of nanoparticles. But the metal particles should still be embedded inside the pores. Then, the perfect design would be a metal-metal oxide hybrid structure, and the metal particle is surrounded by nanostructured metal oxide support—it may be a core-shell structure. However, direct contact of the metal particle and the metal oxide support may block the active surface to limit the reactant and product diffusions. To avoid it, the metal particle should be detached from the metal oxide layer, and the final catalyst design was the metal cores located in the void inside the metal oxide hollow layer [7]. Initially, this structure was called a 'nanorattle' based on its shape. But many researchers preferred to use the term 'yolk-shell', to compare with the core-shell nanoparticles, although some people argued that the structure did not have any egg whites in the void region.

25.3.1   Nanosized Reactors or Nanoreactors

The first use of the yolk-shell design was described in the nanoscale hollow formation by the Kirkendall effect. Alivisatos and Somorjai et al. observed that the oxidation

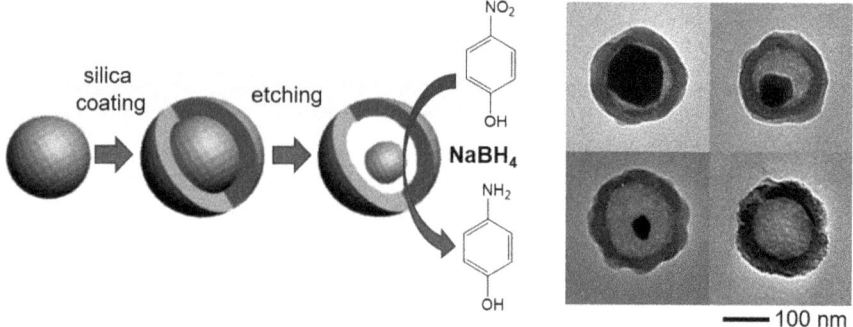

Fig. 25.2 Nanoreactor concepts on Au@SiO$_2$ yolk-shell catalysts with different core sizes formed by the core dissolution with KCN Adapted with permission from Ref. [9]

of Co nanoparticles made a hollow structure due to the difference in diffusion rates inward and outward. When the Pt@Co core-shell nanoparticles were oxidized, a Pt@CoO$_x$ yolk-shell structure was yielded, which was used for ethylene hydrogenation. They used the term 'nanoreactor', because the CoO$_x$ hollow shell structure was similar to that of the reaction chamber with a nanoscale void, and it contained a Pt nanoparticle catalyst inside [8].

The nanoreactor was established as a general concept when the metal@silica yolk-shell nanoparticles were used as a catalyst. Song et al. synthesized Au@SiO$_2$ core-shell nanoparticles, where the Au core was partially dissolved by KCN treatment [9]. The core size was controlled from 104 to 67 and 43 nm by the repetitive treatment (Fig. 25.2). The resulting Au@SiO$_2$ yolk-shell structure was employed as a catalyst for the reduction of p-nitrophenol in the presence of NaBH$_4$. The reactions followed pseudo-first-order kinetics, which was monitored by UV-Vis spectroscopy. Interestingly, the core-shell structure did not progress any reaction, but the yolk-shell structures with different sized Au cores exhibited strong size dependency on the reaction rate and turnover frequency (TOF). In particular, the TOF increased five times from 6.6 to 36 s^{-1} as the particle size decreased from 104 to 43 nm, due to the increase of low coordination sites in smaller nanoparticles.

25.3.2 Enzyme-Mimicking Active Chambers

An enzyme is a biocatalyst in nature and is known to be the most effective catalyst in numerous reactions, even with high activation energies. In organometallics, many researchers tried to mimic the structure of active centers and their environments. The main features near the active centers of metalloenzymes are: the metal center is located in the pockets surrounded by proteins. The ligands coordinating to the metal center alter its electronic property, and the surrounding environment offers steric hindrance as well as functionalities enhancing intermediate stability. The pockets

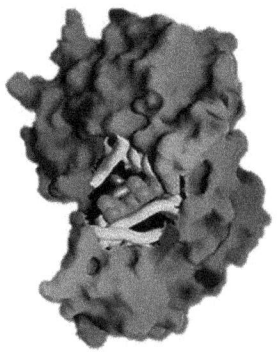 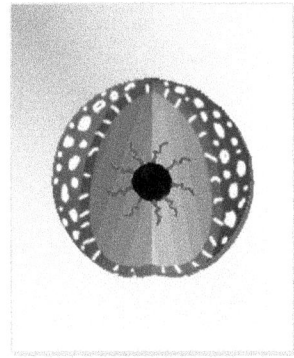

phase	homogeneous	heterogeneous
size range	angstrom + nanometer	nanometer
active site	metal center / ligand	metal particle/surfactant
surroundings	protein shell with a void chamber	metal oxide shell with a void chamber

Fig. 25.3 Striking similarity of yolk-shell nanoparticles to the enzyme structure

have pores and holes to diffuse substrates and products and are sometimes dynamic in structure as the response of binding the substrates. All these factors promote the reaction progress and surprisingly enhance catalytic properties despite neutral and ambient reaction conditions.

The yolk-shell structure shows notable similarity to the enzymes, although the size range is different (Fig. 25.3). The active metal domain is located at the void center, which is surrounded by metal oxide hollow shells. The surfactants and ligands, binding at the metal surface, influence reaction activity. The hollow shells have mesopores which can control the diffusion of reactants and products. Specific functionalities may be introduced either on the metal surface or on the inner side of the hollow structure to assist the reaction property.

In the actual experiments, Song et al. changed the porosity of silica shells by introducing a different amount of porogens, C_{18}TMS (octadecyltrimethoxysilane) [10]. After the thermal treatment, the diffusion coefficient of solvents changed 3.5 fold, and the turnover frequency of o-nitroaniline reduction increased 7 times. For the demonstration of the functionality on the environment, 3-MPA (3-mercaptopropionic acid) was added to the Au@SiO$_2$ yolk-shell particles. At pH 13, the functional groups exposed on the surface were carboxylate anions, which may form a strong hydrogen bonding with the amine groups of o-nitroaniline, and increase the retention time on the gold surface. It resembled cooperative ligands in the active sites of enzymes, and the reaction rate constant increased by 2.4 times compared to the unfunctionalized catalyst. As a result, the combination of porosity and functionality control could tally enhance the reaction rate constant by 13 folds compared to the original Au@SiO$_2$ yolk-shell catalyst.

25.4 Assistive Functions of Shell Layers

The yolk-shell catalysts are very special in structure due to the presence of shell layers. In bifunctional catalysts, metal oxide supports stabilize active metal nanoparticles, and help to transfer the reactants into the metal surface [2]. The functions of the shell layers in the yolk-shell catalysts are basically the same, but their effects are more profoundly presented in chemical and physical properties.

25.4.1 Chambers of Solution- and Solid-Phase Reactions

In general, solution-phase reactions have been widely used for the synthesis of various nanostructures. However, solid-state reactions could not be applied due to their high reaction temperature and severe agglomeration. In this aspect, Lee et al. employed thermally stable silica hollow structures as nanospace-confined media, and proceeded various solution- and solid-phase reactions inside the void space (Fig. 25.4) [11]. In the presence of a Au seed inside the silica hollow, Au, Ag, Ni, and Pt particles were directly grown through the addition of metal precursors and reducing agents. Multiple

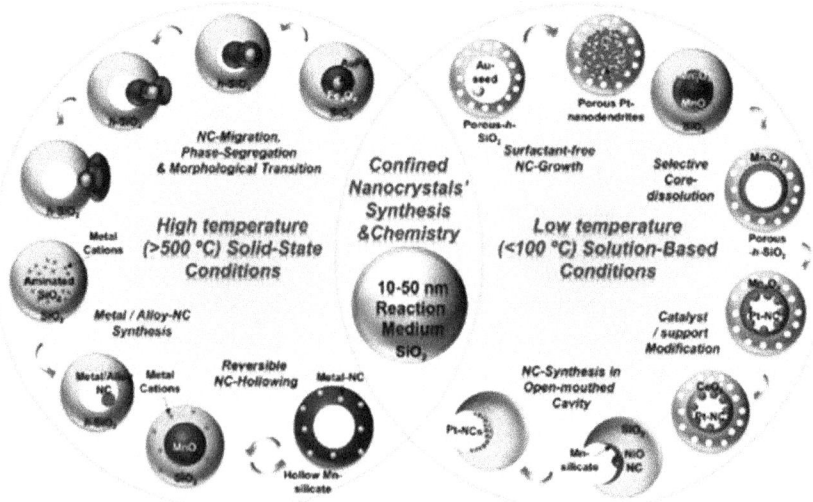

Fig. 25.4 General concepts of the reactions on nano-confined media. Adapted with permission from Ref. [11]

reductions also happened to yield multicomponent hybrid structures. Manganese oxides were unique because they behaved both as sacrificial templates and reductants. By sequential metal deposition, manganese oxides led to form complicated hybrid structures.

The high thermal stability of the silica shells served as excellent reaction chambers for high-temperature solid-state reactions. At high temperatures, metals easily formed alloys at first, but further aging led to phase segregation and migration. When all metals were located at the center of the hollow silica shells, the degree of heating changed the metals into eccentric positions, and the metals finally penetrated the porous silica shells to be exposed outward. Simultaneous phase segregation could generate complex structures containing different metal domains. For instance, Fe/Au/Pd components formed alloys inside the silica, and prolonged heating migrated the metal core from the center to the shell layer and eventually to the outward position. Oxidation in air generated complicated $(AuPd@Fe_3O_4)@SiO_2$ structures [12].

25.4.2 Protective Layers

Metal nanoparticles are readily agglomerated under the harsh reaction environment. However, the yolk-shell structure is very stable due to the existence of highly stable silica layers outside. If the reaction temperature was not as high as a melting point, the neighboring metal cores could not touch each other, and therefore, thermal and chemical stability enormously increased.

Song et al. synthesized $Ni@SiO_2$ yolk-shell catalysts by coating the silica layers and partial dissolution of the nickel cores [13]. The catalysts were employed for steam methane reforming reactions at 700 °C. It was surprising that even a high metal loading exceeding 90% was plausible without particle agglomeration, and the conversion was quantitative for more than 4 h under the severe conditions (Fig. 25.5a).

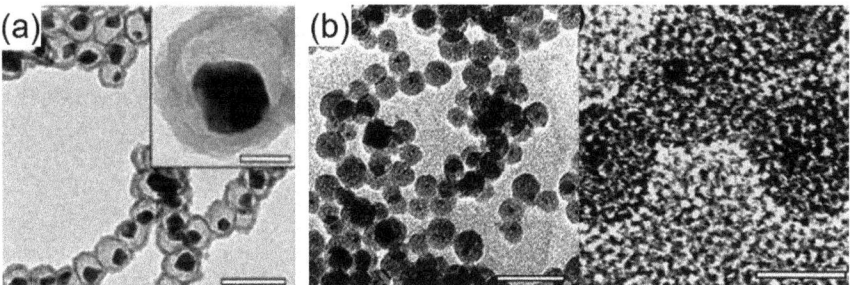

Fig. 25.5 **a** $Ni@SiO_2$ yolk-shell nanocatalysts for steam methane reforming reactions. **b** $Ni@SiO_2$ nanoparticles with tiny Ni cores for hydrogen transfer reactions. Adapted with permission from Ref. [14]

To increase the reactivity under mild conditions, the particle size should be reduced to a few nanometers. It makes the active surface; however, the aggregation generally occurs during the reaction. The yolk-shell structure is another approach to resolve this issue. Tiny Ni@SiO$_2$ yolk-shell nanoparticles were synthesized through a microemulsion method followed by partial core dissolution. The resulting Ni core size was as small as 3 nm (Fig. 25.5b) [14]. Due to the protective silica layer, the catalyst showed high activity on hydrogen transfer of acetophenone with excellent recyclability. Tiny Pd@SiO$_2$ yolk-shell catalysts were also prepared with a maximized pore density and were employed for Suzuki coupling reactions of aryl halides [15]. At room temperature, the reaction activity was similar to that of the freestanding Pd precursors. Still, at 473 K, the catalyst exhibited extremely high activity with a TOF of 78,000 h^{-1}, whereas the other Pd catalysts significantly dropped it. The core-shell structures also exhibited similar high-temperature stability, but could not show any peculiar activity under the mild organic reaction conditions, presumably due to the low diffusion rates of the reactants through the shell layers.

25.4.3 Molecular Sieves for Size Selective Reactions

The high surface area supports containing micropores such as zeolites are known to behave as molecular sieves separating molecules in size. The shells of the yolk-shell structure also played a role to penetrate molecules smaller than their pores into the void, where the metal cores catalyzed the reactions. As a result, the catalysts could exhibit strong size-selectivity.

Schüth et al. synthesized Pd/polymer nanocomposites via emulsion polymerization, and converted them into Pd@microporous carbon yolk-shell particles after pyrolysis at 500 °C [16]. The resulting Pd/C@HCS structure carried out selective hydrogenation of nitrobenzene quantitatively, whereas the conversion of 9-nitroanthracene was less than 5%. It was ascribed that the transport of anthracene was restricted by the small micropores. Tsung et al. used a metal organic framework, ZIF-8, which was promising in gas separation and storage, as a shell material [17]. The resulting yolk-shell Pd@ZIF-8 nanocatalyst successfully proceeded the hydrogenation of ethylene and cyclohexane, but showed no activity for cyclooctene, because the size of cyclooctene (5.5 Å) was much bigger than the pore aperture (3.4 Å).

25.5 Assistive Functions of Core Particles

Metal cores commonly act as active catalysts, but in some cases, shell layers carry out catalytic reactions, and the metal cores assist their functions.

25.5.1 Molecular Sensing Probes

Heterogeneous catalysts comprise an ensemble of distinct surfaces. Therefore, intrinsic heterogeneity of the surfaces makes the catalytic mechanism very hard to understand. If we can monitor the reactions on individual nanoparticles, the reaction process would be precisely resolved. For this purpose, Song et al. synthesized Pt/CdS hollow nanocubes using Galvanic replacement of Ag cubes and subsequent sulfidation and cation exchange reactions [18]. By the irradiation, the catalyst dissociated lactic acid generating hydrogen gas. In an analogous way, Au@Pt/CdS yolk-shell nanocubes were also prepared from Au@Ag nanocubes, where the individual structure had a Au core at the void surrounded by the Pt/CdS hollow nanocube. The Au core at the center behaved as a molecular sensing probe to monitor the reaction progress at a single particle level (Fig. 25.6a). By the irradiation of visible light, hydrogen gas was evolved on the Pt surface, which altered the dielectric constant of the solution media near the Au core. The surface plasmon of the Au nanoparticles is known to be very sensitive to the surface environment. Therefore, the scattering signal of the Au core was significantly changed by the reaction progress, which was precisely monitored by dark-field scattering spectroscopy. As a result, the reaction rate constant and diffusion coefficient on the individual nanocubes were precisely analyzed based on plasmonic scattering shifts during the reaction.

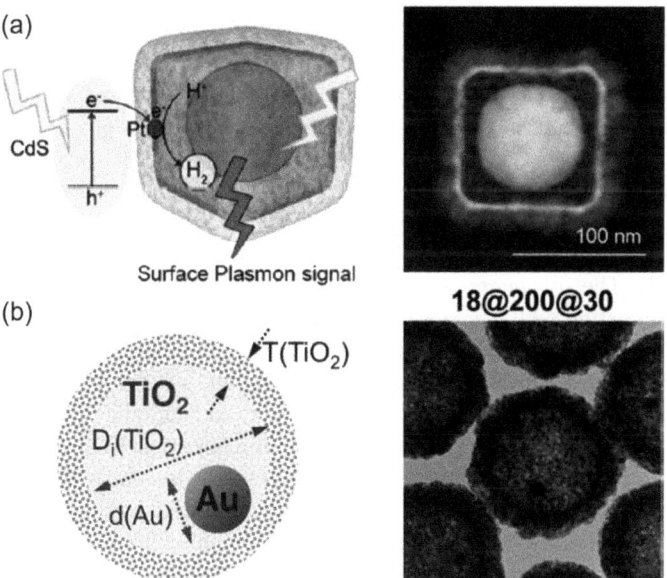

Fig. 25.6 a Au@Pt/CdS yolk-shell nanocubes for monitoring hydrogen evolution from lactic acid at a single particle level. **b** Au@TiO₂ yolk-shell nanostructure to improve photocatalytic hydrogen evolution. Adapted with permission from Refs. [18] and [20]

25.5.2 Enhancer for Photocatalysts

The addition of metal domains on semiconductors can enhance photocatalytic efficiencies by facilitating charge separation generated from the photoexcitation on
the semiconductor, trapping electrons, and promoting the proton reduction to H_2.
Recently, the alternative role of the metal, the recombination of hydrogen atoms
into H_2 on the metal surface, was proposed on Au@TiO$_2$ nanoparticles [19]. Zaera
et al. studied the optimal size of each component on Au@TiO$_2$ yolk-shell nanostructures for photocatalytic hydrogen generation [20]. TiO$_2$ hollow shells were exposed
outside, which were beneficial for absorbing light and proceeding reduction and
oxidation reactions. The hydrogen atoms transferred from the TiO$_2$ combined to
form H_2 on the Au surface in their mechanistic model (Fig. 25.6b). The optimal shell
thickness was evaluated as 28 nm, associated with the light penetration through the
TiO$_2$ shell. The inner diameter of titania was 186 nm at the maximum reactivity,
presumably related to the diffusion of hydrogen atoms to the Au cores. The Au core
size did not influence the photocatalytic activity. In this experiment, the existence of
the Au cores did not affect any geometric feature of the TiO$_2$ hollow shell, but only
played an assistive role in hydrogen evolution.

25.6 Tandem or Multistep Catalysts

In principle, the yolk-shell structure has two distinct domains, which do not have
direct contact with each other. Apparently, this is a perfect platform for a dual catalytic
system within a single nanostructure. Moreover, the reaction process through the
yolk-shell structure is conceptually a sequential tandem type, because the reactants diffuse to the outer shell at first and approach the center core at last. In this
aspect, Liu and Yang cleverly designed the yolk-shell nanoparticles with a basic
core and an acidic shell and employed them for the deacetalization-Henry tandem
reaction [21]. The silica cores were decorated with aminopropyl groups. The additional silica coating with mercaptopropyl groups and organosilane-assisted etching
provided the yolk-shell structure with the distinct void space between the core and
the shell regions. After the oxidation with H_2O_2, the YS-NH$_2$@SO$_3$H yolk-shell
nanoparticles containing amino groups at the cores and sulfonic acid groups on the
shells were yielded. The deacetalization reaction of benzaldehyde dimethyl acetal
occurred on the acidic shell layers to form benzaldehyde, which diffused into the
basic core surface and converted to nitrostyrene. This tandem reaction smoothly
progressed with 100% conversion and selectivity, meaning that the acidic shells and
basic cores were completely separated without neutralization. The physical mixture
of acidic and basic catalysts showed much lower catalytic activity, indicating that
the spatial separation and proximity of the first and second catalytic surfaces were
essential in the tandem reaction. This demonstration opened new possibilities for
designing efficient nanoreactors catalyzing complex multistep reactions.

25.7 Summary and Outlook

The yolk-shell nanostructures are designed as three-dimensional bifunctional model catalysts and have many advantages that arise from their spatial arrangement with a unique core-void-shell structure. As aforementioned, the active catalytic sites can be located at the cores, at the shells, or on both regions, and each arrangement is meaningful in corresponding catalytic reactions [7]. The void is also critical as a reactor space to confine the reactions inside the frame. In this Review, not only the catalysis using the yolk-shell nanoparticles was described, but they are also useful for biomedical and storage applications. As well as biosensing and imaging, the shell structure can also be employed in drug delivery and cancer therapy [22]. The hollow space of the yolk-shell structure prevents the electroactive materials from agglomeration, and also diminishes the stress from huge volume change during the charge-discharge cycling in lithium storage [23]. In any case, the yolk-shell structure is one of the best morphologies to realize the basic concept of nanocatalysts in practical applications, in which each part is adjustable by nano-synthetic techniques, and to explore nanostructure-property relationships [24].

References

1. Somorjai GA (1994) Introduction to surface chemistry and catalysis. Wiley, New York
2. Bell AT (2003) The impact of nanoscience on heterogeneous catalysis. Science 299:1688–1691
3. Somorjai GA, Contreras AM, Montano M, Rioux RM (2006) Clusters, surfaces, and catalysis. Proc Natl Acad Sci USA 103:10577–10583
4. Rioux RM, Song H, Hoefelmeyer JD, Yang P, Somorjai GA (2005) High-surface-area catalyst design: synthesis, characterization, and reaction studies of platinum nanoparticles in mesoporous SBA-15 silica. J Phys Chem B 109:2192–2202
5. Song H, Rioux RM, Hoefelmeyer JD, Komor R, Niez K, Grass M, Yang P, Somorjai GA (2006) Hydrothermal growth of mesoporous SBA-15 silica in the presence of PVP-stabilized Pt nanoparticles: synthesis, characterization, and catalytic properties. J Am Chem Soc 128:3027–3037
6. Zaera F (2013) Nanostructured materials for applications in heterogeneous catalysis. Chem Soc Rev 42:2746–2762
7. Park JC, Song H (2011) Metal@silica yolk-shell nanostructures as versatile bifunctional nanocatalysts. Nano Res 4:33–49
8. Yin Y, Rioux RM, Erdomez CK, Hughes S, Somorjai GA, Alivisatos AP (2004) Formation of hollow nanocrystals through the nanoscale Kirkendall effect. Science 304:711–714
9. Lee J, Park JC, Song H (2008) A nanoreactor framework of a Au@SiO2 yolk/shell structure for catalytic reduction of p-nitrophenol. Adv Mater 20:1523–1528
10. Lee J, Park JC, Bang JU, Song H (2008) Precise tuning of porosity and surface functionality in Au@SiO2 nanoreactors for high catalytic efficiency. Chem Mater 20:5839–5844
11. Kumar A, Jeon K-W, Kumari N, Lee IS (2018) Spatially confined formation and transformation of nanocrystals within nanometer-sized reaction media. Acc Chem Res 51:2867–2879
12. Kim YJ, Choi JK, Lee D-G, Baek KJ, Oh SH, Lee IS (2015) Solid-state conversion chemistry of multicomponent nanocrystals cast in a hollow silica nanosphere: morphology-controlled synthesis of hybrid nanocrystals. ACS Nano 9:10719–10728
13. Park JC, Bang JU, Lee J, Ko CH, Song H (2010) Ni@SiO2 yolk-shell nanoreactor catalysis: High temperature stability and recyclability. J Mater Chem 20:1239–1246

14. Park JC, Lee HJ, Kim JY, Park KH, Song H (2010) Catalytic hydrogen transfer of ketones over Ni@SiO$_2$ yolk-shell nanocatalysts with tiny metal cores. J Phys Chem C 114:6381–6388
15. Park JC, Heo E, Kim A, Kim M, Park KH, Song H (2011) Extremely active Pd@pSiO$_2$ yolk-shell nanocatalysts for Suzuki coupling reactions of aryl halides. J Phys Chem C 115:15772–15777
16. Wang G-H, Chen K, Engelhardt J, Tüysüz H, Bongard H-J, Schmidt W, Schüth F (2018) Scalable one-pot synthesis of yolk-shell carbon nanospheres with yolk-supported Pd nanoparticles for size-selective catalysis. Chem Mater 30:2483–2487
17. Kuo C-H, Tang Y, Chou L-Y, Sneed BT, Brodsky CN, Zhao Z, Tsung C-K (2012) Yolk-shell nanocrystal@ZIF-8 nanostructures for gas-phase heterogeneous catalysis with selectivity control. 134:14345–14348
18. Seo D, Park G, Song H (2012) Plasmonic monitoring of catalytic hydrogen generation by a single nanoparticle probe. J Am Chem Soc 134:1221–1227
19. Joo JB, Dillon R, Lee I, Yin Y, Bardeen CJ, Zaera F (2014) Promotion of atomic hydrogen recombination as an alternative to electron trapping for the role of metals in the photocatalytic production of H$_2$. Proc Natl Acad Sci USA 111:7942–7947
20. Lee YJ, Joo JB, Yin Y, Zaera F (2016) Evaluation of the effective photoexcitation distances in the photocatalytic production of H$_2$ from water using Au@void@TiO$_2$ yolk-shell nanostructures. ACS Energy Lett 1:52–56
21. Yang Y, Liu X, Li X, Zhao J, Bai S, Liu J, Yang Q (2012) A yolk-shell nanoreactor with a basic core and an acidic shell for cascade reactions. Angew Chem Int Ed 51:9164–9168
22. Lin L-S, Song J, Yang H-H, Chen X (2018) Yolk-shell nanostructures: Design, synthesis, and biomedical applications. 30:1704639
23. Liu J, Qiao SZ, Chen JS, Lou XW, Xing X, Lu GQ (2011) Yolk/shell nanoparticles: new platforms for nanoreactors, drug delivery and lithium-ion batteries. Chem Commun 47:12578–12591
24. Song H (2015) Metal hybrid nanoparticles for catalytic organic and photochemical transformations. Acc Chem Soc 48:491–499

Chapter 26
Hollow Carbon Spheres Encapsulating Metal Nanoparticles for CO_2 Hydrogenation Reactions

Guoxiang Yang, Yasutata Kuwahara, Kohsuke Mori, and Hiromi Yamashita

26.1 Introduction

The recent increase in global atmospheric temperature due to greenhouse gas emissions from human activities has been one of the serious challenges facing scientists [1, 2]. This is mainly caused by carbon dioxide (CO_2) emission from the burning of fossil fuels used for transportation of traffic and the production of electric energy [3]. Therefore, it is necessary to develop technologies to reduce the CO_2 generated by the burning of fossil fuels. The method of directly reducing CO_2 emissions by converting CO_2 into chemicals or fuels is one of the promising methods [4, 5]. Since the advent of the field of catalysis, the hydrogenation of CO_2 molecules into value-added products has attracted the attention of chemists. Among them, the generated CH_4 is an important basic step in C1 chemistry and is considered to be one of the strategies that can appropriately reduce the concentration of CO_2 in the atmosphere and at the same time provide the generated carbon fuel [6]. Methanol and formate (formic acid) are also very attractive products [7–9]. They have many applications including being used as fuels, H_2 storage media, fuel additives, energy carriers, etc. [10]. However, due to the chemical inertness of CO_2 molecules and the harsh reaction conditions required to activate CO_2, the design of reliable catalysts with high activity and reusability for CO_2 hydrogenation reactions remain a challenge.

Nowadays, due to high catalytic activity, various metal catalysts (such as Ru, Rh, Pd, Co, and Ni) are effective candidates for heterogeneous CO_2 hydrogenation reactions [11–13]. However, under severe reaction conditions (high temperature and high pressure), it is easy to cause aggregation and loss of metal nanoparticles (NPs), resulting in a gradual decrease in catalytic activity [14–17]. It is well known that the

G. Yang · Y. Kuwahara · K. Mori · H. Yamashita (✉)
Division of Materials and Manufacturing Science, Graduate School of Engineering, Osaka University, Suita, Osaka 565-0871, Japan
e-mail: yamashita@mat.eng.osaka-u.ac.jp

© The Author(s), under exclusive license to Springer Nature Singapore Pte Ltd. 2021   425
H. Yamashita and H. Li (eds.), *Core-Shell and Yolk-Shell Nanocatalysts*,
Nanostructure Science and Technology,
https://doi.org/10.1007/978-981-16-0463-8_26

synthesis and application of carbon materials have a long history because they has excellent characteristics and better stability [18]. For example, discovered fullerenes [19], carbon nanotubes [20], and graphene [21] and those related to valuable carbon materials have always been a hot topic because they are used in catalyst supports, carbon fixation, gas storage, and fuel cells [22–24]. Among them, the hollow carbon spheres (HCS), which are one of the key carbon material families, consist of a carbon shell and internal voids, giving them some special functions, such as low specific density, high surface area, adjustable porosity, and good structural stability [25, 26]. The success of HCS applications in different fields (such as catalysis, adsorption, and energy storage) depends largely on its carefully controlled characteristics including the thickness of the shell, the pore size of the carbon shell, and the dispersion of the active sites [27].

This part reviews recent studies on hollow carbon spheres that supported metal NP catalysts for CO_2 hydrogenation reactions. In the first part of this chapter, the design and the synthesis of various kinds of M(metal)@HCS nanoreactors are summarized. One route is through Stöber templating method [28] to synthesize mesoporous hollow carbon spheres (MHCS) then encapsulating ultrafine metal nanoparticles (M@MHCS). Another route is using MOF-derived porous metal@hollow carbon spheres (M@HCS). In the second part, the applications of M@HCS for the CO_2 hydrogenation reactions are presented. The final part addresses a summary and outlook on M@HCS.

26.2   The Design and the Synthesis of HCS Encapsulating Metal NPs

Although different methods for preparing various M(metal)@HCS have been extensively developed, the synthetic strategies can be roughly divided into two categories: The first type (named "M@void@HCS") is to synthesize the hollow structure carbon sphere (HCS), then introduce seeds into the voids of HCS. The seeds then gradually grow within the void space. Through careful control of seed growth, the yolk-shell structure can be achieved with specific core sizes. Making use of smaller metal particles is ideal for improving the utilization efficiency of metal atoms because catalytic reactions occur on the surface of metal NPs. However, the shape, structure, and size of metal NPs will change under harsh reaction conditions. Encapsulating metal NPs is an effective choice to avoid several reasons for deactivation (sintering, leaching, coking, etc.). Meanwhile, the shell can also be functionalized to further enhance these materials with multiple functions other than physical separation and protection, to affect the catalytic process. When considering the catalyst design that makes full use of unique morphologies to optimize performance, many properties can be controlled and changed to achieve the desired results. In any synthesis strategy, the size and morphology of the metal particles, the thickness and porosity of the shell, and the interaction between the metal particles and the support should be considered. The second type (named "M@HCS" (without void)) is the inner core (M(metal)) which is first formed, and then one or more shell layers are coated through physical and chemical processes. The core–shell structure is constructed by adsorption, self-assembly, or precipitation. Then, by removing some of the core, shell, or any intermediate components to form a space between the core and the shell, a yolk-shell structure can be achieved. Some examples of methods used to achieve this type of structure include Kirkendall diffusion, etching, electroplating replacement, etc. [29, 30].

26.2.1 "M(metal)@Void@HCS" Strategy

The synthesis of HCS is mainly focused on the Stöber template method [31–33]. This method is simple and time-saving. It can synthesize silica primary particles with a polymer shell through a one-step method, which can be mass-produced and practically applied. A core–shell sphere composed of an RF thin layer wrapped with a silica core was successfully prepared and then calcined under N_2 to produce HCS. It should be emphasized that although the hydrolysis condensation of silica alkoxides has many similarities with the polymerization of RF, including ammonia as a catalyst, water/ethanol mixture as a reaction medium, and room temperature reaction, their reaction rate is faster than that of the RF polycondensation reaction. The metal NPs grown in the voids of hollow materials offer more control over particle size and produce narrower particle size distribution because of the adjustable size of the confined space. However, the loading position of the metal NPs remains a substantial challenge. This is because precursors tend to be randomly deposited on the external surface of HCS. Considerable efforts have been made to counter this and to provide targeted deposition of the metal NPs. Qiao et al. reported that the metal ions could be adsorbed and fixed by the amine groups at the surface of the polymer [34].

On this basis, Yang et al. recently reported that the synthetic procedure of PdAg NPs and aminopolymer was confined within mesoporous hollow carbon spheres (PdAg–P@MHCS) composites [35]. Firstly, the MHCS were synthesized through an in situ free assembly of the template and surfactant method to synthesize the SiO_2@RF core–shell structure material by using tetrapropyl orthosilicate (TPOS) as a silica source for SiO_2 particles and resorcinol and formaldehyde as carbon sources (Step 1). After pyrolysis to carbonize the organics and etching in NaOH aqueous solution to remove the SiO_2 template, the hollow structured MHCS were obtained (Step 2). Then amino-polymer dispersed in MHCS (P@MHCS) were synthesized through the polymerization of ethylene diamine and carbon tetrachloride (Step 3). Pd^{2+} and Ag^+ metal ions were bound through the amino-polymer dots. They were reduced by $NaBH_4$ to form PdAg–P@MHCS nanospheres (Step 4) (Fig. 26.1). In this case, the cationic aminopolymer not only interacts electrostatically with the Pd and Ag precursors but also increases the CO_2 adsorption ability, which will affect the catalytic activity.

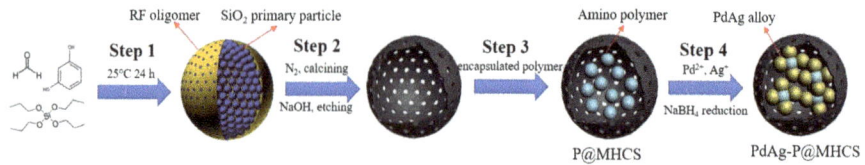

Fig. 26.1 1 Schematic representation of the preparation of PdAg-P@MHCS. Reprinted with permission from Ref. [35]. Copyright 2020, Royal Society of Chemistry

26.2.2 "M(metal)@HCS" (Without Void) Strategy

The direct method of encapsulating the metal NPs in this case is to coat the metal NPs with the desired material. A wide array of sources has been used to produce highly stable shells and has received considerable attention. Recently, the encapsulation of M(metal) into metal-organic frameworks (MOFs) to form M(metal)@HCS composite materials with amazing properties has attracted great interest, because MOFs or coordination polymers are known to be very useful materials in many applications. In general, MOFs have well-developed microporosity or channels accessible to many kinds of molecules or ions, making them suitable for use in gas storage, separation, sensing, catalysis, etc. [36–38]. There are some polymers or surfactant molecules such as polyvinyl pyrrolidone and cetyltrimethylammonium bromide (CTAB) adopted for connecting NPs with MOFs to facilitate the formation of core@shell structured nanocomposites [39]. Recently, methods have been developed to use MOF materials as precursors for preparing customized metal oxides or carbon materials. For instance, Lee et al. prepared well-dispersed hollow porous carbon microspheres using simple pyrolysis of core–shell polystyrene@ZIF-8 MOF [40]. Due to the removal of the polystyrene core evenly, the resulting well-defined HCS has a higher surface area ($1724\ m^2\ g^{-1}$), indicating that the ZIF-8 shell has been transformed into porous carbon, and the polystyrene core is removed at the same time. Due to its excellent dispersibility and high surface areas, it shows great methylene blue adsorption capacity. Once Ag or Au ions are loaded on the polystyrene core and MOF shell, HCS (M@HCS, M = Ag or Au NPs) embedded with Ag or Au NPs can be obtained after one-step pyrolysis. The resulting Ag or Au NPs can be well dispersed in the carbon support without agglomeration. Due to its unique structure, the obtained M@HCS can show effective catalytic activity and also has excellent recyclability [41]. On this basis, Lin et al. recently reported with Ni-MOFs as the precursor, the Ni@C composite was readily produced through thermal carbonation treatment in an N_2 atmosphere [11]. The as-derived hierarchical Ni@C hollow spheres consisting of carbon confined Ni NPs possess high surface area and abundant separated active sites.

26.3 The Applications of HCS Encapsulating Metal NPs for CO_2 Hydrogenation Reactions

26.3.1 Direct Hydrogenation of CO_2 to Formate/Formic Acid

In addition to being a valuable chemical substance commonly used as preservatives and antibacterial agents, formic acid has also become an established hydrogen storage component through decomposition into CO_2 and H_2 and a possible reversible conversion back to regeneration. Besides, formic acid is a low toxicity liquid at atmospheric pressure and room temperature and contains 4.3 wt% hydrogen. Compared with H_2, the transportation and storage of formic acid are safer and more convenient. Therefore, formic acid serves as one of the platforms for chemical energy storage [42]. At present, there are industrial methods such as the oxidation of biomass and the hydrolysis of methyl formate or formamide to produce formic acid [43]. The direct hydrogenation of CO_2 to formic acid has two important differentiating purposes compared with these traditional synthesis methods, such as CO_2 utilization and liquid hydrogen storage.

The conversion of CO_2 and H_2 into formic acid commonly involves a phase change from gaseous reagents into a liquid product. Therefore, when considering gas-phase reactants, the reaction is unfavorable in terms of Gibbs free energy (Eq 26.1) [17]:

$$H_2(g) + CO_2(g) \leftrightarrow HCO_2H(l) \Delta G^0_{298K} = 32.9 \, kJ \, mol^{-1} \tag{26.1}$$

On the other hand, the presence of the solvent changes the thermodynamics of the reaction, and when operating in the aqueous phase, the reaction becomes slightly aggressive (Eq 26.2):

$$H_2(aq) + CO_2(aq) \leftrightarrow HCO_2H(aq) \Delta G^0_{298K} = -4 \, kJ \, mol^{-1} \tag{26.2}$$

The thermodynamic equilibrium can be disrupted through secondary reactions or molecular interactions, making CO_2 easier to convert to formic acid. The usual methods are through esterification, such as the reaction of formic acid/formate with methanol to form formamide, or simply neutralization with a weak base (such as a tertiary amine or alkali/alkaline earth metal bicarbonate) [44].

Su et al. studied the activity of Pd catalysts supported on different materials (such as Al_2O_3, $BaSO_4$, $CaCO_3$, and activated carbon), and Pd supported on activated carbon showed excellent catalytic performance for CO_2 hydrogenation to formic acid/formate [45]. Besides, other researchers have also reported the positive effects of heteroatom modification on carbon carriers. Bi et al. studied a supported Pd catalyst for the reversible (dehydrogenation) reaction between potassium bicarbonate and formic acid as a method of releasing hydrogen into the catalyst solution [46]. Using Pd particles supported on reducing graphite oxide (Pd/r-GO), the turnover number (TON) value was 7088 when the hydrogenation reaction was carried out with 1 wt% Pd loading after 32 h. They concluded that the basic sites of the carrier can stabilize the Pd NPs and promote the interaction with CO_2 and the carrier, thereby promoting the synthesis of formic acid. Furthermore, Masuda et al. investigated highly dispersed PdAg NPs supported on phenylamine-functionalized mesoporous carbon (PdAg/amine-MSC) for CO_2 hydrogenation to formic acid [16]. The activity of PdAg/amine-MSC for CO_2 hydrogenation to formic acid was higher than Pd/amine-MSC. This improved catalytic activity was attributed to the electronic activation of Pd species by charge transfer resulting from the difference in the work functions of the contiguous two metals. However, there are several issues about PdAg alloy catalysts, such as sintering and poor chemical stability. An effective method is to use an ideal carrier to protect the catalysts to solve these problems. Mesoporous hollow carbon sphere (MHCS) is one of the best ideal carriers for immobilizing metal NPs. The main potential benefits of hollow structures in CO_2 methanation are increasing the catalytic activity and stability by increasing metal dispersion, protecting active structures on metal particles, and preventing agglomeration under severe conditions. Furthermore, MHCS is stable in the alkaline solution which is a benefit for increasing CO_2 conversion. Therefore, hollow structured nanomaterials have some unique advantages in terms of catalysis, and also allow flexible combinations of individual functions for specific purposes. Yang et al. designed and synthesized ultrafine and well-dispersed PdAg NPs with an average diameter of 2.8 nm, which were dispersed uniformly within mesoporous hollow carbon spheres (MHCS) of ca. 200–220 nm diameter and a carbon shell thickness of 40 nm containing a polymer (polyethylene diamine) with amino functional groups (Fig. 26.2) [35]. A series of catalysts were used for the hydrogenation of CO_2 to formate, and the catalytic activity was compared with the TON based on the moles of Pd (Table 26.1). The Pd_2Ag_8–P@MHCS was proved to be a successful heterogeneous catalyst with 100% selectivity without the formation of methanol or CO. The total TON was 2680 at 24 h (entry 5). However, P@MHCS (entry 1) and monometallic Ag–P@MHCS (entry 2) showed no reaction for CO_2 hydrogenation to formate, which suggests that Pd atoms are the main active sites. Compared with monometallic Pd–P@MHCS (entry 3), the Pd_2Ag_8–P@MHCS performed significantly better, indicating a positive effect of alloying of Pd with Ag. Besides, compared with the catalyst without aminopolymer, Pd_2Ag_8@MHCS (entry 9), Pd_2Ag_8–P@MHCS showed a significantly better activity, illustrating that the polymer can help to protect the catalyst from agglomeration. The catalyst without SiO_2 being removed, Pd_2Ag_8–P–MHCS@SiO_2, showed low activity (entry 10), indicating that the hollow mesoporous structure helps to protect the catalyst and inhibits its loss. To further illustrate the superiority of the Pd_2Ag_8–P@MHCS, the reaction using the same content of PdAg NPs supported on a mesoporous carbon prepared by the same method as MHCS and those supported on activated carbon was performed, where the TONs were 492 and 95 at 24 h, respectively (entries 11 and 12). These results illustrate that the hollow structure of MHCS is attributed to protecting PdAg NPs and the aminopolymer is promoting PdAg NPs dispersion. Furthermore, the catalytic activity of Pd_2Ag_8–P@MHCS remained almost unvaried at least after 5 cycles (Fig. 26.3(A)). The TEM image and XPS analysis clearly showed that the hollow nanostructured carbon spheres remained and

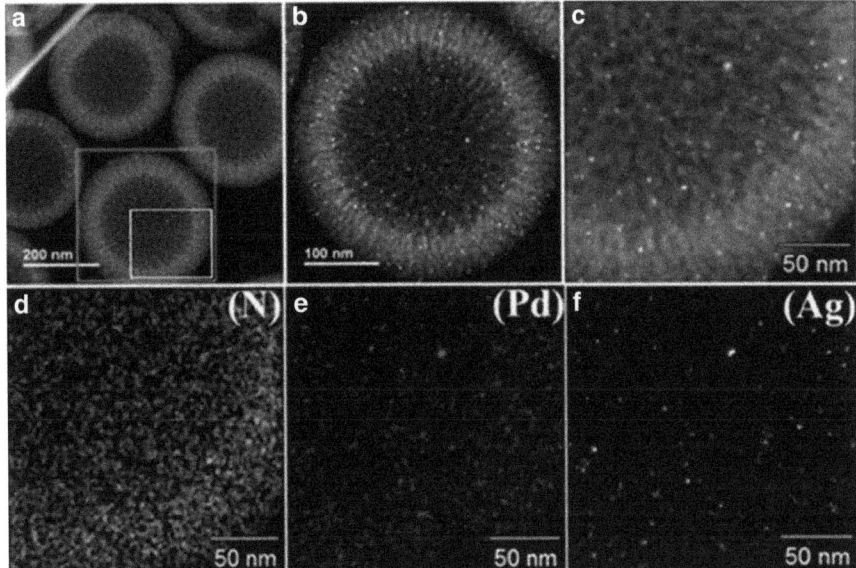

Fig. 26.2 (A) High-angle annular dark-field STEM (HAADF-STEM) image, (B) enlarged high-angle annular dark-field STEM (HAADF-STEM) image of the blue frame in (A), (C) enlarged high-angle annular dark-field STEM (HAADF-STEM) image of the yellow frame in (A), and (D–F) the corresponding STEM elemental maps of (D) N, (E) Pd, and (F) Ag of Pd_2Ag_8–P@MHCS. Reprinted with permission from Ref. [35]. Copyright 2020, Royal Society of Chemistry

the Pd and Ag still existed in Pd (0) and Ag (0) forms after the reaction, respectively (Fig. 26.3(B–D)). These results indicate that the yolk-shell structure of the Pd_2Ag_8–P@MHCS catalyst has good stability.

Combined with the related theoretical studies previously reported, the authors imagined the reaction is initiated by H_2 adsorption into MHCS and activation on the Pd atoms (Step 1). Bicarbonate (HCO_3^-) is adsorbed on the Ag atoms because of the electron transfer from Ag to Pd in PdAg alloy NPs (Step 2), in which CO_2 dissolution in the alkaline aqueous solution allows the further conversion to HCO_3^- (Step 3). A nucleophilic hydrogen atom formed on Pd and HCO_3^- adsorbed on Ag react to form dissociative formate (Step 5). Finally, the hydroxyl groups in the reaction solution react with hydrogen on the Pd to form H_2O, eliminating the remaining active hydrogen (Fig. 26.4).

26.3.2 Direct Hydrogenation of CO_2 to Methane

The process of converting CO_2 into methane through hydrogenation is an industrial process that has environmental protection prospects in coal-based economies that lack natural gas reserves (the Sabatier reaction) [47, 48].

Table 26.1 CO_2 hydrogenation with H_2 over PdAg–P@MHCS and its related catalysts. Reprinted with permission from Ref. [35]. Copyright 2020, Royal Society Chemistry

Entry	Catalyst	The molar ratio of Pd/Ag	TON[b]
1	P@MHCS	none	0
2	Ag–P@MHCS	0:1	0
3	Pd–P@MHCS[c]	1:0	1045
4	Pd_1Ag_9–P@MHCS[c]	1:9	1029
5	Pd_2Ag_8–P@MHCS[c]	2:8	2680
6	Pd_3Ag_7–P@MHCS[c]	3:7	1858
7	Pd_4Ag_6–P@MHCS[c]	4:6	2099
8	Pd_5Ag_5–P@MHCS[c]	5:5	1115
9	Pd_2Ag_8@MHCS[c]	2:8	802
10	Pd_2Ag_8–P@MHCS (without removing SiO_2)[c]	2:8	546
11	Pd_2Ag_8@MCS (without TPOS)[c]	2:8	492
12	Pd_2Ag_8@Activated carbon[c]	2:8	95

[a]Reaction conditions: 10 mg of catalyst, 15 mL of 1.0 M aqueous $NaHCO_3$ solution, 1 MPa H_2 and 1 MPa CO_2, 100 °C, reaction time 24 h. [b]TON = mol formate/mol Pd. [c]Pd theoretical weight was 1 wt%, PdAg alloy were prepared at the different molar ratio, the Pd loadings on the Pd_1Ag_9–P@MHCS, Pd_2Ag_8–P@MHCS, Pd_3Ag_7–P@MHCS, and Pd_4Ag_6–P@MHCS were determined to be 1.15, 0.95, 1.08, and 1.02 wt%, respectively, in the light of ICP data

$$CO_2 + 4H_2 \rightarrow CH_4 + 2H_2O \Delta H^0_{298} = -164.9 \text{ kJ mol}^{-1}$$

Among the possible ways of hydrogenation of CO_2, the conversion of CO_2 to methane has good thermodynamics and can be carried out under atmospheric pressure. Due to the kinetic limitations of catalysts doped with Ni, Rh, or Ru, the CO_2 methanation reaction is carried out at a temperature of about 250–400 °C, rather than at a lower make temperature [49]. The highly exothermic CO_2 methanation is likely to cause catalyst sintering problems.

In recent years, metal-organic frameworks (MOFs) have been widely used as precursors for the design and synthesis of functional materials, which have controlled structures and customized compositions [50]. In particular, the MOF-derived porous metal-carbon (M@C) composite material can make specific metal NPs precisely confined in the carbon shell with high dispersibility [51]. Therefore, the unique configuration of the porous M@C hybrids can not only perform the catalytic function of the dispersed metal NPs but also prevent them from sintering/aggregation at high temperatures.

The main potential benefit of the hollow structure in CO_2 methanation is that by increasing the dispersion of the metal, protecting the active structure on the metal particles, and preventing agglomeration under severe conditions, it can improve the catalytic activity and stability. Furthermore, the hollow structure catalyst can precisely control the catalyst structure and the chemical environment near the

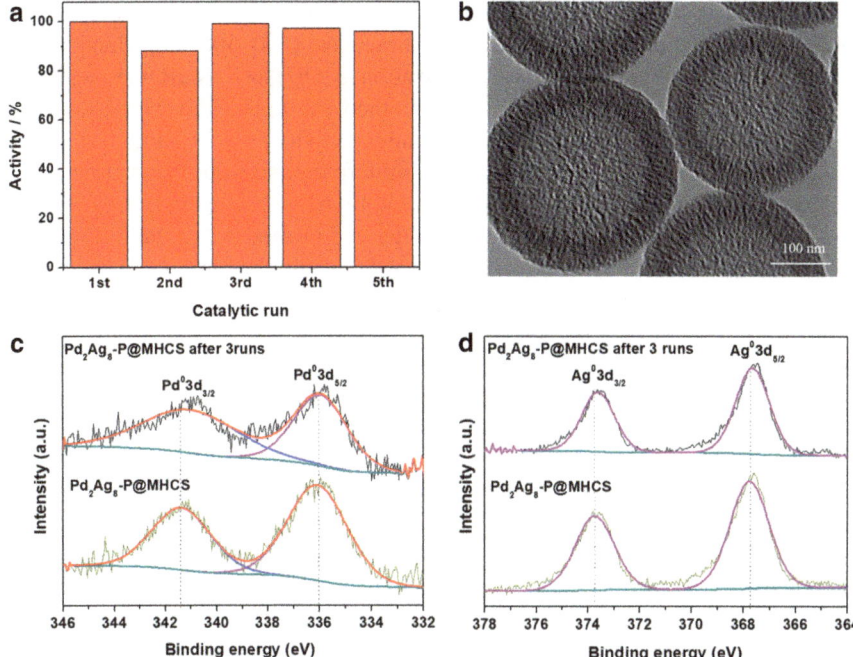

Fig. 26.3 (A) Reusability tests of the Pd$_2$Ag$_8$–P@MHCS catalyst. The TEM image of the Pd$_2$Ag$_8$–P@MHCS catalyst after five catalytic runs. XPS spectra of (C) Pd 3d and (D) Ag 3d for Pd$_2$Ag$_8$–P@MHCS etched after three catalytic runs. Reprinted with permission from Ref. [35]. Copyright 2020, Royal Society of Chemistry

Fig. 26.4 Possible reaction pathway for hydrogenation of CO$_2$ to produce formate over the Pd$_2$Ag$_8$–P@MHCS catalyst. Reprinted with permission from Ref. [35]. Copyright 2020, Royal Society of Chemistry

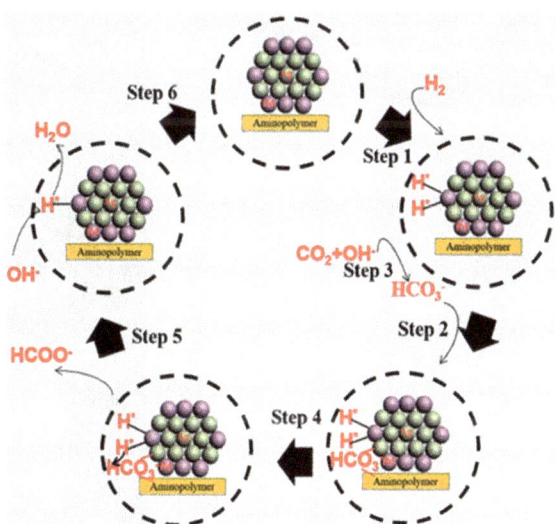

catalytic site to improve the selectivity of the product. Ni-containing catalysts usually require high temperatures to activate CO_2 to methane (400–500 °C), which makes Ni easy to sinter and causes catalyst deactivation. MOF (such as MOF-5 and MIL-101) have been used to encapsulate Ni NPs in the voids of the framework, resulting in high Ni dispersion and high CO_2 methanation activity at low temperatures [52]. As a result, when a MOF precursor with a preferred composition and structure is reasonably manufactured, the derived M@C hybrid may be a favorable CO_2 methanation catalyst. Lin et al. proved that MOF-derived layered Ni@C hollow spheres with unique intra-sphere structure could be easily synthesized and could be used as an effective catalyst for the hydrogenation of CO_2 to CH_4 [11]. With Ni-MOFs as the precursor, Ni@C hollow structure composites could be easily produced by thermal carbonization in the N_2 atmosphere. In this case, Ni-MOFs were synthesized by solvothermal reaction and some modifications were made. The morphologies and microstructures of the samples were characterized by FE-SEM and TEM. The FE-SEM images show that the Ni@C product retains the spherical morphology of the Ni-MOF precursors. The hierarchical surface of the Ni@C particles was identified by the enlarged FE-SEM image (Fig. 26.5a, b). The TEM images reveal that the Ni@C hybrid is composed of many Ni NPs with an average size of about 7 nm and has a hollow ball-in-ball structure (Fig. 26.5c–e). The high-resolution TEM (HR-TEM) image indicated the Ni NPs wrapped by the carbon layer has high crystallinity. The visible lattice fringes with the interlayer distances of 0.34 and 0.20 nm point to the

Fig. 26.5 **a, b** FE-SEM images, **c–e** TEM image, and **f** HR-TEM image of the hierarchical Ni@C hollow spheres. Reprinted with permission from Ref. [11]. Copyright 2019, Royal Society of Chemistry

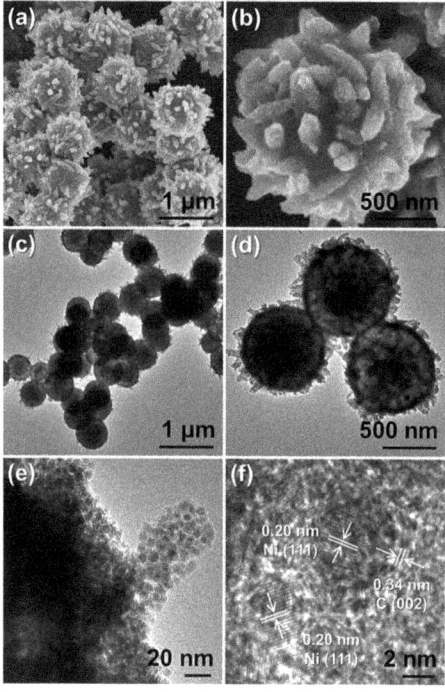

(002) and (111) crystal planes of graphitic carbon and cubic nickel, respectively (Fig. 26.5f). Thus, these results demonstrate the successful synthesis of the Ni@C hollow structure catalyst.

The catalytic performance of the Ni@C hybrid was evaluated by CO_2 methanation reaction in a fixed-bed flow reactor operating at ambient pressure. Figure 26.6(A) shows the CO_2 conversion performance of the Ni@C catalyst at different reaction temperatures. Due to the high chemical stability of CO_2 molecules, the hydrogenation of CO_2 to methane proceeds slowly at the reaction temperature from 150 to 225 °C. However, as the reaction temperature further increases (from 225 to 300 °C), the reaction rate greatly increases. When the reaction temperature rises to 325 °C, all CO_2 gas is converted into CH_4 products at a conversion rate of 100%. It is worth noting that at all the reaction temperatures tested, the Ni@C catalyst showed a fairly high CH_4 selectively, about 99.9%. The by-product was only a trace amount of CO. A series of control experiments were carried out to further prove the remarkable CO_2 hydrogenation activity. Under the same conditions, the CO_2 conversion rate

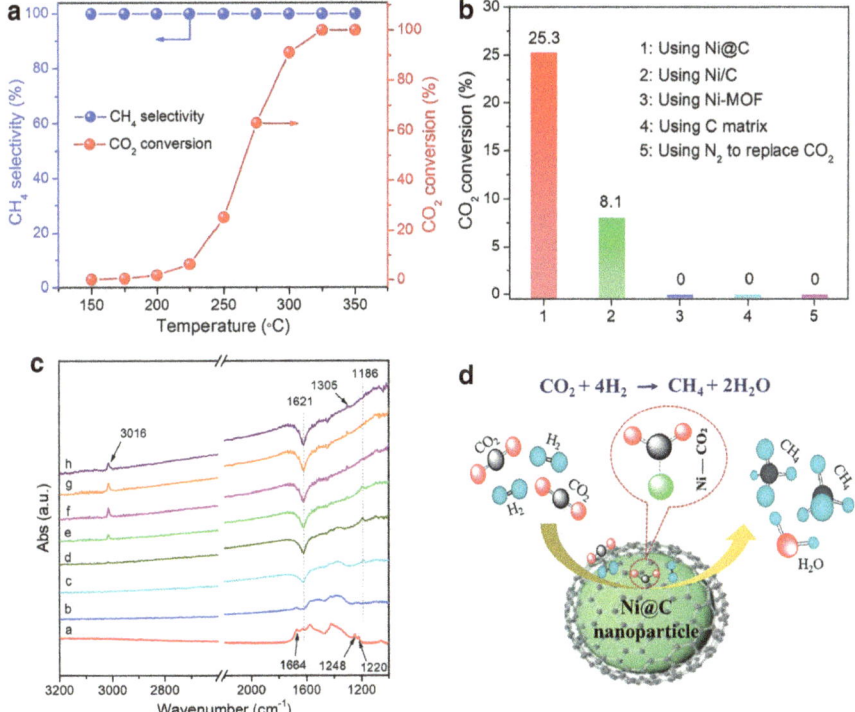

Fig. 26.6 (A) Catalytic CO_2 methanation performance of Ni@C at different reaction temperatures. (B) CO_2 methanation performance of different catalysts at 250 °C. (C) In situ DRIFTS spectra of adsorbed CO_2 and H_2 on the Ni@C catalysts at different temperatures: **a** 50, **b** 100, **c** 150, **d** 200, **e** 250, **f** 300, **g** 350, and **h** 400 °C. (D) Schematic of CO_2 methanation over the Ni@C catalyst. Reprinted with permission from Ref. [11]. Copyright 2019, Royal Society of Chemistry

of the Ni/C solid is 8.1% (entry 2), which is much lower relative to that of Ni@C (entry 1). This observation indicates that the porous structure of the catalyst and the high dispersibility of Ni NPs affect the catalytic performance. When the Ni-MOF (entry 3) or the C matrix (entry 4) was used as the catalyst, CH_4 was not detected, indicating that the metallic Ni NPs are the active sites for the CO_2 hydrogenation to CH_4 (Fig. 26.6(B)). Besides, once Ar is used to replace CO_2 for the reaction, CH_4 or other products will not be generated, which indicates that the source of the generated CH_4 is the CO_2 reactant. Therefore, all these findings emphasize that the excellent CO_2 methanation performance of the MOF-derived Ni@C hybrid can be attributed to the combined catalytic function of Ni NPs and carbon layers that are spatially confined in the unique hierarchical porous architecture.

Generally, there are two proposed reaction pathways and the related intermediates in CO_2 methanation catalysis: (i) the conversion of CO_2 to a CO intermediate and the subsequent hydrogenation of CO to form CH_4, and (ii) the direct hydrogenation of CO_2 to CH_4 through surface carbon species. In situ DRIFTS characterization was performed to identify the surface adsorption state of CO_2 and intermediates on the Ni@C catalyst.

The in situ DRIFTS spectra of the Ni@C sample interacted with CO_2 and H_2 during the temperature-programmed process from 50 to 400 °C (Fig. 26.6(C)). After the adsorption-desorption equilibrium at 50 °C, and several absorption-desorption equilibria at 50 °C, several absorption peaks are detected in the range of 1800–1000 cm^{-1}. The two weak peaks at 1248 and 1220 cm^{-1} are attributed to the carbonate species (CO_3^{2-}), while the band at 1664 cm^{-1} is ascribed to the surface hydrogen carbonate (HCO_3^-). In the beginning, a small peak was observed at 1621 cm^{-1}. This small peak is the bending mode of surface adsorbed H_2O. As the temperature increases from 50 to 150 °C, the peak first decreases and then develops into a negative peak, indicating the heating and surface dehydration during the purging processes. The broadband at 1310–1440 cm^{-1} shows the symmetric stretch mode of carboxylate (CO_2^-), while the band at 1490–1600 cm^{-1} belongs to the CO_2^- group adsorbed on the metallic Ni particle antisymmetric stretch vibration. At the same time, when the temperature reaches 150 °C, a new band emerges at around 1186 cm^{-1}, which corresponds to the stretching vibration of surface C–O(H). These findings indicate that the adsorbed CO_2^- intermediate is reduced to C–H(H) species by the H atoms dissociated from H_2. Besides, the absorption peaks of CH_4 gas (e.g., 3016 and 1305 cm^{-1}) are observed starting from 250 °C, which is accompanied by the decrease of the C–O(H) group (e.g., 1186 cm^{-1}). Therefore, all these DRIFTS results indicate that CO_2 methanation on the Ni@C catalyst is carried out by directly hydrogenating CO_2 without involving CO as an intermediate.

Based on the above results and discussion, the authors proposed a possible reaction mechanism for catalyzing CO_2 methanation over the Ni@C composite catalyst. Firstly, the CO_2 and H_2 gases are chemically adsorbed on the catalyst surface, which leads to the activation of CO_2 molecules and dissociation of H_2 molecules. Due to the abundant d-electrons of metal Ni NPs and the lowest unoccupied molecular orbital (LUMO) of CO_2 molecules, the Ni electrons are beneficial to transfer to the activated CO_2, thus forming the "Ni–CO_2" surface species. Then, the active "Ni–CO_2"

species reacts with the dissociated H atoms to form the "Ni–C–(OH)$_2$" intermediates. Finally, the reaction between Ni–C–(OH)$_2$ and the dissociated H atoms produces the CH$_4$ products and H$_2$O (Fig. 26.6(D)).

26.4 Summary and Outlook

In summary, this chapter deals with two types of new hollow carbon sphere supported metal NPs for CO$_2$ hydrogenation reactions. As for CO$_2$ hydrogenation to formate reaction, thanks to the high surface area and the pore size limitation of the mesoporous hollow carbon sphere(MHCS), the optimized aminopolymer and PdAg NPs are dispersed uniformly within MHCS, which show higher activity than other non-hollow structure carbon materials. As for CO$_2$ hydrogenation to methane reaction, due to the MOF-derived Ni@C distinctive compositions and yolk-shell structures possessing high surface area and more isolated active sites, the catalyst exhibits high activity, superior selectivity, and outstanding stability at atmospheric pressure. The yolk-shell structure of carbon-based nanoreactor catalysts possesses a small size and highly dispersed metal NPs, which are simultaneously prevented from deactivation and loss of NPs during the CO$_2$ hydrogenation reactions. The general synthetic methods for M(metal)@HCS yolk-shell structure and their unique features in regulating catalytic performance were introduced to provide a guide for the design of novel M(metal)@HCS catalysts and to expand their applications into CO$_2$ hydrogenation reactions and other catalytic fields. Although great achievements have been made, we anticipate that the catalytic activity will be further improved by controlling the internal modification of HCS. To this end, the designed M(metal)@HCS catalysts with simple preparation and environmentally friendly methods should be researched, and the mechanism of how the dispersion of M(metal) and the adsorption of CO$_2$ are promoted should be investigated in the future.

References

1. Hoegh-Guldberg O, Jacob D, Taylor M, Bolaños TG, Bindi M, Brown S et al (2019) The human imperative of stabilizing global climate change at 1.5 °C. Science 365:eaaw6974
2. Mora C, Spirandelli D, Franklin EC, Lynham J, Kantar MB, Miles W et al (2018) Broad threat to humanity from cumulative climate hazards intensified by greenhouse gas emissions. Nat Clim Chang 8:1062–1071
3. Schemme S, Samsun RC, Peters R, Stolten D (2017) Power-to-fuel as a key to sustainable transport systems—an analysis of diesel fuels produced from CO$_2$ and renewable electricity. Fuel 205:198–221
4. Wei J, Ge Q, Yao R, Wen Z, Fang C, Guo L et al (2017) Directly converting CO$_2$ into a gasoline fuel. Nat Commun 8:1–9
5. Artz J, Müller TE, Thenert K, Kleinekorte J, Meys R, Sternberg A et al (2018) Sustainable conversion of carbon dioxide: an integrated review of catalysis and life cycle assessment. Chem Rev 118:434–504

6. Jurca B, Bucur C, Primo A, Concepción P, Parvulescu VI, García H (2019) N-doped defective graphene from biomass as catalyst for CO_2 hydrogenation to methane. ChemCatChem 11:985–990

7. Jiang X, Nie X, Guo X, Song C, Chen JG (2020) Recent advances in carbon dioxide hydrogenation to methanol via heterogeneous catalysis. Chem Rev

8. Kuwahara Y, Fujie Y, Mihogi T, Yamashita H (2020) Hollow mesoporous organosilica spheres encapsulating PdAg nanoparticles and poly (ethyleneimine) as reusable catalysts for CO_2 hydrogenation to formate. ACS Catal 10:6356–6366

9. Mori K, Taga T, Yamashita H (2017) Isolated single-atomic Ru catalyst bound on a layered double hydroxide for hydrogenation of CO_2 to formic acid. ACS Catal 7:3147–3151

10. Asefa T, Koh K, Yoon CW (2019) CO2-mediated H2 storage-release with nanostructured catalysts: recent progresses, challenges, and perspectives. Adv Energy Mater 9:1901158

11. Lin X, Wang S, Tu W, Hu Z, Ding Z, Hou Y et al (2019) MOF-derived hierarchical hollow spheres composed of carbon-confined Ni nanoparticles for efficient CO_2 methanation. Catal Sci Technol 9:731–738

12. Mori K, Masuda S, Tanaka H, Yoshizawa K, Che M, Yamashita H (2017) Phenylamine-functionalized mesoporous silica supported PdAg nanoparticles: a dual heterogeneous catalyst for formic acid/CO_2-mediated chemical hydrogen delivery/storage. Chem Comm 53:4677–4680

13. Chen B, Dong M, Liu S, Xie Z, Yang J, Li S et al (2020) CO_2 hydrogenation to formate catalyzed by Ru coordinated with N, P-containing polymer. ACS Catal 10:8557–8566

14. Bian Z, Zhu J, Wen J, Cao F, Huo Y, Qian X et al (2011) Single-crystal-like titania mesocages. Angew Chem Int Ed 123:1137–1140

15. Kuwahara Y, Yamanishi T, Kamegawa T, Mori K, Yamashita H (2011) Enhancement in adsorption and catalytic activity of enzymes immobilized on phosphorus- and calcium-modified MCM-41. J Phys Chem B 115:10335–11345

16. Masuda S, Mori K, Futamura Y, Yamashita H (2018) PdAg nanoparticles supported on functionalized mesoporous carbon: promotional effect of surface amine groups in reversible hydrogen delivery/storage mediated by formic acid/CO_2. ACS Catal 8:2277–2285

17. Mori K, Sano T, Kobayashi H, Yamashita H (2018) Surface engineering of a supported PdAg catalyst for hydrogenation of CO_2 to formic acid: elucidating the active Pd atoms in alloy nanoparticles. J Am Chem Soc 140:8902–8909

18. Benzigar MR, Talapaneni SN, Joseph S, Ramadass K, Singh G, Scaranto J et al (2018) Recent advances in functionalized micro and mesoporous carbon materials: synthesis and applications. Chem Soc Rev 47:2680–2721

19. Cai W, Chen C-H, Chen N, Echegoyen L (2019) Fullerenes as nanocontainers that stabilize unique actinide species inside: structures, formation, and reactivity. Acc Chem Res 52:1824–1833

20. Kinloch IA, Suhr J, Lou J, Young RJ, Ajayan PM (2018) Composites with carbon nanotubes and graphene: an outlook. Science 362:547–553

21. Yu X, Cheng H, Zhang M, Zhao Y, Qu L, Shi G (2017) Graphene-based smart materials. Nat Rev Mater 2:1–13

22. Zhao S, Wang DW, Amal R, Dai L (2019) Carbon-based metal-free catalysts for key reactions involved in energy conversion and storage. Adv Mater 31:1801526

23. Bhanja P, Bhaumik A (2019) Materials with nanoscale porosity: energy and environmental applications. Chem Rec 19:333–346

24. Tian H, Liang J, Liu J (2019) Nanoengineering carbon spheres as nanoreactors for sustainable energy applications. Adv Mater 31:1903886

25. Ding J, Zhang H, Zhou H, Feng J, Zheng X, Zhong C et al (2019) Sulfur-grafted hollow carbon spheres for potassium-ion battery anodes. Adv Mater 31:1900429

26. Liu X, Wu J, Zhang S, Ding C, Sheng G, Alsaedi A et al (2019) Amidoxime-functionalized hollow carbon spheres for efficient removal of uranium from wastewater. ACS Sustain Chem Eng 7:10800–108007

27. Li S, Pasc A, Fierro V, Celzard A (2016) Hollow carbon spheres, synthesis and applications–a review. J Mater Chem A 4:12686–12713

28. Noonan O, Zhang H, Song H, Xu C, Huang X, Yu C (2016) In situ Stöber templating: facile synthesis of hollow mesoporous carbon spheres from silica–polymer composites for ultra-high level in-cavity adsorption. J Mater Chem A 4:9063–9071

29. Lin LS, Song J, Yang HH, Chen X (2018) Yolk–shell nanostructures: design, synthesis, and biomedical applications. Adv Mater 30:1704639

30. Anderson BD, Tracy JB (2014) Nanoparticle conversion chemistry: kirkendall effect, galvanic exchange, and anion exchange. Nanoscale 6:12195–12216

31. Huang S, Gao Y, Zhang Y, Chen S, Xiao Q (2019) Thermal oxidation etching strategy towards mesoporous hollow carbon spheres. Mater Lett 240:253–257

32. Phaahlamohlaka TN, Kumi DO, Dlamini MW, Jewell LL, Coville NJ (2016) Ruthenium nanoparticles encapsulated inside porous hollow carbon spheres: a novel catalyst for fischer–tropsch synthesis. Catal Today 275:76–83

33. Zhang H, Noonan O, Huang X, Yang Y, Xu C, Zhou L et al (2016) Surfactant-free assembly of mesoporous carbon hollow spheres with large tunable pore sizes. ACS Nano 10:4579–4586

34. Qiao ZA, Zhang P, Chai SH, Chi M, Veith GM, Gallego NC et al (2014) Lab-in-a-shell: encapsulating metal clusters for size sieving catalysis. J Am Chem Soc 136:11260–11263

35. Yang G, Kuwahara Y, Masuda S, Mori K, Louis C, Yamashita H (2020) PdAg nanoparticles and aminopolymer confined within mesoporous hollow carbon spheres as an efficient catalyst for hydrogenation of CO_2 to formate. J Mater Chem A 8:4437–4446

36. Wang L, Xu H, Gao J, Yao J, Zhang Q (2019) Recent progress in metal-organic frameworks-based hydrogels and aerogels and their applications. Coord Chem Rev 398:213016

37. Horike S, Nagarkar SS, Ogawa T, Kitagawa S (2020) A new dimension for coordination polymers and metal–organic frameworks: towards functional glasses and liquids. Angew Chem Int Ed 59:6652–6664

38. Nakatsuka K, Yoshii T, Kuwahara Y, Mori K, Yamashita H (2018) Controlled pyrolysis of Ni-MOF-74 as a promising precursor for the creation of highly active Ni nanocatalysts in size-selective hydrogenation. Chem Eur J 24:898–905

39. Zhao M, Deng K, He L, Liu Y, Li G, Zhao H et al (2014) Core–shell palladium nanoparticle@metal–organic frameworks as multifunctional catalysts for cascade reactions. J Am Chem Soc 136:1738–1741

40. Lee HJ, Choi S, Oh M (2014) Well dispersed hollow porous carbon spheres synthesized by direct pyrolysis of core–shell type metal-organic frameworks and their sorption properties. ChemComm 50:4492–4495

41. Choi S, Lee HJ, Oh M (2016) Facile synthesis of Au or Ag nanoparticles-embedded hollow carbon microspheres from metal-organic framework hybrids and their efficient catalytic activities. Small 12:2425–2431

42. Grasemann M, Laurenczy G (2012) Formic acid as a hydrogen source–recent developments and future trends. Energy Environ Sci 5:8171–8181

43. Alvarez A, Bansode A, Urakawa A, Bavykina AV, Wezendonk TA, Makkee M et al (2017) Challenges in the greener production of formates/formic acid, methanol, and DME by heterogeneously catalyzed CO_2 hydrogenation processes. Chem Rev 117:9804–9838

44. Jessop PG, Ikariya T, Noyori R (1999) Homogeneous catalysis in supercritical fluids. Chem Rev 99:475–494

45. Su J, Yang L, Lu M, Lin H (2015) Highly efficient hydrogen storage system based on ammonium bicarbonate/formate redox equilibrium over palladium nanocatalysts. Chemsuschem 8:813–816

46. Bi QY, Lin JD, Liu YM, Du XL, Wang JQ, He HY et al (2014) An aqueous rechargeable formate-based hydrogen battery driven by heterogeneous Pd catalysis. Angew Chem Int Ed 53:13583–13587

47. Vogt C, Monai M, Krämer GJ, Weckhuysen BM (2019) The renaissance of the sabatier reaction and its applications on earth and in space. Nat Catal 2:188–197

48. Sun D, Khan FM, Simakov DS (2017) Heat removal and catalyst deactivation in a sabatier reactor for chemical fixation of CO_2: simulation-based analysis. Chem Eng J 329:165–177
49. Stangeland K, Kalai DY, Li H, Yu Z (2018) Active and stable Ni based catalysts and processes for biogas upgrading: the effect of temperature and initial methane concentration on CO_2 methanation. Appl Energy 227:206–212
50. Sun J-K, Xu Q (2014) Functional materials derived from open framework templates/precursors: synthesis and applications. Energy Environ Sci 7:2071–2100
51. Yi H, Wang H, Jing Y, Peng T, Wang X (2015) Asymmetric supercapacitors based on carbon nanotubes@NiO ultrathin nanosheets core–shell composites and MOF-derived porous carbon polyhedrons with super-long cycle life. J Power Sources 285:281–290
52. Ding M, Flaig RW, Jiang H-L, Yaghi OM (2019) Carbon capture and conversion using metal–organic frameworks and MOF-based materials. Chem Soc Rev 48:2783–2828

Part IV
Yolk-Shell for Photocatalysis and Electrocatalysis

Chapter 27
Effect of Core–Shell Structure of TiO$_2$ on Its Photocatalytic Performance

Yao Chen, Hexing Li, and Zhenfeng Bian

27.1 Introduction

Photocatalysis is an interdisciplinary subject integrating physics, chemistry and materials science. The mechanism of photocatalysis is that semiconductor materials are photoexcited to produce photogenerated electrons (e^-) and holes (h^+), which have strong reduction and oxidation abilities, respectively [1]. They can directly or indirectly participate in the redox reaction on the semiconductor surface. Semiconductor photocatalysis has the advantages of low cost and direct utilization of sunlight [2]. Therefore, photocatalysis has potential applications in energy utilization and environmental protection.

TiO$_2$ is the most studied photocatalytic material by scientists since 1972 [3]. The main reason is that TiO$_2$ has the advantages of light corrosion resistance, good catalytic activity, stable performance, low price and is non-toxic [4]. However, the wide bandgap of TiO$_2$ determines that it can only be activated by ultraviolet light, which greatly limits the practical application of TiO$_2$ [5]. In order to solve the above problems, researchers have adopted a series of modification methods for TiO$_2$, such as noble metal deposition, metal ion doping, non-metallic ion doping, heterojunction structure and self-modification (including particle size, specific surface area, morphology, crystal form, etc.) to improve the visible light catalytic efficiency of TiO$_2$ [6–8].

On the other hand, with the rapid development of nanotechnology, various special structures of TiO$_2$ have been synthesized (nanowires, nanotubes, nanosheets, core–shell structures, petal-like structures, etc.) [9]. It is found that the properties of TiO$_2$ nanomaterials with cavity structure can be significantly improved. TiO$_2$ with special

Y. Chen · H. Li · Z. Bian (✉)
MOE Key Laboratory of Resource Chemistry and Shanghai Key Laboratory of Rare Earth Functional Materials, Shanghai Normal University, Shanghai 200234, China
e-mail: bianzhenfeng@shnu.edu.cn

© The Author(s), under exclusive license to Springer Nature Singapore Pte Ltd. 2021   443
H. Yamashita and H. Li (eds.), *Core-Shell and Yolk-Shell Nanocatalysts*,
Nanostructure Science and Technology,
https://doi.org/10.1007/978-981-16-0463-8_27

core–shell structure can improve the absorption of light on the TiO₂ surface and significantly improve its photocatalytic performance, which has attracted extensive attention [10–12].

This chapter mainly gives a comprehensive overview of the progress of core–shell structured TiO₂ nanomaterials in the field of photocatalysis. Firstly, the controllable synthesis methods of core–shell TiO₂ nanomaterials were summarized. Secondly, the applications of core–shell structure TiO₂ nanomaterials in photocatalysis were reviewed, including pollutant degradation, hydrogen evolution and carbon dioxide reduction. Finally, the development of this field is briefly summarized and prospected.

27.2   Synthetic Methods of Core–Shell TiO₂

It is necessary to reasonably control the change of precursor materials to make TiO₂ preferentially nucleate and grow into a core–shell structure, because different titanium sources and methods can be used to synthesize TiO₂ with different morphologies. The specific formation process of core–shell TiO₂ can be explained by investigating different preparation methods (different temperature, reaction time, reaction environment and other factors). At present, the synthesis methods of core–shell TiO₂ can be roughly divided into sol–gel method, water/solvothermal method, chemical deposition method, template synthesis method and self-assembly strategy.

The sol–gel method uses compounds containing highly chemically active components as precursors. It mixes the raw materials uniformly in the liquid phase, and then undergoes hydrolysis and condensation chemical reactions. The core–shell structure TiO₂ nanomaterials can be prepared by carefully selecting the synthesis parameters. Prashant V. Kamat et al. synthesized a core–shell structure of Ag@TiO₂ by a sol–gel method (Fig. 27.1). This composite material with the metal as the core and the semiconductor as the shell has photocatalytic activity, of which the volume ratio

Fig. 27.1 The formation mechanism of Ag@TiO₂ core–shell clusters. The condensation polymerization of TTEAIP slowly progresses on the surface of Ag particles to yield TiO₂ shell. Reprinted with permission from Ref. [13]. Copyright 2005 American Chemical Society

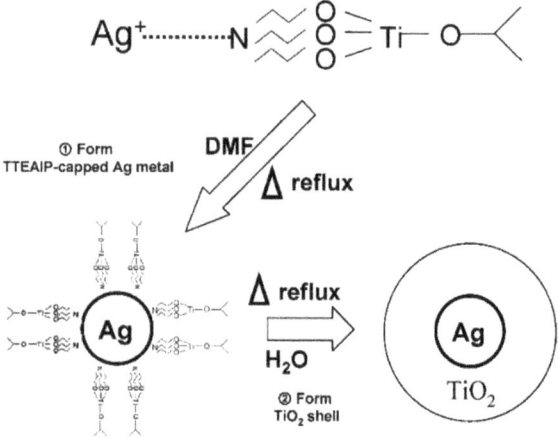

of dimethyl formamide and isopropyl alcohol is an important factor in the preparation of Ag@TiO$_2$ core–shell structure. This method is also suitable for the synthesis of Ag@SiO$_2$ [13]. Masahiko Abe et al. reported core–shell structure Ag@TiO$_2$ nanocapsules with high dispersibility prepared by the sol–gel method [14]. In the TEM image, almost all TiO$_2$ nanocapsules have Ag nanoparticles as the core, which indicates that the precursor of titanium tetraisopropoxide selectively forms a TiO$_2$ shell on the surface of Ag nanoparticles. It is crucial that cetyltrimethyl ammonium bromide (surfactant) as a protective agent can control the thickness of the TiO$_2$ shell layer and the size of Ag nanoparticles. In addition to the core–shell materials prepared with metal nanospheres as the core, Huang et al. precisely controlled the morphology of the metal core, and then synthesized the Au@TiO$_2$ core–shell material with octahedral gold as the core by the sol–gel method (Fig. 27.2) [15]. Compared with spherical nanoparticles, the octahedral gold core exhibits a better plasma effect that random hotspots generated under light are greatly beneficial to the improvement of the activity. The TiO$_2$ shell layer has the function of protecting the metal core structure from changing and preventing the dye molecules from being quenched by the metal surface. In addition, Zhu et al. used sol–gel in situ to assemble a g-C$_3$N$_4$/TiO$_2$ core–shell photocatalyst with TiO$_2$ as the core and wrapped it with an ultra-thin g-C$_3$N$_4$ layer (Fig. 27.3) [16]. The thickness of the g-C$_3$N$_4$ shell can be well controlled by changing the calcination temperature and the concentration of the g-C$_3$N$_4$ colloidal suspension. Generally speaking, ordinary g-C$_3$N$_4$ is difficult to recycle due to its ultra-thin layered structure. This core–shell structure material can effectively recover ultra-thin g-C$_3$N$_4$ while improving the photocatalytic activity.

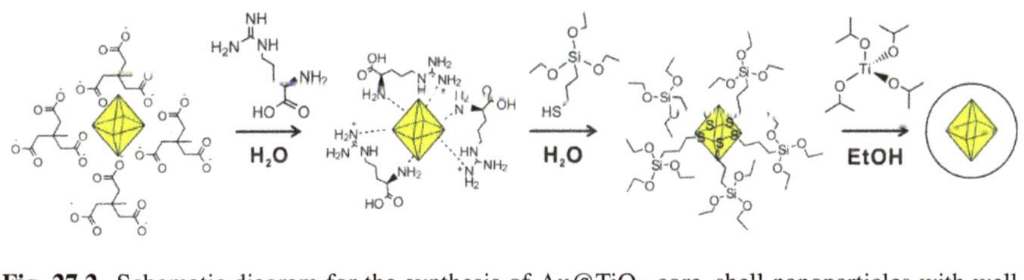

Fig. 27.2 Schematic diagram for the synthesis of Au@TiO$_2$ core–shell nanoparticles with well-controlled core morphology and shell thickness. Reprinted with permission from Ref. [15]. Copyright 2013 Royal Society of Chemistry

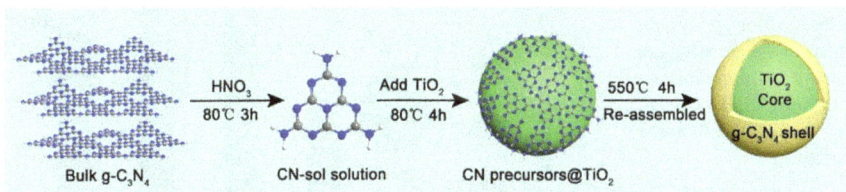

Fig. 27.3 Schematic illustration of preparation of g-C$_3$N$_4$@TiO$_2$ core–shell photocatalyst. Reprinted with permission from Ref. [16]. Copyright 2017 Elsevier B.V. All rights reserved

Hydrothermal and solvothermal syntheses refer to the synthesis of nanomaterials in water or solvent through a specific chemical reaction under certain temperature (373–1273 K) and pressure (1–100 MPa). This method is widely used in the synthesis of core–shell structure TiO_2. Tang et al. used a hydrothermal method to prepare Au@TiO_2 hollow core–shell submicrosphere with TiO_2 as the shell [17]. Different from the traditional hydrothermal synthesis process, this method introduced the Ostwald ripening process technology to prepare a large number of Au@TiO_2 hollow core–shell submicrospheres in a short time, and the chemical waste is greatly reduced. Generally, the hydrolysis rate of titanium source in water is very fast, so it is necessary to add a template to control the morphology and structure of TiO_2 well. It not only makes the operation complicated, but also increases the cost of materials.

Since alcohol can be directly used as a template and alcohol solvents are diversified, the solvothermal alcoholysis method has been rapidly developed in the preparation of metal oxide nanomaterials like TiO_2. Li et al. have done a series of related work in the past 10 years. First, they used the hydroalcoholic method for the first time to prepare core–shell structure mesoporous TiO_2 microspheres with high thermal stability and high surface area (Fig. 27.4) [18]. The size of the TiO_2 core can be effectively controlled by adjusting the reaction time, that is, solid nano-TiO_2 is formed after 12 h of reaction, and hollow TiO_2 nanospheres are formed after 14 days of reaction. The TiO_2 clusters formed by the alcoholysis of titanium oxysulfate ($TiOSO_4$) aggregate and react to form solid spheres containing a large amount of hydrolyzable ligands. As the reaction time increases, water is continuously produced through the etherification reaction and reacts with the spheres, causing the clusters to dissolve and rearrange on the surface, and then dissolve the TiO_2 core. Therefore, solid spheres, core–shell structure and hollow sphere nano-TiO_2 can be observed in sequence. Li et al. considered that the gap between the core–shell structure of TiO_2 core and outer shell could be used as a microreactor, so they prepared Fe-doped TiO_2 microspheres with a core–shell structure by using a similar hydroalcohol method [19]. Through continuous solvothermal and hydrothermal treatment, Li et al. coated Au particles in

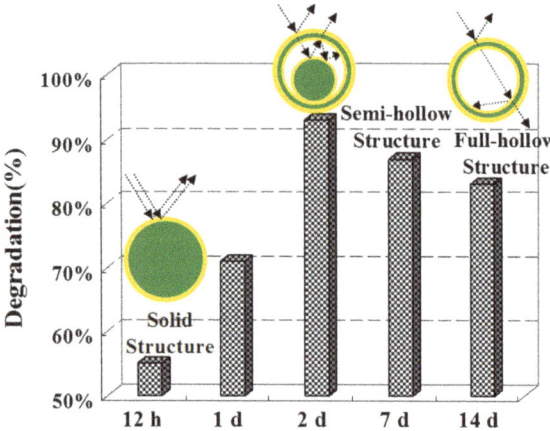

Fig. 27.4 Comparison of photocatalytic activities of the titania spheres with solid, sphere-in-sphere, and hollow structure. Inset shows a schematic illustration of multireflections within the sphere-in-sphere structure. Reprinted with permission from Ref. [18]. Copyright 2007 American Chemical Society

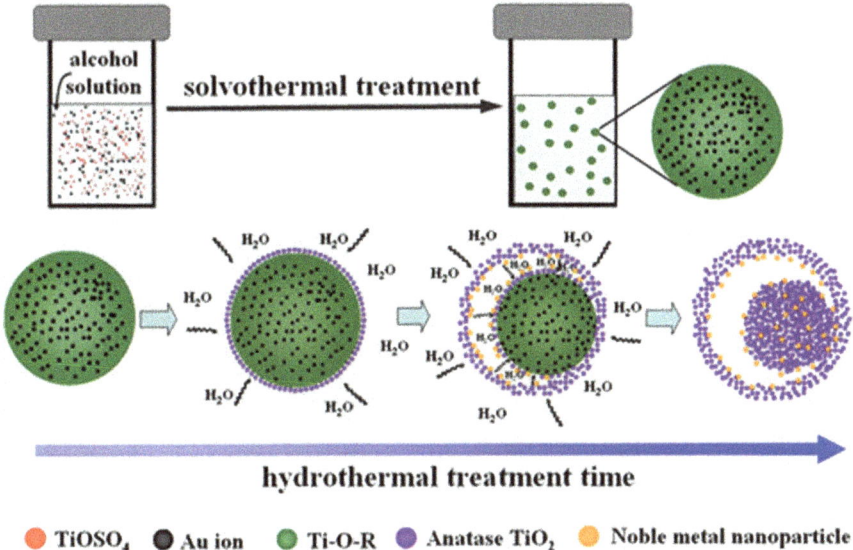

Fig. 27.5 Mechanism for encapsulating noble metal nanoparticles inside the core–shell TiO₂ microspheres. Reprinted with permission from Ref. [20]. Copyright 2009 Royal Society of Chemistry

TiO₂ core–shell spheres in situ to prepare a highly active and durable Au/TiO₂ photo-catalyst (Fig. 27.5) [20]. This method is also applicable to other precious metals. If only the solvent heat treatment is used, it takes 48 h to obtain core–shell Au/TiO₂ microspheres. When only the hydrothermal treatment is used, solid Au/TiO₂ microspheres are only obtained. In addition, Li et al. used glycerol and ethanol as solvents to adjust the morphology and structure of TiO₂ by adjusting the ratio of the two and the reaction time in 2012 [21]. The results showed that the solvothermal alcoholysis of TiOSO₄ in glycerol formed TiO₂ nanoscale plates, while, in the mixture of ethanol and glycerol, the solvothermal alcoholysis of TiOSO₄ produced solid spheres, core–shell spheres and hollow spheres for 1.48 h and 336 h, respectively. Through the same adjustment process, solid balls, nano balls, flower-like particles and hollow balls of TiO₂, TiN, graphitized carbon and MnO₂ materials can also be obtained, respectively (Fig. 27.6).

Chemical deposition is a material preparation method in which materials containing thin-film elements undergo chemical reactions to form thin films on the surface of the substrate. Among them, the atomic deposition method (ALD) can deposit the TiO₂ shell layer more uniformly and controllably than other deposition methods. Yang et al. used the ALD method to coat TiO₂ thin shell on the surface of the ZnO nanowire array [22]. Interestingly, each deposition can form 0.9 Å TiO₂ on the surface of ZnO. When the shell thickness is less than 4 nm, TiO₂ is amorphous and smooth. When the thickness continues to increase, polycrystalline TiO₂ appears and the surface is rough. Therefore, how to prepare the shell uniformly is a key issue for

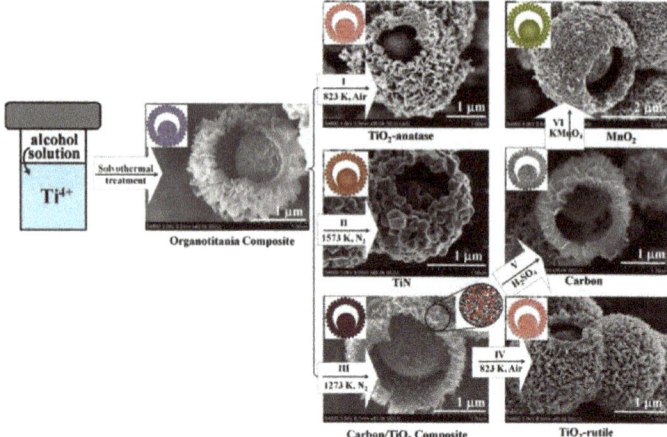

Fig. 27.6 The conversion of the yolk-shell organotitania composite (left) into nanostructured (I) anatase TiO$_2$, (II) TiN, (III) carbon/TiO$_2$ composite, (IV) rutile TiO$_2$, (V) graphitized carbon and (VI) MnO$_2$. Reprinted with permission from Ref. [21]. Copyright 2012 American Chemical Society

researchers. He et al. used the ALD method to prepare one-dimensional core–shell nanowire CNT/TiO$_2$ with metal multi-walled carbon nanotubes (MWNTs) as the core and TiO$_2$ semiconductor as the shell [23]. In order to deposit a flat and uniform TiO$_2$ layer on the surface of MWNTs, the MWNTs array needs to be preheated at a temperature of 150 °C in the ALD reactor. Thus, a uniformly coated TiO$_2$ layer can still be formed on MWNTs after 200 ALD treatments (Fig. 27.7). To obtain a uniformly coated TiO$_2$ shell, researchers often need to use functional processing on

Fig. 27.7 a Schematics of the core–shell nanostructure and the corresponding PD. Inset shows the SEM image of the device. **b** XRD spectra of the core–shell NW before and after the annealing process. **c** TEM image of a core–shell NW. **d** High-resolution TEM image of a core–shell NW at the end. Reprinted with permission from Ref. [23]. Copyright 2012 American Chemical Society

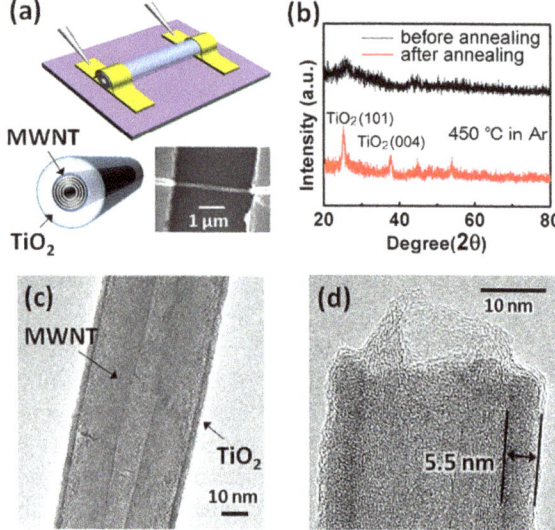

the core, which undoubtedly increases the complexity of the experimental operation. Ivo Utke et al. developed a new strategy that uses the non-functional vertical arrangement of MWCNTs to study the nucleation and growth of TiO₂ shells in the ALD process [24]. On the uneven surface of multi-walled carbon nanotubes, TiO₂ can still be deposited evenly. It is a simple strategy to control the coverage, morphology and crystallinity of the TiO₂ layer. Similarly, Lu et al. also deposited TiO₂ on the unfunctionalized Ge@G surface by the ALD method to synthesize Ge@Graphene@TiO₂ nanofibers with a core–shell structure [25]. Hupp et al. used the ALD method to deposit TiO₂ on the surface of silver, and controllably adjusted the thickness of the TiO₂ shell [26].

The template synthesis method employs materials with nanostructures, easy-to-control shapes, inexpensive and easy-to-obtain as templates. Then, physical or chemical methods are used to deposit raw materials into the pores of the materials template. Finally, the template is removed to obtain the template morphology nanomaterials. The template method is an important method for the synthesis of nanomaterials with specific properties. Niu et al. successfully synthesized C-TiO₂@g-C₃N₄ core–shell hollow nanospheres by the template method [27]. The unique hollow structure improves the efficiency of light energy utilization. First, carbon-doped TiO₂ hollow spheres were prepared using carbon spheres as templates. Then, g-C₃N₄ is grown in situ on the C-TiO₂ hollow sphere to make close contact between the two semiconductors. Furthermore, Zhang et al. synthesized core–shell Ni@TiO₂ and Ni@SiO₂ composites using a two-step template method [28]. First, uniform Ni particles were synthesized by the solvothermal reaction. Then, using tetrabutyl titanate (TBOT) as the precursor, Ni@TiO₂ composite microspheres with core–shell structure were prepared by template deposition (Fig. 27.8). Using ethyl orthosilicate as the precursor, a layer of silicon is coated on the surface of the Ni microspheres by a general sol–gel method. Compared with the original Ni microspheres, the electromagnetic wave absorption properties of Ni-TiO₂ and Ni-SiO₂ are significantly enhanced.

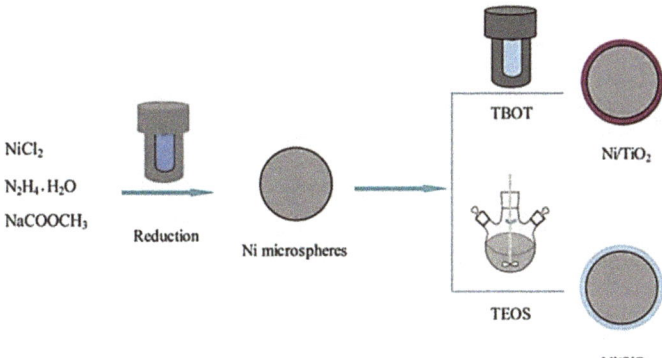

Fig. 27.8 Schematic illustration of the formation process of the Ni@TiO₂ and Ni@SiO₂ core–shell microspheres. Reprinted with permission from Ref. [28]. Copyright 2015 Royal Society of Chemistry

Self-assembly strategy makes the basic structural units (molecules, nanomaterials, micron or larger substances) spontaneously form an ordered structure. In the process of self-assembly, the basic structural units spontaneously organize or aggregate into a stable nanostructure with a certain regular geometric appearance under the interaction based on non-covalent bonds. Bian et al. prepared $Bi_xTi_{1-x}O_2$ mesoporous spheres with core–shell cavities under solvothermal conditions using a self-assembly method [29]. Although Bi atoms are much larger than Ti atoms, the cross-condensation between Ti–OH and Bi–OH ensures that the Bi doping is completely integrated into the TiO_2 lattice. During the reaction, TiO_2 aggregates to form spheroids, and then through dissolution and redeposition to form hollow spheroids with adjustable internal structure. Moreover, Bian et al. used the ultrasound-assisted aerosol-spray method to prepare TiO_2 layered hollow microspheres [30]. This unique structure is assembled from TiO_2 nano wafers. The 3–4 nm ultrafine gold nanoparticles can be uniformly deposited on the surface of TiO_2 hollow microspheres after adding Au^{3+} to the precursor solution. In addition, Ashok K. Ganguli et al. proposed for the first time to design ii-type TiO_2/CuS core–shell nanostructures according to the surface functionalization route [31]. First, they used 3-mercaptopropionic acid (MPA) as a functional ligand to grow on TiO_2 nanorods, and then used different concentrations of copper and sulfur precursors to achieve the best CuS shell thickness on the TiO_2 nanorods. Chen et al. have developed a general method of coating ZnO on various metal nano seeds, including metals, oxides, polymer nanoparticles, graphene oxide and carbon nanotubes [32]. It is necessary to control the nucleation and growth of the shell materials, the wetting of the shell materials of the seeds and the aggregation of the nanoparticles. ZnO is used to encapsulate gold nanoparticles in a case study to understand the multiple effects of polyvinylpyrrolidone (PVP) and its dependence on other factors (Fig. 27.9). The method can be further extended to include Fe_3O_4,

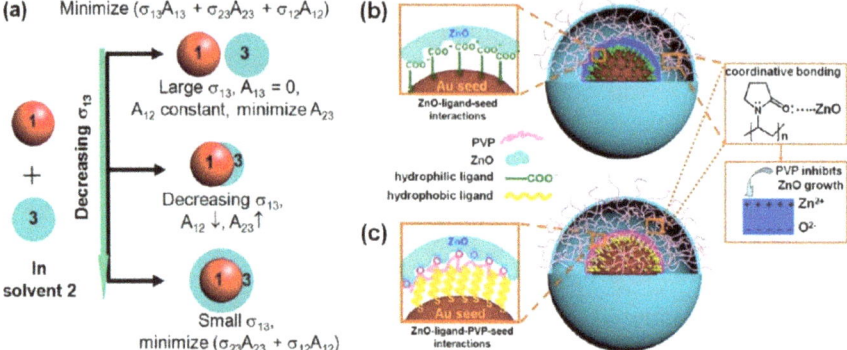

Fig. 27.9 Controlling the "wetting" of ZnO on Au seeds. **a** Equilibrium configurations for two immiscible liquid droplets 1 and 3 in solvent 2; and schematics illustrating **b** the Au–ligand–ZnO and **c** the Au –ligand–PVP–ZnO interactions in Au@ZnO NPs. Reprinted with permission from Ref. [32]. Copyright 2013 American Chemical Society

MnO, Co$_2$O$_3$, TiO$_2$, Eu$_2$O$_3$, Tb$_2$O$_3$, Gd$_2$O$_3$, Fe@Ni(OH)$_2$, ZnS, CdS, etc. as shell materials.

In addition to the above several commonly used methods, there are also some special TiO$_2$ core–shell structure preparation methods. Hu et al. synthesized a unique core–shell structured TiO$_2$ photocatalyst through one-step hydrogen treatment at 400–600 °C [33]. Ti^{3+} is located inside the crystal core, while the unordered Ti^{4+} constitutes the shell layer. Zhao et al. reported a simple extended Stöber's method to synthesize mesoporous core–shell TiO$_2$ nanomaterials [34]. The core–shell p-Si-mesoporous n-TiO$_2$ hybrid material was successfully prepared with silver-assisted p-type silicon nanowires (Si NWs) as the core and mesoporous n-type anatase TiO$_2$ as the outer shell. Further, Zhao et al. reported a multi-purpose dynamic control coating method to construct a uniform porous TiO$_2$ multifunctional core–shell structure [35]. By simply controlling the hydrolysis and condensation kinetics of TBOT in ethanol/ammonia mixtures, uniform porous TiO$_2$ shell-core structures with different diameters, geometries and compositions (for example, Fe-Fe$_2$O$_3$ ellipses) can be prepared. Although this method is very simple, the thickness of the TiO$_2$ shell can be easily controlled.

27.3 Photocatalytic Performance of Core–Shell TiO$_2$

So far, TiO$_2$ is the most widely used photocatalyst. The application of nano-TiO$_2$ photocatalysis is mainly applied to the degradation of pollutants, hydrogen evolution reaction and carbon dioxide reduction. The core–shell structure TiO$_2$ with unique morphology can effectively improve the photocatalytic performance.

Photocatalytic degradation of organic pollutants has the advantages of low energy consumption, high efficiency and direct use of sunlight [36]. Therefore, there are many research reports in wastewater treatment. Hu et al. synthesized a unique core–shell structured TiO$_2$ photocatalyst with Ti^{3+} core and Ti^{4+} outer shell through one-step hydrogen treatment [33]. The inner Ti^{3+} can be stably excited by visible light without changing, while the outer Ti^{4+} has a good ability to adsorb pollutants (Fig. 27.10). This catalyst can effectively degrade pollutants under the visible light environment at room temperature and pressure in which the degradation efficiency of MB is 400% of that of original TiO$_2$. Ashok K. Ganguli and others constructed a CuS/TiO$_2$ heterogeneous core–shell material to degrade toxic organic pollutants in wastewater [31]. Under visible light irradiation, the unique core–shell structure maximizes the interface contact between the core and the shell, thereby reducing the recombination rate of photogenerated carriers and enhancing the transfer of electrons and holes.

In addition, the use of noble metal nanoparticles with strong localized Surface Plasmon Resonance (LSPR) effect is also an effective way to improve the efficiency of photocatalytic carrier separation. However, due to the uneven distribution and rapid agglomeration of precious metal particles, the precious metal photocatalyst is easily deactivated, so researchers focus on the size and stability of precious metals.

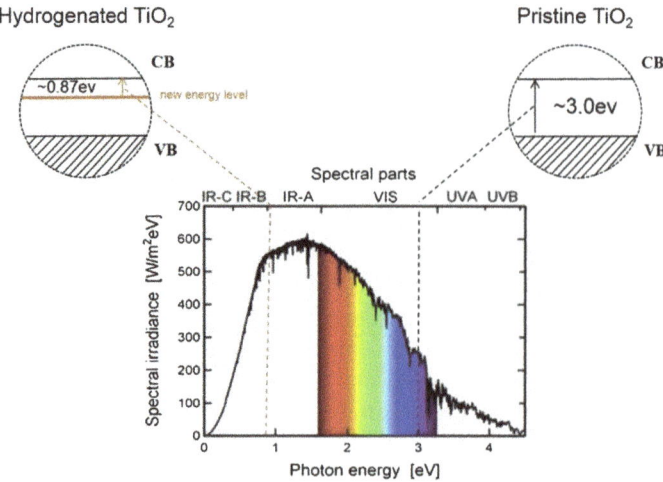

Fig. 27.10 The schematic of energy level transition and solar energy spectrum. Reprinted with permission from Ref. [33]. Copyright 2019 Elsevier B.V. All rights reserved

Li et al. reported a new Au/TiO$_2$ photocatalyst prepared by continuous solvothermal and hydrothermal treatment [20]. They evenly encapsulated the gold nanoparticles with mesoporous core–shell TiO$_2$ microspheres, which can inhibit the agglomeration and shedding of Au, thereby improving its photocatalytic oxidation activity and durability. While protecting Au, the mesoporous pores on the shell can allow reactive molecules to enter the cavity for effective degradation. Other precious metals nanoparticles can also be encapsulated with core–shell TiO$_2$ microspheres, providing a new method for designing high-stability precious metal photocatalysts. Bian et al. first prepared single-crystal mesoporous TiO$_2$ layered hollow microspheres using the ultrasonic-assisted aerosol-spray method [30]. Single crystal mesoporous TiO$_2$ significantly enhances charge separation due to its internal defects and porous characteristics. If gold ion raw materials are added to the TiO$_2$ precursor solution, ultrafine gold nanoparticles can be uniformly deposited on the surface of the TiO$_2$ hollow microspheres, which can effectively degrade pollutants in wastewater. In order to degrade pollutant molecules more efficiently, Bian et al. introduced a piezoelectric field in the photocatalytic system to prevent photoelectron-hole recombination [37]. Generally speaking, the formation of the piezoelectric field requires additional mechanical force or a high-frequency ultrasonic bath, which increases the cost of the reaction. They used a simple coating method to prepare a PZT/TiO$_2$ core–shell catalyst. Only by collecting the mechanical energy of the water, an internal piezoelectric field can be generated (Fig. 27.11). When the stirring speed is 800 rpm/min, the measured transient photocurrent on the PZT/TiO$_2$ electrode is about 1.7 times that of 400 rpm. Therefore, PZT/TiO$_2$ has greatly improved the photocatalytic degradation rate and mineralization efficiency of various pollutants (such as rhodamine B, bisphenol A, phenol and p-chlorophenol).

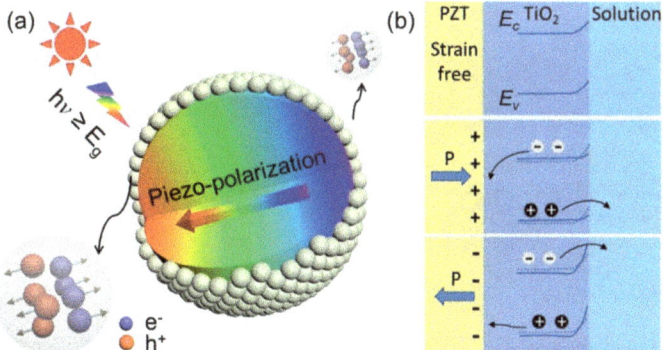

Fig. 27.11 a Diagram illustration on the proposed piezoelectric field enhanced photocatalytic reaction and **b** simplified band structure diagrams of TiO$_2$ at the photocatalyst –solution interface with different PZT polar direction. Reprinted with permission from Ref. [37]. Copyright 2018 American Chemical Society

Solar water splitting is a promising method for producing clean and renewable energy. The decomposition of water into hydrogen fuel and oxygen by photocatalysis provides a particularly attractive solution [38]. Cheng et al. succeeded in activating inert rutile TiO$_2$ for high-performance photocatalytic hydrogen evolution by creating a crystalline Ti^{3+} core and amorphous Ti^{4+} shell structure [39]. This structure can effectively inhibit the recombination of electrons and holes, thereby controllable adjustment of the movement of photogenerated carriers from the inner core and the outer shell. Interestingly, the unique arrangement of the crystalline Ti^{3+} core will make its interface bulge upward without the protection of the amorphous Ti^{4+} shell. This structure can be easily extended to the activation of other oxide/nitrogen-based photocatalysts to achieve efficient solar energy conversion. Patrik Schmuki et al. used a high-pressure hydrogenation method to modify commercial anatase/rutile TiO$_2$ powders, which can generate efficient and stable hydrogen evolution photocatalysts without the use of precious metal promoters [40]. In addition to the self-modification of TiO$_2$, the construction of a heterojunction catalyst is of great benefit to the improvement of water splitting activity. Niu et al. explored a core–shell hollow nanosphere material grown in situ on the surface of C-doped TiO$_2$ with g-C$_3$N$_4$ [27]. Compared with the original C-TiO$_2$ and g-C$_3$N$_4$, the composite heterojunction photocatalyst has significantly enhanced visible light photocatalytic water splitting to generate H$_2$ activity. That is 22.7 times and 10.5 times that of C-TiO$_2$ and g-C$_3$N$_4$, respectively. The performance enhancement is attributed to the formation of a heterojunction between the two semiconductors and the effect of the core–shell hollow structure on the separation of electrons and holes. Li et al. synthesized Ti^{3+} self-doped B-TiO$_2$/g-C$_3$N$_4$ hollow core–shell nano-heterojunctions by continuous hydrothermal deposition and etching reduction methods [41]. The photocatalytic hydrogen evolution activity of B-TiO$_2$/g-C$_3$N$_4$ nano-heterojunction is significantly improved by 18 times and 65 times compared with ordinary TiO$_2$ and

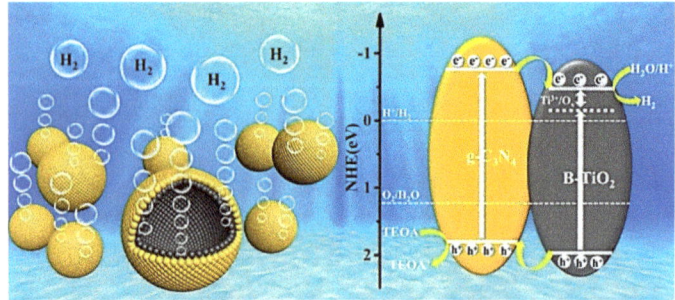

Fig. 27.12 The photocatalytic hydrogen production mechanism of the Ti^{3+} self-doping B-TiO_2/g-C_3N_4 hollow core–shell nano-heterojunction. Reprinted with permission from Ref. [41]. Copyright 2019 Elsevier B.V. All rights reserved

g-C_3N_4. Through continuous electrochemical tests, the author believes that the main mechanism of photocatalytic hydrogen increase is that oxygen vacancies increase visible light response and self-doped Ti^{3+} reduces the activation barrier required for H2 generation (Fig. 27.12). Furthermore, the core–shell nano-heterojunction can promote the separation of photogenerated electrons and holes to improve the efficiency of photocatalytic hydrogen release. More importantly, the hollow core–shell nanostructure improves the specific surface area and stability of the material.

Photocatalytic reduction of carbon dioxide and its conversion into hydrocarbon fuel is a challenging technology. Huang et al. reported a simple low-temperature solvothermal method to prepare a uniform and broad-spectrum hydrogenated blue H-TiO_{2-x} [42]. This material has an obvious crystal nucleus and amorphous shell structure. The amorphous shell contains a large amount of oxygen vacancies which is beneficial to the adsorption and activation of CO_2. The formation of CO_2 reduction intermediate products shows that the defects on the surface of H-TiO_{2-x} can effectively accelerate the adsorption and chemical activation of extremely stable CO_2 molecules, making it easier for CO_2 to CH_4. It is worth noting that the surface electron-modified H-TiO_{2-x} can extend the broad spectrum response, enhancing the separation and transport of electron–hole pairs. Further, the optimized material has a high methane production; the production and the selectivity rate under sunlight irradiation are 16.2 mol g^{-1} h^{-1} and 79%, respectively. Xiong et al. used ultrafast spectroscopy to prove that charge transfer can occur between light-excited inorganic semiconductors and MOFs [43]. To prove this concept, they developed a method to synthesize $Cu_3(BTC)_2$@TiO_2 core–shell structure, and used CO_2 conversion as a model to study the photocatalytic performance. The designed hybrid material uses semiconductor TiO_2 as the shell, which is favorable for light excitation to generate excitons. Moreover, the outer shell has a macroporous structure, which facilitates the capture of gas molecules in the inner core and provides sufficient specific surface area for photocatalysis. Chai et al. demonstrated the synthesis of MWCNT/TiO_2 core–shell nanocomposite and its application in photocatalytic CO_2 reduction under visible light radiation [44]. The photocatalytic mechanism shows that the enhancement of

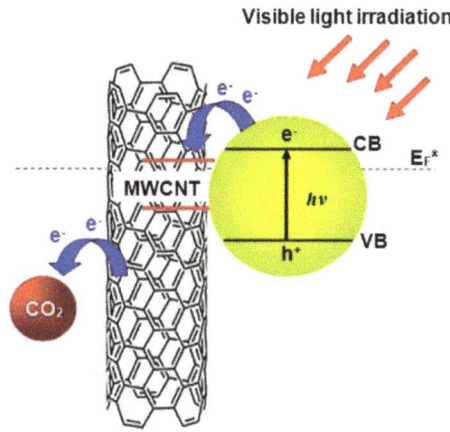

the photoreactivity of the core–shell nanocomposite is achieved by electron transfer between the TiO_2 shell and CNT, which changes the recombination of electron–hole pairs and improves the super-efficiency of photocatalysis (Fig. 27.13).

27.4   Conclusions and Perspectives

In conclusion, the core–shell structured TiO_2 nanomaterials as a new type of photo-catalyst show huge potential advantages [45]. Due to the interaction between the core and shell, the core–shell nano-TiO_2 has unique properties that other TiO_2 morphologies do not have. Compared with traditional TiO_2, its unique size effect, surface effect and cavity structure reflect light making it have higher photocatalytic activity. This chapter reviews the development of core–shell nano-TiO_2 in the field of photocatalysis. Firstly, the synthesis methods of core–shell nano-TiO_2 are introduced, including sol–gel method, water/solvothermal method, chemical deposition method, template synthesis method and self-assembly strategy. We focus on analyzing the basic operations of various methods and the key factors that can be controlled. Secondly, according to the structure–activity relationship, we detailed the application of core–shell nano-TiO_2 in the field of photocatalysis, including pollutant degradation, hydrogen evolution reaction and carbon dioxide reduction.

However, core–shell nanomaterials still have many key scientific and technical issues restricting their industrial applications, and in-depth research is needed. Firstly, based on the diversity of synthesis methods, the position and formation mechanism of TiO_2 in the core-shell structure need to be further studied. Secondly, the performance of core–shell TiO_2 materials can be improved by constructing heterojunctions. There are many elements to choose from, such as metals, non-metals, oxides and organic polymers. How to appropriately select TiO_2 and other materials to form a heterojunction and optimize the ratio of the two in the core–shell structure? This is a problem worthy of systematic research. At present, the preparation conditions of most core–shell TiO_2 composite materials are relatively strict. It is believed that with the deepening of research and the maturity of synthesis methods, core–shell nanomaterials will be further away from the practical application.

References

1. Girish Kumar S, Gomathi Devi L (2011) Review on modified TiO_2 photocatalysis under UV/Visible light: selected results and related mechanisms on interfacial charge carrier transfer dynamics. J Phys Chem a 115:13211–13241
2. Ravelli D, Dondi D, Fagnonia M, Albini A (2008) Photocatalysis. A multi-faceted concept for green chemistry. Chem Soc Rev 38:1999–2011
3. Fujishima A, Honda K (1972) Electrochemical photolysis of water at a semiconductor electrode. Nature 238:37–38
4. Qing G, Zhibo M, Chuanyao Z, Zefeng R, Xueming Y (2019) Single molecule photocatalysis on TiO_2 surfaces. Chem Rev 119:11020–11041
5. Xiaobo C, Samuel SM (2007) Titanium dioxide nanomaterials: synthesis, properties, modifications, and applications. Chem Rev 107:2891–2959
6. Xueqin L, Iocozzia J, Yang W, Xun C, Yihuang C, Shiqiang Z, Zhen L, Zhiqun L (2017) Noble metal–metal oxide nanohybrids with tailored nanostructures for efficient solar energy conversion, photocatalysis and environmental remediation. Energy Environ Sci 10:402
7. Huanli W, Lisha Z, Zhigang C, Junqing H, Shijie L, Zhaohui W, Jianshe L, Xinchen W (2014) Semiconductor heterojunction photocatalysts: design, construction, and Photocatalytic performances. Chem Soc Rev 43:5234
8. Michael D, Yiding L, Yadong Y (2014) Composite titanium dioxide nanomaterials. Chem Rev 114:9853–9889
9. Chunping X, Prasaanth Ravi A, Cyril A, Rafael L, Samuel M (2019) Nanostructured materials for photocatalysis. Chem Soc Rev 48:3868–3902
10. Qiao Z, Ilkeun L, Ji Bong J, Franclsco Z, Yadong Y (2013) Core shell nanostructured catalysts. Accounts Chem Res 46:1816–1824
11. Wei L, Ahmed E, Dhaifallah A, Dongyuan Z (2018) core–shell structured titanium dioxide nanomaterials for solar energy utilization. Chem Soc Rev 47:8203
12. Xin L, Jiaguo Y, Mietek J (2016) Hierarchical photocatalysts. Chem Soc Rev 45:2603
13. Tsutomu H, Prashant VK (2005) Charge separation and catalytic activity of Ag@TiO_2 core–shell composite clusters under UV-irradiation. J Am Chem Soc 127:3928–3934
14. Hideki S, Takashi K, Hirobumi S, Takahiro O, Masahiko A (2005) Preparation of highly dispersed core/shell-type titania nanocapsules containing a single Ag nanoparticle. J Am Chem Soc 128:4944–4945
15. Wei Liang L, Fan Cheng L, Yu Chen Y, Chen Hsien H, Shangjr G, Michael HH, Jer Shing H (2013) The influence of shell thickness of Au@TiO_2 core–shell nanoparticles on the plasmonic enhancement effect in dye-sensitized solar cells. Nanoscale 5:7953
16. Yingying W, Wenjuan Y, Xianjie C, Jun W, Yongfa Z (2018) Photocatalytic activity enhancement of core–shell structure g-C_3N_4@TiO_2 via controlled ultrathin g-C_3N_4 layer. Appl Catal B Environ 220:337–347
17. Jiang D, Jian Q, Dan W, Zhiyong T (2012) Facile synthesis of Au@TiO_2 core–shell hollow spheres for dye-sensitized solar cells with remarkably improved efficiency. Energy Environ Sci 5:6914
18. Hexing L, Zhenfeng B, Jian Z, Dieqing Z, Guisheng L, Yuning H, Hui L, Yunfeng L (2007) Mesoporous titania spheres with tunable chamber stucture and enhanced photocatalytic activity. J Am Chem Soc 129:8406–8407

19. Jingxia L, Jianhua X, Wei-Lin D, Hexing L, Kangnian F (2009) Direct hydro-alcohol thermal synthesis of special core–shell structured Fe-doped titania microspheres with extended visible light response and enhanced photoactivity. Appl Catal B Environ 58:162–170

20. Zhenfeng B, Jian Z, Fenglei C, Yunfeng L, Hexing L (2009) In situ encapsulation of Au nanoparticles in mesoporous core–shell TiO$_2$ microspheres with enhanced activity and durability. Chem Commun 3789–3791

21. Zhenfeng B, Jian Z, Jinguo W, Shengxiong X, Colin N, Hexing L (2012) Multitemplates for the hierarchical synthesis of diverse inorganic materials. J Am Chem Soc 134:2325–2331

22. Lori EG, Matt L, Benjamin DY, Peidong Y (2007) ZnO-TiO$_2$ core–shell nanorod/P3HT solar cells. J Phys Chem C 111:18451–18456

23. Chia Yang H, Der Hsien L, Sheng Yi L, Cheng Ying C, Chen Fang K, Yu Lun C, Wen Kuang H, Jr Hau H (2012) Supersensitive, ultrafast, and broad-band light-harvesting scheme employing carbon nanotube/TiO$_2$ core–shell nanowire geometry. ACS Nano 6:6687–6692

24. Guerra Nuñez C, Yucheng Z, Meng L, Vipin C, Rolf E, Johann M, Hyung Gyu P, Ivo U (2015) Morphology and crystallinity control of ultrathin TiO$_2$ layers deposited on carbon nanotubes by temperature-step atomic layer deposition. Nanoscale 7:10622

25. Xiaoyan W, Ling F, Decai G, Jian Z, Qingfeng Z, Bingan L (2015) Core–shell Ge@Graphene@TiO$_2$ nanofibers as a high-capacity and cycle-stable anode for lithium and sodium ion battery. Adv Funct Mater 26:1104–1111

26. Stacey DS, George CS, Joseph TH (2009) Distance dependence of plasmon-enhanced photocurrent in dye-sensitized solar cells. J Am Chem Soc 131:8407–8409

27. Yajun Z, Jian-Wen S, Dandan M, Zhaoyang F, Lu L, Chunming N (2017) In-situ synthesis of C-doped TiO$_2$@g-C$_3$N$_4$ core–shell hollow nanospheres with enhanced visible-light photocatalytic activity for H$_2$ evolution. Chem Eng J 322:435–444

28. Biao Z, Gang S, Bingbing F, Wanyu Z, Rui Z (2015) Investigation of the electromagnetic absorption properties of Ni@TiO$_2$ and Ni@SiO$_2$ composite microspheres with core–shell structure. Phys Chem Chem Phys 17:2531

29. Zhenfeng B, Jie R, Jian Z, Shaohua W, Yunfeng L, Hexing L (2009) Self-assembly of Bi$_x$Ti$_{1-x}$O$_2$ visible photocatalyst with core–shell structure and enhanced activity. Appl Catal B Environ 89:577–582

30. Chao T, Longfei L, Yali L, Zhenfeng B (2017) Aerosol spray assisted assembly of TiO$_2$ mesocrystals into hierarchicalhollow microspheres with enhanced photocatalytic performance. Appl Catal B Environ 201:41–47

31. Sunita K, Sandeep K, Ashok KG (2016) Comparative study of TiO$_2$/CuS core/shell and composite nanostructures for efficient visible light photocatalysis. ACS Sustain Chem Eng 4:1487–1499

32. Hang S, Jiating H, Jiangyan W, Shuang Yuan Z, Cuicui L, Thirumany S, Subodh M, Ming Yong H, Dan W, Hongyu C (2013) Investigating the multiple roles of polyvinylpyrrolidone for a general methodology of oxide encapsulation. J Am Chem Soc 135:9099–9110

33. Haiyang H, Yan L, Yun Hang H (2020) core–shell structured TiO$_2$ as highly efficient visible light photocatalyst for dye degradation. Catal Today 341:90–95

34. Manas P, Hao W, Yunke J, Xiaomin L, Hongwei Z, Changyao W, Shuai W, Abdullah MA, Yonghui D, Gengfeng Z, Dongyuan Z (2016) Core–shell silicon@mesoporous TiO$_2$ heterostructure: towards solar-powered photoelectrochemical conversion. ChemNanoMat 2:647–651

35. Wei L, Jianping Y, Zhangxiong W, Jinxiu W, Bin L, Shanshan F, Yonghui D, Fan Z, Dongyuan Z (2012) A versatile kinetics-controlled coating method to construct uniform porous TiO$_2$ shells for multifunctional core–shell structures. J Am Chem Soc 134:11864–11867

36. Malato S, Fernández-Ibáñez P, Maldonado MI, Blanco J, Gernjak W (2009) Decontamination and disinfection of water by solar photocatalysis: Recent overview and trends. Catal Today 147:1–59

37. Yawei F, Hao L, Lili L, Sa Y, Donglai P, Hao G, Hexing L, Zhenfeng B (2018) Enhanced photocatalytic degradation performance by fluid-oinduced piezoelectric field. Environ Sci Technol 52:7842–7848

38. Akihiko K, Yugo M (2009) Heterogeneous photocatalyst materials for water splitting. Chem Soc Rev 38:253–278
39. Yongqiang Y, Gang L, John TSI, Hui-Ming C (2016) Enhanced photocatalytic H_2 production in core–shell engineered rutile TiO_2. Adv Mater 28:5850–5856
40. Ning L, Christopher S, Detlef F, Umamaheswari V, V. R. Reddy M, Martin Hn, Benjamin W, Erdmann S Andres O, Eva M. Z, Karsten M, Tomohiko N, Xuemei Z, Patrik S, (2014) Hydrogenated anatase: strong photocatalytic dihydrogen evolution without the use of a co-catalyst. Angew Chem Int Ed 53:14201–14205
41. Jiaqi P, Zongjun D, Beibei W, Ziyuan J, Chuang Z, Jingjing W, Changsheng S, Yingying Z, Chaorong L (2019) The enhancement of photocatalytic hydrogen production via Ti^{3+} self-doping black TiO_2/g-C_3N_4 hollow core–shell nano-heterojunction. Appl Catal B Environ 242:92–99
42. Guoheng Y, Xieyi H, Tianyuan C, Wei Z, Qingyuan B, Jing X, Yifan H, Fuqiang H (2018) Hydrogenated blue titania for efficient solar to chemical conversions: preparation, pharacterization, and reaction mechanism of CO_2 reduction. ACS Catal 8:1009–1017
43. Rui L, Jiahua H, Mingsen D, Helin W, Xijun W, Yingli H, Hai-Long J, Jun J, Qun Z, Xie Yi, Yujie X (2014) Integration of an inorganic semiconductor with a metal–organic framework: a platform for enhanced gaseous photocatalytic reactions. Adv Mater 26:4783–4788
44. Meei Mei G, Siang Piao C, Bo Qing X, Abdul Rahman M (2014) Enhanced visible light responsive MWCNT/TiO_2 core–shell nanocomposites as the potential photocatalyst for reduction of CO_2 into methane. Sol Energy Mater Sol Cells 122:183–189
45. Hua T, Shuxin O, Yingpu B, Naoto U, Mitsutake O, Jinhua Y (2012) Nano-photocatalytic materials: possibilities and challenges. Adv Mater 24:229–251

Chapter 28
Synthesis of Yolk-Shell Structured Fe$_3$O$_4$@Void@CdS Nanoparticles: A General and Effective Structure Design for Photo-Fenton Reaction

Lingzhi Wang and Jinlong Zhang

28.1 Introduction

The photo-Fenton reaction, which combines Fenton reagents that generate highly aggressive ·OH from H$_2$O$_2$ and light irradiation that gives rise to extra ·OH radicals is considered to be one of the most effective ways to decontaminate [1, 2]. However, the generation of ferric-hydroxide sludge and quick loss of ferrous in the Fe^{2+}–H$_2$O$_2$ system have severely hindered its practical applications [3]. Compared with a homogeneous photo-Fenton reaction, a heterogeneous photo-Fenton reaction in which Fe is stabilized within the catalyst structure partly overcomes the aforementioned problems [4, 5]. Considering the higher content of structural Fe^{2+}, Fe$_3$O$_4$ is extensively reported to be the most effective and widely used heterogeneous photo-Fenton catalyst among Fe-based materials [6, 7]. Moreover, the merit of inherent magnetic property contributes to not only the fast separation but also the controllable synthesis [8]. Nevertheless, bare Fe$_3$O$_4$ still has some inevitable problems such as limited adsorption ability, spontaneous aggregation, narrow suitable pH range and inferior cyclic performance [4].

On the one hand, in order to gain better enrichment ability for pollutants as well as avoiding aggregation, inert materials such as carbon and, SiO$_2$ are introduced to fabricate loading- or coating-type catalyst [9–11]. Among them, yolk-shell structured Fe$_3$O$_4$@void@shell composites composed of a movable Fe$_3$O$_4$ core and a layer of permeable shell have recently gained increasing attention in the fields of catalysis [12, 13], biology [14, 15] and energy storage [16, 17], where the interior core is efficiently protected from agglomeration and is accessible for small molecules. In the context

L. Wang · J. Zhang (✉)
Key Laboratory for Advanced Materials and Institute of Fine Chemicals, School of Chemistry and Molecular Engineering, East China University of Science and Technology, 130 Meilong Road, Shanghai 200237, People's Republic of China
e-mail: jlzhang@ecust.edu.cn

of catalysis, the cavity between the core and shell with a flexibly tunable volume provides a nanoreactor for a variety of reactions, in which the confined reactants may result in improved reaction rate or altered synthetic route [18–20].

On the other hand, the limited working pH range around 3 commonly encountered by most Fenton agents has significantly obstructed the wide applications [21, 22]. Although the heterogeneous Fenton agent generally can be applied in less acidic conditions, the regeneration efficiency of Fe^{2+} from Fe^{3+} and the oxidation potential of hydroxyl radicals have actually both been decreased. Recently, the combination of semiconductor and Fe-containing compound has proven to be effective on reducing Fe^{3+} to Fe^{2+} by the photogenerated electron from a semiconductor [23, 24]. Considering the priority of the yolk-shell structured composite as a nanoreactor, it is desirable to apply it in a photo-Fenton process, which, however, has been retarded by the tedious preparation process including layer-by-layer coating and removal of sacrificing layer, as well as the underdeveloped techniques on the component-tailoring [25–27]. As such, current progress is far from the application requirements covering the fast degradation rate, efficient utilization of solar light and reasonable recyclability. To this end, a simple and general synthesis technique for yolk-shell structured composite is extremely desired to put forward its application in the photo-Fenton treatment of sewage water [28–30].

Owing to the excellent absorption ability to visible light, CdS has been recently integrated into versatile yolk-shell structured nanocomposites and applied in photocatalysis. For example, Han et al. achieved an improved photocatalytic activity for hydrogen evolution using yolk-shell structured AuNR-CdS [31]. Xiao et al. fabricated the yolk-shell structured $CdS@void@TiO_2$ for H_2 evolution and dye degradation; the TiO_2 shell can both provide more active sites and protect CdS from oxidation [32]. Lou et al. reported a sulfidation, and cation-exchange strategy for preparing multi-shelled ZnS-CdS rhombic dodecahedral cages (RDCs) by the yolk-shell Zn-based ZIF-8 RDCs [33]. Zhong et al. synthesized yolk-shell CdS–graphene composite with a hollow core for dye degradation [34]. Zhou et al. revealed the heterostructures of NH_2-MIL-125/TiO_2/CdS yolk-shell can bring superior photocatalytic performances [35]. Considering the unique cavity for promoting the light absorption and the possible coupling effect between a core and shell, a yolk-shell structured catalyst with CdS is also expected to be functional in the Fenton reaction. Herein, using hydrophilic $Fe_3O_4@SiO_2$ particle as the cores, we report a simple one-pot strategy for the rational design and successful synthesis of $Fe_3O_4@void@$semiconductor[36]. As demonstrated by $Fe_3O_4@void@CdS$, the inner SiO_2 layer can be gradually etched with the coating of the CdS shell through a CBD process. The obtained yolk-shell NPs are monodisperse and the CdS shells are highly uniform. This catalyst shows excellent activity toward the degradation of methylene blue in a wide pH range from 4.5 to 11. Control experiments are carried out to fully understand the possible mechanism of this heteroconjuction. Further studies show similar enhanced activities can also be achieved on yolk-shell structure containing different semiconductor shells including TiO_2 and CeO_2, demonstrating the general feasibility of this kind of photo-Fenton agent in efficient water decontamination.

28.2 Experimental Section

Synthesis of yolk-shell structured Fe$_3$O$_4$@void@CdS NPs
First, the magnetic Fe$_3$O$_4$ cores were synthesized according to a solvothermal method described previously [37]. And then, Fe$_3$O$_4$@SiO$_2$ core-shell NPs were prepared through a modified Stöber method [38]. Finally, the synthesis of Fe$_3$O$_4$@void@CdS NPs adopts the method in our previous work with some modifications [12]. Typically, 50 mg of Fe$_3$O$_4$@SiO$_2$ NPs was dispersed into 100 mL distilled water under vigorous stirring. After 30 min, CdCl$_2$ (10 mL, 0.05 M), trisodium citrate (3 mL, 0.2 M) and thiourea (2.5 mL, 0.2 M) were added into the above aqueous solution orderly. An appropriate amount of concentrated ammonia solution was then introduced to the suspension drop by drop until the final pH value reached 11. Afterward, the resultant solution was kept at 60 °C under ultrasonic conditions for 2 h. Fe$_3$O$_4$@void@CdS were magnetically collected, washed and finally dried before use.

Synthesis of Fe$_3$O$_4$@SiO$_2$@CdS NPs
Fe$_3$O$_4$@SiO$_2$@CdS NPs were synthesized according to our previous work [12].

Synthesis of SiO$_2$@void@CdS NPs
SiO$_2$@void@CdS NPs were synthesized just taking SiO$_2$ NPs (ca. 220 nm) as cores with other experimental reagents and method unchanged. Because of the similar SiO$_2$ surface compared with Fe$_3$O$_4$@SiO$_2$, the right size of hollow space could be got as long as the right mass of SiO$_2$ NPs is applied.

Synthesis of Fe$_3$O$_4$@CdS NPs
Fe$_3$O$_4$@CdS NPs were synthesized by coating CdS on Fe$_3$O$_4$ cores directly.

Synthesis of CdS NPs
CdS NPs were synthesized just without adding the Fe$_3$O$_4$@SiO$_2$ cores.

Synthesis of Fe$_3$O$_4$@void@TiO$_2$ NPs
The core-shell structured Fe$_3$O$_4$@SiO$_2$@TiO$_2$ NPs were synthesized based on the hydrolysis and condensation of TBOT [39]. The yolk-shell structured Fe$_3$O$_4$@void@TiO$_2$ NPs were obtained after etching of the SiO$_2$ layer by concentrated ammonia solution in a water bath at 60 °C.

Synthesis of Fe$_3$O$_4$@void@CeO$_2$ NPs
The core-shell structured Fe$_3$O$_4$@SiO$_2$@CeO$_2$ NPs were synthesized according to the method reported previously [40]. The yolk-shell structured Fe$_3$O$_4$@void@CeO$_2$ NPs were obtained after etching of the SiO$_2$ layer by concentrated ammonia solution in a water bath at 60 °C.

Characterizations
Scanning electron microscopy (SEM) images were taken on a field emission scanning electron microscope (FESEM, S4800, Hitachi) equipped with an energy-dispersive

X-ray spectrum (EDS, JEOLJXA-840). The transmission electron microscopy (TEM JEOL JEM-2100EX) was used to characterize the samples' morphologies. The crystal phases of the obtained samples were analyzed by X-ray diffraction (XRD) on a SHIMADZU XRD-7000 XRD diffractometer using Cu Kα radiation. Raman spectra were recorded on a Renishaw inVia-Reflex Raman microprobe system. The Brunauer-Emmett-Teller (BET) surface area of the sample was determined through nitrogen adsorption at 77 K (Micromeritics ASAP2010). A vibrating sample magnetometer (VSM, LakeShore7407) was used to measure the magnetic hysteresis loop of the obtained sample at room temperature. The photoluminescence (PL) spectra were measured with a spectrofluorophotometer (Shimadzu, RF-5301). UV–Vis spectra of methylene blue were recorded using an ultraviolet visible spectrophotometer (Cary 100).

Photo-Fenton Experiments
The photo-Fenton activities of the as-prepared samples were tested by the degradation experiment under visible light irradiation using methylene blue as a model pollutant. The light source was a 1000 W tungsten-halide lamp equipped with wavelength cutoff filters ($\lambda \geq 420$ nm). In a typical experiment, a certain amount of the samples was added into methylene blue aqueous solution. Before turning on the light, the suspension was vigorously stirred in dark for 40 min to establish the adsorption/desorption equilibrium. In the beginning, a certain amount of H_2O_2 and NaOH was added into the solution together with turning on the lamp. At fixed time intervals during the degradation process, 2 mL of the solution was taken out and immediately magnetic separated for 30 s to get the upper clear solution. The concentration of methylene blue was analyzed by recording the characteristic absorption peak at 664 nm using a UV-Vis spectrophotometer. The photo-Fenton experiments of Fe_3O_4@void@TiO_2 and Fe_3O_4@void@CeO_2 were similar to that of Fe_3O_4@void@CdS only using a Xe lamp (300 W) as the light source.

28.3   Results and Discussion

The synthesis method of the yolk-shell structured Fe_3O_4@void@CdS NPs is explained in Scheme 28.1. First, magnetic Fe_3O_4 NP as a core was synthesized by a solvothermal process. Then a SiO_2 layer was coated above Fe_3O_4 through a modified Stöber process, which serves as the hard template for the void space. The yolk-shell structured Fe_3O_4@void@CdS NPs were gained by a one-pot ultrasound-assisted CBD method, where SiO_2 was gradually etched during the in situ growth process of the CdS layer.

The FESEM images of Fe_3O_4@SiO_2 and Fe_3O_4@void@CdS NPs show these two composites are monodisperse and have uniform sizes of ca. 170 and 200 nm, respectively (Fig. 28.1a, b). The TEM image of Fe_3O_4@SiO_2 indicates that the diameter of the Fe_3O_4 core is ca. 100 nm and the thickness of the SiO_2 shell is ca. 35 nm (Fig. 28.1c). Fe_3O_4@void@CdS has a coarse CdS shell with the thickness

Scheme 28.1 Schematic illustration for the formation of the Fe$_3$O$_4$@void@CdS NPs. Reprinted with permission from Ref. [36]. Copyright 2016 American Chemical Society

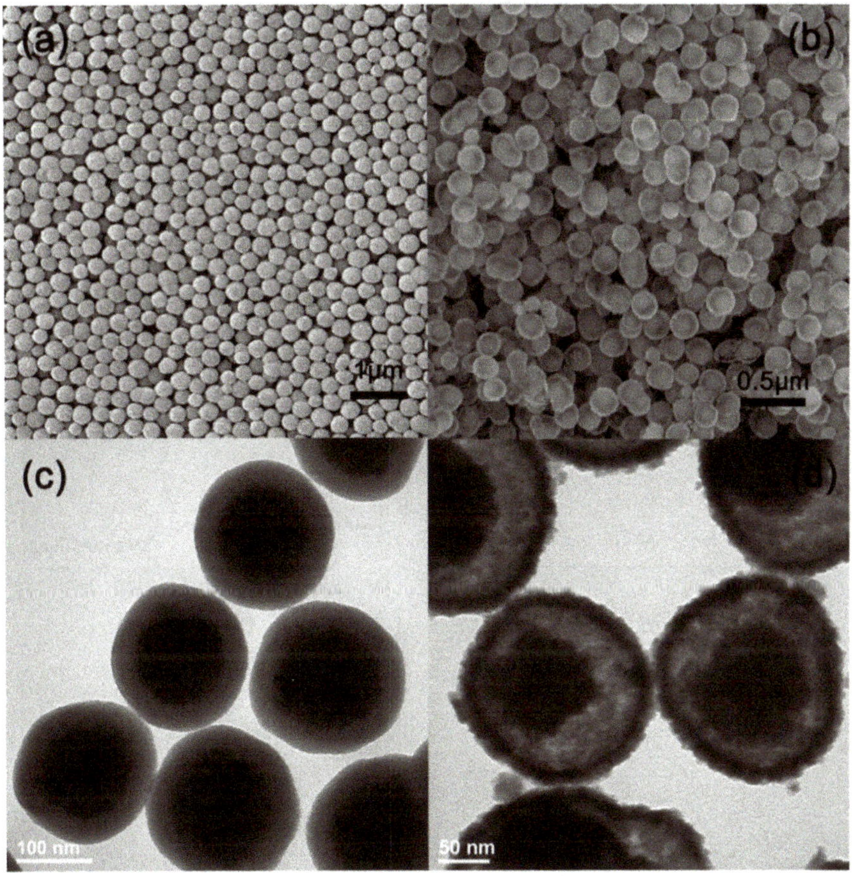

Fig. 28.1 FESEM images of **a** Fe$_3$O$_4$@SiO$_2$ and **b** Fe$_3$O$_4$@void@CdS NPs; TEM images of **c** Fe$_3$O$_4$@SiO$_2$ and **d** Fe$_3$O$_4$@void@CdS NPs. Reprinted with permission from Ref. [36]. Copyright 2016 American Chemical Society

of ca. 15 nm and the void space with a radius of ca. 35 nm between the Fe_3O_4 core and the CdS shell, which is consistent with the thickness of the original SiO_2 layer and indicates the successful etching of SiO_2 during the coating of the CdS layer (Fig. 28.1d).

The EDS result (Fig. 28.2a) shows S, O, Fe and Cd elements and negligible Si, further verifying the effective etching of SiO_2. The wide-angle XRD patterns of Fe_3O_4@void@CdS NPs are displayed in Fig. 28.2b. Obvious peaks at 2θ values of 25.3° (100), 26.6° (002), 28.3° (101), 44.2° (110), 48.0° (103) and 51.8° (112) can be indexed to the planes of hexagonal phase CdS (JCPDS Card No. 80-0006). In addition, the peaks located at the 2θ values of 30.1° (220), 35.4° (311), 57.0° (511) and 62.5° (440) correspond to the characteristic peaks of orthorhombic phase Fe_3O_4 (JCPDS Card No. 19-0629) very well. The Raman spectrum is further used to confirm the presence of CdS and Fe_3O_4. As shown in Fig. 28.2c, two prominent peaks located at 301 and 601 cm^{-1} originate from 1 LO (longitudinal optical) and 2 LO vibrational modes of CdS, respectively [41]. Two peaks located at 480 and 667 cm^{-1} can also be found in the spectrum, which is accordant with the T_{2g} and A_{1g} bands of Fe_3O_4, respectively [42]. The magnetic property of the Fe_3O_4@void@CdS

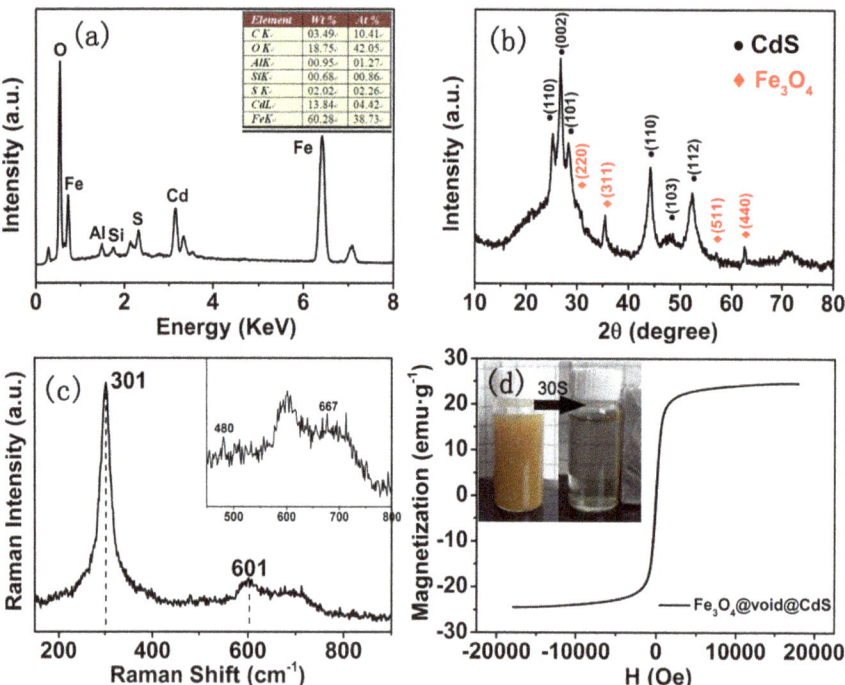

Fig. 28.2 **a** EDS patterns of Fe_3O_4@void@CdS NPs; **b** XRD patterns of Fe_3O_4@void@CdS NPs; **c** Raman spectra of Fe_3O_4@void@CdS NPs; **d** Field-dependent magnetization curve of Fe_3O_4@void@CdS at room temperature. Reprinted with permission from Ref. [36]. Copyright 2016 American Chemical Society

NPs was quantified with VSM at room temperature. As illustrated in Fig. 28.2d, the hysteresis loop shows the ferromagnetism of the Fe$_3$O$_4$@void@CdS NPs. The saturation magnetization for Fe$_3$O$_4$@void@CdS NPs is about 24.56 emu·g^{-1}, which can be quickly separated from the solution by applying an external magnetic field within 30 s (inset in Fig. 28.2d).

Formation Mechanism of Fe$_3$O$_4$@void@CdS NPs

To reveal the formation mechanism of Fe$_3$O$_4$@void@CdS NPs, the influence of pH values in the CBD process was further studied. Tiny and disconnected CdS nanoclusters are deposited on the surface of Fe$_3$O$_4$@SiO$_2$ and the SiO$_2$ layer is well preserved at pH < 9 (Fig. 28.3a). A uniform CdS shell is formed around the SiO$_2$ layer when the pH value increases to 10 (Fig. 28.3b). The yolk-shell structured Fe$_3$O$_4$@void@CdS is finally formed at pH = 11 with the assistance of ultrasonic treatment. In contrast, the CdS shell is broken at a higher pH value of 12 and no shell can be formed at pH = 13 (Fig. 28.3c, d), where only bulk CdS NPs detached from Fe$_3$O$_4$ are observed. It is known that S^{2-} gradually releases from thiourea in alkaline conditions (Eqs. 28.1–28.3). Therefore, Cd^{2+} can hardly be adsorbed on the SiO$_2$ surface when the releasing speed of S^{2-} is too fast in a strong alkaline system, resulting in the homogeneous nucleation of CdS in the solution. In a mild reaction system with a pH of 11, Cd^{2+} can be firstly adsorbed onto the surface of SiO$_2$ uniformly in the case of retarded decomposition of thiourea. With the continuous generation of S^{2-} and gradual etching of SiO$_2$, a uniform CdS shell is finally formed

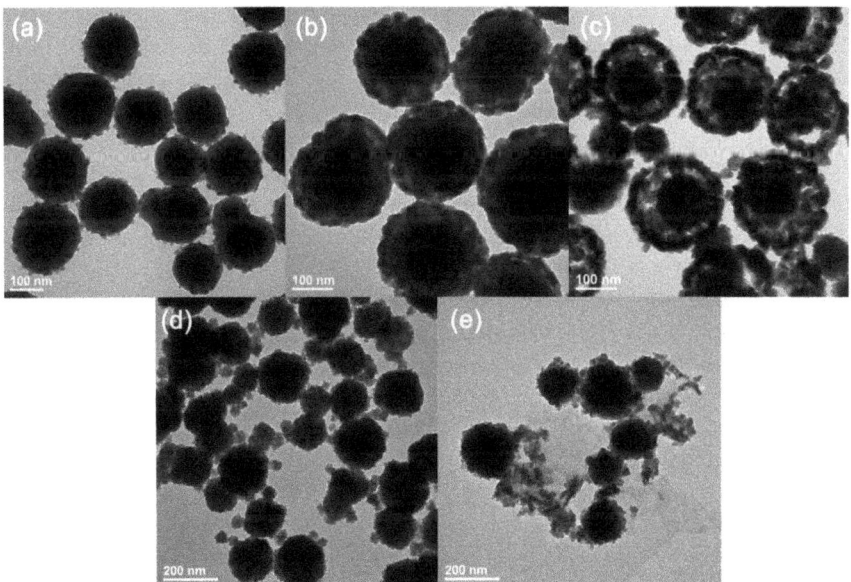

Fig. 28.3 TEM images of the samples **a** pH = 9; **b** pH = 10; **c** pH = 12; **d** pH = 13; and **e** without trisodium citrate. Reprinted with permission from Ref. [36]. Copyright 2016 American Chemical Society

(Eq. 28.5), with the formation of a hollow cavity. Moreover, trisodium citrate here acts as a complex agent to control the concentration of Cd^{2+} (Eq. 28.4) in the case of the nucleation of CdS in the solution as mentioned above because of its low solubility product constant ($K_{sp} = 10^{-28}$) [43]. In the absence of trisodium citrate, CdS nanoclusters generate irregularly and of course, no shells can be formed even if the SiO_2 layer has been successfully etched (Fig. 28.3e).

$$NH_3 + H_2O \rightarrow NH + 4 + OH^- \tag{28.1}$$

$$(NH_2)_2CS + OH^- \rightarrow CH_2N_2 + H_2O + HS^- \tag{28.2}$$

$$HS^- + OH^- \rightarrow S^{2-} + H_2O \tag{28.3}$$

$$[CdCitr]^- \leftrightarrow Cd^{2+} + Citr^{3-} \tag{28.4}$$

$$Cd^{2+} + S^{2-} \rightarrow CdS \downarrow \tag{28.5}$$

Photo-Fenton Degradation Activities

The photo-Fenton activity of as-prepared Fe_3O_4@void@CdS NPs was studied using methylene blue as a model pollutant. As shown in Fig. 28.4a, under visible light irradiation ($\lambda \geq 420$ nm), about 90% of methylene blue is degraded just within 5 min and no methylene blue can be detected after 10 min in the presence of 0.2 g·L^{-1} of Fe_3O_4@void@CdS and 300 mM of H_2O_2, exhibiting an extremely excellent degradation activity. The self-degradation of methylene blue can be excluded because no obvious degradation is observed under light irradiation. Meanwhile, without Fe_3O_4@void@CdS, only 10% of methylene blue is degraded in the presence of H_2O_2 after 10 min irradiation. In the absence of H_2O_2, 18% and 22% removal efficiencies are achieved within 10 and 20 min, which should be caused by the photocatalytic activity of Fe_3O_4@void@CdS. For the Fenton reaction in the dark condition, a relatively fast degradation rate is observed within the first 10 min (16%) but significantly slows in 20 min. The above results well demonstrate the high degradation efficiency is mainly caused by the photo-Fenton reaction on Fe_3O_4@void@CdS.

The influence of H_2O_2 amount on the degradation efficiency is further studied in a wide concentration range of 20–300 mM (Fig. 28.4b). About 88% of methylene blue can be degraded within 20 min in the presence of 20 mM of H_2O_2. Complete degradation is achieved only within 7.5 min when the concentration increases to 200 mM. Further increasing of the H_2O_2 concentration to 300 mM enables a faster degradation rate in the first 2.5 min. In addition, the effect of pH value was further explored in the range of 4.5–11 (Fig. 28.4c). More than 90% removal of methylene blue can be achieved within 7.5 min at pH = 4.5 and 7, which is slightly lower than those without pH adjusting and with 100% efficiency at basic conditions. Further extending the time to 10 min results in a nearly complete elimination, suggesting an

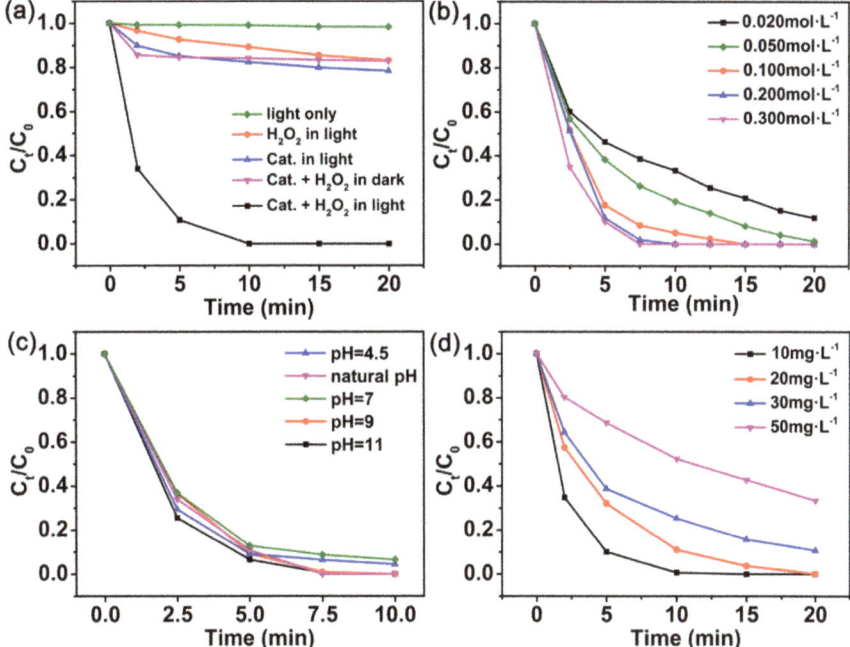

Fig. 28.4 a Time profiles of degradation performances under different conditions at room temperature (Cat. 0.2 g·L^{-1}, MB, 10 mg·L^{-1}, H$_2$O$_2$, 300 mM, Vol., 50 mL). Photo-Fenton performances of Fe$_3$O$_4$@void@CdS with different **b** H$_2$O$_2$ dosages, **c** pH values and **d** concentrations of MB. Reprinted with permission from Ref. [36]. Copyright 2016 American Chemical Society

excellent performance of Fe$_3$O$_4$@void@CdS in a wide pH range. Notably, a higher pH value (e.g. 11) even facilitates the degradation and the possible reasons will be discussed later. The degradation abilities of Fe$_3$O$_4$@void@CdS to methylene blue with different concentrations were also studied (10–50 mg·L^{-1}). As shown in Fig. 28.4d, methylene blue can be completely degraded at a concentration of 20 mg·L^{-1} in 20 min. 90% of methylene blue is eliminated within the same time when the concentration increases to 30 mg·L^{-1}. An extremely high degradation ratio of ca. 67% still can be achieved even at a higher concentration of 50 mg·L^{-1}, indicating the extremely high decontamination efficiency of this yolk-shell structured photo-Fenton agent.

Moreover, to understand the possible photosensitization effect of methylene blue during the photo-Fenton process, tetracycline (TC), a very refractory antibiotic with the maximum absorption peak at 357 nm, was further adopted as the pollutant model (Fig. 28.5) [44]. 20 mg·L^{-1} of TC can be completely degraded within 20 min under the visible light irradiation, which clearly excludes the contribution from the dye-sensitization [45].

Study on Reaction Mechanism and Morphology

To explore the mechanism of the photo-Fenton reaction, several control experiments were carried out using Fe_3O_4, CdS, $Fe_3O_4@CdS$, $Fe_3O_4@SiO_2@CdS$, $SiO_2@void@CdS$ and the physical mixture of Fe_3O_4 and $SiO_2@void@CdS$ NPs as catalysts (Fig. 28.6). All the reference samples are synthesized via similar methods with the size and morphology precisely controlled (Fig. 28.7). Obviously, methylene blue can be more efficiently preadsorbed by yolk-shell structured

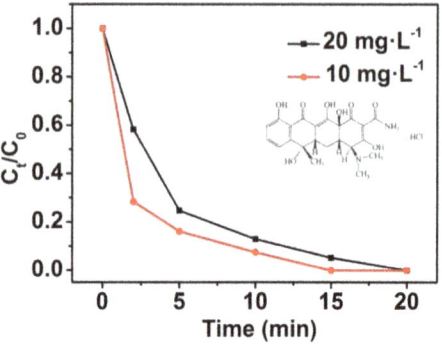

Fig. 28.5 Photo-Fenton degradation performances of $Fe_3O_4@void@CdS$ toward TC at room temperature under natural pH value (Cat. 0.2 $g·L^{-1}$, H_2O_2, 0.300 $mol·L^{-1}$, Vol., 50 mL, visible light irradiation). Reprinted with permission from Ref. [36]. Copyright 2016 American Chemical Society

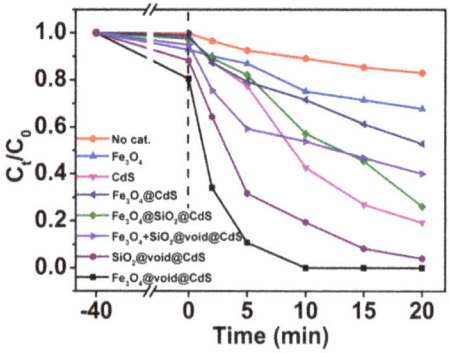

Fig. 28.6 Photo-Fenton degradation of MB by different catalysts (no cat., Fe_3O_4, CdS, $Fe_3O_4@CdS$, $Fe_3O_4@SiO_2@CdS$, physical mixture of Fe_3O_4 and $SiO_2@void@CdS$, $SiO_2@void@CdS$ and $Fe_3O_4@void@CdS$, Cat., 0.2 $g·L^{-1}$, MB, 10 $mg·L^{-1}$, H_2O_2, 300 mM, Vol., 50 mL). Reprinted with permission from Ref. [36]. Copyright 2016 American Chemical Society

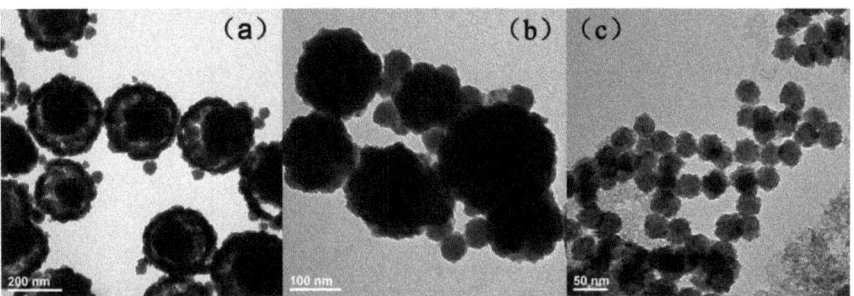

Fig. 28.7 TEM images of the as-prepared **a** SiO$_2$@void@CdS, **b** Fe$_3$O$_4$@CdS and **c** CdS NPs

samples Fe$_3$O$_4$@void@CdS (ca. 20%) and SiO$_2$@void@CdS (ca. 12%), demonstrating more adsorption sites result from the open inner surface of the yolk-shell structure. Compared with SiO$_2$@void@CdS, the improved adsorption efficiency of Fe$_3$O$_4$@void@CdS should be attributed to the rough surface of Fe$_3$O$_4$ core. In fact, a considerable adsorption efficiency of ca. 8% is revealed from naked Fe$_3$O$_4$ NPs, in accordance with the difference between the adsorption ability of SiO$_2$@void@CdS and Fe$_3$O$_4$@void@CdS. Fe$_3$O$_4$@void@CdS also shows the best degradation performance, where no methylene blue can be detected after light irradiation for 10 min. The tests on Fe$_3$O$_4$ and CdS NPs suggest that they are both active components in the photo-Fenton reaction, but both of the activities are inferior to Fe$_3$O$_4$@void@CdS. In addition, only around 40% and 30% of methylene blue can be degraded by Fe$_3$O$_4$@SiO$_2$@CdS and Fe$_3$O$_4$@CdS within 10 min, indicative of the important role of the yolk-shell structure. It is interesting that SiO$_2$@void@CdS NPs show better activities toward the degradation of methylene blue than CdS NPs, further implying the priority of the unique shell structure and cavity. A physical mixture of Fe$_3$O$_4$ (8 mg) and SiO$_2$@void@CdS (2 mg) NPs is also used to confirm the possible synergic effect between Fe$_3$O$_4$ and CdS (the ratio is calculated according to EDS to simulate the composition of Fe$_3$O$_4$@void@CdS NPs), which also exhibits an inferior performance to Fe$_3$O$_4$@void@CdS suggesting the importance of the synergistic effect between Fe$_3$O$_4$ and CdS in Fe$_3$O$_4$@void@CdS.

The possible photo-Fenton mechanism of Fe$_3$O$_4$@void@CdS was further explored by the radical-capture strategy. The addition of AgNO$_3$ as an electron scavenger significantly decreases the degradation efficiency (Fig. 28.8). Moreover, stronger photoluminescence is observed from Fe$_3$O$_4$@SiO$_2$@CdS compared with Fe$_3$O$_4$@void@CdS (Fig. 28.9), which should be attributed to the electrons transferring from CdS to Fe$_3$O$_4$ without the insulating SiO$_2$ layer. In contrast, a decrease of the degradation rate is hardly observed in the presence of hole scavenger Na$_2$C$_2$O$_4$ (Fig. 28.8), suggesting holes are not the key factor in this photo-Fenton system.

Based on the above results, a possible mechanism is proposed as illustrated in Fig. 28.10. Under the visible light irradiation, electrons and holes are generated on the CdS shell simultaneously (Eq. 28.6). On the one hand, electrons on the conduction band of CdS can initiate the direct decomposition of H$_2$O$_2$ to ·OH (Eq. 28.7) as shown

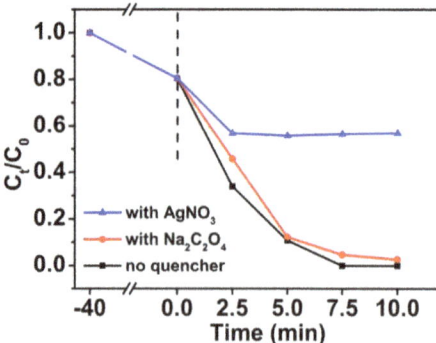

Fig. 28.8 Comparison of photo-Fenton activities of the yolk-shell structured Fe_3O_4@void@CdS NPs with different scavengers. Reprinted with permission from Ref. [36]. Copyright 2016 American Chemical Society

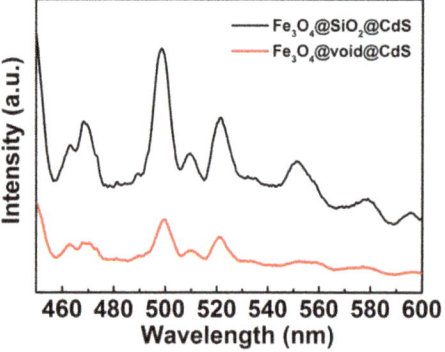

Fig. 28.9 PL spectra of the Fe_3O_4@void@CdS and Fe_3O_4@SiO_2@CdS NPs. Reprinted with permission from Ref. [36]. Copyright 2016 American Chemical Society

in pathway 1 [46]. The electron may also transfer to the Fe_3O_4 cores, reducing Fe^{3+} to Fe^{2+} (Eq. 28.8). Because Fe^{2+} is more active in the Fenton process [4], this will greatly speed up the generation of $\cdot OH$ (pathway 2). On the other hand, the residue holes in the value band of CdS may be able to react with H_2O to generate active $\cdot OH$ (Eq. 28.10) or directly with methylene blue (Eq. 28.11) as illustrated in pathway 3. Another important factor for the enhanced photo-Fenton activity is related to the yolk-shell structure. In the photo-Fenton process, hydroxyl radical with great oxidative potential ($E^0 = 2.8$ V vs. SHE) is considered to be the predominant active species for the degradation of dye [47, 48]. However, the lifetime of hydroxyl radicals in water is on the scale of a few nanoseconds [49]. Dye molecules adsorbed on the surface can capture the short-lived hydroxyl radicals to initiate the dye degradation process [50, 51]. Therefore, the adsorption of dye molecules on the surface of the catalysts may largely affect the catalytic efficiency [52]. As a consequence, yolk-shell

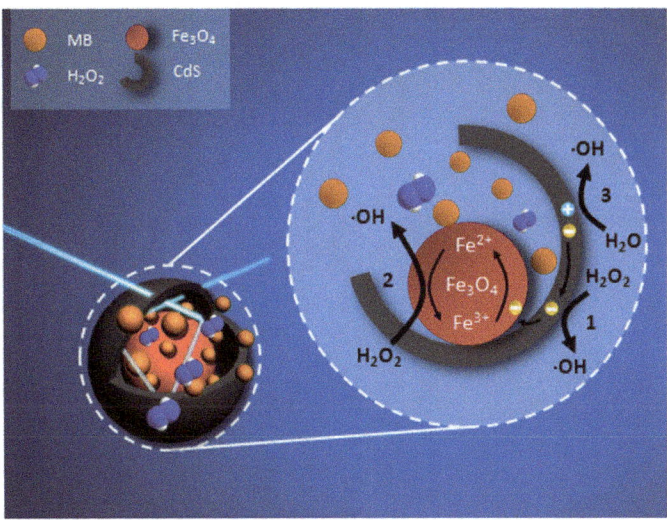

Fig. 28.10 Possible mechanism for the photo-Fenton degradation of MB by Fe$_3$O$_4$@void@CdS. Reprinted with permission from Ref. [36]. Copyright 2016 American Chemical Society

structure maximizes the BET surface (Fig. 28.11) to obtain better adsorption ability and exposes more active sites on the inner and outer surfaces. Another advantage of the yolk-shell structure is that the incident light may be reflected in the hollow space, resulting in better utilization of light energy [53, 54]. This is in accordance with the stronger absorbance of Fe$_3$O$_4$@void@CdS as shown in UV-Vis diffuse reflectance spectra (Fig. 28.12).

$$CdS \longrightarrow hue^-(CdS) + h^+(CdS) \tag{28.6}$$

$$H_2O_2 + e^- \rightarrow OH^- + \cdot OH \tag{28.7}$$

Fig. 28.11 Nitrogen adsorption-desorption measurements and the estimated BET surface areas of the Fe$_3$O$_4$@void@CdS, Fe$_3$O$_4$@SiO$_2$@CdS and Fe$_3$O$_4$@CdS NPs. Reprinted with permission from Ref. [36]. Copyright 2016 American Chemical Society

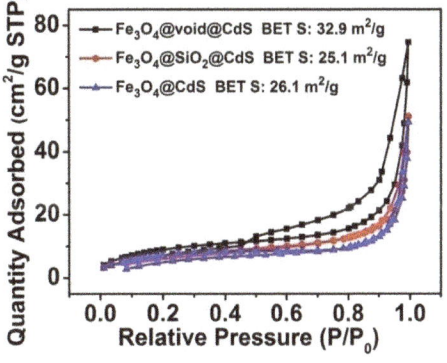

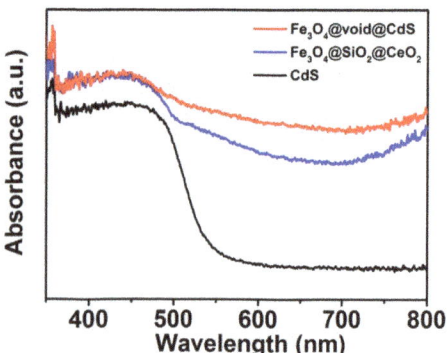

$$Fe^{3+} + e^- \rightarrow Fe^{2+} \qquad (28.8)$$

$$Fe^{2+} + H_2O_2 \rightarrow Fe^{3+} + OH^- + \cdot OH \qquad (28.9)$$

$$H_2O + h^+ \rightarrow H^+ + \cdot OH \qquad (28.10)$$

$$Dye + h^+ \rightarrow dye^+ \rightarrow \text{ degradation product} \qquad (28.11)$$

As mentioned above, Fe_3O_4@void@CdS can still exhibit good performance under basic conditions, but bare Fe_3O_4 is generally regarded as a poor Fenton reagent in high pH value [55]. The activity of SiO_2@void@CdS was thus investigated at different pH values to explore the possible reason. As seen from Fig. 28.13, SiO_2@void@CdS shows a faster degradation rate when the pH value increases, which indicates OH^- is favorable to the photo-Fenton degradation of methylene blue by CdS. There are two possible reasons: one is the large amount of OH^- in the solution that may react with holes directly to generate active $\cdot OH$ (Eq. 28.12) and the other is that the enhanced

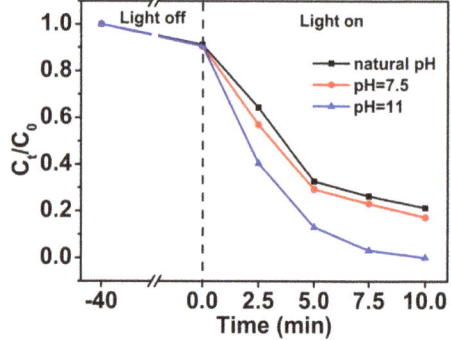

adsorption ability toward methylene blue in a basic environment [56]. Benefiting from the enhanced activity of CdS as well as the synergic effect between CdS and Fe$_3$O$_4$, Fe$_3$O$_4$@void@CdS keeps high activity under basic conditions.

$$OH^- + h^+ \rightarrow \cdot OH \tag{28.12}$$

Moreover, to verify the possible general feasibility of the above strategy in the photo-Fenton process, yolk-shell structured Fe$_3$O$_4$@void@TiO$_2$ and Fe$_3$O$_4$@void@CeO$_2$ are further prepared and applied to the methylene blue degradation. The morphologies and XRD patterns can be seen in Figs. 28.14, 28.15.

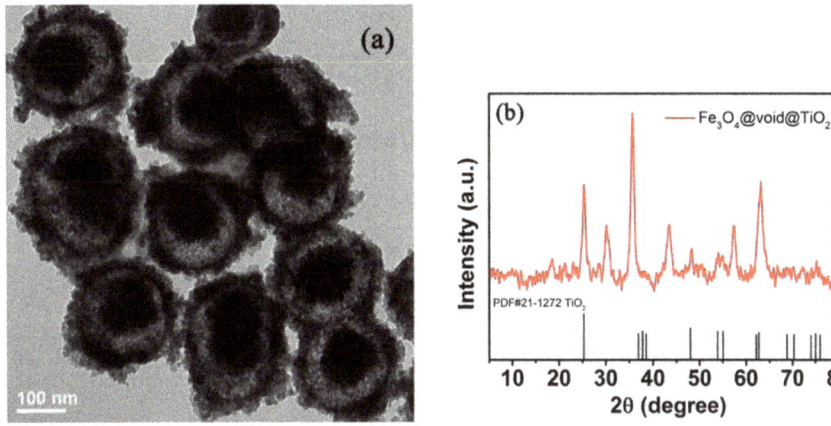

Fig. 28.14 **a** TEM image of Fe$_3$O$_4$@void@TiO$_2$ NPs; **b** XRD pattern of Fe$_3$O$_4$@void@TiO$_2$ NPs. Reprinted with permission from Ref. [36]. Copyright 2016 American Chemical Society

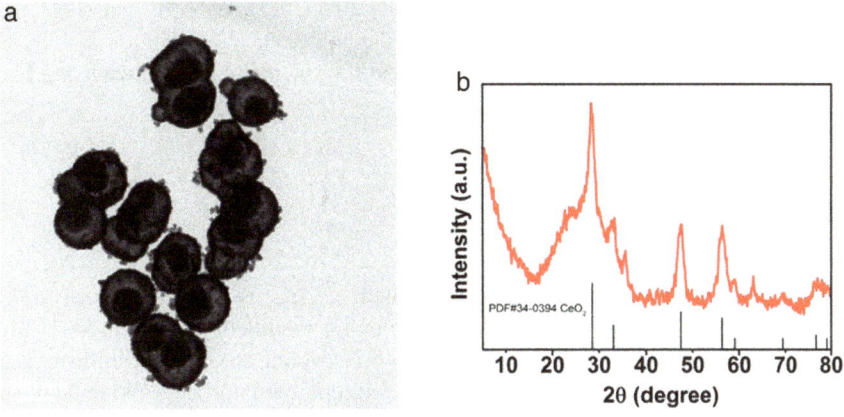

Fig. 28.15 **a** TEM image of Fe$_3$O$_4$@void@CeO$_2$ NPs; **b** XRD patterns of Fe$_3$O$_4$@void@CeO$_2$ NPs. Reprinted with permission from Ref. [36]. Copyright 2016 American Chemical Society

Fig. 28.16 Photo-Fenton
performances of
$Fe_3O_4@SiO_2@TiO_2$,
$Fe_3O_4@TiO_2$ and
$Fe_3O_4@void@TiO_2$ at room
temperature under natural
pH value (Cat., 0.2 g·L^{-1},
MB, 10 mg·L^{-1}, H_2O_2,
300 mM, Vol., 50 mL, no
filter). Reprinted with
permission from Ref. [36].
Copyright 2016 American
Chemical Society

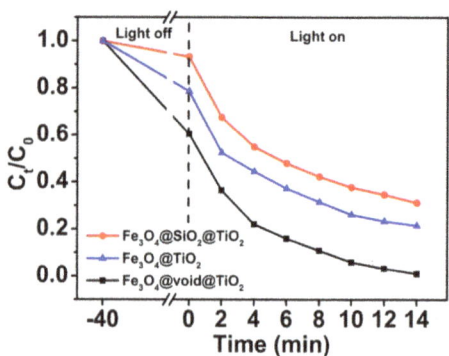

Fig. 28.17 Photo-Fenton
performances of
$Fe_3O_4@SiO_2@CeO_2$,
$Fe_3O_4@CeO_2$ and
$Fe_3O_4@void@CeO_2$ at
room temperature under
natural pH value (Cat.,
0.2 g·L^{-1}, MB, 10 mg·L^{-1},
H_2O_2, 300 mM, Vol., 50 mL,
AM1.5 filter). Reprinted
with permission from Ref.
[36]. Copyright 2016
American Chemical Society

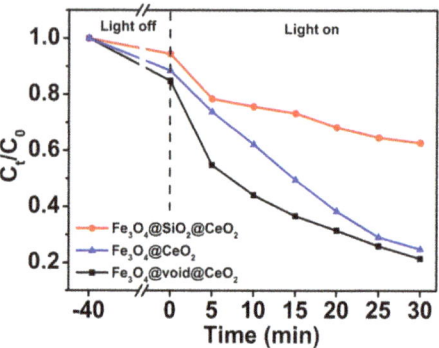

Both $Fe_3O_4@void@TiO_2$ and $Fe_3O_4@void@CeO_2$ show better performance in dark reaction and degradation stages than the counterparts without a cavity structure (Figs. 28.16, 28.17), which is consistent with that of $Fe_3O_4@void@CdS$. These results further confirm the cooperative effect between core and shell and the priority of the cavity as a nanoreactor. The enhanced photo-Fenton activity is well repeated demonstrating the wide applicability of photo-Fenton catalysts like $Fe_3O_4@void@semiconductor$.

28.4 Conclusions

In summary, we have designed and illustrated the successful synthesis of yolk-shell structured $Fe_3O_4@void@CdS$ NPs through a modified CBD method. Using $Fe_3O_4@SiO_2$ as cores, $Fe_3O_4@void@CdS$ NPs which are highly uniform and monodisperse can be obtained by a one-pot coating-etching process. When applied to photo-Fenton reaction, this yolk-shell structured catalysis shows excellent performance due to the inherent synergic effect between CdS and Fe_3O_4. Meanwhile, the yolk-shell structure also plays a key role: one is it functions as a nanoreactor to enrich the pollutant and provide a suitable reaction place; the other one is it max-utilizes light irradiation owning to light reflection in the hollow place. A general structure design is further proposed and verified by replacing the shell with TiO_2 or CeO_2. The enhanced photo-Fenton activity is well repeated confirming the wide applicability of photo-Fenton catalysis like Fe_3O_4@void@semiconductor. Benefitting from the magnetic property of Fe_3O_4, this kind of catalyst could also be fast separated by an external magnetic field for a secondary use. It is assumed the yolk-shell structured Fe_3O_4@void@semiconductor catalyst which combines photocatalytic degradation and Fenton reaction processes together by utilizing the optimally spatial arrangement has a great potential for practical water treatment in the future.

References

1. Navalon S, de Miguel M, Martin R, Alvaro M, Garcia H (2011) Enhancement of the catalytic activity of supported gold nanoparticles for the Fenton reaction by light. J Am Chem Soc 133(7):2218–2226
2. Navalon S, Martin R, Alvaro M, Garcia H (2010) Gold on diamond nanoparticles as a highly efficient Fenton catalyst. Angew Chem 122(45):8581–8585
3. Pouran SR, Aziz AA, Daud WMAW (2015) Review on the main advances in photo-Fenton oxidation system for recalcitrant wastewaters. J Ind Eng Chem 21:53–69
4. Munoz M, de Pedro ZM, Casas JA, Rodriguez JJ (2015) Preparation of magnetite-based catalysts and their application in heterogeneous Fenton oxidation–A review. Appl Catal B 176:249–265
5. Hammouda SB, Adhoum N, Monser L (2015) Synthesis of magnetic alginate beads based on Fe_3O_4 nanoparticles for the removal of 3-methylindole from aqueous solution using Fenton process. J Hazard Mater 294:128–136
6. Xu L, Wang J (2012) Fenton-like degradation of 2, 4-dichlorophenol using Fe_3O_4 magnetic nanoparticles. Appl Catal B 123:117–126
7. Pastrana-Martínez LM, Pereira N, Lima R, Faria JL, Gomes HT, Silva AM (2015) Degradation of diphenhydramine by photo-Fenton using magnetically recoverable iron oxide nanoparticles as catalyst. Chem Eng J 261:45–52
8. Minella M, Marchetti G, De Laurentiis E, Malandrino M, Maurino V, Minero C, Vione D, Hanna K (2014) Photo-Fenton oxidation of phenol with magnetite as iron source. Appl Catal B 154:102–109
9. Cleveland V, Bingham J-P, Kan E (2014) Heterogeneous Fenton degradation of bisphenol A by carbon nanotube-supported Fe 3 O 4. Sep Purif Technol 133:388–395
10. Yang S-T, Zhang W, Xie J, Liao R, Zhang X, Yu B, Wu R, Liu X, Li H, Guo Z (2015) Fe_3O_4@SiO_2 nanoparticles as a high-performance Fenton-like catalyst in a neutral environment. RSC Adv 5(7):5458–5463
11. Zhou L, Shao Y, Liu J, Ye Z, Zhang H, Ma J, Jia Y, Gao W, Li Y (2014) Preparation and characterization of magnetic porous carbon microspheres for removal of methylene blue by a heterogeneous Fenton reaction. ACS Appl Mater Interfaces 6(10):7275–7285
12. Shi W, Lu D, Wang L, Teng F, Zhang J (2015) Core–shell structured Fe_3O_4@SiO_2@ CdS nanoparticles with enhanced visible-light photocatalytic activities. RSC Adv 5(128):106038–106043
13. Zeng T, Zhang X, Wang S, Ma Y, Niu H, Cai Y (2013) A double-shelled yolk-like structure as an ideal magnetic support of tiny gold nanoparticles for nitrophenol reduction. J Mater Chem A 1(38):11641–11647

14. Wang Y, Wang G, Xiao Y, Yang Y, Tang R (2014) Yolk-Shell nanostructured Fe_3O_4@ $NiSiO_3$ for selective affinity and magnetic separation of his-tagged proteins. ACS Appl Mater Interfaces 6(21):19092–19099
15. Zhang L, Wang T, Li L, Wang C, Su Z, Li J (2012) Multifunctional fluorescent-magnetic polyethyleneimine functionalized Fe_3O_4–mesoporous silica yolk–shell nanocapsules for siRNA delivery. Chem Commun 48(69):8706–8708
16. Li L, Wang T, Zhang L, Su Z, Wang C, Wang R (2012) Selected-control synthesis of monodisperse Fe_3O_4@C core-shell spheres, chains, and rings as high-performance anode materials for lithium-ion batteries. Chemistry 18(36):11417–11422. https://doi.org/10.1002/chem.201 200791
17. Zhang J, Wang K, Xu Q, Zhou Y, Guo S (2015) Beyond yolk-shell nanoparticles: Fe_3O_4@Fe_3C Core@Shell nanoparticles as yolks; carbon nanospindles as shells for efficient lithium ion storage. ACS Nano 9(3):3369–3376
18. Liu J, Qiao SZ, Budi Hartono S, Lu GQM (2010) Monodisperse yolk–shell nanoparticles with a hierarchical porous structure for delivery vehicles and nanoreactors. Angew Chem 122(29):5101–5105
19. Liu J, Yang HQ, Kleitz F, Chen ZG, Yang T, Strounina E, Lu GQM, Qiao SZ (2012) Yolk–shell hybrid materials with a periodic mesoporous organosilica shell: ideal nanoreactors for selective alcohol oxidation. Adv Funct Mater 22(3):591–599
20. Liu C, Li J, Qi J, Wang J, Luo R, Shen J, Sun X, Han W, Wang L (2014) Yolk-Shell FeO@ SiO_2 nanoparticles as nanoreactors for fenton-like catalytic reaction. ACS Appl Mater Interfaces 6(15):13167–13173
21. Xu L, Wang J (2012) Magnetic nanoscaled Fe_3O_4/CeO_2 composite as an efficient Fenton-like heterogeneous catalyst for degradation of 4-chlorophenol. Environ Sci Technol 46(18):10145–10153
22. Qiu B, Li Q, Shen B, Xing M, Zhang J (2016) Stöber-like method to synthesize ultradispersed Fe_3O_4 nanoparticles on graphene with excellent Photo-Fenton reaction and high-performance lithium storage. Appl Catal B 183:216–223
23. Yang X, Chen W, Huang J, Zhou Y, Zhu Y, Li C (2015) Rapid degradation of methylene blue in a novel heterogeneous Fe_3O_4@rGO@TiO_2-catalyzed photo-Fenton system. Sci Rep 5
24. Abbas M, Rao BP, Reddy V, Kim C (2014) Fe_3O_4/TiO_2 core/shell nanocubes: single-batch surfactantless synthesis, characterization and efficient catalysts for methylene blue degradation. Ceram Int 40(7):11177–11186
25. Zeng T, Zhang X, Wang S, Ma Y, Niu H, Cai Y (2014) Assembly of a nanoreactor system with confined magnetite core and shell for enhanced fenton-like catalysis. Chem Eur J 20(21):6474–6481
26. Liang X, Li J, Joo JB, Gutiérrez A, Tillekaratne A, Lee I, Yin Y, Zaera F (2012) Diffusion through the shells of yolk–shell and core–shell nanostructures in the liquid phase. Angew Chem 124(32):8158–8160
27. Liu J, Cheng J, Che R, Xu J, Liu M, Liu Z (2013) Synthesis and microwave absorption properties of Yolk-Shell microspheres with magnetic iron oxide cores and hierarchical copper silicate shells. ACS Appl Mater Interfaces 5(7):2503–2509
28. Do QC, Kim DG, Ko SO (2018) Catalytic activity enhancement of a Fe_3O_4@SiO_2 yolk-shell structure for oxidative degradation of acetaminophen by decoration with copper. J Cleaner Prod 172 (pt.2):1243–1253
29. Zhuang Y, Yuan S, Liu J, Zhang Y, Pei Y (2019) Synergistic effect and mechanism of mass transfer and catalytic oxidation of octane degradation in yolk-shell Fe_3O_4@C/fenton system. Chem Eng J 379:122262
30. Niu H, Zheng Y, Wang S, Zhao L, Cai Y (2017) Continuous generation of hydroxyl radicals for highly efficient elimination of chlorophenols and phenols catalyzed by heterogeneous Fenton-like catalysts yolk/shell Pd@Fe_3O_4@metal organic frameworks. J Hazard Mater 346:174
31. Lee S-U, Jung H, Wi DH, Hong JW, Sung J, Choi S-I, Han SW (2018) Metal–semiconductor yolk–shell heteronanostructures for plasmon-enhanced photocatalytic hydrogen evolution. J Mater Chem A 6(9):4068–4078. https://doi.org/10.1039/c7ta09953c

32. Zhao J, Li W, Liu H, Shi H, Xiao C (2019) Yolk-shell CdS@void@TiO$_2$ composite particles with photocorrosion resistance for enhanced dye removal and hydrogen evolution. Adv Powder Technol 30(9):1965–1975. https://doi.org/10.1016/j.apt.2019.06.015

33. Zhang P, Guan BY, Yu L, Lou XW (2018) Facile synthesis of multi-shelled ZnS-CdS cages with enhanced photoelectrochemical performance for solar energy conversion. Chem 4(1):162–173. https://doi.org/10.1016/j.chempr.2017.10.018

34. Wang H, Zhu C, Xu L, Ren Z, Zhong C (2020) Layer-by-layer assembled synthesis of hollow yolk-shell CdS–graphene nanocomposites and their high photocatalytic activity and photostability. J Nanoparticle Res 22(4). https://doi.org/10.1007/s11051-020-04826-6

35. Bibi R, Huang H, Kalulu M, Shen Q, Wei L, Oderinde O, Li N, Zhou J (2018) Synthesis of amino-functionalized Ti-MOF derived yolk-shell and hollow heterostructures for enhanced photocatalytic hydrogen production under visible light. ACS Sustain Chem Eng

36. Shi W, Du D, Shen B, Cui C, Lu L, Wang L, Zhang J (2016) Synthesis of yolk-shell structured Fe$_3$O$_4$@void@CdS nanoparticles: a general and effective structure design for photo-fenton reaction. ACS Appl Mater Interfaces 8(32):20831–20838. https://doi.org/10.1021/acsami.6b0 7644

37. Liu J, Sun Z, Deng Y, Zou Y, Li C, Guo X, Xiong L, Gao Y, Li F, Zhao D (2009) Highly water-dispersible biocompatible magnetite particles with low cytotoxicity stabilized by citrate groups. Angew Chem 121(32):5989–5993

38. Stöber W, Fink A, Bohn E (1968) Controlled growth of monodisperse silica spheres in the micron size range. J Colloid Interface Sci 26(1):62–69

39. Li W, Yang J, Wu Z, Wang J, Li B, Feng S, Deng Y, Zhang F, Zhao D (2012) A versatile kinetics-controlled coating method to construct uniform porous TiO$_2$ shells for multifunctional core-shell structures. J Am Chem Soc 134(29):11864–11867. https://doi.org/10.1021/ja3037146

40. Cheng G, Zhang J-L, Liu Y-L, Sun D-H, Ni J-Z (2011) Synthesis of novel Fe$_3$O$_4$@ SiO$_2$@ CeO$_2$ microspheres with mesoporous shell for phosphopeptide capturing and labeling. Chem Commun 47(20):5732–5734

41. Prabhu RR, Khadar MA (2008) Study of optical phonon modes of CdS nanoparticles using Raman spectroscopy. Bull Mater Sci 31(3):511–515

42. Shebanova ON, Lazor P (2003) Raman spectroscopic study of magnetite (FeFe$_2$O$_4$): a new assignment for the vibrational spectrum. J Solid State Chem 174(2):424–430

43. Zhou J, Wu X, Teeter G, To B, Yan Y, Dhere R, Gessert T (2004) CBD-Cd1-xZnxS thin films and their application in CdTe solar cells. Physica Status Solidi 241(3):775–778

44. Figueroa RA, Leonard A, MacKay AA (2004) Modeling tetracycline antibiotic sorption to clays. Environ Sci Technol 38(2):476–483

45. Liu Y, Ohko Y, Zhang R, Yang Y, Zhang Z (2010) Degradation of malachite green on Pd/WO 3 photocatalysts under simulated solar light. J Hazard Mater 184(1):386–391

46. Xu W, Zhu S, Liang Y, Li Z, Cui Z, Yang X, Inoue A (2015) Nanoporous CuS with excellent photocatalytic property. Sci Rep 5

47. Liu W, Wang Y, Ai Z, Zhang L (2015) Hydrothermal synthesis of FeS$_2$ as a high-efficiency fenton reagent to degrade alachlor via superoxide-mediated Fe(II)/Fe(III) cycle. ACS Appl Mater Interfaces 7(51):28534–28544

48. Voinov MA, Pagán JOS, Morrison E, Smirnova TI, Smirnov AI (2010) Surface-mediated production of hydroxyl radicals as a mechanism of iron oxide nanoparticle biotoxicity. J Am Chem Soc 133(1):35–41

49. Buxton GV, Greenstock CL, Helman WP, Ross AB (1988) Critical review of rate constants for reactions of hydrated electrons, hydrogen atoms and hydroxyl radicals (·OH/·O-in aqueous solution. J Phys Chem Ref Data 17(2):513–886

50. Stefan MI, Mack J, Bolton JR (2000) Degradation pathways during the treatment of methyl tert-butyl ether by the UV/H$_2$O$_2$ process. Environ Sci Technol 34(4):650–658

51. Chen C, Ma W, Zhao J (2010) Semiconductor-mediated photodegradation of pollutants under visible-light irradiation. Chem Soc Rev 39(11):4206–4219

52. Zhao Y, Pan F, Li H, Niu T, Xu G, Chen W (2013) Facile synthesis of uniform α-Fe$_2$O$_3$ crystals and their facet-dependent catalytic performance in the photo-Fenton reaction. J Mater Chem A 1(24):7242–7246

53. Dong W, Zhu Y, Huang H, Jiang L, Zhu H, Li C, Chen B, Shi Z, Wang G (2013) A performance study of enhanced visible-light-driven photocatalysis and magnetical protein separation of multifunctional yolk–shell nanostructures. J Mater Chem A 1(34):10030–10036
54. Wang J, Li X, Li X, Zhu J, Li H (2013) Mesoporous yolk–shell SnS 2–TiO$_2$ visible photocatalysts with enhanced activity and durability in Cr (vi) reduction. Nanoscale 5(5):1876–1881
55. Dutta K, Mukhopadhyay S, Bhattacharjee S, Chaudhuri B (2001) Chemical oxidation of methylene blue using a Fenton-like reaction. J Hazard Mater 84(1):57–71
56. Houas A, Lachheb H, Ksibi M, Elaloui E, Guillard C, Herrmann J-M (2001) Photocatalytic degradation pathway of methylene blue in water. Appl Catal B 31(2):145–157

Chapter 29
Organic Transformations Enabled by Yolk–Shell and Core–Shell Structured Catalysts

He Huang, Weixiao Ji, and Wei Wang⊙

29.1 Introduction

Heterogeneous catalysis plays an indispensable role in organic synthesis and a wide range of chemicals is produced by this catalytic power. It has been widely used in every aspect of industrial process. Among all heterogeneous catalysts, nanomaterial-based catalysts have emerged as a new horizon in recent years. The nanoeffects enable catalysts to exhibit significantly different activity and efficiency compared to traditional heterogeneous catalysts. However, one main issue of nanocatalysts is that they tend to agglomerate during the reaction process, which can deteriorate the catalytic performance. In the past few years, the design of core–shell and yolk–shell nanostructures has offered a promising solution to minimize the problem. Furthermore, distinct from homogeneous and other nanocatalytic machinery, these nanocatalysts can achieve the selectivity of substrate size by regulating the hole size of the shell structure. In this chapter, we will focus on the developments of organic transformations enabled by yolk–shell or core–shell structured catalysts categorized by the reaction type.

H. Huang
Department of Chemistry and Chemical Biology, Cornell University, Ithaca, NY 14853, USA

W. Ji
Department of Mechanical Engineering, University of Wisconsin Milwaukee, Milwaukee, WI 53211, USA

W. Wang (✉)
Department of Pharmacology and Toxicology, University of Arizona, Tucson, AZ 85721, USA

© The Author(s), under exclusive license to Springer Nature Singapore Pte Ltd. 2021   479
H. Yamashita and H. Li (eds.), *Core-Shell and Yolk-Shell Nanocatalysts*,
Nanostructure Science and Technology,
https://doi.org/10.1007/978-981-16-0463-8_29

29.2 Reduction Reactions

Selective hydrogenation of the carbonyl group in unsaturated aldehydes can provide
an efficient route for the production of unsaturated alcohols, which are broadly used in
perfume, flavoring, and pharmaceutical industries. However, developing an effective
catalyst for this transformation without affecting the carbon–carbon double bond is
challenging because the hydrogenation is a thermodynamically favored process for
C=C bond. In 2016, Li, Zhao, and Tang reported a novel core–shell nanocatalyst
with a sandwich nanostructure [1]. The MIL-101 contains metal nodes of Fe^{3+}, Cr^{3+},
or both and connected with 1,4-benzenedicarboxylate (BDC) linkers (Fig. 29.1).
By sandwiching platinum between a metal–organic framework (MOF) MIL-101
inner core and outer shell, the catalyst displays high stability and robustness in the
hydrogenation catalysis. Notably, the catalytic system can selectively convert a broad
range of α,β-unsaturated aldehydes to unsaturated alcohols with high efficiency and
significantly improve chemoselectivity over C=C double bond.

In the study, five typical sandwich nanostructures were designed and synthesized. Among them, two MIL-101(Fe)@Pt@MIL-101(Fe) with shell thicknesses of
about 9.2 nm and 22.0 nm were synthesized, denoted as MIL-101(Fe)@Pt@MIL-
$101(Fe)^{9.2}$ and MIL-101(Fe)@Pt@MIL-101(Fe)$^{22.0}$. Other metal-containing nanostructures MIL-101(Cr)@Pt@MIL-101(Cr) and MIL-101(Cr)@Pt@MIL-101(Cr)
were also prepared for performance comparison study. Cinnamaldehyde was chosen
as the standard substrate to test their efficiency and selectivity. The regular Pt
nanoparticles delivered double bond hydrogenation product as the major product
(Table 29.1, entry 1). By contrast, those catalysts with sandwich nanostructure
were capable of selective reduction of C=O over C=C bond (entries 2–5). In addition to cinnamaldehyde, they functioned smoothly for other unsaturated aldehydes
hydrogenation reaction with high chemoselectivity as well (entries 6–8).

In 2016, Jiang reported a Pd nanocubes@ZIF-8 composite material for the efficient
hydrogenation of olefins (Fig. 29.2) [2]. The reaction was performed under mild

Synthetic route

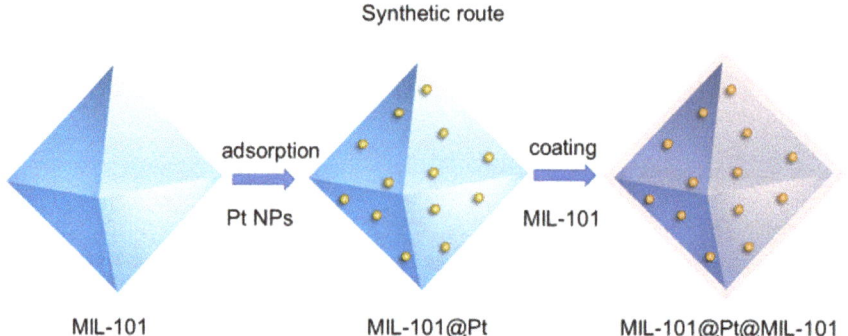

Fig. 29.1 Sandwich MIL-101@Pt@MIL-101 nanostructures. Reprinted with permission from Ref.
[1]. Copyright 2016 Springer Nature

Table 29.1 Selective hydrogenation of different α,β-unsaturated aldehydes by different catalysts

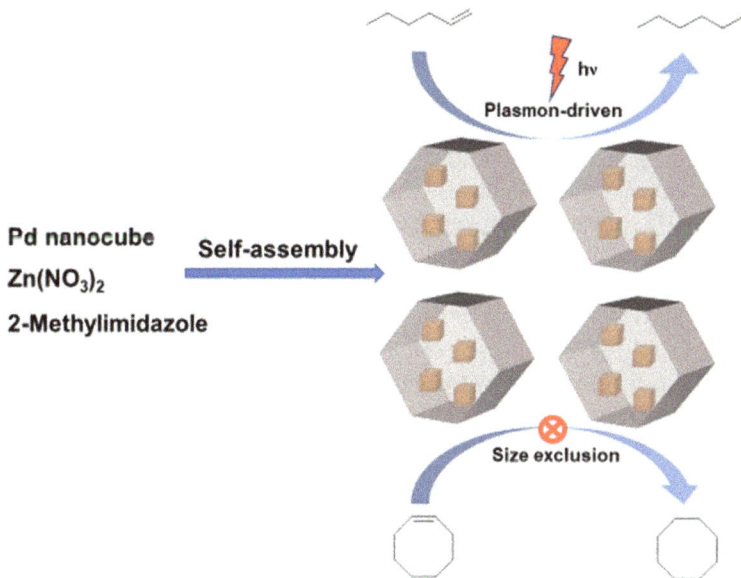

Entry	Catalyst	A%	B%	C%
Substrate: cinnamaldehyde				
1	Pt NPs	18.3	60.6	21.1
2	MIL-101(Fe)@Pt@MIL-101(Fe)$^{22.0}$	96.3	1.9	1.8
3	MIL-101(Cr)@Pt@MIL-101(Fe)$^{2.9}$	94.2	4.7	1.1
4	MIL-101(Cr)@Pt@MIL-101(Cr)$^{5.1}$	79.2	14.6	6.2
5	MIL-101(Fe)@Pt@MIL-101(Fe)$^{9.2}$	94.1	2.9	3.0
Substrate: furfural				
6	MIL-101(Cr)@Pt@MIL-101(Cr)$^{5.1}$	99.8	0	0.2
Substrate: 3-methyl-2-butenal				
7	MIL-101(Fe)@Pt@MIL-101(Fe)$^{22.0}$	92.5	2.8	4.7
Substrate: acrolein				
8	MIL-101(Fe)@Pt@MIL-101(Fe)$^{22.0}$	97.3	2.7	0

Reprinted with permission from Ref. [1]. Copyright 2016 Springer Nature

Fig. 29.2 Synthesis of Pd nanocubes@ZIF-8 and catalysis of the hydrogenation of olefins. Reprinted with permission from Ref. [2]. Copyright 2016 John Wiley and Sons

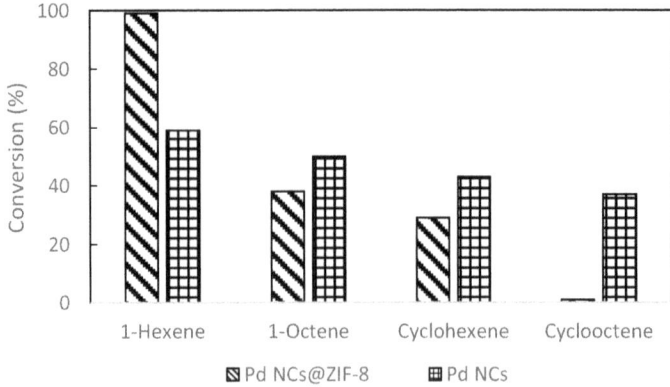

Fig. 29.3 Pd nanocubes@ZIF-8-catalyzed hydrogenation reaction. Reprinted with permission from Ref. [2]. Copyright 2016 John Wiley and Sons

reaction conditions with 1 atm H_2 and light irradiation at room temperature. The reaction was benefited from plasmonic photothermal effects of the Pd nanocube core. The catalytic structures exhibited the unique performance by accelerating the reaction by H_2 enrichment and the high stability of the Pd core. Most importantly, it displayed the selectivity on olefins with specific sizes (Fig. 29.3).

The stability of the Pd nanocubes@ZIF-8 catalyst was much higher than the Pd nanocubes without a shell. After three cycles of reaction run, Pd nanocubes@ZIF-8 maintained the same catalytic ability while a significant drop in performance was observed for the Pd nanocubes without a shell. Better substrate size control was also achieved by the Pd nanocubes@ZIF-8.

Hydrogenation of nitroaromatics is a widely used process for the synthesis of anilines. In 2018, Ma and Cheng reported a new Cu_2O@ZIF-8 yolk–shell structure. By employing a template protection-sacrifice (TPS) technique, Cu_2O nanocubes can be successfully encapsulated into a MOF ZIF-8 and form a composite material Cu_2O@ZIF-8 [3].

The Cu_2O@ZIF-8 composite exhibits an excellent catalytic efficiency and a good recycling stability in the hydrogenation of 4-nitrophenol by using $NaBH_4$ as a reducing reagent (Fig. 29.4). The reduction process was monitored by UV/Vis spectroscopy. Without the addition of Cu_2O@ZIF-8, reaction did not take place. Upon the addition of Cu_2O@ZIF-8, the reduction happened immediately and finished within 14 min with a conversion rate of nearly 99%. After five testing cycles, this catalyst

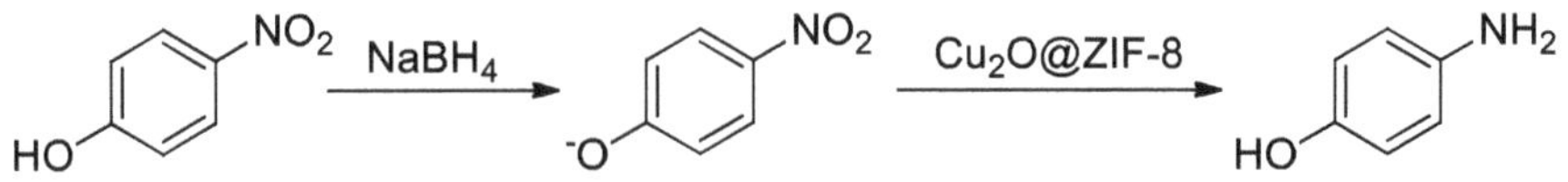

Fig. 29.4 Catalytic reduction reaction of 4-nitrophenol to 4-aminophenol with Cu_2O@ZIF-8. Reprinted with permission from Ref. [3]. Copyright 2018 John Wiley and Sons

was still highly active with 98% conversion in 20 min, indicating that the MOF shell provides solid protection for the catalyst core structure throughout the whole reaction course.

Pd@SiO$_2$ core–shell nanocatalyst [4] and silver nanoparticle (Ag NPs) decorated on tannic acid (TA)-covered magnetite (Fe$_3$O$_4$) (TA@Fe$_3$O$_4$-Ag NPs) nanohybrid [5] were also exploited as catalysts for the hydrogenation of nitrobenzenes.

Besides MOF, covalent organic framework (COF) can also function as a shell in the core–shell structure. Benefited from the large pore size and ordered π-columnar structures of COF, the internal core is more accessible compared to the common MOF shell. In 2018, Kim reported a Pd-doped MOFs@COFs hybrid as a photocatalytic platform. A Pd-doped NH$_2$-MIL-125(Ti) (TiATA) was encapsulated into the LZU1 COF shell to form the Pd-doped MOFs@COF hybrid (Fig. 29.5) [6]. As demonstrated, the nanostructure showed a great photocatalytic performance for tandem dehydrogenation of ammonia borane and hydrogenation of olefins reaction in both continuous-flow microreactor and batch reactor.

Pd/TiATA@LZU1 was tested in the photocatalytic hydrogenation of styrene. As shown in Table 29.2, almost 100% conversion of styrene to ethylbenzene was

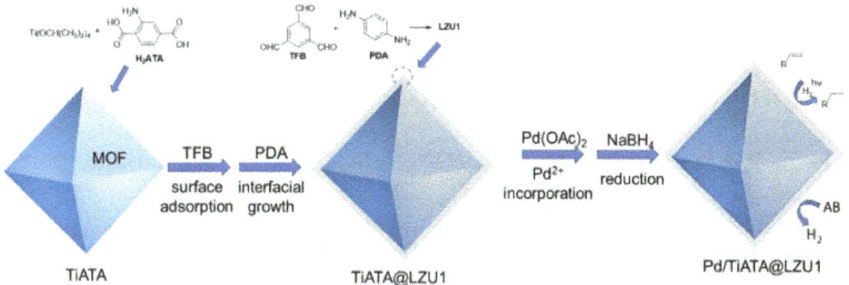

Synthesis of Pd doped MOF@COF core-shell hybrid

Fig. 29.5 Preparation of the Pd-doped TiATA@LZU1 core–shell and the photocatalytic applications. Reprinted with permission from Ref. [6]. Copyright 2018 John Wiley and Sons

Table 29.2 Photocatalytic hydrogenation of styrene

Entry	Photocatalyst	Conversion (%)	Selectivity (%)
1	Pd/TiATA@LZU1	>99	>99
2	Pd/LZU1	42	>99
3	Pd/TiATA	83	>99

Reprinted with permission from Ref. [6]. Copyright 2018 John Wiley and Sons

achieved with >99% selectivity (Table 29.2, entry 1). In contrast, without the core–shell structure, Pd/LZU1 and Pd/ TiATA can only deliver 42% and 83% conversion, respectively (entries 2–3).

Besides MOF and COF as the shells, other inorganic materials were also employed as shells. For example, Kaneda reported a core–shell structure nanocomposite catalyst for the reduction of unsaturated aldehydes with high chemoselectivity [7].

29.3 Oxidation Reactions

Oxidation reaction represents another important type of transformation in organic synthesis. Shen and Li developed a yolk–shell Co@C-N nanoreactor as a highly efficient and recyclable catalyst for the oxidation of alcohols to carbonyls (Fig. 29.6) [8]. The traditional Co nanoparticles suffer from serious aggregation during the catalytic process which led to reduced surface area and thus catalyst performance. In contrast, the yolk–shell structure could overcome the problem. The hollow yolk–shell structure offered a higher surface area even better than solid core–shell Co@C-N.

This unique yolk–shell Co-C-N nanoreactor displayed excellent catalytic activity in aerobic oxidation of alcohols. The reaction proceeded smoothly in water without base. Air was used as the oxidant. This protocol can be compatible with various functional groups to produce diverse functionalized ketones and aldehydes (Fig. 29.7).

In 2019, Li reported a hydrophobic MOFs@COFs core–shell structure for the oxidation of styrene with high selectivity [9]. The heterogeneous composites were fabricated by employing MOFs NH_2-MIL-101(Fe) as the core and COFs NUT-COF-1(NTU) as the shell via a covalent linking process (Fig. 29.8). The NH_2-MIL-101(Fe)@NTU composite showed a large enhanced catalytic conversion and good selectivity toward aldehyde products. This MOFs@COFs structure could create a hydrophobic environment, which is favor for enriching the hydrophobic substrate styrene.

The styrene oxidizing reaction was performed in the mixture of acetonitrile and TBHP with styrene and catalyst at 80 °C (Fig. 29.9). Without a shell, MOFs NH_2-MIL-101(Fe) delivered a very limited selectivity by forming a mixture of

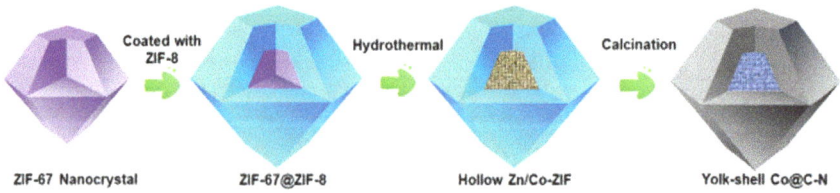

Fig. 29.6 Preparation of the yolk–shell Co@C-N. Reprinted with permission from Ref. [8]. Copyright 2018 American Chemical Society

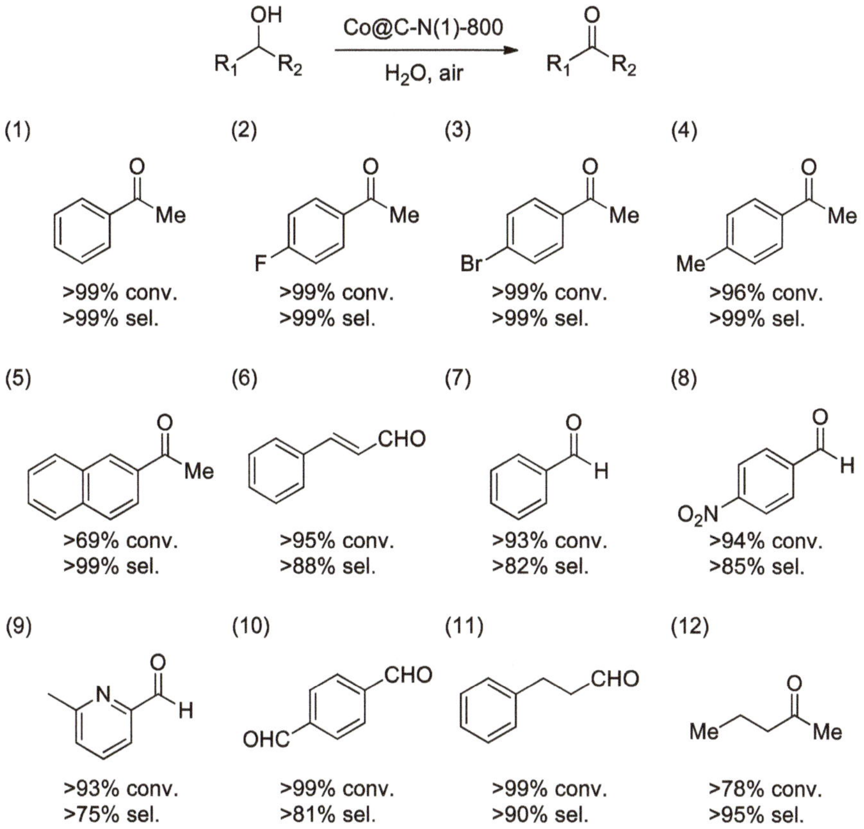

OH
R1 R2
Co@C-N(1)-800
H2O, air
O
R1 R2
(1)
O
Me
>99% conv.
>99% sel.
(2)
O
Me
F
>99% conv.
>99% sel.
(3)
O
Me
Br
>99% conv.
>99% sel.
(4)
O
Me
Me
>96% conv.
>99% sel.
(5)
O
Me
>69% conv.
>99% sel.
(6)
CHO
>95% conv.
>88% sel.
(7)
O
H
>93% conv.
>82% sel.
(8)
O
H
O2N
>94% conv.
>85% sel.
(9)
O
N
H
>93% conv.
>75% sel.
(10)
CHO
OHC
>99% conv.
>81% sel.
(11)
CHO
>99% conv.
>90% sel.
(12)
O
Me
Me
>78% conv.
>95% sel.

Fig. 29.7 Oxidation reaction by Co@C-N-800. Reprinted with permission from Ref. [8]. Copyright 2018 American Chemical Society

Hydrophilic
One step
Hydrophobic
NH2-MIL-101(Fe)
NH2-MIL-101(Fe)@NTU-COF

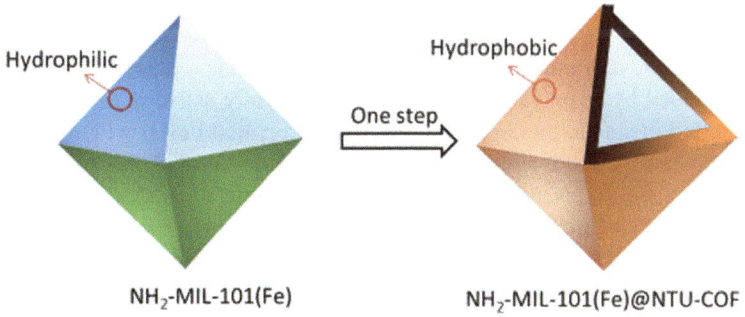

Fig. 29.8 Fabrication of NH_2-MIL-101(Fe)@NTU-COF. Reprinted with permission from Ref. [9]. Copyright 2019 John Wiley and Sons

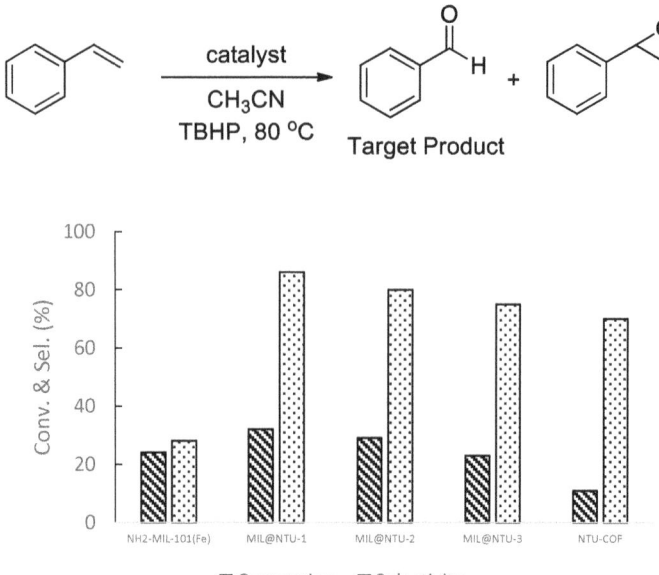

Fig. 29.9 Styrene oxidation reaction with different catalysts. Reprinted with permission from Ref. [9]. Copyright 2019 John Wiley and Sons

benzaldehyde and styrene oxide. However, the core–shell catalyst machinery offered a significant enhanced selectivity.

29.4 Cross-Coupling Reactions

Suzuki cross-coupling reaction is one of the most widely used reactions in synthesis in both academic and industrial sectors, especially in many pharmaceutical industries, demanded by large various drugs and biologically active materials. In 2018, Bian reported magnetic hierarchical core–shell structured $Fe_3O_4@PDA-Pd@MOF$ nanocomposites for catalyzing Suzuki coupling reaction [10]. This catalyst showed great activity and functional group tolerance. Magnetic separation of the catalyst from the reaction solution offered another advantage of this structure.

Two years later Liu reported a series of nanoreactors, including core–shell (CS), yolk–shell (YS), hollow (HS), and multilayer yolk–shell structures (AYS and DYS) [11]. By adjusting reaction–diffusion process, these structures could be produced in various controlled size (Fig. 29.10).

The COF shells in those core–shell and yolk–shell structures offered a significant size selectivity. For example, A3 + B3 COF shells have small pore size ranging from 1.1 to 1.5 nm, which only allowed reactants <1.5 nm to diffuse into the reaction vessel. For example, the coupling reaction between phenylboronic acid and iodobenzene

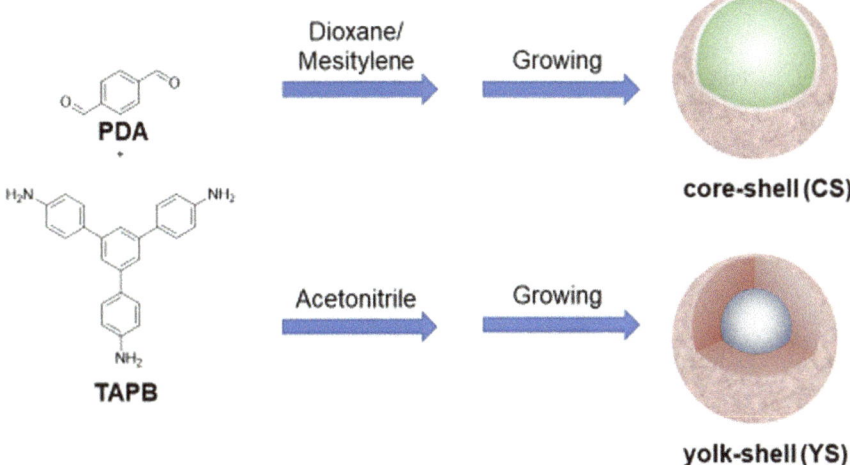

Fig. 29.10 COF morphology design. Reprinted with permission from Ref. [11]. Copyright 2020 Elsevier

occurred smoothly in 99% yield. Comparing with iodobenzene with a size of 0.67 nm, 1,3,5-tris(4-iodophenyl)benzene has a larger size of 1.7 nm. Notably, under the same coupling reaction conditions (in the presence of the same catalyst), no conversion was observed. In a control experiment, Pd@core with no COF shells was also applied in those two reactions. Both of those coupling reactions happened with 99% and 58% yields, respectively. This result confirmed the size selectivity of the COF shells structures offered (Fig. 29.11).

Besides the size selectivity, another advantage of those catalysts endowed was the high stability for repetitive use. After five cycling runs, Pd@core–shell and Pd@yolk–shell catalyst could still be highly effective while Pd@core exhibited a decrease in catalytic activity (Fig. 29.12).

Cross-dehydrogenative coupling (CDC) is a powerful approach for C–C bond formation. Han and Chen developed a nanocatalyst for promoting CDC of tetrahydroisoquinoline with indole (Fig. 29.13). The catalyst is prepared by in situ anchoring of Cu_2O nanoparticles on an N-doped porous C yolk−shell cuboctahedral (CNPC) framework. This CNPC nanoparticles showed excellent photocatalytic efficiency and great stability for the photocatalytic CDC reaction to produce functionalized indole. Without the yolk–shell structure, Cu_2O could only give very low yields. What is more, the CNPC nanoparticles could be reused for five times without significant loss of photocatalytic efficiency [12].

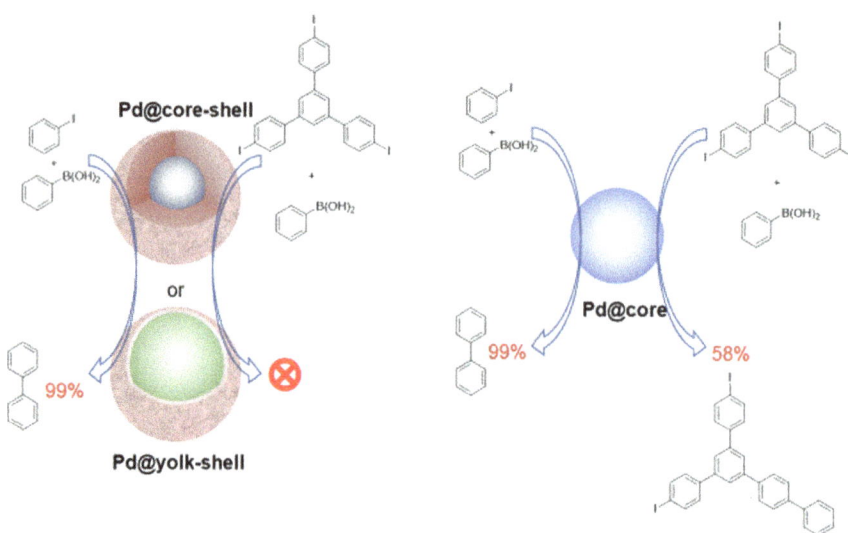

Fig. 29.11 Size-selective controlled cross-coupling reactions. Reprinted with permission from Ref. [11]. Copyright 2020 Elsevier

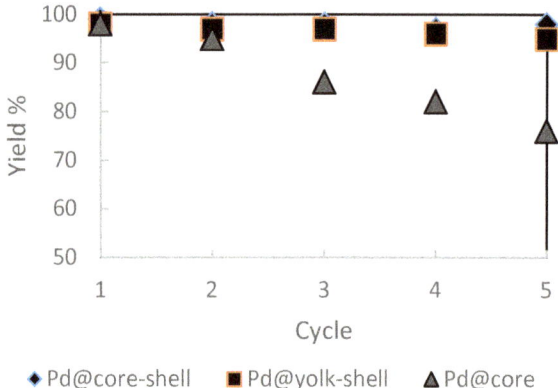

Fig. 29.12 Cyclic catalysis performances. Reprinted with permission from Ref. [11]. Copyright 2020 Elsevier

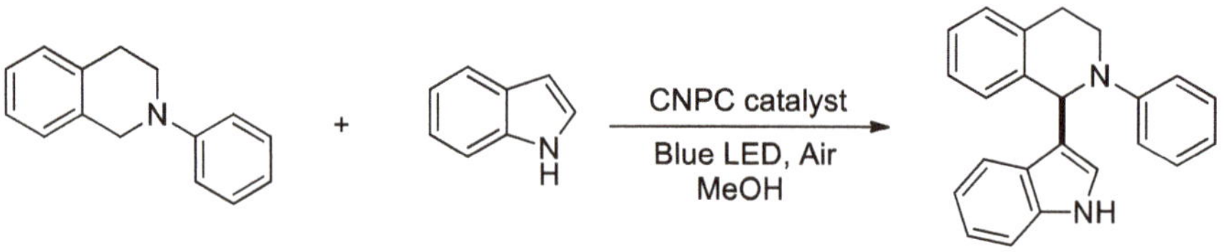

Fig. 29.13 Cross-dehydrogenative coupling catalyzed by yolk–shell catalysts. Reprinted with permission from Ref. [12]. Copyright 2018 American Chemical Society

29.5 Cascade Reactions

A cascade reaction or domino reaction is a powerful manifold for facile construction of complex molecular architectures, powered by the capacity of at least two consecutive reactions occurring in one operation. Therefore, the features of high atom and step economy and reduction of waste generated have made the strategy particularly appealing in modern synthesis.

Li and Tang reported a core–shell Pd nanoparticles@MOF as multifunctional catalyst for Knoevenagel condensation–reduction cascade reaction [13]. Pd nanoparticles core was encapsulated by amine-functionalized IRMOF-3 shell (Fig. 29.14a). The base-transition metal bifunctional catalyst could be used to catalyze Knoevenagel condensation reaction of 4-nitrobenzaldehyde and malononitrile and subsequent reduction of $-NO_2$ group in one-pot operation (Fig. 29.14b).

Recently, a template-assisted protocol by incorporating small molecular catalysts into the hollow porous capsule was reported by Jiang [14]. The soluble yolk has a homogeneous environment while the outside shell possesses many properties of a heterogeneous catalyst, such as selective permeability, substrate enrichment, and size selectivity. Therefore, the yolk–shell MOF capsule offered the catalytic properties of both homogeneous and heterogeneous catalysts.

As a showcase of the size-selective heterogeneous catalysis, metalloporphyrin (e.g., co-centered porphyrin, co-TCPP) was prepared. It was encapsulated into H-LDH@ZIF-8 to produce yolk–shell Co-TCPP@H-LDH@ZIF-8. The nanocatalyst was the effective promoter for the carbonylation of epoxides with CO_2 (Fig. 29.15). It gave a highest conversion (up to 94%) and size selectivity (up to 99%) toward small-sized epoxides compared to Co-TCPP and Co-TCPP in situ embedded in ZIF-8 (Co-TCPP@ZIF-8).

Moreover, an oxidation-Knoevenagel condensation cascade reaction was successfully demonstrated by the yolk–shell catalytic system. An amine-functionalized porphyrin (TAPP) was encapsulated in H-LDH@ZIF-8 to form yolk–shell TAPP@H-LDH@MOF-74. The MOF shell served as the catalyst for oxidation and the TAPP

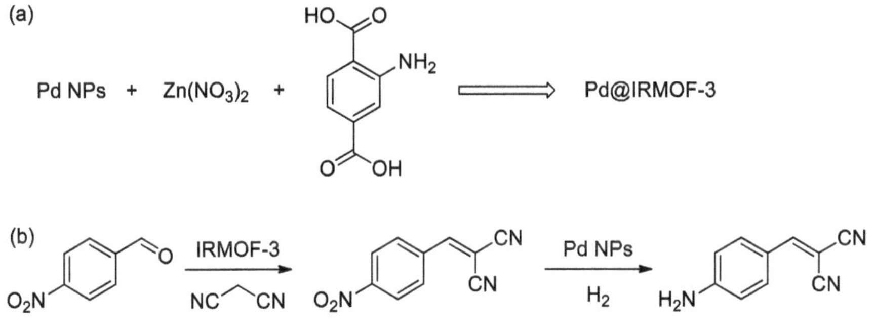

Fig. 29.14 Amine-functionalized IRMOF-3 shell for Knoevenagel condensation reduction cascade. Reprinted with permission from Ref. [13]. Copyright 2014 American Chemical Society

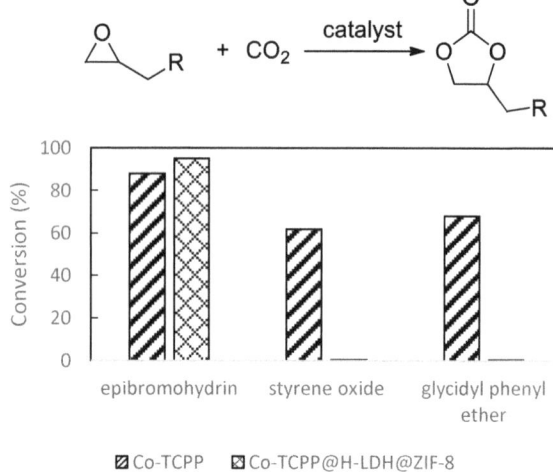

Fig. 29.15 Size selectivity. Reprinted with permission from Ref. [14]. Copyright 2019 Oxford University Press

yolk promoted the Knoevenagel condensation. This cascade process efficiently delivered an α,β-unsaturated substance using benzyl alcohol as substrate. Notably, TAPP@H-LDH@MOF-74 catalyst exhibited higher activity than the homogeneous catalyst TAPP (Fig. 29.16).

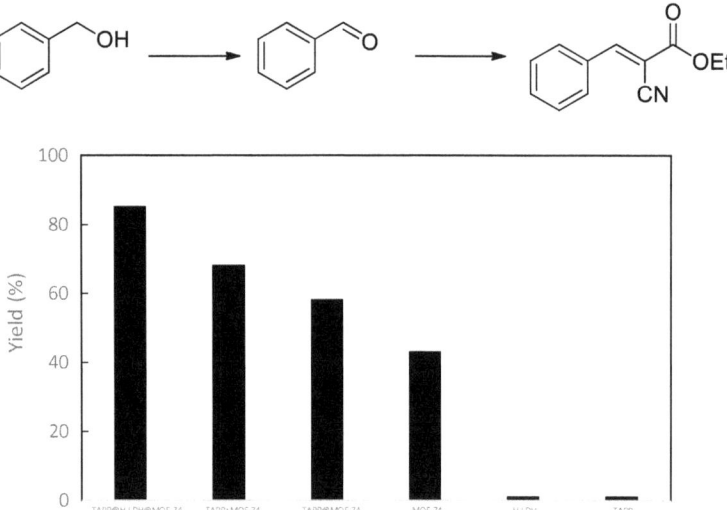

Fig. 29.16 An oxidation-Knoevenagel condensation cascade reaction. Reprinted with permission from Ref. [14]. Copyright 2019 Oxford University Press

29.6   Conclusion

In summary, this chapter summarizes the recent development of core–shell or yolk–shell structured catalysts for organic transformations including reduction, oxidation, cross-coupling, and cascade reactions. These nanocatalysts offer the merits of reducing agglomeration while enhancing the catalyst performance in terms of size and chemoselectivity and catalyst stability and robustness for recycling and reuse. It is recognized that still a very limited number of reactions have been explored by the nanocatalytic machinery. Furthermore, enantioselective processes have not been developed with the strategy. Therefore, the field holds great potential for developing new and practically useful reactions, which may be difficult or impossible to achieve by homogenous and other heterogeneous catalytic systems.

Acknowledgements Financial support for this work provided by the NIH (5R01GM125920-04) is gratefully acknowledged.

References

1. Zhao M, Yuan K, Wang Y et al (2016) Metal–organic frameworks as selectivity regulators for hydrogenation reactions. Nature 539:76–80
2. Yang Q, Xu Q, Yu S-H, Jiang H-L (2016) Pd nanocubes@ZIF-8: integration of plasmon-driven photothermal conversion with a metal-organic framework for efficient and selective catalysis. Angew Chem Int Ed 55:3685–3689
3. Li B, Ma J-G, Cheng P (2018) Silica-protection-assisted encapsulation of Cu_2O nanocubes into a metal-organic framework (ZIF-8) to provide a composite catalyst. Angew Chem Int Ed 57:6834–6837
4. Hu Y, Tao K, Wu C et al (2013) Size-controlled synthesis of highly stable and active pd@SiO_2 core–shell nanocatalysts for hydrogenation of nitrobenzene. J Phys Chem C 117:8974–8982
5. Sangili A, Annalakshmi M, Chen S-M et al (2019) Synthesis of silver nanoparticles decorated on core-shell structured tannic acid-coated iron oxide nanospheres for excellent electrochemical detection and efficient catalytic reduction of hazardous 4-nitrophenol. Compos B Eng 162:33–42
6. Sun D, Jang S, Yim S-J et al (2018) Metal doped core-shell metal-organic frameworks@covalent organic frameworks (MOFs@COFs) hybrids as a novel photocatalytic platform. Adv Funct Mater 28:1707110
7. Mitsudome T, Matoba M, Mizugaki T et al (2013) Core-shell AgNP@CeO_2 nanocomposite catalyst for highly chemoselective reductions of unsaturated aldehydes. Chem Eur J 19:5255–5258
8. Chen H, Shen K, Mao Q et al (2018) Nanoreactor of MOF-derived yolk–shell Co@C–N: precisely controllable structure and enhanced catalytic activity. ACS Catal 8:1417–1426
9. Cai M, Li Y, Liu Q et al (2019) One-step construction of hydrophobic MOFs@COFs core–shell composites for heterogeneous selective catalysis. Adv Sci 6:1802365
10. Ma R, Yang P, Ma Y, Bian F (2018) Facile synthesis of magnetic hierarchical core-shell structured Fe_3O_4@PDA-Pd@MOF nanocomposites: highly integrated multifunctional catalysts. ChemCatChem 10:1446–1454
11. Wang S, Yang Y, Liu P et al (2020) Core-shell and yolk-shell covalent organic framework nanostructures with size-selective permeability. Cell Rep Phys Sci 1:100062

12. Han X, He X, Sun L et al (2018) Increasing effectiveness of photogenerated carriers by in situ anchoring of Cu_2O nanoparticles on a nitrogen-doped porous carbon yolk–shell cuboctahedral framework. ACS Catal 8:3348–3356

13. Zhao M, Deng K, He L et al (2014) Core–shell palladium nanoparticle@metal–organic frameworks as multifunctional catalysts for cascade reactions. J Am Chem Soc 136:1738–1741

14. Cai G, Ding M, Wu Q, Jiang H-L (2020) Encapsulating soluble active species into hollow crystalline porous capsules beyond integration of homogeneous and heterogeneous catalysis. Natl Sci Rev 7:37–45

Chapter 30
Yolk-Shell Nanostructures: Charge Transfer and Photocatalysis

Yi-An Chen, Yu-Ting Wang, and Yung-Jung Hsu

30.1 Introduction

Due to the excessive consumption of finite fossil fuels today, the development of alternatively sustainable energy sources has become an urgent problem. Among renewable sources, solar energy attracts much attention. Taken from the sun, solar energy is a nearly inexhaustible source of energy, while photocatalysts provide a route to transform solar energy into chemical energy. Such a potential solution has prompted many researchers to devote time into further investigation. At present, hydrogen and hydrocarbons are considered as the most promising solar fuels, which are produced from photocatalyst-induced water splitting and carbon dioxide reduction; no matter how the energy is generated, the products after use are all clean and eco-friendly. In addition, with the booming economy, wastewater treatment in industries and farms has also received much attention. Photocatalysts are used to decompose organic pollutants in water such as antibiotics [1–5], bacteria [6, 7], dyes [8], and other organic compounds hazardous to the environment. Moreover, many high value-added organic compounds can be synthesized by photocatalytic reactions, which can refrain from the high cost incurred through traditional manufacture [9, 10].

When one irradiates photocatalysts with a light source with appropriate energy, electrons will be excited to the conduction band (CB) leaving holes at the valance band (VB). A fraction of the electron-hole pairs will be consumed through recombination, and the remaining electrons and holes will transfer to the catalyst's surface. After that,

Y.-A. Chen · Y.-T. Wang · Y.-J. Hsu (✉)
Department of Materials Science and Engineering, National Chiao Tung University, Hsinchu 30010, Taiwan
e-mail: yhsu@cc.nctu.edu.tw

Y.-J. Hsu
Center for Emergent Functional Matter Science, National Chiao Tung University, Hsinchu 30010, Taiwan

© The Author(s), under exclusive license to Springer Nature Singapore Pte Ltd. 2021
H. Yamashita and H. Li (eds.), *Core-Shell and Yolk-Shell Nanocatalysts*,
Nanostructure Science and Technology,
https://doi.org/10.1007/978-981-16-0463-8_30

electrons and holes undergo reduction and oxidation reactions respectively. However, the most serious bottleneck is the high probability of electron-hole recombination, which results in only a few carriers actually participating in the reaction. In addition, poor light absorption efficiency also leads to a low solar energy conversion rate. To conquer these obstacles, a great deal of photocatalysts with heterostructures have been proposed in which charge separation can be easily achieved via a difference in band structure, for example, at metal-semiconductor [11], type-II semiconductor [12], p-n semiconductor [13], and Z-scheme semiconductor interfaces [14].

The yolk-shell nanostructure consists of a yolk, hollow space, and a permeable shell, and the number of yolks and shells can be single or multiple. The fascinating features of this structure include a confined space, which can facilitate the diffusion of reacting species, abundant active sites for driving chemical reactions, which are offered by both the inner and outer surfaces, and multi-reflection of light inside the void space, which improves the efficiency of light absorption [15, 16]. As stated above, the overall catalytic performance can be effectively promoted. In addition, due to the distinct band structure between yolk and shell, photoexcited charges can be efficiently separated and can carry out the catalytic reaction, respectively [17, 18]. In this chapter, we mainly focus on synthetic methods for fabricating yolk-shell nanostructures, including the Kirkendall effect, Galvanic replacement, Ostwald ripening, chemical etching, and thermal treatment methods. Moreover, photocatalytic applications with yolk-shell nanostructures such as dye degradation, decomposition of toxins, carbon dioxide reduction, water splitting, and fine chemical synthesis are also reported. A future outlook for the advancement of the field will also be discussed.

30.2 Synthetic Approach

As the yolk-shell nanostructure has received considerable attention, many researchers have devoted effort into developing the fabrication methods for this amazing hollow structure. Up to now, a wide variety of chemical compositions in the form of yolk-shell nanostructures have been successfully fabricated. The different synthetic systems can be roughly divided into five approaches: (1) Kirkendall effect, in which the void space is formed via the difference between the diffusion rate of various kinds of ions [19]. This method is widely used in the synthesis of metal-semiconductor yolk-shell heterostructures, for example, FePt-CoS_2 [20], Ag-Fe_2O_3 [21]; (2) Galvanic replacement, for which the underlying principle for the formation of a yolk-shell structure is also due to a diversity in diffusion rate, but the driving force for ion exchange is the redox potential difference between ions; the cases include Au-AuAg [22], Au/PdAg [23], Ni-SnO_2/Ni_3Sn_2 [24]; (3) Ostwald ripening refers to a process involving dissolution and re-deposition of smaller colloidal particles onto larger colloidal particles, and this mechanism is also often adopted to synthesize yolk-shell heterostructures, such as Au-TiO_2 [25], $CoSe_2$-(NiCo)Se_2 [26], Pt-CeO_2 [27]; (4) Chemical etching is a relatively intuitive preparation method, which involves coating a layer of sacrificial template between the core and shell, and then performing selective etching. As a result, hollow space or void is created between the core and shell. With this method, as long as the shells can be successfully grown onto the template, complex compositions can be achieved, for instance, Fe_3O_4-CdS [28], $ZnFe_2O_4$-reduced graphene oxide-TiO_2 [29]. Moreover, even multi-shell structures can also be simply produced, for example, TiO_2-TiO_2-TiO_2 [30], Si-C-C [31], and SiO_2-SiO_2-SiO_2 [32]; (5) Thermal treatment can be used to generate a hollow space through volatilization or oxidation of chemicals in solution, and most of them are manufactured via pyrolysis (e.g., Co-C-Co_9S_8 [33], $MnFe_2O_4$-SnO_2 [34]) or calcination (e.g., C-TiO_2 [35], Pd-In_2O_3 [36]), which are quite suitable for large-scale production.

30.3   Photocatalytic Applications

30.3.1   Dye Degradation

In the past ten years, environmental scientists from various countries have conducted a large number of studies for conducting dye degradation by using semiconductor photocatalysts. From the research results and current situation, such a method is effective for treating dyeing wastewater, which is mainly due to the strong oxidizing ability of photocatalytic oxidation that can completely oxidize and decompose organic pollutants. Additionally, the photocatalytic reaction mechanism for dye degradation is a relatively simple dynamic system that can achieve the effect of environmental cleanliness.

Many different types of yolk-shell nanostructures have demonstrated remarkable photocatalytic dye degradation, which include C@TiO_2 [37], SiO_2-void-Ag/TiO_2 [38], $MnFe_2O_4$-SnO_2 [34]. J.-B. Joo et al. reported the successful synthesis of C@TiO_2 yolk-shell nanostructures by thermal treatment [37]. The structures were composed of a conducting carbon core and a photoactive TiO_2 shell. The resultant C@TiO_2 particles displayed favorable properties, e.g., the presence of conducting carbon in the void space, good microstructural integrity, regular size distribution, and mesoporous shell with great crystallinity and controlled crystal phase. The yolk-shell particles were prepared by consecutively depositing an amorphous TiO_2 and a SiO_2 protecting layer on the surface of resin spheres. Further calcining in Ar conditions at high temperatures can convert resin core into carbon core with smaller size, accompanied by the crystallization of TiO_2 shell. After removing SiO_2 layer using NaOH, C@TiO_2 yolk-shell nanostructures can be obtained. Further post-treatment by means of acid-treatment and re-calcination can enhance the photocatalytic efficiency of TiO_2 shell as a result of the improved crystallinity [39, 40]. In this work, various re-calcination temperatures (700, 750, 800 °C) were used to compare the performance of different samples (C@TiO_2-HCl-700, C@TiO_2-HCl-750, C@TiO_2-HCl-800).

As shown in Fig. 30.1a, dark carbon cores and porous TiO_2 shells were evident for the sample with post-treatment, indicating the successful formation of yolk-shell nanostructures. The photocatalytic efficiency of the samples for rhodamine

Fig. 30.1 **a** TEM image and the sketch of C@TiO$_2$ yolk-shell particles prepared with post-treatment. **b** Results of RhB degradation on different samples under UV illumination. **c** Proposed photocatalytic mechanism for C@TiO$_2$ toward RhB degradation. Reprinted with permission [37]. Copyright 2016, Elsevier

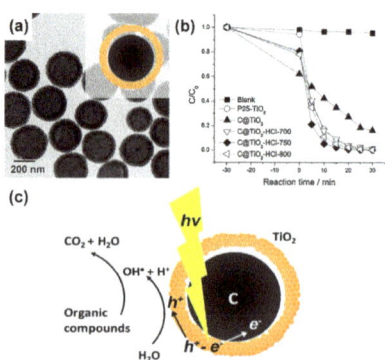

B (RhB) degradation under UV illumination, as measured by monitoring the peak intensity at a wavelength of 554 nm vs. time, is compared in Fig. 30.1b. Prior to UV illumination, the photocatalyst powder and RhB solution were mixed and stirred in the dark for 30 min in order to assure that the adsorption of RhB onto the catalyst surface was saturated. All the four C@TiO$_2$ were found to display a better RhB adsorption capacity (18–38% of RhB) compared to P25-TiO$_2$ (6% of RhB). Due to hydrophobicity and a microporous carbon core, the sample without post-treatment (C@TiO$_2$) showed the highest adsorption capacity. In a blank test without adding photocatalyst, RhB was found to degrade by 4.5% after 30 min of UV radiation, while in the presence of a catalyst, the degradation was significantly enhanced. The photocatalytic activity of the samples for RhB degradation followed the sequence of C@TiO$_2$-HCl-750 > C@TiO$_2$-HCl-800 > P25-TiO$_2$ > C@TiO$_2$-HCl-700 > C@TiO$_2$. Upon acid-treatment and re-calcination operation, the photocatalytic efficiency of C@TiO$_2$ was greatly improved.

Figure 30.1c shows that under UV light irradiation, photogenerated charge carriers are separated during the photocatalytic process. Electrons tend to transfer to the conducting carbon core and holes stay on the TiO$_2$ shell, thereby reducing the charge carrier recombination and extending the carrier lifetime. The holes at the TiO$_2$ shell can oxidize water to produce hydroxyl radicals, which leads to the oxidative decomposition of organic molecules. In addition, the conducting carbon core can act as an electron acceptor, promoting charge separation efficiency for TiO$_2$ to enhance the overall photocatalytic performance.

30.3.2 *Decomposition of Toxics*

With the rapid development of modern industry, toxic substances become consequential environmental pollutants at the globe scale. Owing to their extensive use in daily life as sanitizers, valid solutions are demanded to dissipate such pollution from the environment. Hereinto, photocatalysis using semiconductor and solar irradiation is regarded as an effective and environmental-friendly approach toward the decomposition of toxic substances. In particular for yolk-shell nanostructures such as Bi-void-SnO_2 [41], SnO_2-$LaFeO_3$ [42], $NiFe_2O_4$-C [43], many studies have demonstrated notable photocatalytic activity for decomposition of toxics.

In the past few years, the typical colorless antibiotic, tetracycline (TC), has been widely used in the treatment of human disease and animal husbandry [44, 45]. Excessive usage of antibiotics, such as TC, has led to widespread water pollution and affected the sustainable development of human society [46]. Recently, the use of two different materials to construct the yolk-shell nanostructure has attracted widespread attention in the photocatalytic decomposition of TC.

In a study by Wu et al. [41], Bi spheres were used as core to synthesize yolk-shell Bi@void@SnO_2 spheres with a chemical etching approach. The photocatalytic performance was evaluated by using TC as a test toxic under visible light irradiation ($\lambda > 420$ nm, light intensity $= 300$ W/cm^2). During the photocatalytic degradation process, three different scavengers were employed to recognize the types of the reactive species responsible for TC decomposition. The photocatalytic mechanism was then studied. Figure 30.2a is the TEM image for Bi@void@SnO_2 showing the size of Bi core in a range from 50 nm to 200 nm. As shown in Fig. 30.2b, Bi spheres show a noticeable activity with the TC decomposition amount of 16.33%,

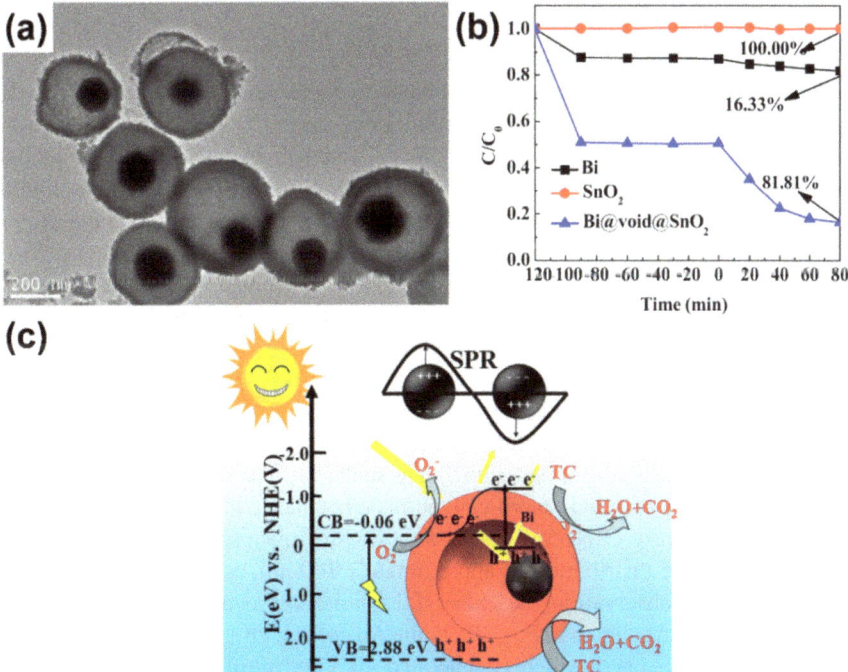

Fig. 30.2 **a** TEM image of Bi@void@SnO_2 spheres. **b** Results of TC degradation on different samples under visible light illumination. **c** Proposed photocatalytic mechanism for Bi@void@SnO_2 toward TC degradation. Reprinted with permission [41]. Copyright 2019, Springer Nature

whereas SnO_2 displays considerably low activity of TC decomposition. Significantly, Bi@void@SnO_2 shows considerably high performance with the decomposition efficiency reaching 81.81%. A photocatalytic mechanism for Bi@void@SnO_2 for TC decomposition is proposed in Fig. 30.2c. The electrons at the VB of SnO_2 are excited, which then migrate to the CB of SnO_2; meanwhile, holes remain at the VB of SnO_2. Note that Bi spheres have surface plasmon resonance (SPR) property that can generate hot electrons under visible light illumination [47]. Because the SPR level of Bi is higher than the CB of SnO_2, the hot electrons in Bi can be injected into the CB of SnO_2. Simultaneously, the potential at Bi is decreased rapidly, which promotes the capture of electrons from the VB of SnO_2 to restore the original state. On the other hand, the electromagnetic field induced by the SPR of Bi reduces charge carrier recombination in SnO_2 [48], which can extend the carrier lifetime and enhance the photocatalytic efficiency as well. The separated electrons may react with the dissolved O_2 to form $\cdot O_2^-$, and such a reactive substance together with the holes at the SnO_2 surface can significantly decompose TC. Compared with pure SnO_2, the Bi@void@SnO_2 has better photocatalytic efficiency due to a higher specific surface area and more pronounced charge separation caused by the SPR effect of Bi [48].

30.3.3   Carbon Dioxide Reduction

As human society becomes more prosperous, a large amount of carbon dioxide is produced by modern society in transportation, electricity generation, industrial manufacture, and other activities, which can lead to the greenhouse effect and global warming. To alleviate this phenomenon, many scientists have devoted research into the topic of carbon dioxide reduction using photocatalysts [49, 50]. The feature of this reaction is the employment of carbon dioxide as feedstock to obtain other carbon materials, including CH_4, C_2H_6, and CH_3OH, which can also be used as fuel. The ultimate goal of carbon dioxide reduction is to mimic natural photosynthesis and simultaneously balance the global carbon dioxide footprint. Many different yolk-shell nanostructures have demonstrated remarkable photocatalytic carbon dioxide reduction activity, which include Au-TiO_2 [17], Ni-SiO_2 [51], and Au-Cu_2O [52].

One of the studies of Guo et al. fabricated a yolk-shell structure, which is composed of Au as core, and C_3N_4 decorated with SnS as shell [53]. The samples were prepared by selective etching of a SiO_2 sacrificial template between a Au core and C_3N_4 shell with NH_4HF_2. Au@g-C_3N_4/SnS (41.2 wt% of SnS) nanospheres were prepared with a uniform size distribution of approximately 100 nm, as shown in Fig. 30.3a. Through the TEM observation, the successful synthesis of a yolk-shell structure was further confirmed. As shown in Fig. 30.3b, the Au cores with a size of approximately 5 nm and a hollow space, which is generated via removal of SiO_2, are apparently observed; meanwhile, SnS nanosheets adhered onto the g-C_3N_4 shell may also be easily recognized from the uneven appearance of the shell surface. Figure 30.3c shows the result of carbon dioxide reduction under visible light irradiation ($\lambda > 420$ nm). The prepared samples include Au@SnS, Au@g-$C_3N_{4,}$ and Au@g-C_3N_4/SnS with

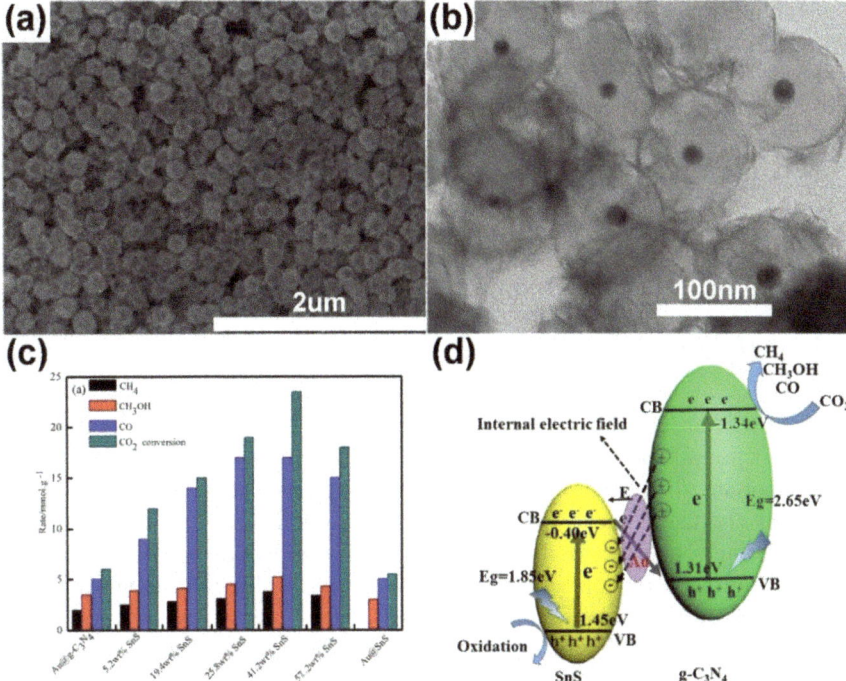

Fig. 30.3 **a** SEM and **b** TEM images of Au@g-C$_3$N$_4$/SnS nanospheres. **c** Results of photocatalytic carbon dioxide reduction on different samples upon 4 h of visible light illumination. **d** Proposed photocatalytic mechanism for Au@g-C$_3$N$_4$/SnS toward carbon dioxide reduction. Reprinted with permission [53]. Copyright 2018, American Chemical Society

varying SnS content. Compared with Au@g-C$_3$N$_4$, Au@SnS displays a worse carbon dioxide conversion efficiency despite larger surface area, stronger carbon dioxide adsorption ability and higher visible light absorption. Moreover, no CH$_4$ product is observed and only a few CO and CH$_3$OH are detected on Au-SnS. This can be attributed to the unsuitable potential of the CB and fast carrier recombination in this sample. For the different Au@g-C$_3$N$_4$/SnS, the optimal SnS content for maximizing carbon dioxide reduction efficiency was determined to be 41.2 wt%, which gives rise to the production of 3.8 μmol g^{-1} CH$_4$, 5.3 μmol g^{-1} CH$_3$OH and 17.1 μmol g^{-1} CO.

Figure 30.3d depicts a plausible photocatalytic mechanism for Au@g-C$_3$N$_4$/SnS toward carbon dioxide reduction. In this heterojunction, if the movement of photoexcited carriers follows a type-II route, then the electrons can transfer to the CB of SnS. Consequently, there would be no sign of CH$_4$ production because the CB level of SnS is too low to trigger carbon dioxide reduction. However, the results reveal that the electrons in Au@g-C$_3$N$_4$/SnS can efficiently conduct carbon dioxide reduction. This means that these electrons are mainly separated at the CB of g-C$_3$N$_4$, the level of which is sufficiently high. Consequently, a Z-scheme pathway is proposed. Under visible light irradiation, both SnS and C_3N_4 will generate electron and hole pairs. Here, Au serves as charge-transfer mediator, promoting the transportation of electrons from the CB of SnS to the VB of g-C_3N_4, followed by the recombination with holes. The electrons left at the CB of g-C_3N_4 and the holes left at the VB of SnS then participate in carbon dioxide reduction and oxidation reaction, respectively. With this Z-scheme mechanism as well as the advantages of the yolk-shell structure such as large surface area and effective photon absorption, Au@g-C_3N_4/SnS demonstrates the promotion of photocatalytic performance toward carbon dioxide reduction.

30.3.4 Hydrogen Production

Currently, how to efficiently convert solar energy to other energies, such as chemical energy, is an important subject. From an environmental point of view, solar energy is an important alternative energy source. To overcome the crisis of petrochemical energy shortages, hydrogen is considered as a significant green energy source. Ever since Fujishima and Honda first observed photocatalytic activity of TiO_2 in 1972 [54], direct water splitting to produce hydrogen under light illumination has been the Holy Grail of solar energy harvesting and conversion. Therefore, photocatalytic water splitting for hydrogen production has become a topic of much attention. Many different yolk-shell nanostructures have demonstrated remarkable photocatalytic hydrogen production activity, which include Au-CdS [55], Au-TiO_2 [56], and CdS-SiO_2 [57].

Compared to traditional core-shell structure [58], the void space between core and shell in a yolk-shell system allows more interactions of core and shell components, which can promote the resultant synergy. For instance, for yolk-shell Au-CdS system, charge separation of CdS shell can be improved by the constantly contacting Au core, thereby increasing the photocatalytic efficiency. In addition, the SPR of Au particles can be used to enable hot electron injection [59], resonant energy transfer [60, 61], and electromagnetic field intensification [62]. These features would increase photon harvesting and thus carrier generation for CdS. Chiu et al. successfully synthesized Au-CdS yolk-shell nanostructures and investigated the photocatalytic performance toward hydrogen generation [55]. The samples were prepared by employing the Kirkendall effect with two consecutive ion-exchange procedures. Briefly, the Au-CdS yolk-shell sample is synthesized through consecutively ion-exchange processes from Au-Cu_2O to Au-Cu_7S_4, and finally to Au-CdS. By varying the amount of Au, the shell thickness of Cu_2O of Au-Cu_2O can be controlled, thereby generating different void sizes for Au-CdS with the same shell thickness (11.5 ± 0.4 nm). Four Au-CdS yolk-shell nanoparticles along with pure CdS hollow particles were prepared, and the void size was determined to be 40.2 ± 3.3 nm, 47.3 ± 3.3 nm, 52.1 ± 3.3 nm, 82.7 ± 3.9 nm, and 64.0 ± 6.3 nm for Au-CdS-1, Au-CdS-2, Au-CdS-3, Au-CdS-4, and pure CdS, respectively. Figure 30.4a shows the TEM image for Au-CdS-4. The hydrogen production activity was estimated under visible light illumination ($\lambda = 400$–700 nm, light intensity $= 690$ mW/cm^2) by using Na_2S/Na_2SO_3 as electrolyte. As shown in

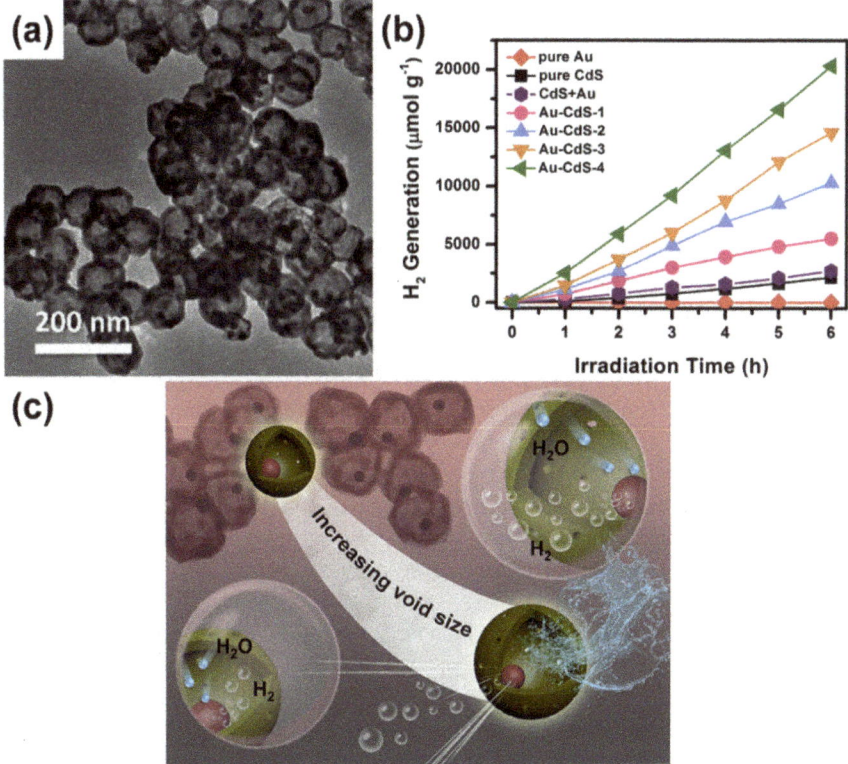

Fig. 30.4 **a** TEM image of Au-CdS-4 nanostructures. **b** Results of photocatalytic hydrogen production on different samples under visible light illumination. **c** Proposed photocatalytic mechanism for Au-CdS toward hydrogen production. Reprinted with permission [55]. Copyright 2019, Elsevier

Fig. 30.4b, the yolk-shell nanoparticles show better performance compared to pure CdS and mixture of CdS and Au (CdS + Au), suggesting that the yolk-shell structure is important for increasing the photocatalytic efficiency. Among the four Au-CdS samples, Au-CdS-4 has the highest hydrogen generation performance. This can be ascribed to the synergy of effective charge separation, the SPR effect, and improved mass transport due to large void size. The photocatalytic hydrogen production mechanism for Au-CdS is further shown in Fig. 30.4c. The findings from this work illustrate that the void size is a significant cause for optimizing the photocatalytic activity of yolk-shell nanostructures toward hydrogen production.

30.3.5 Oxygen Production

In comparison with hydrogen production, only a handful of photocatalysts show effective oxygen production from water splitting. The reason why oxygen production is more difficult is that water oxidation needs four equivalent holes at the same time to generate an oxygen molecule. Conceptually, this is much more complicated than reducing water to produce hydrogen, which requires two electrons. Although oxygen is an important resource, the oxygen production reaction is difficult to achieve. However, it can provide guidelines for photocatalytic system optimization. Many different yolk-shell nanostructures have demonstrated remarkable photocatalytic oxygen production activity, which include Fe_2TiO_5-TiO_2 [63], $Pt@TiO_2@In_2O_3@MnO_x$ [64], Co_3O_4/g-C_3N_4/Pt [65], and Ta_3N_5/Pt [66].

TiO_2 and α-Fe_2O_3 meet several basic requirements for oxygen production from water splitting [67, 68]. The disadvantages of TiO_2 are a large bandgap (~3.2 eV), which can only harvest 4% of solar spectrum, and the fast charge recombination induced by trap states [69]. The narrow bandgap (~2.1 eV) of α-Fe_2O_3 can enable visible light absorption, which covers 46% of solar spectrum. Moreover, α-Fe_2O_3 is abundant, nontoxic, and low-cost and is stable in aqueous solution [68]. However, due to the rapid charge recombination, there is a large amount of energy loss due to the low electrical conductivity of α-Fe_2O_3, which deteriorates the process of charge generation, transportation, collection, and injection [70, 71]. With a sufficiently anodic VB level, TiO_2 (E_{VB} = +2.91 eV vs. NHE [72]) and α-Fe_2O_3 (E_{VB} = +2.48 eV vs. NHE [72]) meet the standards for efficient water oxidation. To further increase oxygen production on α-Fe_2O_3 and TiO_2, a heterojunction based on Ti–Fe–O structures has been constructed. Fe_2TiO_5 is an n-type semiconductor with a suitable bandgap (~2.1 eV) for visible light absorption [73]. In addition, the band structure of Fe_2TiO_5 matches well with that of TiO_2 [74]. Therefore, by forming a heterostructure between Fe_2TiO_5 and TiO_2, the water oxidation performance can be improved.

In the work by M. Waqas et al., pristine Fe_2TiO_5, pristine TiO_2, and Fe_2TiO_5-TiO_2 yolk-shell hollow spheres (FTYS-HS) were prepared to demonstrate oxygen production under light illumination [63]. These samples were prepared through a simple sacrificial hard template strategy. The benefits of the designed FTYS-HS include the hollow cavity, porous shell, and the heterostructure interface. First, the hollow cavity increases the surface area, so there is an increased number of active sites ready for redox reactions. Second, the porous shell allows the reacting species to access the minority carriers at the inner core, thereby effectively utilizing all of the charge carriers. Finally, the multi-reflection of light inside the yolk-shell structure increases the lifetime of incident photons. As shown in Fig. 30.5a, uniformly thick FTYS-HS were obtained. Figure 30.5b compares the photocatalytic performance of four relevant samples toward oxygen production under white light illumination (light intensity = 300 W/cm^2) by using $AgNO_3$ as the electrolyte. The results show that oxygen produced within the first 5 h is stable. At a prolonged time of 5.5 h, needless reduction of Ag^+ occurred to form Ag, thereby blocking light absorption to decrease the activity. FTYS-HS shows a substantially higher oxygen evolution rate

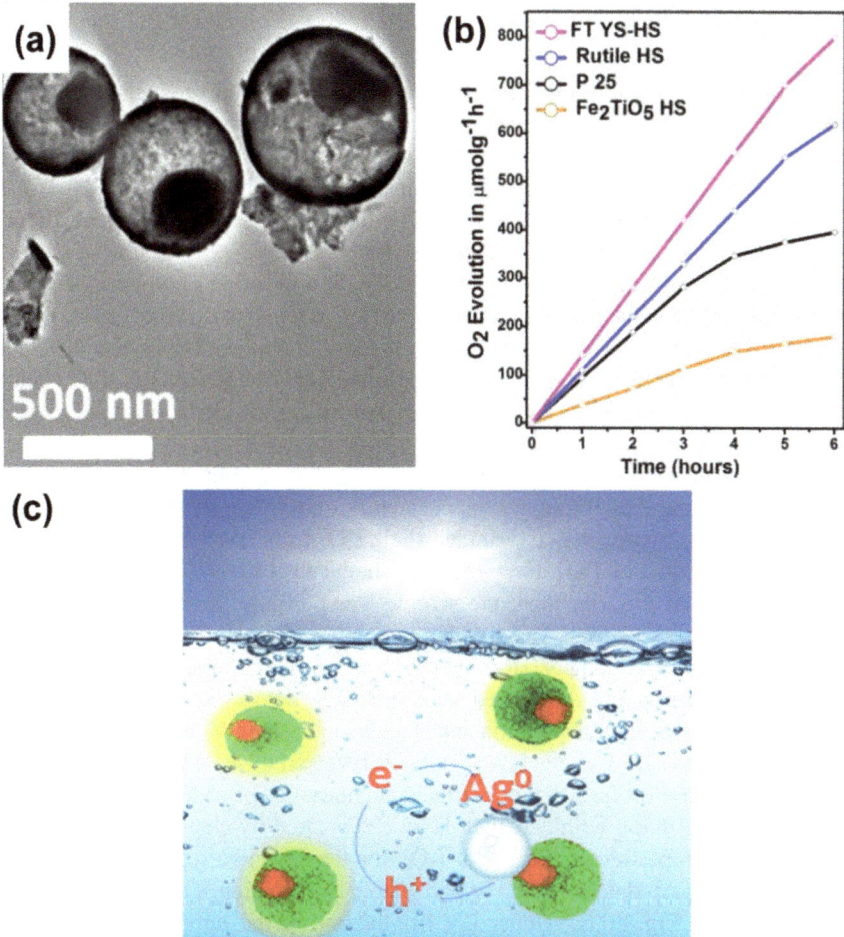

Fig. 30.5 a TEM of FTYS-HS. **b** Results of photocatalytic oxygen production on different samples under white light illumination. **c** Proposed photocatalytic mechanism for FTYS-HS toward oxygen production. Reprinted with permission [63]. Copyright 2017, Elsevier

compared to the rutile TiO_2 hollow sphere (Rutile HS), P25 TiO_2 (P25), and Fe_2TiO_5 hollow sphere (Fe_2TiO_5 HS). The better performance of FTYS-HS is attributed to the enhanced charge separation at the Fe_2TiO_5/TiO_2 heterojunction. Moreover, charge carriers can transfer between the two systems because of the higher E_{VB} and E_{CB} of Fe_2TiO_5 compared to TiO_2. [75] The mechanism for oxygen production for FTYS-HS is shown in Fig. 30.5c. The incident light can be harvested effectively due to the multiple reflections in the yolk-shell structure, which can prolong the carrier lifetime. Moreover, the heterojunction formed at the TiO_2 hollow shell and Fe_2TiO_5 core can separate the electrons and holes, allowing holes to participate in oxygen production without rapid charge recombination.

30.3.6   Fine Chemical Synthesis

Apart from the applications mentioned above, organic molecular conversion through photocatalysts is a topic that attracts less attention. Traditionally, organic chemicals are mostly manufactured under high temperature with an oxidation agent. Although high selectivity and conversion can be achieved, this method is also expensive and the reagents used during the reaction may be harmful to the environment. Therefore, the use of photocatalysts to produce organic chemicals is desirable. A few works have investigated organic chemical production by using photocatalysts [76–78], producing the value-added products including aldehydes and biphenyls, which are used in drugs, fragrances, and precursors for further applications. Many different yolk-shell nanostructures have demonstrated remarkable photocatalytic fine chemical synthesis, which include TiO_2-TiO_2 [79], Au-TiO_2 [80], Fe_3O_4-nitrogen-doped carbon [81], and WO_3-WO_3 [82].

Ouyang et al. successfully synthesized a TiO_2-based yolk-shell structure and further modified the shell through chemical etching and hydrogen treatment, creating an advanced yolk-hydrogenated wrinkled shell TiO_2 (Y@HWS-TiO_2) for selective benzyl alcohol oxidation [83]. First, the yolk-shell TiO_2 (Y@S-TiO_2) was prepared via a solvothermal process; then, the sample was partially etched and hydrogen-treated to produce Y@HWS-TiO_2, which is composed of an inner TiO_2 core and a hydrogenated wrinkled outer TiO_2 shell. It is worth mentioning that Ti^{3+} species and oxygen vacancies are introduced after hydrogen treatment, both of which are beneficial for visible light absorption and charge separation [84–86]. The yolk-shell structure of Y@S-TiO_2 and Y@HWS-TiO_2 are revealed by the TEM images shown in Fig. 30.6a, b. In addition, a distinct surface roughness can be observed in Y@HWS-TiO_2, confirming that a wrinkled shell is indeed created by chemical etching. The photocatalytic properties of Y@HWS-TiO_2 for the selective oxidation of benzyl alcohol were evaluated under both UV and visible illumination. The results obtained are compared in Fig. 30.6c. Under UV irradiation, all the samples including the commercial P25 TiO_2 (C-TiO_2), Y@S-TiO_2, yolk-wrinkled shell TiO_2 (Y@WS-TiO_2) and Y@HWS-TiO_2 demonstrate vital activity for benzyl alcohol oxidation. However, under visible light illumination, only Y@HWS-TiO_2 exhibits noticeable photocatalytic activity. Figure 30.6d illustrates the advantages of the yolk-shell structure: (1) Hollow space may induce multiple reflections and scattering of incident light, rendering light absorption to be more effective; (2) Organic substances including alcohol (blue spots) and aldehyde (red spots) can be confined between the core and shell, which may also facilitate benzyl alcohol oxidation activity. The band structure of Y@HWS-TiO_2 shows sublevel states created by the introduction of Ti^{3+} species, which results in a decrease in the TiO_2 bandgap to allow for visible light absorption. Therefore, under visible light irradiation, the electrons at the hydrogenated wrinkled shell will be excited to the sublevel and then take part in oxygen reduction to produce $\cdot O_2^-$ radicals. On the other hand, free $\cdot OH$ radicals may be created by the holes left behind in the VB. Both of these two reactive species can greatly improve the photocatalytic alcohol oxidation capability.

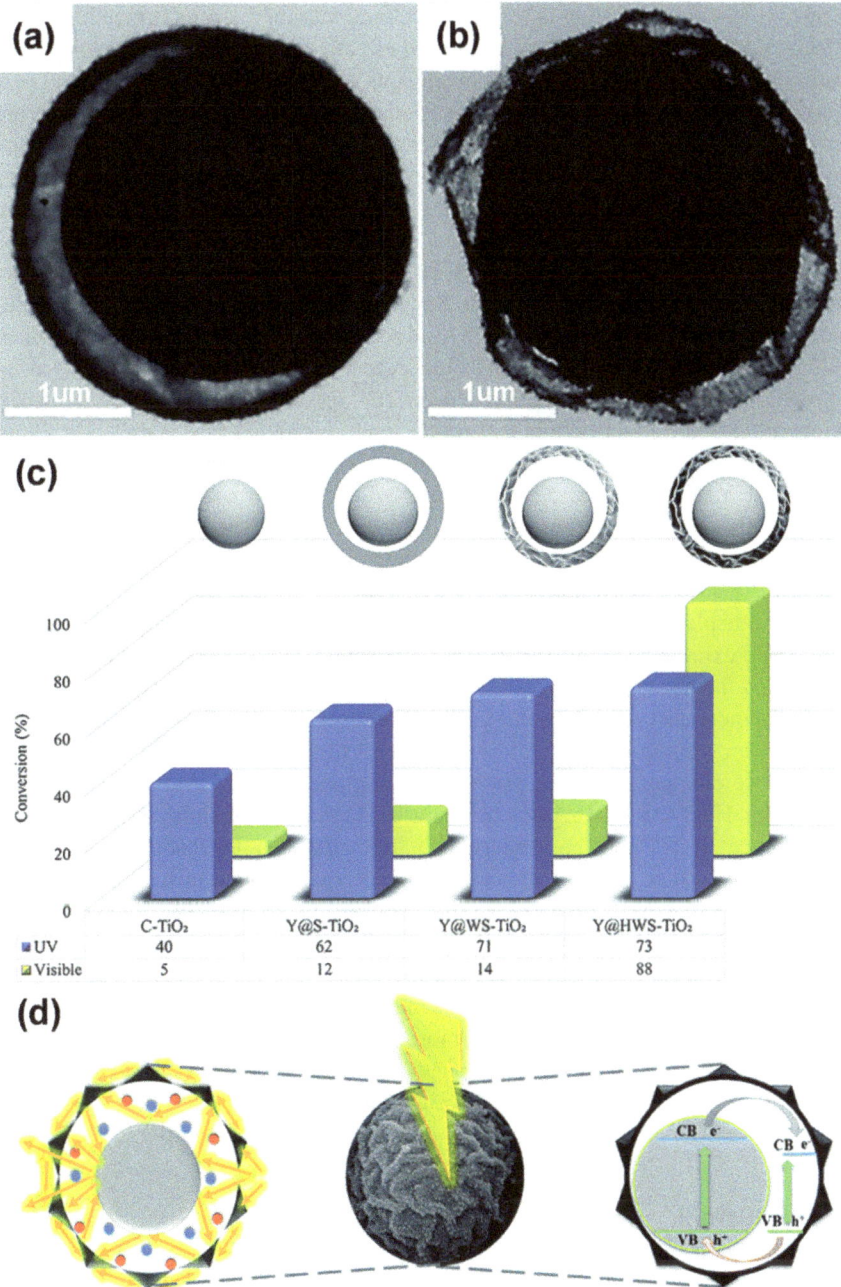

Fig. 30.6 TEM images of **a** Y@S-TiO$_2$, **b** Y@HWS-TiO$_2$. **c** Results of photocatalytic selective benzyl alcohol oxidation on different samples under UV (blue bars) and visible light (yellow bars) illumination. **d** Proposed photocatalytic mechanism for Y@HWS-TiO$_2$ toward benzyl alcohol oxidation. Reprinted with permission [83]. Copyright 2018, The Royal Society of Chemistry

Suzuki-Miyaura coupling is a cross-coupling reaction that was first published in 1979 by Akira Suzuki [87]. By adopting Pd as catalyst, a punch of organics can be synthesized for C-C bond formation in aqueous solution through this reaction, such as styrenes and biphenyls. At present, various palladium hybrid photocatalysts are also used to carry out the coupling reaction of organics [88–90]. With the effort of researchers, even Pd-free photocatalysts are capable of exhibiting Suzuki-Miyaura coupling [91]. Rohani et al. designed a novel yolk-shell architecture composed of a hydrogenated urchin-like TiO_2 shell and a TiO_2 yolk with Au-Pd core-shell particles decorated onto the shells (HUY@S-TOH/AuPd) [92]. This study investigated the photocatalytic activity for the famous Suzuki–Miyaura coupling. The TiO_2 yolk-shell structure was first prepared with a solvothermal method. Next, a hierarchical urchin-like shell was created by a dissolution-recrystallization process, followed by hydrogen treatment, which also introduced Ti^{3+} species. Finally, Au/Pd core/shell nanoparticles were attached onto the surface to produce HUY@S-TOH/AuPd. From the TEM images shown in Fig. 30.7a, b, the solid cores and hierarchical urchin-like shell of HUY@S-TOH/AuPd can be observed. The 5 nm-sized Au/Pd nanoparticles were also obviously decorated onto the surface of the shell. The photocatalytic activity was evaluated for the coupling reaction of 4-iodotoluene and phenylboronic acid under visible light illumination. The expected products are biphenyls and the results are displayed in Fig. 30.7c. Compared with the relevant samples, HUY@S-TOH/AuPd exhibits the highest yields for biphenyls. P25 TiO_2 does not display any activity and the need to employ Au/Pd nanoparticles was confirmed by the observed significant increased yield from P25/AuPd. Moreover, owing to the larger surface area and improved photon harvesting from yolk-shell structure, Y@S-TO/AuPd demonstrates more efficient photocatalytic activity compared to P25/AuPd. The superior activity for HUY@S-TOH/AuPd over Y@S-TO/AuPd is mainly associated with the presence of Ti^{3+}, which renders TiO_2 the capability to absorb visible light. Figure 30.7d further reveals the possible photocatalytic mechanism for HUY@S-TOH/AuPd for Suzuki-Miyaura coupling. Hydrogenated TiO_2 can harvest visible photons to generate photoexcited charge carriers. The electrons may move onto Au/Pd. On the other hand, hot electrons can also be induced on account of the SPR of Au, which are subsequently injected into Pd to promote the activation of the $C-X$ bond of aryl halides [93]. Furthermore, the holes can oxidize the protic organic solvent or cleave the $C-B$ bonds of phenylboronic acid, which may further react with activated aryl halides by reductive elimination.

30.4 Outlook

In this chapter, a concise introduction to the five different methods used for the preparation of yolk-shell nanostructures is provided. The underlying mechanism for Kirkendall effect and galvanic replacement is based on a difference in the ion diffusion rate and redox potential, respectively. Ostwald ripening can be used to generate hollow structures through dissolution and re-deposition of nanoparticles.

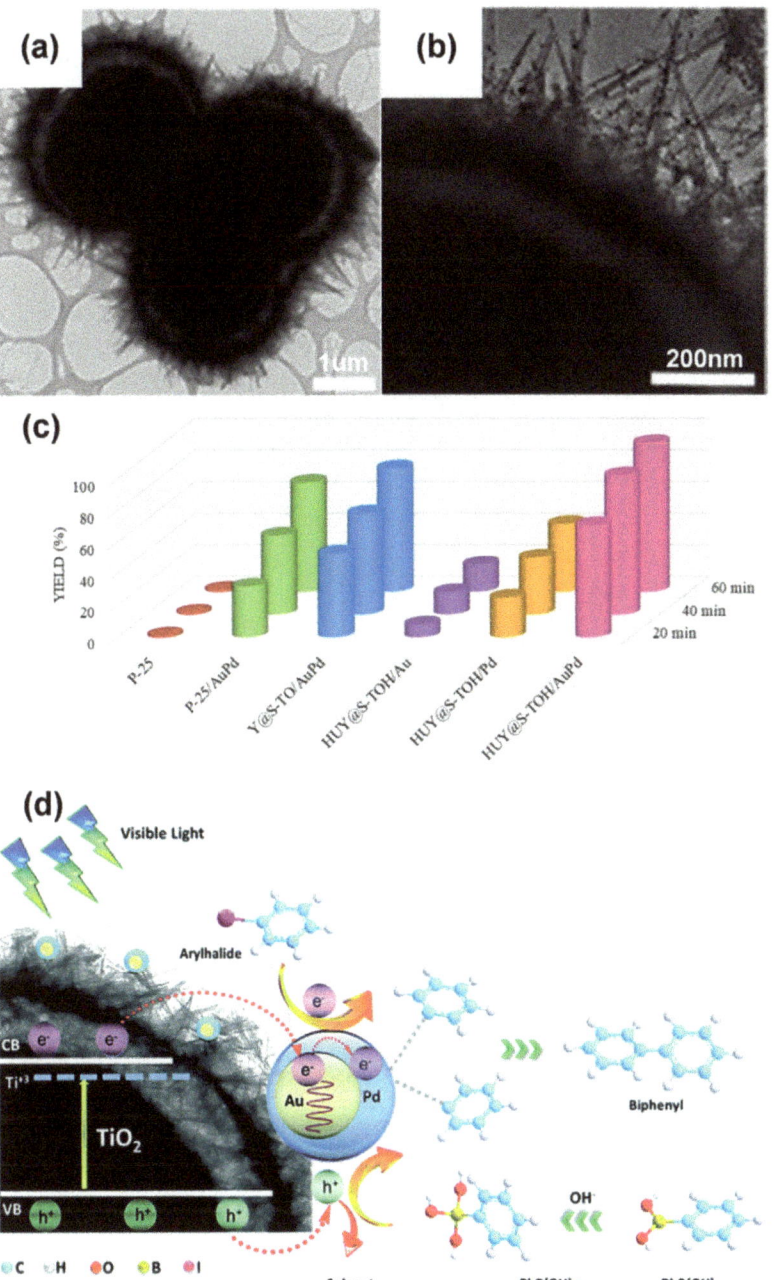

Fig. 30.7 **a, b** TEM images of HUY@S-TOH/AuPd. **c** Results of photocatalytic Suzuki-Miyaura coupling reaction on different samples under visible light illumination. **d** Proposed photocatalytic mechanism for HUY@S-TOH/AuPd toward the Suzuki-Miyaura coupling reaction. Reprinted with permission [92]. Copyright 2019, The Royal Society of Chemistry

Chemical etching and thermal treatment methods employ sacrificial template removal and chemical volatilization to create void structures. The yolk-shell structure has the advantage of a high reaction surface area, so it has a greater advantage compared with other nanostructure paradigms such as core-shell structures. In addition, the separated electrons and holes at the yolk and shell are both accessible by reactive species, allowing for the concurrent occurrence of reduction and oxidation reactions, which is particularly significant for overall water splitting and artificial photosynthesis. Although the photocatalytic use of yolk-shell structure has been well developed, it faces the problem of an ambiguous charge-transfer mechanism as a result of the movable yolk. Unlike the stationary core in core-shell structures, the yolk materials in a yolk-shell structure are constantly moving during the photocatalytic process. This can complicate the charge carrier transfer and separation behavior. How to intuitively monitor the charge-transfer dynamics of a yolk-shell structure to allow for the understanding of a realistic charge-transfer mechanism is the future direction of this field.

References

1. Yan M, Hua Y, Zhu F, Gu W, Jiang J, Shen H, Shi W (2017) Fabrication of nitrogen doped graphene quantum dots-BiOI/MnNb$_2$O$_6$ p-n junction photocatalysts with enhanced visible light efficiency inphotocatalytic degradation of antibiotics. Appl Catal B 202:518–527
2. Wang W, Fang J, Shao S, Lai M, Lu C (2017) Compact and uniform TiO$_2$@g-C$_3$N$_4$ core-shell quantum heterojunction for photocatalytic degradation of tetracycline antibiotics. Appl Catal B 217:57–64
3. Shi W, Guo F, Yuan S (2017) In situ synthesis of Z-scheme Ag$_3$PO$_4$/CuBi$_2$O$_4$ photocatalysts and enhanced photocatalytic performance for the degradation of tetracycline under visible light irradiation. Appl Catal B 209:720–728
4. Hong Y, Li C, Zhang G, Meng Y, Yin B, Zhao Y, Shi W (2016) Efficient and stable Nb$_2$O$_5$ modified g-C$_3$N$_4$ photocatalyst for removal of antibiotic pollutant. Chem Eng J 299:74–84
5. Wang K, Zhang G, Li J, Li Y, Wu X (2017) 0D/2D Z-scheme heterojunctions of bismuth tantalite quantum dots/ultrathin g-C$_3$N$_4$ nanosheets for highly efficient visible light photocatalytic degradation of antibiotics. ACS Appl Mater Interfaces 9:43704–43715
6. Wu TS, Wang KX, Li GD, Sun SY, Sun J, Chen JS (2010) Montmorillonite-supported Ag/TiO$_2$ nanoparticles: an efficient visible-light bacteria photodegradation material. ACS Appl Mater Interfaces 2:544–550
7. Kuo MY, Hsiao CF, Chiu YH, Lai TH, Fang MJ, Wu JY, Chen JW, Wu CL, Wei KH, Lin HC, Hsu YJ (2018) Au@Cu$_2$O core@shell nanocrystals as dual-functional catalysts for sustainable environmental applications. Appl Catal B 242:499–506
8. Zhang H, Lv X, Li Y, Wang Y, Li J (2010) P25-graphene composite as a high performance photocatalyst. ACS Nano 4:380–386
9. Chen W, Liu TY, Huang T, Liu XH, Yang XJ (2016) Novel mesoporous P-doped graphitic carbon nitride nanosheets coupled with ZnIn$_2$S$_4$ nanosheets as efficient visible light driven heterostructures with remarkable enhanced photo-reduction activity. Nanoscale 8:3711–3719
10. Hu Z, Quan H, Chen Z, Shao Y, Li D (2018) New insight into efficient visible-light-driven photocatalytic organic transformation over CdS/TiO$_2$ photocatalysts. Photochem Photobiol Sci 17:51–59
11. Liu X, Iocozzia J, Wang Y, Cui X, Chen Y, Zhao S, Li Z, Lin Z (2017) Noble metal–metal oxide nanohybrids with tailored nanostructures for efficient solar energy conversion, photocatalysis and environmental remediation. Energy Environ Sci 10:402–434

12. Wang Y, Wang Q, Zhan X, Wang F, Safdar M, He J (2013) Visible light driven type II heterostructures and their enhanced photocatalysis properties: a review. Nanoscale 5:8326–8339

13. Pirhashemi M, Habibi-Yangjeh A, Pouran SR (2017) Review on the criteria anticipated for the fabrication of highly efficient ZnO-based visible-light-driven photocatalysts. J Ind Eng Chem 62:1–25

14. Li H, Tu W, Zhou Y, Zou Z (2016) Z-scheme photocatalytic systems for promoting photocatalytic performance: recent progress and future challenges. Adv Sci 3:1500389

15. Jiang Z, Zhu C, Wan W, Qian K, Xie J (2016) Constructing graphite-like carbon nitride modified hierarchical yolk-shell TiO_2 sphere for water pollution treatment and hydrogen production. J Mater Chem A 4:1806–1818

16. Sun H, Yip HY, Jiang Z, Ye L, Lo IMC, Wong PK (2018) Facile synthesis of oxygen defective yolk-shell BiO_{2-x} for visible light-driven photocatalytic inactivation of Escherichia coli. J Mater Chem A 6:4997–5005

17. Tu W, Zhou Y, Li H, Li P, Zou Z (2015) Au@TiO_2 yolk–shell hollow spheres for plasmon-induced photocatalytic reduction of CO_2 to solar fuel via a local electromagnetic field. Nanoscale 7:14232–14236

18. Wang W, Zhu S, Cao Y, Tao Y, Li X, Pan D, Phillips DL, Zhang D, Chen M, Li G, Li H (2019) Edge-enriched ultrathin MoS_2 embedded yolk-shell TiO_2 with boosted charge transfer for superior photocatalytic H_2 evolution. Adv Funct Mater 29:1901958

19. Yang Z, Yang N, Pileni MP (2015) Nano kirkendall effect related to nanocrystallinity of metal nanocrystals: influence of the outward and inward atomic diffusion on the final nanoparticle structure. J Phys Chem C 119:22249–22260

20. Gao J, Liang G, Zhang B, Kuang Y, Zhang X, Xu B (2007) FePt@CoS_2 yolk-shell nanocrystals as a potent agent to kill HeLa cells. J Am Chem Soc 129:1428–1433

21. Wei Z, Zhou Z, Yang M, Lin C, Zhao Z, Huang D, Chen Z, Gao J (2011) Multifunctional Ag@Fe_2O_3 yolk–shell nanoparticles for simultaneous capture, kill, and removal of pathogen. J Mater Chem 21:16344–16348

22. Zhu J, Zhang S, Weng GJ, Li JJ, Zhao JW (2020) The morphology regulation and plasmonic spectral properties of Au@AuAg yolk–shell nanorods with controlled interior gap. Spectrochim Acta A 236:118343

23. Fang C, Zhao G, Zhang Z, Zheng J, Ding Q, Xu X, Shao L, Geng B (2018) Morphology engineering of Au/(PdAg alloy) nanostructures for enhanced electrocatalytic ethanol oxidation. Part Part Syst Char 35:1800258

24. Zhao B, Guo X, Zhao W, Deng J, Fan B, Shao G, Bai Z, Zhang R (2017) Facile synthesis of yolk–shell Ni@void@SnO_2(Ni_3Sn_2) ternary composites via galvanic replacement/Kirkendall effect and their enhanced microwave absorption properties. Nano Res 10:331–343

25. Sun H, He Q, Zeng S, She P, Zhang X, Li J, Liu Z (2017) Controllable growth of Au@TiO_2 yolk–shell nanoparticles and their geometry parameter effects on photocatalytic activity. New J Chem 41:7244–7252

26. Park SK, Kima JK, Kang YC (2017) Metal–organic framework-derived $CoSe_2$/(NiCo)Se_2 box-in-box hollow nanocubes with enhanced electrochemical properties for sodium-ion storage and hydrogen evolution. J Mater Chem A 5:18823–18830

27. Zhang N, Fu X, Xu YJ (2011) A facile and green approach to synthesize Pt@CeO_2 nanocomposite with tunable core-shell and yolk-shell structure and its application as a visible light photocatalyst. J Mater Chem 21:8152–8158

28. Shi W, Du D, Shen B, Cui C, Lu L, Wang L, Zhang J (2016) Synthesis of yolk−shell structured Fe_3O_4@void@CdS nanoparticles: a general and effective structure design for photo-fenton reaction. ACS Appl Mater Interfaces 8:20831–20838

29. Feng J, Wang Y, Hou Y, Li L (2017) Tunable design of yolk-shell $ZnFe_2O_4$@RGO@TiO_2 microspheres for enhanced high-frequency microwave absorption. Inorg Chem Front 4:935–945

30. Hwang SH, Yun J, Jang J (2014) Multi-shell porous TiO_2 hollow nanoparticles for enhanced light harvesting in dye-sensitized solar cells. Adv Funct Mater 24:7619–7626

31. Hu L, Luo B, Wu C, Hu P, Wang L, Zhang H (2019) Yolk-shell Si/C composites with multiple Si nanoparticles encapsulated into double carbon shells as lithium-ion battery anodes. J Energy Chem 32:124–130

32. Liu X, Qian G, Jiao Z, Wu M, Zhang H (2017) The transformation of hybrid silica nanoparticles from solid to hollow or yolk-shell nanostructures. Chem Eur J 23:8066–8072

33. Liu X, Hao C, He L, Yang C, Chen Y, Jiang C, Yu R (2018) Yolk–shell structured Co-C/Void/Co_9S_8 composites with a tunable cavity for ultrabroadband and efficient low-frequency microwave absorption. Nano Res 11:4169–4182

34. Li Y, Li L, Hu J, Yan L (2017) A spray pyrolysis synthesis of $MnFe_2O_4/SnO_2$ yolk/shell composites for magnetically recyclable photocatalyst. Mater Lett 199:135–138

35. Li Y, Shen Q, Guan R, Xue J, Liu X, Jia H, Xua B, Wu Y (2020) A $C@TiO_2$ yolk–shell heterostructure for synchronous photothermal–photocatalytic degradation of organic pollutants. J Mater Chem C 8:1025–1040

36. Rai P, Yoon JW, Kwak CH, Lee JH (2016) Role of Pd nanoparticles on gas sensing behaviour of $Pd@In_2O_3$ yolkshell nanoreactors. J Mater Chem A 4:264–269

37. Joo JB, Liu HY, Lee YJ, Dahl M, Yu HX, Zaera F, Yin YD (2016) Tailored synthesis of $C@TiO_2$ yolk–shell nanostructures for highly efficient photocatalysis. Catal Today 264:261–269

38. Zhao J, Li WJ, Fan LP, Quan Q, Wang JF, Xiao CF (2019) Yolk-porous shell nanospheres from silver-decorated titanium dioxide and silicon dioxide as an enhanced visible-light photocatalyst with guaranteed shielding for organic carrier. J Colloid Interface Sci 534:480–489

39. Joo JB, Dahl M, Li N, Zaera F, Yin YD (2013) Tailored synthesis of mesoporous TiO_2 hollow nanostructures for catalytic applications. Energy Environ Sci 6:2082–2092

40. Joo JB, Lee I, Dahl M, Moon GD, Zaera F, Yin YD (2013) Controllable synthesis of mesoporous TiO_2 hollow shells: toward an efficient photocatalyst. Adv Funct Mater 23:4246–4254

41. Wu XF, Wang YJ, Song LJ, Su JZ, Zhang JR, Jia YN, Shang JL, Nian XW, Zhang CY, Sun XG (2019) A yolk–shell $Bi@void@SnO_2$ photocatalyst with enhanced tetracycline degradation. J Mater Sci: Mater Electron 30:14987–14994

42. Khan I, Sun N, Wang Y, Li ZJ, Qu Y, Jing LQ (2020) Synthesis of SnO_2/yolk-shell $LaFeO_3$ nanocomposites as efficient visible-light photocatalysts for 2,4-dichlorophenol degradation. Mater Res Bull 127:110857

43. Chen Z, Gao YT, Mu DZ, Shi HF, Lou DW, Liu SY (2019) Recyclable magnetic $NiFe_2O_4/C$ yolk–shell nanospheres with excellent visible-light-Fenton degradation performance of tetra-cycline hydrochloride. Dalton Trans 48:3038–3044

44. Zheng NC, Ouyang T, Chen YB, Wang Z, Chen DY, Liu ZQ (2019) Ultrathin CdS shell-sensitized hollow S-doped CeO_2 spheres for efficient visible-light photocatalysis. Catal Sci Technol 9:1357–1364

45. Liu S, Zhao MY, He ZT, Zhong Y, Ding H, Chen DM (2019) Preparation of a p-n hetero-junction 2D BiOI nanosheet/1DBiPO$_4$ nanorod composite electrode for enhanced visible light photoelectrocatalysis. Chin J Catal 40:446–457

46. Liu ZF, Song QG, Zhou M, Guo ZG, Kang JH, Yan HY (2019) Synergistic enhancement of charge management and surface reaction kinetics by spatially separated cocatalysts and p-n heterojunctions in $Pt/CuWO_4/Co_3O_4$ photoanode. Chem Eng J 374:554–563

47. Gao YX, Huang Y, Li Y, Zhang Q, Cao JJ, Ho WK, Lee SC (2016) Plasmonic $Bi/ZnWO_4$ microspheres with improved photocatalytic activity on NO removal under visible light. ACS Sustain Chem Eng 4:6912–6920

48. Dong F, Li QY, Sun YJ, Ho WK (2014) Noble metal-like behavior of plasmonic Bi particles as a cocatalyst deposited on $(BiO)_2CO_3$ microspheres for efficient visible light photocatalysis. ACS Catal 4:4341–4350

49. Bo Y, Gao C, Xiong Y (2020) Recent advances in engineering active sites for photocatalytic CO_2 reduction. Nanoscale 12:12196–12209

50. Wang Z, Monny SA, Wang L (2020) Hollow structure for photocatalytic CO_2 reduction. ChemNanoMat 6:881–888

51. Liu H, Meng X, Dao TD, Liu L, Li P, Zhao G, Nagao T, Yang L, Ye J (2017) Light assisted CO_2 reduction with methane over SiO_2 encapsulated Ni nanocatalysts for boosted activity and stability. J Mater Chem A 5:10567–10573

52. Zhang BB, Wang YH, Xu SM, Chen K, Yang YG, Kong QH (2020) Tuning nanocavities of Au@Cu$_2$O yolk–shell nanoparticles for highly selective electroreduction of CO$_2$ to ethanol at low potential. RSC Adv 10:19192–19198

53. Liang M, Borjigin T, Zhang Y, Liu H, Liu B, Guo H (2018) Z-scheme Au@Void@g-C$_3$N$_4$/SnS yolk–shell heterostructures for superior photocatalytic CO$_2$ reduction under visible light. ACS Appl Mater Interfaces 10:34123–34131

54. Fujishima A, Honda K (1972) Electrochemical photolysis of water at a semiconductor electrode. Nature 238:37–38

55. Chiu YH, Naghadeh SB, Lindley SA, Lai TH, Kuo MY, Chang KD, Zhang JZ, Hsu YJ (2019) Yolk-shell nanostructures as an emerging photocatalyst paradigm for solar hydrogen generation. Nano Energy 62:289–298

56. Shi XW, Lou ZZ, Zhang P, Fujitsuka M, Majima T (2016) 3D-Array of Au–TiO$_2$ yolk–shell as plasmonic photocatalyst boosting multi-scattering with enhanced hydrogen evolution. ACS Appl Mater Interfaces 8:31738–31745

57. Zhao J, Tian SN, Shi HT, Quan Q, Xiao CF (2019) Encapsulated cadmium sulfide in silicon dioxide porous shells for enhanced photocatalytic sustainability and commendable protection of organic carriers. Adv Mater Interfaces 6:1801933

58. Yang TT, Chen WT, Hsu YJ, Wei KH, Lin TY, Lin TW (2010) Interfacial charge carrier dynamics in core-shell Au-CdS nanocrystals. J Phys Chem C 114:11414–11420

59. Naya SI, Kume T, Akashi R, Fujishima M, Tada H (2018) Red-light-driven water splitting by Au(Core)–CdS(Shell) half-cut nanoegg with heteroepitaxial junction. J Am Chem Soc 140:1251–1254

60. Cushing SK, Li JT, Meng F, Senty TR, Suri S, Zhi MJ, Li M, Bristow AD, Wu NQ (2012) Photocatalytic activity enhanced by plasmonic resonant energy transfer from metal to semiconductor. J Am Chem Soc 134:15033–15041

61. Zhou N, Vanesa LP, Wang Q, Polavarapu L, Isabel PS, Xu QH (2015) Plasmon-enhanced light harvesting: applications in enhanced photocatalysis, photodynamic therapy and photovoltaics. RSC Adv 5:29076–29097

62. Zhao Q, Ji MW, Qian HM, Dai BS, Weng L, Gui J, Zhang JT, Ouyang M, Zhu H. Controlling structural symmetry of a hybrid nanostructure and its effect on efficient photocatalytic hydrogen evolution. Adv Mater 26:1387–1392

63. Waqas M, Iqbal S, Bahadur A, Saeed A, Raheel M, Javed M (2017) Designing of a spatially separated hetero-junction pseudobrookite (Fe$_2$TiO$_5$-TiO$_2$) yolk-shell hollow spheres as efficient photocatalyst for water oxidation reaction. Appl Catal B 219:30–35

64. Li A, Chang XX, Huang ZQ, Li CC, Wei YJ, Zhang L, Wang T, Gong JL (2016) Thin heterojunctions and spatially separated cocatalysts to simultaneously reduce bulk and surface recombination in photocatalysts. Angew Chem Int Ed Engl 55:13734–13738

65. Zheng DD, Cao X-N, Wang XC (2016) Precise formation of a hollow carbon nitride structure with a Janus surface to promote water splitting by photoredox catalysis. Angew Chemie 128:11684–11688

66. Wang D, Hisatomi T, Takata T, Pan CS, Katayama M, Kubota J, Domen K (2013) Core/shell photocatalyst with spatially separated co-catalysts for efficient reduction and oxidation of water. Angew Chem Int Ed Engl 52:11252–11256

67. Cowan AJ, Tang JW, Leng WH, Durrant JR, Klug DR (2010) Water splitting by nanocrystalline TiO$_2$ in a complete photoelectrochemical cell exhibits efficiencies limited by charge recombination. J Phys Chem C 114:4208–4214

68. Sivula K, Formal FL, Grätzel M (2011) Solar water splitting: progress using hematite (α-Fe$_2$O$_3$) photoelectrodes. Chemsuschem 4:432–449

69. Ni M, Leung MKH, Leung DYC, Sumathy K (2007) A review and recent developments in photocatalytic water-splitting using TiO$_2$ for hydrogen production. Renew Sustain Energy Rev 11:401–425

70. Dotan H, Kfir O, Sharlin E, Blank O, Gross M, Dumchin I, Ankonina G, Rothschild A (2013) Resonant light trapping in ultrathin films for water splitting. Nat Mater 12:158–164

71. Petit S, Melissen STAG, Duclaux L, Sougrati MT, Bahers TL, Sautet P, Dambournet D, Borkiewicz O, Laberty-Robert C, Durupthy O. How should iron and titanium be combined in oxides to improve photoelectrochemical properties? J Phys Chem C 120:24521–24532

72. Guo XH, Liu JY, Guo GB (2017) Photocatalytic removal of dye and reaction mechanism analysis over Y_2O_3 composite nanomaterials. Matec Web Conf 88:02003

73. Deng JJ, Lv XX, Liu JY, Zhang H, Nie KQ, Hong CH, Wang J, Sun XH, Zhong J, Lee ST (2015) Thin-layer Fe_2TiO_5 on hematite for efficient solar water oxidation. ACS Nano 9:5348–5356

74. Bassi PS, Chiam SY, Gurudayal Barber J, Wong LH (2014) Hydrothermal grown nanoporous iron based titanate, Fe_2TiO_5 for light driven water splitting. ACS Appl Mater Interfaces 6:22490–22495

75. Liu QH, He JF, Yao T, Sun ZH, Cheng WR, He S, Xie Y, Peng YH, Cheng H, Sun YF, Jiang Y, Hu FC, Xie Z, Yan WS, Pan ZY, Wu ZY, Wei SQ (2014) Aligned Fe_2TiO_5-containing nanotube arrays with low onset potential for visible-light water oxidation. Nat Commun 5:5122

76. Yoon TP, Ischay MA, Du J (2010) Visible light photocatalysis as a greener approach to photochemical synthesis. Nat Chem 2:527–532

77. Zhao Y, Zhao B, Liu J, Chen G, Gao R, Yao S, Li M, Zhang Q, Gu L, Xie J, Wen X, Wu LZ, Tung CH, Ma D, Zhang T (2016) Oxide-modified nickel photocatalysts for the production of hydrocarbons in visible light. Angew Chem Int 55:4215–4219

78. Liu S, Yang MQ, Tang ZR, Xu YJ (2014) A nanotree-like CdS/ZnO nanocomposite with spatially branched hierarchical structure for photocatalytic fine-chemical synthesis. Nanoscale 6:7193–7198

79. Ziarati A, Badiei A, Luque R (2019) Engineered bi-functional hydrophilic/hydrophobic yolk@shell architectures: a rational strategy for non-time dependent ultra selective photocatalytic oxidation. Appl Catal B 240:72–78

80. Li A, Zhang P, Chang X, Cai W, Wang T, Gong J (2015) Gold nanorod@TiO_2 yolk-shell nanostructures for visible-light-driven photocatalytic oxidation of benzyl alcohol. Small 11:1892–1899

81. Movahed SK, Lehi NF, Dabiri M (2018) Palladium nanoparticles supported on core-shell and yolk-shell Fe_3O_4@nitrogen doped carbon cubes as a highly efficient, magnetically separable catalyst for the reduction of nitroarenes and the oxidation of alcohols. J Catal 364:69–79

82. Chen Z, Wang J, Zhai G, An W, Men Y (2017) Hierarchical yolk-shell WO_3 microspheres with highly enhanced photoactivity for selective alcohol oxidations. Appl Catal B 218:825–832

83. Ziarati A, Badiei A, Luque R, Ouyang W (2018) Designer hydrogenated wrinkled yolk@shell TiO_2 architectures towards advanced visible light photocatalysts for selective alcohol oxidation. J Mater Chem A 6:8962–8968

84. Yan Y, Han M, Konkin A, Koppe T, Wang D, Andreu T, Chen G, Vetter U, Morantee JR, Schaaf P (2014) Slightly hydrogenated TiO_2 with enhanced photocatalytic performance. J Mater Chem A 2:12708–12716

85. Wang Z, Yang C, Lin T, Yin H, Chen P, Wan D, Xu F, Huang F, Lin J, Xie X, Jiang M (2013) H-doped black titania with very high solar absorption and excellent photocatalysis enhanced by localized surface plasmon resonance. Adv Funct Mater 23:5444–5450

86. Yu X, Kim B, Kim YK (2013) Highly enhanced photoactivity of anatase TiO_2 nanocrystals by controlled hydrogenation-induced surface defects. ACS Catal 3:2479–2486

87. Miyaura N, Yamada K, Suzuki A (1979) A new stereospecific cross-coupling by the palladium-catalyzed reaction of 1-alkenylboranes with 1-alkenyl or 1-alkynyl halides. Tetrahedron Lett 20:3437–3440

88. Gao S, Shang N, Feng C, Wang C, Wang Z (2014) Graphene oxide–palladium modified Ag–AgBr: a visible-light-responsive photocatalyst for the Suzuki coupling reaction. RSC Adv 4:39242–39247

89. Hosseini-Sarvari M, Bazyar Z (2018) Visible light driven photocatalytic cross-coupling reactions on nano Pd/ZnO photocatalyst at room-temperature. ChemistrySelect 3:1898–1907

90. Wang N, Ma L, Wang J, Zhang Y, Jiang R (2019) Graphitic carbon nitride (g-C_3N_4) supported palladium species: an efficient heterogeneous photocatalyst surpassing homogeneous thermal heating systems for Suzuki coupling. ChemPlusChem 84:1164–1168

91. Sharma K, Kumar M, Bhalla V (2015) Aggregates of the pentacenequinone derivative as reactors for the preparation of Ag@Cu_2O core–shell NPs: an active photocatalyst for Suzuki and Suzuki type coupling reactions. Chem Commun 51:12529–12532
92. Rohani S, Ziarati A, Ziarani GM, Badiei A, Burgi T (2019) Engineering of highly active Au/Pd supported on hydrogenated urchin-like yolk@shell TiO_2 for visible light photocatalytic Suzuki coupling. Catal Sc. Technol 9:3820–3827
93. Farina V, Krishnan B (1991) Large rate accelerations in the stille reaction with tri-2-furylphosphine and triphenylarsine as palladium ligands: mechanistic and synthetic implications. J Am Chem Soc 113:9585–9595

Chapter 31
Yolk–Shell Materials for Photo and Electrocatalysis

Yulin Min

31.1 Introduction

31.1.1 Introduction to the Core–Shell Structure

In recent years, with the development of nanomaterial synthesis technology, the design of specific morphological structures through physical or chemical means is crucial for conversion into particular functions. At the same time, multi-component nanocatalysts not only have a more complex structure and composition than single-component materials in the field of catalysis but also have new characteristics that single-component materials lack. As the simplest model in the two-component system, the core–shell structures with ordered assembly have become one of the most promising materials in the field of catalysis.

The core–shell nanoparticles are a type of materials that encapsulate the inner "guest" particles of one component in one or more outer shells of another element through a specific interaction force, usually expressed as core@shell. The synergy between multiple ingredients can make them exert the effect of "1 + 1 > 2." When the core is used as the active center, the unique geometry can improve the reactivity, thermal stability, or oxidation stability of the core. The cheaper cores can also support more expensive thin shells to reduce the synthesis cost of composite materials. The broad definition of a core–shell structure also includes elements such as hollow spheres and microcapsules. In addition to the field of catalysis, core–shell structural materials that are easy to design are also commonly used in biology, sensors, and other areas.

Y. Min (✉)
Shanghai Key Laboratory of Materials Protection and Advanced Materials in Electric Power, Shanghai University of Electric Power, Shanghai 200090, People's Republic of China
e-mail: minyulin@shiep.edu.cn

© The Author(s), under exclusive license to Springer Nature Singapore Pte Ltd. 2021
H. Yamashita and H. Li (eds.), *Core-Shell and Yolk-Shell Nanocatalysts*,
Nanostructure Science and Technology,
https://doi.org/10.1007/978-981-16-0463-8_31

31.1.1.1 Classification of the Core–Shell Structure

The core–shell nanomaterials are classified from the perspectives of composition and morphology.

Composition

According to the composition of core–shell nanomaterials, it can be roughly divided into inorganic–inorganic materials, organic–organic materials, and inorganic–organic materials [1].

(1) Inorganic–inorganic core–shell structured materials: They are the most widely studied materials, usually with nanoparticles such as metals, metal oxides, metal sulfides as cores, and metals and metal oxides as shells. Guo et al. [2] combined the template method and two-step hydrothermal method to prepare the hollow cubic Co_3S_4@MoS_2 heterostructure. Due to the strong interfacial coupling between the two phases, it can be assembled into a two-electrode setup in an alkaline environment, which is an excellent water splitting electrocatalyst.

(2) Organic–organic core–shell structure materials: They are mainly polymers composed of three-dimensional network structured polymer materials or other carbon-containing materials. Wang et al. [3] treated two MOF materials through hydrothermal and carbonization: ZIF-8 and ZIF-67. Nitrogen-doped carbon (NC, carbonized from ZIF-8) and high graphitic carbon (GC, carbonized from ZIF-67) (NC@GC) were obtained. The core–shell catalysts combined the advantages of the two materials. It can exhibit high activity in both electrocatalytic oxygen reduction and oxygen evolution reactions due to high conductivity and stability.

(3) Inorganic–organic core–shell structured materials: It usually uses metal or metal oxide as the core and covers the surface by the polymer shell stably by electrostatic or chemical action. Li et al. [4] synthesized the core–shell structured upconversion nanoparticles (UCNPs)-Pt@MOF using the layer-by-layer growth method. Then they prepared the (UCNPs)-Pt@MOF/Au photocatalyst by microwave synthesis method loading plasma gold nanoparticles. Through a series of light absorption and emission cycles of MOF, gold nanoparticles, and UCNPs-Pt, the broadband spectral response of the catalyst from the ultraviolet to the near-infrared (NIR) region was realized, and the excellent photocatalytic hydrogen production activity was achieved.

Morphology

Depending on the shape of the core–shell structured nanomaterials, it can be divided into core–shell structure, and yolk–shell structure, hollow structure, and sandwich structure (Fig. 31.1).

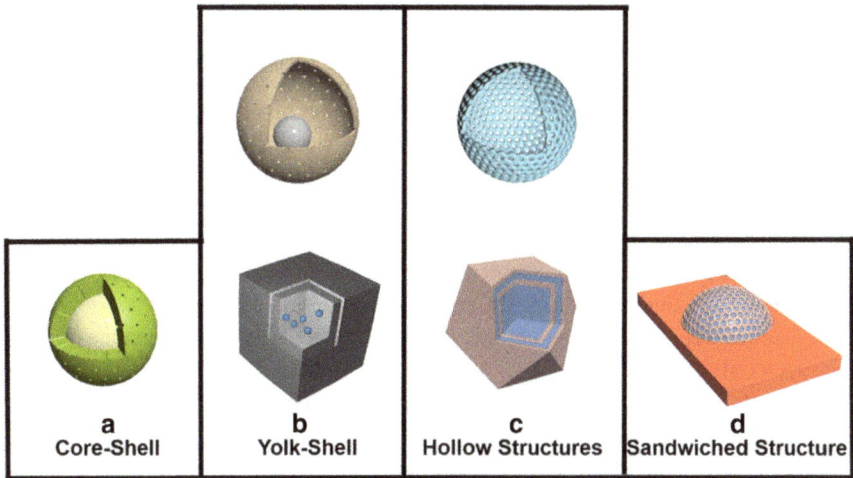

Core-Shell Structures

Fig. 31.1 Classification of core–shell structures according to morphology. **a** Core–shell structure: a structure with continuity between core and shell; **b** Yolk–shell structure: similar to the structure of egg yolk and eggshell, expressed as the core@void@shell. The sacrificial layer is often removed by selective etching to create a void, which will be discussed in detail in the second section; **c** Hollow structure: a special case of yolk–shell structures with the same formation mechanism, which can be synthesized by template induction method, self-assembly method, microemulsion method, etc.; **d** Sandwich structure: the active component core is embedded between the same or different shell materials. The core and the two carriers cooperate to catalyze. The shell is often deposited on the core by the sol–gel method, hydrothermal method, etc

31.1.1.2   Mechanism of the Core–Shell Synergy

It is very important to have stable and highly dispersed cores for the core–shell structure catalysts. The key point is to minimize interfacial tension [5]. After ligand exchange on the surface of the inner core, a wetting layer is formed with the shell. Therefore, the core nanoparticles are often modified with suitable ligands or shells such as surfactants, amphiphilic block copolymers, etc. with specific linking groups.

31.1.1.3   Advantages of the Core–Shell Structure in the Catalytic Process

(1)   The most noticeable feature of the core–shell structure is the protective effect of the outer shell. Due to the high surface energy of nanoparticles, the particles tend to be in a small specific surface state with low energy under the action of surface tension. They are accessible to agglomerate/sinter to reduce the specific surface area and lose active sites. In this case, the outer shell of the core–shell structure acts as a physical barrier to improve the stability of the composite

material. For shells with low thermal stability, it is also an excellent method to coat mesoporous silica to enhance sintering resistance.

(2) Integrated multi-component functions. In materials with a clear structure, the determination of the position of different components will greatly facilitate the selectivity of the catalytic reaction. In the integrated catalyst system, other specific functional components, such as photosensitizers, can also be introduced to enhance the synergy between the components.

(3) For catalytic sites with multiple reactive activities, bifunctional catalysis can sometimes be achieved through the optimization of materials. By adjusting the size of components, the time difference of bifunctional catalysis can be reduced as much as possible to avoid undesirable side effects reaction, enhance product selectivity.

(4) Screen the reactants by adjusting the size of the shell holes to reduce side reactions.

(5) The formation of core–shell heterostructures may be accompanied by lattice strain, which changes the electronic properties of the material. Strain and ligand effects change the center of the metal d-band, thereby changing the adsorption capacity of the reactants and intermediates on the metal and other activity indicators of the catalysts.

(6) The spherical core–shell structure can provide a maximized active interface in bifunctional reactions or other reactions with inconsistent active sites. Due to its stability, the structure can also be maintained well.

Re-optimizing the porosity and combining the advantages of (1), (3), (4), and (6) can significantly improve the catalytic reaction activity, selectivity, and stability.

31.1.2 Introduction of the Yolk–Shell Structure

The yolk–shell structure is one of the subsets of the core–shell structure divided by morphology. It is based on the synthesis of the core–shell structure and then introduces voids through selective etching and other methods. Its structure is presented as core@void@shell, which is composed of a hollow shell and a movable core. The hollow shell is often used as a space-restricted nanoreactor. Under the effect of space-limiting capacity and adjustable porosity, the integrated yolk–shell structure nanoreactor has excellent catalytic performance and the utilization rate.

At the end of 2002, Kim et al. [6] used the mesoporous silica template to prepare hollow spherical polymers and carbon capsules (Fig. 31.2a). For the first time, they encapsulated gold nanoparticles and named the structure of this core encapsulated in the cavity as "Nanorattles." In the second year, Kamata et al. [7] synthesized a spherical hollow colloid of poly(benzyl methacrylate) (PBzMA) containing a movable gold core and defined as "core–shell spherical colloids with movable cores" (Fig. 31.2b).

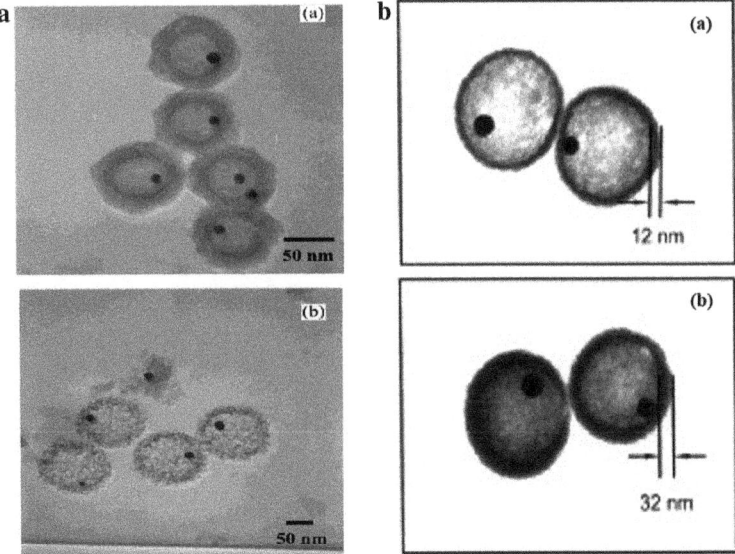

Fig. 31.2 TEM imagines of **a** Au@HCMs: **a** polymer, **b** carbon capsules. Reproduced with permission [6]. Copyright © 2002 American Chemical Society. **b** Au@PBzMA with different aggregation times: 3 h (**a**), 6 h (**b**). Reproduced with permission [7]. Copyright © 2003 American Chemical Society

Until 2004, Yin et al. [8] synthesized Pt@CoO yolk–shell structure through a mechanism analogous to the Kirkendall effect. This is the first time to use the "yolk–shell " concept to describe this structure.

31.1.2.1 Classification of the Yolk–Shell Structure

According to the number of cores and shells, the structure of yolk–shell can be classified as six forms (Fig. 31.3) [9–14]: (1) single-core@single-shell; (2) multiple-cores@single-shell; (3) void@single-shell; (4) single-core@multiple-shells; (5) multiple-cores@multiple-shells; and (6) void@multiple-shells. Others are simply divided into "spherical" and "non-spherical" based on whether the core–shell material has a spherical structure.

The structure determines performance. When the yolk–shell structure is used for photo/photoelectric catalysis, we summarize the following rules:

(1) The void space provides a wide space for light reflection and scattering in the shell, so the larger the void, the more photon energy can be used.

(2) The thinner the shell thickness, the shorter the charge diffusion distance, and the easier the charge migration.

(3) The presence of both internal and external surfaces leads to a relatively large specific surface area, which enhances the surface reaction.

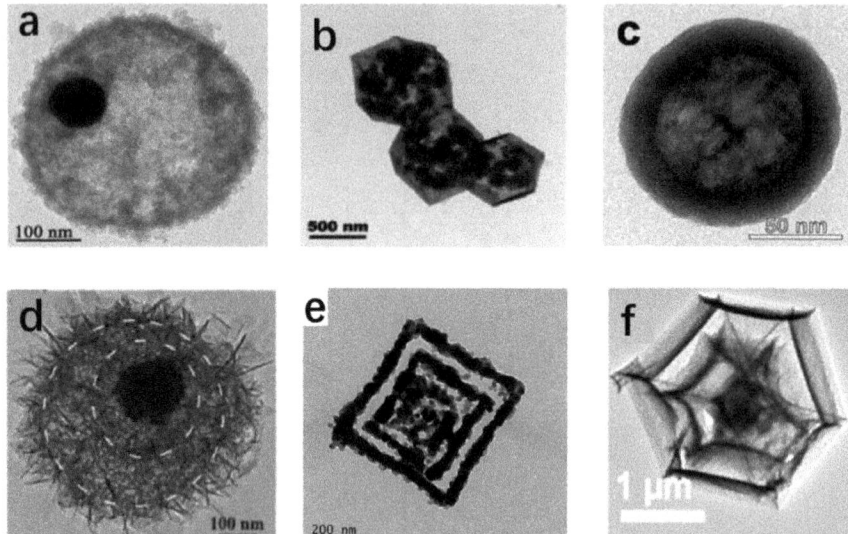

Fig. 31.3 Classify the structure of yolk–shell according to the number of cores and shells: **a** single-core@single-shell. Reproduced with permission [9]. Copyright © 2020 Elsevier Ltd; **b** multiple-cores@single-shell. Reproduced with permission [10]. Copyright © 2020 Elsevier B.V; **c** void@single-shell. Reproduced with permission [11]. Copyright © 2019 American Chemical Society; **d** single-core@multiple-shells. Reproduced with permission [12]. Copyright © 2011 American Chemical Society; **e** multiple-cores@multiple-shells. Reproduced with permission [13]. Copyright © 2014, American Chemical Society; **f** void@multiple-shells. Reproduced with permission [14]. Copyright © 2019, American Chemical Society

(4) The hollow structure is the basis of all yolk–shell structures. The large specific surface area improves the contact between the catalyst and the reactants, which is beneficial to the surface reaction. The cavity extends the light reflection path and enhances light absorption.

(5) Multi-cores highly dispersed in the shell will increase the active site.

(6) Each void can reflect and scatter light for multiple shells to enhance light capture and increase the generation of photogenerated charges, thereby photo/photoelectric catalytic activity will improve.

(7) For the multi-core@multi-shell structure, the active center is enriched, and the light scattering is enhanced. The generation of photogenerated charges and the surface reaction work synergistically to improve the catalytic activity.

31.1.2.2 Advantages of the Yolk–Shell Structure

(1) The presence of cavities leads to a low density of materials.

(2) The existence of both the inner and outer surfaces of the shell leads to a high specific surface area.

(3) The void space helps the movement of fluid reactants.

(4)   It can effectively isolate the catalytically active substances to improve the stability of the catalysts.
(5)   The selectivity of the catalysts can be improved by precisely controlling the pore size on the shell.

31.1.3   The Similarities and Differences Between Core–Shell and Yolk–Shell Structures

31.1.3.1   Differences

The yolk–shell structure is a kind of core–shell structure. The most essential difference is that the yolk–shell structure has a cavity for the core to move. However, how did this cavity come about?

(1)   The synthesis method of the two materials is different. The yolk–shell structure can be obtained by removing the sacrificial layer on the core–shell structure. This is the origin of the cavity.
(2)   Therefore, the yolk–shell structure has the advantages of both the core–shell structure and the hollow structure compared with the core–shell structure. It has a low density and high specific surface area. This structure can be used as a "nanoreactor" to allow reactants to shuttle more freely and improve the mass transfer rate and adsorption capacity. The reaction particles also increase the probability of collision to improve the utilization of active sites. Therefore, the catalytic activity will also increase. At the same time, light reflection and diffraction can be enhanced in photo/photocatalysis to improve light utilization.
(3)   In addition to adjusting the thickness of the shell, the yolk–shell structure can also achieve a better catalytic effect by adjusting the size of the cavity. In 2019, Chiu et al. [15] used $Au–Cu_2O$ core–shell nanostructures as growth templates to form $Au–Cu_7S_4$ yolk–shell nanostructures by sulfidation. $Au–Cu_7S_4$ was then converted to Au–CdS by cation exchange. By changing the reagent volume, the void size is changed. The study found that Au–CdS nanostructures with larger voids have higher hydrogen decomposition and hydrogen production performance.
(4)   In summary, the field of application of the yolk–shell structure is also different from the core–shell structure. In addition to the field of catalysis, it also provides a buffering space for the electroactive core material during lithium insertion and extraction [16]. In biomedicine, the void in the yolk–shell structure is also conducive to targeted drug/gene delivery, controlled release, bioimaging, diagnosis, therapeutic agents, biosensors, antibacterial activity, etc.

31.1.3.2 Similarities

(1) The core and shell have the same function: usually, the core is used as the active site when the shell is used as the protective layer to improve the stability of the catalysis.
(2) The means of characterizing the morphology are the same (mainly through transmission electron microscopy).

31.2 Preparation

31.2.1 Template Method

The template method is generally used to prepare core–shell materials with a hollow structure. The process usually refers to a three-step synthesis that the template materials are deposited into the pores or surface of the solid substances through physical or chemical methods. The first step is the preparation of template nanomaterials with controlled size and morphology, while the second step involves the incorporation of the templates into hollow structures [17]. Thirdly, the template materials are removed to yield void spaces. There are two main ways to remove the template body of hollow materials:

(1) Dissolution method: It is suitable for removing inorganic substances and polymer cores, which can maintain the original shape of the particles.
(2) Calcination method: It is suitable for removing the polymer core, but it will cause local structure collapse during calcination.

The size, morphology, and structure of the nanomaterials can be controlled easily based on the spatial confinement of the template [18]. According to the attribute diversities, the templates can be generally divided into soft template, hard template, and self-template [19, 20].

31.2.1.1 Soft Template

Soft templating is a flexible, simple, and effective method, especially for the fabrication of yolk–shell nanomaterials. Soft templates are surfactant and polymer with a hydrophilic head and a hydrophobic chain [21], including surfactants (CTAB [22, 23], PVP [24–26], MPN [27], SDS [28, 29], etc.), microemulsions [30–32], and vesicles [33, 34]. These materials can provide cavities in dynamic equilibrium, and substances can diffuse in and out through the walls of the cavities. When the nuclear precursors are uniformly dispersed in the soft-template mixture, the template agents react with the precursors in the nanoscale interlayer gap or micropores under the action of non-covalent bonds. Based on the spatial confinement, the precursors can be assembled regularly to form an encapsulation structure. The hydrolysis and condensation of the core–shell nanoparticles then remove the hydrophobic chains and form voids between the core and shell. Finally, the porous core–shell structure is obtained by calcination.

For example, Lee et al. [35] synthesized the Sn@C of yolk–shell structure using CTAB as the soft template. The precursors of Sn and C were adsorbed on the lipophilic parts inside the micelle by CTAB micelles, and then the templates were removed by calcination at 700 °C. Liu and Qiao [36] described a kind of fluorocarbon surfactant to prepare a series of yolk–shell nanoparticles with different cores. Fluorocarbon surfactants FC4 and F127 are adsorbed on the core through electrostatic interaction to form a vesicle complex. Then ethyl orthosilicate was added and deposited on the surface of the nuclear vesicle complex by vesicle template method. Besides, the orthosilicate was polymerized by sol–gel reaction to form a silicate layer. The yolk–shell structure was obtained through the shrinkage of the silicate shell and the maturation of TEOS in the condensation and hydrolysis process. Finally, calcining removes surfactants and forms a mesoporous SiO_2 shell. Through this method, a variety of nanoparticles, such as magnetic particles, silicon spheres, Au NPs, mesoporous SiO_2 NPs, can be encapsulated into the SiO_2 shell.

31.2.1.2 Hard Template

The hard template mainly refers to a substance with a relatively rigid structure and a rigid shape maintained by covalent bonds, such as carbon [37–39], silica [40–42], metals [43, 44], metal oxides [45, 46], and oligomers [47] (Fig. 31.4) [48]. This substance provides static pores in which the core precursors are encapsulated inside, and then the required shell materials are coated on the external surface of the hard-template agents to form core–shell nanoparticles with sandwich structure. Finally, the etchant is introduced to remove the hard template and create voids between the core and shell.

In recent years, Fe-based materials have been widely used in photoelectrocatalysis due to their rich content and suitable binding energy. DeKrafft et al. [49] reported a metal–organic framework (MOF)-templated strategy for the synthesis of a mixed metal oxide nanocomposite with useful photophysical properties. Fe-containing nanoscale MOFs were coated with amorphous titania, and then calcined to produce crystalline Fe_2O_3@TiO_2 core–shell nanoparticles. He et al. [50] successfully synthesized Ce@Fe hollow nanotubes with core–shell structure by combining electrochemical deposition method and thermal polymerization strategy (Fig. 31.5a). Using ZnO nanowire arrays (NRAs) with smooth surfaces as templates, CeO_2 was evenly wrapped on ZnO. After electrodeposition, ZnO was etched to form rough nanotube arrays (NTAs). By controlling the deposition time, different Ce@Fe samples with varied thickness of α-Fe_2O_3 shell were prepared. Besides, with Fe_3O_4 as the core, Du et al. [51] demonstrated Fe_3O_4@TiO_2 yolk–shell structure generation process (Fig. 31.5b). First, the pre-synthesized magnetic Fe_3O_4 NPs were coated with a nonporous silica layer through a sol–gel approach in the presence of tetraethyl

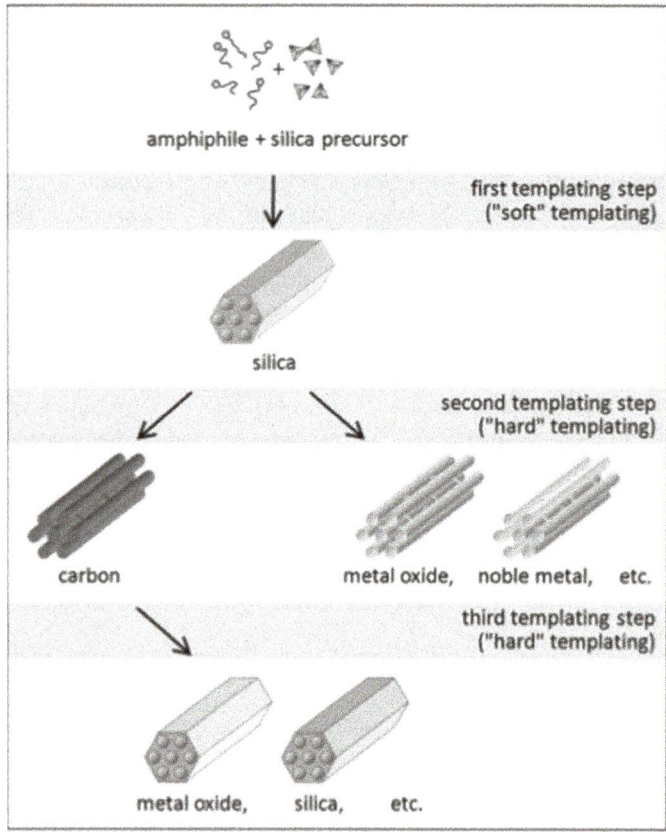

Fig. 31.4 Schematic drawing of hard templates for the synthesis of ordered mesoporous materials. Reproduced with permission [48]. Copyright © 2008 American Chemical Society

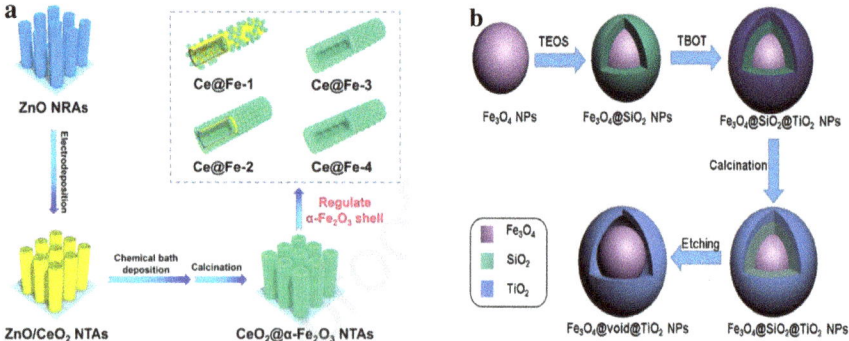

Fig. 31.5 Schematic illustration by the hard-template method of **a** fabrication process of the Ce@Fe NTAs heterojunction. Reproduced with permission [50]. Copyright © 2020 Elsevier B.V. **b** The formation procedure of the yolk–shell structured Fe_3O_4@void@TiO_2 NPs. Reproduced with permission [51]. Copyright © 2016 Elsevier B.V

orthosilicate (TEOS). Then, a further sol–gel coating process was utilized to deposit a porous TiO_2 shell onto the nonporous silica layer using tetrabutyl titanate (TBOT) as the precursor, followed by a calcination at 450 °C in N_2 atmosphere. Finally, an ultrasound-assisted etching method was used to etch off the nonporous silica layer in a weak alkaline media to the resultant Fe_3O_4@void@TiO_2.

Meanwhile, because of the advantages of stability, non-toxicity, and low price, TiO_2 has been extensively studied in photocatalysis. But its wide bandgap limits the effective use of light. Therefore, the strategy of preparing composites by coating TiO_2 with other materials has attracted great attention. Li et al. [52] prepared Pt@TiO_2 hollow spheres (Pt-HSs) using SiO_2 as a template. The SiO_2 nanospheres prepared by the modified Stöber method were used as a template to adsorb H_2PtCl_6 on the surface, and then they were calcined at 500 °C for 2 h under H_2 atmosphere to form Pt particles anchored on SiO_2. Subsequently, the coating of the amorphous TiO_2 shell was carried out by hydrolysis of titanium tert-butoxide (TBOT). The crystallinity of TiO_2 was improved by calcination. It is worth noting that direct calcination will destroy the structure of the TiO_2 shell. Therefore, in order to maintain the morphology, the outermost protective layer was formed by coating another layer of SiO_2. After calcination, the inner and outer layers of SiO_2 were etched in NaOH solution at 70 °C for 8 h to form Pt @ TiO_2 hollow spheres (PT-HSs). After that, Li et al. [53] continued to use Pt nanoparticles and MnO_X to selectively modify the inner and outer surfaces of the TiO_2–In_2O_3 double-layer shell to synthesize Pt@TiO_2@In_2O_3@MnO_X mesoporous hollow spheres (PTIM-MSs) as photocatalyst.

31.2.1.3   Self-template

Both the soft-template method and the hard-template method must remove the template to form a void, which inevitably causes material loss and waste. Meanwhile, the choice of template agents, affinity with the precursors, solution concentration, surfactant types, etc. have a significant influence on the coating effect. In recent years, the introduction of self-template to construct hollow nanostructures has attracted extensive attention. Professor David Lou, Yadong Yin et al. [20, 54, 55] have reviewed the synthesis strategy and application of the self-template method. Compared with the traditional soft and hard templating method, the template materials in the self-templating approach not only serve as supporting frames as traditional templates, but also can directly participate in the formation process of the hollow shell layer: the template materials are directly converted into a shell layer or as precursors to the shell. Therefore, the self-templating method has the characteristics of fewer reaction steps and no need for additional templates, and has significant advantages in the design and optimization of hollow nanostructures.

The key to the self-templating method is the formation of the internal spatial structure of the composite material. According to the forming mechanism of the internal hollow structure, the self-templating method can be roughly divided into four synthetic mechanisms, namely, selective etching method, outward diffusion method, heterogeneous shrinkage method, and electrochemical replacement method.

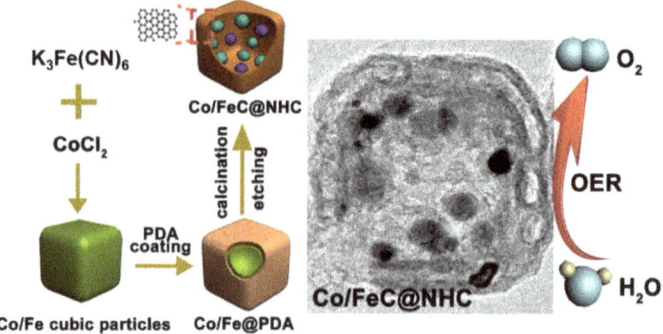

Fig. 31.6 Schematic illustration by selective etching method of Co/FeC core–nitrogen-doped hollow carbon shell structure for highly efficient oxygen evolution reaction. Reproduced with permission [56]. Copyright © 2020 Elsevier Inc

Selective Etching Method

The formation of the hollow structure comes from the selective etching of the internal region of the templates. In general, selective etching requires the use of capping agents, such as PVP, to maintain nanoparticles' stability, so that the hollow structure obtained based on this mechanism will maintain the original template dimensions and crystal crystallinities. Zhang et al. [56] reported a multi-core–shell structure Co/FeC@N-doped hollow carbon (Co/FeC@NHC) with tunable carbon shell thickness (Fig. 31.6). A distinctive core–shell structure consisting of a bimetal-based MOF (Co/Fe) core and polydopamine shell was crafted, followed by high-temperature pyrolysis and subsequent hydrochloric acid etching process to synthesize multiple Co/FeC nanoparticles encapsulated in N-doped hollow carbon. Polydopamine (PDA) is used as a morphology regulator playing a critical role in formation of core–shell structure, as well as a carbon source and nitrogen source decomposing at high temperature and in situ forming the carbon matrix. Notably, multi-core–shell structure composites have adjustable shell diameters ranging from 4.8 to 47.8 nm through simply adjusting the amounts of PDA.

Outward Diffusion Method

The reaction mechanism of the outward diffusion method can be divided into three types: the Kirkendall effect on the nanometer scale, Ostwald ripening, and ion-exchange reaction.

(1) Kirkendall effect: When two metals with different diffusion rates diffuse to each other, the vacancy defect will be formed on the side of the metal with a faster diffusion rate, and then holes will be generated. He et al. [57] demonstrated a nanoscale Kirkendall cavitation process that can transform solid palladium nanocrystals into hollow palladium nanocrystals through the insertion and

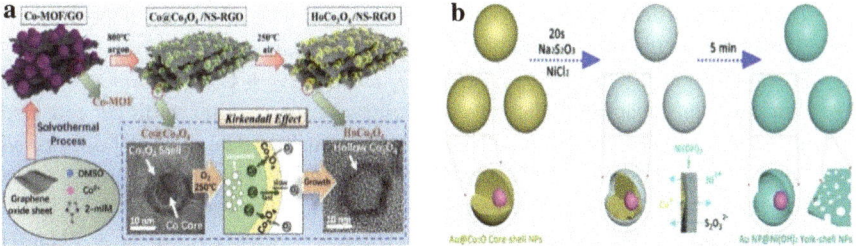

Fig. 31.7 Schematic illustration by outward diffusion method of **a** the synthetic route of Co$_3$O$_4$/NS-RGO framework via nanoscale Kirkendall effect. Reproduced with permission [58]. Copyright © 2020, American Chemical Society. **b** Au NPs@ Ni(OH)$_2$ yolk–shell NPs via ion-exchange reaction. Reproduced with permission [9]. Copyright © 2020 Elsevier Ltd

extraction of phosphorus. Solid metal nanocrystals of Pd, which became solid nanocrystals of PdP$_2$ after reacting with species P due to the faster diffusion of P than Pd, were eventually converted to hollow metal nanocrystals with an increased outer diameter through an artful application of the Kirkendall effect by extracting the species P. The key to success in producing monometallic hollow nanocrystals was the effective extraction of P through an oxidation reaction, which promoted the outward diffusion of P from the compound nanocrystals of PdP$_2$ and consequently the inward diffusion of vacancies and their coalescence into larger voids. Zhu et al. [58] utilized a two-step calcination strategy to fabricate hollow Co$_3$O$_4$ nanoparticles embedded in a N,S-co-doped reduced graphene oxide framework (Fig. 31.7a). In the first step, core–shell-like Co@Co$_3$O$_4$ embedded in N,S-co-doped reduced graphene oxide was synthesized by pyrolysis of Co-based metal–organic framework/graphene oxide precursors in an inert atmosphere at 800 °C. The designed hollow Co$_3$O$_4$ nanoparticles with an average particle size of 25 nm and a wall thickness of about 4–5 nm are formed by a further calcination process in air at 250 °C via the nanoscale Kirkendall effect.

(2) Ostwald ripening: The process is generally divided into two steps: the dissolution of small crystals and the redeposition of dissolved substances on the surface of large particles to induce secondary crystal growth. Nagaraju et al. [59] composed a yolk–shell structured Mn$_3$O$_4$ nanosphere (Mn$_3$O$_4$ NSs). The formation process of yolk–shell structured Mn$_3$O$_4$ NSs included nucleation, aggregation, and Ostwald ripening process from inside to outside. During the high-temperature solvothermal process, the initially formed cores with precursors became nanoparticles, which were further aggregated into spherical shapes to reduce surface energy. Subsequently, under continuous autoclave treatment, the spherical materials underwent the Ostwald ripening process from the inside to the outside, dissolved, and then deposited to form yolk–shell-shaped nanospheres. It is worth noting that the addition of polar protic solvent H$_2$O will cause changes in the properties (polarity, viscosity, and solubility) of alcohol solvents, which may change the shape and morphology of metal

oxide nanostructures. Therefore, H_2O promoted the continuous deposition of $Mn(OH)_2$ nucleation and reunion to form porous nanospheres ($Mn(CH_3COO)_2$ + $2H_2O \rightarrow Mn(OH)_2 + 2CH_3COOH$), $2Mn(OH)_2 + O_2 \rightarrow 2MnO(OH)_2\downarrow$).

(3) Ion exchange reaction: The outward migration of internal particles can also be achieved by ion exchange on the template surface. Cai et al. [9] developed a general method for preparing Au NPs@Ni(OH)$_2$ yolk–shell nanoparticles based on the etching of Cu_2O nanoparticles and the formation of $Ni(OH)_2$ shell ($Cu_2O + Na_2S_2O_3 + H_2O \rightarrow Cu_2S_2O_3 + 2NaOH$; $NiCl_2 + 2NaOH \rightarrow Ni(OH)_2\downarrow + 2NaCl$) (Fig. 31.7b). The yolk–shell structure was the result of the interdiffusion of multiple ions, including Cu^{2+} outward diffusion and Ni^{2+} inward diffusion through vacancy exchange, forming a unique hollow inner shell and porous shell. Huang et al. [60] synthesized the electrocatalyst $CoS_2/NC@Co\text{-}WS_2$ with a yolk–shell structure by the ion-exchange method. ZIF-67 was the self-template and Co sources. Co^{2+} diffused out of ZIF-67 and formed a Co-WS$_X$ layer with WS_4^{2-} diffused from $(NH_4)_2WS_4$. As the process is going, Co^{2+} was further released from the core, and Co-WS$_X$ simultaneously grew on the shell, resulting in the formation of yolk–shell. At this stage, due to the combination of Co^{2+} and WS_4^{2-}, the shell became concave and rough, which could provide more exposed active sites and shorten the charge transfer paths.

Heterogeneous Shrinkage Method

In the heterogeneous shrinkage mechanism, the formation of the hollow structure comes from the huge mass or volume difference between the template and the product during the heat treatment process, which leads to the volume shrinkage of the template from outside to inside. Many thermally degradable substances can be used as template precursors in this process, such as metal carbonates/alkoxides and metal–organic framework structures. For example, in the process of MOF derivation of the hollow structure, the MOF precursors can be gradually decomposed from the outside to the inside caused by the temperature gradient. The coordination bonds will be cracked, the linkers are carbonized, and the dense shell is separated from the in situ generated core to form void spaces (Fig. 31.8a) [61].

Electrochemical Replacement Method

It is based on the electrochemical oxidation–reduction reaction between the templates and the solution metal ions, which guides the metal ions in the solution to nucleate and grow on the surfaces of the templates, while causing the internal template etched to form a hollow structure. Specifically, there is an electrode potential difference between the two metals, using one metal as the reducing agent (anode) and the other as the oxidizing agent (cathode). After the anode template materials are synthesized, the cathode metal salts will be added, and under the action of the electrode potential

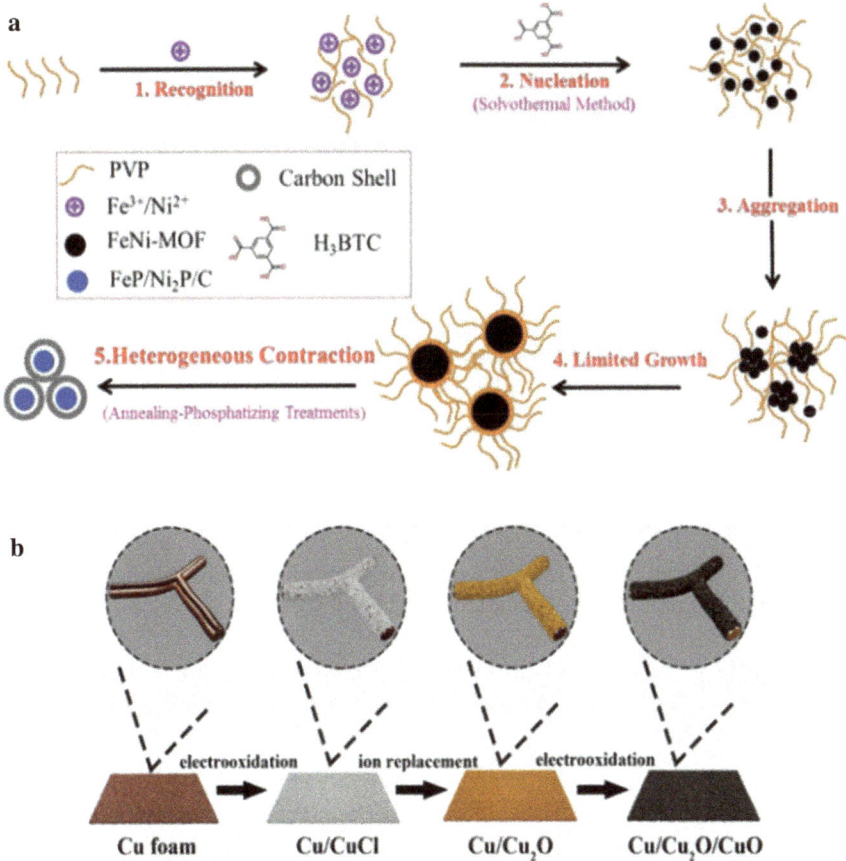

Fig. 31.8 Schematic illustration of **a** the process of heterogeneous shrinkage by MOF derivation. Reproduced with permission [61]. Copyright © 2020 Elsevier B.V. **b** the synthesis of three dimensional (3D) Cu/Cu$_2$O/CuO hybrid foams. Reproduced with permission [61]. Copyright © 2020 Elsevier B.V

difference, they will spontaneously undergo a redox reaction. Li et al. [62] reported a facile halides-assisted electrooxidation method for the synthesis of three-dimensional (3D) Cu/Cu$_2$O/CuO hybrid foams as free-standing OER electrodes (Fig. 31.8b). With the assistance of halide ions (Cl, Br, I), nanostructured cuprous phase is facilely constructed via a potentiostatic electrooxidation process. Then, active Cu$_2$O/CuO shell with nanoarray architectures is achieved after ion replacement and a further electrooxidation process. Ghasemian et al. [63] introduced a "liquid metal" reaction process. In aqueous solution, MnO^{4-} on the metal surface of the liquid gallium–indium alloy (EGaIn) underwent an electrochemical displacement reaction with EGaIn, leading MnO^{4-} to be reduced to an oxide layer at the interface. Thus, a single-layer hydrated MnO$_2$ nanosheet shell was obtained, named EGaIn/Mn core–shell structure. Hou et al. [64] reported for the first time a self-template strategy, through pyrolysis of core@shell MOFs, to construct a nanoflower-like carbon cage. The key was attributed to the induction of Fe^{3+}. Generally, carbonized MOFs would produce nanocarbon particles with structural collapse or isolation. However, with Fe^{3+} doped into the Zn@Co-MOF precursors, the Fe–Co nanoalloy produced by Fe^{3+} induced the continuous growth of carbon tubes on the surface of the carbon cage. The carbon tubes of adjacent carbon cages were connected to each other, making the carbon cages assemble spontaneously into a complete hydrangea flower-shaped nanosuperstructure.

31.2.2 Seed Deposition Method

The seed deposition method uses nanoparticles as seeds. By adding metal ion solution and reducing agent, and controlling the deposition process through chemical or physical actions, the shell precursors grow on the surfaces of seeds.

Generally, the seed deposition method consists of two steps:

(1) Nuclear formation: By reducing the precursors of the core, core nanoparticles with different morphologies are obtained and used as seeds for the next shell coating.

(2) Epitaxial growth of the shell: By adding a reducing agent, blocking agent, and inorganic ions to the precursor of the shell, the shell is evenly coated on the surface of the core.

Therefore, the main factors affecting the size and morphology of core–shell nanoparticles are the type and concentration of the precursors, the type of blocking agent, the reducing agent, etc.

Zhou et al. [65] synthesized the electrocatalyst Fe_3O_4@CoO with a core–shell structure by seed-mediated method. The Fe_3O_4 nanoparticles were first synthesized by the thermal decomposition of $Fe(acac)_3$ in a mixed solution of hydrophobic monomer octadecyl acrylamide (OAm) and oleic acid (OA). The obtained Fe_3O_4 nanoparticles were then employed as the seeds for the growth of the CoO shell by the thermal decomposition of $Co(acac)_2$. It is notable that the high-temperature organic colloidal environment can regulate the nucleation size and shell thickness. Aitbekova et al. [66] used the seed-mediated colloid method to synthesize the heterodimer Ru@Fe_2O_3. The transformation of heterodimers into core–shell structures resulted from the encapsulation of ruthenium by iron oxide upon reductive pretreatment. Specifically, Ru nanoparticle seeds were obtained by thermally decomposing $Ru_3(CO)_{12}$ in oleylamine. Then, by adding $Fe(CO)_5$, Fe_2O_3 nanoparticles were grown onto Ru seeds and decomposed at 300 °C in a mixture of 1-octadecene, oleic acid (OA) and octadecyl acrylamide (OAm). The contact between Ru and Fe_2O_3 in the ruthenium-iron oxide colloidal heterodimer promoted the reduction of Fe_2O_3, leading to the formation of a ruthenium-iron core–shell structure through the hydrogen overflow effect.

Tang et al. [67] designed a MOF@MOF (ZIF-8@ZIF-67) core–shell structure using ZIF-8 as seeds, and derivatized a new type of functionalized nanoporous materials NC@GC with nitrogen-doped carbon (NC) as the core and high graphitized carbon (GC) as the shell through high-temperature pyrolysis. Nanoporous NC prepared from ZIF-8 crystals had a relatively high N content and a large surface area, and nanoporous GC prepared from ZIF-67 crystals possessed highly graphitic walls with good conductivity due to the catalytic graphitization effect of well-dispersed Co species in the parent ZIF-67 crystals. Furthermore, the NC@GC materials possess interconnected hierarchically micro/mesoporous structure originated from the core–shell ZIF-8@ZIF-67. Similar to Tang, Wu et al. [68] used ZIF-67 as seeds, deposited Co_3O_4 nanoshells, and derivatized graphite phase r-Go by calcination to prepare r-Go@Co_3O_4 yolk-shell structure [63]. The high conductivity of r-GO and the highly porous Co_3O_4 shells were organically combined.

Ahmad et al. [69] took Ag@Au as an example to study the growth and evolution of nanostructures during the seed-mediated process by adjusting the combination of end-capping agent and coordination complex (Fig. 31.9a). The capping agent, dimethyl-amine (DMA), and the coordinating complex, potassium iodide (KI), in an organic solvent (methanol) could slow down the reaction kinetics to observe mechanistic insights into the overgrowth process and shift the growth regime from galvanic-replacement mode to direct synthesis mode resulting in the conventional synthesis of Ag@Au core–shell structures. Specifically, applying DMA and KI could completely block the current displacement reaction. Under this growth condition, the equilibrium concentration of the Au monomer and the reduction potential was significantly reduced. By changing the concentration of DMA, the growth rate of Au deposited on Ag could be directly adjusted: under the same electron dose conditions, a higher DMA concentration resulted in a lower Au deposition rate. On the other hand, the use of DMA itself did not completely stop the electrical reaction at the interface, but it could cause the formation of a hollow structure and the diffusion of Ag into the shell region. The addition of KI inhibited the electrochemical reaction of Au–Ag alloy at the interface. The combined use of this end-capping agent and coordination complex is the best strategy to form a stable Ag@Au nanostructure (Fig. 31.9b).

In addition, alloys have attracted extensive attention in the field of catalysis due to their bimetallic synergy. Zhao et al. [70] combined core–shell structure with alloy design to prepare electrocatalyst Pd@$Pt_{1.8}$Ni, containing bimetallic nanoshells. With octahedral Pd as seeds, about four atomic layers of PtNi alloy were deposited on their crystal plane. By simply changing the concentration of Pt and Ni precursors added to the system, the Pt:Ni atomic ratio, controlled from 0.6 to 1.8, could be successfully adjusted. The ultra-thin PtNi alloy shell with only a few atomic layers greatly improved the utilization efficiency, thereby reducing the use of Pt. Zhang et al. [71] reported a promisingly dendritic core–shell nickel–iron–copper metal/metal oxide electrode CS-NiFeCu, prepared via dealloying with an electrodeposited nickel–iron–copper alloy as a precursor. Dealloying is a commonly used top-down nanosynthesis technique where one or more chemically active elements are selectively oxidized and removed from parent alloy by chemical and/or electrochemical methods. This keeps

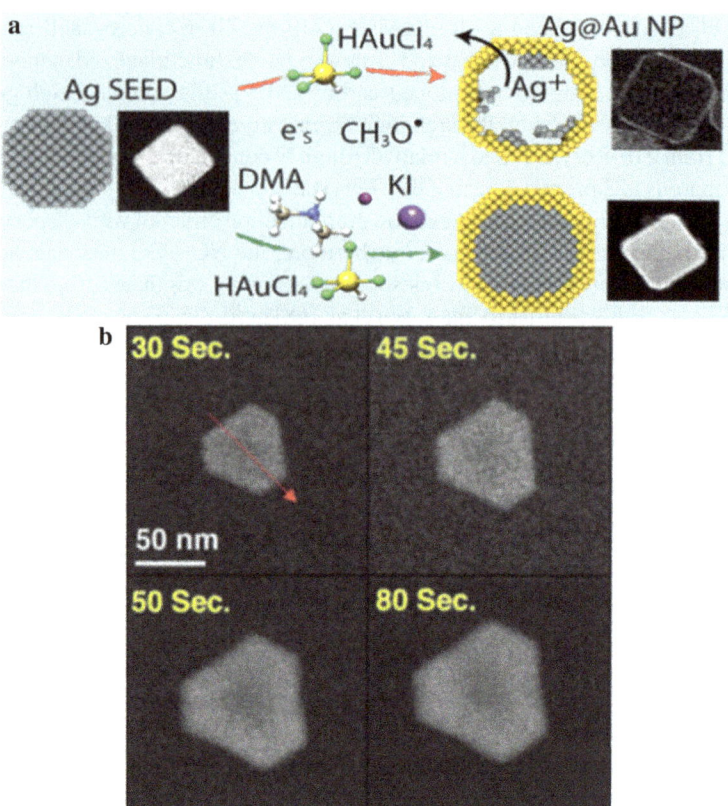

Fig. 31.9 Schematic illustration by seed deposition method of **a** the effect of the combination of end-capping agent and coordination complex; **b** shape evolution of bimetallic Ag@Au morphologies. Reproduced with permission [69]. Copyright © 2019, American Chemical Society

the metallic core at high electrical conductivity, which enhances the facile electron transfer during the catalytic process.

31.2.3 Self-assembly Method

Self-assembly is also a common method of core–shell structure preparation. It is a process in which the components spontaneously form thermodynamically stable, structurally determined aggregates by biological affinity or bonding between groups. Once this process starts, it will automatically proceed to a certain expected endpoint, and the structural units such as molecules will be automatically arranged in an orderly manner.

There are two common forms of self-assembly method: (1) amphiphilic copolymerization and (2) layer-by-layer deposition.

(1) Amphiphilic copolymerization: Its principle is that the copolymer contains both hydrophilic and hydrophobic functional ends. Therefore, the self-assembly of the amphiphilic block copolymer in a selective solvent can form a polymer micelle with a hydrophobic core-hydrophilic shell. Li et al. [72] reviewed the research progress of the preparation of spherical nanoparticles by the block polymer template method, and the templates were divided into two categories: micelles self-assembled from linear block copolymers and unimolecular star-shaped block copolymers. The article pointed out that changing the molecular weight of the block polymer could adjust the size of the block polymer assembly, and further control the size of the nanoparticles. Fei et al. [73] used a new type of block copolymer molecular brush to synthesize polydimethylsiloxane block-polyethylene oxide and phenolic resin to obtain an ordered precursor film, and then carbonized to prepare nanometers (Fig. 31.10a). Porous carbon provided new ideas for subsequent core–shell structure preparation. Through hypercrosslinking-induced self-assembly strategy, Xu et al. [74] prepared a porous polymer composite with a functional hollow structure, and then encapsulated Pd nanoparticles in the hollow cavity of hypercrosslinked polymer (HCPs) nanospheres by in situ impregnation-reduction method. Thus, a selective catalyst with yolk–shell structure was prepared (Fig. 31.10b). HCPs exhibited different intrinsic swelling behaviors in solvents of different polarities. Therefore, by changing the polarity of the solvent medium, the pore sizes of micropores in HCPs-based catalysts could be precisely adjusted, so that the catalyst had selectivities for substrates of different molecular sizes.

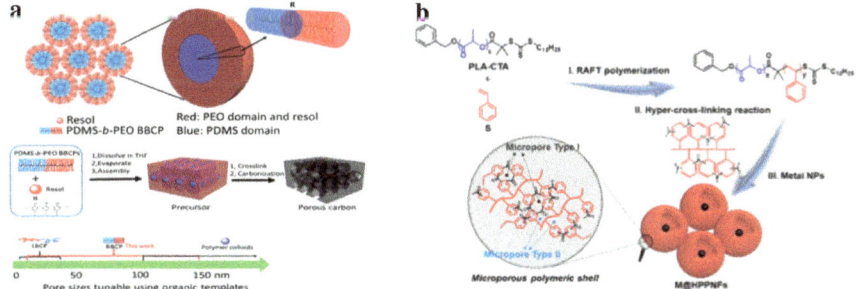

Fig. 31.10 Schematic illustration by amphiphilic copolymerization of **a** a block copolymer molecular brush to synthesize polydimethylsiloxane block-polyethylene oxide and phenolic resin. Reproduced with permission [73]. Copyright © 2019, American Chemical Society. **b** In-situ impregnation-reduction method to encapsulate Pd nanoparticles in the hollow cavity of hypercrosslinked polymer (HCPs) nanospheres. Reproduced with permission [74]. Copyright © 2019, American Chemical Society

(2) Layer-by-layer deposition (LBL): The process of alternate layer-by-layer deposition and the weak interaction between the molecules of each layer can result in that the layers and layers spontaneously associate to form a stably coating structure with specific functions. LbL deposition is suitable for depositing many kinds of materials containing nanoparticles, polymers, proteins, dye molecules, and lipids on various substrates. During the deposition process, the matching of the surface properties of the two layers has an essential effect on the uniformity of the coating. In addition to the surface charge involved in electrostatic interactions, hydrophobic interactions, hydrogen bonding, complementary base pairs, and covalent bonding should also be considered. The thickness, function, and composition of the final material prepared by LbL can be easily adjusted by changing the number of deposited layers, the type of adsorption, and the conditions used in the assembly process. Caruso et al. [75] reported the preparation of monodisperse hollow titania spheres with defined diameter, wall thickness, and crystal phase by the layer-by-layer templating of colloid particles and subsequent calcination. Specifically, using LbL technology, with titanium(IV) bis(ammonium lactato) dihydroxide (TALH) as precursor and poly(dimethyldiallyl ammonium chloride) (PDADMAC) as polyelectrolyte, a uniform TiO_2 layer was deposited on various colloidal nanoparticles. The ability to form core–shell nanoparticles uniformly depended on the water stability of TALH: it could be assembled with PDADMAC through electrostatic interaction at room temperature, and then subjected to constrained hydrolysis and condensation to form a TiO_2 shell. The calcined PS nanoparticles coated with TALH/PDADMAC formed hollow TiO_2 spheres. By adjusting the number of deposited TALH/PDADMAC layers, the thickness of the shell could be easily adjusted from 5 nm to 50 nm. Wang et al. [76] synthesized the $ZnO@void@SiO_2$ yolk–shell structure by the LbL method. The hydrothermal treatment of glucose was used to form a carbon layer on the surface of ZnO, and then a silicon shell was prepared on the surface by the sol–gel method. Finally, glucose decomposed under high-temperature calcination, resulting in voids, namely, $ZnO@void@SiO_2$. Glucose acted as a C source and end-capping agent. During the SiO_2 coating process, according to the Ostwald ripening principle, SiO_2 nanoparticles were adsorbed on the specific C surfaces to form a shell, resulting in an increase in the surface roughness of the particles. In addition, Wang et al. [77] achieved a marigold-like $SiC@MoS_2$ nanoflower with a unique Z-scheme structure through electrostatic assembly technology (Fig. 31.11a). The positively charged 3D-SiC nanocrystals and the negatively charged 2D-MoS_2 nanosheets underwent electrostatic self-assembly in aqueous solution. The electrostatic repulsion between the negatively charged MoS_2 nanosheets suppressed their stacking on the SiC surface, thereby leading to the formation of nanoflower morphology. Yun et al. [78] successfully constructed ZIF-8@Au$_{25}$@ZIF-67 with a sandwich structure through coordination-induced self-assembly strategy. The composite structure had an atomically accurate surface interface structure and could be used as a heterogeneous catalyst (Fig. 31.11b).

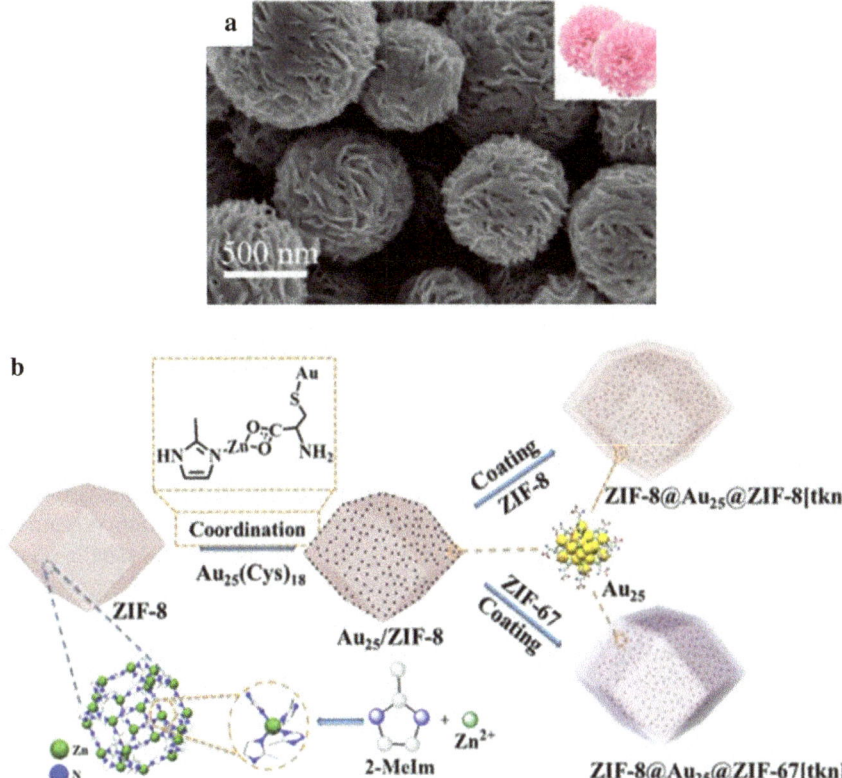

Fig. 31.11 Schematic illustration by LbL method of **a** SEM images of SiC@MoS₂ nanoflower. Reproduced with permission [77]. Copyright © 2018, American Chemical Society. **b** The sandwich structures of ZIF-8@Au₂₅@ZIF67. Reproduced with permission [78]. Copyright © 2020, American Chemical Society

31.2.4 Sol–Gel Method

The sol usually refers to the colloidal solution in which solids are dispersed in a liquid. The gel is the frozen substance formed between a solid state and a liquid state under specific conditions during the coagulation process of the sol. The sol–gel method refers to dispersing the particles to be coated in the prepared sol, and then performing gelation under certain reaction conditions, so that the desired shell layer can be coated on the surface of the core particles. By adjusting the synthesis parameters, such as capping agents, precursors, and solvents, not only can organic or inorganic cores of various sizes (nanoscale, micro-scale, and sub-micro-scale) be coated, but also multiple shells can be prepared to core–shell material with layer structure.

Ullmann et al. [79] entrapped Cp_2ZrCl_2 in dual-layered silica by two iterated nonhydrolytic sol–gel steps with organosilanes (methyltriethoxysilane, isopropyltriethoxysilane, octyltriethoxysilane, octadecyltrimethoxysilane) and WCl_6 as modifiers (Fig. 31.12a). In the case of dual systems, the presence of two distinct hybrid silica layer effects, an internal silica layer modified with C18-Si, and an external layer with C8-Si moieties impinged on the broadening of polydispersity (Fig. 31.12b).

Afterward, Lan et al. [80] continued to explore a confined interfacial monomicelle assembly approach for accurately coating ordered monolayered TiO_2 mesopores on diverse surfaces (Fig. 31.13a). They used the triblock copolymer F127 as the template and glycerol as the restricted solvent to achieve the control of the ordered mesostructure and thickness of the TiO_2 shell layer at the level of single micelles. First, titania monomicelle hydrogels were prepared after preferential evaporation of tetrahydrofuran (THF). A sol–gel process was then initiated by subsequently mixing the monomicelles and solid silica cores into ethanol/glycerol solvents under stirring. The high-viscosity glycerol was chosen as a co-solvent, because it enabled induction of monomicellar self-assembly in spatially confined directions as well as simultaneously retarding hydrolysis and the condensation rate of titanium oligomers by firmly adhering to titania monomicelles. As a result, the coated TiO_2 shells posed a monolayer of mesopores (Fig. 31.13b–d). Furthermore, by tuning the amount of swelling agent, the accurate controllability of such a confined assembly process achieved the formation of TiO_2 shells from mono- to multilayers (up to five layers) of mesopores, and the mesopore size also can be manipulated from 4.7 to 18.4 nm (Fig. 31.13e–n).

Stober et al. [81] reported that under the catalysis of ammonia water, alkyl silicate was hydrolyzed in an alcohol solution to obtain SiO_2 spherical core particles with good monodispersity and controllable size. This is the well-known Stober method. Recently, Salinas et al. [82] linked the Stober principle and sol–gel method to prepare Ca^{2+}-doped core–shell structured catalyst $SiO_2@ZrO_2$-CaO. Experiments showed that adding calcium to the $SiO_2@ZrO_2$ core–shell would produce calcium effect: (1) CaO could partially stabilize ZrO_2, causing the ZrO_2 crystal phase to change from cubic phase to the monoclinic phase. The cubic phase and the orthorhombic phase had similar XRD spectra, that is to say, no XRD spectrum can detect the crystalline phase of zirconia. However, due to the broad-spectrum characteristics of c-ZrO_2, after adding calcium, the Raman spectra of the cubic phase and the orthorhombic phase showed a significant difference. (2) CaO increased the total number of basic sites, which were related to the strong basic sites of CaO and the presence of monoclinic ZrO_2. In the Stober system, Wei et al. [83] used the sol–gel method and carbothermal reduction method to prepare hollow silicon carbide balls Ni@HSS with complete shape and uniform dispersion (Fig. 31.14). Specifically, yeast was applied as a template, PVP was used to disperse and pretreat the agglomerated colonies. Under alkaline conditions, TEOS hydrolyzed and condensed to generate SiO_2, and nano-SiO_2 would gradually coat condensates on the surface of the yeast through electrostatic interaction between particles. With the control of PVP, uniform and complete coating layers were formed onto yeast surface. By roasting at high temperature in the air, the material inside yeast was removed, leaving only a layer of SiO_2 shell outside,

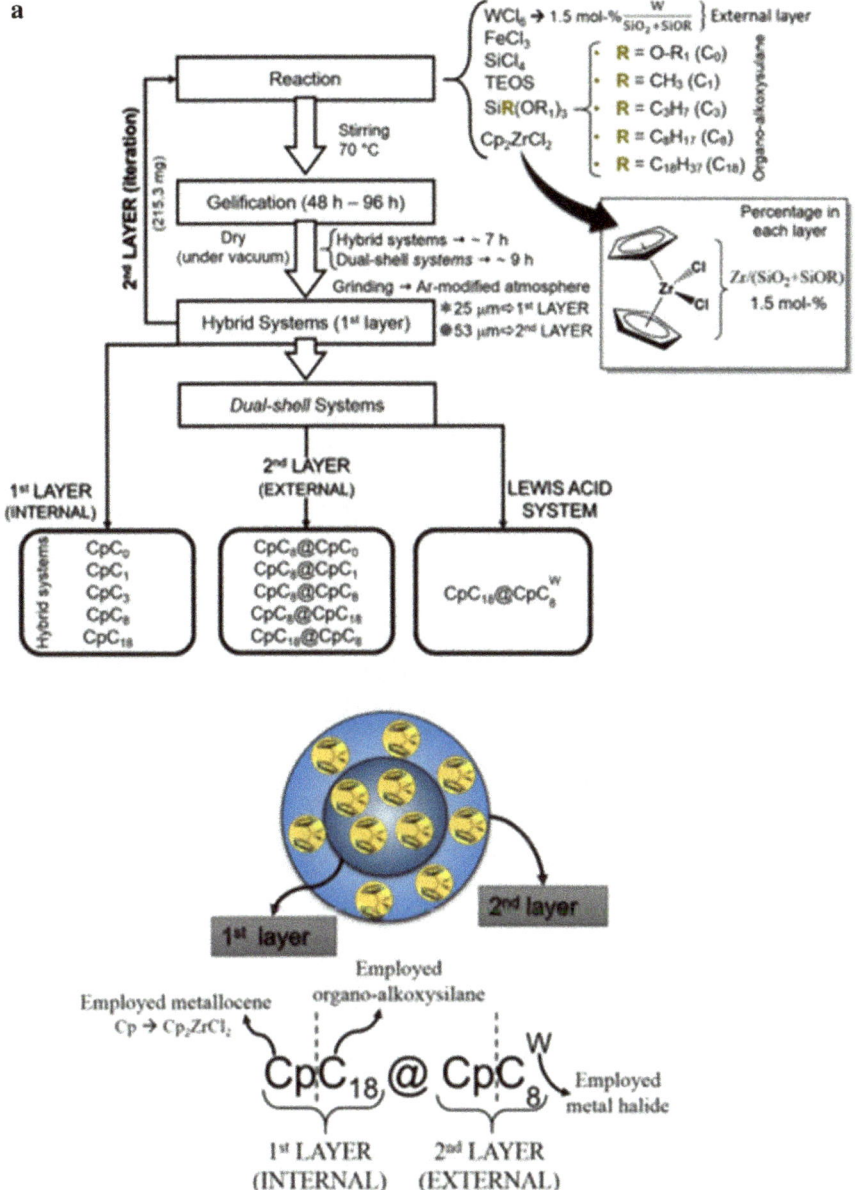

Fig. 31.12 Schematic illustration by the sol–gel method of **a** hybrid dual-shell silica-based catalysts synthesis route; **b** labeling example for dual-shell system. Reproduced with permission [79]. Copyright © 2020 Elsevier Inc

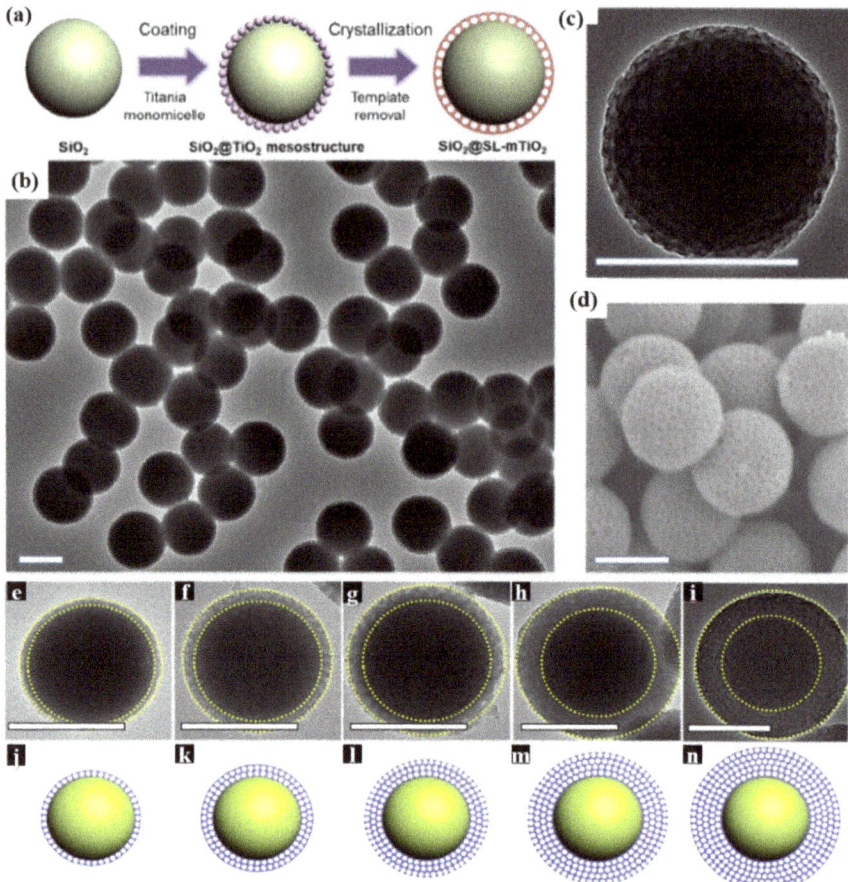

Fig. 31.13 a Schematic illustration of the preparation of single-layer TiO$_2$ mesopore-coated core–shell structures via sol–gel method; **b–d** Morphological characterization images of the SiO$_2$@SL-mTiO$_2$ core–shell nanostructures; **e–n** SiO$_2$@mTiO$_2$ core–shell structures with highly tunable coated TiO$_2$ layers from one to five layers of mesopores. Reproduced with permission [80]. Copyright © 2019 Elsevier Inc

which was the hollow silicon template that can be used as the silicon source of HSS. After the sol–gel reaction, binary phenols with different molecular structures were formed on the surface of a hollow silicon template, named RF@hollow SiO$_2$. HSS was finally achieved after calcination in Ar and air atmosphere separately.

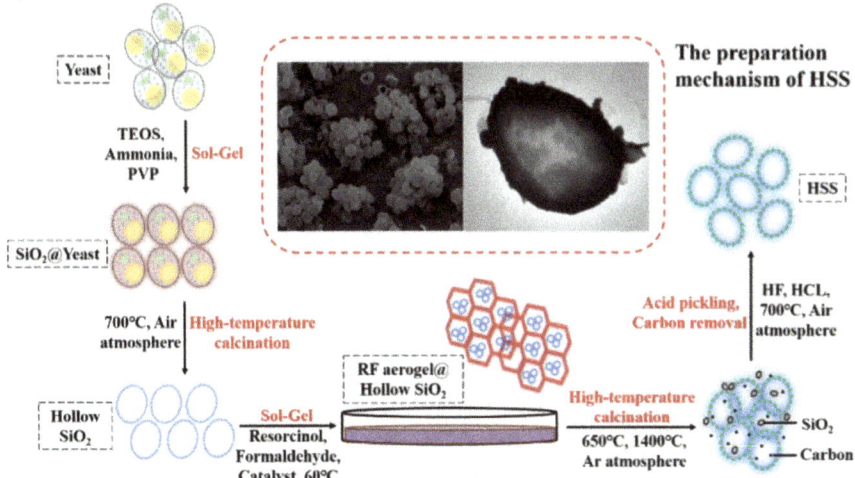

Fig. 31.14 Schematic illustration of preparation process of HSS. Reproduced with permission [83]. Copyright © 2020 Elsevier B.V

31.2.5 Microemulsion Method

The microemulsion method is due to the formation of nanoparticles after nucleation, coalescence, agglomeration, and heat treatment in the emulsion microbubbles formed by two incompatible solvents under the action of surfactants. The characteristic particles have good monodispersity and good interfacial property, and the semiconductor nanoparticles of group II–VI are mostly prepared by this method. According to the difference of surfactant, chemical composition, and continuous phase, it is mainly divided into three structures: oil-in-water (O/W), water-in-oil (W/O), and oil-water bi-continuous. In recent years, there have been more and more reports on the preparation of core/yolk–shell structure using microemulsion technique at home and abroad, and it has become one of the main techniques for preparing this kind of material.

Wang et al. [84] used $PdCl_4^{2-}$ ions as precursors of Pd nanoparticles to combine $P123$/sodium oleate/$PdCl_4^{2-}$ solution with polymer precursors (2,4-dihydroxybenzoic acid (DA) and hexamethylenetetramine (HMT)) acidic solution to prepare an emulsion solution containing Pd precursor (Fig. 31.15a). During hydrothermal treatment, the formaldehyde produced by HMT can not only reduce Pd precursors to Pd nanoparticles, but also polymerize with DA on the surface of emulsion droplets. The Pd nanoparticles formed in situ first induced the formation of Pd/polymer nanocomposites due to strong interaction with polymer carboxyl groups. As the polymerization proceeds further, the polymer-yolk-supported Pd nanoparticles produce the final egg yolk shell polymer nanospheres.

Luo [85] and others have demonstrated for the first time that metal complexes are beneficial to the formation of water–scCO$_2$ microemulsion. Compared with

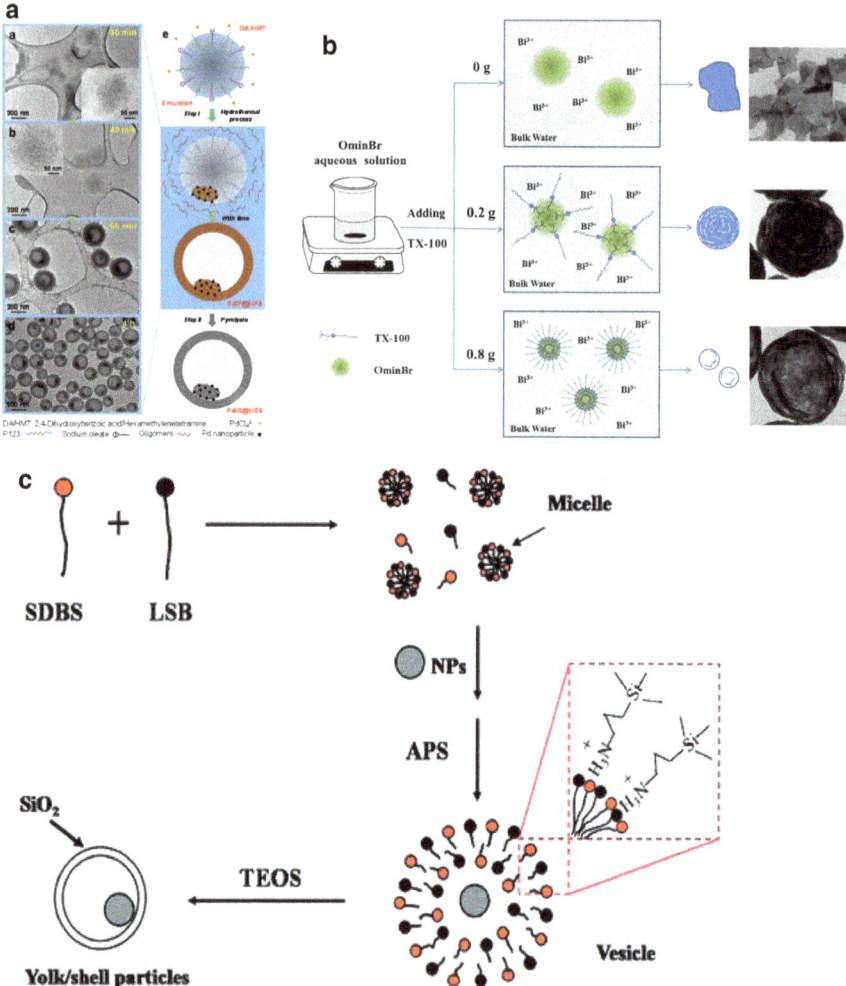

Fig. 31.15 a Using PdCl$_4^{2-}$ as a precursor, the TEM image of the sample after hydrothermal treatment at 160 °C for different times. And a schematic diagram of the formation process of Pd/C @ HCS. Reproduced with permission [84]. Copyright © 2018, American Chemical Society; **b** Schematic illustration for the proposed formation process of Bi-rich Bi$_4$O$_5$Br$_2$ with different morphologies. Reproduced with permission [86]. Copyright © 2017 Elsevier B.V. **c** Yolk/SiO$_2$ shell particles generation containing movable NP cores. Reproduced with permission [90]. Copyright © 2009, American Chemical Society

conventional hydrocarbon emulsifiers, the water solubility of metal complex stabilized microemulsion is greatly improved. At the same time, this microemulsion provides photocatalytic in situ reduction of carbon dioxide by the metal complex at the water/scCO$_2$ interface and a new approach is introduced. This method promotes mass transfer because the metal complex increases the interface area between the two phases.

Mao et al. [86] successfully synthesized $Bi_4O_5Br_2$ with two-dimensional nanoflakes, three-dimensional monodispersed layered microspheres, and hollow microspheres with adjustable morphology using ionized water-liquid microemulsion method. Surfactant TX-100 as a stabilizer of IL/W microemulsion, its concentration significantly affects the morphology and size of the obtained $Bi_4O_5Br_2$ crystals (Fig. 31.15b). The prepared $Bi_4O_5Br_2$ HSs and LMs have higher photocatalytic efficiency than $Bi_4O_5Br_2$ and BiOBr NSs, due to the hollow/layered spherical structure and the large amount of Bi.

Kuwahara et al. [87] synthesized $TiO_2 @ SiO_2$ photocatalyst with yolk–shell structure using oil/water (O/W) emulsion as a template. The oil/water system prepared by adding oleic acid (oil phase) to water (aqueous phase) modifies the outer surface of TiO_2 NPs to form a Ti-COOH coordination bond, resulting in a lipophilic surface. Under vigorous stirring, TEOS and 3-aminopropyltriethoxysilane (APTES) were added as SiO_2 precursors. The carboxyl group of oleic acid is neutralized with the amino group of APTES, resulting in the deposition of APTES at the interface between the continuous phase and the emulsion droplets of TiO_2 NPs. At the same time, the amino group of APTES provides an alkaline environment, causing the silicon alkoxide to polymerize, thereby further forming a SiO_2 shell.

Wu and Xu [88] reported an early soft-template method to synthesize yolk–shell structure, successfully prepared materials with different particle cores ($SiO_2 @ SiO_2$, Au@ SiO_2, and $Fe_2O_3 @ SiO_2$), and provides a standard process for this method (Fig. 31.5c). The synthesis procedure includes three main steps: (1) Disperse the nanoparticles (usually SiO_2 and Au) into dodecyl sulfonate betaine (LSB), zwitterionic surfactant, and sodium dodecyl benzene sulfonate (SDBS) (anionic surfactant) in an aqueous mixture with a molar ratio of 1:1. (2) Add 3-amino-propyltriethoxysilane (APS) to induce vesicles with movable nanoparticle cores formation. At the same time, some protonated APS used as structure guiding agents adhere to the surface of the vesicles by electrostatic attraction. (3) Sol–gel method was used to prepare SiO_2 shell by hydrolyzing APS and tetraethyl orthosilicate (TEOS). Wu et al. [89] used oil/water (O/W) emulsion as a template to prepare Au nanocatalysts (2.8–4.5 nm) with controllable size in hollow SiO_2 nanospheres. Adjusting the Au precursor and the concentration of $HAuCl_4$ used in the synthesis process can easily control the size of the Au nanocatalyst. Liu et al. [90] described a simple and general strategy to prepare a series of yolk–shell structures with different cores: fluorocarbon surfactants FC4 and F127 adsorbed on the core through electrostatic interaction to form a vesicle complex. Then add ethyl orthosilicate and deposit it on the surface of the nuclear vesicle complex by vesicle template method. The orthosilicate is polymerized by a sol–gel reaction to form a silicate shell. Then the yolk–shell structure is obtained through the shrinkage of the silicate shell and the maturation process of TEOS in the condensation and hydrolysis process. The surfactant is calcined to remove mesoporous SiO_2 shell. Through this method, a variety of nanoparticles such as magnetic particles, silicon spheres, Au NPs, mesoporous SiO_2 NPs, etc. can be encapsulated into the SiO_2 shell.

The advantages of preparing nanoparticles by microemulsion method are simple experimental device, low energy consumption, easy operation, and controllable particle size; the particles are not easy to coalesce and have good stability; and the surface active agent encapsulates the nanoparticle surface. The coating improves the interface properties of the nanomaterials and significantly improves its catalytic performance.

31.2.6 Some Other Methods

Electrodeposition technology is a new method for preparing core–shell structure through the redox reaction on the electrode to form a coating through the migration of positive and negative ions in the electrolyte solution under the action of an external electric field. The electrodeposited alloy nanospheres can grow tightly on the electrode without any polymer binder. Chen [91] and others used a two-step electrochemical strategy to fabricate core–shell nanospheres with highly active OER electrocatalysts. A simple and fast electrodeposition method prepared $NiFe_xSn$ alloy nanospheres and the surface of $NiFe_xSn$ alloy nanospheres. It will undergo electrooxidation to produce an amorphous shell of NiFe (hydroxyl) hydroxide, thereby forming a core–shell structure. The metal core of $NiFe_xSn$ helps electron transfer to the shell of amorphous NiFe (oxygen) hydroxide, which in turn prevents further oxidation of the metal core. The selective electrochemical etching of tin in alkaline solution produces a large surface area, which can expose a large number of active sites, which is beneficial to the diffusion and transport of substances. By using electrodeposition adjustment, it is easy to optimize the Ni/Fe ratio to obtain better OER activity. The core–shell structured catalyst prepared by the electrodeposition method has stronger stability, and at the same time can modify the original substrate in many ways, and has more and more extensive applications in the construction of the catalyst.

Atomic layer deposition (ALD) technology is a method that can deposit substances layer by layer in the form of a monoatomic film. Atomic layer deposition is similar to ordinary chemical deposition. However, in the process of atomic layer deposition, the chemical reaction of a new layer of atomic film is directly related to the previous layer. This way allows only one layer of atoms to be deposited per reaction. Ma [92] and others used atomic layer deposition (ALD) to precisely control the thickness of the ultra-thin ZnO shell, adjust the absorption of light by the CdS core, and designed a CdS @ ZnO core–shell structure catalyst to achieve a close contact between the core and shell to accelerate the migration of photogenerated carriers, and at the same time, the ZnO shell can also effectively prevent the photocorrosion of the CdS core, thus showing good photocatalytic stability (Fig. 31.16a). Ma et al. [93] used a one-step co-deposition process of atomic layer deposition (ALD) to deposit highly uniform GaON films. This material only exhibits high-quality epitaxial growth on ZnO nanowires (NWs) at 200 °C (Fig. 31.16b). Behavior to construct ZnO–GaON

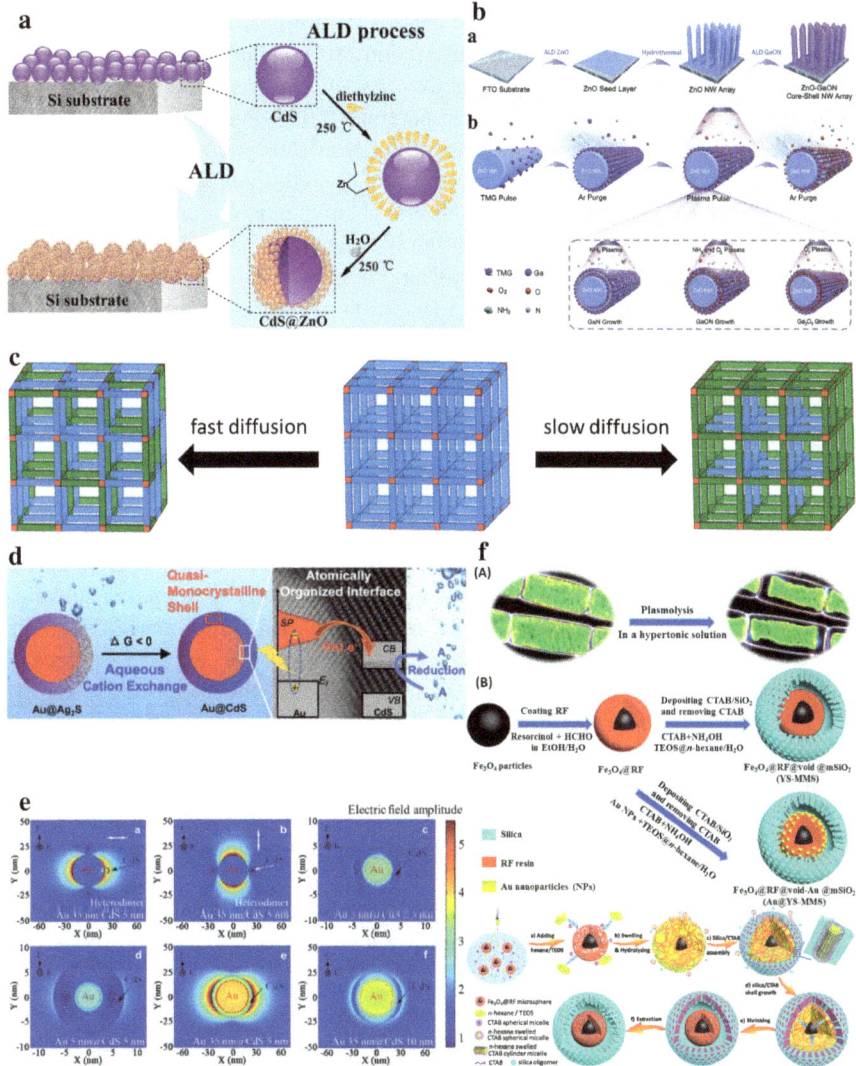

Fig. 31.16 a Schematic procedure of CdS @ ZnO core–shell structure prepared by an ALD method. Reproduced with permission [92]. © 2017 Elsevier Ltd; **b** A schematic representation specifying the fabrication of the ZnO–GaON composite array (**a**). Schematic representation of the one-step ALD growth of the GaON film on the surface of the ZnO NWs (**b**). Reproduced with permission [93]. © 2019 Elsevier Ltd; **c** Two different ligand incorporation models: (left) a uniform distribution associated with fast diffusion relative to the exchange process and (right) a core–shell distribution associated with slow diffusion relative to the exchange process. Reproduced with permission [95]. Copyright © 2017, American Chemical Society; **d** The scheme of transition metal cation replacement; **e** FEM simulations of the electric field amplitude of Au 35/CdS 5 heterodimer (35-nm-sized Au domain and 5-nm-sized CdS domain) NCs with different polarization directions. Reproduced with permission [99]. Copyright © 2018 Elsevier Ltd; **f** Schematic illustration of the proposed formation mechanism for YS-MMS microspheres and Plasmolysis process of plant cells upon immersion in a hypertonic solution (**a**); and synthesis procedure for the YS-MMS microspheres (**b**). Reproduced with permission [100]. Copyright © 2017, American Chemical Society

core–shell NW with different shell thicknesses (5–60 nm). It is found that ZnO–GaON NW with the best shell thickness (about 40 nm) has the greatest electric field enhancement and light trapping ability. Atomic layer deposition technology due to its highly controllable deposition parameters (thickness, composition, and structure), excellent deposition uniformity, and consistency makes it have a wide range of application potential in the construction of core–shell structures.

Glow discharge is a traditional cold plasma phenomenon with high-energy electrons. Wang [94] and others used argon glow discharge as the electron source to prepare a highly active and stable carbon-supported Pt–Pd alloy catalyst by the room temperature electron reduction method. In addition, when the precursor is a mixture of different metal salts, electron reduction can also form alloy NPs with smaller particles. The core–shell nanostructure prepared by glow discharge has the characteristics of strong interaction between carriers and high catalytic performance. This method does not require chemical reducing agents, protective agents, and dispersants. The preparation process minimizes environmental damage and efficiency also relatively impressive.

Ligand exchange (substitution of ligands) refers to the ligand exchange reaction: the ligand in the coordination compound can be replaced by other ligands, called ligand exchange reaction, the general reaction mechanism is nucleophilic substitution reaction. Boissonnault et al. [95] studied the microstructure of metal–organic frameworks (MOFs) after ligand exchange (PSE), and found that the exchanged ligands were concentrated at the edge of the crystal. As the depth of the crystal increased, the ligand concentration gradually lowers, forming a core–shell structure. Diffusion studies of carboxylic acid ligands in MOF-5 indicate that this is because the carboxylic acid ligands slowly diffuse into the pores of the MOF, causing exchange to occur faster than diffusion, resulting in the formation of a core–shell structure (Fig. 31.16c). The study of PSE in UMCM-8 and UiO-66 single crystals shows a similar trend, indicating the applicability of PSE as a method for creating core–shell MOFs. The core–shell structure catalyst prepared by ligand exchange has a wide range of modification applications for MOF-type materials, greatly broadening its catalytic capacity.

Zhang et al. [96], based on the sequential chemistry of the solubility product constant (Ksp), can sequentially precipitate metal sulfides to construct a uniform core–shell structure. The utility model is composed of a metal sulfide absorbent core and a metal sulfide auxiliary catalyst shell. The results show that the solubility product constant (Ksp) of metal sulfides can drive the formation of a unique core–shell structure. In a uniform core–shell structure, the instability of CdS nanoparticles under irradiation can be well overcome.

Ion exchange is the function or phenomenon of the exchange of ions in solution with ions on a certain ion exchanger. It is the exchange of ions in a solid ion exchanger with ions in a dilute solution. Ghasemian [97] and others introduced a "liquid metal" reaction process in which a single layer of hydrated manganese dioxide nanoparticles was prepared in aqueous solution based on the reduction of permanganate ions on the surface of the liquid EGaIn metal by galvanic-replacement EGaIn/Mn was obtained for the chip case. Under simulated sunlight, the photocatalytic degradation performance of droplets as heterogeneous catalysts on Congo red model dyes was studied, and it was found that EGaIn/Mn-2.5 is superior to MnO_2 photocatalysts reported in previous literatures.

Chen [98] and others used sonochemical precipitation and hydrogen ion exchange to synthesize WO_3 microspheres with controllable internal structure, and heat-treated them. Under full-wavelength light irradiation, aromatic alcohols are selectively photocatalyzed and oxidized to the corresponding aldehyde in aqueous solution. The photocatalytic activity of egg yolk–shell WO_3 microspheres is significantly higher than solid structure WO_3 microspheres and hollow structure WO_3 microspheres. Liu et al. [99] achieved the replacement of transition metal cations in aqueous solution by the aqueous cation exchange method, and synthesized the transition metal coating of different metal elements (Fig. 31.16d, e) by adjusting the core size and shell thickness, different morphological adjustments: spherical, rod-like. Using the ion-exchange method, by changing the concentration of metal ions, the particle size of the particles can be controlled, and a new method for the overall construction of the core–shell structure is proposed.

Yue [100] and others proposed a brand-new "quasi-wall separation" interface nanoengineering idea, successfully constructed a yolk–shell magnetic mesoporous composite microsphere material with uniform and controllable morphology, and in the synthesis process the synchronous coating of the functional nanocatalyst in the cavity is realized in (Fig. 31.16f). The team mimicked the process of plasmolysis of plant cells in hypertonic solutions in nature using magnetic particles coated with polymer resin as the core and using interface nanoengineering strategies to controllably deposit surfactants on the swollen polymer resin (CTAB)-SiO_2 composite coating, during the subsequent ethanol extraction to remove CTAB and organic swelling agent, the polymer resin shrinks, and the inorganic SiO_2 shell forms a large number of radial mesopores due to the removal of CTAB, thereby forming a yolk–shell composite mesoporous micro-ball material.

31.2.7 Use of Photocatalysts for Yolk–Shell and Core–Shell Materials

For current photocatalysts, the main factors restricting their development are the excessively low light utilization efficiency, the corrosion dissolution of precious metals, and the excessively low specific surface area, which cannot provide enough active sites. Because yolk–shell and core–shell have the properties of shell and cavity or form a heterojunction at the interface of the two materials, this can greatly improve the catalytic performance and can also avoid photocorrosion caused by light when using precious metals, and significantly improve the life and stability of the catalyst. These advantages make yolk–shell and core–shell structure catalysts have important applications in photocatalysis [101, 102].

31.2.7.1   Use of Yolk–Shell Photocatalyst

Chiu [15] and others used Au–Cu$_2$O core–shell nanostructures as growth templates to form Au–Cu$_7$S$_4$ yolk–shell nanostructures by sulfidation. Au–Cu$_7$S$_4$ was then converted to Au–CdS by cation exchange and by changing the reagent capacity, by changing the void size, the prepared Au–CdS nanostructures showed a faster hydrogen generation rate. Besides, the study found that Au–CdS nanostructures with larger cavities have a faster hydrogen production rate, which is mainly due to their increased diffusion rate on the CdS shell. Jiang et al. [103] prepared Pt/porous CeO$_2$ nanostructured photocatalysts by heterogeneous growth of CeO$_2$ on porous metal nanoparticles and then calcination to induce shrinkage of the nanoparticles. The resulting (Pt yolk)/(porous CeO$_2$ shell) nanostructures exhibited stronger broadband absorption in the visible region than their pure core/shell CeO$_2$ and Pt NPs. The thin CeO$_2$ shell and sufficient void space in the PtAg-yolked nanostructure is the key factor for its high photocatalytic activity. The selective oxidation of benzyl alcohol under visible light proves that the nanostructures obtained have excellent photocatalytic activity. Its performance is attributed to the synergistic effect of multiple components on light absorption and electron–hole separation and effective mass transfer. Cai et al. [104] successfully encapsulated gold cores of different diameters and shapes into Ni(OH)$_2$ shells, and found that the synergistic effect of Au and Ni(OH)$_2$ makes the heterogeneous structure of egg yolk shells have enhanced catalysis active. Li et al. [105] can adjust the gap between the yolk–shell ZnFe$_2$O$_4$ nanostructure (CN–ZnFe$_2$O$_4$) egg yolk, and the shell through the heating rate during the calcination of the precursor, so that it has excellent separation efficiency of photogenerated electron–hole pairs And the transfer efficiency of photogenerated electrons, the obtained CN–ZnFe$_2$O$_4$ sample showed a stronger visible light response than ZnFe$_2$O$_4$ (Fig. 31.17a).

Wang et al. [106] prepared upconversion nanoparticles (UCNPs) and Zn$_x$Cd$_{1-x}$S yolk–shell nanoparticles by a simple template-assisted hydrothermal method (Fig. 31.17b). The results of steady-state and dynamic fluorescence spectrum analysis show that the prepared UCNPs/Zn$_x$Cd$_{1-x}$S egg yolk–shell nanoparticles have higher energy transfer efficiency and can be used as an efficient nanosensor for near-infrared light, and the prepared egg yolk–shell nanoparticles are the dye which has good photocatalytic degradation performance, good biocompatibility, and strong ability to generate hydroxyl radical (OH) and monooxygen (^{1}O$_2$). Waqas et al. [107] controlled the distribution of TiO$_2$ in hollow spheres by loading aqueous solutions of TiCl$_4$ precursors onto carbonaceous templates, and then annealed, and Fe^{3+} ions radially penetrated the hydrophobic core of the carbonaceous templates (Fig. 31.17c). For the first time, photocatalytic water oxidation was carried out with a controlled pseudo-brookite phase as yolk and titanium dioxide as a hollow spherical shell. Due to the geometry of the hollow spheres, the light collection effect is better. Fe$_2$TiO$_5$-TiO$_2$ yolk–shell hollow spheres show high oxygen evolution reaction (OER) rate of up to 148 μmol g^{-1} h^{-1} under ultraviolet-visible light. Through the thin-shell–yolk heterojunction, the reaction solvent is brought close to the cavity of the reaction

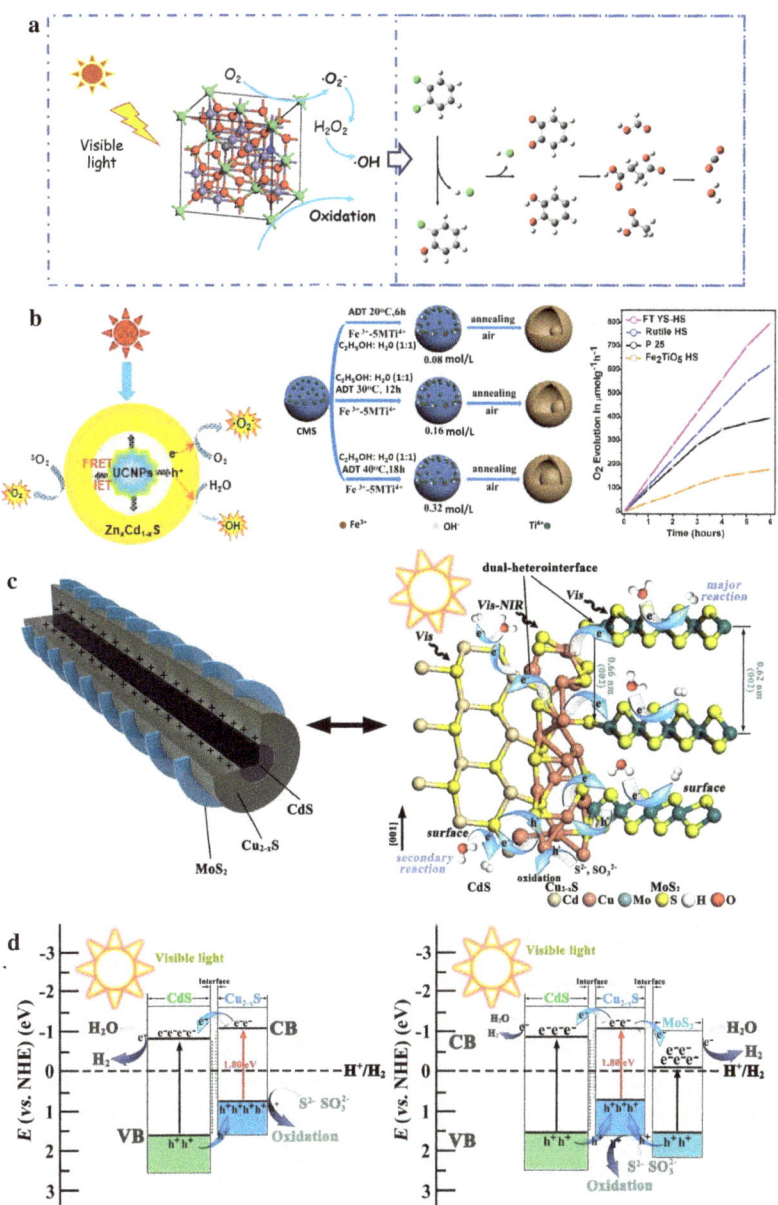

Fig. 31.17 a The proposed photocatalysis degradation path of o-DCB on the surface of CN–ZnFe$_2$O$_4$. Reproduced with permission [105]. Copyright © 2017 Elsevier B.V; **b** The photocatalytic mechanism of Zn$_x$Cd$_{1-x}$S under near-infrared light. Reproduced with permission [106]. Copyright © 2017 Elsevier B.V; **c** The synthesis process of Fe$_2$TiO$_5$-TiO$_2$ yolk–shell hollow sphere and analysis of its oxygen evolution performance. Reproduced with permission [107]. Copyright © 2017 Elsevier B.V; **d** Schematic diagram of the crystal structure of the heterogeneous interface of photocatalytic H$_2$ under visible light (λ > 400 nm) and the proposed mechanism of binary CdS–Cu$_{2-x}$S and ternary layered CdS–Cu$_{2-x}$S/MoS$_2$. Reproduced with permission [108]. Copyright © 2020, American Chemical Society

site and the Fe–O–Ti heterojunction in the hollow structure is subjected to charge separation.

31.2.7.2 Use of Core–Shell Photocatalyst

Liu [108] and others used Cu^+ migration to design layered $CdS–Cu_{2-x}S/MoS_2$ photocatalysts (Fig. 31.17d) as migrated and embedded into the adjacent MoS_2 surface to create CuI @ MoS_2 (CuI is embedded in the MoS_2 base surface). The epitaxial growth of $CuI@MoS_2$ nanosheets on the surface of the one-dimensional core–shell $CdS–Cu_{2-x}S$ nanorods forms a catalytic layer and a protective layer to enhance both catalytic activity and stability. Wang [109] and others studied the effects of proton irradiation on the structure and physical properties of giant CdSe/CdS core–shell quantum dots (g-CS QDs). These experiments reveal the delocalization of photoelectrons in g-CS quantum dots, where the strong changes in the current link and light emission are caused by the spatial expansion of the photoelectron wave function over the conduction bands of CdSe and CdS. Monte Carlo simulations of ionic-substance interactions indicate that the rate of destruction can be set by the energy of impacting protons, thereby promoting the formation of structural defects in the core or shell. After an irradiation dose higher than about $1017 H^+ cm^{-2}$, the formation of a nanocavity was confirmed, and a continuous decrease in luminescence intensity was observed, thereby increasing the proton flux. This marks an increase in non-radiation phenomena and a more photocarrier transfer between CdS and CdSe.

Xiao et al. [110] synthesized copper nanowires/ZnS (CuNWs/ZnS) with core–shell structure in situ through microwave-induced metal dissolution. Cu NW is used as a microwave antenna to create a local superheated surface to further induce ZnS crystallization and completely cover the Cu NW. With S^{2-}, the molten iron surface further causes the dissolution of Cu NWs and promotes the diffusion of Cu^+ into the ZnS lattice. By the narrow ZnS bandgap and the strong coupling interface formed between Cu NWs and ZnS by microwaves, the prepared hybrid composites showed excellent activity under the visible light and the stability of photocatalytic hydrogen production. Wang et al. [111] polymerized quantum thick graphite carbon nitride (g-C_3N_4) onto the surface of anatase titanium dioxide (TiO_2) nanoflakes, and at the same time exposed the $TiO_2\{001\}$ surface to form $TiO_2@g-C_3N_4$ (TCN) core–shell quantum heterojunction, the decisive factor for improving the photocatalytic activity of TCN is its unique structural advantages: the g-C_3N_4 shell is dense and the contact interface is uniform, the reaction sites are rich, and the surface adsorbed hydroxyl (OH) groups There are many groups, so it can greatly improve the photocatalytic tetracycline degradation activity. The degradation rate of tetracycline of 100 mg TCN photocatalyst is 36% higher than that of the $TiO_2/g-C_3N_4$ random mixture (TCN(mix)), 2 times higher than TiO_2, and 2.3 times higher than bulk g-C_3N_4. Li [112] and others used a layer-by-layer growth method to prepare core–shell structured upconversion nanoparticles (UCNPs)-Pt@MOF, and then used microwave synthesis method to load plasma gold nanoparticles to prepare (UCNPs)-Pt @ MOF/Au. In this material, MOF has a small amount of visible light response to ultraviolet light.

Plasma gold nanoparticles (NPs) absorb visible light, while UCNPs-Pt absorbs NIR light and emits ultraviolet and visible light. These lights are then captured by MOF and Au to achieve a broadband spectral response to the near-infrared (NIR) region. Besides, MOF achieves spatial separation of Au and Pt and also provides a way for the adsorption and desorption of reactants and products. Sun et al. [113] prepared the core–shell structure TiATA@LZU1 by using MOFs containing NH_2 groups directly grown in the COF shell, and synthesized Pd/TiATA@LZU1 photocatalyst by doping metal Pd. The experiments show the three components. There is a synergy between the metal as the active center, the MOF core as an electron donor, and the COF shell as an intermediary for electron transfer, forming a "donor-mediator-acceptor" system. The photocatalytic hydrogenation of olefins and the photocatalytic dehydrogenation of ammonia borane (NH_3BH_3, AB) under visible light proved the effectiveness of Pd doping.

Yang et al. [114] obtained a Ti^{3+} core/amorphous Ti^{4+} shell structure through a two-step hierarchical dispersion modification combining thermal reduction and hydrothermal fluorination. The formed core/shell structure is adjusted by the energy band structure. The rutile titanium dioxide crystallites are transformed into an efficient hydrogen evolution photocatalyst. The average precipitation rate of hydrogen in the mixture of water and methanol increased by more than 100 times from 1.7 to 268.3 μmol h^{-1}. Li [115] and others prepared Pt@TiO$_2$ hollow spheres (Pt-HSs) using silica as a template and then deposited MnO_x on the outer surface by light deposition to form Pt@TiO$_2$ @ MnO$_x$ hollow spheres (PTM-HSs), Pt presence is conducive to electron trapping, while MnO_x tends to collect holes. When TiO$_2$ is generated, electrons and holes flow inside and outside the spherical photocatalyst, gather on the corresponding cocatalyst, and participate in the redox reaction. Experiments show that the composite material has a higher photocatalytic oxidation efficiency for water and benzyl alcohol solution. Simultaneously with this work, the team of Li et al. [116] prepared Pt @ TiO$_2$ @ In$_2$O$_3$ @ MnO$_x$ mesoporous hollow spheres (PTIM-MSs) using the hard-template method. The inner and outer surfaces of TiO$_2$–In$_2$O$_3$ double-layer shells were selectively modified with Pt nanoparticles and MnO_x, respectively. The space-separated promoter promotes the flow of electrons and holes near the surface in opposite directions, while the thin heterogeneous shell separates the charges generated in the phase. The synergy between thin heterojunctions and spatially separated cocatalysts can simultaneously reduce bulk and surface/subsurface recombination. Indium oxide can also be used as a sensitizer to enhance light absorption. PTIM-MSs show high photocatalytic activity for the oxidation of water and alcohol.

31.2.8 Application of Yolk–Shell and Core–Shell Materials in Electrochemical Catalysis

The role of electrocatalysts in electrochemical reactions is to catalyze reactions by forming reactive intermediates to reduce overpotentials and increase current density. A good electrocatalyst should have high catalytic activity, stable corrosion resistance, good electronic conductivity, simple synthesis method, low cost, etc. The materials of yolk–shell and core–shell structure have low density, large surface area, and adjustable pore structure. Through structural design and synthesis, not only can the advantages of the internal and external materials be integrated, but also the deficiencies of the two materials can be overcome to achieve the improvement and optimization of the overall performance of the materials. In recent years, the yolk–shell and core–shell structures have become an important direction for the synthesis and application of nanomaterials. And it has broad application prospects in the field of electrocatalysis.

Wu et al. [117] designed a simple method to realize a new type of reduced graphene oxide-coated porous Co_3O_4 yolk–shell nanocage composite material. This special composite structure achieves enhanced OER activity and stability by combining high-conductivity r-GO with highly porous Co_3O_4 YSNCs. Cai et al. [104] obtained Au NP@Ni(OH)$_2$ yolk–shell structure through multi-ion interdiffusion (Fig. 31.18a, b). Au yolk and Ni(OH)$_2$ shell layers have abundant porosity and synergy effect, thus showing excellent OER performance. Zhao et al. [118] obtained Pd@Pt Ni/C core–shell catalysts by depositing about four PtNi alloy atomic layers on the surface of Pd nanocrystal seeds (Fig. 31.18c). They changed the number of platinum and nickel precursors added to the system to control the atomic ratio of platinum/nickel in the shell from 0.6 to 1.8. The ultra-thin PtNi alloy surface layer with only a few atomic layers can greatly improve the utilization efficiency and thus minimizes the use of platinum. Hou et al. [119] synthesized ultra-long Fe(OH)$_3$:Cu(OH)$_2$ core–shell nanowires in situ on a three-dimensional foamed copper (CF) electrode by a simple,

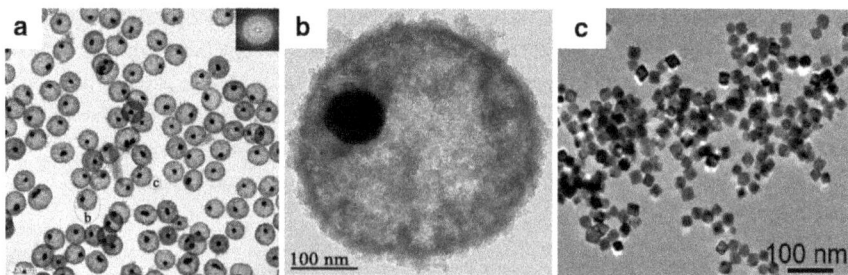

Fig. 31.18 **a** and **b** TEM images of Au NP@Ni(OH)$_2$ yolk–shell NPs. Low-magnification image and SAED (inset) and high-resolution images. Reproduced with permission [104]. Copyright © 2020 Elsevier Ltd. **c** Bright-field TEM image of octahedral Pd@Pt$_{1.8}$Ni core–shell nanocrystals. Reproduced with permission [118]. Copyright © 2015, American Chemical Society

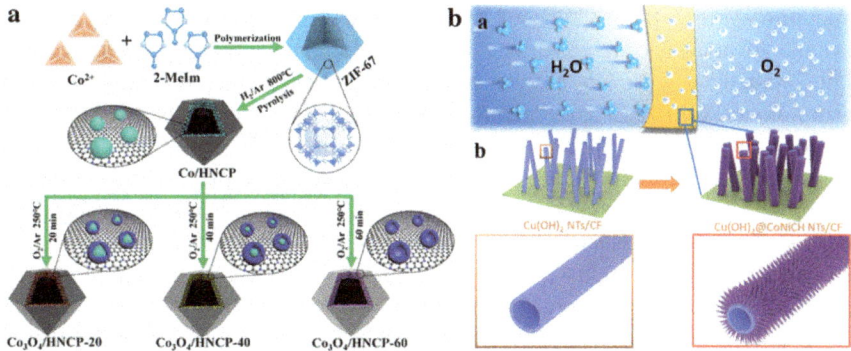

Fig. 31.19 **a** Illustrative expression of the procedures in the synthesis of ZIF-8 and NC, ZIF-67 and GC, and ZIF-8@ZIF-67 and NC@GC. Reproduced with permission [3]. Copyright © 2016 Elsevier Ltd. **b** Fabrication process of Cu@CoFe LDH core–shell nanostructure electrocatalysts. Reproduced with permission [120]. Copyright © 2017 Elsevier Ltd

scalable, and rapid synthesis method. The 3D Fe(OH)$_3$: Cu(OH)$_2$/CF electrode has high efficiency OER catalytic performance, the overpotential at 10 mA cm^{-2} is 365 mV, and Tafel slope is 42 mV dec^{-1}.

Wang et al. [3] designed a dual-function electrocatalyst with a core–shell structure synthesized with ZIF-8 and ZIF-67 organic framework structures (Fig. 31.19a). Although ZIF-8 and ZIF-67 have similar structures, they will form nitrogen-doped carbon (NC) and highly graphitized carbon (GC) with unique properties after carbonization. The core–shell structure of NC@GC integrates the advantages of these two carbon materials, high conductivity and good stability show good electrocatalytic activity for ORR and OER. Yu et al. [120] combined the structure of 1D Cu nanowires (NWs) and 2D CoFe LDH nanosheets (NSs) to prepare a new layered core–shell nanostructure on Cu foam (Fig. 31.19b) CoFe–LDH NSs have a large surface area and active site accessibility is conducive to the adsorption and catalytic reaction of water molecules. The unique layered structure is conducive to the diffusion of water and the release of gas products, ensuring the close contact between the catalyst and the electroactive substances. The channel provided by the Cu NWs core for electron transport reduces the electron transport distance and the barrier layer, which is beneficial to the reaction kinetics. Taking advantage of these advantages, Cu@CoFe LDH core–shell catalysts exhibit excellent overall water decomposition performance in alkaline media.

Wang et al. [121] used argon glow discharge as the electron source to prepare a highly active and stable carbon-supported Pt–Pd alloy catalyst using room temperature electron reduction method. A glow discharge is a traditional plasma phenomenon with high-energy electrons. The reduced metal nanoparticles have the characteristics of small size, steady interaction with the carrier, and high catalytic performance. Ding et al. [122] adopted a two-step pyrolysis-oxidation method. First, ZIF-67 was simply pyrolyzed under the H$_2$/Ar atmosphere. Secondly, Co NPs undergo controlled oxidation in O$_2$/Ar to obtain a series of Co-based nanoparticles with different multilayer structures. Among these Co-based hybrids, hollow N-doped carbon polyhedrons have the advantages of large surface area, rich hierarchical pores, good dispersion, high conductivity, sufficient oxygen vacancies and tetrahedral Co^{2+} can be used as adsorption sites for H_2O or oxygen ions, etc. The catalyst has the best bifunctional electrocatalytic performance for ORR and OER, in which activity and durability can exceed commercial Pt/C and IrO_2. It has become a potential high-performance non-noble metal bifunctional catalyst for sustainable energy applications. Kang et al. [123] used a three-dimensional open-hole structure $Cu(OH)_2$@CoNiCH core–shell nanotubes (NTs) as a highly efficient OER electrocatalyst. It is through a simple, highly controllable two-step method using copper foam (CF) as a carrier. At room temperature, a large-sized tubular $Cu(OH)_2$ array was directly grown on the copper foam (CF). High-porosity nano spiny cones were grown on the wall surfaces of these $Cu(OH)_2$ NTs by low-temperature hydrothermal method. This method simplifies the manufacturing process and reduces the inherent resistance of the electrode. This unique 3D layered core–shell structure has a hollow tubular $Cu(OH)_2$ core and a porous CoNiCH nanospine shell. This structure provides a large number of active sites, abundant defects, a large specific surface area, and fast electron transport, thereby enhancing the electrocatalytic performance and durability of OER.

Hwang et al. [124] used a one-pot method to synthesize hexagonal, sandwich-structured M@Ru (M = Ni, NiCo) core–shell nanoparticles. Ni core particles having a hexagonal plate-like morphology are generated from Ni and Ru precursors. The Ru shell layer grows in a regioselective manner around the top and bottom and the center edge of the Ni nanoplate. Therefore, the Ni@Ru core–shell hexagonal nanosandwich structure is generated to accelerate the slow kinetics of water oxidation. Zhang et al. [125] proposed a very promising dendritic core–shell nickel–iron–copper metal/metal oxide electrode. It is a core–shell nickel–iron–copper electrode prepared from electrodeposited nickel–iron–copper alloy as a template precursor with a porous oxide shell and metal core. This trimetallic core–shell nickel–iron–copper electrode has significant activity on water oxidation in alkaline media. At a current density of $10 \, mA \, cm^{-2}$, the overpotential is only $180 \, mV$. It is one of the most promising OER catalysts. Lu et al. [126] encapsulated Ni, Co, and Fe alloy cores in a graphite carbon shell (NiCoFe@C) to obtain a multifunctional catalyst material. It has good activity on oxygen evolution reaction (OER), oxygen reduction reaction (ORR), and hydrogen evolution reaction (HER). The encapsulation of the nickel–cobalt alloy core causes changes in the electronic structure of the graphitized shell layer, which adjusts the bonding strength of the reaction intermediates and improves the catalytic activity.

Zhang et al. [127] used wet tissues (WTs) as templates, and carbon nanotubes and ultra-thin $NiCo_2O_4$ nanosheet arrays as inner and outer microtubes to synthesize double microtube structures. Carbon in WTs is an excellent defect inducer, which induces a large number of defects into $NiCo_2O_4$ nanosheets. In addition, the corrugated structure and microtubular structure of the nickel oxide nanosheets can buffer the strain caused by the catalyst phase transition in the electrocatalytic reaction. In addition, the folded structure and microtubules of the nickel oxide nanocrystals can buffer the strain caused by the phase transition of the catalyst in the electro-catalytic reaction. The dual microtubule structure also provides abundant diffusion channels for oxygen and electrolytes with significant electrocatalytic activity, rapid reaction kinetics, and excellent OER stability. Dai et al. [128] used a wet chemical reduction method to prepare a ternary Co-Pd-Pt nanocatalyst supported by carbon nanotubes. The new ternary Co-Pd-Pt catalyst is composed of the core–shell structure of Co@Pd and the surface modification of Pt trimer (Pt_3). This three-way catalyst shows ORR quality activity, which is 30.6 times higher than that of commercial Pt catalyst. The Co-Pd-Pt catalyst showed excellent durability due to the survival and dispersion of the modified Pt_3 species during the ORR. It maintains a high activity of more than 322,000 potential cycles in alkaline electrolytes. Zhou et al. [129] colloidally synthesized Fe_3O_4@CoO NCs core–shell structure in a high-temperature organic solvent through seed-mediated growth. The core size and shell thickness are controlled by high-temperature colloidal synthesis to systematically study the effect of shell thickness on the catalytic activity of Fe_3O_4@CoO NCs.

Liu et al. [130] used an iron-based crystalline organic framework with nanorod morphology as the starting material, and treated it in a 2-methylimidazole solution to induce amorphization to generate MIL-88B nanorods. A MOF-NC is obtained by adding Co^{2+} to the reaction solution, including iron-rich Fe–Co–aMOF core and cobalt-rich Fe–Co–MOFs nanorods, and amorphous $Co(OH)_2$ nanosheets as the outer layer. Gohl et al. [131] synthesized a thin layer of Pt on TiWC and TiWN by a general high-temperature self-assembly method, and prepared a stable atomic-grade ultra-thin platinum shell titanium tungsten carbide Pt/TiWC catalyst. The material has stability at high oxidation potential and better catalytic durability of ORR than Pt/C. Xie et al. [132] used the Fe–PBA cage as a precursor to synthesize the open-cage bimetallic core–shell structure of Fe–CoP through a simple hydrothermal method and phosphating treatment. There is a hole in the center of each face of the Fe–CoP cage, which exposes more active parts of the OER. Due to the doping of iron and the unique open-cage core–shell structure, the Fe–CoP cage can withstand a current density of 10 mA cm^{-2} at a low overpotential (300mv), which is superior to the performance of commercial RuO_2.

At present, the better explanation of the catalytic performance of the core/yolk structure catalyst is almost limited to the restrictive effect of preventing the sintering of active sites. In addition to advanced characterization methods, a deeper under-standing of the sintering mechanism at the atomic level is required. For the design of core–shell catalysts, almost all designs are based on the knowledge of the supported catalyst or the performance of the material itself. When different materials are combined to form a core–shell structure, the characteristics of individual materials may change, especially in the interface area. Therefore, when the multifunctional design is required, material compatibility, electronic performance, the composition, location, and the number of different active sites can be rationally designed by computer simulation.

31.2.9 Application of Yolk–Shell and Core–Shell Materials in Other Efficient Catalysis Catalysts

Generally, yolk–shell metal–carbon nanostructures are constructed by template-assisted selective etching. The core nanoparticles coat the intermediate template layer and the carbon precursor shell layer in sequence. Then selectively remove the inter-layer template after carbonization. However, from the point of view of green chemistry, this method has a low atomic economy. Although template-free methods have been reported for yolk–shell nanostructures without carbon shells, they are still rarely explored. Xu et al. [133] used the method of micellar interfacial copolymerization to directly prepare polymers and carbon nanospheres with uniform hollow morphology. This new method is a template-free and unconstrained interfacial copolymerization reaction and a carbonization reaction without a sacrificial template reaction. In the presence of gold nanospheres, yolk–shell Au@C nanospheres were prepared by copolymerization of aniline and pyrrole on the interface of TritonX-100.

Since the production of fuels and chemicals can reduce the emission of greenhouse gas CO_2, hydrogenation of CO_2 into hydrocarbons in the scientific community has received widespread attention. Aitbekova et al. [134] used a seed-mediated colloidal method to synthesize heterodimers. Ru nanoparticle seeds (average particle size 4.8 nm) were obtained by thermal decomposition $Ru_3 (CO)_{12}$ in oleylamine, and they share the crystal plane and direct contact with iron oxide to obtain a ruthenium-iron oxide colloidal heterostructure. They demonstrated the synergistic effect in the ruthenium-iron oxide heterodimer catalyst. Among them, the hydrogenation of CO_2 to form hydrocarbons is realized by hydrogen activation and overflow from the proximal ruthenium phase to iron components.

With the continuous improvement of the performance of the photoelectrode for solar water splitting, improving the long-term stability of the photoelectrode has become an increasingly important issue to meet the needs of the factory-scale operation. Guo et al. [135] developed a core–shell structure of Au@CdS nanoparticles immobilized on zinc oxide nanowires (Au@CdS–ZnO) to promote charge transfer of zinc oxide-based photoanodes (Fig. 31.20). The core–shell structure forms a nano-junction between CdS and ZnO, which greatly promotes charge transfer, thereby significantly improving the IPCE efficiency of the PEC battery in a two-electrode configuration. He et al. [136] prepared bionic Au@TiO$_2$ yolk–shell nanostructures by growing TiO$_2$ shells in situ on pre-formed AuNPs. The Au@TiO$_2$ yolk–shell NPs photoelectrode prepared by self-assembly not only maintains the yolk–shell characteristics, but also has a good three-dimensional structure and photocatalytic activity. Due to the unique hollow nanostructure and the efficient charge separation provided by AuNPs, Au@TiO$_2$ yolk–shell NPs are superior to commercial TiO$_2$ and Au@TiO$_2$ core–shell NPs in both photocurrent and photocatalytic H_2 generation. Wang et al. [137] used aluminum-doped ZnO nanorod arrays (NRs) as an effective electron transport layer and CdS as a light collection layer. Atomic layer deposition (ALD) is used to deposit the TiO$_2$ protective layer on the Al–ZnO/CdS photoanode to improve the light stability. Compared with untreated Al–ZnO/CdS photoanode, the

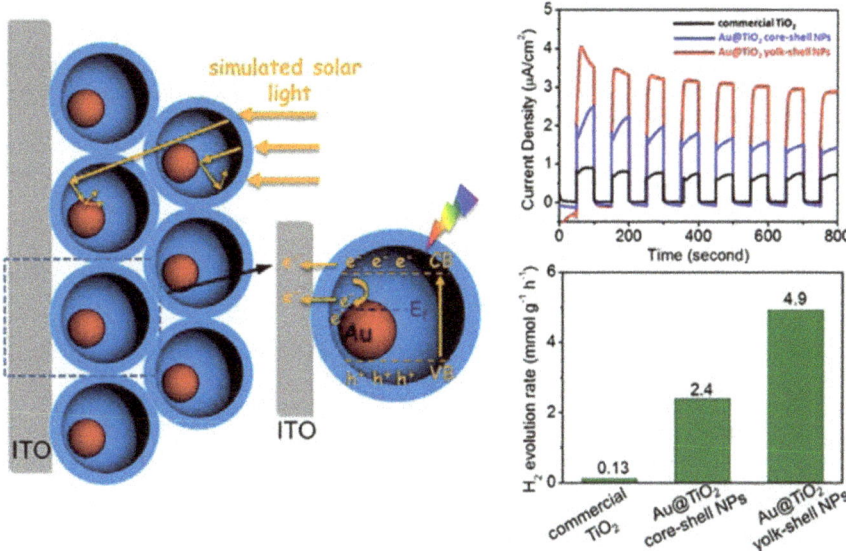

Fig. 31.20 Schematic illustration for the photocurrent generation with Au@TiO₂ yolk–shell NPs under irradiation [135]. Copyright © 2018 Elsevier B.V

light stability of Al–ZnO/CdS/TiO₂ photoanode is greatly improved. The enhanced stability is due to the triple function of the ALD TiO₂ layer, which is to isolate the direct contact between the photoanode and the surrounding liquid environment, passivate the surface state of CdS, and capture and store photogenerated holes. Jin et al. [138] first prepared PbWO₄ spheres by solvothermal method. First, two pickling was used to prepare a yolk–shell structure WO₃ film. Then the surface and inside of the WO₃ shell are coated with a layer of BiVO₄ film with a smaller bandgap to obtain better electrolyte accessibility. The core–shell-type PEC photoanode not only has good light absorption performance, but also facilitates the electron collection of BiVO₄ due to the increase of the contact area. Under the light of 100 mW/cm², the Fe–Ni promoter is added at a bias of 1.23 V versus RHE, and the photocurrent density can reach 5.0 mA/cm².

We believe that the core/yolk–shell nanoreactor has great potential for improving chemical reactions. Multidisciplinary research and cooperation are needed to fully realize the potential of core/yolk–shell nanoreactors. Although significant progress has been made in the synthesis and application of the core/yolk–shell nanoreactor, the core/yolk–shell structure will continue to be popular in the future.

References

1. Das S, Pérez-Ramírez J, Gong J (2020) Core-shell structured catalysts for thermocatalytic, photocatalytic, and electrocatalytic conversion of CO_2. Chem Soc Rev 49:2937–3004
2. Guo Y, Tang J, Wang ZL (2018) Elaborately assembled metal sulfides as a bifunctional core-shell structured catalyst for highly efficient electrochemical overall water splitting. Nano Energy 47:494–502
3. Wang ZL, Lu YZ, Yan Y (2016) Core-shell carbon materials derived from metal-organic frameworks as an efficient oxygen bifunctional electrocatalyst. Nano Energy 30:368–378
4. Li D, Yu SH, Jiang HL (2018) From UV to near-infrared light-responsive metal-organic framework composites: plasmon and upconversion enhanced photocatalysis. Adv Mater 30:1707377
5. Wang H, Chen L, Feng Y, Chen H (2013) Exploiting coreshell synergy for nanosynthesis and mechanistic investigation. Acc Chem Res 46:1636–1646
6. Kim M, Sohn K, Na HB (2002) Synthesis of nanorattles composed of gold nanoparticles encapsulated in mesoporous carbon and polymer shells. Nano Lett 2:1383–1387
7. Kamata K, Lu Y, Xia Y (2003) Synthesis and characterization of monodispersed core-shell spherical colloids with movable cores. J Am Chem Soc 125:2384–2385
8. Yin YD, Rioux RM, Erdonmez CK (2004) Formation of hollow nanocrystals through the nanoscale Kirkendall effect. Science 304:711–714
9. Cai R, Jin H, Yang D (2020) Generalized preparation of Au NP @ $Ni(OH)_2$ yolk-shell NPs and their enhanced catalytic activity. Nano Energy 71:104542
10. Shi J, Qiu F, Yuan W (2020) Nitrogen-doped carbon-decorated yolk-shell CoP@FeCoP micro-polyhedra derived from MOF for efficient overall water splitting. Chem Eng J 403:126312
11. Su X, Tang Y, Li Y (2019) Facile synthesis of monodisperse hollow mesoporous organosilica/silica nanospheres by an in-situ dissolution and reassembly approach. ACS Appl Mater Interfaces 11:12063–12069
12. Li W, Deng Y, Wu Z (2011) Hydrothermal etching assisted crystallization: a facile route to functional yolk-shell titanate microspheres with ultrathin nanosheets-assembled double shells. J Am Chem Soc 133:15830–15833
13. Ghosh S, Roy M, Naskar MK (2014) A facile soft-chemical synthesis of cube-shaped mesoporous CuO with microcarpet-like interior. Cryst Growth Des 14:2977–2984
14. Wang L, Wan J, Zhao Y (2019) Hollow multi-shelled structures of Co_3O_4 dodecahedron with unique crystal orientation for enhanced photocatalytic CO_2 reduction. J Am Chem Soc 141:2238–2241
15. Chiu YH, Naghadeh SB, Lai SAL (2019) Yolk-shell nanostructures as an emerging photocatalyst paradigm for solar hydrogen generation. Nano Energy 62:289–298
16. Purbia R, Paria S (2015) Yolk/shell nanoparticles: classifications, synthesis, properties and applications. Nanoscale 7:19789–19873
17. Zhang Q, Wang W, Goebl J, Yin Y (2009) Self-templated synthesis of hollow nanostructures. Nano Today 4:494
18. Xuan CY, Qing GQ, Yang S, Jie L (2016) General space-confined on-substrate fabrication of thickness-adjustable hybrid perovskite single-crystalline thin films. J Am Chem Soc 138:16196–16199
19. Yu L, Wu HB, David Lou (2017) Self-templated formation of hollow structures for electrochemical energy applications. Acc Chem Res 50:293–301
20. Feng J, Yin YD (2019) Self-templating approaches to hollow nanostructures. Adv Mater 31:1802349
21. Wang MW, Boyjoo Y, Pan J (2017) Advanced yolk-shell nanoparticles as nanoreactors for energy conversion. Chin J Catal 38:970–990
22. Saha A, Bharmoria P, Mondal A (2015) Generalized synthesis and evaluation of for-mation mechanism of metal oxide/sulphide@C hollow spheres. J Mater Chem A 3:20297–20304
23. Sheng Y, Zeng HC (2015) Monodisperse aluminosilicate spheres with tunable Al/Si ratio and hierarchical macro-meso-microporous structure. ACS Appl Mater Interfaces 7:13578–13589

24. Chen J, Wu X, Hou XD (2014) Shape-tunable hollow silica nanomaterials based on a soft-templating method and their application as a drug carrier. ACS Appl Mater Interfaces 6:21921–21930

25. Shao J, Qu QT, Wan ZM (2015) From dispersed microspheres to interconnected nanospheres: carbon-sandwiched monolayered MoS_2 as high-performance anode of Li-Ion batteries. ACS Appl Mater Interfaces 7:22927–22934

26. Dassanayake AC, Wickramaratne NP, Hossain M (2019) Prussian blue-assisted one-pot synthesis of nitrogen-doped mesoporous graphitic carbon spheres for super-capacitors. J Mater Chem A 7:22092–22102

27. Long YK, Xiao L, Cao QH (2017) Efficient incorporation of diverse components into metal organic frameworks via metal phenolic networks. Chem Commun 53:10831–10834

28. Zheng JN, Lv JJ, Li SS (2014) One-pot synthesis of reduced graphene oxide sup-ported hollow Ag@Pt core–shell nanospheres with enhanced electrocatalytic activity for ethylene glycol oxidation. J Mater Chem A 2:3445–3451

29. Cai WR, Zhang GY, Lu KK (2017) Enhanced electrochemiluminescence of one-dimensional self-assembled porphyrin hexagonal nanoprisms. ACS Appl Mater Interfaces 9:20904–20912

30. Jang J, Bae J (2005) Fabrication of mesoporous polymer using soft template method. Chem Commun 1200–1202

31. Chen HY, Zhang TL, Fan J (2013) Electrospun hierarchical TiO_2 nanorods with high porosity for efficient dye-sensitized solar cells. ACS Appl Mater Interfaces 5:9205–9211

32. Zhang J, Li JY, Wang WP (2018) Microemulsion assisted assembly of 3D porous S/Graphene@g-C_3N_4 hybrid sponge as free-standing cathodes for high energy density Li–S batteries. Adv Energy Mater 8:1702839

33. Dong RH, Liu WM, Hao JC (2012) Soft vesicles in the synthesis of hard materials. Acc Chem Res 45:504–513

34. Northcutta RG, Sundaresan VB (2014) Phospholipid vesicles as soft templates for electropoly-merization of nanostructured polypyrrole membranes with long range order. J Mater Chem A 2:11784–11791

35. Lee KT, Jung YS, Oh SM (2003) Synthesis of tin-encapsulated spherical hollow carbon for anode material in lithium secondary batteries. J Am Chem Soc 125:5652–5653

36. Liu J, Qiao SZ (2010) Monodisperse yolk-shell nanoparticles with a hierarchical porous structure for delivery vehicles and nanoreactors. Angew Chem Int Ed 49:4981–4985

37 Wu ZX, Li Q, Feng D (2010) Ordered mesoporous crystalline γ-Al2O3 with variable architecture and porosity from a single hard template. J Am Chem Soc 132:12042–12050

38. Cui ZM, Chen Z, Cao CY (2013) A yolk–shell structured Fe_2O_3@mesoporous SiO_2 nanore-actor for enhanced activity as a Fenton catalyst in total oxidation of dyes. Chem Commun 49:2332

39. Lee J, Oh J, Jeon Y (2018) Multi-heteroatom-doped hollow carbon attached on graphene using $LiFePO_4$ nanoparticles as hard templates for high-performance lithium–sulfur batteries. ACS Appl Mater Interfaces 10:26485–26493

40. Cai Z, Xu L, Yan M (2015) Manganese oxide/carbon yolk-shell nanorod anodes for high capacity lithium batteries. Nano Lett 15:738

41. Baddour FG, Nash CP, Schaidle JA (2016) Angew Chem Int Ed 55:9026

42. Anibal J, Romero HG, Leonard ND (2016) Effect of silica morphology on the structure of hard-templated, non-precious metal catalysts for oxygen reduction. Appl Catal B 198:32–37

43. Shi QR, Song Y, Zhu CZ (2015) Mesoporous Pt nanotubes as a novel sensing platform for sensitive detection of intracellular hydrogen peroxide. ACS Appl Mater Interfaces 7:24288–24295

44. Song SY, Liu XC, Li JQ (2017) Confining the nucleation of Pt to in situ form (Pt-Enriched Cage)@CeO_2 Core@Shell nanostructure as excellent catalysts for hydrogenation reactions. Adv Mater 29:1700495

45. Kuo CH, Hua TE, Huang MH (2009) Au nanocrystal-directed growth of Au–Cu_2O core–shell heterostructures with precise morphological control. J Am Chem Soc 131:17871

46. Niu W, Zhang L, Xu G (1987) Shape-controlled synthesis of single-crystalline palladium nanocrystals. ACS Nano 2010:4
47. Sun H, Shen X, Yao L (2012) Measuring the unusually slow ionic diffusion in poly-aniline via study of yolk-shell nanostructures. J Am Chem Soc 134:11243
48. Tiemann M (2008) Repeated templating. Chem Mater 20:961–971
49. DeKrafft KE, Wang C, Lin WB (2012) Metal-organic framework templated synthesis of Fe_2O_3/TiO_2 nanocomposite for hydrogen production. Adv Mater 24:2014–2018
50. He S, Yan C, Chen XZ (2020) Construction of core-shell heterojunction regulating α-Fe_2O_3 layer on CeO_2 nanotube arrays enables highly efficient Z-scheme photoelectrocatalysis. Appl Catal B 276:119138
51. Du D, Shi W, Wang LZ (2017) Yolk-shell structured Fe_3O_4@void@TiO_2 as a photo-Fenton-like catalyst for the extremely efficient elimination of tetracycline. Appl Catal B 200:484–492
52. Li A, Wang T, Chang XX (2016) Spatial separation of oxidation and reduction co-catalysts for efficient charge separation: Pt@TiO_2@MnOx hollow spheres for pho-tocatalytic reactions. Chem Sci 7:890–895
53. Li A, Chang X, Huang Z (2016) Thin heterojunctions and spatially separated cocatalysts to simultaneously reduce bulk and surface recombination in photocatalysts. Angew Chem Int Ed 55:13734
54. Yu L, Yu XY, David Lou (2018) The design and synthesis of hollow micro-/nanostructures: present and future trends. Adv Mater 30:1800939
55. Zong L, Wang Z, Yu R (2019) Lanthanide-doped photoluminescence hollow structures: recent advances and applications. Small 15:1804510
56. Zhang Q, Fu MX, Ning GY (2020) Co/FeC core–nitrogen doped hollow carbon shell structure with tunable shell-thickness for oxygen evolution reaction. J Coll Interface Sci 580:794–802
57. He TN, Wang WC, Yang XL (2017) Inflating hollow nanocrystals through a repeated Kirkendall cavitation process. Nat Commun 8:1261
58. Zhu JK, Tu WM, Pan HF (2020) Self-templating synthesis of hollow Co_3O_4 nanoparticles embedded in N, S-dual-doped reduced graphene oxide for lithium ion batteries. ACS Nano 14:5780–5787
59. Nagaraju G, Sekhar SC, Raju GSR (2017) Designed construction of yolk–shell structured trimanganese tetraoxide nanospheres via polar solvent-assisted etching and biomass-derived activated porous carbon materials for high-performance asymmetric supercapacitors. J Mater Chem A 5:15808–15821
60. Huang J, Qian X, Yang JH (2020) Construction of Pt-free electrocatalysts based on hierarchical CoS_2/N-doped C@Co-WS_2 yolk-shell nano-polyhedrons for dye-sensitized solar cells. Electrochim Acta 340:135949
61. Yao CL, Zha JW, Li CP (2020) Yolk-shelled FeP/Ni_2P/C@C nanospheres with void: controllable synthesis and excellent performance as the anode for lithium-ion batteries. Colloids Surf, A 602:125103
62. Li R, Xu JS, Zeng RG (2020) Halides-assisted electrochemical synthesis of Cu/Cu_2O/CuO core-shell electrocatalyst for oxygen evolution reaction. J Power Sources 457:228058
63. Ghasemian MB, Mayyas M, Idrus-Saidi SA (2019) Self-limiting galvanic growth of MnO_2 monolayers on a liquid metal—applied to photocatalysis. Adv Funct Mater 29:1901649
64. Hou CC, Zou L, Xu Q (2019) A hydrangea-like superstructure of open carbon cages with hierarchical porosity and highly active metal sites. Adv Mater 31:1904689
65. Zhou LS, Deng BL, Jiang ZQ (2019) Shell thickness controlled core–shell Fe_3O_4@CoO nanocrystals as efficient bifunctional catalysts for the oxygen reduction and evolution reactions. Chem Commun 55:525–528
66. Aitbekova A, Goodman ED, Wu LH (2019) Innentitelbild: Fein-Tuning der Porengröße in versteiften ZIF-8_Cm-Gerüsten durch eine Mixed-Linker-Strategie für verbesserte permeative CO_2/CH_4-Trennung. Angew Chem 131:2–9
67. Tang J, Salunkhe RR, Liu J (2015) Thermal conversion of core-shell metal–organic frameworks: a new method for selectively functionalized nanoporous hybrid carbon. J Am Chem Soc 137(4):1572–1580

68. Wu ZL, Sun LP, Yang M (2016) Facile synthesis and excellent electrochemical performance of reduced graphene oxide–Co_3O_4 yolk-shell nanocages as a catalyst for oxygen evolution reaction. J Mater Chem A 4:13534–13542

69. Ahmad N, Bon M, Passerone D (2019) Template-assisted in situ synthesis of Ag@Au bimetallic nanostructures employing liquid-phase transmission electron microscopy. ACS Nano 13:13333–13342

70. Zhao X, Chen S, Fang ZC (2015) Octahedral Pd@Pt1.8Ni core–shell nanocrystals with ultra-thin PtNi alloy shells as active catalysts for oxygen reduction reaction. J Am Chem Soc 137:2804–2807

71. Zhang PL, Li L, Nordlund D (2018) Dendritic core-shell nickel-iron-copper met-al/metal oxide electrode for efficient electrocatalytic water oxidation. Nat Commun 9:381

72. Li X, Iocozzia J, Chen Y (2018) From precision synthesis of block copolymers to properties and applications of nanoparticles. Angew Chem Int Ed 57:2046

73. Fei HF, Li WH, Bhardwaj A (2019) Ordered nanoporous carbons with broadly tunable pore size using bottlebrush block copolymer templates. J Am Chem Soc 141:17006–17014

74. Xu Y, Yao YX, Yu HT (2019) Nanoparticle-encapsulated hollow porous polymeric nanosphere frameworks as highly active and tunable size-selective catalysts. ACS Macro Lett 8:1263–1267

75. Caruso F, Shi X, Caruso RA (2001) Hollow Titania spheres from layered precursor deposition on sacrificial colloidal core particles. Adv Mater 13:740–744

76. Wang XM, Song JD, Chen HL (2016) Preparation of ZnO@void@SiO_2 rattle type core-shell nanoparticles via layer-by-layer method. NANO 11:1650103

77. Wang Y, Zhang ZZ, Zhang LN (2018) visible-light driven overall conversion of CO_2 and H_2O to CH_4 and O_2 on 3D-SiC@2D-MoS_2 heterostructure. J Am Chem Soc 140:14595–14598

78. Yun YP, Sheng HZ, Bao K (2020) Design and remarkable efficiency of the robust sandwich cluster composite nanocatalysts ZIF-8@Au25@ZIF-67. J Am Chem Soc 142:4126–4130

79. Ullmann MA, dos Santos JHZ (2020) Zirconocene immobilization into organic-inorganic dual-shell silicas prepared by the nonhydrolytic sol-gel method for poly-ethylene production. J Catalysis 385:30–43

80. Lan K, Xia Y, Wang RC (2019) Confined interfacial monomicelle assembly for precisely controlled coating of single-layered Titania mesopores. Matter. 1:527–538

81. Stober W, Fink A, Bohn E (1968) Controlled growth of monodisperse silica spheres in the micron size range. J Colloid Interface Sci 26(1):62–69

82. Salinas D, Guerrero S, Campos CH (2020) The effect of the ZrO_2 loading in SiO_2@ZrO_2-CaO catalysts for transesterification reaction. Materials 13(1):221

83. Wei B, Zhou JT, Yao ZJ (2020) Excellent microwave absorption property of nano-Ni coated hollow silicon carbide core-shell spheres. Appl Surf Sci 508:145261

84. Wang GH, Chen K, Engelhardt J, Scalable one-pot synthesis of yolk-shell carbon nanospheres with yolk-supported Pd nanoparticles for size-selective catalysis

85. Luo T, Zhang J, Tan X (2016) Water-in-supercritical CO_2 microemulsion stabilized by a metal complex. Angew Chemie Int Ed 55(43):13533–13537

86. Mao D, Ding S, Meng L, One-pot microemulsion-mediated synthesis of Bi-rich $Bi_4O_5Br_2$ with controllable

87. Kuwahara Y, Sumida Y, Fujiwara K (2016) Facile synthesis of yolk-shell nanostructured photocatalyst with improved adsorption properties and molecular-sieving properties. Chemcatchem 8(17):2781–2788

88. Wu XJ, Xu D (2009) Formation of yolk/SiO_2 shell structures using surfactant mixtures as template. J Am Chem Soc 131(8):2774–2775

89. Wu SH, Tseng CT, Lin YS (2011) Catalytic nano-rattle of Au @ hollow silica: towards a poison-resistant nanocatalyst. J Mater Chem 21(3):789–794

90. Liu J, Qiao SZ, Hartono SB (2010) Monodisperse yolk-shell nanoparticles with a hierarchical porous structure for delivery vehicles and nanoreactors. Angew Chemie Int Ed 49(29):4981–4985

91. Chen M, Lu S, Fu XZ (2020) Core-shell structured NiFeSn@NiFe (oxy)hydroxide nanospheres from an electrochemical strategy for electrocatalytic oxygen evolution reaction. Adv Sci 7(10)

92. Ma D, Shi JW, Zou Y (2017) Rational design of CdS@ZnO core-shell structure via atomic layer deposition for drastically enhanced photocatalytic H-2 evolution with excellent photostability. Nano Energy 39:183–191

93. Ma HP, Yang JH, Tao JJ (2019) Low-temperature epitaxial growth of high-quality GaON films on ZnO nanowires for superior photoelectrochemical water splitting. Nano Energy 66

94. Wang W, Wang ZY, Wang JJ (2017) Highly active and stable Pt-Pd alloy catalysts synthesized by room-temperature electron reduction for oxygen reduction reaction. Adv Sci 4(4)

95. Boissonnault JA, Wong-Foy AG, Matzger AJ (2017) Core-shell structures arise naturally during ligand exchange in metal-organic frameworks. J Am Chem Soc 139(42):14841–14844

96. Zhang XM, Liang HC, Li H (2020) Sequential chemistry toward core-shell structured metal sulfides as stable and highly efficient visible-light photocatalysts. Angewandte Chemie Int Ed

97. Ghasemian MB, Mayyas M, Idrus-Saidi SA (2019) Self-limiting galvanic growth of MnO_2 monolayers on a liquid metal-applied to photocatalysis. Adv Function Mater 29(36)

98. Chen Z, Wang J, Zhai G (2017) Hierarchical yolk-shell WO_3 microspheres with highly enhanced photoactivity for selective alcohol oxidations. Appl Cataly B-Environ 218:825–832

99. Liu J, Feng J, Gui J (2018) Metal@semiconductor core-shell nanocrystals with atomically organized interfaces for efficient hot electron-mediated photocatalysis. Nano Energy 48:44–52

100. Yue Q, Li JL, Zhang Y (2017) Plasmolysis-Inspired nanoengineering of functional yolk-shell microspheres with magnetic core and mesoporous silica shell. J Am Chem Soc 139(43):15486–15493

101. Zhang N, Fu X, Xu YJ (2011) A facile and green approach to synthesize Pt@CeO_2 nanocomposite with tunable core-shell and yolk-shell structure and its application as a visible light photocatalyst. J Mater Chem 21(22):8152–8158

102. Marschall R (2014) Semiconductor composites: strategies for enhancing charge carrier separation to improve photocatalytic activity. Adv Func Mater 24(17):2421–2440

103. Jiang N, Li D, Liang L (2020) (Metal yolk)/(porous ceria shell) nanostructures for high-performance plasmonic photocatalysis under visible light. Nano Research 13(5):1354–1362

104. Cai R, Jin H, Yang D (2020) Generalized preparation of Au NP @ Ni(OH)(2) yolk-shell NPs and their enhanced catalytic activity. Nano Energy 71

105. Li JA, Li XY, Chen X (2019) In situ construction of yolk-shell zinc ferrite with carbon and nitrogen co-doping for highly efficient solar light harvesting and improved catalytic performance. J Colloid Interface Sci 554:91–102

106. Wang WN, Huang CX, Zhang CY (2018) Controlled synthesis of upconverting nanoparticles/ZnxCd1-xS yolk-shell nanoparticles for efficient photocatalysis driven by NIR light. Appl Cataly B-Environ 224:854–862

107. Waqas M, Iqbal S, Bahadur A (2017) Designing of a spatially separated hetero-junction pseudobrookite (Fe_2TiO_5-TiO_2) yolk-shell hollow spheres as efficient photocatalyst for water oxidation reaction. Appl Cataly B-Environ 219:30–35

108. Liu G, Kolodziej C, Jin R (2020) MoS_2-stratified CdS-Cu_2-xS core-shell nanorods for highly efficient photocatalytic hydrogen production. ACS Nano 14(5):5468–5479

109. Wang C, Barba D, Selopal GS (2019) Enhanced photocurrent generation in proton-irradiated "giant" CdSe/CdS core/shell quantum dots. Adv Function Mater 29(46)

110. Xiao S, Dai W, Liu X (2019) Microwave-induced metal dissolution synthesis of core-shell copper nanowires/ZnS for visible light photocatalytic H-2 evolution. Adv Energy Mater 9(22)

111. Wang W, Fang J, Shao S (2017) Compact and uniform TiO_2@g-C_3N_4 core-shell quantum heterojunction for photocatalytic degradation of tetracycline antibiotics. Appl Cataly B-Environ 217:57–64

112. Li D, Yu SH, Jiang HL (2018) From UV to near-infrared light-responsive metal-organic framework composites: plasmon and upconversion enhanced photocatalysis. Adv Mater 30(27)

113. Sun D, Jang S, Yim SJ (2018) Metal doped core-shell metal-organic frameworks @ covalent organic frameworks (MOFs @ COFs) hybrids as a novel photocatalytic platform. Adv Function Mater 28(13)
114. Yang Y, Liu G, Irvine JTS (2016) Enhanced photocatalytic H-2 production in core-shell engineered rutile TiO_2. Adv Mater 28(28), 5850–+
115. Li A, Wang T, Gong J (2016) Spatial separation of oxidation and reduction cocatalysts for efficient charge separation: $Pt@TiO_2@MnOx$ hollow spheres for photocatalytic reactions. Chem Sci 7(2):890–895
116. Li A, Chang X, Gong J (2016) Thin heterojunctions and spatially separated cocatalysts to simultaneously reduce bulk and surface recombination in photocatalysts. Angewandte Chemie Int Ed 55(44):13734–13738
117. Wu Z, Sun LP, Yang M, Grenier (2016) Facile synthesis and excellent electrochemical performance of reduced graphene oxide–Co_3O_4 yolk-shell nanocages as a catalyst for oxygen evolution reaction. J Mater Chem A 4(35):13534–13542
118. Zhao X, Chen S, Fang Z (2015) Octahedral $Pd@Pt1.8Ni$ core-shell nanocrystals with ultrathin PtNi alloy shells as active catalysts for oxygen reduction reaction. J Am Chem Soc 137(8):2804–2807
119. Hou CC, Wang CJ, Chen QQ (2016) Rapid synthesis of ultralong $Fe(OH)_3:Cu(OH)_2$ coreshell nanowires self-supported on copper foam as a highly efficient 3D electrode for water oxidation. Chem Commun (Camb) 52(100):14470–14473
120. Yu L, Zhou H, Sun J (2017) Hierarchical Cu@CoFe layered double hydroxide core-shell nanoarchitectures as bifunctional electrocatalysts for efficient overall water splitting. Nano Energy 41:327–336
121. Wang W, Wang Z, Wang J (2017) Highly active and stable Pt-Pd alloy catalysts synthesized by room-temperature electron reduction for oxygen reduction reaction. Adv Sci (Weinh) 4(4):1600486
122. Ding D, Shen K, Chen X (2018) Multi-level architecture optimization of MOF-templated co-based nanoparticles embedded in hollow N-Doped carbon polyhedra for efficient OER and ORR. ACS Catal 8(9):7879–7888
123. Kang J, Sheng J, Xie J (2018) Tubular $Cu(OH)_2$ arrays decorated with nanothorny Co–Ni bimetallic carbonate hydroxide supported on Cu foam: a 3D hierarchical core–shell efficient electrocatalyst for the oxygen evolution reaction. J Mater Chem A 6(21):10064–10073
124. Hwang H, Kwon T, Kim HY (2018) Ni@Ru and NiCo@Ru core-shell hexagonal nanosandwiches with a compositionally tunable core and a regioselectively grown shell. Small 14(3):1702353
125. Zhang P, Li L, Nordlund D (2018) Dendritic core-shell nickel-iron-copper metal/metal oxide electrode for efficient electrocatalytic water oxidation. Nat Commun 9(1):381
126. Lu X, Tan X, Zhang Q (2019) Versatile electrocatalytic processes realized by Ni, Co and Fe alloyed core coordinated carbon shells. J Mater Chem A 7(19):12154–12165
127. Zhang X, Li X, Li R (2019) Highly active core-shell carbon/$NiCo_2O_4$ double microtubes for efficient oxygen evolution reaction: ultralow overpotential and superior cycling stability. Small 15(42):e1903297
128. Dai S, Chou JP, Wang KW (2019) Platinum-trimer decorated cobalt-palladium core-shell nanocatalyst with promising performance for oxygen reduction reaction. Nat Commun 10(1):440
129. Zhou L, Deng B, Jiang Z (2019) Shell thickness controlled core-shell $Fe_3O_4@CoO$ nanocrystals as efficient bifunctional catalysts for the oxygen reduction and evolution reactions. Chem Commun (Camb) 55(4):525–528
130. Liu C, Wang J, Wan J (2020) Amorphous metal-organic framework-dominated nanocomposites with both compositional and structural heterogeneity for oxygen evolution. Angew Chem Int Ed Engl 59(9):3630–3637
131. Gohl D, Garg A, Paciok P (2020) Engineering stable electrocatalysts by synergistic stabilization between carbide cores and Pt shells. Nat Mater 19(3):287–291

132. Xie JY, Liu ZZ, Li J (2020) Fe-doped CoP core–shell structure with open cages as efficient electrocatalyst for oxygen evolution. J Energy Chem 48:328–333

133. Xu F, Lu Y, Ma J (2017) Facile, general and template-free construction of monodisperse yolk-shell metal@carbon nanospheres. Chem Commun (Camb) 53(89):12136–12139

134. Aitbekova A, Goodman ED, Wu L (2019) Engineering of ruthenium-iron oxide colloidal heterostructures: improved yields in CO_2 hydrogenation to hydrocarbons. Angew Chem Int Ed Engl 58(48):17451–17457

135. Guo CX, Xie J, Yang H (2015) Au@CdS core-shell nanoparticles-modified ZnO nanowires photoanode for efficient photoelectrochemical water splitting. Adv Sci (Weinh) 2(12):1500135

136. He Q, Sun H, Shang Y (2018) Au@TiO_2 yolk-shell nanostructures for enhanced performance in both photoelectric and photocatalytic solar conversion. Appl Surf Sci 441:458–465

137. Wang R, Wang L, Zhou Y (2019) Al-ZnO/CdS photoanode modified with a triple functions conformal TiO_2 film for enhanced photoelectrochemical efficiency and stability. Appl Catal B 255:117738

138. Jin B, Jung E, Ma M (2018) Solution-processed yolk-shell-shaped WO_3/$BiVO_4$ heterojunction photoelectrodes for efficient solar water splitting. J Mater Chem A 6(6):2585–2592

Chapter 32
Conclusion

Yasutaka Kuwahara and Hiromi Yamashita

This book made a comprehensive review on recent progress in the construction of core–shell/yolk–shell nanostructures and their applications in catalysis, photocatalysis, and electrocatalysis. With the rapid development of synthetic chemistry, colloid, and interfacial science, a numerous number of core–shell and yolk–shell nanostructured catalysts with diverse core and shell combinations have been developed. Core–shell nanostructures have been regarded as valuable and versatile nanomaterials for catalysis because of their enhanced catalytic properties compared to their single-component counterparts. For example, the synergism (e.g., ligand effect, ensemble effect, and geometric effect) at the interface between the core and the surrounding shell can be exploited for tailoring the catalytic activity and selectivity as well as stability of core–shell nanocatalysts. Recent developments in the preparation of core–shell nanocatalysts and their catalytic applications, including chemoselective transformation reactions, hydrogenation reactions, selective synthesis of valuable chemicals by syngas conversion and CO_2 hydrogenation, etc. have been summarized in Part I. As reviewed in some chapters in Part I, tuning the compositions in the core/shell regions and the thickness of the shell in the core–shell nanocatalysts allows high chemoselectivity in specific organic transformations to produce fine chemicals, and zeolite-based core–shell structure appears to be an effective configuration for selective transformation reactions of C1-3 feedstock chemicals to produce value-added chemicals and fuels involving gas phase reactors. Moreover, core–shell catalysts have become promising materials for photocatalysis and electrocatalysis. As summarized in Part II, manipulation of core/shell characteristics,

Y. Kuwahara · H. Yamashita (✉)
Division of Materials and Manufacturing, Science, Graduate School of Engineering, Osaka University, Suita, Osaka 565-0871, Japan
e-mail: yamashita@mat.eng.osaka-u.ac.jp

Y. Kuwahara
e-mail: kuwahara@mat.eng.osaka-u.ac.jp

© The Author(s), under exclusive license to Springer Nature Singapore Pte Ltd. 2021   563
H. Yamashita and H. Li (eds.), *Core-Shell and Yolk-Shell Nanocatalysts*,
Nanostructure Science and Technology,
https://doi.org/10.1007/978-981-16-0463-8_32

core/shell interfaces, as well as the growth of multiple shells have provided many opportunities for improving catalytic performances and adding new functionalities in photo/electro-catalysis. Thus, the integration of two different components into a single core–shell structure has led to the emergence of a new heterogeneous catalyst that can be explored for a number of important chemical reactions.

Furthermore, yolk–shell nanostructures with unique core@void@shell nano-configuration have been emerged as valuable and versatile nanocatalysts for the last decade, because of their many unique advantages, as follows: (1) encapsulation of free, movable cores inside the hollow shells; (2) encapsulation and compartmentation of large guest molecules inside the interior void space; (3) protection of the encapsulated core particles from leaching, sintering, and aggregation by the outer shell; (4) molecular-sieving effect for reactant molecules; (5) adsorption and condensation property toward reactant molecules inside the cavity space; (6) easy tunability in the core and the shell regions to add a number of functionalities. With these advantages, yolk–shell nanostructure can be used as a promising platform for the design and development of heterogeneous catalysis. In this book, recent developments in the design and fabrication of yolk–shell nanocatalysts and their catalytic and photo-/electro-catalytic applications have been summarized in Part III and Part IV, respectively. The design and creation of yolk–shell structure composites with multiple components and well-controlled configurations are likely to bring new benefits to many catalytic reactions, with keeping the encapsulated components stable during the catalytic reactions. In particular, for the design and construction of efficient photocatalysts, which requires multiple components to fulfill various functions, including light absorption, photo-excited charge separation and transportation, and redox reactions, yolk–shell nanocatalysts with controllable structures exert a significant influence.

As introduced in each chapter, a variety of a type of core/yolk–shell nanostructures with different core/shell combinations have been successfully fabricated via different synthetic strategies (e.g., seed-based growth method, selective etching method, hard/soft-templating method, and ship-in-bottle approach) to construct nano-configurations suitable for target chemical reactions. However, one-step/pot strategy in the preparation of core/yolk–shell nanostructures is still in its infancy, and many challenges remain to be solved. Multiple synthetic processes, preparation cost, and difficulty in large-scale production of these nanostructures have limited their ultimate applications in industrial applications. To this end, the development of simpler one-step/pot synthetic strategies is necessary for the green, scalable, and cost/energy-effective production of core/yolk–shell NPs, and extended application of the core/yolk–shell nanocatalysts to industrially important catalytic applications, which produce high-value-added fine and specialty chemicals, is highly desirable.

In comparison with traditionally supported catalysts, the unique collective and synergism between the core and the shell materials in the core/yolk–shell nanostructures have not fully been understood yet, mainly due to the difficulty in precise control of the nanostructures, the diffusion of reactants and products within confined nano-space, and the limited characterization technique available to understand the core/shell interface and the nano-environments. For future development of core/yolk–shell catalysts, it is extremely important to develop new synthetic methods to construct well-defined core/yolk–shell nanostructures, in which the sizes and shapes of the metal cores as well as the thicknesses and structures of the shells can be tuned precisely, to make a deep understanding of the structure–activity relationships and the interfacial synergism between the cores and the shells. It is also highly desirable to develop advanced characterization techniques and theory calculations to gain insights into core/yolk–shell nanostructures, for example, the diffusion and transfer behaviors of both reactants and products in the core/yolk–shell NPs, the charge transfer/distribution between the cores and the shells and the Schottky barrier at the core/shell interface. Indeed, further exploration of new, undiscovered functionalities is necessary to expand the possibilities of the core/yolk–shell nanocatalysts to a wide range of different catalytic reactions, especially those important for industrial applications. In particular, core/yolk–shell nanostructures containing multiple catalytic components with controllable configurations/arrangements are believed to bring new benefits to realize one-pot catalytic reactions (namely, cascade or domino reactions), which can lead to energy-saving and cost-saving catalytic processes.

We believe that core/yolk–shell nanostructures have great potential as platforms for designing and fabricating high-performance heterogeneous catalysts in the field of catalysis, photocatalysis, and electrocatalysis, which will eventually contribute to the development of efficient catalytic processes and sustainable industrial applications in the future.

Index